CRC SERIES IN AGRICULTURE

Editor-in-Chief

Angus A. Hanson, Ph.D.
Vice President
and
Director of Research
W-L Research, Inc.
Highland, Maryland

HANDBOOK OF SOILS AND CLIMATE IN AGRICULTURE

Editor
Victor J. Kilmer (Retired)
Chief
Soils and Fertilizer Research Branch
National Fertilizer Development Center
Tennessee Valley Authority
Muscle Shoals, Alabama

HANDBOOK OF PLANT SCIENCE IN AGRICULTURE

Editor
B. R. Christie, Ph.D.
Professor
Department of Crop Science
Ontario Agricultural College
University of Guelph
Guelph, Ontario, Canada

HANDBOOK OF PEST MANAGEMENT IN AGRICULTURE

Editor
David Pimentel, Ph.D.
Professor
Department of Entomology
New York College of Agricultural
and Life Sciences
Cornell University
Ithaca, New York

HANDBOOK OF ENGINEERING IN AGRICULTURE

Editor
R. H. Brown, Ph.D.
Chairman
Division of Agricultural Engineering
Agricultural Engineering Center
University of Georgia
Athens, Georgia

HANDBOOK OF TRANSPORTATION AND MARKETING IN AGRICULTURE

Editor
Essex E. Finney, Jr., Ph.D.
Assistant Center Director
Agricultural Research Center
U.S. Department of Agriculture
Beltsville, Maryland

HANDBOOK OF PROCESSING AND UTILIZATION IN AGRICULTURE

Editor
Ivan A. Wolff, Ph.D.
Director
Eastern Regional Research Center
Science and Education Administration
U.S. Department of Agriculture
Philadelphia, Pennsylvania

HANDBOOK OF ANIMAL SCIENCE IN AGRICULTURE

CRC Handbook of Pest Management in Agriculture

Volume I

Editor
David Pimentel
New York State College of Agriculture and Life Sciences
Cornell University
Ithaca, New York

CRC Series in Agriculture

A. A. Hanson, Editor-in-Chief
Vice President and
Director of Research
W-L Research, Inc.
Highland, Maryland

CRC Press, Inc.
Boca Raton, Florida

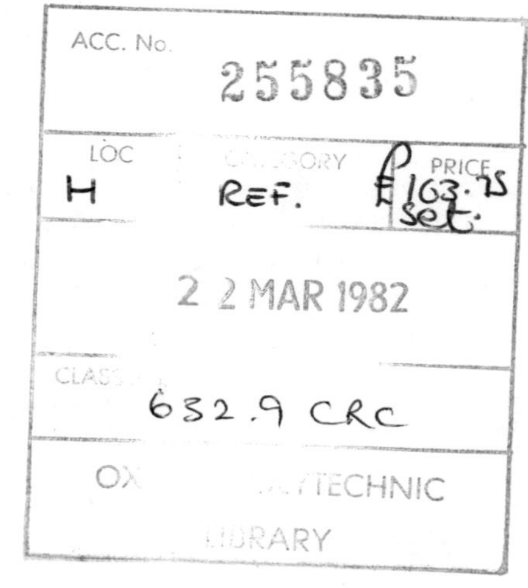

Library of Congress Cataloging in Publication Data

Main entry under title:

Handbook of pest management in agriculture.

(CRC series in agriculture)

Bibliography: p.
Includes index.
1. Pest control—Handbooks, manuals, etc.
I. Pimentel, David, 1925- II. Hanson, Angus Alexander. III. Series.
SB950.H33 632 80-15635
ISBN 0-8493-3841-7 (v. 1)

This book represents information obtained from authentic and highly regarded sources. Reprinted material is quoted with permission, and sources are indicated. A wide variety of references are listed. Every reasonable effort has been made to give reliable data and information, but the author and the publisher cannot assume responsibility for the validity of all materials or for the consequences of their use.

Direct all inquiries to CRC Press, Inc., 2000 N.W. 24th Street, Boca Raton, Florida 33431.

International Standard Book Number 0-8493-3841-7 (Volume I)
International Standard Book Number 0-8493-3842-5 (Volume II)
International Standard Book Number 0-8493-3843-3 (Volume III)

Library of Congress Card Number 80-15635
Printed in the United States

PREFACE
CRC SERIES IN AGRICULTURE

Agriculture, because of its pivotal role in the development of civilized societies, contributed much to the development of various scientific disciplines. Thus, agricultural pursuits led to the practical application of chemistry, and gave rise to such major disciplines as economics and statistics. The expansion of scientific frontiers, and the concomitant specialization within disciplines, has proceeded to the point where agricultural scientists classify themselves in an array of disciplines and subdisciplines, i.e., nematologist, geneticist, physicist, virologist, and so forth. Nevertheless, within the framework of these various disciplines and mission-oriented agricultural research, information of primary interest and concern in the solution of agriculturally oriented problems is generated. Although some of the basic information finds its way into the plethora of reference books available within most disciplines, no attempt has been made to develop a comprehensive handbook series for the agricultural sciences.

It is recognized that there are serious difficulties in developing a meaningful handbook series in agriculture because of the range and complexity of agricultural enterprises. In fact, the single common denominator that applies to all agricultural scientists is their universal concern with at least some aspect of the production and utilization of farm products. The disciplines and resources that are called for in a specific investigation are either the same or similar to those utilized in any area of biological research, or in any one of several fields of scientific endeavor.

The sections in this handbook series reflect the input of different Editors and Advisory Boards, and as a consequence, there is considerable variation in both the depth and coverage offered within a given area. However, an attempt has been made throughout to bring together pertinent information that will serve the needs of nonspecialists, provide a quick reference to material that might otherwise be difficult to locate, and furnish a starting point for further study.

The project was undertaken with the realization that the initial volumes in the series could have some obvious deficiencies that will necessitate subsequent revisions. In the meantime, it is felt that the primary objectives of the Section Editors and their Advisory Boards have been met in this first edition.

A. A. Hanson
Editor-in-Chief

EDITOR-IN-CHIEF

Angus A. Hanson, Ph.D., is Vice President and Director of Research, W-L Research, Inc., Highland, Maryland, and has had broad experience in agricultural research and development. He is a graduate of the University of British Columbia, Vancouver, and McGill University, Quebec, and received the Ph.D. degree from the Pennsylvania State University, University Park, in 1951.

An employee of the U.S. Department of Agriculture from 1949 to 1979, Dr. Hanson worked as a Research Geneticist at University Park, Pennsylvania (1949 to 1952) and at Beltsville, Maryland, serving succesively as Research Leader for Grass and Turf Investigations (1953 to 1965), Chief of the Forage and Range Research Branch (1965 to 1972), and Director of the Beltsville Agricultural Research Center (1972 to 1979). He has been appointed to a number of national and regional task forces charged with assessing research needs and priorities, and has participated in reviewing agricultural needs and research programs in various foreign countries. As Director at Beltsville, he was directly responsible for programs that included most dimensions of agricultural research.

In his personal research, he has emphasized the improvement of forage crops, breeding and management of turfgrasses, and the breeding of alfalfa for multiple pest resistance, persistence, quality, and sustained yield. He is the author of over one hundred technical and popular articles on forage crops and turfgrasses, and has served as Editor of *Crop Science* and the *Journal of Environmental Quality*.

PREFACE

The aim of these volumes is to assemble the data currently available on pest management. Over time pest control technologies have grown in variety and complexity. The intensive use of the new synthetic chemical pesticides that started about 1945 has benefited agriculture while at the same time creating problems for agriculture, the environment, and society.

The public has demanded, and rightly so, to see the pest control "balance sheet" to determine the risks versus the benefits of chemical and nonchemical pest controls relative to public health, food quantity and quality, and the environment. In these volumes an attempt is made to examine several major questions related to chemical and nonchemical pest controls:

1. How serious are the agricultural losses to pests in this nation?
2. What chemical and nonchemical control strategies are employed and what are their relative efficiencies?
3. What are the benefits and risks of the various pest control technologies to society?
4. What are the benefits and risks of the new pest management strategies that agriculturalists are developing?

The authors of this Handbook have assembled the best information that is currently available, but fully recognize the limitations of any such effort as this. We look forward to receiving any comments and suggestions from researchers in helping strengthen any further editions.

Despite the limitations, we hope that the information assembled by the numerous authors will prove valuable to agriculturalists as they strive to make more effective use of the land, water, and energy resources needed for successful crop and livestock production in the U.S. and elsewhere in the world.

All the authors and advisory board to this Handbook devoted considerable time to assembling this Handbook. The successful completion of this Handbook is due to their cooperation and hard work.

I am also indebted to Ms. Susan Pohl and Nancy Goodman Sorrells for their valuable assistance in compiling this volume. I also wish to thank Terrill Evon of CRC for her technical assistance in editing this book.

David Pimentel
Section Editor

THE EDITOR

David Pimentel is Professor of Insect Ecology and Agricultural Sciences in the Department of Entomology, the Section of Ecology and Systematics, and the Field of Natural Resources, College of Agriculture and Life Sciences, Cornell University. Dr. Pimentel received his B.S. in 1948 from the University of Massachusetts and his Ph.D. in 1951 from Cornell University. From 1951 to 1954 he was Chief of the Tropical Research Laboratory, U.S. Public Health Service, San Juan, Puerto Rico. From 1954 to 1955 he was post-doctoral research fellow at the University of Chicago; 1960 to 1961 OEEC Fellow at Oxford University (England); and 1961, Research Scholar at Massachusetts Institute of Technology. He was appointed an Assistant Professor of Insect Ecology at Cornell University in 1955 and Associate Professor in 1961. In 1963 he was appointed Professor and Head of the Department of Entomology and Limnology. He served as Head until 1969 when he returned to full time research and teaching.

Nationally he has served on numerous Presidential Committees and National Academy of Sciences Committees and Boards. In the recent World Food and Nutrition Study of the NAS he chaired the Panel on *Interdependencies of Population, Food, Health, Energy, and the Environment*. He is chairman of the Biomass Panel of Energy Research Advisory Board of the U.S. Department of Energy. He is a Fellow of the American Association for the Advancement of Science and Fellow of the Entomological Society of Canada.

Dr. Pimentel has authored nearly 200 scientific publications, written two books, and edited five books.

ADVISORY BOARD

CONTRIBUTORS

VOLUME I

John R. Abernathy
Associate Professor
Texas Agricultural Experiment Station
Texas A & M University Research and Extension Center
Lubbock, Texas

Keith A. Arnold
Professor, Department of Wildlife and Fisheries Sciences
Texas A & M University
College Station, Texas

D. E. Bay
Associate Professor
Department of Entomology
Texas A & M University
College Station, Texas

M. G. Boosalis
Professor and Head
Department of Plant Pathology
University of Nebraska-IANR
Lincoln, Nebraska

J. M. Chandler
Research Agronomist
SEA-USDA
Southern Weed Science Laboratory
Stoneville, Mississippi

William J. Cromartie, Jr.
Associate Professor of Entomology
ENVL
Stockton State College
Pomona, New Jersey

David E. Davis
Professor Emeritus
North Carolina State University
Raleigh, North Carolina

Ben Doupnik
Professor of Plant Pathology
Department of Plant Pathology
South Central Station
University of Nebraska
Clay Center, Nebraska

Roger O. Drummond
Laboratory Director
U.S. Livestock Insects Laboratory
AR/SEA-USDA
Kerrville, Texas

Don M. Huber
Plant Pathologist and Associate Professor
Department of Botany and Plant Pathology
Purdue University
West Lafayette, Indiana

W. Clive James
Agricultural Specialist
Canadian International Development Agency
Hull, P.Q., Canada

Waldemar Klassen
National Research Program Leader
Crop Production Staff
NPS/AR/SEA/USDA
Beltsville, Maryland

Thor Kommedahl
Professor, Department of Plant Pathology
University of Minnesota
St. Paul, Minnesota

George Lambert
Assistant Director
National Animal Disease Center
AR/SEA-USDA
Ames, Iowa

S. Leach
Research Plant Pathologist
AR/SEA-USDA
N.E. Plant, Soil and Water Laboratory
University of Maine
Orono, Maine

Robert D. Lumsden
Research Plant Pathologist
Plant Protection Institute
Soilborne Diseases Laboratory
Agricultural Research Center
Beltsville, Maryland

Gaylord I. Mink
Professor and Plant Pathologist
Irrigated Agriculture Research and Extension Center
Washington State University
Prosser, Washington

Neal O. Morgan
Research Entomologist
Livestock Insects Laboratory
Beltsville Agricultural Research Center
AR/SEA-USDA
Beltsville, Maryland

G. N. Odvody
Assitant Professor of Plant Pathology
Texas A & M Research and Extension Center
Texas A & M University
Corpus Christi, Texas

Einar W. Palm
Professor of Plant Pathology
Department of Plant Pathology
University of Missouri
Columbia, Missouri

G. Gregor Rohwer
Director, National Program Planning Staff
Plant Protection and Quarantine
Animal and Plant Health Inspection Service
U.S. Department of Agriculture
Hyattsville, Maryland

Leon H. Russell, Jr.
Professor
Department of Veterinary Public Health
Texas A & M University
College Station, Texas

Paul H. Schwartz, Jr.
National Research Program Leader
Crop Producton Staff
NPS/AR/SEA/USDA
Beltsville, Maryland

Gregory Shaner
Associate Professor of Plant Pathology
Department of Botany and Plant Pathology
Purdue University
West Lafayette, Indiana

Fred W. Slife
Professor — Weed Science
Agronomy Department
University of Illinois
Urbana, Illinois

Harry E. Smalley
Consultant in Veterinary Toxicology
Laboratory Director (retired)
Veterinary Toxicology and Entomology Research Laboratory
AR/SEA-USDA
College Station, Texas

C. Dayton Steelman
Professor of Entomology and Assistant Director
Louisiana Agricultural Experiment Station
Louisiana State University and A & M College
Baton Rouge, Louisiana

Vernon Stern
Professor of Entomology
Department of Entomology
University of California
Riverside, California

Pete D. Teel
Research Scientist and Entomologist
Department of Entomology
Texas A & M University
College Station, Texas

George L. Teetes
Professor
Department of Entomology
Texas A & M University
College Station, Texas

C. E. Terrill
Staff Scientist
National Program Staff
AR/SEA-USDA
Beltsville, Maryland

Ward M. Tingey
Assistant Professor of Entomology
Department of Entomology
Cornell University
Ithaca, New York

Frederic H. Wagner
Associate Dean, College of Natural Resources
Director of Ecology Center
Utah State University
Logan, Utah

Howard E. Waterworth
Chief, Germplasm Resources Laboratory
Beltsville Agricultural Research Center
Beltsville, Md.

TABLE OF CONTENTS
CRC HANDBOOK OF PEST MANAGEMENT IN AGRICULTURE

Volume I

Volume II

Volume III

TABLE OF CONTENTS

Volume I

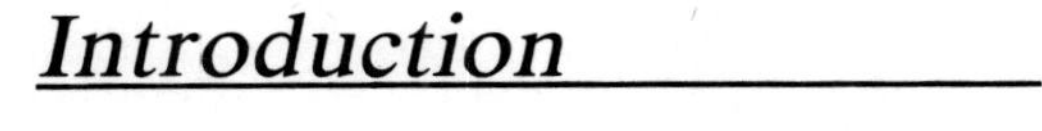

Introduction

INTRODUCTION

David Pimentel

Pest management, the science and technology of pest control, has as its goal the control of agricultural pests to benefit society as a whole. Therefore, successful pest control strategies must take into account the complexities of society and the entire ecosystem. Recognizing this, then, pest control strategies are influenced by socioeconomics, land and water resources, fossil energy resources, human population density, and the environment, as well as the ecology of the pests.

An examination of the interdependency of pest mangement strategies and this complex set of factors provides a valuable perspective of the wide range of topics included in this Pest Management Handbook. Some of the interdependencies of pest management and society and the environment are examined below.

Food Losses to Pests

All crops and livestock are attacked by pests and the resulting losses are high. World crop losses to pests (insects, pathogens, and weeds) are currently estimated to be about 35%.[1] Mammal and bird losses appear to be more severe in the tropics and subtropics than in the temperate region, but are still low compared with losses to the three major pest groups of insects, pathogens, and weeds.

Losses to pests may have been intensified by the use of the new intensive crop production technology introduced with the green revolution. The new high-yield crop varieties in use today are often more susceptible to pests than the old varieties were. Before the green revolution farmers usually selected seeds from individual plants that survived best under native cultural conditions. The plants surviving and yielding best under these environmental conditions contained alleles resistant to insects and pathogens. These plants also competed successfully with weeds.[2,3]

Worldwide postharvest losses to pests are estimated to range from 10% to 20%.[4] The major pests of harvested foods are microorganisms, insects, and rodents. When postharvest losses are added to preharvest losses, worldwide food losses due to pests are estimated to be about 45%. This significant loss of valuable food occurs in spite of all of the methods used to control pests.

In the U.S. preharvest losses to pests are estimated to be about 37% even with the use of modern pest control technology. Insects account for 13% of the losses,[5] plant pathogens 12%,[5] and weeds, 12%.[6] It should be noted that U.S. food losses are based upon our high-quality food standards.

Postharvest losses to pests are estimated to be about 9%.[5] Thus, total losses to pests in the U.S. are estimated to be more than 40%.

Numerous nonchemical or bioenvironmental methods are used throughout the U.S. and the world in an attempt to control pests and reduce crop losses. Bioenvironmental control is described as any method used to reduce pest populations "by manipulation of the pest's environment and ecology or by altering the pest's physiology, genetics, and behavior or by a combination of these".[7]

In the U.S., nonchemical controls are employed on more cropland (about 90%) than pesticide controls (about 20%).[8] Although pesticides are often considered to be the most important control of pests, nonchemical controls are more important than pesticides. The extent of nonchemical pest control use worldwide in agriculture is believed to be substantial but no published estimates are available.

Over 5 billion lb (1 lb = 0.45 kg) of pesticides are applied annually in the world today.[9] About 34% of the total is applied in North America, about 45% in western and eastern Europe, and the remaining 21% in the rest of the world (primarily devel-

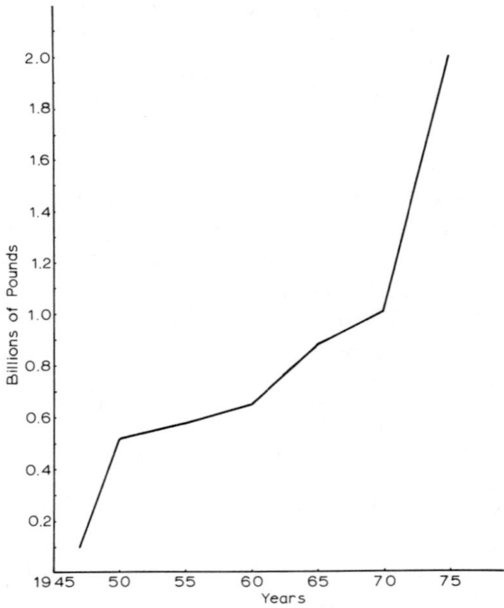

FIGURE 1. Estimated amount of pesticide produced in the U.S.[36,37]

oping countries). Estimates are that by 1985 the total quantity of pesticide applied annually in the world will total more than 6 billion lb. Of the 5 billion lb of pesticide used in the world, one fifth or 1 billion lb are used in the U.S.[10] (Figure 1).

Human Populations

One of the major reasons for the increased need of chemical and nonchemical controls of pests is the growing need for food to feed a rapidly growing world population. At no previous time in history have humans so dominated their environment by the sheer numbers of their species. This phenomenon, however, is a fairly recent event. For the first 990,000 years of the more than 1 million years that humans have inhabited the earth, the maximum world population numbered about 200,000 (about one fourth the current population of Boston). For most of that period population growth was only about 0.01% per year (Figure 2). That comparatively slow growth rate accounts for why the world population took 990,000 years to reach about 200,000. During that long period of time people were hunter-gatherers and depended exclusively on the natural ecosystem for their food.

Then about 10,000 years ago humans began to settle and cultivate food crops. With established agriculture came an increased and stable food supply. This contributed to the increased growth rate of the world population (Figure 2).

The next major increase in the population growth rate coincided with the discovery and use of stored fossil energy resources such as coal, oil, and gas. Since that time, a mere 300 to 400 years ago, rapid population growth closely paralleled the increased use of fossil energy (Figures 2 and 3) because humans have been able to use energy to manipulate and manage their environment. In particular, energy has been used to increase yields from food production and to improve public health. Improvements in both food and public health have been the prime contributors to rapid population growth in the world.

At present the world population stands at more than 4 billion, but what is more alarming is the high growth rate of nearly 2%. This is 200 times greater than the growth

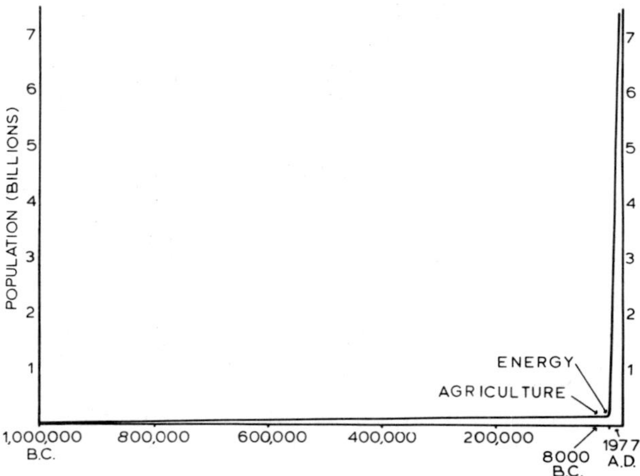

FIGURE 2. Human population trends over the last million years.

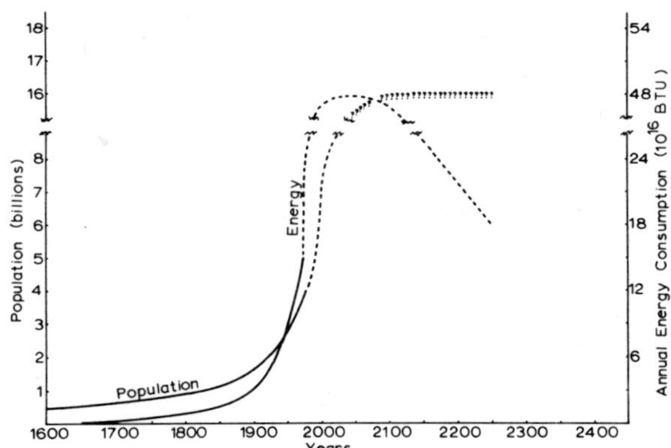

FIGURE 3. Estimated world population numbers (—) from 1600 to 1975 and projected numbers (- - - -) (?????) to the year 2250. Estimated fossil fuel consumption (—) from 1650 to 1975 and projected (- - - -) to the year 2250.[16]

rate for the entire first 990,000 years of human existence on earth. This rate of increase contributes about 200,000 people to the world population daily.

Demographers project that the world population will reach 6 to 7 billion by the year 2000 and by 2100 will reach 10 to 16 billion.[11] At present there appears to be no generally accepted way to limit this rate of growth.

The population in the U.S. is also projected to increase (Figure 4). The U.S. population is growing at about half the rate of the world population or about 0.9% per year.[11] This rate is about 90 times the growth rate of the world population during the first 990,000-year period (Figures 2 and 4). During the next 25 years the U.S. population is projected to increase 24% from 212 million to 262 million. Although about two thirds of the increase will be due to immigration, one third will be the result of the U.S. birth rate.

Contributing to the high growth rate of the world population is the young age-structure of many populations in many nations. For example, 52% of the population in

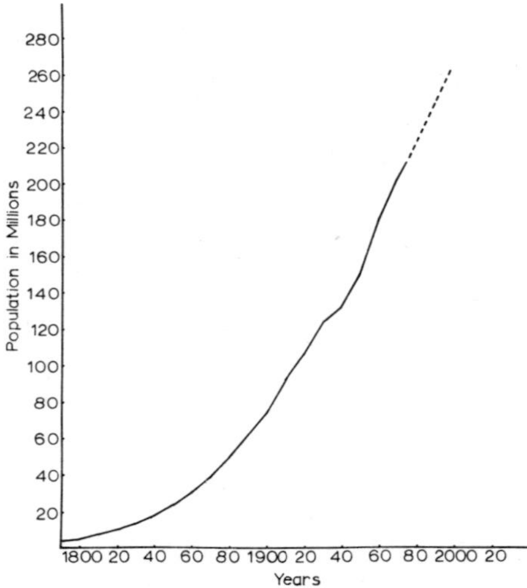

FIGURE 4. Population growth in the U.S., actual (solid line) and projected (broken line).[38]

Honduras is 15 years old or younger.[12] This means that a large proportion of the population is within childbearing age. This age structure has long-term implications for human population growth.

For example, in India where population density is already high, if all Indian families were to begin now to limit their family size to just two children, the population would continue to increase for the next 70 years.[13] As a result the Indian population would increase from its present 600 million to more than 900 million. The same would be true of all countries with a high proportion of young in their populations and which restricted birth rates to just two replacement children per family.

Food is a vital resource that must supply the basic nutritionl needs of the U.S. and world population. Therefore, it is relevant to examine the environmental resources utilized in food production, because these resources also influence pest control strategies.

Environmental Resources Utilized in Food Production

To gain some idea of the challenges of feeding a rapidly growing world population, estimates are made of the animal and vegetable matter production relative to land, water, and fossil energy constraints. This analysis assumes a present population of 4 billion, 6 billion in the year 2000, and 16 billion for the year 2100.

Most peoples of the world desire to eat and live as the people of the U.S. Hence, in the first analysis, we calculate land and energy required to feed a population of 4 billion a U.S. high protein-calorie diet produced with U.S. agricultural technology.

In the U.S. about 160 million ha are planted to crops.[14] With a U.S. population of about 215 million, this averages about 0.7 ha planted to crops per capita. Since about 20% of our crop yield is exported, the estimated arable land per person is about 0.62 per hectare.[14] The world arable land resources are about 1.5 billion ha.[15] With 4 billion humans in the world today, the per capita land available is only 0.38 ha. In the U.S., 0.62 ha of land plus a high-energy agricultural technology are necessary to produce the high protein-calorie diet that is consumed. Hence, in the world today, there is insufficient arable land (even assuming that energy resources and other technology are also available) to feed the current world population a diet similar to and produced in the same manner as that consumed in the U.S.

In this analysis of land resources, fossil energy was assumed to be unlimited. Unfortunately, fossil energy is also in limited supply for food production. This can be put into perspective with the following analysis. If petroleum were the only source of energy for food production, and if all petroleum reserves were used solely to feed the world population, the 66,000-billion-ℓ oil reserve in the world would last a mere 13 years.[16] Both the land and energy estimates indicate that the human population has already reached a density too great for the arable land and energy resources that are required to feed the world population a U.S. diet utilizing U.S. technology. Both estimates were made from known arable land and petroleum resources. If we include potential arable land and potential petroleum resources, the situation appears to be improved. It should be pointed out that a population of only 4 billion was used in the calculations. The world population is now 4.3 billion and is projected to reach more than 6 billion in the next 21 years.

We will make another analysis recognizing the constraints of land and energy resources while food demand increases with a rapidly growing world population. The focus is on both the animal and vegetable foods and their availability to the human population. Total animal protein consumption by man today amounts to about 25% of the total protein supply consumed by the world population. Cereals contribute nearly half of the total protein supply consumed by man. There would be sufficient foods for all people throughout the world if pests and other production factors were under control. Even with pest losses and other types of loss, there should be adequate amounts of food available if it were equitably distributed to all peoples.

Livestock production may be increased 30% by the year 2000 through reduced overgrazing and the use of better pasture plant species and application of limited amounts of fertilizer under certain advantageous conditions. But to hold the per capita food supply in the year 2000 at 1975 levels will require a 66% increase in legumes, a 100% increase in other vegetables, and a 75% increase in cereals.[16] This 75% increase in the next 25 years is technically feasible; the 55% increase for legumes and the 100% increase for vegetables appears less likely.

To feed the 16 billion humans predicted for the year 2100, utilizing a similar diet to that consumed by the world population in 1975, will require significant increases in food production. For example, legumes must be increased by 173%, vegetables 233%, and cereals 330%.[16] With the resources of land, energy, and water that are available, these increases appear to be doubtful.

One means of increasing the total amount of food available to man would be to reduce the amount of vegetable and other animal products that are currently fed to livestock. An estimated 51 million metric tons of protein suitable for man's use were fed to the world's livestock in 1975.[16] This 51 million metric tons that is fed livestock is nearly equal to the total cereal protein available to man for 1975. Therefore, if man could switch from consuming quite as large an amount of animal products to consuming more plant products there would be more food available for the world's population.

With careful management of land, water, energy, and human resources, and cooperation among nations of the world, we believe that it is possible to maintain current per capita levels of food supplies for the next 21 years, as the world increases to more than 6 billion humans. Serious malnutrition already exists with some half billion humans. Efforts are needed to eliminate this deficiency by better food production and distribution.

Science and technology in pest control can help man to overcome future food crises that face him as his numbers rapidly increase, but the necessary solution to the well-being of mankind will likely also require a more equitable distribution of resources and effective population control. The problems associated with pest control are inextricably bound to complex biological and environmental aspects of crop and livestock management as well as to other aspects of the ecosystem of man. In the next section, we will briefly examine the ecological basis for pest outbreaks.

Ecological Basis for Pest Outbreaks

The ecological basis for insect pests, pathogens, and weed problems is complex. Pest outbreaks are often the result of a combination of ecological factors. One of these cases is the monoculture of crops. Natural ecosystems tend to evolve toward stable climax communities for each particular habitat in the world. For agricultural production, however, the natural plant community is removed and destroyed, and is replaced by a single crop species. As soon as the land is cleared of the natural vegetation, man's battle with what he terms pests begins. The seeds that are planted germinate, but so do hundreds of seeds of other plant species that lay in the soil, some of which may have remained dormant for many years. In addition, various microorganisms are present in the soil or may drift in the wind. Insects may be present or fly in from other locations, with all of these insects tending to attack the crop.

One aspect of the monoculture problem is that the larger the area that is planted to a single crop, the greater the potential for pest problems. This also relates to the problem of continuous culture or a monoculture of one crop in one location for several years. When crops are maintained in the same area year after year, pests associated with the crop tend to increase in number and in severity. This is true of cole crops. For example, if they are cultured for several years in the same soil, club root *(Plasmodiophora brassicae)* organisms increase rapidly and can totally ruin production.[17] The same is true of the corn rootworm that is now the primary pest of corn. When corn is planted following soybeans or small grains, corn rootworms are not a problem. Crop rotations, however, can sometimes increase a problem with black cutworms and wireworms. The use of crop rotations is not, therefore, a complete answer to insect problems in corn.[18]

Some pest problems occur when crops are introduced into new biotic communities. For example, when the potato, which originated in Bolivia and Peru[19] was introduced into the southwestern U.S., it acquired a serious pest, the Colorado potato beetle. Native to the U.S., this beetle had originally coevolved with and fed on wild sand bur.[20] When the potato was introduced into the southwest, the beetle spread onto it. Because the potato had never been exposed to the beetle, it lacked any natural resistance to it. Since then, this insect has become the most serious pest to potato in the world and has accompanied the plant as its cultivation spread to other areas.

In addition to introducing crops into new locations and having pest problems develop, the introduction of pest species is one of the most important causes of pest problems. A few examples of newly introduced species that became pests are the European rabbit that was introduced into Australia, the Japanese beetle that was introduced into the U.S., the gypsy moth, Dutch elm disease, and water hyacinth.

One of the most important ecological factors involved in pest problems is the breeding of susceptible crop genotypes.[21] An example of this is sorghum. On a susceptible strain of commercial sorghum, the mean rate of oviposition (eggs per generation) of the chinchbug was about 100. On a resistant strain, however, the mean oviposition was less than 1.[22] In this instance, the animal feeding was reduced by 99% on the resistant plants. This has had dramatic effects on the population dynamics of chinchbugs.

Much has been written about diversity and its influence on the stability of pest populations. Frequently, outbreaks of insect pests in agriculture have been attributed to crop monocultures. For example, Marchal[23] wrote that when man plants a vast extent of the country with certain crops, while excluding others, he offers to the insects feeding on those plants favorable conditions for their explosive increase. This has been documented with experimental studies on the plant *Brassica oleracea* by Pimentel,[24] Tahvanainen and Root,[25] Cromartie,[26] and Root.[27]

Based on numerous examples, it is clear that many parasites have the genetic variability to evolve and overcome what might be termed single-factor resistance in their host. For example, parasitic stem rust and crown rust have been found to overcome genetic resistance bred into their oat hosts. Since 1940, oat varieties have been changed in the Corn Belt region every 5 years to counter the changes in the rises of stem rust and crown rust.[29,30] In experiments with an animal and simulated plant model, genetic stability has been demonstrated in an animal-plant relationship when six resistant factors (genes) were present in the plant population.[31]

Another factor contributing to pest outbreaks is plant spacing. In cultivated fields, crop plant densities are carefully controlled to obtain the maximum population possible for optimal growth, resulting in maximum economic yield. Seldom are the spacings of such plants similar to those in the wild. The new plant spacings often result in an ecological situation that encourages pest outbreaks.[32]

All plant-feeding insects have specific nutrient requirements. Altering the nutrient level in the soil and then in the host plant influences the pests that are feeding on the plant. An improvement in nutrients often results in the parasite increasing, and a decline in the nutrients results in the reverse. For example, Haseman[33] reported that the grain aphid, feeding on small grain plants with high nitrogen, produced an average of 33 progeny per aphid, whereas on plants with a low level of nitrogen progeny production averaged only 13 per aphid.

Another factor contributing to pest outbreaks is the particular host plant associations that are employed by growers. Some pests, for example, can attack and feed on several species of host plants. These pests can move from one host plant to another when one of the host plant populations declines in abundance for some reason. For example, plant bugs feed on alfalfa and cotton, and when alfalfa is mowed for hay and eliminated as a food source the bugs will move onto cotton in large numbers.[34]

Some pesticides may alter the physiology of crop plants and therefore make them more susceptible to pest attack. For example, herbicides have been found to increase insect pest and pathogen problems associated with corn.[35]

Knowledge of the ecological causes of pest outbreaks is an essential component of pest management strategies that are briefly discussed next.

Pest Management Strategies

In developing any strategy for a pest management program, information on the following components is essential: (1) The ecological basis for the pest problem, (2) monitoring of the pest and natural enemy populations, and (3) an analysis of the benefits and costs of the pest control technique. Knowledge of the ecological basis of the pest problem will suggest ways and means of altering the crop or livestock environment to reduce the attack of the pest. Such environmental manipulations have several advantages and should be the prime means of attacking the pest problem.

Not all pest problems, however, will be solved by manipulating the environment of the pest. The second line of defense is pesticides. When pesticides are needed then integrated pest control technology (a combination of pesticides and natural enemies) should be employed.

Successful integrated pest control or management depends upon monitoring the pest populations and their predator and parasite populations. With adequate information concerning beneficial and pest populations the pest control specialist can determine the best pesticide to use and when to apply it for maximum effectiveness.

Limiting pesticide applications to a minimum has environmental and public health advantages while at the same time being important to the farmer. First of all, reducing pesticide use reduces crop production costs. Second, and equally important, is the fact that reducing pesticide use lessens the chance for pesticide resistance to develop in pest populations. This then extends the useful life of the pesticide and reduces the costs of pest control to society.

The final yet most important step in developing any pest management strategy entails a careful benefit analysis of the technique, including measuring the environmental and social costs of the particular technique. This is essential if the implementation of any pest control program is to benefit agriculture and society as a whole.

The specific aims of this Pest Management Handbook are to identify and describe the numerous pest control technologies that might be employed against a wide array of pests including insects, pathogens, weeds, mammals, and birds; plus analyze the environmental and social implications of some of these technologies.

REFERENCES

1. **Cramer, H. H.,** Plant protection and world crop production, *Pflanzenschutznachrichten,* 20, 1, 1967.
2. **Oka, I. N.,** personal communication, 1975.
3. **Robinson, R. A.,** Crop resistance may be our best crop protection, *World Crops Livestock,* 29, 104, 1977.
4. National Academy of Sciences, *Postharvest Food Losses in Developing Countries,* Board on Science and Technology for International Development, National Academy of Sciences, Washington, D.C., 1978.
5. U.S. Department of Agriculture, Losses in Agriculture, U.S. Dep. Agric. Agric. Handb. 291, 1965.
6. **Zimdahl, R. L.,** Extent of mechanical, cultural and other nonchemical methods of weed control, in *Pest Management,* CRC Handbook Series in Agriculture, CRC Press, Boca Raton, Florida, 1980.
7. President's Scientific Advisory Committee, Restoring the Quality of our Environment, Rep. Environmental Pollution Panel, President's Scientific Advisory Committee, Washington, D.C., 1965.
8. **Pimentel, D., Ed.,** *CRC Handbook of Pest Management,* CRC Press, Boca Raton, Florida, 1981.
9. United States Agency for International Development, Environmental Impact Statement on the AID Pest Management Program, Vols. 1 and 2, United States Agency for International Development, Department of State, Washington, D.C., 1977.
10. **Berry, J. H.,** Pesticides and energy utilization, in *Pesticides: Role in Agriculture, Health, and Environment,* Sheets, T. J. and Pimentel, D., Eds., Humana Press, Clifton, N.J., 1979, chap. 4.
11. National Academy of Sciences, *Population and Food,* Committee on World Food, Health and Population, National Academy of Sciences, Washington, D.C., 1975.
12. National Academy of Sciences, *Rapid Population Growth,* Vols. 1 and 2, National Academy of Sciences, Johns Hopkins Press, Baltimore, 1971.
13. **Gulhati, K.,** Compulsory sterilization: the change in India's population policy, *Science,* 195, 1300, 1977.
14. U.S. Department of Agriculture, *Agricultural Statistics 1977,* U.S. Government Printing Office, Washington, D.C., 1977.
15. Food and Agriculture Organization, *Production Yearbook 1972,* Vol. 26, Food and Agriculture Organization, Rome, 1973.
16. **Pimentel, D., Dritschilo, W., Krummel, J., and Kutzman, J.,** Energy and land constraints in food-protein production, *Science,* 190, 754, 1975.
17. **Walker, J. C., Larson, R. H., and Taylor, A. L.,** Diseases of cabbage and related plants, *U.S. Dep. Agric. Agric. Handb.* No. 144, 1958.
18. National Academy of Sciences, *Pest Control: An Assessment of Present and Alternative Technologies,* Vol. 2, National Academy of Sciences, Washington, D.C., 1975.
19. **Hawkes, J. G.,** *Potato-Collecting Expeditions in Mexico and South America,* School of Agriculture, Cambridge, England, 1944.
20. **Elton, C. S.,** *The Ecology of Invasions by Animals and Plants,* Methuen, London, 1958.

21. **Lupton, F. G. H.,** The plant breeders' contribution to the origin and solution of pest and disease problems, in *Origins of Pest, Parasite, Disease and Weed Problems,* Cherrett, J. M. and Sagar, G. R., Eds., Blackwell Scientific Publications, Oxford, 1977.
22. **Dahms, R. G.,** Effect of different varieties and ages of sorghum on the biology of the chinch bug, *J. Agric. Res.,* 76, 271, 1948.
23. **Marchal, P.,** The utilization of auxiliary entomophagous insects in the struggle against insects injurious to agriculture, *Pop. Sci. Monthly,* 72, 352, 1908.
24. **Pimentel, D.,** Animal population regulation by the genetic feedback mechanism, *Am. Nat.,* 95, 79, 1961.
25. **Tahvanainen, J. O. and Root, R. B.,** The influence of vegetational diversity on the population ecology of a specialized herbivore, *Phyllotreta cruciferae* (Coleoptera: Chrysomelidae), *Oecologia (Berl.),* 10, 321, 1972.
26. **Cromartie, W. J., Jr.,** The effect of stand size and vegetational background in the colonization of cruciferous plants by herbivorous insects, *J. Appl. Ecol.,* 12, 517, 1975.
27. **Root, R. B.,** Organization of a plant-arthropod association in simple and diverse habitats: the fauna of collards *(B. oleracea), Ecol. Monogr.,* 43, 95, 1973.
28. **Root, R. B.,** Some consequences of ecosystem texture, in *Ecosystem Analysis and Prediction,* Levin, S. A., Ed., Society for Industrial and Applied Mathematics, Philadelphia, 83, 1975.
29. **Stevens, N. E. and Scott, W. O.,** How long will present spring oat varieties last in the central corn belt?, *Agron. J.,* 42, 307, 1950.
30. **van der Plank, J. E.,** *Disease Resistance in Plants,* Academic Press, New York, 1968.
31. **Pimentel, D. and Bellotti, A. C.,** Parasite-host population systems and genetic stability, *Am. Nat.,* 110, 877, 1976.
32. **Pimentel, D.,** Species diversity and insect population outbreaks, *Entomol. Soc. Am.,* 54, 76, 1961.
33. **Haseman, L.,** Influence of soil mineral on insects, *J. Econ. Entomol.,* 39, 8, 1946.
34. **Stern, V. M.,** Interplanting alfalfa in cotton to control lygus bugs and other insect pests, *Proc. Tall Timbers Conf. Ecol. Anim. Contr. Habit. Mgmt.,* 1, 55, 1969.
35. **Oka, I. N. and Pimentel, D.,** Herbicide (2,4-D) increases insect and pathogen pests on corn, *Science,* 193, 239, 1976.

Estimated Losses of Crops and Livestock to Pests

ESTIMATE OF LOSSES CAUSED BY INSECTS AND MITES TO AGRICULTURAL CROPS

P. H. Schwartz and W. Klassen

INTRODUCTION

It is generally considered that there are more than 10,000 species of pest insects that cause losses. About 600 species are serious enough to warrant control measures each year, and which, together with diseases, nematodes, and weeds, cause losses that amount to about 33% of potential agricultural production each year.[227] Approximately 10% to 12% of the 33% losses are attributed to insect and mite pests. The authors' objective was to bring together the information available in the literature on losses caused by insects and mites (Table 1). This information was limited to losses caused by either yield or quality and the cost of controlling the pest. This paper will not deal with losses to society in general or losses caused by pest control because of adverse environmental impacts.

Several studies have estimated the losses caused by insects and mites to agricultural commodities in the U.S. The earliest of these was Marlatt.[176] The U.S. Department of Agriculture, at various times, has developed loss estimates.[10,11,12,146,274-6] Other scientists have developed estimates of losses to crops by insects and mites.[50,87,157,186,208] There have also been a number of studies concerning losses to individual crops or by pests or groups of pests reported in the literature. For use in our calculations we have defined loss as the difference in yield resulting from a pest infestation as measured by insecticide-treated and untreated plot studies. Where we cite estimates of losses by other workers, the loss data may include a reduction in yield or quality or both, unless stated otherwise.

Surveys concerning the use of insecticides on various agricultural commodities have been conducted by the Economics, Statistics, and Cooperative Service (ESCS) of the U.S. Department of Agriculture for the following years: 1964,[7] 1966,[83] 1971,[84] and 1976.[85] These surveys are a continuing function of that agency and appear periodically. The Census of Agriculture also develops information concerning the amount and use of insecticides for various commodities and the cost of chemicals for control of insects on commodities for various types of farms. The latest census was taken in 1974.[270]

METHODS

The *Journal of Economic Entomology* from the years 1942 to 1978 and the Insecticide and Acaricide Tests published by the Entomological Society of America for the years 1976, 1977, and 1978 were reviewed for studies containing yield data on all commodities except cotton, which was done for 1965 and subsequent years. In this study the authors utilized the experiments where yield data were taken in conjunction with pest density data and in which an untreated control was included in tests to evaluate the effects of insecticides and acaricides.

Various investigators have used different approaches to measuring the densities of a given pest. Also, they often used different units in measuring yield of a given commodity. Therefore, a common basis was needed for the data in all studies involving a given pest on a given commodity. This was accomplished by the equation $I = 100T/U$ where I is equal to the index number for the yield or pest density; T is equal to the yield or pest density in the treated plot, and U is equal to the yield or pest density in

Table 1
LOSSES OF VARIOUS AGRICULTURAL COMMODITIES CAUSED BY VARIOUS ARTHROPOD PESTS[a]

	Calculated yield loss (%)		Estimated loss by various workers (%)						
Crop/pest	Without control	With control	Marlatt 1904	Hyslop 1938	USDA 1946	USDA 1954	USDA 1965	Metcalf, et al. 1962	ARS 1976(a,b,c)
Alfalfa (hay)							15.0		
Alfalfa caterpillar				1—2			0.1		
Alfalfa weevil	27.0 ± 22.1	0.53 ± 0.2		5—50			2.5		
Aphids	48.8 ± 27.0	1.2 ± 0.8			5.0	4.1	2.4—4.9		
Armyworm									
Clover root curculio	6.0 ± 6.0	2.5 ± 2.5							
Cutworm									
Grasshoppers							1.0		
Leafhoppers	26.3 ± 15.5	2.6 ± 1.9					3.1		
Meadow spittlebug	8	1				5.2	0.8		
Mirid bugs	19.5 ± 12.5	4.5 ± 0.5							
Velvet bean caterpillar									
Webworm							0.2		
Alfalfa (seed)							38.0	10.0	
Alfalfa seed chalcid							10.5		
Alfalfa plant bug							2.1		
Aphids							0.3		
Grasshoppers							0.7		
Leafhoppers	14.0 ± 8.6	6.8 ± 4.3							
Leafminers	4.2 ± 2.4	1.4 ± 1.4							
Lygus bugs	43.5 ± 8.5	3.5 ± 0.5				35.0	24.4		
Mirids	0	0							
Stinkbugs									
Thrips	38	12							
Apples							13.0	20.0	23.0
Aphids			5 (wooly)						
Apple and thorn skeletonizer									1

Apple maggot						
Apple red bug					3.0	2
Bagworm						
Borer — round headed apple tree						
Borer — shot hole						
Cankerworms						
Casebearers						
Codling moth	100	16	12.0	15.0	11.0	1.0
Cutworms — climbing						
Eye-spotted budmoth						
Fall webworm						
Fruitworms						
Grasshoppers						
Japanese beetle			43.0			
Leafhoppers						
Leafroller — fruit tree						
Leafroller — red-banded	0	0				
Mites — apple rust						
Mites — brown						
Mites — European red						
Mites — two spotted						
Mites — McDaniel						
Mites — pear leaf blister						
Periodical cicadas						
Plum curculio			3.0		1.2	
Scales — Forbes						
Scales — Putnam						

[a] All arthropod species are included for which the U.S. Department of Agriculture has issued control guidelines.

Table 1 (continued)
LOSSES OF VARIOUS AGRICULTURAL COMMODITIES CAUSED BY VARIOUS ANTHROPOD PESTS[a]

Crop pest	Calculated yield loss (%)		Estimated loss by various workers						
	Without control	With control	Marlatt 1904	Hyslop 1938	USDA 1946	USDA 1954	USDA 1965	Metcalf, et al. 1962	ARS 1976 (a,b,c)
Apples (cont.)									
Scales — oystershell									
Scales — San Jose									
Scales — scurfy									1
Tent caterpillars									
Treehoppers									
Tentiform leafminers									
Yellow-necked and red humped caterpillars									
Asparagus									
Asparagus beetle									
Cutworms									
Thrips									
Wireworms									
Barley — (see also wheat)							5.0	8.0	
Brown wheat mite							2.8		
Cutworms							1.9		
Green bug	84	7					0.3		
Beans (lima and snap)									
Aphids									
Bean leaf beetle									
Bean thrips	0	0							
Corn earworm	37.0 ± 31.7	6.0 ± 3.5							
Cowpea cuculio									
Cucumber beetle									
Cutworms									
European corn borer	5	0							
Flea beetle									

Garden symphylan						
Japanese beetle						
Leafhoppers	3.0 ± 1.0	1 ± 0				
Lesser cornstalk borer						
Lima bean pod borer			4.0			
Lygus bugs	0 ± 0	0 ± 0				
Mexican bean beetle (lima)	97.5 ± 2.5	5.0 ± 3.0	10	9	8.0	
Mexican bean beetle (snap)	20.0 ± 20.0	9.9 ± 9.4	10	9	7.0	
Potato leaf beetle						
Potato leafhopper	41.0 ± 33.0	0 ± 0				
Root maggots						
Salt-marsh caterpillars						
Spider mites	23.5 ± 8.8	1.5 ± 1.0			3.0	
Stink bugs						
Western bean cutworm						
Wireworms	47.0 ± 39.0	3.0 ± 0.0				
Tarnished plant bug	0	0				
Beet, table						
Aphids						
Beet webworm						
Blister beetle						
Cutworms						
Flea beetles						
Leafminers						
Wireworms						
Beet, sugar						
Alfalfa looper						
Aphids	0	0				3.0
Armyworm						2
Beet leafhopper			6—25			4.0
Blister beetles						
Cutworms						
Flea beetles						1

Table 1 (continued)
LOSSES OF VARIOUS AGRICULTURAL COMMODITIES CAUSED BY VARIOUS ANTHROPOD PESTS[a]

Crop pest	Calculated yield loss (%)		Estimated loss by various workers						
	Without control	With control	Marlatt 1904	Hyslop 1938	USDA 1946	USDA 1954	USDA 1965	Metcalf, et al. 1962	ARS 1976 (a,b,c)
Beet, sugar (cont.)									
Garden symphylan									
Grasshoppers									
Leafminers									
Lygus bugs									1
Root maggot	22.7 ± 6.8	8.2 ± 2.9							2
Spider mites									1
Stink bugs									
Webworms	0	0							
Wireworms				8.0					
Blackberry							31.0		31.0
Aphids									
Japanese beetle									
Orange tortrix									
Raspberry crown borer									
Raspberry fruitworms				8.0					
Raspberry sawfly									
Redberry mite									
Rednecked cane borer									
Rose chafer									
Rose leafhopper									
Rose scale									
Spider mites									
Stink bugs									
Strawberry weevil									
White grubs									
Blueberry									
Blueberry maggot									
Fruit worms									

Plum curculio		
Broccoli		
Aphids		
Cabbage looper		
Caterpillars		
Cutworms		
Flea beetles		
Harlequin bug		
Imported cabbageworm		
Mole crickets		
Root maggots		
Stink bugs		
Vegetable weeviles		
Wireworms		
Brussels sprouts		
Aphids		
Cabbage looper		
Caterpillars		
Cutworms		
Flea beetles		
Harlequin bug		
Imported cabbageworm		
Mole crickets		
Root maggots		
Stink bugs		
Vegetable weevils		
Wireworms		
Cabbage		
Aphids		
Cabbage looper	9.5 ± 5.7	0.8 ± 0.8
Caterpillars		
Cutworms		
Flea beetles		
Harlequin bug		
Imported cabbageworm	41.0 ± 41.0	0 ± 0
Mole crickets		

Table 1 (continued)
LOSSES OF VARIOUS AGRICULTURAL COMMODITIES CAUSED BY VARIOUS ANTHROPOD PESTS[a]

Crop pest	Calculated yield loss (%)		Estimated loss by various workers						
	Without control	With control	Marlatt 1904	Hyslop 1938	USDA 1946	USDA 1954	USDA 1965	Metcalf, et al. 1962	ARS 1976 (a,b,c)
Root maggots									
Stink bugs									
Vegetable weevils									
Wireworms									
Cantaloupe									
Aphids									
Beet leafhopper									
Cucumber beetles									
Leafhoppers									
Leafminers									
Pickleworm									
Spider mites									
Thrips									
Wireworms									
Carrot									
Carrot rust fly	97	2							
Leafhoppers									
Vegetable weevils									
Wireworms				20					
Cauliflower									
Aphids									
Cabbage looper									
Caterpillars									
Cutworms									
Flea beetles									
Harlequin bug									
Imported cabbageworm									

Mole crickets					
Root maggots					
Stink bugs					
Vegetable weevils					
Wireworms					
Celery					
Aphids			5.0		
Celery leaf tier			2.0		
Cutworms (soil)			1.0		
Leafminers					
Loopers			2.0		
Lygus bugs					
Spider mites			1.0		
Wireworms					
Cherries			3.0	10.0	3.0
Black cherry aphid					
Borer, lesser peach tree					
peach tree					
shot-hole					
Cherry fruitworm					
Fruit flies					
Japanese beetle					
Mites, European red					
McDaniel					
plum rust					
two-spotted spider					
Pear slug					
Pear thrip					
Periodical cicadas					
Plum curculio					
Rose chafer					
Scales, Forbes					
San Jose					
Cole crops					
Aphids	56.0 ± 23.0	1.5 ± 0.5			
Cabbage looper	0	0			
Caterpillars	32.0 ± 19.0	11.5 ± 11.5			

Table 1 (continued)
LOSSES OF VARIOUS AGRICULTURAL COMMODITIES CAUSED BY VARIOUS ANTHROPOD PESTS[a]

	Calculated yield loss (%)		Estimated loss by various workers						
Crop pest	Without control	With control	Marlatt 1904	Hyslop 1938	USDA 1946	USDA 1954	USDA 1965	Metcalf, et al. 1962	ARS 1976 (a,b,c)
Cole crops (cont.)									
Cutworms (soil)									
Diamondback moth	0.5 ± 0.5	0.5 ± 0.5							
Flea beetles									
Harlequin bug									
Loopers	25	3							
Mole crickets									
Root maggots									
Vegetable weevils									
Wireworms									
Collards									
Aphids									
Cabbage looper	49	0							
Caterpillars									
Cutworms									
Flea beetles									
Harlequin bug									
Imported cabbage worm									
Mole crickets									
Root maggots									
Stink bugs									
Vegetable weevils									
Webworm, cabbage	21	0							
Wireworms									
Corn, field							12.0	9.0	
Armyworm									
Chinch bug			2.0				0.2		
Corn earworm			2.0	1—6.0	4.0	1.2	4.0		2.5
Cornfield ant	20	3							

Corn leaf aphid							0.4		0.2
Corn rootworms	15.7 ± 4.6	5.0 ± 1.3	2.0				2.1		2.0
Cutworms									0.2
European corn borer						1.9	3.5		2.0
Fall armyworm							0.5		2.0
Flea beetles									
Garden symphalid	14	7							
Grasshoppers							0.6		
Japanese beetle				35—80					
Leafhoppers	74.7 ± 15.7	38.3 ± 14.8							
Mormon cricket									
Southwestern corn borer	34.4 ± 10.4	9.9 ± 2.7				0.2	0.2		1.0
Spider mites									
Sugarcane beetle									
Sugarcane borer	0	0		20					
White fringed beetle									1.0
White grubs	43	11							
Wireworms	48.3	18.4		16					0.2
Corn, sweet									
Corn earworm				10—50			13.8		
Corn sapbeetle									
Cutworms	22	7							
European corn borer	4.3 ± 4.3	0.3 ± 0.3					4.1		
Fall armyworm	67.5 ± 1.5	27.0 ± 1.0							
Japanese beetle				35.0					
Rootworms	45.5 ± 45.5	0 ± 0							
Spider mites	0	0							
Wireworms	29.0 ± 19.0	4.0 ± 2.0							
Cotton							19.0	15.0	19.0
Aphids	22.3 ± 22.3	0.3 ± 0.3						0.05[a]	
Armyworm, beet									
Bandwing white fly								0.18[a]	
Boll weevil	30.9 ± 8.6	19 ± 6.7	6.7	10.9	8.7	10.1	8.0	2.2[a]	8.0
Cabbage looper	35	20							
Cotton aphid	4.8 ± 4.8	0 ± 0							
Cotton flea hopper	39.0 ± 11.6	12.5 ± 3.9		2—50				0.2	

Table 1 (continued)
LOSSES OF VARIOUS AGRICULTURAL COMMODITIES CAUSED BY VARIOUS ANTHROPOD PESTS[a]

	Calculated yield loss (%)		Estimated loss by various workers						
Crop pest	Without control	With control	Marlatt 1904	Hyslop 1938	USDA 1946	USDA 1954	USDA 1965	Metcalf, et al. 1962	ARS 1976 (a,b,c)
Cotton (cont.)									
Cotton leaf perforator	0	0							
Cotton leafworm									
Cutworms									
Darkling ground beetle									
Fall armyworm									
Garden webworm									
Grasshoppers									
Heliothis	90.8 ± 14.2	12.1 ± 2.2	4.0	2.0			4.0	3.3[b]	4.0
Lygus bugs								0.67[b,c]	
Pink bollworms	35.5 ± 13.5	10.0 ± 6						0.32[b]	
Salt-marsh caterpillar									
Spider mites	0 ± 0	0 ± 0						0.2[b]	
Stink bugs				7.5				0.67[b,c]	
Thrips	67.8 ± 22.4	18.0 ± 6.1		13				0.2[b]	
Yellow-striped armyworm									
Cucumber									
Aphids				3.0					
Armyworms									
Cabbage looper									
Cucumber beetle				9.0					
Earwigs									

[b] Based on bale losses in cotton-producing states reporting data, 1974—76 acreage. Source: The Cotton Foundation, Dec. 1977.
[c] All plant bugs including lygus.

Garden flea hopper					
Garden symphylan					
Greenhouse whitefly					
Leafminers					
Mealy bugs					
Mushroom mite					
Pickleworm	25.0				
Spider mites					
Thrips					
Wireworms					
Currant and Gooseberry					
Currant aphid					
Gooseberry fruitworm					
Imported currant worm					
San Jose scale					
Spider mites					
Eggplant					
Aphids					
Colorado potato beetle					
Cutworms					
Eggplant lace bug					
Flea beetles					
Hornworms					
Spider mites					
Whiteflies					
Wireworms					
Fig					
Grape			4.0	15.0	20.0
Climbing cutworms					
European fruit lecanium					
Grape berrymoth		8.7			

Table 1 (continued)
LOSSES OF VARIOUS AGRICULTURAL COMMODITIES CAUSED BY VARIOUS ANTHROPOD PESTS[a]

Crop pest	Calculated yield loss (%)		Estimated loss by various workers						
	Without control	With control	Marlatt 1904	Hyslop 1938	USDA 1946	USDA 1954	USDA 1965	Metcalf, et al. 1962	ARS 1976 (a,b,c)
Grape cane gallmaker									
Grape flea beetle									
Grape leaf folder									
Grape leafhopper									
Grape mealy bug									
Grape phylloxera	6.5 ± 2.5	1.5 ± 0.5							
Grape rootworm									
Grapevine aphid									
Japanese beetle									
Red-banded leafroller									
Rose chafer									
Two-spotted spider mite									
Grapefruit (citrus)							5.0	10.0	
Aphids									
Citrus thrips									
Mite, citrus red							2.4 (all)		
Texas citrus									
six-spotted									
false spider									
citrus rust									
Scales, black									
brown soft							2.4 (all)		
California red									
purple									
yellow									
Glover									
Florida red									

chaff			
citrus snow			
Whiteflies			
Guar			33.0
Alfalfa stem borer			1.0
Gall midge			5.0
Guar midge			25.0
Three-cornered alfalfa hopper			1.0
Hops	15.0	12.0	
Aphids	4.0		
Spider mites	6.0		
Kale			
Aphids			
Cabbage looper			
Caterpillars			
Cutworms			
Flea beetles			
Harlequin bug			
Imported cabbageworm			
Mole crickets			
Root maggots			
Stink bugs			
Vegetable weevils			
Wireworms			
Kohlrabi			
Aphids			
Armyworms			
Cabbage looper			
Cutworms			
Flea beetles			
Harlequin bug			
Imported cabbageworm			
Root maggots			
Stink bugs			
Lemons	6.0	10.0	
Mites	2.4		
Scale insects	2.0		

Table 1 (continued)
LOSSES OF VARIOUS AGRICULTURAL COMMODITIES CAUSED BY VARIOUS ANTHROPOD PESTS[a]

Crop pest	Calculated yield loss (%)		Estimated loss by various workers						
	Without control	With control	Marlatt 1904	Hyslop 1938	USDA 1946	USDA 1954	USDA 1965	Metcalf, et al. 1962	ARS 1976 (a,b,c)
Lettuce									
Aphids							0.8		
Armyworm									
Cabbage looper	81	2		10.0			3.9		
Corn earworm							2.3		
Cutworms									
Grasshoppers									
Greenhouse whitefly									
Imported cabbageworm									
Leafhoppers									
Mole crickets									
Sow bugs									
Wireworms				10.0					
Mint							15.0		
Aphids									
Caterpillars							4.7		
Flea beetles							4.7		
Garden symphylan									
Loopers									
Root weevils									
Spider mites							7.0		
Mustard greens									
Aphids									
Cabbage looper									
Caterpillars									
Flea beetles									
Harlequin bug									

Imported cabbageworm							
Stink bugs							
Vegetable weevils							
Oats						4.0	5.0
Brown wheat mite						0.3	
Cutworms						1.4	
Green bug				0.6		2.3	
Okra							
Aphids							
Blister beetles							
Corn earworm							
Cucumber beetles							
Flea beetles							
Japanese beetles							
Leafminers							
Stink bugs							
Wireworms							
Onion							
Leafminers							
Onion maggot	31	15				8.0	
Onion thrips	54.7 ± 11.8	7.5 ± 3.6			17.0	10.0	
Wireworms			12.0				
Oranges						6.0	10.0
Mites						2.7	
Scale insects						2.5	
Pasture and Range							
Armyworm							
Capsids	26	3					
Fall armyworm	0	0					
Grasshoppers			20—75				
Leafhoppers	100	33					
Mormon cricket							
Range caterpillar							
Peas, garden							
Alfalfa looper						0.4	
Aphids	48.7 ± 22.3	4.0 ± 2.3	5—40			3.6	
Celery looper							

Table 1 (continued)
LOSSES OF VARIOUS AGRICULTURAL COMMODITIES CAUSED BY VARIOUS ANTHROPOD PESTS[a]

Crop pest	Calculated yield loss (%)		Estimated loss by various workers						
	Without control	With control	Marlatt 1904	Hyslop 1938	USDA 1946	USDA 1954	USDA 1965	Metcalf, et al. 1962	ARS 1976 (a,b,c)
Pea weevil				6.0					
Wireworms									
Peas, southern									
Aphids									
Beet armyworm	54	18							
Cowpea curculio									
Lygus bugs									
Pea moth	9.0 ± 9.0	2.5 ± 2.5							
Thrips	58	20							
Wireworms									
Peaches							4.0	20.0	4
Aphids									
Borer, American plum									
lesser peach tree									
peach tree							0.7		
peach twig									
shot-hole									
Climbing cutworms									
Earwigs									
Grasshoppers									
Green June beetles									
Japanese beetles				27.0					
Leaf rollers									
Mites									
Oriental fruit moth						4.0			
Pandemis moth									
Periodical cicada									
Plum curculio				15.0		5.0			

Rose chafer					
Scales					
Stink bugs					
Tarnished plant bug					
Peanut			3.0	3.0	3.0
Corn earworm					0.4
Cornstalk borer	7	2			0.5
Cutworms					
Fall armyworm					0.3
Leafhoppers					
Peanut worm	0	0			
Potato leafhopper					
Southern corn root-worm	15.5 ± 1.5	1.0 ± 0	0.8		0.8
Thrips	7.8 ± 6.3	0.4 ± 0.2			
Velvet bean caterpil-lar					
White-fringed beetle					
Pear			6.0	8.0	23.0
Aphids					
Borer, shot-hole					
Codling moth					
Fruit tree leafroller					1
Fruitworms					
Mite, brown					
European red					2 (all)
McDaniel					
pear leaf blister					
pear rust					
Pear midge					
Pear psylla					
Pear slug					1
Pear thrips					
Periodical cicadas					
Plum curculio					
Scales, Forbes					
San Jose					
Scurfy					

Table 1 (continued)
LOSSES OF VARIOUS AGRICULTURAL COMMODITIES CAUSED BY VARIOUS ANTHROPOD PESTS[a]

Crop pest	Calculated yield loss (%)		Estimated loss by various workers						
	Without control	With control	Marlatt 1904	Hyslop 1938	USDA 1946	USDA 1954	USDA 1965	Metcalf, et al. 1962	ARS 1976 (a,b,c)
Treehoppers									
Peppers									
Aphids	24	0							
Corn earworm	48.5 ± 48.5	0.5 ± 0.5					4.1		
Cutworms									
European corn borer	54.5 ± 19.9	0.0 ± 0.0							
Fall armyworm	0 ± 0	0 ± 0							
Flea beetle									
Hornworms									
Leafminers									
Pepper maggot									
Pepper weevil	37	4		33.3			0.8		
Spider mites									
Wireworms									
Plums and Prunes								8.0	6.0
Aphids									
Borer, American									
plum									
lesser peach tree									
peach tree							0.7		
peach twig									
shot-hole									
Eye-spotted bud moth									
Mite, European red									
McDaniel									
plum rust									
two-spotted spider									

Plum curculio				
Scales, Forbes				
oystershell				
San Jose				
Potatoes				
Aphids	14.5 ± 8.1	0.6 ± 0.4		
Armyworms	59.0 ± 17.0	1.3 ± 1.0		
Blister beetles				
Colorado potato beetles	46.6 ± 10.4	1.0 ± 0.6	8.0	
Cucumber beetles				
Cutworms				
European corn borer	54.2 ± 26.6	1.5 ± 0.6		
False chinch bug				
Flea beetles	43.3 ± 10.3	0.7 ± 0.4		
Garden symphylan				
Grasshoppers				
Green peach aphid	3.7 ± 3.7	0 ± 0		
Leafhoppers	36.0 ± 26.6	0 ± 0		
Leafminers				
Millipedes				
Mole crickets				
Plant bugs				
Potato aphid	5	0		
Potato leafhopper	43.2 ± 18.6	0.4 ± 0.2	14.0	
Potato psyllid	32.0 ± 17.2	1.0 ± 1.0	0.6	25—30
Potato tuberworm	91	1		
Slugs				
Wireworms			20.0	
Spider mites	5	4		
Three-lined potato beetle				
Thrips				
Vegetable weevils				
Whiteflies				
White fringed beetles				
White grubs				
Wireworms	4.9 ± 3.5	0.2 ± 0.2		

Table 1 (continued)
LOSSES OF VARIOUS AGRICULTURAL COMMODITIES CAUSED BY VARIOUS ANTHROPOD PESTS[a]

Crop pest	Calculated yield loss (%)		Estimated loss by various workers						
	Without control	With control	Marlatt 1904	Hyslop 1938	USDA 1946	USDA 1954	USDA 1965	Metcalf, et al. 1962	ARS 1976 (a,b,c)
Pumpkin									
Aphids									
Cucumber beetles									
Cutworms									
Squash bug									
Squash vine borer									
Radish									
Root maggots									
Rape (seed)									
Cabbage aphids	100	0							
Raspberry							23.0		23.0
Aphids									
Japanese beetle									
Oblique-banded leafroller									
Orange tortrix									
Raspberry crown boror									
Raspberry fruitworm									
Raspberry sawfly									
Rednecked cane borer									
Rose chafer									
Rose scale									
Spider mites									
Stink bugs									
Strawberry weevil									
White grubs									
Rice							4.0	8.0	
Grape colaspis									
Leafhoppers	42	9							

Rice stink bug				3.1		
Rice water weevil	5.2 ± 4.6	1.5 ± 1.5		1.0		
Sugarcane borer			2—6			
Rutabaga						
Aphids						
Cabbage looper						
Cutworms						
Flea beetles						
Wireworms						
Safflower						
Lygus	21	3				
Thrips	16 ± 16	3.5 ± 3.5				
Sorghum				9.0	7.0	9.0
Armyworm				0.5		
Chinchbug	0	0		1.3		
Corn earworm				4.1		
Cornleaf aphid	0 ± 0	0 ± 0		0.4		
Cutworm				0.5		
Fall armyworm						
Green bug	19.4 ± 7.2					
Mites	0 ± 0	0 ± 0				
Sorghum midge	5.0 ± 5.0	0.2 ± 0.2	1.0	1.1		
Sorghum webworm	3	0		1.4		
Southwestern corn borer	24	4		0.2		
Whitegrubs	39.0 ± 26.3	9.3 ± 4.7				
Soybean				3.0	5.0	3.0
Caterpillar complex	46	0				
Corn earworm	1.5 ± 0.5	0.6 ± 0.4				1.0
Green cloverworm	8.8 ± 6.6	3.0 ± 3.0				
Loopers	25.5 ± 5.5	10.5 ± 3.5				
Mexican bean beetle	26.0 ± 13.3	0.4 ± 0.3				
Mites	89	0				
Soybean looper	15.7 ± 7.1	4.8 ± 1.9				0.2
Stink bugs	15.0 ± 2.0	8.5 ± 7.5				
Velvetbean caterpillar	16.6 ± 3.7	2.4 ± 0.9				0.7

Table 1 (continued)
LOSSES OF VARIOUS AGRICULTURAL COMMODITIES CAUSED BY VARIOUS ANTHROPOD PESTS[a]

	Calculated yield loss (%)		Estimated loss by various workers						
Crop pest	Without control	With control	Marlatt 1904	Hyslop 1938	USDA 1946	USDA 1954	USDA 1965	Metcalf, et al. 1962	ARS 1976 (a,b,c)
Spinach									
Alfalfa looper									
Aphids									
Beet webworm									
Spinach leafminer									
Wireworms									
Squash									
Aphids									
Cucumber beetles									
Cutworms									
Leafminers									
Pickleworm									
Squash bug									
Squash vine borer	0	0							
Strawberry							25.0	10.0	25.0
Ants									
Aphids							5.0		
Cutworms									
Cyclamen mite							2.7		
Earwigs									
Field crickets									
Flea beetles									
Lygus bugs									
Mole crickets									
Omnivorous leaftier									
Potato leafhopper									
Spider mites	23.0 ± 8.0	2.0 ± 1.0					10.0		
Spittle bugs	0	0							

Strawberry crown borer							
Strawberry leaf beetle							
Strawberry leaf roller							
Strawberry root weevil							
Strawberry weevil			6.0				
Sugarcane beetle							
Thrips							
Whiteflies					5.0		
White grubs							
Wireworms							
Sugarbeets					12.0	15.0	15.0
Alfalfa looper							
Aphids							3.0
Armyworms							2
Beet leafhopper			6—25				
Blister beetles							4.0
Cutworms							1.0
Flea beetles							
Garden symphylan							
Grasshoppers							
Leafminers							
Lygus bugs							
Root aphids	0	0					1.0
Root maggots	22.7 ± 6.8	8.2 ± 2.9					
Spider mites							2.0
Stink bugs							1.0
Webworms	0	0					
Wireworms			8.0				
Sugarcane					15.0	20.0	10—15
Aphids					1.1		1.1
Armyworm							
Fall armyworm							
Sugarcane borer	28.6 ± 12.4	5.7 ± 2.5	10 15	9.4	12.1		8.0
White grubs							1.0
Wireworms							0.8
Sunflowers							8.1
Carrot beetle							0.2

Table 1 (continued)
LOSSES OF VARIOUS AGRICULTURAL COMMODITIES CAUSED BY VARIOUS ANTHROPOD PESTS[a]

Crop pest	Calculated yield loss (%)		Estimated loss by various workers						
	Without control	With control	Marlatt 1904	Hyslop 1938	USDA 1946	USDA 1954	USDA 1965	Metcalf, et al. 1962	ARS 1976 (a,b,c)
Girdlers									0.1
Seed midge									1.0
Seed weevils									0.2
Suleima									0.1
Sunflower beetle									0.5
Sunflower moth	40.3 ± 8.7	7.1 ± 2.1							5.0
Sweet potatoes									
Flea beetle	0	0							
Sweet potato weevil	29.7 ± 3.7	1.7 ± 1.2		13.0		1.7			
Wireworms	38	16							
Swiss chard									
Aphids									
Flea beetles									
Tobacco						11.0	11.0	10.0	2—11
Aphids	0.2 ± 0.2	0.3 ± 0.3							
Budworms	13.0 ± 6.7	1.3 ± 0.9		3.0					
Cutworms				2.0					
Flea beetle	0.24	0		2.0					
Grasshoppers									
Green June beetle									
Hornworms				5.0					
Loopers	10	4							
Mole crickets									
Tobacco suckfly									
Vegetable weevil									
Tomato									
Aphids	6.5 ± 4.5	0 ± 0							
Armyworm	0	0							

Beet leafhopper			7	
Blister beetle				
Cabbage looper	15	0		
Colorado potato beetle	93	0		
Cutworms	16	0.3		
Drosophila				
Flea beetles				
Garden symphylan				
Hornworms				
Leafminers	38	10		
Spider mites				
Stalk borer				
Tomato fruitworm			13.0	4.5
Tomato pinworm	50.0 ± 50.0	5.0 ± 5.0	20—25	
Tomato psyllid				4.0
Tomato russet mite				
White flies	38	12		
Wireworms				
Worms	0	0		
Turnips				
Aphids			40	
Cabbage looper				
Cabbage webworm				
Flea beetle				
Root maggots				
Vegetable weevils				
White-fringed beetle				
Wireworms				
Walnuts				
Curculio				
Fall webworm				
Walnut caterpillar				
Walnut husk maggot				
Walnut lace bug				
Watermelons				
Aphids				

Table 1 (continued)
LOSSES OF VARIOUS AGRICULTURAL COMMODITIES CAUSED BY VARIOUS ANTHROPOD PESTS[a]

	Calculated yield loss (%)		Estimated loss by various workers						
Crop pest	Without control	With control	Marlatt 1904	Hyslop 1938	USDA 1946	USDA 1954	USDA 1965	Metcalf, et al. 1962	ARS 1976 (a,b,c)
Watermelons (cont.)									
Cucumber beetle									
Cutworms									
Leafhoppers									
Leafminers									
Spider mites									
Thrips									
Wireworms									
Wheat and other small grains							6.0	9.0	
Armyworm									
Banks grass mite	61	18					0.6		
Brown wheat mite	100	21							
Cereal leaf beetle							0.5		
Corn leaf aphid									
Cutworms	54.7 ± 12.4	7.7 ± 3.2					0.9		
Fall armyworm									
Grasshoppers							0.4		
Green bugs	0	0	2—3			0.9	1.4		
Hessian fly	0	0	10—50	6—15	5—0	0.9	1.2		
Morman cricket									
Red harvester ant									
Wheat stem sawfly						0.2	1.0		
Winter grain mite									

the untreated plot. In using this equation, the untreated plot yield or pest population density would equal 100. If the yield in the treated plot is greater than in the untreated control, the yield index is greater than 100, and if the yield is less than the untreated control the index is less than 100. If the pest population densities in the treated plots are greater than in untreated plots, the pest population index is greater than 100. If the pest population densities in the treated plots are less than in the untreated plots, the pest population index is less than 100. The transformed data (i.e., indices) were then used in subsequent analyses to calculate loss estimates. The transformation did not affect the outcome of these estimates, but allowed direct comparision of analyzed data across commodities and pests (Table 2).

For the purposes of this study we assumed that a given pest population density would cause a corresponding loss in yield. We also assumed that adequately reliable loss estimates could be made by treating this relationship as though it is linear. Other workers have used this relationship.[20,124,129,195,218,251,264] Thus, as pest population densities increase, the losses in yield increase proportionately if the plant is attacked by the pest in a susceptible stage of growth. However, we recognize that under some circumstances there could be no cause and effect relationship or the relationship could be sigmoidal or curvilinear as indicated, for example, by Hartstack et al.,[122] for damage to cotton by *Heliothis*.

The relationship of pest density and yield or quality was calculated by means of the linear equation: $Y = Y_0 + kP$, where Y is the yield index for any pest population density index P. The regression coefficients were Y_0, which is the yield index where there is no pest infestation, and k reflects the change in Y per unit of P. Also the coefficient of determination (r^2) was calculated. The coefficients (Y_0 and k) were used to calculate the losses when no treatments were applied (the untreated infestation) and those losses incurred when the most effective treatment in the study was used as follows:

1. % loss without controls $= ((Y_0 - Y_{100})/Y_0) \times 100$
2. % loss with best treatment $= ((Y_0 - Y_x)/Y_0) \times 100$

where Y_0 is the yield index without any pest infesation; Y_{100} is the yield index when no control was applied (untreated check plot) and Y_x is the yield index for the most effective treatment.

When the computation of percent loss was less than 0, it was rounded off to 0%; when greater than 100, it was rounded off to 100%. A standard error of the mean was computed when two or more data sets were available for each crop/pest situation (Table 2).

A summary of the uses of insecticides on various commodities in the U.S. was developed from the survey information obtained by the Economics, Statistics, and Cooperative Service, U.S. Department of Agriculture (Table 3). These data were used to project estimates for 1980 and 1990 (Table 4). The estimates were developed by the regression equation: $Y = a + bx$, where a and b are constants, x is the year, and Y is the amount of insecticide (lb) or acre-treatments, or pounds per acre-treatment. The coefficient of determination (r^2) was also calculated. In those instances where negative values were obtained for Y they were rounded off to 0. Only data with three or more observations were used for each of the commodities to make the projections to 1980 and 1990. The projections for 1976 were made from 1964, 1966, and 1971 data, if available, to see how well the projection compared with the actual data for 1976.

The estimated losses caused by insects and mites (Table 5) were developed from the overall cost of pest control obtained by the State of California Department of Food

Table 2
SUMMARY OF DATA ON THE EFFECT OF INSECT DENSITY AND OF CONTROL MEASURES ON YIELDS OF VARIOUS COMMODITIES

Crop/pest	No. of data sets	Calculated losses (%) With best control	Calculated losses (%) Without control (untreated check)	Yield index (%) with no pest infestation Y_0	Slope k	Coefficient of determination r^2	Ref.
Alfalfa							
Alfalfa weevil	3	0.53 ± 0.24	27.0 ± 22.1	151.0 ± 32.7	−0.82 ± 0.43	0.32 ± 0.15	9, 88, 248
Aphids	4	1.25 ± 0.75	48.75 ± 27.0	119.76 ± 28.0	−1.25 ± 0.79	0.55 ± 0.21	91, 94, 268, 269
Clover root curculio	2	2.5 ± 2.5	6.0 ± 6.0	108.0 ± 1.0	−0.05 ± 0.09	0.19 ± 0.18	193
Leafhoppers	4	2.59 ± 1.86	26.3 ± 15.5	155.8 ± 48.2	−0.36 ± 0.25	0.10 ± 0.01	94, 102, 103, 184
Meadow spittlebug	1	1.0	8.0	111	−0.09	0.03	94
Mirid bugs	2	4.5 ± 0.5	19.5 ± 12.5	192.5 ± 72.5	−0.28 ± 0.10	0.12 ± 0.11	94, 184
Alfalfa (seed)							
Leafhoppers	5	6.80 ± 4.27	14.00 ± 8.58	127.60 ± 17.69	−0.20 ± 0.16	0.23 ± 0.10	34, 55, 143, 231
Leafminers	5	1.45 ± 1.39	4.2 ± 2.35	112 ± 4.51	0.06 ± 0.03	0.05 ± 0.03	93, 94, 102, 103, 143
Lygus bugs	2	3.5 ± 0.5	43.5 ± 8.50	323.5 ± 204.5	−1.57 ± 1.16	0.48 ± 0.10	55, 158
Mirids	1	0	0	186	.00026	0.000	183
Thrips	1	12	38	135	−0.51	0.77	55
Apples							
Codling moth	1	16	100	134	−5.32	0.79	114
Red-banded leafroller	2	0	0	84.5 ± 10.5	1.06	0.38 ± 0.01	153, 154
Barley							
Greenbug	1	7	84	1741.63	−14.55	0.04	69
Beans							
Bean thrips	2	0	0	85.5 ± 8.5	0.24 ± 0.16	0.24 ± 0.12	110
Corn earworm	3	6.0 ± 3.5	37.0 ± 31.7	446.3 ± 300.9	−6.01 ± 6.29	0.36 ± 0.19	169, 214, 236
European corn borer	1	0	5	102	−0.05	0.03	165
Leafhoppers	2	1 ± 0	3.0 ± 1.0	103 ± 1.0	−0.04 ± 0.01	0.03 ± 0.01	110

Lygus bugs	4	0 ± 0	0 ± 0	86.8 ± 3.86	0.33 ± 0.12	0.29 ± 0.14	109, 110, 245
Mexican bean beetle (lima)	2	5.0 ± 3.0	97.5 ± 2.50	164.50 ± 64.50	−2.13 ± 0.06	0.27 ± 0.16	73, 220
Mexican bean beetle (snap)	8	8.75 ± 5.86	29.00 ± 15.98	171.13 ± 26.02	−0.41 ± 0.85	0.32 ± 0.12	26, 72, 214, 236, 237, 294, 295
Potato leafhopper	2	0 ± 0	41.0 ± 33.0	100.5 ± 9.5	−0.38 ± 0.29	0.53 ± 0.15	82, 220
Spider mites	4	1.51 ± 0.95	23.5 ± 8.81	199.8 ± 63.0	−0.04 ± 0.77	0.29 ± 0.20	225, 236, 241, 287
Tarnished plant bugs	1	0	0	239	7.51	0.32	180
Wireworms	2	3 ± 0	47.0 ± 39.0	115.0 ± 8.0	−0.57 ± 0.49	0.48 ± 0.37	163, 244
Cabbage							
Cabbage looper	3	0.75 ± 0.75	9.50 ± 5.72	234.0 ± 50.38	0.12 ± 0.51	0.09 ± 0.03	147
Imported cabbageworm	2	0 ± 0	41.0 ± 41.0	297.50 ± 29.50	−0.96 ± 1.23	0.14 ± 0.04	147
Carrot							
Carrot rust fly	1	2	97	2761	−26.71	0.9	144
Celery							
Leafminers	1	0	0	84	0.17	0.07	62
Cole crops							
Aphids	2	1.5 ± 0.5	56.0 ± 23.0	288.5 ± 96.5	−1.84 ± 1.21	0.58 ± 0.29	80, 116
Cabbage looper	1	0	0	141	0.46	0.44	246
Caterpillars	7	11.5 ± 11.5	32.0 ± 19.0	383.0 ± 97.0	−1.41 ± 1.04	0.12 ± 0.10	106, 246
Diamond back moth	2	0.5 ± 0.5	0.5 ± 0.5	139.0 ± 16.0	−0.01 ± 0.01	0.04 ± 0.04	246
Looper and cabbageworm	1	3	25	132	−0.34	0.08	148
Collards							
Cabbage looper	1	0	49	664	−3.29	0.22	118
Cabbage webworm	1	0	21	650	−1.39	0.5	118
Corn, field							
Ant	1	3	20	138	−0.28	0.61	224
Corn rootworms	12	4.75 ± 1.23	16.17 ± 4.18	0.42 ± 10.58	−0.27 ± 0	0.25 ± 0.08	95, 117, 178, 192, 222
Garden symphylid	1	7	14	133	−0.19	0.31	223

Table 2 (continued)
SUMMARY OF DATA ON THE EFFECT OF INSECT DENSITY AND OF CONTROL MEASURES ON YIELDS OF VARIOUS COMMODITIES

Crop/pest	No. of data sets	Calculated losses (%)		Yield index (%) with no pest infestation	Slope	Coefficient of determination	Ref.
		With best control	Without control (untreated check)	Y_0	k	r^2	
Southwestern corn borer	11	9.91 ± 2.71	34.45 ± 10.40	146.64 ± 17.91	−0.48 ± 0.49	0.60 ± 0.10	125, 126, 152
Sugarcane borer	1	0	0	109	0.35	0.42	194
White grubs	1	11	43	149	−0.65	0.86	224
Wireworms	1	18.36	48.31	296.0	−1.43	0.34	76
Leafhopper on silage corn	3	38.33 ± 14.81	74.67 ± 15.72	308.67 ± 100.65	−2.47 ± 1.00	0.69 ± 0.17	3, 33
Corn, sweet							
Cutworms	1	7	22	132	−0.29	0.92	111
European corn-borer	3	0.33 ± 0.33	4.33 ± 4.33	107.33 ± 3.18	0.14 ± 0.19	0.42 ± 0.30	107, 247, 290
Fall armyworm	2	27.0 ± 1.0	67.5 ± 1.5	1130.0 ± 898.0	−7.81 ± 6.28	0.62 ± 0.06	1, 2
Rootworms	2	0 ± 0	45.5 ± 45.5	145.5 ± 26.5	−0.66 ± 0.92	0.21 ± 0.15	8, 78
Spider mites	1	0	0	95	0.10	0.03	159
Wireworms	2	4.0 ± 2.0	29.0 ± 19.0	137.0 ± 17.0	−0.36 ± 0.21	0.51 ± 0.50	151, 238
Cotton							
Aphids	3	0.33 ± 0.33	22.3 ± 22.3	133.7 ± 22.2	0.07 ± 0.68	0.36 ± 0.25	57, 58, 60
Boll weevil	17	19.00 ± 6.69	30.94 ± 8.63	176.42 ± 50.68	−0.40 ± 1.01	0.37 ± 0.08	49, 142, 181, 182, 205, 215, 249, 291
Cabbage looper	1	20	35	152	−0.53	0.63	144
Cotton aphid	4	0.0005 ± 0.0005	4.75 ± 4.75	121.0 ± 59.36	0.98 ± 0.80	0.25 ± 0.20	142, 215, 249
Cotton fleahopper	4	12.5 ± 3.88	39.00 ± 11.57	185.75 ± 30.1	−0.81 ± 0.37	0.59 ± 0.18	57, 58, 59, 60
Cotton leaf perforator	1	0	0	122	0.1	0.08	284

Heliothis	29	12.07 ± 2.17	90.82 ± 14.21	315.76 ± 55.15	−2.00 ± 0.58	0.59 ± 0.06	141, 156, 181, 182, 205, 206, 207, 211, 216, 228, 249
Pink bollworm	2	10.0 ± 6.0	35.5 ± 13.5	169.0 ± 30.0	−0.64 ± 0.34	0.50 ± 0.25	285, 291
Spider mites	2	0 ± 0	0 ± 0	84.0 ± 3.0	4.70 ± 4.10	0.60 ± 0.30	142
Thrips	4	18.01 ± 6.08	67.75 ± 22.40	167.25 ± 23.14	−1.29 ± 0.50	0.46 ± 0.21	57, 58, 142
Grapes							
Grape phylloxera	2	1.50 ± 0.50	6.50 ± 2.50	108.50 ± 4.50	−0.07 ± 0.02	0.04 ± 0.02	240
Lettuce							
Cabbage looper	1	2	81	107.56	−87.56	0.83	97
Oats							
Cereal leaf beetle	2	0.65 ± 0.35	26.00 ± 21.00	163.00 ± 35.00	−0.50 ± 0.43	0.52 ± 0.48	185, 289
Onion							
Onion maggot	1	15	31	164	−0.51	0.56	229
Onion thrips	6	7.50 ± 3.59	54.67 ± 11.76	822.83 ± 659.80	−19.43 ± 18.66	0.63 ± 0.14	42, 52, 79, 204, 230, 288
Pasture and range							
Capsids on bluegrass	1	3	26	126	−0.33	0.70	203
Fall armyworm	1	0	0	102	0.19	0.25	35
Leafhoppers	1	33	100	170	−1.93	0.90	35
Peanuts							
Cornstalk borer	1	2	7	117	−0.08	0.1	18
Peanut worm	1	0	0	74	0.23	0.17	14
Southern corn rootworm	1	1.0 ± 0	15.50 ± 1.50	119.0 ± 3.0	−0.18 ± 0.01	0.28 ± 0.05	233
Thrips	6	0.35 ± 0.21	7.83 ± 6.31	100.17 ± 4.56	0.01 ± 0.09	0.24 ± 0.10	13, 14, 190, 232
Peas							
Aphids	3	4.0 ± 2.31	48.67 ± 22.33	123.67 ± 10.90	−0.45 ± 0.25	0.12 ± 0.06	56, 74, 297
Pea moth on field peas	2	2.5 ± 2.5	9.0 ± 9.0	107.50 ± 1.50	−0.13 ± 0.06	0.06 ± 0.01	259
Peas (southern)							
Beet armyworm	1	18	54	207	−1.13	0.52	48
Thrips	1	20	58	273	−1.58	0.76	48
Peppers							
Aphids	1	0	24	151	−0.36	0.11	30

Table 2 (continued)
SUMMARY OF DATA ON THE EFFECT OF INSECT DENSITY AND OF CONTROL MEASURES ON YIELDS OF VARIOUS COMMODITIES

Crop/pest	No. of data sets	Calculated losses (%) With best control	Calculated losses (%) Without control (untreated check)	Yield index (%) with no pest infestation Y_0	Slope k	Coefficient of determination r^2	Ref.
Corn earworm	2	0.5 ± 0.5	48.50 ± 48.50	228.0 ± 100.0	−0.14 ± 0.18	0.38 ± 0.38	29
European corn borer	6	0.0 ± 0.0	54.50 ± 19.89	171.33 ± 36.12	−3.27 ± 2.18	0.25 ± 0.15	29, 30, 31, 139
Fall armyworm	2	0 ± 0	0 ± 0	124.50 ± 19.50	0.16 ± 0.08	0.02 ± 0.02	29, 30
Pepper weevil	1	4	37	145	−0.54	0.2	86
Potatoes							
Aphids	13	0.54 ± 0.37	15.00 ± 7.45	124.00 ± 6.44	−0.17 ± 0.13	0.14 ± 0.05	27, 36, 99, 115, 187, 209, 226
Armyworms	2	1.28 ± 0.98	59.0 ± 17.0	745.0 ± 197	−4.74 ± 2.42	0.45 ± 0.31	
Colorado potato beetle	13	1.04 ± 0.58	46.62 ± 10.38	450.85 ± 101.19	−3.05 ± 1.13	0.48 ± 0.08	130, 131, 136, 137, 138, 187, 200, 220, 242
European corn borer	5	1.50 ± 0.65	54.25 ± 26.64	598.20 ± 251.35	−3.41 ± 1.69	0.34 ± 0.08	136, 138, 234, 235
Flea beetles	7	0.74 ± 0.35	43.29 ± 10.30	175.29 ± 46.10	−0.81 ± 0.52	0.44 ± 0.08	115, 135, 138, 187, 200
Green peach aphid	3	0 ± 0	3.67 ± 3.67	120.00 ± 4.04	−0.08 ± 0.04	0.16 ± 0.04	36, 37, 51
Leafhoppers	3	0 ± 0	36.00 ± 26.63	139.00 ± 9.50	−10.19 ± 9.56	0.44 ± 0.22	36, 134, 188
Potato aphid	1	0.001	5	118	−0.06	0.03	36
Potato leafhopper	6	0.37 ± 0.20	43.17 ± 18.62	162.50 ± 17.98	−0.85 ± 0.43	0.39 ± 0.13	36, 37, 132, 138, 166, 173
Potato psyllid	3	1.0 ± 1.0	32.0 ± 17.2	271.7 ± 124.6	−0.52 ± 0.17	0.12 ± 0.08	98, 99, 100
Potato tuberworm	1	1	91	560	−5.1	0.39	136
Spider mites	1	4	5	84	−0.04	0.09	160
Wireworms	14	0.22 ± 0.21	4.86 ± 3.47	111.07 ± 4.70	−0.08 ± 0.06	0.11 ± 0.03	32, 167, 197, 212, 221, 243

Rape (seed)							
Cabbage aphid	1	0	100	146	−46	0.75	39
Rice							
Leafhopper	1	9	42	133	−0.56	1.0	23
Rice water weevil	8	1.51 ± 1.50	5.25 ± 4.55	107.50 ± 3.97	−0.01 ± 0.08	0.18 ± 0.10	21, 22, 23, 23a, 24, 25, 105
Safflower							
Lygus	1	3	21	109	−0.23	0.47	68
Thrips	2	3.5 ± 3.5	16 ± 16	122 ± 17.0	−0.06 ± 0.39	0.62 ± 0.29	43, 45
Sorghum							
Chinch bug	1	0	0	139	+1.29	1.0	281
Cornleaf aphid	2	0 ± 0	0 ± 0	126.50 ± 9.50	0.18 ± 0.17	0.21 ± 0.03	253
Greenbug	18	2.80 ± 1.55	19.39 ± 7.16	119.94 ± 4.51	−0.24 ± 0.17	0.35 ± 0.07	43, 53, 54, 64, 65, 66, 123, 282
Mites	3	0 ± 0	0 ± 0	97.33 ± 4.26	0.22 ± 0.12	0.54 ± 0.02	283
Sorghum midge	2	0.2 ± 0.2	5.0 ± 5.0	136.00 ± 31.00	+0.11 ± 0.27	0.16 ± 0.14	75, 213
Sorghum web-worm	1	0	3	5184	−1.44	0.8	155
Southwestern corn borer	1	4	24	186	−0.45	0.60	101
White grubs	3	9.33 ± 4.70	39.00 ± 26.27	592.33 ± 446.64	−4.54 ± 4.37	0.59 ± 0.20	252
Soybeans							
Caterpillar complex	1	0	46	198	−0.91	0.98	161
Corn earworm	2	0.65 ± 0.35	1.50 ± 0.50	100.00 ± 8.00	−0.02 ± 0.01	0.42 ± 0.41	262, 263
Green cloverworm	4	3.00 ± 2.98	8.75 ± 6.55	117.75 ± 9.98	−0.12 ± 0.10	0.26 ± 0.21	262, 265
Looper	3	10.50 ± 3.50	25.50 ± 5.50	130.67 ± 11.62	−0.24 ± 0.13	0.34 ± 0.17	262
Mexican bean beetle	7	0.41 ± 0.28	26.00 ± 13.33	121.57 ± 16.59	−0.77 ± 0.73	0.44 ± 0.13	133, 262, 263, 265
Mites	1	0	89	221	−1.97	0.93	133
Soybean looper	6	4.83 ± 1.92	15.67 ± 7.06	138.67 ± 21.45	−0.28 ± 0.15	0.5 ± 0.14	263, 265, 266
Stink bugs	2	8.5 ± 7.5	15.00 ± 2.00	120.50 ± 0.50	−0.19 ± 0.03	0.12 ± 0.04	265
Velvet bean caterpillar	10	2.35 ± 0.94	16.64 ± 3.73	143.64 ± 13.23	−0.30 ± 0.09	0.51 ± 0.09	262, 265, 266

Table 2 (continued)
SUMMARY OF DATA ON THE EFFECT OF INSECT DENSITY AND OF CONTROL MEASURES ON YIELDS OF VARIOUS COMMODITIES

Crop/pest	No. of data sets	Calculated losses (%) With best control	Calculated losses (%) Without control (untreated check)	Yield index (%) with no pest infestation Y_0	Slope k	Coefficient of determination r^2	Ref.
Squash							
Squash vineborer	1	0	0	94	0.02	0.0005	46
Strawberry							
Spider mites	2	2.0 ± 1.0	23.00 ± 8.00	191.00 ± 63.00	−0.49 ± 0.30	0.34 ± 0.27	140, 216
Spittlebugs	1	0	0	113	0.11	0.06	216
Sugar beets							
Root aphids	1	0	0	105	0.4	0.03	117
Root maggot	6	8.17 ± 2.89	22.67 ± 6.82	173.80 ± 22.29	−0.44 ± 0.17	0.37 ± 0.09	4, 5, 6, 119, 201
Webworms	1	0	0	101	1.0	0.02	112
Sugarcane							
Sugarcane borer	5	5.74 ± 2.51	28.6 ± 12.36	127.80 ± 5.77	−0.45 ± 0.16	0.39 ± 0.17	128, 171, 172
Sunflowers							
Sunflower moth	15	7.13 ± 2.07	40.27 ± 8.66	222.40 ± 30.96	−1.83 ± 0.76	0.58 ± 0.08	40, 41, 255, 256, 257
Sweet potatoes							
Flea beetle	1	0	0	56	0.46	0.02	15
Sweet potato weevil	3	1.67 ± 1.20	29.67 ± 3.71	149.00 ± 21.66	−0.46 ± 0.13	0.34 ± 0.05	89, 162, 279
Wireworms	1	16	38	130	−0.50	0.08	15
Tobacco							
Aphids	3	0.33 ± 0.33	0.20 ± 0.20	112.00 ± 7.64	14.40 ± 14.30	0.67 ± 0.21	17, 108, 121
Budworm	3	1.33 ± 0.88	13.00 ± 6.66	128.00 ± 15.04	−0.11 ± 0.18	0.61 ± 0.09	96, 149
Flea beetle	1	0	0.24	137	−0.32	0.23	96
Green peach aphid	6	0.04 ± 0.03	22.17 ± 13.45	184.67 ± 48.13	0.56 ± 1.39	0.43 ± 0.18	20, 77, 250
Loopers	1	4	10	143	−0.15	0.12	96
Tomato							
Aphids	2	0 ± 0	6.50 ± 4.50	105.00 ± 13.00	−0.08 ± 0.06	0.09 ± 0.01	127, 296

Armyworms	1	0	0	85	0.08	0.01	278
Cabbage looper	1	0	15	125	−0.19	0.36	127
Colorado potato beetle	1	0	93	805	−7.48	0.81	170
Cutworms	1	0.3	16	123	−0.2	0.36	127
Leafminers	1	10	38	124	−0.47	0.59	219
Tomato pinworm	2	5.0 ± 5.0	50 ± 50	201.50 ± 118.50	−0.96 ± 0.97	0.42 ± 0.39	278, 280
Whiteflies	1	12	38	143	−0.55	0.99	168
Worms (hornworms and fruitworms)	1	0	0	89	0.06	0.02	296
Wheat							
Banks grass mite	1	18	61	120	−0.73	0.02	71
Brown wheat mite	1	21	100	193	−6.47	0.33	70
Cutworms	3	7.67 ± 3.18	54.67 ± 12.44	164.00 ± 48.00	−0.73	0.52 ± 0.11	63, 67
Greenbug	2	0 ± 0	0 ± 0	266.50 ± 147.50	1.51 ± 1.47	0.01 ± 0.004	53, 189
Hessian fly	3	0 ± 0	0 ± 0	121.00 ± 7.02	0.13 ± 0.07	0.04 ± 0.02	28, 191

Note: Regression equations were calculated to establish the mathematical relationship among yield, pest density, and losses. The regression equations have the form $Y = Y_0 + kP$, where Y = yield index, P = pest population index, Y_0 and k are constants, and r^2 is the coefficient of determination.

Source: *J. Econ. Entomol.* 1942—1978 and *Insecticide and Acaricide Tests* 1976—1978.

Table 3
SUMMARY OF THE USAGE OF INSECTICIDES ON VARIOUS COMMODITIES IN THE U.S.

Commodity	1964[a]			1966[b]			1971[c]			1976[d]		
	1,000 lbs	Treatment — 1,000 acres	Treatment — lbs/acre	1,000 lbs	Treatment — 1,000 acres	Treatment — lbs/acre	1,000 lbs	Treatment — 1,000 acres	Treatment — lbs/acre	1,000 lbs	Treatment — 1,000 acres	Treatment — lbs/acre
Corn	15,668	18,156	0.86	23,656	24,308	0.97	27,315	28,987	0.94	31,979	33,988	0.94
Cotton	78,022	22,650	3.44	65,368	18,516	3.53	73,365	17,991	4.08	64,139	20,635	3.11
Wheat				922	1,584	0.58	1,712	3,820	0.45	7,236	15,694	0.46
Sorghum										4,604	5,827	0.79
Rice										508	237	2.14
Other grain				1,352	1,130	1.20	7,551	11,847	0.64	1,823	1,458	1.25
Soybeans	4,997	4,109	1.22	3,737	2,678	1.40	5,749	5,052	1.14	7,866	5,859	1.34
Tobacco	5,471	2,566	2.13	4,203	1,911	2.20	4,143	1,378	3.01	3,240	1,593	2.03
Peanuts				5,529	1,412	3.92	5,993	2,043	2.93	2,439	1,371	1.78
Alfalfa				3,615	3,147	1.15	2,317	2,776	0.83	5,391	6,173	0.87
Other hay and forage				487	463	1.05	333	760	0.44	959	907	1.06
Pasture and rangeland										114	132	0.86
Other field crops	12,551	9,187	1.37	916	1,013	0.90	2,285	2,065	1.12			
Irish potatoes	1,456	1,210	1.20	2,972	1,610	1.85	2,889	1,737	1.66			
Other vegetables	8,290	3,827	2.17	8,184	3,242	2.52	8,494	3,258	2.61			
Citrus	1,425	853	1.67	9,435	1,387	6.80	43,131	2,102	20.52			
Apples	10,828	2,545	4.25	10,884	2,180	4.99	10,521	1,686	6.24			
All other fruits and vegetables	4,476	2,913	1.54	7,307	2,629	2.78	18,518	3,746	4.94			
Total	143,184	71,016	2.02	148,753	67,210	2.21	214,977	89,248	2.41	130,298	93,871	1.39

[a] Ref. 85.
[b] Ref. 84.
[c] Ref. 7.
[d] Ref. 83

Table 4A
LINEAR REGRESSION ESTIMATES FOR USAGE OF INSECTICIDES ON SELECTED CROPS BY 1980 AND 1990

Commodtiy	1976[a]			1980			1990		
	Lbs	Treatment (acre)	Treatment (lbs/acre)	Lbs	Treatment (acre)	Treatment (lbs/acre)	Lbs	Treatment (acre)	Treatment (lbs/acre)
Corn	35,574	36,684	1.0	37,607	39,474	0.97	49,655	51,673	1.00
Cotton	70,184	14,699	4.09	62,697	18,917	3.47	55,696	17,958	3.42
Wheat	—	—	—	8,973	19,732	0.39	15,287	33,842	0.27
Other grain	—	—	—	3,999	5,107	1.08	4,470	5,435	1.13
Soybeans	6,305	5,747	1.08	8,658	6,669	1.29	11,514	8,757	1.30
Tobacco	3,206	533	3.64	2,621	1,054	2.42	1,092	303	2.49
Peanuts	—	—	—	1,873	1,498	0.95	0	1,397	0
Alfalfa	—	—	—	5,373	6,755	0.70	7,149	9,781	0.42
Other hay and forage	—	—	—	1,018	1,110	0.86	1,490	1,554	0.87
Other field crops	—	—	—	0 (−9464)	0 (−6069)	0.87	0 (−20,783)	0 (−13,882)	0.66
Irish potatoes	—	—	—	4,547	2,373	2.16	6,169	3,030	2.62
Other vegetables	—	—	—	8,781	2,597	2.09	9,135	1,946	1.94
Citrus	—	—	—	97,404	3,678	44.80	158,486	5,394	71.83
Apples	—	—	—	10,102	601	8.77	9,608	0 (−580)	11.55
All other fruits and vegetables	—	—	—	36,769	4,904	9.27	57,283	6,925	14.02
Total	266,449	102,386	2.69	161,851	104,940	1.49	164,220	127,827	1.01

Table 4B
LINEAR REGRESSION ESTIMATES FOR USAGE OF INSECTICIDES ON SELECTED CROPS BY 1980 AND 1990

	Coefficients[b]								
	Total lbs			Treatments (acre)			Treatment (lbs/acre)		
Commodity	a	b	r^2	a	b	r^2	a	b	r^2
Corn	−2347990.97	1204.85	0.88	−2375925.49	1219.90	0.95	−6.39	0	0.18
Cotton	1448914.15	−700.11	0.32	208791.70	−95.90	0.06	13.73	−0.01	0.01
Wheat	−1241199.40	631.4	0.84	−2774048.33	1411.00	0.87	24.15	−0.01	0.69
Other grain	−89258.77	47.10	0	−59837.11	32.80	0	−8.83	0.01	0.01
Soybeans	−556852.11	285.61	0.79	−406757.17	208.80	0.68	−0.65	0	0
Tobacco	305321.15	−152.88	0.80	149822.48	−75.14	0.61	−12.13	0.01	0.01
Peanuts	613692.67	−309.00	0.64	21495.77	−10.10	0.02	424.67	−0.21	1.00
Alfalfa	−346275.27	177.60	0.33	−592392.60	302.60	0.66	56.14	−0.03	0.64
Other hay and forage	−92438.20	47.20	0.52	−86802.40	44.40	0.96	−1.12	0	0
Other field crops	2231667.70	−1131.88	0.41	1540920.56	−781.31	0.40	41.23	−0.02	0.1
Irish potatoes	−316517.62	162.15	0.47	−127697.77	65.69	0.74	−88.46	0.05	0.24
Other vegetables	−61430.18	35.46	0.66	131372.99	−65.04	0.50	31.79	−0.02	0.07
Citrus	−11996892.92	6108.23	0.99	−336120.13	171.62	0.97	−5306.53	2.70	1.00
Apples	107883.87	−49.38	0.83	234469.96	−118.12	0.98	−541.06	0.28	0.99
All other fruits and vegetables	−4025048.86	2051.42	0.99	−270468.31	139.08	0.75	−931.99	0.48	0.99
Total	−307369.52	236.98	0	−4426731.06	2288.72	0.87	97.06	−0.05	0.35

[a] 1976 data estimated from 1964, 1966, and 1971 data to compare to actual 1976 data. See Refs. 7, 83, 84, and 85.

[b] a = intercept where x = 0, b = slope, and r^2 = coefficient of determination.

Table 5
ESTIMATED LOSSES CAUSED BY INSECTS AND MITES TO SELECTED CROPS IN THE U.S. IN 1977 (YIELD LOSS + CONTROL COSTS)

Crop	Acres harvested[a] (×1000)	Production value (×1000)	Yield loss and pest control costs (×1000)		% Crop grown in:		Estimated national overall costs of pest control (×1000)		Estimated % loss based on:	
			CA[b]	GA[c]	CA	GA	CA	GA	CA	GA
Alfalfa (hay)	27,075[d]									
Alfalfa (seed)	392.5	103,318								
Apples	—	692,532	2,864	690	5.41[e]	0.28[e]	52,976	200,996	7.65	35.71
Asparagus	85.8	81,445								
Barley	9,490	746,199	2,513		10.01		16,022		2.15	
Beans (lima)	60.3	25,426	2,186		44.61		3,351		13.18	
Bean (snap)	335.3	155,654								
Beet, table	14.1	7,981								
Beet, sugar	1,217.8	662,103	8,351		17.82		27,697		4.18	
Blackberry	2.8	8.842								
Blueberry (bushberries)	0.8	2,552								
Broccoli	59.1	88,059	3,297		27.75		3,432		3.90	
Brussels sprouts	5.4	12,765	1,195							
Cabbage	91.8	175,985	1,066		8.82		11,651		6.62	
Cantaloupe	79.5	113,963	3,586		50.82		6,549		5.75	
Carrot	72.7	151,733	2,091		45.67		4,996		3.29	
Cauliflower	34.4	64,236	2,687		78.78		3,369		5.25	
Celery	33.6	148,193	6,053		60.12		9,320		6.29	
Cherries	—	135,447	1,001				8,643		6.38	
Collards	10[f]									
Corn, field (grain)	70,006	12,886,621	8,856	14,355	0.35	3.24	1,659,622	1,170,542	12.88	8.10
Corn, sweet	618.3	220,340								
Cotton (lint)	13,279	3,568,682	139,420	18,820	19.28	2.2	26,875	974,630	0.19	29.54
Cucumber	175.3	136,500	410		3.96		3,688		2.70	
Currant and gooseberry	0.14	75								
Eggplant	3.1	7,241								

Table 5 (continued)
ESTIMATED LOSSES CAUSED BY INSECTS AND MITES TO SELECTED CROPS IN THE U.S. IN 1977 (YIELD LOSS + CONTROL COSTS)

Crop	Acres harvested[a] (×1000)	Production value (×1000)	Yield loss and pest control costs (×1000)		% Crop grown in:		Estimated national overall costs of pest control (×1000)		Estimated % loss based on:	
			CA[b]	GA[c]	CA	GA	CA	GA	CA	GA
Fig (fresh)	—	11,081								
Grape	—	828,695	115,492		91.39[e]					
Grapefruit	—	167,528								
Guar	100[f]									
Hops	30.5	49,095	78		4.92		1,748		3.56	
Kale										
Kohlrabi	0.1									
Lemons	—	87,538	14,168		81.89[e]		17,300		19.76	
Lettuce	228.7	438,390	14,441		68.52		20,760		4.74	
Mint (spearmint and peppermint)	120.1	91,788								
Mustard greens	9.5[f]									
Oats	13,447	853,434	36		0.77		3,995		0.47	
Okra	19.8[f]									
Onion	105.2	227,094	381		26.81		1.825		0.80	
Oranges	—	648,552								
Pasture and range										
Peas, garden	352.2	100,842	18		2.02		874		0.87	
Peas, southern	300									
Peach	—	282,222	32,450	10,271	52.25[e]	5.95[e]	62,110	155,856	22.01	61.14
Peanut (nuts)	1,516	783,302		5,570		34.57		13,426		1.79
Pear	—	114,233	6,201		39.54[e]		15,682			13.73
Pecans		136,459		9,981		50.37		19,816		
Peppers	55.6	90,313	416		14.03		1,979		2.19	
Plums and prunes	—	135,318	31,207		93.17[e]		33,496		24.75	
Potatoes	1,349	1,275,261	1,848		4.5		43,905		3.44	
Pumpkin	16[f]									

Radish	33.2[f]									
Rape (seed)										
Raspberry	6.1	13,904								
Rice	2,249	935,673	4,375		13.69		24,234		2.59	
Rutabaga	1.9[f]									
Safflower			2,040							
Sorghum (grain)	14,065	1,357,262	430	722	1.07	0.3	25,773	287,850	1.90	19.37
Soybean (beans)	57,911	9,944,975		31,247		1.9		1,883,500		20.35
Spinach	30.3	24,008	192		38.61		401		1.67	
Squash (summer and winter)	95.6[f]		2.7							
Strawberry	34.3	214,947	21,521		3.82		44,701		20.80	
Sugarcane (sugar and seed)	758.2	328,364								
Sweet potatoes	112.9	119,280								
Swiss chard	0.3									
Sunflowers	2,205[d]	2,760,470[d]								
Tobacco	957.7	2,235,773		1,514		6.51		26,599		1.11
Tomato	471.2	914,121	26,856		6.28		159,453		17.44	
Turnips (roots)	22.2[f]									
Walnuts		139,555	14,524		99.75[d]		14,561		10.43	
Watermelons	227.4	89,413								
Wheat	66,216	4,677,339	87		1.38		3,682		0.08	

[a] Ref. 272.
[b] Ref. 271.
[c] Ref. 175.
[d] Ref. 239.
[e] Ref. 261. All data reported for Georgia is for 1976, from Ref. 272.
[f] Percentage of value of U.S. production for the state — all other figures are a percentage of acres harvested.

and Agriculture (1979) for 1977, and by the state of Georgia for 1976.[261] These national estimates were developed from the state data by means of the following simple proportion:

$$\frac{\text{California pest control costs}}{\text{Value of California production for commodity}} = \frac{\text{National pest control costs (x)}}{\text{Production value of commodity for U.S.}}$$

RESULTS AND DISCUSSION

We believe that superior means of obtaining information on crop losses is that of Judenko.[150] This method is based on the comparison of the yields from two sets of plants that are precisely the same and, as far as is known, are growing under identical conditions in all respects, except that one is uninfested and the other is infested by the specific pest being considered. Judenko provides a series of equations through which can be obtained an economic loss and an expected yield in the absence of pests. Unfortunately, most of the data on the evaluation of pest control methods, insect-induced losses, and economic thresholds do not lend themselves to an anlysis of this type.

Generally, insect and mite losses expressed in yield or quality are approximately proportional to the density of the pest population when a critical stage or part of the plant is under attack by the pest. For instance, Rogers[217] demonstrated a strong correlation between bud destruction by midge larvae when guar plants were attacked between 45 to 90 days after emergence from the soil and the subsequent yield loss at harvest. Birdsfoot trefoil must be protected from mirid species through bud development, bloom, and early pod growth.[174]

Spring generations of cabbage maggots attacking plants during the seed production year have no effect on yield, but maggots attacking seedlings in August and September cause reductions in seed yields.[104] On the other hand, if a noncritical stage or part of the plant is attacked by the same pest population densities that cause economic losses under the above assumption, little or no economic losses in yield or quality will occur. Several workers have shown that the greater the density of infestation, the more damage is inflicted on the crop and the greater is the loss of yield or quality.[145,195,196,264,292]

We present specific information obtained by Waddill[280] on tomatoes for tomato pinworm to demonstrate the concept of the data transformation and its use. The transformed data were plotted (Figure 1) to show estimates for losses from no pest infestation where the Y axis crosses P at 0; the yield index for the most effective insecticide treatment, P_{16} Y'_{289}; and the yield index for the untreated check where Y′ crosses P at 100. The difference on the line between P_{16} and P_0 (289.28 − 320 = 30.72) is the yield lost from the maximum potential yield (if no insects were present), which is 10.4 lb per plot (0.3072 × 33.8 = 10.38). If the infestation were not treated (P_{100}) the yield index was calculated to be 128 (where the line crosses Y on the P_{100} axis). The difference between P_0 and P_{100} on the line (320 − 128 = 192) is the amount of potential yield lost due to the pest population with no treatment, which is 64.9 lb per plot (1.92 × 33.8 = 64.9). The percent of loss with control is 9.6% and without control it is 60%. These data suggest that a population of pinworms causing 90% damaged fruit are reducing total yield by 60%.

The equations which we developed and presented in Table 2 may be used to a limited extent to project the probable yield differences when only the pest population densities are given in studies involving an untreated plot and one or more treatments. The untreated plot would be P_{100} and the relative magnitude of yield gain or loss in the treatments would be determined.

This method may be useful to those workers attempting to determine the benefits of a particular pesticide such as is now occurring in the rebuttable presumption against registration process (RPAR) in the U.S. Environmental Protection Agency and the National Agricultural Pesticide Impact Assessment Program of the U.S. Department of Agriculture (NAPIAP).

To illustrate this point, again we can use the example of tomato pinworm.[280] The equation for tomato pinworm on tomato (Table 2) is $Y' = 201.5 - 0.96P$. This can then be applied to the data of Waddill[280] and compared to the actual data as follows:

Treatment	% Damaged fruit	P	Y′	Calculated yield ((Y′/100) × 33.8)	Actual yield	Differences between treatments: Calculated yield	Differences between treatments: Actual yield
FMC 33297	21	23.3	179.1	60.5	84.4	3.2	−5.1
SD 41706	30	33.3	169.5	57.3	89.5	16.6	26.4
Bay NTN 9306	76	84.4	120.5	40.7	63.1	0	0.3
Methomyl 1.8L	76	84.4	120.5	40.7	62.8	5	29.0
Untreated check	90	100.0	105.5	35.7	33.8	3.7	6.4
Orthene 75S	100	11.1	94.8	32.0	27.4	—	—

As expected, the calculated yields are not the same as the actual yields, but this is not too important. It is the relationship of the treatment yields to each other that is important. Starting with the first treatment yield, we compared it to the preceding one in the table. These are the differences between treatments columns. An analysis of this data indicated the following:

Observation	Mean	Sum of square
Calculated yield difference	5.7	162.04
Actual yield difference	11.4	955.22

An F-test performed on these data indicated that there was no significant difference between the relative treatments for actual and observed data ($F = 0.58$) at the 1% level of signifance.

Since these equations can be used to project yield differences at various population densities, the data in the literature where only pest densities are reported may be used by economists and others who wish to compare insecticide benefits. Actual yields need not be known since Y′ is reported as a percentage increase or decrease of the untreated plot yield. For instance, in the above-cited example, the FMC compound produced about a 69% increase in total yield over the check and about an 89% increase in total yield over the Orthene® treatment for the calculated data.

At high pest densities where a substantial portion of the crop is destroyed, there is a point of no return and the entire crop may be lost. The state of Georgia reported severe losses to many crops in 1977.[260] There was an extended drought which favored insect development, making the infestations more severe than normal. Cotton and corn were heavily damaged (Table 6). The costs of control (yield loss plus cost of insecticide application) exceeded the value of the cotton crop by almost 44%. *Heliothis* spp. (bollworm and tobacco budworm) were observed to occur earlier and in higher numbers than usual in cotton fields. Heavy infestations of beet armyworms and cabbage loopers also were pest problems on cotton in Georgia in 1977.[260]

The fall armyworm caused a substantial amount of damage to Georgia corn and in combination with the severe drought caused a yield reduction from 62 bu/acre in 1976

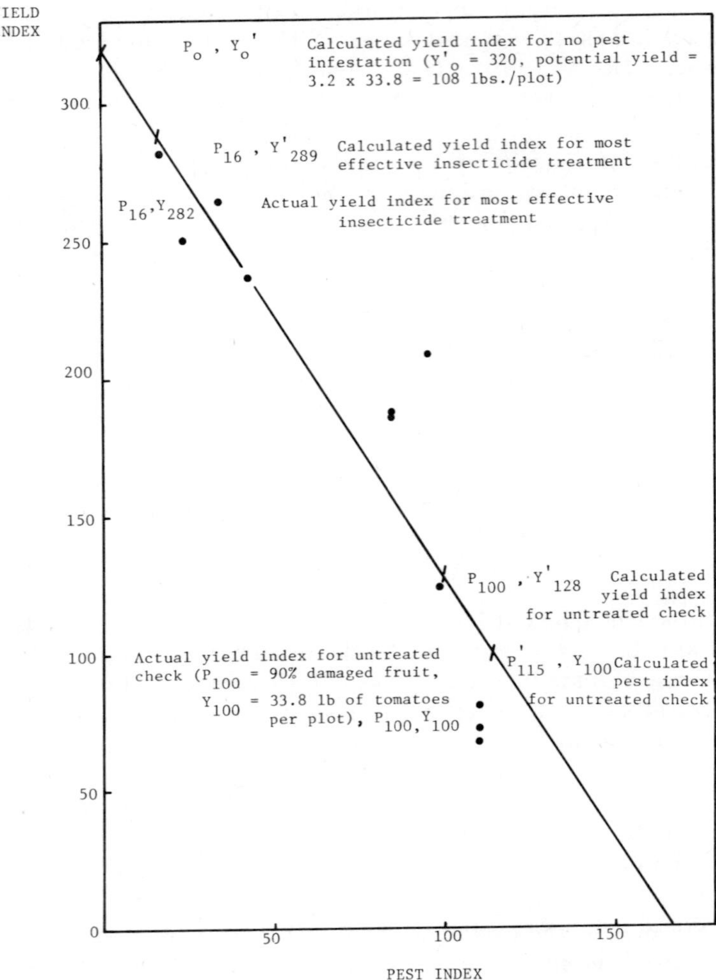

FIGURE 1. Effect of tomato pinworm population densities on tomato yield.[280]

to 24 bu/acre in 1977. Cost of control for the fall armyworm in Georgia was over $3 million on about a million acres of corn which was harvested in 1977 (Table 6). Even with our present pest control technology these data indicate that insect infestations can still severely damage crops and cause serious reductions in yield.

Pest damage can be significant only at certain stages and on certain parts of the plant. For instance, when the soybean plant was defoliated when the pod was just visible at one of the four uppermost nodes, there was no significant effect on seed size.[258] When ⅔ of the plant was defoliated at this stage, the number of pods was significantly reduced. However, if ⅓ of depodding occurred when the pod contained full-size green beans at one of the four uppermost nodes, seed size was significantly increased. When the pod contained full-size green beans at one of the four uppermost nodes, then only complete defoliation significantly reduced the number of pods per plant at harvest.

These data show that soybean plants can compensate for losses in pods at the earlier stages of pod formation but cannot fully replace lost pods when defoliation or depodding occurs after the early budding stages.[210,258]

Table 6
LOSSES ON SELECTED CROPS ATTRIBUTED TO INSECT DAMAGE IN GEORGIA IN 1977[260]

Crop	Harvested acres (× 1,000)	Production value (× 1,000)	Control and yield reduction costs (× 1,000)	% Loss
Cotton	170	18,552	26,712	143.98
Corn	1,000	49,200	46,287	94.08
Sorghum	24	1,411	664	47.06
Peanuts	519	312,256	27,738	8.88
Soybeans	1,090	124,260	20,236	16.29
Vegetable				
Tomato	2.9	7,830	4,011	51.23
Pepper	3.67	5,726	2,024	35.35
Sweet potato	7.5	11,250	1,978	17.58
Sweet corn	3.0	2,250	554	24.62
Cabbage	3.0	1,080	308	28.52
Squash	5.0	2,000	188	9.40
Okra	5.0	1,650	83	5.03
S. peas	55.0	8,250	165	2.00
Fruit				
Apple	—	2,746	463	16.86
Peach	—	18,359	9,173	49.96
Pecan	—	70,425	12,155	17.26

When two leaves are removed from the sorghum plant just prior to boot or a later stage, no effect on yield occurs.[253] Loss of four leaves affected yield if the loss occurred just before boot or at bloom. Greenbug population levels of about 1300 to 1500 resulting in leaf loss of more than three per plant at about the bloom stage were required to cause significant yield loss.[253]

In another example, Bowling[19] simulated insect damage to rice by removing 25% and 50% of the leaf in the various stages of rice. He found that, in the seedling stage, yields were reduced 3% and 8%; in the tillering stage, yields were reduced 5% and 12%. Yield loss from leaf removal was also greater on rice planted in April then on rice planted in May.

Some plant species are more susceptible to insect damage than others. Snap beans are more susceptible to seed maggot damage than Fordhook lima beans or California light red kidney beans.[277] When up to 25% loss of the first pair of unifoliate leaves of snap bean seedlings occur there is a yield loss of 11 to 48%. Yield loss occurred in lima and kidney beans when seedlings lost more than 75% of the first pair of unfoliated leaves.

Several workers have studied losses to crops over a period of several years. A study of the value of soil insect control in Iowa corn for the period 1951 to 1970 by Peters[202] showed that plots receiving a soil insecticide treatment had an 8 bu/acre increase in yield where corn followed corn, and an 8.7 bu/acre increase where corn followed a crop other than corn. Corn yields during these periods averaged about 100 bu/acre in the untreated plots. Unfortunately, these data are not correlated with pest population data; therefore, the actual losses caused by soil insects may not be reflected from these studies because the untreated acreage in Iowa may not have had soil insect populations of sufficient density to justify insecticidal treatment. Our data (Table 2) indicate the soil insect populations on field corn (rootworms, white grubs, and wireworms) can cause yield reductions of 16% to 48% in untreated plots. The loss in Peters'[202] study was 8%. A study from 1972 to 1974 of soil insecticide usage on corn production in

Indiana by Turpin and Thieme[267] showed that corn rootworm caused economic loss in only 3.4% of the fields and was present in only 34% of the untreated fields. The average yields from insecticide-treated fields averaged 2 bu/acre higher than fields that were not treated.

We anticipate that soil insects will cause greater losses to crops in the future as the chlorinated hydrocarbon insecticide residues dissipate from the soil. The less persistant insecticides now in use allow the buildup of greater populations of soil insects. For instance, recent studies have shown that corn rootworms were held to about one emerging adult per corn plant in chlorinated hydrocarbon insecticide-treated fields. In fields treated with less persistant insecticides about ten adults are emerging per plant without any difference in yield loss. However, the adult populations of rootworms are spreading to more fields and plants.[146a]

One of the reasons for the increased population without a difference in loss is that the rootworm larvae are killed off early in the season when the plant is young and more susceptible to damage. Larvae from eggs hatched later in the season are not killed as they would have been by the persistant insecticides.[146a] We believe that this pattern can develop with other soil insects pests such as wireworms, white grubs, grasshoppers, cutworms, and the white fringed beetle. This is potentially a very explosive situation which could lead to some disastrous results when the right environmental conditions occur because we have no insecticides currently capable of keeping heavy population densities of these pests under control.

Data by Parencia and Cowan[199] show that an average of 6.1 applications of insecticides gave an increase in yield of 366 lb of seed cotton per acre or a 54.3% increase during the 32-year period, 1939 to 1970, in central Texas. The insects of primary importance in the study were the boll weevil, bollworm, and tobacco budworm. This is consistant with our calculations (Table 2) which show that these pests are capable of causing losses of 30% to 90% in the untreated plots. Other pests in the study which contributed to yield reduction of cotton were cotton fleahopper, cotton aphid, carmine spider mite, thrips, and pink bollworm.

The pear leaf rust mite in unsprayed orchards caused damage to the fruit amounting to 30% or more of the harvested crop, while pear psylla did not.[286] In an unsprayed orchard a proportion of fruit infested with San Jose scale increased during the first 4 years and reached 92% in 1967. By 1969 the infestation had decreased to about 4% and remained below 12% through 1971. Codling moth infestation levels ranged from 21% to 81% with a mean of 42.4% in the unsprayed orchard.[286] Small losses (2% to 3%) from codling moth early in the season will result in a harvest infestation of 40% or more if left untreated.[113] These studies showed that the first year of growing the crop without the use of pesticides reduced yield and quality by 45% and 85%, respectively, at harvest. In the second year of pesticide absence the development of crop injury was more rapid and three of the four varieties had complete loss of marketable fruit.

Barnes and Moffitt[16] conducted a 5-year study of the effects of walnut aphid and European red mite on walnuts and concluded that these pests had a significant impact on the productivity of staminate flowers, which subsequently resulted in loss of both yield and quality. Yield reductions by aphids averaged 25% over the first 3 years and continued at 28% and 43% during the fourth and fifth years after control was reestablished. Mite infestations had no effect on yield until the third year, when there was a 40% loss. Losses of 22% and 31% persisted after control was reestablished. After the second and third year of infestation of these pests, there was an average reduction in number of staminate flowers of 68% attributable to aphids; 61% to mites; and 85% when both species were present.

Unfortunately, there is very little information on yield losses caused by insects on fruit crops. As indicated in Table 1 we were unable to find data to calculate loss estimates for nearly all of the pests attacking fruit crops.

Our study of the literature indicates that there are a few basic minimal standards that researchers should adhere to so that their data are of maximal use to others who may seek information on losses, insecticide benefit, etc. as follows:

1. Measure the correct parameters such as pest density, yield, and/or quality at the critical times when the pest can cause a yield or quality loss.
2. Deal with an economically important insect in the development of control measures.
3. Develop valid information on losses with no pest control vs. the best agricultural practice in pest control.

Failure of scientists to adhere to these minimal standards results in data that has very limited usefulness to those who seek to use it in developing loss information, evaluating insecticide benefits, and comparing the effectiveness of insecticides.

Estimates of crop losses can be relatively simple, such as the estimates of knowledgeable experts in the field who assess pest damage and provide estimates of losses after application of the best agricultural practices. These estimates have been documented in publications for the period 1951 to 1960.[274] These estimates have not been comprehensively updated for the U.S. since 1960.

We present the estimates from various knowledgeable experts for each of the commodity/pest combinations in Table 1. This table also provides a summary of our calculated loss information. There are 734 pests on 73 crops in Table 1. There are some duplications of pests on crops, particularly for the commodities within the general class of cole crops. Therefore, the number of actual pests is slightly less than the total indicated. Unfortunately there is little information concerning losses caused by many of these pests. This void is particularly noticeable for fruit insects. The pests in Table 1 are there because either data exists concerning losses from yield or because they are pests for which the U.S. Department of Agriculture has guidelines for the use of insecticides.[273] There are 142 pests on 35 crops for which yield data exists, but there are 20 crops for which no loss data are available for any of the pests listed under that crop.

Several reviewers of the manuscript have indicated that the information in Table 1 may give the wrong impression of the relative importance of pests on some crops. For instance, the European corn borer and corn rootworm are generally considered to be the most important pests of field corn throughout the corn-growing region. However, leafhoppers show a greater loss, which probably reflects a localized situation where leafhoppers were a severe problem on silage corn.[202a]

Greenbugs and Hessian flies are usually considered the most important pests of wheat, but mites appear to cause greater losses. We believe this is similar to the corn example cited above. On sorghum, the sorghum midge generally causes far greater losses than the Southwestern corn borer and white grubs, especially if sorghum is planted 3 to 4 weeks after the normal planting date.[257a]

The above examples point out some of the deficiencies in developing loss estimates. Losses vary widely in different growing regions. In some areas some insect species are major pests, whereas in other areas they are of minor concern. Noticeable differences in losses also occur from season to season. Loss estimates may vary widely from year to year. The authors hope that the information in this text will further stimulate workers in entomology to document losses when and where they occur. The following discussion and Table 1 may help point out where the critical needs are in documenting insect loss information.

For purposes of discussion we arbitrarily selected those pests that we calculated caused a loss of over 5%, even with the best treatment, in the studies. The treatment may be a registered insecticide or an experimental compound which was not available for practical use. The following commodity/insect combinations will cause losses of over 5% even with the best control:

Alfalfa seed/leaf hopper, thrips
Apples/codling moths
Beans/corn earworm, Mexican bean beetle
Beets/root maggot
Cole crops/caterpillars
Corn, field/corn rootworms, leafhoppers, Southwestern corn borer, white grubs
Corn, sweet/cutworms, fall armyworm
Cotton/boll weevil, cabbage looper, cotton fleahopper, *Heliothis,* pink bollworms, thrips
Onion/onion maggot, onion thrips
Pasture and range/leafhoppers
Peas, Southern/beet armyworm, wireworm
Rice/rice stinkbug
Soybeans/loopers, stinkbugs
Sunflowers/sunflower moth
Tomato/leafminers, tomato pinworm, whiteflies
Wheat/brown wheat mite, cutworms, white grubs

The data in Table 1 revealed some interesting information concerning the potential for pests to cause losses where there is no chemical control. A cutoff figure of over 50% was arbitrarily selected for the purposes of discussion in this paper. The following commodity/pest combination had over a 50% loss in the untreated plots:

Apples/codling moth
Beans, lima/Mexican bean beetle
Carrot/carrot rust fly
Cole crops/aphids
Corn, field/leafhoppers
Corn, sweet/fall armyworm
Cotton/ *Heliothis,* thrips
Onion/onion thrips
Pasture and range/leafhoppers
Peas, Southern/thrips
Peppers/European cornborer
Potatoes/armyworm, European cornborer, potato tuberworm
Rape (seed)/cabbage aphids
Tomatoes/Colorado potato beetle, tomato pinworm
Wheat/brown wheat mite, cutworms

We recognize that this information has limited value because of the conditions under which the information is obtained (i.e., the research plots were most likely picked where heavy pest population densities occurred for evaluation of the candidate insecticides, and a great number of other pests and commodities are not represented with any yield data). However, the information does indicate some general trends and possible areas of potential problems. It also may be useful in some respects for planning research programs and evaluating current control methods on pests. The loss calculations developed for the pests are not additive for the pests under a given commodity. The figures should be used for the comparative purposes of the destructive potential of each pest. We have not attempted to estimate an overall loss figure for the commodity using these calculations.

These crop/pest combinations need a more careful scrutiny because it is possible that yield reductions did not occur because sufficient pest population densities did not develop.

On the other hand, the Y_0 statistic reveals that there are some pests that are capable of causing heavy yield losses. We can accept these data with agreater degree of confidence than the preceding data because we know that the pest populations did develop and they did cause heavy yield losses. Some examples are:

Alfalfa (seed)/lygus bugs
Beans/corn earworm
Carrot/carrot rust fly
Cole crops/caterpillars
Collards/cabbage looper and cabbage webworm
Corn, sweet/fall armyworm
Corn, field/leafhopper on silage corn
Cotton/*Heliothis*
Onion/onion thrips
Pasture and range/*Thymelicus lineola* on timothy
Potatoes/armyworms, Colorado potato beetle, European cornborer, potato tuberworm
Sorghum/white grubs
Tomato/Colorado potato beetle

These data substantiate what economic entomologists already know — nearly all of these examples are major pests of primary importance on the crop.

We have already discussed how the regression coefficients may be utilized for comparing the benefits of insecticides as estimated by protection from yield loss when only pest density data is provided. There is another use that we see for the data in Table 2. The parameter Y_0 is an interesting statistic because it gives the percentage change in yield over an untreated plot when no pest population is present. Any value of Y_0 that is 100 or less indicates that, potentially, the pest does not cause any yield reduction. Of course, this information must be handled very cautiously because it is possible that a population of insects of sufficient density to cause damage in the untreated plots was not present.

We do need to raise the question: if the insect pest does not reduce yield directly or through an adverse change in quality, are control measures warranted? We may need to examine more critically the following pests in light of this question:

Apples/red-banded leafroller
Beans/bean thrips
Beans/lygus bugs
Beans/potato leafhopper
Celery/leafminers
Corn, sweet/spider mites
Cotton/spider mites
Peanuts/peanut worm
Potatoes/spider mites
Sorghum/spider mites
Soybeans/corn earworm
Squash/squash vine borer

We have reached the general conclusion that insect loss data are dynamic; they change from year to year and in geographical areas. We have also found that losses from insect damage continue to occur at magnitudes that have not decreased appreciably over the past 30 years. While the pests causing the losses may have changed and some crops may now fare better than others, the magnitude of the losses, in general, has not shifted downward to any extent. To us, this means that we are barely staying abreast of the insect pests in our research, development, and implementation of pest management programs. There are times when the insects can tip the scales in their favor, as evidenced by the atypical year Georgia had in 1977.

The information concerning the uses of insecticides on selected crops (Table 4) provides some interesting information on the general trends of insecticide treatments. This is generally downward on cotton, tobacco, other field crops, and apples in the total pounds of insecticides used on these crops. Corn, wheat, citrus, and all other fruits and vegetables show substantial increases in the total amounts of insecticides used on these crops. We have made the 1980 and 1990 predictions from these data only to illustrate the general trend of insecticide usage, and we do not suggest they be used as an actual prediction of use.

California and Georgia have developed data concerning losses attributed to insects and mites for the various commodities in their states[239,261] These losses were used in Table 5 to project losses on a national basis. We recognize the limitations in this type of projection and that they do not provide a completely adequate picture of the losses for the nation. California does not have some of the pests in the East, whereas Georgia represents the South, where insects are generally more damaging. We have production values for 56 of the 73 commodities presented in Tables 1 and 5. Of these, California and Georgia had estimates for 38 commodities. This represented 90% of the total production value of the crops listed in Table 2. The percentage-loss figure calculated for national overall costs of pest control based on the California data was 7.35%, and on Georgia data, 14.84%. These figures compare fairly closely with the generally accepted range of estimated losses caused by insect pests of about 10% to 12%.

REFERENCES

1. **All, J. N.,** Field corn, fall armyworm control, 1977, in *Insecticide and Acaricide Tests,* Vol. 3, Rep. No. 164, Entomological Society of America, College Park, Md., 1977, 118.
2. **All, J. N.,** Field corn, fall armyworm control, 1977, in *Insecticide and Acaricide Tests,* Vol. 3, Rep. No. 165, Entomological Society of America, College Park, Md., 1977, 119.
3. **All, J. N., Kuhn, C. W., Gallaher, R. N., Jellum, M. D., and Jussey, R. S.,** Influence of no-tillage-cropping, carbofuran, and hybrid resistance on dynamics of maze chlorotic dwarf and maze dwarf mosaic diseases of corn, *J. Econ. Entomol.,* 70, 221, 1977.
4. **Allen, W. R., Askew, W. L., and Schreiber, K.,** Insecticidal control of the sugarbeet root maggot and yield of sugarbeets, *J. Econ. Entomol.,* 54, 178, 1961.
5. **Anderson, A. W., Carlson, R. B., and Dregseth, R.,** Evaluation of in-furrow applications of insecticides for sugarbeet root maggot control, 1975, in *Insecticide and Acaricide Tests,* Vol. 2, Rep. No. 86, Entomological Society of America, College Park, Md., 1975, 69.
6. **Anderson, A. W., Carlson, R. B., and Dregseth, R.,** Evaluation of insecticides to control the sugarbeet root maggot, 1975, in *Insecticide and Acaricide Tests,* Vol. 2, Rep. No. 87, Entomological Society of America, College Park, Md., 1975, 70.
7. **Andrilenas, P. A.,** Farmers' use of pesticides in 1971 — quantities, Agric. Econ. Rep. No. 252, Economic Research Service, U.S. Department of Agriculture, Washington, D.C., 1971.
8. **Apple, J. W.,** Reduced dosages of insecticides for corn rootworm control, *J. Econ. Entomol.,* 50, 28, 1957.

9. **Armbrust, E. J., Wilson, M. C., and Hintz, T. R.,** Chemical control of the alfalfa weevil in Illinois and Indiana. I. Comparison of registered and experimental materials, *J. Econ. Entomol.,* 61, 1050, 1968.
10. Insect control — fruits, vegetables, nut trees, and nursery stock, ARS-NRP 20220, Agricultural Research, U.S. Department of Agriculture, Beltsville, Md., 1976.
11. Cotton and tobacco insect control, ARS-NRP 20230, Agricultural Research, U.S. Department of Agriculture, Beltsville, Md., 1976.
12. Insect control—grains, forages, sugar crops and oilseeds, ARS-NRP 20240, Agricultural Research, U.S. Department of Agriculture, Beltsville, Md., 1976.
13. **Arthur, B. W. and Arant, F. S.,** Effect of systemic insecticides upon certain peanut insects and upon peanuts, *J. Econ. Entomol.,* 47; 1111, 1954.
14. **Arthur, B. W. and Hyche, L. L.,** Soil applications of insecticides for control of tobacco thrips on peanuts, *J. Econ. Entomol.,* 52, 451, 1959.
15. **Averre, C. W., Barker, K. R., and Sorensen, K. R.,** Evaluation of nematicides and insecticides for sweetpotatoes, 1975, in *Insecticide and Acaricide Tests,* Vol. 1, Rep. No. 100, Entomological Society of America, College Park, Md., 1976.
16. **Barnes, M. N. and Moffitt, H. R.,** A five-year study of the effects of the walnut aphid and the European redmite on Persian walnut productivity in coastal orchards, *J. Econ. Entomol.,* 71, 1978.
17. **Barney, W. P. and Reagan, T. E.,** Aphid control on tobacco, 1976, in *Insecticide and Acaricide Tests,* Vol. 2, Rep. No. 137, Entomological Society of America, College Park, Md., 1977.
18. **Berberet, R. C. and Rawl, R.,** *E. lignosellus,* control in dry land peanuts, Oklahoma, 1975, in *Insecticide and Acaricide Tests,* Vol. 1, Rep. No. 127, Entomological Society of America College Park, Md., 1976.
19. **Bowling, C. C.,** Simulated insect damage to rice: effects of leaf removal, *J. Econ. Entomol.,* 71, 377, 1978.
20. **Bowling, C. C.,** "Crop Loss Assessment Methods," Chiarappa, L., Ed., Food and Agriculture Organization, Rome, 1971.
21. **Bowling, C. C.,** Rice water weevil control with granular insecticides, *J. Econ. Entomol.,* 69, 680, 1976.
22. **Bowling, C. C.,** Comparison of three formulations of Aldrin for use as seed treatments for rice water weevil control, *J. Econ. Entomol.,* 59, 1008, 1966.
23. **Bowling, C. C.,** Chemical control of the rice water weevil, *J. Econ. Entomol.,* 54, 710, 1961.
23a. **Bowling, C. C.,** Tests with systemic insecticides on rice, *J. Econ. Entomol.,* 54, 937, 1961.
24. **Bowling, C. C.,** Seed treatment for control of the rice water weevil, *J. Econ. Entomol.,* 50, 450, 1957.
25. **Bowling, C. C. and Flinchum, W. T.,** Interaction of Propanil with insecticides applied as seed treatments on rice, *J. Econ. Entomol.,* 61, 67, 1968.
26. **Brannon, L. W.,** Tests of some new insecticides to control Mexican bean beetle, *J. Econ. Entomol.,* 42, 928, 1949.
27. **Bray, D. F.,** European cornborer control in potatoes, *J. Econ. Entomol.,* 54, 782, 1961.
28. **Brown, H. E.,** Insecticidal control of the Hessian fly, *J. Econ. Entomol.,* 53, 501, 1960.
29. **Burbutis, P. P.,** Insect control in sweet peppers, Delaware, 1977, in *Insecticide and Acaricide Tests,* Vol. 3, Rep. No. 105, Entomological Society of America, College Park, Md., 1978.
30. **Burbutis, P. P.,** Insect control on sweet peppers, Delaware, 1975, in *Insecticide and Acaricide Tests,* Vol. 1, Rep. No. 67, Entomological Society of America, College Park, Md., 1976.
31. **Burbutis, P. P. and Lesiewicz, D. S.,** Exclusion as a means of control of the European cornborer in sweet peppers, *J. Econ. Entomol.,* 67, 97, 1974.
32. **Burrage, R. H., Menzies, J. A., and Zirk, E.,** Soil treatments with broadcast or band applications of organophosphorus or carbamate insecticides for prevention of wireworm damage to potatoes, *J. Econ. Entomol.,* 60, 1489, 1967.
33. **Bushing, R. W. and Burton, V. E.,** Leafhopper damage to silage corn in California, *J. Econ. Entomol.,* 67, 656, 1974.
34. **Byers, R. A., Neal, J. W., Jr., Elgin, J. H., Jr., Hill, K. R., McMurtrey, J. E., III, and Feldmesser, J.,** Systemic insecticides with spring-seeded alfalfa for control of potato leafhopper, *J. Econ. Entomol.,* 70, 337, 1977.
35. **Byers, R. A.,** Increased yields of coastal Bermuda grass after application of insecticides to control insect complex, *J. Econ. Entomol.,* 60, 315, 1967.
36. **Cancellado, R. and Radcliffe, E. B.,** Potato pest control, Rosemount, Minnesota, 1976, in *Insecticide and Acaricide Tests,* Vol. 2, Rep. No. 68, Entomological Society of America, College Park, Md., 1977.

37. **Cancellado, R. and Radcliffe, E. B.,** Potato leafhopper and greenpeach aphid control on potato with soil applied systemic insecticides, 1975, in *Insecticide and Acaricide Tests,* Vol. 1, Rep. No. 71, Entomological Society of America, College Park, Md., 1976.
38. **Cancellado, R. and Radcliffe, E. B.,** Potato leafhopper and greenpeach aphid control with foliar insecticidal sprays, 1975, in *Insecticide and Acaricide Tests,* Vol. 1, Rep. No. 72, Entomological Society of America, College Park, Md., 1976.
39. **Carlson, E. C.,** Cabbage and turnip aphids and their control and damage on rape and mustard, *J. Econ. Entomol.,* 66, 1303, 1973.
40. **Carlson, E. C.,** New insecticides to control sunflower moth, *J. Econ. Entomol.,* 64, 208, 1971.
41. **Carlson, E. C.,** Control of sunflower moth larvae and their damage to sunflower seeds, *J. Econ. Entomol.,* 60, 1068, 1967.
42. **Carlson, E. C.,** Effect of flower thrips on onion seed plants and a study of their control, *J. Econ. Entomol.,* 57(5), 735, 1964.
43. **Carlson, E. C.,** Damage to safflower plants by thrips and lygus bugs and a study of their control, *J. Econ. Entomol.,* 57, 140, 1964.
44. **Carlson, E. C.,** New insecticides for lygus bug control on vegetable seed crops, *J. Econ. Entomol.,* 53, 767, 1960.
45. **Carlson, E. C. and Witt, R. L.,** Insecticides for *Frankliniella occidentalis* and *lygus hesperus* on safflower plants, *J. Econ. Entomol.,* 70, 460, 1977.
46. **Carruth, L. A. and Howe, W. L.,** Factors affecting use and phytotoxicity of DDT and other insecticides for squash borer control, *J. Econ. Entomol.,* 41, 352, 1948.
47. **Cate, J. R., Jr., Bottrell, D. G., and Teetes, G. L.,** Management of the greenbug on grain sorghum. I. Testing foliar treatments of insecticides against greenbugs and corn leaf aphids, *J. Econ. Entomol.,* 66, 945, 1973.
48. **Chalfant, R. B. and Johnson, A. W.,** Field evaluation of pesticides applied to the soil for control of insects and nematodes affecting southern peas in Georgia, *J. Econ. Entomol.,* 65, 1711, 1972.
49. **Cowan, C. B., Jr. and Davis, J. W.,** Field tests for the conventional low volume or ultra-low volume sprays for control of the boll weevil, bollworm, and tobacco budworm on cotton in 1967, *J. Econ. Entomol.,* 61, 1115, 1968.
50. **Cramer, H. H.,** Plant protection and world crop production *Pflanzenschutz Nachr.,* 20(1), 7, 1967.
51. **Cranshaw, W. S. and Radcliffe, E. B.,** Potato insect control, Grand Forks, North Dakota, 1977, in *Insecticide and Acaricide Tests,* Vol. 3, p. 85, Rep. No. 114, Entomological Society of America, College Park, Md., 1978.
52. **Crowell, H. H.,** Onion thrips control, 1975, in *Insecticice and Acaricide Tests,* Vol. 1, Rep. No. 64, Entomological Society of America, College Park, Md., 1976.
53. **Dahms, R. G. and Wood, E. A., Jr.,** Evaluation of greenbug damage to small grains, *J. Econ. Entomol.,* 50, 443, 1957.
54. **Daniels, N. E.,** Insecticidal control of greenbugs in grain sorghum, *J. Econ Entomol.,* 65, 235, 1972.
55. **Daniels, N. E.,** Insects affecting alfalfa seed production, *J. Econ. Entomol.,* 48, 339, 1955.
56. **Davis, A. C., McEwen, F. L., and Schroeder, W. T.,** Control of pea enation mosaic in peas with insecticides, *J. Econ. Entomol.,* 54, 161, 1961.
57. **Davis, J. W. and Cowan, C. B., Jr.,** Early season insects on cotton: control with two systemic insecticides, *J. Econ. Entomol.,* 67, 130, 1974.
58. **Davis, J. W. and Cowan, C. B., Jr.,** Field evaluation of three formulations of aldicarb for control of cotton insects, *J. Econ. Entomol.,* 65, 231, 1972.
59. **Davis, J. W., Cowan, C. B., Jr., Watkins, W. C., Jr., Lindgren, P. D., and Ridgway, R. L.,** Experimental insecticides applied as sprays to control thrips and the cotton fleahopper, *J. Econ. Entomol.,* 59, 980, 1966.
60. **Davis, J. W., Watkins, W. C., Jr., Cowan, C. B., Jr., Ridgway, R. L., and Lindquist, D. A.,** Control of several cotton pests with systemic insecticides, *J. Econ. Entomol.,* 59, 159, 1966.
61. **DeBord, D. V.,** Cotton insect and weed loss analysis, The Cotton Foundation, unpublished information, 1977.
62. **Denton, W. H., White, J. M., Musgrave, C. A., and Bennett, D. R.,** Celery, vegetable leafminer control, 1977, in *Insecticide and Acaricide Tests,* Vol. 3, Rep. No. 91, Entomological Society of America, College Park, Md., 1978.
63. **DePew, L. J.,** Field evaluation of insecticides to control pale western cutworm in winter wheat, *J. Econ. Entomol.,* 68, 85, 1975.
64. **DePew, L. J.,** Controlling greenbugs in grain sorghum with foliar and soil insecticides, *J. Econ. Entomol.,* 67, 533, 1974.
65. **DePew, L. J.,** Further evaluation of insecticides for greenbug control on grain sorghum in Kansas, *J. Econ. Entomol.,* 65, 1095, 1972.

66. **DePew, L. J.,** Evaluation of foliar and soil treatments for greenbug control on sorghum, *J. Econ. Entomol.,* 64, 169, 1971.
67. **DePew, L. J.,** Further studies on pale western cutworm control in Kansas, *J. Econ. Entomol.,* 63, 1842, 1970.
68. **DePew, L. J.,** Field studies on control of lygus bugs and onion thrips infesting safflower, *J. Econ. Entomol.,* 60, 1224, 1967.
69. **DePew, L. J.,** Systemic insecticides to control greenbugs on spring-planted barley, *J. Econ. Entomol.,* 57(2), 250, 1964.
70. **Depew, L. J.,** Evaluation of brown wheat mite control on yield of winter wheat in Kansas, *J. Econ. Entomol.,* 55, 1010, 1962.
71. **DePew, L. J.,** Control of *Oligonychus Pratnsis* attacking winter wheat in western Kansas, *J. Econ. Entomol.,* 53, 1061, 1960.
72. **Ditman, L. P. and Wiley, R. C.,** The effectiveness of several insecticides for control of insects on snap beans, *J. Econ. Entomol.,* 51, 258, 1958.
73. **Ditman, L. P., Owens, H. B., and Harrison, F. P.,** Experiments with sprays, dusts, and aerosols for the home garden, *J. Econ. Entomol.,* 50, 324, 1957.
74. **Ditman, L. P. and Burkhardt, G.,** Further experiments on pea aphid control, *J. Econ. Entomol.,* 45, 880, 1952.
75. **Doering, G. W. and Randolph, N. M.,** Habits and control of the sorghum midge, *Contarinia sorghicola,* on grain sorghum, *J. Econ. Entomol.,* 56, 454, 1963.
76. **Dogger, J. R. and Lilly, J. H.** Seed treatment as a means of reducing wireworm damage to corn, *J. Econ. Entomol.,* 42, 663, 1949.
77. **Dominick, C. B.,** Evaluation of systemic insecticides for greenpeach aphid control on tobacco, *J. Econ. Entomol.,* 64, 1565, 1971.
78. **Dominick, C. B.,** Control of the cornroot worm, *J. Econ. Entomol.,* 53, 670, 1960.
79. **Douglass, J. R. and Shirck, F. H.,** Experiments for control of onion thrips, *J. Econ. Entomol.,* 42, 68, 1949.
80. **Dowdy, A. C. and Sleesman, J. P.,** Systemic poisons on vegetable crops, *J. Econ. Entomol.,* 45, 640, 1952.
81. **DuRant, J. A.,** Methomyl on cotton: evaluation of use patterns for phytotoxicity and efficacy against the bollworm and tobacco budworm, *J. Econ. Entomol.,* 70, 641, 1977.
82. **Eckenrode, C. J. and Ditman, L. P.,** An evaluation of potato leafhopper damage to lima beans, *J. Econ. Entomol.,* 56, 551, 1963.
83. **Eichers, T. R., Andrilenas, P., and Anderson, T. W.,** Farmer's Use of Pesticides in 1976, Agric. Econ. Rep. No. 418, Economics, Statistics, and Cooperatives Service, U.S. Department of Agriculture, Washington, D.C., 1978.
84. **Eichers, T. R., Andrilenas, P., Blake, H., et al.,** Quantities of pesticides used by farmers in 1966, Agric. Econ. Rep. No. 179, Economics, Statistics and Cooperatives Service, U.S. Department of Agriculture, Washington, D.C., 1970.
85. **Eichers, T. R., Andrilenas, P., Jenkins, R., and Fox, A.,** Quantities of pesticides used by farmers in 1964, Agric. Econ. Rep. No. 131, Economics, Statistics, and Cooperatives Service, U.S. Department of Agriculture, 1968.
86. **Elmore, J. C. and Campbell, R. E.,** Control of a pepper weevil, *J. Econ. Entomol.,* 47, 1141, 1954.
87. **Ennis, W. B., Jr., Klassen, W., and Dowler, W. M.,** Crop protection to increase food supplies, *Science,* 188, 593, 1975.
88. **Eshbaubh, E. L. and Sorensen, E. E.,** Effects of insecticides on alfalfa weevil, 1976, in *Insecticide and Acaricide Tests,* Vol. 2, Rep. No. 103, Entomological Society of America, College Park, Md., 1977.
89. **Floyd, E. H.,** Control of the sweet potato weevil and several insects attacking roots of sweet potatoes in the field, *J. Econ. Entomol.,* 48, 644, 1955.
90. Food and Agriculture Organization of the United Nations, *Crop Loss Assessment Methods,* Suppl. 2, Sec. 3.3, Food and Agriculture Organization of the United Nations, Rome, 1977, 120.
91. **Franklin, W. W.,** Insecticidal control plot test for pea aphids in relation to alfalfa hay yields, *J. Econ. Entomol.,* 46, 462, 1953.
92. **Gauthier, N. L.,** Northern corn rootworm larval control, 1976, in *Insecticide and Acaricide Tests,* Vol. 3, Rep. No. 168, Entomological Society of America, College Park, Md., 1978.
93. **Gauthier, N. C.,** Third cutting alfalfa insect control study, 1976, in *Insecticide and Acaricide Tests,* Vol. 3, Rep. No. 155, Entomological Society of America, College Park, Md., 1978.
94. **Gauthier, N. L.,** Second cutting alfalfa insect control study, 1976, in *Insecticide and Acaricide Tests,* Vol. 3, Rep. No. 154, Entomological Society of America, College Park, Md., 1978.

95. **Gauthier, N. L., Hower, A. A., Jr. and Gesell, S. G.,** Field corn, northern corn rootworm larval control, 1977, in *Insecticide and Acaricide Tests,* Vol. 3, Rep. No. 169, Entomological Society of America, College Park, Md., 1978.
96. **Gentry, C. R., Kincaid, R. R., and Bowman, M. C.,** Soil treatment with disulfoton against certain insect pests of cigar-wrapper tobacco, *J. Econ. Entomol.,* 63, 1139, 1970.
97. **Gerhardt, P. D.,** Cabbage looper control on head lettuce, 1977, in *Insecticide and Acaricide tests,* Vol. 3, Rep. No. 100, Entomological Society of America, College Park, Md., 1978.
98. **Gerhardt, P. D.,** Potato psyllid control with synthetic pyrethroids, 1976, in *Insecticide and Acaricide Tests,* Vol. 2, Rep. No. 70, Entomological Society of America, College Park, Md., 1977.
99. **Gerhardt, P. D.,** Potato psyllid and green peach aphid control on Kennebec potatoes with temik and other insecticides, *J. Econ. Entomol.,* 59, 9, 1966.
100. **Gerhardt, P. D. and Bennett, L. H.,** *P. cockerelli* control by seed piece treatment with systemic insecticide, 1975, in *Insecticide and Acaricide Tests,* Vol. 1, Rep. No. 73, Entomological Society of America, College Park, Md., 1976.
101. **Gerhardt, P. D., Moore, L., Armstrong, J. F. and Kaspersen, L. J.,** Southwestern cornborer control in grain sorghum., *J. Econ. Entomol.,* 65, 491, 1972.
102. **Gerhardt, P. D., Moore, L., Armstrong, J. F., and Kaspersen, L. J.,** Cost/benefits of candidate insecticides for control of potato leafhopper and alfalfa blotch leafminer, 1976, in *Insecticide and Acaricide Tests,* Vol. 2, Rep. No. 104, Entomological Society of America, College Park, Md., 1977.
103. **Gerhardt, P. D., Moore, L., Armstrong, J. F., and Kasperson, L. J.,** Cost/benefits of potato leafhopper and blotch leafminer control on new seeding of alfalfa, 1976, in *Insecticide and Acaricide Tests,* Vol. 2, Rep. No. 105, Entomological Society of America, College Park, Md., 1977.
104. **Getzin, L. W.,** Effect of cabbage maggot damage on yield and quality of hybrid cabbage seed, *J. Econ. Entomol.,* 71, 528, 1978.
105. **Gifford, J. R., Oliver, B. F., and Trahan, G. B.** Rice water weevil with Pirimiphos-ethyl seed treatment, *J. Econ. Entomol.,* 68, 79, 1975.
106. **Gould, G. E.,** Effectiveness of DDT for cabbage caterpillar control in Indiana, 1945 to 1960, *J. Econ. Entomol.,* 54, 475, 1961.
107. **Gould, G. E. and Wilson, M. C.,** Granulated insecticides for European cornborer control, *J. Econ. Entomol.,* 50, 510, 1957.
108. **Guthrie, F. E., Rabb, R. L., and VanMiddelem, C. H.,** Control of aphids on cigar wrapper and flue cured tobacco, *J. Econ. Entomol.,* 49, 602, 1956.
109. **Hagel, G. T.,** *Lygus* spp.: Damage to beans by reducing yields, seed pitting, and control by varietal resistance and chemical sprays, *J. Econ. Entomol.,* 71, 613, 1978.
110. **Hagel, G. T.,** Systemic insecticides and control of insects and mites on beans, *J. Econ. Entomol.,* 63, 1486, 1970.
111. **Hagen, A. F.,** The biology and control of the western bean cutworm in dent corn in Nebraska, *J. Econ. Entomol.,* 55, 628, 1962.
112. **Hagen, A. F.,** Evaluation of Thiodan and Sevin for control of webworms in sugarbeets, *J. Econ. Entomol.,* 54, 799, 1961.
113. **Hall, F. R.,** Bioeconomics of apple pests: cost appraisal of crop injury data, *J. Econ. Entomol.,* 67, 517, 1974.
114. **Hamilton, D. W., McAlister, H. J., Summerland, S. A., and Fahey, J. E.,** Control of pests attacking apples, peaches and pears with Nitroparaffin compounds, *J. Econ. Entomol.,* 45, 463, 1952.
115. **Harding, J. A.,** Tests with systemic insecticides for control of insects and certain diseases on potatoes, *J. Econ. Entomol.,* 55, 62, 1962.
116. **Harding, J. A. and Wolfenbarger, D. A.,** Granulated systemic insecticides for vegetable insect control in south Texas, *J. Econ. Entomol.,* 56, 687, 1963.
117. **Harper, A. M.,** Effect of insecticides on the sugarbeet root aphid, *Pemphigus betae, J. Econ. Entomol.,* 54, 1151, 1961.
118. **Harper, A. M., Lilly, C. E., and Bergen, P.,** Effect of insecticides on emergence of sugarbeet seedlings and on control of the sugarbeet root maggot, *J. Econ. Entomol.,* 54, 895, 1961.
119. **Harper, J. D., Kouskolekas, C. A., and Hollingsworth, M. H.,** Cabbage looper and cabbage webworm control on collars (sic), 1977, in *Insecticide and Acaricide Tests,* Vol. 3, Rep. No. 92, Entomological Society of America, College Park, Md., 1978.
120. **Harrison, F. P. and Wooldridge, A. W.,** Green peach aphid control on tobacco with systemic insecticides, *J. Econ. Entomol.,* 59, 270, 1966.
121. **Harrison, F. P. and Osgood, C. E.,** Insecticide tests for tobacco flea beetle and greenpeach aphid control in Maryland tobacco, *J. Econ. Entomol.,* 53, 963, 1960.
122. **Hartsack, A. W., Jr., Ridgway, R. L., and Jones, S. L.,** Damage to cotton by the bollworm and tobacco budworm, *J. Econ. Entomol.,* 71, 239, 1978.
123. **Harvey, T. L. and Hackerott, H. L.,** Chemical control of a greenbug on sorghum and infestation effects on yields, *J. Econ. Entomol.,* 63, 1536, 1970.

124. **Headley, J. C.,** Defining the economic threshold, in Pest Control Strategies for the Future, National Academy of Science, Washington, D.C., p. 100, 1972.
125. **Henderson, C. A. and Davis, F. M.,** Four insecticides tested in the field for control of *Diatraea grandiosella, J. Econ. Entomol.,* 63, 1495, 1970.
126. **Henderson, C. A. and Davis, F. M.,** Insecticidal control of the southwestern cornborer, *J. Econ. Entomol.,* 629, 1967.
127. **Henne, R. C.,** Control of aphids, cabbage looper and variegated cutworms on tomatoes, in *Insecticide and Acaricide Tests,* Vol. 3, Rep. No. 142, Entomological Society of America, College Park, Md., 1978.
128. **Hensley, S. D., McCormick, W. J., Long, W. H., and Concienne, E. J.,** Field tests with new insecticides for control of the sugarcane borer in Louisiana in 1959, *J. Econ. Entomol.,* 54, 1153, 1961.
129. **Hintz, T. R., Wilson, M. C., and Armbrust, E. J.,** Impact of alfalfa weevil larval feeding on the quality and yield of first cutting alfalfa, *J. Econ. Entomol.,* 69, 749, 1976.
130. **Hofmaster, R. N.,** Colorado potato beetle control with foliage application of insecticides, 1976, in *Insecticide and Acaricide Tests,* Vol. 2, Rep. No. 71, Entomological Society of America, College Park, Md., 1977.
131. **Hofmaster, R. N.,** Colorado potato beetle control with systemic insecticides applied to the soil, 1976, in *Insecticide and Acaricide Tests,* Vol. 2, Rep. No. 72, Entomological Society of America, College Park, Md., 1977.
132. **Hofmaster, R. N.,** Potato tuberworm and potato aphid control on potato with foliar insecticidal sprays, 1976, in *Insecticide and Acaricide Tests,* Vol. 2, Rep. No. 73, Entomological Society of America, College Park, Md., 1977.
133. **Hofmaster, R. N.,** Systemic granulars to control Mexican bean beetle and Atlantic spider mite in soybeans, 1976, in *Insecticide and Acaricide Tests,* Vol. 2, Rep. No. 132, Entomological Society of America, College Park, Md., 1977.
134. **Hofmaster, R. N.,** Effectiveness of new insecticides on the potato leafhopper and the influence of leafhopper control and potato variety on tuber worm infestations, *J. Econ. Entomol.,* 52, 908, 1959.
135. **Hofmaster, R. N.,** Flea beetle control on Irish potatoes in eastern Virginia, *J. Econ. Entomol.,* 49, 530, 1956.
136. **Hofmaster, R. N. and Francis, J.,** Potato insect control with foliar sprays, 1977, in *Insecticide and Acaricide Tests,* Vol. 3, Rep. No. 116, Entomological Society of America, College Park, Md., 1978.
137. **Hofmaster, R. N. and Waterfield, R. L.,** Insecticides applied to the soil for control of the Colorado potato beetle in Virgina, *J. Econ. Entomol.,* 65, 1672, 1972.
138. **Hofmaster, R. N., Waterfield, R. L., and Boyd, J. C.,** Insecticides applied to the soil for control of eight species of insects on Irish potatoes in Virginia, *J. Econ. Entomol.,* 60, 1311, 1967.
139. **Hofmaster, R. L., Bray, D. F., and Ditman, L. P.,** Effectiveness of insecticides against the European cornborer and green peach aphid on peppers, *J. Econ. Entomol.,* 53, 624, 1960.
140. **Hofmaster, R. L. and Greenwood, D. E.,** Control of the 2-spotted mite on strawberries, *J. Econ. Entomol.,* 44, 514, 1951.
141. **Hopkins, A. R. and Taft, H. M.,** Control of cotton pests by aerial application of ultra-low-volume (undiluted) technical insecticides, *J. Econ. Entomol.,* 60, 561, 1967.
142. **Hopkins, A. R. and Taft, H. M.,** Control of certain cotton pests with a new systemic insecticide, UC-21149, *J. Econ. Entomol.,* 58, 746, 1965.
143. **Hower, A. A.,** Potato leafhopper and alfalfa blotch leafminer control, 1977, in *Insecticide and Acaricide Tests,* Vol. 3, Rep. No. 156, Entomological Society of America, College Park, Md., 1978.
144. **Howitt, A. J. and Cole, S. G.,** Chemical control of the carrot rustfly, *Psila Rosila Rosea* (F)., in western Washington, *J. Econ. Entomol.,* 52, 963, 1959.
145. **Huddleston, E. W., Dressel, E. M., and Watts, J. G.,** Economic threshold for range caterpillar larvae on blue grama pasture in northeastern Lincoln County, New Mexico, in 1975, *N. M. Agric. Expt. Stn. Res. Rep.,* 314, 1, 1976.
146. **Hyslop, J. A.,** Losses occasioned by insects, mites, and ticks in the United States, E-444, Bureau of Entomology and Plant Quarantine, U.S. Department of Agriculture, Beltsville, Md., 1938.
146a. **Jackson, R. D.,** personal communication, 1979.
147. **Jaques, R. P.,** Field efficacy of viruses infectious to the cabbage looper and imported cabbage worm on late cabbage, *J. Econ. Entomol.,* 70, 111, 1977.
148. **Jaques, R. P.,** Control of the cabbage looper and the imported cabbage worm by viruses and bacteria, *J. Econ. Entomol.,* 65, 757, 1972.
149. **Johnson, A. W.,** Yield and value of flue-cured tobacco treated with carbofuran in certain foliar insecticides for tobacco budworm control, *J. Econ. Entomol.,* 69, 715, 1976.
150. **Judenko, E.,** The assessment of economic losses in yield of annual crops caused by pests, and the problem of the economic threshold, *Pest Art. News Summ.,* 18(2), 186, 1972.

151. **Keaster, A. J. and Fairchid, M. L.,** Occurrence and control of sandwire worm in Missouri, *J. Econ. Entomol.,* 53, 963, 1960.
152. **Keaster, A. J. and Fairchild, M. L.,** Reduction of a corn virus disease incidence and control of southwestern cornborer with systemic insecticides, *J. Econ. Entomol.,* 61, 367, 1968.
153. **King, H. L. and Huston, R.,** Further studies on control of red-banded leafroller with parathion, *J. Econ. Entomol.,* 42, 398, 1949.
154. **King, H. L., Hutson, R., and Farr, T. H.,** The control of red-banded leaf roller with parathion, *J. Econ. Entomol.,* 41, 976, 1948.
155. **Kinzer, H. G. and Henderson, C. F.,** Effect of sorghum webworm on yield of grain sorghum in Oklahoma, *J. Econ. Entomol.,* 60., 118, 1967.
156. **Kinzer, R. E., Meariola, L. A., Ridgway, R. L., and Jones, S. L.,** Insecticides in a nuclear polyhedrosis virus for control of the budworm and tobacco budworm on cotton, *J. Econ. Entomol.,* 69, 697, 1976.
157. **Klassen, W.,** Pest management: organization and resources for implementation, in *Insects, Science, and Society,* Pimentel, D., Ed., Academic Press, New York, 1975.
158. **Klostermeyer, E. C.,** The relationship among pea aphids, lygus bugs, and alfalfa seed yields, *J. Econ. Entomol.,* 55, 462, 1962.
159. **Klostermeyer, E. C.,** Effect of mite control on corn yield, *J. Econ. Entomol.,* 54, 608, 1961.
160. **Klostermeyer, E. C. and Rasmussen, W. B.,** The effect of soil insecticide treatments on mite population and damage, *J. Econ. Entomol.,* 46, 910, 1953.
161. **Kulash, W. M.,** Benzene hexachloride, DDT, and *Ryanex* to control soybean caterpillars, *J. Econ. Entomol.,* 40, 927, 1947.
162. **Kung, S. P., Su, C. Y., and Rose, R. I.,** Control of sweet potato weevil, in *Insecticide and Acaricide Tests,* Vol. 1, Rep. No. 101, Entomological Society of American, College Park, Md., 1976.
163. **Lange, W. H., Jr., Carlson, E. C., and Leach, L. D.,** Seed treatments for wireworm control with particular reference to the use of lindane, *J. Econ. Entomol.,* 42, 942, 1949.
164. **Libby, J. L.,** European cornborer control in yield of Yolo wonder peppers, in *Insecticide and Acaricide Tests,* Vol. 1, Rep. No. 68, Entomological Society of America, College Park, Md., 1976.
165. **Libby, J. L.,** Potato leafhopper control, European cornborer damage, yield, and grade of Tenderette vari. snap beans, in *Insecticide and Acaricide Tests,* Vol. 1, Rep. No. 36, Entomological Society of America, College Park, Md., 1976.
166. **Libby, J. L. and Longridge, J. L.,** Soil applied systemic insecticides for potato leafhopper control and yield of potatoes, 1977, in *Insecticide and Acaricide Tests,* Vol. 3, Rep. No. 117, Entomological Society of America, College Park, Md., 1978.
167. **Lilly, C. E.,** Wireworms: efficacy of various insecticides for protection of potatoes in southern Alberta, *J. Econ. Entomol.,* 66, 1205, 1973.
168. **Lindquist, R. K., Bauerle, W. L., and Spadafora, R. R.,** Effect of the greenhouse whitefly on yields of greenhouse tomatoes, *J. Econ. Entomol.,* 65, 1406, 1972.
169. **Linduska, J. J.,** Lima beans, corn earworm control, 1977, in *Insecticide and Acaracide Tests,* Vol. 3, Rep. No. 74, Entomological Society of America, College Park, Md., 1978.
170. **Lunduska, J. J.,** Control of Colorado potato beetle in direct seeded tomatoes with systemic insecticides, 1976, in *Insecticide and Acaricide Tests,* Vol. 2, Rep. No. 99, Entomological Society of America, College Park, Md., 1977.
171. **Long, W. H., Concienne, E. J., Hensley, S. D., McCormick, W. J., and Newson, L. D.,** Control of the sugarcane borer with insecticides, *J. Econ. Entomol.,* 52, 821, 1959.
172. **Long, W. H., Hensley, S. D., Concienne, E. J., and McCormick, W. J.,** Field tests with new insecticides for sugarcane borer control in Louisiana in 1960, *J. Econ. Entomol.,* 54, 1155, 1961.
173. **Longridge, J. L. and Libby, J. L.,** Potato leafhopper control and potato yields, 10-day interval spray program, 1977, in *Insecticide and Acaricide Tests,* Vol. 3, Rep. No. 118, Entomological Society of America, College Park, Md., 1978.
174. **MacCollom, G. B.,** Control of *Madidae* spp. in birdsfoot trefoil seed fields, *J. Econ. Entomol.,* 60, 1116, 1967.
175. **Magness, J. R., Markle, G. M., and Compton, C. C.,** Food and feed crops of the United States, a descriptive list classified according to potentials for pesticide residues, Bull. No. 882, *N.J. Agric. Exp. Stn.,* New Brunswick, N.J.
176. **Marlatt, C. L.,** U.S. Department of Agriculture Yearbook, The annual loss occasioned by destructive insects in the United States, U.S. Department of Agriculture, Washington, D.C., 1904, 461.
177. **Mayo, Z. B.,** Emergency post planting applications of insecticides to control larvae of the western and northern corn rootworm in Nebraska, *J. Econ. Entomol.,* 69, 600, 1976.
178. **Mayo, Z. B., Peters, L. L., Campbell, J. B., Hagen, A. F., and Witkowski, J. F.,** Field corn, corn rootworm larvae control, 1977, in *Insecticide and Acaricide Tests,* Vol. 3, Rep. No. 170, Entomological Society of America, College Park, Md., 1978.

179. **Mayo, Z. B., Peters, L. L., Campbell, J. B., and Hagen, A. F., and Witkowski, J. F.,** Evaluation of insecticides to control larvae of the corn rootworm in Nebraska, 1975, in *Insecticide and Acaricide Tests,* Vol. 1, Rep. No. 113, Entomological Society of America, College Park, Md., 1976.
180. **McEwen, F. L. and Hervey, G. E.,** The effect of lygus bug control on the yield of lima beans, *J. Econ. Entomol.,* 53, 513, 1960.
181. **McGarr, R. L., Dulmage, H. T., and Wolfenbarger, D. A.,** Field tests with HD-1 Delta Endotoxin of *Bacillus thuringiensis* and with chemical insecticides for control of the tobacco budworm and the bollworm in 1970, *J. Econ. Entomol.,* 65, 897, 1972.
182. **McGarr, R. L. and Wolfenbarger, D. A.,** Insecticides for control of four cotton insects in 1968, *J. Econ. Entomol.,* 63, 1324, 1970.
183. **Medler, J. T. and Brooks, G. N.,** Insect control in relation to alfalfa seed production in central Wisconsin, *J. Econ. Entomol.,* 50, 336, 1957.
184. **Medler, J. T. and Fisher, E. H.** Leafhopper control with Methoxychlor and Parathion to increase alfalfa hay production, *J. Econ. Entomol.,* 56, 511, 1953.
185. **Merritt, D. L. and Apple, J. W.,** Yield reduction of oats caused by the cereal leaf beetle, *J. Econ. Entomol.,* 62, 298, 1969.
186. **Metcalf, C. L., Flint, W. P., and Metcalf, R. L.,** *Destructive and Useful Insects, Their Habits and Control,* 4th ed., McGraw-Hill, New York, 1962.
187. **Miller, P. M. and Kring, J. B.,** Reduction of nematode and insect damage to potatoes by band application of systemic insecticides and soil fumigation, *J. Econ. Entomol.,* 63, 186, 1970.
188. **Moore, D. H.,** Field evaluation of phiodan as an insecticide for potatoes, *J. Econ. Entomol.,* 52, 564, 1959.
189. **Moore, C. R., Owens, J. C., Ashdown, D., Huddleston, E. W., and Turner, W. F.,** Greenbug control on wheat in 1967—69, *J. Econ. Entomol.,* 65, 764, 1972.
190. **Morgan, L. W., Snow, J. W., and Peach, M. J.,** Chemical thrips control; effects on growth and yield of peanuts in Georgia, *J. Econ. Entomol.,* 63, 1253, 1970.
191. **Morrill, W. L. and Nelson, L. R.,** Hessian fly control with carbofuran, *J. Econ. Entomol.,* 69, 123, 1976.
192. **Muma, M. H., Hill, R. E., and Hixson, E.,** Soil treatments for corn rootworm control, *J. Econ. Entomol.,* 42, 822, 1949.
193. **Neal, J. W., Jr. and Ratcliffe, R. H.,** Clover root curculio; control with granular carbofuran as measured by alfalfa regrowth, yield, and root damage, *J. Econ. Entomol.,* 68, 829, 1975.
194. **Negm, A. A., Hensley, S. D., and Concienne, E. J.,** Insecticidal control of the sugarcane in corn, *J. Econ. Entomol.,* 62, 245, 1969.
195. **Ogunlana, M. O. and Pedigo, L. P.,** Economic injury levels of the potato leafhopper on soybeans in Iowa, *J. Econ. Entomol.,* 67, 29, 1974.
196. **Onsager, J. A.,** Pacific coast wireworm: relationship between injury and damage to potatoes, *J. Econ. Entomol.,* 68, 203, 1975.
197. **Onsager, J. A. and Foiles, L. L.,** Control of wireworms on summer potatoes in eastern Washington, *J. Econ. Entomol.,* 63, 1883, 1970.
198. **Owens, J. C., Witkowski, J. F., Tollefson, J. J., Rogers, R. R., and Peters, D. C.,** Greenhouse evaluation of soil insecticides for western corn root worm control, *J. Econ. Entomol.,* 67, 772, 1974.
199. **Parencia, C. R., Jr. and Cowan, C. B., Jr.,** Comparative yield of cotton in treated and untreated plots in insect-control experiments in central Texas, 1939—1970, *J. Econ. Entomol.,* 65, 480, 1972.
200. **Parker, W. L., Warren, J, C., and Stearns, L. A.,** Chlorinated camphene on potatoes, *J. Econ. Entomol.,* 41, 275, 1948.
201. **Peay, W. E., Stanger, C. E., and Swenson, A. A.,** Preliminary evaluation of soil insecticides for sugarbeet root maggot control, *J. Econ. Entomol.,* 61, 19, 1968.
202. **Peters, D. C.,** The value of soil insect control in Iowa corn, 1951-70, *J. Econ. Entomol.,* 68, 483, 1975.
202a. **Peters, D. C.,** personal communication, 1979.
203. **Peterson, A. G. and Vea, E. V.,** Silvertop of bluegrass in Minnesota, *J. Econ. Entomol.,* 64, 247, 1971.
204. **Peterson, A. G., Silberman, M. S., and Meade, A. B.,** Development of resistance to insecticides by the onion maggot, *Hulemya antiqua,* in Minnesota, *J. Econ. Entomol.,* 56, 580, 1963.
205. **Pfrimmer, T. R., Furr, R. E., and Stadelbacher, E. A.,** Materials for control of boll weevils, bollworms, and tobacco budworms on cotton at Stoneville, Mississippi, *J. Econ. Entomol.,* 64, 475, 1971.
206. **Pieters, E. P., Mullins, W., and Halford, W. T.,** Chemical control of *Heltiohis* (sic) Spp. on cotton, 1977, in *Insecticide and Acaricide Tests,* Vol. 3, Rep. No. 176, Entomological Society of America, College Park, Md., 1978.

207. **Pieters, E. P. and Boyette, J. D.,** Evaluation of insecticides to control *Heliothis* spp. on cotton in Mississippi, 1976, in *Insecticide and Acaricide Tests,* Vol. 2, Rep. No. 121, Entomological Society of America, College Park, Md., 1977.
208. **Pimentel, D., Drummel, J., Gallahan, D., Hough, J., Merrill, A., Schreiner, I., Vittum, P., Koziol, F., Back, E., Yen, D., and Fiance, S.,** Benefits and costs of pesticide use in US food production, *BioScience,* 28(12), 778, 1978.
209. **Pond, D. D.,** Field evaluation of insecticides for the control of aphids on potatoes, *J. Econ. Entomol.,* 60, 1203, 1967.
210. **Poston, F. L. and Pedigo, L. P.,** Simulation of painted lady and green clover worm damage to soybeans, *J. Econ. Entomol.,* 69, 423, 1976.
211. **Price, R. G., Young, J. H., Bogel, D., and Ree, B.,** Cotton bollworm and tabacco (sic) budworm control, 1977, in *Insecticide and Acaricide Tests,* Vol. 3, Rep. No. 177, Entomological Society of America, College Park, Md., 1978.
212. **Radcliffe, E. B.,** Wireworm control on potatoes grown on sod, 1973, in *Insecticide and Acaricide Tests,* Vol. 1, Rep. No. 77, Entomological Society of America, College Park, Md., 1976.
213. **Randolph, N. M., Meisch, M. V., and Teetes, G. L.,** Effectiveness of certain insecticides against the sorghum midge based on a new method of determining infestation, *J. Econ. Entomol.,* 64, 87, 1971.
214. **Ratcliffe, R. H., Ditman, L. P., and Young, J. R.,** Field experiments on the insecticidal control of insects attacking peas, snap, and lima beans, *J. Econ. Entomol.,* 53, 18, 1960.
215. **Ridgway, R. L., Walker, H. J., Hanna, R. L., and Owen, W. L.,** Fertilizers impregnated with systemic insecticides for control of cotton insects, *J. Econ. Entomol.,* 60, 592, 1967.
216. **Rodriguez, J. G. and Chaplin, C. E.,** Pesticide performance on strawberries, *J. Econ. Entomol.,* 55, 184, 1962.
217. **Rogers, C. E.,** Economic injury level for *Contarnia texana* on guar, *J. Econ. Entomol.,* 69, 693, 1976.
218. **Rogers, C. E.,** Economic impact of midges on guar seed production in the Texas rolling plains, *J. Econ. Entomol.,* 65, 1689, 1972.
219. **Schuster, D. J.,** Vegetable leafminer control on tomato, 1977, in *Insecticide and Acaricide Tests,* Vol. 3, Rep. No. 147, Entomological Society of America, College Park, Md., 1978.
220. **Schwartz, P. H., Jr., Osgood, C. E., and Ditman, L. P.,** Experiments with granulated systemic insecticides for control of insects on potatoes, lima beans, and sweet corn, *J. Econ. Entomol.,* 54, 663, 1961.
221. **Scott, D. R. and Carpenter, G. P.,** Wireworm control on potatoes in Idaho with side-dressed and broadcast insecticides, *J. Econ. Entomol.,* 64, 945, 1971.
222. **Sechriest, R. E.,** Suppression of corn rootworm larvae in Illinois, 1975, in *Insecticide and Acaricide Tests,* Vol. 1, Rep. No. 118, Entomological Society of America, College Park, Md., 1976.
223. **Sechriest, R. E.,** Control of the garden snyphylan in Illinois cornfields, *J. Econ. Entomol.,* 65, 599, 1972.
224. **Sechriest, R. E. and Paullus., J. H.,** Control of Phyllophaga larvae and cornfield ants in field corn in northern Illinois, 1975, in *Insecticide and Acaricide Tests,* Vol. 2, Rep. No. 113, Entomological Society of America, College Park, Md., 1977.
225. **Semel, M.,** Tests for the control of *Tetranychus Telarius* (L.) on lima beans, *J. Econ. Entomol.,* 51, 735, 1958.
226. **Shands, W. A., Simpson, G. W., and Gordon, C. C.,** Insect predators for controlling aphids on potatoes. V. Numbers of eggs and schedules for introducing them in large field cages, *J. Econ. Entomol.,* 65, 810, 1972.
227. **Shaw, W. C. and Jansen, L. L.,** Chemical weed control strategies for the future, in *Pest Control Strategies for the Future,* National Academy of Sciences, Washington, D.C., 1972.
228. **Shipp, O. E. and Earhart, R. W.,** Comparative seasonal abundance of *Heliothis* larvae in cotton fields in Texas and Arkansas during summer 1964, *J. Econ. Entomol.,* 60, 893, 1967.
229. **Shirck, F. H.,** Experiments for the control of the onion maggot, *J. Econ. Entomol.,* 50, 577, 1957.
230. **Shirck, F. H. and Douglass, J. R.,** Experiments on control of the onion thrips in Idaho, *J. Econ. Entomol.,* 49, 526, 1956.
231. **Simonet, D. E., Martinez, D. G., and Pienkowski, R. L.,** Alfalfa foliar sprays for potato leafhopper control, 1977, in *Insecticide and Acaricide Tests,* Vol. 3, Rep. No. 160, Entomological Society of America, College Park, Md., 1978.
232. **Smith, J. C.,** Tobacco thrips-nematode control on Virginia-type peanuts, *J. Econ. Entomol.,* 65, 1700, 1972.
233. **Smith, J. C.,** Field evaluation of candidate insecticides for control of the southern corn rootworm on peanuts in Virginia, *J. Econ. Entomol.,* 64, 280, 1971.
234. **Sorensen, K. A.,** Irish potato, European corn borer control, 1975, in *Insecticide and Acaricide Tests,* Vol. 3, Rep. No. 131, Entomological Society of America, College Park, Md., 1978.

235. **Sorensen, K. A.,** Irish potato, European corn borer control, 1976, in *Insecticide and Acaricide Tests,* Vol. 3, Rep. No. 132, Entomological Society of America, College Park, Md., 1978.
236. **Sorensen, K. A.,** Snap beans, insect control, 1976, in *Insecticide and Acaricide Tests,* Vol. 3, Rep. No. 76, Entomological Society of America, College Park, Md., 1978.
237. **Sorensen, K. A.,** Snap beans, Mexican bean beetle control, 1974, in *Insecticide and Acaricide Tests,* Vol. 2, Rep. No. 50, Entomological Society of America, College Park, Md., 1976.
238. **Starks, K. J. and Lilly, J. H.,** Some effects of insecticide seed treatment on dent corn, *J. Econ. Entomol.,* 48, 549, 1955.
239. California Department of Food and Agriculture, Estimated Damage and Crop Loss Caused by Insects and Mite Pests, 1977, California Department of Food and Agriculture, Sacramento, 1979.
240. **Stevenson, A. P.,** Soil treatments with insecticides to control the root form of grape Phylloxera, *J. Econ. Entomol.,* 61, 1168, 1968.
241. **Stoltz, R. L.,** Two-spotted spider mite control on beans with a soil-applied systemic insecticide, 1977, in *Insecticide and Acaricide Tests,* Vol. 3, Rep. No. 77, Entomological Society of America, College Park, Md., 1978.
242. **Stoltz, R. L., Bishop, G. W., and Sandvol, L.,** Control of Colorado potato beetle and green peach aphid with soil applied systemic insecticides, 1977, in *Insecticide and Acaricide Tests,* Vol. 3, Rep. No. 133, Entomological Society of America, College Park, Md., 1978.
243. **Stone, M. W. and Foley, F. B.,** Field experiments with insecticides for the control of wireworms in irrigated lands, *J. Econ. Entomol.,* 46, 1075, 1953.
244. **Stone, M. W. and Foley, F. B.,** Field tests with ethylene chlorobromide on wireworms, *J. Econ. Entomol.,* 44, 711, 1951.
245. **Stone, M. W., Foley, F. B., and Campbell, R. E.,** Field tests with various insecticides for control of lygus bugs and the corn earworm on lima beans, *J. Econ. Entomol.,* 53, 397, 1960.
246. **Su, C. Y., Kung, S. P., and Rose, R. I.,** Control of cruciferous insect pests, 1974 and 1975, in *Insecticide and Acaricide Tests,* Vol. 1, Rep. No. 49, Entomological Society of America, College Park, Md., 1976.
247. **Su, C. Y., Kung, S. P., Rose, R. I.,** Control of cornborers in corn, 1974 and 1975, in *Insecticide and Acaricide Tests,* Vol. 1, Rep. No. 119, Entomological Society of America, College Park, Md., 1976.
248. **Summers, C. G. and McClellan, W. D.,** Interaction between Egyptian alfalfa weevil feeding and foliar disease: impact on yield and quality in alfalfa, *J. Econ. Entomol.,* 68, 487, 1975.
249. **Taft, H. M. and Hopkins, A. R.,** Control of cotton pests with low-volume insecticides applied with a low-volume mist sprayer, *J. Econ. Entomol.,* 60, 607, 1967.
250. **Tappan, W. B.,** Insecticides for the control of the green peach aphid on shade-grown tobacco, *J. Econ. Entomol.,* 56, 34, 1963.
251. **Teetes, G. L.,** in Crops Loss Assessment Methods, Chiarappa, L., Ed., Food and Agriculture Organization, Rome, 1977.
252. **Teetes, G. L.,** *Phyllophaga crinita:* damage assessment and control in grain sorghum and wheat, *J. Econ. Entomol.,* 66, 773, 1973.
253. **Teetes, G. L. and Johnson, J. W.,** Assessment of damage by the greenbug in grain sorghum hybrids of different maturities, *J. Econ. Entomol.,* 67, 514, 1974.
254. **Teetes, G. L., and Johnson, J. W.,** Damage assessment of the greenbug on sorghum, *J. Econ. Entomol.,* 66, 1181, 1973.
255. **Teetes, G. L. and Randolph, N. M.,** Effects of pesticides and dates of planting sunflowers on the sunflower moth, *J. Econ. Entomol.,* 64, 124, 1971.
256. **Teetes, G. L. and Randolph, N. M.,** Chemical and cultural control of the sunflower moth in Texas, *J. Econ. Entomol.,* 62, 1444, 1969.
257. **Teetes, G. L. and Randolph, N. M.,** Chemical control of sunflower moth on sunflowers, *J. Econ. Entomol.,* 61, 1344, 1968.
257a. **Teetes, G. L.,** personal communication, 1979.
258. **Thomas, G. D., Ignoffo, C. M., and Smith, D. B.,** Influence of defoliation and depodding on quality of soybeans, *J. Econ. Entomol.,* 69, 737, 1976.
259. **Thompson, L. S. and Sanderson, J. B.,** Pea moth control in field peas with insecticides and the effect on crop yield, *J. Econ. Entomol.,* 70, 518, 1977.
260. **Todd, J. W. and Suber, E. F.,** *Losses due to insect damage and costs of control in Georgia in 1977,* Research Report, Division of Entomology, University of Georgia, Athens, 1979.
261. **Todd, J. W. and Suber, E. F.,** *Economic losses due to insect damage and costs of control in Georgia for the years 1971—1976,* Division of Entomology, University of Georgia, Athens, 1978.
262. **Todd, J. W. and Canerday, T. D.,** Control of soybean-insect pest with certain systemic insecticides, *J. Econ. Entomol.,* 65, 501, 1972.

263. **Todd, J. W., Minton, N. A., and Dukes, P. D.,** Infestations of phytophagous insects on soybeans following applications of DuPont 1410 foliar sprays and other insecticides applied to the soil, *J. Econ. Entomol.,* 65, 295, 1972.
264. **Toscanono, N. C. and Stern, V. N.,** Cotton yield and quality loss caused by various levels of stinkbug infestations, *J. Econ. Entomol.,* 69, 53, 1976.
265. **Turnipseed, S. G.,** Systemic insecticides for control of soybean insects in South Carolina, *J. Econ. Entomol.,* 60, 1054, 1967.
266. **Turnipseed, S. G., Heinrichs, E. A., Dasalva, R. F., and Todd, J. W.,** Response of soybean insects to foliar applications of chitin synthesis inhibitor TH-6040, *J. Econ. Entomol.,* 67, 760, 1974.
267. **Turpin, F. T. and Thieme, J. M.,** Impact of soil insecticide uses on corn production in Indiana: 1972—1974, *J. Econ. Entomol.,* 71, 83, 1978.
268. **Tuttle, D. M., and Arvizo, G. L.,** Alfalfa, blue alfalfa aphid control, 1977, in *Insecticide and Acaricide Tests,* Vol. 3, Rep. No. 161, Entomological Society of America, College Park, Md., 1978.
269. **Tuttle, D. M. and Butler, G. D., Jr.,** The yellow clover aphid — a new alfalfa pest in the Southwest, *J. Econ. Entomol.,* 47, 1157, 1954.
270. United States Bureau of the Census, Census of Agriculture, Vol. 2, Part IV., U.S. Department of Commerce, Washington, D.C., IV-37 to IV-40, 1979.
271. United States Department of Agriculture, Crop production 1978 annual summary acreage yield production ESCS, CrPr 2-1, Crop Reporting Board, U.S. Department of Agriculture, Washington, D.C., 1979.
272. U.S. Department of Agriculture, Agricultural Statistics, U.S. Government Printing Office, Washington, D.C., 1978.
273. U.S. Department of Agriculture, Guidelines for the use of insecticides to control insects affecting crops, livestock, households, stored products, forests, and forest products. *U.S. Dept Agric. Agric. Handb.* 452, 1974.
274. U.S. Department of Agriculture, Losses in agriculture, *U.S. Dep. Agric. Agric. Handb.* 291, 1965.
275. U.S. Deptartment of Agriculture, Losses in agriculture — A preliminary appraisal for review, ARS 20-1, Agricultural Research Service, U.S. Department of Agriculture, 1954.
276. U.S. Department of Agriculture, Examples of Estimated Losses Caused by Insects, Bureau of Entomology and Plant Quarantine, U.S. Department of Agriculture, Washington, D.C., 1946.
277. **Vea, E. V. and Eckenrode, C. J.,** Seed maggot injury on surviving bean seedlings influences yield, *J. Econ. Entomol.,* 69, 545, 1976.
278. **Waddill, V.,** Evaluation of insecticides for control of insect pests of tomatoes, 1977, in *Insecticide and Acaricide Tests,* Vol. 3, Rep. No. 149, Entomological Society of America, College Park, Md., 1978.
279. **Waddill, V.,** Sweet potato weevil control, 1976, in *Insecticide and Acaricide Tests,* Vol. 2, Rep. No. 97, Entomological Society of America, College Park, Md., 1978.
280. **Waddill, V.,** Tomato pinworm control, 1975, in *Insecticide and Acaricide Tests,* Vol. 1, Rep. No. 105, Entomological Society of America, College Park, Md., 1976.
281. **Walton, R. R.,** Effects of chlorinated hydrocarbons and Sabadilla on insects of plants, *J. Econ. Entomol.,* 40, 389, 1947.
282. **Ward, C. R., Huddleston, E. W., Ashdown, D., Owens, J. C., and Polk, K. L.,** Greenbug control on grain sorghum and the effects of tested insecticides on other insects, *J. Econ. Entomol.,* 63, 1929, 1970.
283. **Ward, C. R., Huddleston, E. W., Owens, J. C., Hills, T. M., Richardson, L. G., and Ashdown, D.,** Control of the banks grass mite attacking grain sorghum and corn in west Texas, *J. Econ. Entomol.,* 65, 523, 1972.
284. **Watson, T. F.,** *B. thurberiella* control in cotton, 1975, in *Insecticide and Acaricide Tests,* Vol. 1, Rep. No. 124, Entomological Society of America, College Park, Md., 1976.
285. **Watson, T. F.,** *P. gossypiella* control in cotton, 1975, in *Insecticide and Acaricide Tests,* Vol. 1, Rep. No. 125, Entomological Society of America, College Park, Md., 1976.
286. **Westigard, P. H.,** Pest status of insects and mites on pear in southern Oregon, *J. Econ. Entomol.,* 66, 227, 1973.
287. **Wilcox, J. and Howland, A. F.,** Experiments for the control of spider mites on lima beans, *J. Econ. Entomol.,* 50, 129, 1957.
288. **Wilcox, J. and Howland, A. F.,** New insecticides for control of onion thrips in southern California, *J. Econ. Entomol.,* 43, 690, 1950.
289. **Wilson, M. D., Treece, R. E., Shade, R. E., Day, K. M., and Stivers, R. K.,** Impact of cereal leaf beetle larvae on yields of oats, *J. Econ. Entomol.,* 62, 699, 1969.
290. **Witkowski, J. F.,** Evaluation of five insecticides to control a natural infestation of first brood larvae of European cornborer in northeast Nebraska, 1975, in *Insecticide and Acaricide Tests,* Vol. 1, Rep. No. 121, Entomological Society of America, College Park, Md., 1976.

291. **Wolfenbarger, D. A., McGarr, R. L., Longoria, R. R., and Nosky, J. B.,** Toxicity of EPN, Accothion, and certain chlorinated hydrocarbons to certain cotton insects, *J. Econ. Entomol.,* 63, 1568, 1970.

292. **Yeargan, K. V.,** Effects of green stinkbug damage on yield and quality of soybeans, *J. Econ. Entomol.,* 70, 619, 1977.

293. **York, A. C.,** Dry beans, Mexican bean beetle control, 1975, in *Insecticide and Acaricide Tests,* Vol. 1, Rep. No. 37, Entomological Society of America, College Park, Md., 1976.

294. **York, A. C.,** Snap beans, *E. varivestis* control, in *Insecticide and Acaricide Tests,* Vol. 1, Rep. No. 38, Entomological Society of America, College Park, Md., 1976.

295. **York, A. C.,** Snap beans, *E. varivestis* control, in *Insecticide and Acaricide Tests,* Vol. 1, Rep. No. 39, Entomological Society of America, College Park, Md., 1976.

296. **York, A. C.,** Tomato miscellaneous pests, in, *Insecticide and Acaricide Tests,* Vol. 1, Rep. No. 107, Entomological Society of America, College Park, Md., 1976.

297. **Young, J. R. and Ditman, L. P.,** Effectiveness of some newer insecticides for control of *Macrosiphum Pisi* (Harris) and *Epilachna Varivestis* Muls., *J. Econ. Entomol.,* 52, 541, 1959.

ESTIMATED LOSSES OF CROPS FROM PLANT PATHOGENS

W. Clive James

INTRODUCTION

The destruction of food and fiber crops by plant pathogens has been a consistently reported feature throughout the annals of history. During the last 200 years, there have been some notable examples of epidemics which have had a profound effect on the social development of man. The most devastating epidemic occurred in Ireland in 1845, when famine resulted after the potato crop was completely destroyed by the late blight fungus *Phytophthora infestans.* Twenty-five years later the coffee rust disease, caused by *Hemileia vastatrix,* was responsible for the complete destruction of the coffee industry in Ceylon. A century later, in 1970, the same fungus spread for the first time into the western hemisphere and caused substantial losses in coffee production in Brazil. In 1975, the coffee rust was observed in Nicaragua and unless the current eradication campaign is completely successful it now threatens coffee production in the whole of Central America.

The evolvement of crop monocultures with increasing genetic homogeneity in modern agriculture has considerably increased the vulnerability of crops to disease. In 1970 southern corn leaf blight epidemic in the U.S. dramatically demonstrated that gains from high yielding cultivars can be ephemeral and that diseases can still seriously disrupt world food supplies and trade patterns.

Although it is generally recognized that plant diseases cause significant losses in production there are actually very few reliable loss data available, with the exception of epidemics which lead to complete destruction. However, estimates of loss are a prerequisite to the rational development of any plant protection program. Plant pathology is basically a practical exercise in problem solving, aimed at the economic control of plant diseases. There are two basic and sequential processes involved in the resolution of any problem — the definition and solution — and within the context of plant pathology, estimation of loss is the definition process. Since it is difficult, illogical, and inefficient to attempt to solve a problem that has not been adequately defined, estimation of losses due to disease is considered to be a primary function in any plant protection program. Reliable loss estimates facilitate the objective identification of needs and, consequently, limited resources can be assigned on a priority basis to optimize the benefits from a given effort and increase the returns from investments in plant pathology.

The estimation of losses in food production is particularly important in a world that, on the one hand, is conscious of food deficits and on the other, of the pollution problems arising from fungicides produced from nonrenewable resources. In 1975 the food deficit of the developing countries of the world was estimated at 37 million tons and this deficit is expected to increase to between 120 to 145 million tons by 1990.[1] The current world fungicide market has been estimated at $1 billion with substantial increased usage on food crops.[2] One way to attempt a partial resolution of the dilemma is to rationalize fungicide usage using objective estimates of losses due to disease to determine plant protection needs.

In this chapter information from selected references has been collated to provide a brief overview of the impact of diseases on production. A global perspective will be accompanied by a summary of losses affecting major crops in the U.S., as well as information from disease surveys conducted on major crops in different countries.

Table 1
GLOBAL ESTIMATES OF LOSSES DUE TO PLANT DISEASES[3]

Crop	Loss in production (%)	Actual loss in production (millions of tons)	Value of loss ($ billions)	Corrected value of loss for 1976 ($ billions)
Wheat	9.1	33.3	2.2	4.4
Rice	8.9	39.4	3.2	6.4
Maize	9.4	32.7	1.6	3.2
Other cereals	8.6	29.9	1.7	3.4
Potatoes	21.8	88.9	3.4	6.8
Sugar beet/cane	16.5	232.3	2.3	4.6
Vegetables	10.1	31.1	2.3	4.6
Fruit/citrus/grapes	16.4	32.6	3.3	6.6
Stimulants	14.9	2.6	1.7	3.4
Oil crops	10.2	13.5	1.6	3.2
Fiber crops/rubber	11.8	3.1	1.5	3.0
Total		539.4	24.8	49.6

The quality of the data is variable and qualifications are often necessary when interpreting the information. Space will not allow consideration of all published material on crop losses due to disease, but the body of information presented is a representative sample of the data currently available.

GLOBAL ESTIMATES OF PLANT DISEASE LOSSES

The data collated and published by Cramer[3] is the only attempt to generate a comprehensive estimate of losses due to disease on a worldwide basis. The global loss due to diseases was estimated as 12% of the potential production, and at current day prices this would be equivalent to an annual monetary loss of approximately $50 billion* at the producer level. This estimate is based on the data published by Cramer[3] in 1967, and summarized in Table 1, with a correction to allow for the twofold increase in crop prices between 1967 and 1976, reported by the Food and Agriculture Organization.[4] In the calculation of such estimates many factors have to be considered, and the original reference should be consulted for any necessary qualifications and to facilitate better interpretation of the data. The estimated tonnage lost to disease on a global basis is approximately 550 million tons/annum (Table 1). Of this total approximately 135 million tons are cereals, which is considerably more than the 37-million-ton global food deficit in 1975 and approximately equivalent to the 120 to 145-million-ton deficit projected by 1990.

The data in Table 1 shows that, of the major crops, the highest percentage loss is 21.8% for potatoes, followed by 16.5 and 16.4% for sugar beet/cane and fruit crops, respectively. The high losses in potatoes are due to a complex of virus diseases, late blight, and some bacterial diseases, which are particularly important in developing countries without certified seed programs.

In terms of value, the highest losses are reported for potatoes, fruits, and rice, estimated at $6.8, $6.6, and $6.4 billion per annum, respectively. These substantial losses due to plant diseases occur despite the fact that a considerable and increasing quantity of fungicides is used annually, and currently estimated at $1 billion. The latter cost can be considered an additional loss factor to which should also be added the application costs sustained by the farmer. On a worldwide basis the cost of fungicide ($1 billion) is approximately 2% of the estimated total losses ($50 billion) due to disease.

* All numbers in U.S. dollars.

Table 2
GLOBAL DISTRIBUTION OF ESTIMATED PLANT DISEASE LOSSES[3]

Region	Value of losses ($ billions)		Loss/region expressed as % of total global loss	% Loss due to disease within each region
	1967	1976		
Asia	7.1	14.2	29	11.3
Europe	6.2	12.4	25	13.1
North/Central America	3.9	7.8	16	11.3
USSR/People's Republic of China	2.6	5.2	11	9.1
Africa	2.4	4.8	10	12.9
South America	2.1	4.2	9	15.2
Oceania	0.2	0.4	<1	12.6
Total	24.5	49.0	100	—

The distribution of disease losses in the different regions of the world is shown in Table 2, and is reported on an absolute value and percentage basis. Of the total global loss due to disease, approximately 30% occurs in the developing countries of Asia, with 25% in Europe and 15% in North and Central America. The figures reported by Cramer for percentage loss in Asia (11.3%) may be an underestimate because the International Rice Research Institute suggests that a conservative estimate for rice disease losses is 15%. Since losses from perishable crops would be higher than for rice, the average percentage loss for Asia is likely to be higher than 15%. The higher percentage loss for Europe (13.1%) compared with North and Central America (11.3%) is probably due to many factors including the more intensive agriculture of Europe compared with the extensive methods of production in North America.

LOSSES IN THE U.S.

The U.S. is the only country which has attempted to produce publications listing estimates of disease loss for all major crops. Three publications in 1937, 1954, and 1965[5] list total estimated losses due to plant diseases at $298 million, $2,194 million, and $3,251 million, respectively. The losses are expressed in terms of quantity (percent loss) and value, with the latter including quality losses. The reliability of loss estimates varies considerably, with the better data generated from surveys and the more subjective estimates based on judgment. The potential production, in the absence of disease and other constraint factors, has been used as the reference for calculating percentage loss. Furthermore, the estimators have assumed that losses from different diseases are additive, and this may overestimate the total loss. Within the last 50 years the U.S. has suffered some devastating epidemics which have resulted in significant losses in production. In 1935 stem rust of wheat resulted in a loss of 100 million bushels of wheat worth approximately $50 million. The development of race 15B of stem rust destroyed 65% of the durum wheat in 1953 and 75% in 1954, in addition to 25% of bread wheat production in the 2 years. A sudden outbreak of *Phytophthora* blight of tomatoes in 1946 caused an estimated loss of $40 million. In 1970, a new race of *Helminthosporium maydis,* which was highly virulent on corn with T-type cytoplasm, was responsible for decreasing yields in the U.S. by 710 million bushels, estimated to be worth approximately $1 billion.[6] Such a significant loss in production showed that the normal reserves were not large relative to the magnitude of this loss, and the epidemic

Table 3
ESTIMATED ANNUAL LOSSES CAUSED BY PLANT DISEASES FOR THE MAJOR CROP GROUPS IN U.S.[5]

Crop	Annual loss based on 1951—60 costs ($ billions)	Loss expressed as % of total	Annual loss corrected for 1976 costs ($ billions)
Field crops	1.9	58	3.8
Forage/pasture	0.8	26	1.6
Fruit/vegetables/ ornamentals	0.5	16	1.0
Total	3.2	100	6.4

had a major impact on prices of domestic food and on the international market. The above examples are exceptions, but disease does vary significantly from year to year depending on weather, cultural conditions, etc. It is for this reason that published figures represent an average over a period of years. The data summarized here are from the most recent report on disease losses in the U.S., published in 1965, and the data collected during the period 1951 to 1960. For convenience, losses are summarized for three crop groups: field, forage/pasture, and fruit/vegetables/ornamentals.

Table 3 shows that the total loss due to diseases was estimated at $3.25 billion in 1965. Food and Agriculture Organization statistics[4] show that crop prices received by U.S. farmers increased from an index of 225 in 1967 to 405 in 1976, i.e., a factor of 2, and on this basis the current estimated annual loss due to disease would be a minimum of $6 to $7 billion. Approximately 60% of the losses are sustained in field crops, with 25% and 15% in pasture/forage and fruit/vegetables/ornamental, respectively. It is noteworthy that total cost of controlling diseases in the U.S. in 1965 was estimated at $115 million or 3% of the total loss due to disease.

Field Crops

This represents the largest of the three groups of crops, and almost two thirds of the total losses occur in the field crops listed in Table 4. In 1965, the total disease losses on a value basis for the groups were estimated at almost $2 billion, and the major crops affected by diseases were corn, wheat, soybeans, cotton, tobacco, and oats. Subsequent to 1965 the acreages of soybeans, wheat, and corn have substantially increased and, consequently, given that their current disease susceptibility is the same, the value of losses from these crops will have substantially increased. The soybean hectarage has increased markedly from 12 million ha in 1965 to 20 million ha in 1976, and the value per unit of production has also increased, thus making the value of losses significantly greater. The estimated annual losses caused by specific diseases of the six major crops in the group are shown in Table 5. No comprehensive data on crop losses have been published since 1965, and, therefore, the data do not reflect significant changes that have occurred in the interim period, e.g., the 1970 southern corn leaf blight epidemic. However, in some cases it has been possible to check more current data for specific crops, and in these cases there has been little change in the relative importance of diseases and their total impact on crop production. The Cotton Disease Loss Estimate Committee of the Cotton Disease Council publishes annual loss estimates and the data for 1973[7] are very similar to those reported for cotton in Table 5. A total disease loss of 13.9% was reported for cotton in 1973 with verticillium wilt, seedling diseases, and boll rots causing losses of 2.9%, 2.7%, and 2.5%, respectively.

Table 4
ESTIMATED LOSSES DUE TO DISEASES AFFECTING FIELD CROPS IN U.S.[5]

Crop	Loss from potential production	
	% Loss	Value of loss ($ millions)[a]
Corn	12	527
Wheat	14	331
Cotton	12	301
Oats	21	198
Soybeans	14	143
Tobacco	11	132
Barley	14	55
Peanuts	28	54
Sorghum	9	43
Sugarbeets	16	27
Beans, dry	17	25
Rice	7	19
Sugar cane	23	14
Flax (seed)	10	12
Others	3—14	8

[a] Based on 1951—60 values.

Table 5
ESTIMATION OF PERCENTAGE LOSSES DUE TO DISEASES OF SELECTED MAJOR FIELD CROPS IN THE U.S.[a]

Crop/disease	% Loss
Corn	
Stalk rots (fungus)	3.0
Helminthosporium leaf blights	2.3
Seedling blights	1.6
Root rots	1.3
Ear rots	1.2
Smut	.7
Bacterial leaf blight	.5
Physoderma brown spot	.2
Others	1.2
Total	12.0
Wheat	
Stem rust	4.0
Leaf rust	2.5
Root rots	1.0
Septoria leaf and glume blotch	1.0
Wheat streak mosaic	1.0
Loose smut	.8
Cercosporella foot rot	.6
Scab	.5
Soilborne mosaics	.5
Common bunt	.4
Powdery mildew	.4
Take-all	.2
Bacterial diseases	.1
Dwarf bunt	.1
Miscellaneous leaf and head blights	.1
Miscellaneous virus diseases	.1
Fusarium and Typhula snow molds	.1
Stripe rust	.1
Others	.5
Total	14.0
Soybeans	
Phytophthora	2.3
Bacterial blight	2.2
Downy mildew	1.6
Bacterial pustule	1.4
Pod and stem blight	1.2
Stem canker	1.1
Bud blight	.8
Brown spot	.8
Brown stem rot	.8
Purple stain	.7
Fusarium root rot	.6
Frogeye leaf spot	.1
Others	.4
Total	14.0

Table 5 (continued)
ESTIMATION OF PERCENTAGE LOSSES DUE TO DISEASES OF SELECTED MAJOR FIELD CROPS IN THE U.S.[a]

Crop / Disease	%
Cotton	
Seedling diseases	2.6
Verticillium wilt	2.3
Boll rots	2.2
Bacterial blight	1.7
Fusarium wilt	1.1
Phymatotrichum root rot	.9
Ascochyta blight	.2
Others	1.0
Total	12.0
Tobacco	
Leaf decay during curing	2.7
Tobacco mosaic	1.4
Wildfire	1.3
Miscellaneous stalk and root diseases	1.1
Black shank	1.0
Black root rot	.9
Brown spot	.8
Blue mold	.6
Blackfire	.3
Miscellaneous leaf diseases	.3
Frogeye	.2
Fusarium wilt	.2
Granville wilt	.2
Total	11.0
Oats	
Yellow dwarf	3.8
Crown rust	3.7
Blast (sterility)	2.5
Root necrosis	2.4
Stem rust	2.3
Septoria foliage blight	1.8
Scab	.9
Helminthosporium leaf blotch	.7
Smuts	.6
Soilborne mosaic	.5
Bacterial stripe blight	.4
Victoria blight	.4
Blue dwarf	.3
Halo blight	.3
Helminthosporium culm rot	.2
Physiologic leaf spot	.1
Others	.1
Total	21.0

[a] Estimates from "Losses in Agriculture"[5] during the period 1951—60.

On a percentage loss basis the most susceptible crops within the group are peanuts and sugar cane (Table 4). Peanuts are particularly susceptible to *Cerospora* leaf spots and stem rot/southern blight, while many pathogens, particularly viruses, result in a decrease in production of sugar and sucrose content. No data are available to estimate the current value of losses due to diseases of field crops, but an estimate of $4 billion would be conservative considering the substantial increase in acreage of the crops and the twofold increase of general crop prices since 1965.

Forage/Pasture Crops

Approximately 50% of the total land area of the U.S. is utilized for pastures and this provides half of all the feed for domestic livestock. The estimation of losses due to disease in perennial grasses and legumes is extremely difficult. Losses due to forage plants grown for seed are also included in this group. The total losses for the group, estimated at $834 million, represent about one quarter of the total losses due to disease on all crops. Almost half of the total losses are associated with alfalfa cultivated for hay (Table 6) and one quarter associated with other hay crops. It is noteworthy that the percentage loss due to diseases of alfalfa is high at 24%, and is the result of multiple infections by foliar pathogens, wilts, crown and root rots, and virus diseases. Many of the clovers also suffer a high percentage of loss and are infected by a similar spectrum of diseases. The crops in this group differ from the field crops in that the value of the crop/acre is substantially lower. It follows that much of the loss sustained by forage crops is not preventable because of economic reasons, whereas many of the higher value field crops can be economically treated with fungicides.

Table 6
ESTIMATED LOSSES DUE TO DISEASES AFFECTING FORAGE/ PASTURE CROPS IN U.S.[5]

Crop	Loss from potential production	
	Loss (%)	Value of loss ($ millions)[a]
Alfalfa (hay)	24	389
All other hay plants	15	226
Forage plants for seed	4—52	24
Cropland pastures	9	77
Forestland pastures	3	16
Grassland	5	102
Total		834

[a] Based on 1951—60 values.

Table 7
ESTIMATED LOSSES DUE TO DISEASES AFFECTING FRUIT/VEGETABLE/ ORNAMENTAL CROPS[5]

Crop	Loss from potential production	
	Loss (%)	Value of loss ($ millions)[a]
Potatoes	19	90
Tomatoes	21	60
Grapes	27	48
Oranges	12	39
Strawberries	26	27
Beans (green)	20	20
Peaches	14	20
Apples	8	18
Lettuce	12	17
Onions	20	14
Ornamental plants and shade trees	1—22	14
Cherries	24	12
Peas	23	12
Celery	17	11
Lemons	25	11
Pears	17	10
Other vegetables	2—21	69
Other fruits and nuts	2—38	39
Total		531

[a] Based on 1951—60 values.

Fruit/Vegetable/Ornamental Crops

The annual losses due to diseases for this group were estimated at approximately $500 million in 1965. Vegetables constitute about 50% of the losses estimated at $300 million, and fruit and nuts at $225 million, with the remaining 8% equivalent to approximately $15 million, associated with ornamental plants. Of the crops listed in Table 7, the major losses occur in potatoes, tomatoes, grapes, and oranges. Among the

Table 8
ESTIMATION OF PERCENTAGE LOSSES DUE TO DISEASES OF SELECTED MAJOR FRUIT AND VEGETABLE CROPS IN THE U.S.[a]

Crop/disease	% Loss	Crop/disease	% Loss
Potatoes		Grapes	
Late blight	4.0	Leaf roll	8.4
Leaf roll	3.0	Black measles	5.0
Verticillium wilt	2.7	Fanleaf, yellow mosaic, and vein banding	5.0
Scab	1.5	Summer bunch rot	2.2
Early blight	1.0	Dead arm	2.0
Latent mosaic	1.0	Powdery mildew	1.8
Mild mosaic	1.0	Fruit rots	1.0
Rhizoctonia black scurf	1.0	Pierce's disease	.5
Rugose mosaic	1.0	Armillaria root rot	.2
Blackleg	.5	Black rot	.2
Fusarium wilt	.5	Yellow vein	.2
Ring rot	.5	Anthracnose	.1
Spindle tuber	.2	Downy mildew	.1
Bacterial brown rot	.1	Others	.3
Others	1.0	**Total**	27.0
Total	19.0		
Tomatoes		Oranges	
Tobacco mosaic	5.0	Root and foot rot	3.2
Gray leaf spot	3.0	Psorosis	2.0
Verticillium wilt	3.0	Exocortis	1.9
Bacterial spot	1.0	Tristeza	1.9
Blossom-end rot	1.0	Fruit and leaf spots	1.3
Curly top	1.0	Wood and heart rot	1.3
Early blight (alternaria leaf spot)	1.0	Blossom blight	.1
Fusarium wilt	1.0	"Stubborn" disease	.1
Bacterial wilt	.5	Twig blight	.1
Late blight	.5	Xyloporosis	.1
Leaf mold	.5	**Total**	12.0
Septoria leaf spot	.5		
Others	3.0		
Total	21.0		

[a] Estimates from "Losses in Agriculture"[5] during the period 1951—60.

major crops, the highest percentage loss occurs in grapes and strawberries, both of which sustain more than 25% loss. Unlike the crops in the forage and pasture group, the crops in this group have a high value and are often subject to intensive fungicide programs. In the absence of these plant protection programs, the losses reported for this group of crops would be substantially higher. A list of the specific diseases and corresponding estimates of loss for selected major crops in the group are tabulated in Table 8. Virus diseases are a critical problem in both grapes and oranges, and foot and root rots are particularly serious in the latter crop. For the potato crops, late blight is still considered to be the most important disease, despite the fact that many fungicide sprays are applied. Virus diseases are problems for both potatoes and tomatoes, and as a complex they are responsible for 25% of the total losses in both crops. Assuming that the current levels of loss are approximately the same as those reported here, the current total estimated value of loss due to diseases for this group is $1 billion/annum.

Losses Due to Nematodes

The estimates of loss for the three crop groups already discussed do not include loss estimates for nematodes because they are not available for all crops. For this reason

Table 9
ESTIMATED LOSSES DUE TO NEMATODES ON SELECTED CROPS IN THE U.S.[5]

Crop	% Loss	Value of loss ($ millions)[a]
Corn	3	132
Cotton	3	50
Alfalfa	3	48
Tobacco	3	36
Tomatoes	8	21
Soybeans	2	20
Potatoes	4	19
Oranges	4	13
Sugarbeet	4	7
Peanuts	3	6
Peaches	4	6
Beans	5	5
Raspberries	4	4
Cantaloupes	5	3
Total		370

[a] Based on 1951—60 values.

they are reported separately in this review, and it follows that the figures are underestimates because all crops are not represented. The losses for the crops listed in Table 9 total $370 million, with the major losses occurring in corn, cotton, alfalfa, and tobacco. It is much more difficult to estimate losses from nematodes than from other diseases, and the reported values of loss are to a great extent a function of the limitations of appraisal techniques.

LOSSES IN SELECTED MAJOR CROPS

The estimates of losses published by the U.S. and discussed previously are not a product of a specific investigation utilizing crop loss models and surveys to assess precisely the level of loss. On the contrary, the estimates of loss discussed above basically represent a consensus of opinion which may or may not be supported by experimental or survey evidence. Another approach, which has been followed by some pathologists in the U.S. and other countries, is to generate more reliable estimates for a few major crops, rather than an indication of loss for all crops. This approach usually requires the development of carefully tested disease-loss models which accurately describe the relationship between disease and yield loss (production/quality). The second stage consists of large scale surveys to estimate disease severity, and the crop loss model is then used to interpret the survey data in terms of losses in production/quality. Much of this type of work has been developed in the United Kingdom and Canada where permanent national programs are specifically operated for this purpose. The strategy and techniques utilized in disease loss appraisal programs have been the subject of regular reviews and the reader is referred to them for detailed discussions.[2,8-10] Selected examples will be discussed to facilitate comment on the programs and the data generated therefrom. Most of the programs feature cereal crops and this discussion will reflect this emphasis.

Table 10
SEVERITY OF FOLIAGE DISEASES[a] AFFECTING BARLEY IN ENGLAND AND WALES 1967—1975[12]

Disease	1967	1968	1969	1970	1972	1973	1974	1975
Powdery mildew	17	19	8	11	10	15	15	7
Leaf blotch	2	2	3	tr	3	2	1	<1
Brown rust	5	3	1	20	3	4	2	1
Yellow rust	1	<1	<1	0	tr	1	2	2
Halo spot	<1	<1	<1	0	tr	<1	tr	tr
Net blotch	0	<1	<1	0	tr	tr	tr	tr
Septoria	—	—	—	—	3	1	<1	tr
Total diseases	25	24	13	30	19	23	20	10
Samples estimated	293	250	282	272	217	231	275	243

Note: tr = trace.

[a] Estimates of percentage of area diseased, made on the lamina of the 2nd youngest leaf (penultimate) at a growth stage when grain was milky ripe.

Surveys of Cereal Diseases in England and Wales

Barley is the major crop in England and Wales and occupies approximately 5.5 million acres, whereas wheat occupies 3.0 million acres. Crop loss models have been developed for transforming the severity of the major foliage diseases of cereals into yield loss,[10] and the first national survey was initiated by James[11] in 1967 to assess the relative importance of foliage diseases of barley. Three hundred randomly selected farms were sampled in proportion to the acreage of barley in each of nine regions. One barley field on each farm was selected and an identical sample of plants, at a critical growth stage, was removed from each field and transmitted to a central laboratory for assessment. Trained disease assessors objectively estimated the percentage leaf area affected by different diseases on the top two leaves, and the disease data with corresponding crop cultural information were accessed directly to a computer to facilitate processing and analysis of results. The 1967 barley survey demonstrated for the first time the significance of powdery mildew and the need for chemical control measures which were hitherto considered unnecessary, but are currently applied to over 2 million acres. The total loss due to barley foliage diseases in 1967 was estimated at 20 to 25%, worth approximately $70 to 90 million, and mildew, the major disease, decreased yield by 13 to 18%.[10] The success of the 1967 barley survey not only ensured its annual repetition but also prompted the development of a similar survey on winter wheat in 1970. Both surveys are now conducted annually with continuous modifications to provide more comprehensive information on foliage and other diseases. Such surveys generate an indispensable data base which can be used for estimating losses. They may also be used for monitoring regional or cultivar variations in disease from year to year or differences due to cultural practices.

The data in Table 10 show the variation in barley leaf diseases during the period 1967 to 1975. Powdery mildew was the most severe disease every year with the exception of 1970, when a severe epidemic of rust was observed. The complex of foliage diseases was least severe in 1975 when only a total of 19% severity was recorded, and most severe in 1970 when 30% severity was observed. The tabulated data represent national averages, but the data base from which they are derived has been categorized in many ways to detect trends or differences due to region, varieties, cultural practices, etc.[11] A similar set of data has been accumulated for winter wheat since 1970. Estimates of yield loss are generated from the survey data to determine the relative importance of diseases, and the estimates for 1974 and 1975 for spring barley and winter wheat are given in Table 11.[12,14] In 1974 and 1975, loss from barley powdery mildew was estimated at 6.9% and 6.0%, at a cost of $67 and $60 million, respectively.

Approximately $11 million was spent on powdery mildew fungicide in 1974, which, in turn, increased production by $35 million to give a net profit, resulting from fungicide usage, of $24 million. A similar calculation for 1975 shows that the actual loss due to mildew was 6.0%, worth $60 million. The fungicide costs were the same as 1974 and resulted in an increase in production worth $27 million and a net profit of $16 million.[14] It was further calculated that, if treated with powdery mildew fungicides, approximately 60% of the unsprayed crops in 1974 (46% of total) and 1975 (41% of total) would have resulted in an economic return. Losses from the other foliage diseases of barley in 1974 and 1975 totalled 2.5% and 1.8% at a cost of $25 and $8 million respectively. The total losses due to foliage diseases of barley in 1974 and 1975 were 9.4% and 7.8%, equivalent to $92 million and $68 million, respectively (Table 11).

The total losses due to foliage diseases of winter wheat in England and Wales in 1974 and 1975 were estimated at approximately 4% at a cost of $25 million and $20 million, respectively (Table 11). In 1975 only about 10% of crops were treated with fungicide: 5% for mildew and 5% for other diseases. Wheat powdery mildew caused the highest losses in 1974 and 1975, and the data of King[13] show that it also caused the highest average loss (2.8%) during the period 1970 to 1975 (Table 12). The equivalent average loss for Septoria diseases was 2.2%, but the range (0.6 to 7.4%) was greater than for mildew. The total losses due to winter wheat foliage diseases ranged from 2.9% in 1970 to 11.0% in 1972.

The series of disease surveys on barley and winter wheat in England and Wales probably provides the most convincing demonstration of the usefulness of disease-loss surveys. It also provides an excellent example of the relevance of estimating losses within the context of policy making and implementation of a plant protection program. The desirability of utilizing reliable crop-loss models is evident, as is the need to continuously update the methods as requirements demand. The reader is referred to the frequently published reports[11-14] for further details.

Cereal Disease Surveys in Canada

Disease-loss surveys have also been conducted in Canada on wheat, oats, and barley, but the techniques differ from those used in England and Wales because the nature of agriculture in Canada is more extensive. Two types of surveys have been conducted. The objective of the first type of survey has been to estimate losses, mainly from all major foliage diseases affecting cereals in selected prairie and other provinces. The second type of survey has estimated the losses due to a single disease complex, common root rot, on barley and wheat. For convenience, the two types of survey will be discussed separately.

Surveys to Estimate Losses Mainly from Foliage Diseases

Two surveys were conducted in Manitoba and eastern and northern Saskatchewan in 1969[15] and 1970,[16] but only results from the latter survey are quoted here. Approximately 150 farms were visited for each of three crops, wheat, oats, and barley, and assessments of the major foliage diseases were recorded in the fields. Local experimental data and a crop-loss model developed elsewhere, but reported to be applicable in Canada,[15] were used to interpret the disease survey data in terms of percentage losses. The data in Table 13 show that the total losses due to the diseases assessed in the survey of wheat, oats, and barley in Manitoba and eastern and northern Saskatchewan

Table 11
ESTIMATED LOSSES DUE TO BARLEY AND WHEAT FOLIAGE DISEASE IN ENGLAND AND WALES[14]

	Barley				Wheat			
	1974		1975		1974		1975	
Disease	% Loss	Value ($ millions)	% Loss	Value ($ millions)	% Loss	Value ($ millions)	% Loss	Value ($ millions)
Powdery mildew	6.9	67	6.0	60	2.0	13	1.7	9
Brown rust	1.3	13	0.5	5	0.2	1	0.8	4
Rhynchosporium	0.3	3	0.1	1	—	—	—	—
Yellow rust	0.9	9	1.2	2	0.2	1	0.4	2
Septoria	—	—	—	—	1.5	10	1.0	5
Total	9.4	92	7.8	68	3.9	25	3.9	20

Table 12
ESTIMATED ANNUAL PERCENTAGE LOSS IN YIELD OF WINTER WHEAT CAUSED BY DISEASES IN ENGLAND AND WALES[13]

Disease	1970	1971	1972	1973	1974	1975	Average 1970—75
Powdery mildew	2.3	3.9	2.4	4.4	2.2	2.1	2.8
Septoria	0.6	1.8	7.4	2.5	0.8	0.1	2.2
Yellow rust	t	0.2	1.2	0.1	0.2	0.3	0.3
Eyespot	—	—	—	—	—	0.9	
Total	2.9	5.9	11.0	7.0	3.2	3.4	

Note: t = <0.05.

Table 13
VALUE OF LOSSES DUE TO DISEASE IN WHEAT, OATS, AND BARLEY IN THE PRAIRIE PROVINCES OF CANADA IN 1970
($ millions)

Crop	Manitoba	Saskatchewan (N. and E.)
Wheat	3.9	5.9
Oats	5.0	0.4
Barley	2.3	2.0
Total	11.2	8.3

in 1970 were equivalent to approximately $20 million. In Saskatchewan, the loss of $6 million for wheat was the highest for the three crops, and in Manitoba the highest was $5 million for oats. The authors considered that the losses were conservative and did not take into account losses in terms of quality that are known to occur in the presence of severe diseases. The data in Table 14 show the losses due to specific diseases both in terms of percentage loss and in bushels. The total loss of wheat in Manitoba was 2.8 million bu, equivalent to 8% loss, and the major disease was rust (5% loss). For Saskatchewan the percentage loss for wheat was approximately the same as Manitoba (8%) but this resulted in a loss of 4.2 million bu because the acreage was substantially larger. Total losses due to disease in oats in Manitoba were substantially higher than Saskatchewan (15.8% cf. 1.4%); crown rust and stem rust caused almost all the loss and were equally important, each reducing yield by approximately 7%. Barley diseases caused losses of 5.6 and 3.3% in Manitoba and Saskatchewan, respectively, with leaf spots and viruses accounting for most of the loss. Additional data on disease severity were recorded on susceptible varieties to calculate the value of rust and smut resistance in current varieties grown in Manitoba. This economic gain due to rust and smut resistance was valued at $5.6 million, and clearly indicated the value of research aimed at controlling plant diseases. Similar surveys have been conducted on winter wheat in Ontario in 1969 and 1970[17] where losses due to mildew, leaf rust, and virus diseases were estimated at approximately 5%/year, representing an annual loss of $1.25 million.

Surveys to Estimate Losses due to Common Root Rot of Cereals

Surveys were conducted on wheat in 1969 to 1971,[18] and on barley in 1970 to 1972[19] in Saskatchewan, Alberta, and Manitoba. The technique for estimating losses differed

Table 14
ESTIMATED LOSSES DUE TO CEREAL DISEASES IN MANITOBA AND SASKATCHEWAN IN 1970[15]

	Manitoba		Saskatchewan (N. and E.)	
Disease	1000 bu	% loss	1000 bu	% loss
	Wheat Losses			
Leaf rust	1680	5.1	1911	3.6
Leaf spots	1061	3.2	2198	4.2
Virus	16	<0.1	12	<0.1
Root rot	80	0.2	160	0.3
Total	2837	8.6	4281	8.2
	Oat Losses			
Crown rust	4614	7.3	130	0.2
Stem rust	4638	7.4	—	—
Virus	235	0.4	564	0.9
Leaf spots	354	0.6	81	0.1
Blast	66	0.1	68	0.1
Smut	—	—	20	<0.1
Total	9907	15.8	863	1.4
	Barley Losses			
Virus	1177	2.1	441	0.5
Leaf spots	1193	2.2	1449	1.8
Leaf rust	299	0.5	47	<0.1
Stem rust	107	0.2	—	—
Smut	162	0.3	449	0.6
Root rot	163	0.3	250	0.3
Total	3101	5.6	2636	3.3

substantially from that used in the other Canadian surveys because in addition to disease assessments, yield was also assessed from the sample plants and used in conjunction with a formula to calculate loss in yield.[19] The estimated average annual loss in wheat during the period 1969 to 1971 was 5.7%, which was equivalent to 30 million bu/annum. At a cost of $1.40/bu, the annual loss was equivalent to $42 million/annum. Similar surveys conducted on common root rot of barley showed that the losses in 1970, 1971, and 1972 were equivalent to 37, 42, and 84 million bu, respectively. The average annual loss was 10.3%, equivalent to 54 million bu. At an average cost of $1.00/bu this loss is equivalent to $54 million/annum. Over the 3-year period the losses were highest in Saskatchewan (6 to 20%), followed by Alberta (8 to 11%), and Manitoba had the lowest loss (0 to 14%), with no loss recorded for common root rot in Manitoba in 1970. The latter result is in agreement with the results of the surveys that concentrated on foliage diseases,[16] which also concluded that root rot caused minimal damage. Negligible losses were also recorded for common root rot of wheat in Manitoba in 1970 for both surveys.[16,18]

THE GENERATION OF CROP-LOSS ESTIMATES IN THE FUTURE

In 1967 the Food and Agriculture Organization (FAO) convened a symposium in Rome which recognized the need for more reliable data on crop losses due to diseases, and also the need to standardize methodology. Accordingly, a crop loss manual[20] was published which includes general guidelines and detailed examples of specific methods for estimating losses due to particular diseases. A series of workshops has also been organized in different countries to provide practicing plant pathologists with the necessary training to enable them to initiate and develop crop loss programs. Disease-assessment manuals[21,22] have also been published in different languages to facilitate the standardization of methods for appraising diseases. Plant pathologists are encouraged to employ disease-loss models for reliably assessing disease, rather than guestimate losses which could be more misleading than the complete absence of loss information. Many crop loss models have now been published[2] and pathologists should test models developed elsewhere in order to accelerate the employment and acceptance of standardized techniques. Although almost all the crop-loss models to date only feature a single disease, the need to develop multiple-disease models to cater for interaction is evident and appropriate techniques have already been developed.[2] Indeed, the arguments for quantifying losses or production constraints associated with plant diseases apply equally well to pests, weeds, and any other production constraint. Therefore, the objective should be to quantify production constraints due to diseases within the context of the total production system so that a crop loss profile can be generated to establish the relative importance of all constraint factors within the system. In a recent review, James and Teng[2] have discussed how this objective can be achieved, and outlined the role of a crop-loss program for estimating pre- and post-harvest losses due to diseases, within the context of an integrated plant protection program. These authors considered that a disease-loss program should be a permanent feature of any national integrated plant-protection program to monitor changes in a dynamic system resulting from varieties becoming susceptible, the development of resistance to fungicides, or a change in cultural practice leading to an increase or decrease in disease. Such a system, if implemented in the majority of national plant protection programs, would allow the economic importance of diseases in major crops throughout the world to be monitored continuously and priorities revised to optimize benefits from our limited resources.

REFERENCES

1. International Food Policy Research Institute, Food needs of developing countries: projections of production and consumption to 1990, *Res. Rep.*, 3, International Food Policy Research Institute, Washington, D.C., 1977.
2. **James, W. C. and Teng, P. S.,** The quantification of production constraints associated with plant diseases, *Appl. Biol.*, 4, 201, 1979.
3. **Cramer, H. H.,** Plant Protection and World Crop Production, Bayer, Leverkusen, West Germany, 1967.
4. Food and Agriculture Organization of the United Nations, *Production Year Book*, Food and Agriculture Organization of the United Nations, Rome, 1977, 30.
5. Losses in Agriculture, Agricultural Handbook No. 291, Agricultural Research Service, U.S. Department of Agriculture, Washington, D.C., 1965.
6. **Tatum, L. A.,** The southern corn leaf blight epidemic, *Science*, 171, 1113, 1971.
7. Cotton Disease Council, Reduction in yield of cotton caused by parasitic diseases in 1973, *Plant Dis. Rep.*, 58, 373, 1974.
8. **Chester, K. S.,** Plant disease losses: their appraisal and interpretation, *Plant Dis. Rep., Suppl.*, 193, 189, 1950.

9. **Large, E. C.,** Measuring plant disease, *Ann. Rev. Phytopathol.,* 4, 9, 1966.
10. **James, W. C.,** Assessment of plant diseases and losses, *Ann. Rev. Phytopathol.,* 12, 27, 1974.
11. **James, W. C.,** A survey of foliar diseases of spring barley in England and Wales in 1967, *Ann. Appl. Biol.,* 63, 253, 1969.
12. **King, J. E.,** Surveys of foliar diseases of spring barley in England and Wales, 1972-75, *Plant Pathol.,* 26, 21, 1977.
13. **King, J. E.,** Surveys of diseases of winter wheat in England and Wales, 1970-75, *Plant Pathol.,* 26, 8, 1977.
14. **Cock, L. J.,** The control of cereal diseases in the U.K., in *Proc. 8th Brit. Insecticide Fungicide Conf.,* Brighton, U.K., 1975, 859.
15. **McDonald, W. C., Martens, J. W., Green, G. J., Samborski, D. J., Fleischmann, G., and Gill, C. G.,** Losses from cereal diseases and value of disease resistance in Manitoba in 1969, *Can. Plant Dis. Surv.,* 49, 114, 1969.
16. **McDonald, W. C., Martens, J. W., Nielsen, J., Green, G. J., Sambroski, D. J., Fleischmann, G., Gill, C. G., Chiko, A. W., and Baker, R. J.,** Losses from cereal diseases and value of disease resistance in Manitoba and Eastern and Northern Saskatchewan in 1970, *Can. Plant Dis. Surv.,* 51, 105, 1971.
17. **James, W. C.,** Importance of foliage diseases of winter wheat in Ontario in 1969 and 1970, *Can. Plant Dis. Surv.,* 51, 24, 1971.
18. **Ledingham, R. J., Atkinson, T. G., Horricks, J. S., Mills, J. T., Piening, L. J., and Tinline, R. D.,** Wheat losses due to common root rot in the Prairie Provinces of Canada, 1969-71, *Can. Plant Dis. Surv.,* 53, 113, 1973.
19. **Piening, L. J., Atkinson, T. G., Horricks, J. S., Ledingham, R. J., Mills, J. T., and Tinline, R. D.,** Barley losses due to common root rot in the Prairie Provinces of Canada, 1970-72, *Can. Plant Dis. Surv.,* 56, 41, 1976.
20. **Chiarappa, L., Ed.,** Crop loss assessment methods: FAO manual on the evaluation and prevention of losses by pests, diseases, and weeds, Commonwealth Agricultural Bureau, Food and Agriculture Organization, Slough, U.K., 1971.
21. **James, W. C.,** An illustrated series of assessment keys for plant diseases, their preparation and usage, *Can. Plant Dis. Surv.,* 51, 39, 1971.
22. **James, W. C.,** A Manual of Disease Assessment Keys for Plant Diseases, Pub. 1458, Canada Department of Agriculture, Ottawa, 1971.

ESTIMATED LOSSES OF CROPS TO WEEDS

J. M. Chandler

NEED FOR DOCUMENTATION OF LOSSES CAUSED BY WEEDS

The diverse influences and burdens that weeds inflict on the total human race are especially felt by the producers of our raw agricultural products. An estimate of losses caused by specific weeds in monetary terms could offer considerable insight and help to producers of agricultural products in planning their production programs. This information would also be very useful to developers of herbicides, in identification of those areas of greatest need and areas of greatest potential monetary returns. Government officials at all levels could use weed loss estimates to better evaluate needs and use of public funds for meaningful research and development of improved weed control practices and regulation of these practices. This chapter represents an attempt to estimate the monetary losses resulting from weeds during the period 1972 to 1976 in major field crops, vegetable crops, fruit and nut crops, forage seed crops, and pastures and rangelands in the U.S. The seven most frequently reported weeds are listed for each crop or group of crops.

DEFINITION OF A WEED

Plant species are considered weeds when they interfere with man's activities or his welfare. In simple terms, weeds are unwanted plants or plants out of place. A particular plant is a weed only in terms of a human attitude. Before man tilled the soil, the concept of plants out of place had not been conceived; as man developed plants for higher production, weeds seem to have followed so that now we have weeds that are adapted to the same environment as our major crops. Thus weeds have become an important part of our environment. Losses from weeds are of several kinds because weeds are undesirable for various reasons.

EFFECT OF WEEDS ON THE PUBLIC WELFARE

Weeds directly alter the health and well-being of many people among the general public. Weeds such as poison sumac, poison oak, and poison ivy cause much suffering and loss in efficiency among our labor force. In the early 1960s the loss sustained annually by the Department of Industrial Relations of the state of California for time lost because of poison oak totaled $160,000.[3] The adverse effects on health from allergies to weed pollen cause many persons to travel to noninfested areas during the hayfever seasons and others spend large sums of money for medical attention. Most cases of hayfever east of the Mississippi River are caused by the pollen from ragweed, while in the Rocky Mountain states it is caused mainly by the pollen of sagebrush, bursage and sumpweed. On occasion death occurs (mainly among children) from eating poisonous plants, berries, seeds, or tubers.

Losses caused by weeds on nonagricultural sites are numerous. Large sums of money are spent annually to control weeds on industrial sites, on ditchbanks, on highway, railroad, and utility rights-of-way, on parking lots, storage areas, and on recreational areas such as lawns, parks, golf courses, playgrounds, and athletic fields. In 1975, the direct losses to agriculture caused by aquatic and ditchbank weeds amounted to $175 million. An estimated 2 million acre-feet of water are lost each year in the 17 western

Table 1
THE WATER REQUIREMENTS OF CROPS AND WEEDS, AKRON, COLORADO[13]

Crop	Water requirement[a]	Weed	Water requirement[a]
Alfalfa	844	Annual sunflower	577
Barley	518	Cocklebur	415
Bromegrass	977	Gumweed	585
Buckwheat	540	Knotweed	678
Corn	349	Lambsquarter	658
Flax	783	Nightshade	487
Millet	285	Pigweed, prostrate	260
Oats	583	Pigweed, rough	305
Potatoes	575	Purslane	281
Red clover	759	Ragweed	912
Rye	634	Russian thistle	314
Sorghum	305		
Soybeans	646		
Sugar beets	377		
Sweet clover	731		
Wheat	545		

[a] Pounds of water per pound of dry matter produced.

states because of weeds. The net productive value of lost water amounts to more than $110 million per year.[5]

Weeds have adverse effects on livestock production. In extreme cases, death of animals can occur from eating poisonous weeds. Weeds such as bitterweed, wild onion, and wild garlic cause off-flavors in milk and other dairy products. Weeds often damage wool, hides, and carcasses, thereby lowering the quality of animal products. Thorns, awns, spines, and burs from weeds can cause losses through irritational effect.

WEED COMPETITION WITH CROPS

Weed competition with crop plants and the costs of their control constitute some of the highest costs in the production of food, feed, and fiber. The losses from weeds to farmers and those in agribusiness are many and occur at various stages in the crop production and processing cycle. Weeds directly compete with the crop for water, mineral nutrients, light, and other growth factors. Weeds (1) reduce the yield of crops, (2) impair the quality of crops, (3) reduce the value and productivity of land and the efficiency with which it is used, (4) increase the costs of hand tillage, mechanical tillage, fertilizer, crop drying, and storage and transportation, (5) prevent efficient irrigation and water management, (6) increase both the acreage needed for crop production and the amount of labor required, (7) cause large amounts of fuel and other types of energy to be expended for control, and (8) serve as hosts and habitats for insects, nematodes, disease-causing organisms, and rodents.[14]

Weeds are strong competitors for soil moisture. Competition for water begins when root systems overlap in their search for water and nutrients.[12] Sunflower requires almost twice as much water as corn to produce the same amount of dry matter, ragweed requires about three times that of millet or corn, and common mustard requires four times as much water as oats.[3] Cocklebur requires almost three times as much water as sorghum but Russian thistle requires about 20% less water than sorghum.[4] The water requirements of several crop and weed species have been determined and are compared in Table 1.[13]

Shading of weeds drastically influences their ability to compete with crops. Several weeds have been classified as to their response to shading.[10] Rough pigweed and green foxtail are intolerant of shading. Prairie ground cherry, clasping dogbane, prairie rose, spotted spurge, and field bindweed are tolerant to shade. Lambsquarters, common sunflower, Pennsylvania smartweed, common ragweed, wild buckwheat, common peppergrass, and woodsorrel are suppressed by corn. Cocklebur, ground cherry, and hedge bindweed are suppressed by winter wheat. Purple nutsedge, classified as the world's worst weed, is intolerant to shade and forms 10 to 57% fewer tubers and bulbs when shaded continuously.[7,9]

Soil nutrients influence the competitive interactions of weeds with various crops. Blackman and Templeman reported that the competition between barley and weeds is principally for nitrogen in a year of normal rainfall. In Iowa, mature foxtail infestations reduced corn yields approximately 14, 10, and 5 bu/acre with applications of 0, 70, and 140 lb, respectively, of elemental nitrogen.[17] Nitrogen did not affect the competitive relationship between cotton and weeds in some years but in others the duration of competition tolerated was increased by the addition of nitrogen.[2] The application of nitrogen to weedy upland rice fields benefits purple nutsedge more than the rice. Purple nutsedge and upland rice compete extensively for moisture and the competition is much more serious with increased nitrogen fertilization.[11] In West Texas, cocklebur, Russian thistle, and puncturevine utilized twice the amount of nitrogen as sorghum during the growing season, but crabgrass, buffalobur, and barnyardgrass used only half as much on a per acre basis.[16]

TAXONOMIC RELATIONSHIP OF WEEDS AND CROPS

Relatively few weed species are responsible for the major portion of the losses sustained in our crops. Holm states that across the world about 200 weed species are involved in 95% of man's weed problems as related to food production.[8] He lists 80 weed species as primary weeds and 120 weed species as secondary weeds. To date, man has named about 250,000 flowering plants; thus, these 200 weed species comprise less than 1% of all the world's species. On a global basis, one third of the worst weeds are grasses and sedges and two thirds are broadleaf weeds. Within the 80 weeds of primary importance, 35% are grasses, 56% are broadleaf weeds, 6% are sedges and two species are ferns. Of these weeds, 44% are perennials and 56% are annuals. Within the 120 species of secondary importance, 13% are grasses, 79% are broadleaf weeds, 8% are sedges and one species is a fern. Annuals comprise 70% of this group, whereas the remaining 30% are perennials. In the future many of our emerging weed problems will probably come from this secondary group. The U.S. has 70% of the world's important weeds: 24% of these are grasses, 67% broadleaf weeds, and 7% sedges.

The important families of the world's worst weeds and number of weed species in each are Poaceae (44 species), Cyperaceae (12 species), Asteraceae (32 species), Polygonaceae (8 species), Amaranthaceae (7 species), Brassicaceae (7 species), Leguminosae (6 species), Convolvulaceae (5 species), Euphorbiaceae (5 species), Chenopodiaceae (4 species), Malvaceae (4 species), Solonaceae, and 47 other families with 3 species or less. These 12 families provide 68% of the world's main weed problems and the Poaceae, Cyperaceae, and Asteraceae contain 43% of the main weed problems.[8]

On a global basis, five plant families containing 12 crop species provide 75% of our food. The families and associated species are Poaceae (barley, maize, millet, oats, rice, sorghum, sugarcane, and wheat), Solanaceae (white potato), Convolvulaceae (sweet potato), Euphorbiaceae (cassava), and Leguminosae (soybean, peas, and beans). These five plant families also supply us with many of our weeds. Correlation between crops and weeds belonging to the same botanical family are numerous and support the idea that both domesticated plants and weeds are adapted to the same habitat. Thus, practices that tend to favor domesticates also tend to favor weeds. Although this relationship exists it is not the rule, because 25% of the world's worst weeds are found in the Compositae and Cyperaceae families, which contain no major crops.

WEED LOSSES IN CROP PRODUCTION — HISTORICAL PERSPECTIVE

The universal occurrence of weeds as constant components of agricultural environments as opposed to the epidemic nature of other pests has delayed recognition of the importance of weed control in crop production. Crop production consists largely of growing crop plants in pure culture, in contrast to nature's mixed plant communities. To accomplish this, man started controlling weeds by hand-pulling and using simple hand tools. The horse-drawn wheel hoe brought about a revolution in row-crop agriculture during the early 1700s and with the gasoline tractor in the early 1900s a rapid expansion in crop production was observed. Mechanical cutivation for weed control played an essential role in the rapid expansion in productivity.

Over the past 3 decades, U.S. agriculture has had to make dramatic changes in order to meet the food and fiber requirements for the American people and to supply export markets. During this period the American farmer has been able to shift from labor-intensive farming to a more efficient capital-intensive production. Examples of this shift are the utilization of newer, bigger, and better machines, increased use of fertilizers, utilization of selective herbicides, improved crop varieties, etc. In the late 1950s only 2% of the cotton grown in the southeastern U.S. was harvested mechanically and the primary means of weed control in the cotton drill-row was hand hoeing.[6,18] In fields with heavy weed growth, up to 50 hr of hoe labor per acre were required to remove weeds. Currently, cotton producers use about 4 hr of hoe labor per acre to remove weeds. In the past 30 years, wheat, rice, and potato yields in the U.S. have doubled and corn yields have tripled. Improved weed control practices in these crops alone have reduced labor requirements by 30 to 50%. Across all crops, the use of mechanical power has increased 30% and herbicides have increased sevenfold since 1950, while manual labor decreased 40%.[15]

MONETARY LOSSES FROM WEEDS IN CROPS

Even with the shift in weed control practices from hand hoeing to selective herbicides, we still have weeds as major pests in our crops. With the introduction of herbicides, the degree of crop losses resulting from weeds has been reduced over the past 3 decades.[15] In Tables 2, 4, 6, 8, 9, and 10 we have presented, by crop, the estimated average annual losses due to weeds during 1972 through 1976 in the U.S. For a discussion of how the monetary losses due to weeds were tabulated see the footnotes in Table 2. In Tables 3, 5, and 7, the seven weeds most frequently reported and order of frequency are presented by crop or crop grouping.

Losses in Field Crops

The total monetary loss due to weeds in the 13 field crops listed in Table 2 was $4,986,388,000 per year for the period of 1972 to 1976. Losses in soybeans and corn were 33 and 30%, respectively, of the total during this time interval. Monetary losses in the four small grain crops account for 19% of the total loss but the largest portion of this loss occurred in wheat. Grain sorghum, cotton, and rice losses were approxi-

Table 2
FIELD CROPS: ESTIMATED AVERAGE ANNUAL LOSSES DUE TO WEEDS, 1972—1976

		Loss from potential production[a]		
Commodity	Production unit	Reduction[b] (%)	Quantity[c] (1,000 units)	Value[c] ($1,000)
Row crops				
Corn	Bushel	10	619,929	1,487,831
Cotton[d]	Bale	8	992	219,033
Peanuts	Pound	15	636,148	111,962
Grain sorghum	Bushel	13	115,124	246,366
Soybeans[e]	Bushel	17	280,340	1,620,365
Sugarbeets	Ton	8	2,333	65,230
Sugarcane	Ton	13	4,072	59,461
Drill crops				
Barley	Bushel	12	52,098	113,053
Flax (seed)[f]	Bushel	12	1,797	12,183
Oats[f]	Bushel	17	13,078	16,739
Rice[f]	Hundredweight	17	21,939	204,910
Rye	Bushel	10	2,428	4,954
Wheat[f]	Bushel	12	254,414	824,301
Total				4,986,388

[a] Estimate is based on full production with the cause eliminated. The percentages were applied to the data of actual farm production to obtain estimates of loss of farm production in terms of quantity and value. Loss in quality is included in the loss in value. Value loss is computed upon the assumption that market outlets would be available for reduced production with no change from average farm prices.

[b] Estimated percent reduction due to weeds for each crop was adapted from Agricultural Handbook No. 291, U. S. Department of Agriculture, Washington, D. C., 1965.

[c] The basic data used to calculate the loss in quantity and value represent the averages for the period 1972 to 1976 as estimated by the Statistical Reporting Service.

[d] Percent reduction includes a 2% loss in fiber quality due to lowered grades caused by grass, weeds, and associated trash.

[e] Percent reduction includes a 3% loss due to weed seed dockage, damage in cleaning to remove weed seed, and off-flavors.

[f] Percent reduction includes a 3% loss in quality in flax and oats, 4% in rice, and 1% in wheat due to weed seed dockage, delayed maturity, cleaning losses, and off-flavors.

mately 5, 4, and 4%, respectively. A monetary loss of 1% or less occurred in sugarbeets, sugarcane, and flax.

In these field crops the seven most frequently reported weeds out of the 31 listed were cocklebur, crabgrass, foxtails, pigweeds, johnsongrass, nutsedge, and morningglories (Table 3). In soybeans and corn, respectively, where 63% of the monetary loss occurred, the most frequently listed weeds were annual broadleaves (cocklebur and pigweed) and the second most frequently listed were annual grasses (foxtails and crabgrass). The perennials, nutsedge and johnsongrass, were reported less frequently than the annual grasses and broadleaves but yet are very important weed species in these crops. In small grains the most frequently reported weeds were mustard followed by wild oat, bromes, and wild garlic. A variety of broadleaves and grasses competed with the other seven crops. It is of interest to note that each crop usually has a weed that is unique to it due to the environment and/or the cultural practices utilized to

Table 3
SEVEN WEEDS REPORTED MOST FREQUENTLY IN INDIVIDUAL FIELD CROPS[a]

Weed species[a]	Corn	Cotton	Peanuts	Grain sorghum	Soybeans	Sugarbeets	Sugarcane	Rice	Wheat	Other small grains
Alexandergrass							1			
Barnyardgrass				7		4		1		
Bindweed									7	6
Bromes									2	
Buckwheat, wild									4	
Bulrushes								4		
Cocklebur		2	3	5	1					
Crabgrass	2	6	2	3			2			
Ducksalad								2		
Foxtails	3			4	2	3				
Henbit									5	7
Garlic, wild									3	2
Johnsongrass	7	1		2	5		3			
Kochia						5				
Lambsquarter						2				5
Morning glories		4	7	6	4		4			
Mustards						6			1	1
Napiergrass							5			
Nutsedge	5	3	1		7					
Oats, wild						7			6	3
Panicum	4		6				6			
Paragrass							7			
Pigweed	1	7	5	1	3	1				4
Quackgrass	6									
Rice, red								5		
Sidas		5								
Sesbania, hemp								3		
Sicklepod			4							
Signalgrass								6		
Sprangletop								7		
Velvetleaf					6					

[a] Weed frequency data constructed from information given in "Extent and cost of weed control with herbicides and an evaluation of important weeds, 1968, ARS-H-1.

[b] Within a crop the order of weed frequency is in descending order from one being reported most frequently.

produce that crop. Examples are: quackgrass in corn, prickly sida in cotton, sicklepod in peanuts, red rice in rice, velvetleaf in soybeans, kochia in sugarbeets, and alexandergrass in sugarcane.

Losses in Vegetable Crops

The estimated monetary loss due to weeds in the 22 vegetable crops listed in Table 4 was $312,548,000 annually during the period 1972 to 1976. Approximately one fourth of this loss occurred in vegetable legumes such as beans and peas. Substantial losses were encountered in tomatoes, with 17% of the total loss. In potatoes and lettuce, monetary losses from weeds were 12 and 9% of the total loss, respectively. Seven percent of the total loss occurred in sweet corn. Monetary losses due to weed competition ranged from 1 to 5% of the total in the other vegetable crops.

The most frequently reported weeds in the above vegetable crops were pigweed, crabgrass, lambsquarter, and nutsedge (Table 5). In dry beans, green lima beans, green

Table 4
VEGETABLES: ESTIMATED AVERAGE ANNUAL LOSSES DUE TO WEEDS, 1972—1976

Commodity	Production unit	Loss from potential production[a] Reduction[b] (%)	Quantity[c] (1,000 units)	Value[c] ($1,000)
Cole crops				
Broccoli	Hundredweight	7	286	3,890
Cabbage	Hundredweight	7	1,782	7,680
Greens				
Spinach[d]	Hundredweight	13	589	2,858
Root and bulb crops				
Carrots[d]	Hundredweight	9	2,040	11,424
Onions[d]	Hundredweight	7	2,322	15,650
Potatoes	Hundredweight	3	9,962	39,349
Sweet potatoes	Hundredweight	4	551	3,961
Beets[d]	Hundredweight	6	254	429
Salad crop				
Lettuce	Hundredweight	7	3,876	27,713
Solanaceous				
Peppers[e]	Hundredweight	7	373	5,580
Tomatoes (fresh)[e]	Hundredweight	7	1,532	26,503
(processed)[e]	Hundredweight	7	10,156	26,701
Vegetable legumes				
Beans, dry[c]	Hundredweight	15	21,055	59,117
Green lima[c]	Hundredweight	8	151	2,027
Green snap[c]	Hundredweight	9	1,626	14,071
Peas, green[d]	Hundredweight	13	1,570	13,237
Vegetable, others				
Corn, sweet	Hundredweight	10	6,352	21,494
Hops	Hundredweight	10	615	4,854
Peppermint	Hundredweight	12	39	5,015
Spearmint	Hundredweight	12	17	1,912
Vine crops				
Cucumbers (fresh)	Hundredweight	6	299	2,864
(pickled)	Hundredweight	6	785	4,547
Cantalopes	Hundredweight	6	684	6,402
Watermelon	Hundredweight	6	1,597	5,270
Total				312,548

[a] Estimate is based on full production with the cause eliminated. The percentages were applied to the data of actual farm production to obtain estimates of loss of farm production in terms of quantity and value. Loss in quality is included in the loss in value. Value loss is computed upon the assumption that market outlets would be available for reduced production with no change from average farm prices.

[b] Estimated percent reduction due to weeds for each crop was adapted from Agricultural Handbook No. 291, U. S. Department of Agriculture, Washington, D. C., 1965.

[c] The basic data used to calculate the loss in quantity and value represent the averages for the period 1972 to 1976 as estimated by the Statistical Reporting Service.

[d] Percent reduction includes a 2% loss in quality.

[e] Percent reduction includes a 3% loss in quality.

Table 5
SEVEN WEEDS REPORTED MOST FREQUENTLY IN INDIVIDUAL AND GROUPS OF VEGETABLE CROPS

Weed species	Vegetable crops[a]									
	Cole crops	Greens	Root and bulb crops	Potatoes	Salad crops	Solanaceous fruits	Vegetable legume	Vegetable, others	Corn, sweet	Vine crops
Barnyardgrass			7	6	5	7			6	5
Chickweed	5	2								
Cocklebur								6		
Crabgrass	3	1	3	3	4	2	2	1	2	3
Foxtails			6	5			6		3	6
Henbit		5			6					
Johnsongrass								5		
Lambsquarter	2	3	2	2	1	5	3	3	7	2
Nightshade						6	7			
Nutsedge	7		4	4		3	4	4	4	7
Pigweed	1	4	1	1	2	1	1	2	1	1
Purslane	4	6	5		3					
Quackgrass				7				7	5	
Ragweed	6	7			7	4	5			4

[a] Within a crop the order of weed frequency is in descending order from one being reported most frequently.

[b] Weed frequency data constructed from information given in "Extent and cost of weed control with herbicides and an evaluation of important weeds," 1968, ARS-H-1.

Table 6
FRUITS AND NUTS: ESTIMATED AVERAGE ANNUAL LOSSES DUE TO WEEDS, 1972—1976

Commodity	Production unit	Loss from potential production[a]		
		Reduction[b] %	Quantity[c] (1,000 units)	Value[c] ($1,000)
Citrus fruits				
Grapefruit	Ton	5	131	8,705
Lemons	Ton	5	43	5,366
Oranges	Ton	5	492	33,142
Tangerines	Ton	5	11	1,220
Pome fruits				
Apples	Ton	3	99	15,418
Pears	Ton	5	38	5,476
Stone fruits				
Avocadoes	Ton	6	6	4,565
Olives	Ton	4	3	945
Peaches	Ton	6	86	15,404
Small fruits				
Bushberries	Ton	20	7	4,387
Cranberries	Barrel	10	242	3,075
Grapes (fresh)	Ton	15	682	104,073
Strawberries	Ton	25	86	49,427
Deciduous tree nuts				
Almonds	Ton	7	12	12,064
Pecans	Ton	9	9	9,354
Walnuts	Ton	5	9	4,707
Total				277,328

[a] Estimate is based on full production with the cause eliminated. The percentages were applied to the data of actual farm production to obtain estimates of loss of farm production in terms of quantity and value. Loss in quality is included in the loss in value. Value loss is computed upon the assumption that market outlets would be available for reduced production with no change from average farm prices.

[b] Estimated percent reduction due to weeds for each crop was adapted from Agricultural Handbook No. 291, U. S. Department of Agriculture, Washington, D. C., 1965.

[c] The basic data used to calculate the loss in quantity and value represent the averages for the period 1972 to 1976 as estimated by the Statistical Reporting Service.

snap beans, green peas, peppers, and tomatoes, where 47% of the losses due to weed competition occurred, the most frequently observed weeds were pigweed and crabgrass. Losses due to weeds in sweet corn, potatoes, broccoli, cabbage, lettuce, and spinach add up to one third of the total for vegetable crops, and the predominant weeds were broadleaf annuals, especially pigweed and lambsquarter. As with the field crops, certain weed species tend to be more numerous in specific vegetable crops.

Losses in Fruit and Nut Crops

The estimated annual monetary loss due to weeds in the 16 fruit and nut crops listed in Table 6 was $277,328,000 for the period 1972 to 1976. Over two thirds of this total loss occurred in grapes, strawberries, and oranges with 38, 18, and 12%, respectively. Among the pome fruits (apples and pears) and the stone fruits (avocados, olives, and peaches) a loss of 7% of the total resulted from weed competition in each group. In almonds, pecans, and walnuts, monetary losses from weeds totaled $26 million.

Table 7
SEVEN WEEDS REPORTED MOST FREQUENTLY IN GROUPS OF FRUIT AND NUT CROPS

Weed species[b]	Fruit and nut crops[a]				
	Citrus fruits	Pome fruits	Stone fruits	Small fruits	Deciduous tree nuts
Barnyardgrass	5			7	4
Bermudagrass	1	6	5	5	3
Bindweed		2	3		7
Chickweed				3	
Crabgrass	6	3	1	1	5
Ivy, poison		1	7		6
Johnsongrass	3	4	4	6	1
Milkweed		7			
Nightshade, silverleaf	7				
Nutsedge	2				
Pigweed	4	5	6	4	2
Quackgrass			2	2	

[a] Within a crop the order of weed frequency is in descending order from one being reported most frequently.

[b] Weed frequency data constructed from information given in "Extent and cost of weed control with herbicides and an evaluation of important weeds," 1968, ARS-H-1.

Crabgrass, bermudagrass, johnsongrass, and pigweed were the most frequently reported weeds in the above fruit and nut crops (Table 7). Five of the seven weed species listed most frequently in the small fruits were annual or perennial grasses with crabgrass and quackgrass being number 1 and 2, respectively. About two thirds of the weed species listed as those occurring most frequently in fruit and nut crops were perennials and within each crop grouping at least 50% of the most frequently reported weeds were perennials.

Losses in Forage Seed Crops

In forage seed crops $38,897,000 is the estimated annual loss due to weeds during 1972 to 1976 (Table 8). In general, weeds caused a 1% loss in seed quality and a 5% loss can be attributed to the seed-cleaning process. Monetary losses due to weeds in alfalfa seed production account for 43% of the total for all forage seed crops. For other legume seed crops and grass seed crops, the percentage of total monetary loss was 19 and 38%, respectively.

In the grass seed crops the seven most frequently reported weeds, in descending order, were bromes, quackgrass, foxtails, pigweeds, annual bluegrass, crabgrass, and wild garlic. In legume seed crops the order was dodders, plantains, thistles, quackgrass, pigweed, docks, and crabgrass. In the grass seed crops the weeds are mainly monocots and in the legume seed corps the weeds are mainly dicots.

Losses in Hay Crops, Pastures, and Rangelands

An annual monetary loss of $676,221,000 due to weeds has been calculated for all hay crops during the period 1972 to 1976 (Table 9). The most frequently reported weeds in all hay crops were quackgrass, thistles, chickweed, dandelions, bromes, pigweeds, and ragweed. Eastern pastures and western rangelands are best divided at the eastern borders of North Dakota, South Dakota, Nebraska, Kansas, Oklahoma, and

Table 8
FORAGE SEED CROPS: ESTIMATED AVERAGE ANNUAL LOSSES DUE TO WEEDS, 1972—1976

Commodity[b]	Production unit	Loss from potential production[a] Reduction[c] (%)	Quantity[d] (1,000 units)	Value[d] ($1,000)
Grass seed crops				
Bluegrass, Kentucky	Pounds	16	8,476	4,317
Fescue, tall	Pounds	15	19,586	3,035
Orchardgrass	Pounds	14	2,727	790
Ryegrass	Pounds	13	31,711	4,268
Timothy	Pounds	14	3,099	681
Turf grasses	Pounds	17	4,246	1,827
Legume seed crops				
Alfalfa	Pounds	18	21,519	16,578
Clover				
Crimson	Pounds	20	652	273
Ladina	Pounds	17	736	806
Red	Pounds	22	9,005	4,913
Lespedeza	Pounds	19	3,440	1,173
Vetch, hairy	Pounds	17	1,272	236
Total				38,897

[a] Estimate is based on full production with the cause eliminated. The percentages were applied to the data of actual farm production to obtain estimates of loss of farm production in terms of quantity and value. Loss in quality is included in the loss in value. Value loss is computed upon the assumption that market outlets would be available for reduced production with no change from average farm prices.

[b] Percent reduction for each crop contains a 1% loss in seed quality and a 5% loss attributed to the seed cleaning process.

[c] Estimated percent reduction due to weeds for each crop was adapted from Agricultural Handbook No. 291, U.S. Department of Agriculture, Washington, D.C., 1965.

[d] The basic data used to calculate the loss in quantity and value represent the averages for the period 1972 to 1976 as estimated by the Statistical Reporting Service.

Texas. The estimated annual monetary losses due to weeds in these two geographic regions were $410,896,000 and $377,909,000, respectively (Table 9). The most frequently reported weeds in the eastern pastures were thistles, ragweeds, docks, pigweeds, horsenettle, wild barley, and crabgrass. In the western states, the most frequently reported weeds were sagebrush, weedy bromes, larkspurs, thistles, pricklypear, rabbitbrush, and spurges. It is on the western ranges where biocontrol methods are most extensively practiced. Successful biocontrol techniques have been developed for St. Johnswort, tansy ragwort, and puncturevine. Research is underway to develop biocontrol methods for several of the thistles and spurges.

In summary, an average annual monetary loss of $7 billion can be attributed to the presence of weeds in our crops (Table 10). The total potential value of the crops discussed in this chapter to the producer was estimated to be $60,088,402,000; thus, an 11.7% loss can be attributed to weeds. Approximately 70% of this loss occurs in field crops and 20% in those crops utilized as forage by the livestock industry. The remaining 10% accounts for losses in vegetable, fruit, and nut crops. The American farmers are now producing 70% more crops and 40% more livestock than they were 28 years

Table 9
PASTURES, RANGELANDS, AND HAY: ESTIMATED AVERAGE ANNUAL LOSSES DUE TO WEEDS, 1972—1976

Kind of land and location[b]	Loss from potential production[a] Reduction[c] (%)	Value[d] ($1,000)
Pastures and rangelands in 31 eastern states	20	410,896
Rangelands in 17 western states	13	377,909
Hay in 48 states[e]	10	676,221
Total		1,465,026

[a] Estimate is based on full production with the cause eliminated. The percentages were applied to the data of actual farm production to obtain estimates of loss of farm production in terms of quantity and value. Loss in quality is included in the loss in value. Value loss is computed upon the assumption that market outlets would be available for reduced production with no change from average farm prices.

[b] Excludes cropland used for pasture.

[c] Estimated percent reduction due to weeds was adapted from Agricultural Handbook No. 291, U.S. Department of Agriculture, Washington, D.C., 1965.

[d] The basic data used to calculate the monetary losses was obtained from "Major uses of land in the United States," USDA-SEA-ER, 1974 (in press).

[e] All hay with 60% of the total being alfalfa.

Table 10
ALL CROPS: ESTIMATED AVERAGE ANNUAL LOSSES DUE TO WEEDS, 1972—1976

Commodity group	Average annual monetary losses ($1,000)
Field crops	4,986,388
Vegetables	312,548
Fruits and nuts	277,328
Forage seed crops	38,897
Hay	676,221
Pasture and rangelands	788,805
Total	7,080,187

ago. For similar gains in the future, the management of weeds will be one of the essential ingredients necessary for maximum productivity.

APPENDIX

Crops

Common name	Scientific name
Alfalfa	*Medicago sativia* L.
Apple	*Pyrus malus* L.
Avocado	*Persea americana* Mill.
Barley	*Hordeum vulgare* L.
Bean, green lima	*Phaseolus limensis* Macf.
Bean, green snap	*Phaseolus vulgaris* L.
Bluegrass, Kentucky	*Poa pratensis* L.
Broccoli	*Brassica oleracea*, var. *italic* Plenck.
Buffalograss	*Buchloe dactyloides* (Nutt.) Engelm.
Cabbage	*Brassica oleracea*, var. *capitata* L.
Canteloupe	*Cucumis melo* L.
Clover, Ladino	*Trifolium repens* L.
Clover, crimson	*Trifolium incarnatum* L.
Clover, red	*Trifolium pratense* L.
Corn	*Zea mays* L.
Corn, sweet	*Zea saccharata* L.
Cotton	*Gossypium hirsutum* L.
Cucumber	*Cucumis sativus* L.
Fescue, tall	*Festuca elatior* L.
Flax	*Linum usitatissimum* L.
Grapes	*Vitis vinifera* L.
Hops	*Humulus lupulus* L.
Lespedeza	*Lespedeza striata* (Thund. ex murr.) Hook. and Arn.
Lettuce	*Lactuca sativa* L.
Oats	*Avena sativa* L.
Olive	*Olea europaea* L.
Orange	*Citrus aurantium* L.
Orchardgrass	*Dactylis glomerata* L.
Pea, green	*Pisum sativum* L.
Pear	*Pyrus communis* L.
Peach	*Prunus persica* (L.) Batsch
Peanut	*Abrus hypogaea* L.
Pecan	*Carya illinoensis* (Wangern.) K. Koch
Pepper	*Capsicum annuum* L.
Peppermint	*Mentha piperita* L.
Potato	*Solanum tuberosum* L.
Rice	*Oryza sativa* L.
Rye	*Secale cereale* L.
Ryegrass, perennial	*Lolium perenne* L.
Spinach	*Spinacia oleracea* L.
Sorghum, grain	*Sorghum bicolor* (L.) Moench.
Soybean	*Glycine max* (L.) Merr.
Spearmint	*Mentha spicata* L.
Strawberry	*Fragaria chiloensis* Duch.
Sugarbeet	*Beta vulgaris* L.
Sugarcane	*Saccharum officinarum* L.
Timothy	*Phleum pratense* L.
Tomato	*Lycopersicum esculentum* Mill.
Vetch, hairy	*Vicia villosa* Roth
Walnut	*Juglans regia* L.
Watermelon	*Citrullus vulgaris* Schrad.
Wheat	*Triticum aestivum* L.

APPENDIX (continued)

Weeds

Common name	Scientific name
Alexandergrass	*Brachiaria plantaginea* (Link) A. Hitchc.
Barley, wild	*Hordeum leporinum* Link
Barnyardgrass	*Echinochloa crusgalli* (L.) Beauv.
Bermudagrass	*Cynodon dactylon* (L.) Pers.
Bindweed, field	*Convolvulus arvensis* L.
Bindweed, hedge	*Convolvulus sepium* L.
Bitterweed	*Ambrosia artemisiifolia* L.
Bluegrass, annual	*Poa annua* L.
Bromes	*Bromus* spp.
Buckwheat, wild	*Polygonum convolvulus* L.
Bursage	*Franseria* spp.
Buttercup	*Ranunculus* spp.
Bulrush	*Scirpus* spp.
Chickweed, common	*Stellaria media* (L.) Cyrill.
Cocklebur	*Xanthium* spp.
Crabgrass	*Digitaria* spp.
Dandelion	*Taraxacum* spp.
Dodders	*Cuscuta* spp.
Docks	*Rumex* spp.

REFERENCES

1. **Blackman, G. E. and Templeman, W. C.,** The nature of the competition between cereal crops and annual weeds, *J. Agric. Sci.,* 28, 247, 1938.
2. **Buchanan, G. A. and McLanghlin, R. D.,** Influence of nitrogen on weed competition in cotton, *Weed Sci.,* 23, 324, 1975.
3. **Crafts, A. S. and Robbins, W. W.,** *Weed Control,* McGraw-Hill, New York, 1967, 19.
4. **Davis, R. G. and Wiese, A. F.,** Weed root growth patterns in the field, *Proc. South. Weed Sci. Soc.,* 17, 367, 1964.
5. **Frank, P. A.,** Report of the SEA research planning conference on aquatic weed control, U.S. Government Printing Office, Washington, D.C., 1978, 1.
6. **Grissom, P. H. and Spurgeon, W. I.,** Crop systems and weed control, *Miss. Agric. Exp. Stn. Inf. Sheet,* 799, 1963.
7. **Hauser, E. W.,** Establishment of nutsedge from space-planted tubers, *Weed Sci.,* 10, 209, 1962.
8. **Holm, L. G.,** Some characteristics of weed problems in two worlds, *Proc. West. Soc. Weed Sci.,* 31, 3, 1978.
9. **Holm, L. G., Plucknett, D. L., Pancho, J. V., and Herberger, J. P.,** *The World's Worst Weeds, Distribution and Biology,* University Press of Hawaii, Honolulu, 1977, 8.
10. **Norris, E. L.,** Competing ability of different plants, *University of Nebraska Studies,* Vol. 39, 1939, 2.
11. **Okafor, L. I. and De Datta, S. K.,** Competition between upland rice and purple nutsedge for nitrogen, moisture and light, *Weed Sci.,* 24, 43, 1976.
12. **Pavlychenko, T. K. and Harrington, J. B.,** Root development of weeds and crops in competition under dry farming, *Sci. Agric.,* 16, 151, 1935.
13. **Shantz, H. L. and Piemeisel, L. N.,** The water requirement of plants at Akron, Colorado, *J. Agric. Res.,* 34, 1093, 1927.
14. **Shaw, W. C.,** Weed science — revolution in agricultural technology, *Weed Sci.,* 12, 153, 1964.
15. **Shaw, W. C.,** Herbicides: The cost/benefit ratio — the public view, *Proc. South. Weed Sci. Soc.,* 31, 28, 1978.

16. **Shipley, J. L. and Wiese, A. F.,** Economics of weed control in sorghum and wheat, *Tex. Agric. Exp. St. Misc. Publ.*, 909, 8, 1969.
17. **Staniforth, D. W.,** Effect of annual grass weeds on the yield of corn, *Agron. J.*, 49, 551, 1957.
18. **Street, J. H.,** *The New Revolution in the Cotton Economy,* University of North Carolina, Chapel Hill, 1957.

ESTIMATED LOSSES OF LIVESTOCK TO PESTS

R. O. Drummond, G. Lambert, H. E. Smalley, Jr., and C. E. Terrill

INTRODUCTION

There is no doubt that the increasing human population of the world will create an ever-increasing need for the animal proteins that are essential for human nutrition. Because of these increasing demands and because the natural resources of the earth are finite, it is necessary to maximize the production of livestock, especially that of ruminants, that convert plant proteins into animal proteins. Unfortunately, the efficient production of livestock is prevented by many factors, including "pests". In this chapter we have defined pests as any noxious or destructive organism that affects livestock. These organisms include arthropods, diseases, internal parasites, weeds, poisonous plants, and predators. Although our knowledge of the magnitude of losses caused by these pests is, at best, fragmentary and incomplete, the estimates presented herein can serve as a measure of the impact these organisms cause in efficiency of livestock production and should stimulate efforts to more accurately define the impact of these pests on animal production and to find solutions to these problems.

ARTHROPOD PESTS

Production of livestock in the U.S. is affected by a large number of arthropod pests that suck blood, transmit diseases, cause downgrading at market, make holes and other blemishes in hides, decrease production of wool, mohair, milk, and eggs, decrease weight gains, predispose animals to secondary infections, irritate and annoy animals, inject toxins, cause paralysis, create unthriftiness, and may even cause the death of livestock.

Efforts have been made for many years to determine monetary losses resulting from these pests; yet it is extremely difficult to establish satisfactory experiments that eliminate all variables except the presence or absence of a specific arthropod. Nevertheless, a number of excellent trials have produced valid data concerning the effect of arthropod pests on the efficiency of production of livestock in the U.S.

The first organized report on losses in animal production caused by arthropod pests was published by Hyslop,[1] who listed total annual losses of $168.8 million in the late 1930s. This report was followed in 1954 by the U.S. Department of Agriculture publication ARS-20-1,[2] which was updated in 1965 as U.S. Department of Agriculture Handbook No. 291,[3] in which annual losses in livestock and poultry production to arthropods were estimated at $877.8 million. Recently, Steelman[4] reviewed the data available on the effects of arthropod pests on livestock production and called for more accurate data on economics in order to establish the basis for pest management programs.

In this section on arthropod pests, figures for losses in animal production, in terms of amount or value of product, such as meat, milk, eggs, and wool, in the absence of control measures, were taken from original publications (not cited). The losses obtained for limited tests with small numbers of animals in specific areas were extrapolated to include the entire U.S.; the values for production, sales, and numbers of livestock as presented for 1977 in Agricultural Statistics[5] were used.

Beef and Dairy Cattle

Beef and dairy cattle are parasitized by insects, ticks, and mites. Bloodsucking flies (often called "biting" flies) that attack cattle interfere with normal feeding activities and cause loss of blood, reduced weight gains, increased susceptibility to diseases, and death of cattle, and may spread anaplasmosis, anthrax, tularemia, bluetongue, and other diseases. The most important biting flies are the horn fly, *Haematobia irritans* (L.), found on cattle in pastures, and the stable fly, *Stomoxys calcitrans* (L.), found around cattle in pens and other confinements. Other biting flies include horse flies, deer flies, black flies or buffalo gnats, biting gnats, and mosquitoes.

Other flies that attack cattle do not suck blood. One important species, the face fly, *Musca autumnalis* De Geer, which frequents the moist areas, especially eyes and nostrils, of cattle, is difficult to control and can contribute to the spread of pinkeye and eye worms (*Thelazia* spp.). Cattle are infested with northern cattle grub, *Hypoderma bovis* (L.), and the common cattle grub, *H. lineatum* (de Villers), which cause decreased weight gains and milk flow, excessive trimming of the carcass, and downgrading of hides. Cattle are also attacked by the screwworm fly, *Cochliomyia hominivorax* (Coquerel), but a highly successful federal-state eradication campaign has virtually eliminated this pest from the U.S.

Cattle are infested with the cattle biting louse, *Bovicola bovis* (L.), the shortnosed cattle louse, *Haematopinus eurysternus* (Nitzsch), the longnosed cattle louse, *Linognathus vituli* (L.), the little blue cattle louse, *Solenopotes capillatus* Enderlein, and the cattle tail louse, *H. quadripertusus* Fahrenholz. Lice that cause unthriftiness and anemia are usually most numerous in the cooler times of the year.

Cattle are infested with five species of scabies and mange mites that live on or in the skin and cause severe dermatitis. Because of its pathogenicity and contagious nature, the sheep scab mite, *Psoroptes ovis* (Hering), is the subject of a state-federal eradication campaign that includes compulsory treatment and quarantine. Cattle are also infested with the cattle itch mite, *Sarcoptes bovis* Robin; a chorioptic scab mite, *Chorioptes bovis* (Hering); *Psorergates bos* Johnson; and the cattle follicle mite, *Demodex bovis* Stiles.

Cattle are parasitized by several species of ticks that suck blood, inject toxins, cause paralysis, damage hides, cause "tick worry", and transmit diseases. The most important ticks found on cattle are the lone star tick, *Amblyomma americanum* (L.); the Gulf Coast tick, *A. maculatum* Koch; the winter tick, *Dermacentor albipictus* (Packard); the Rocky Mountain wood tick, *D. andersoni* Stiles; the Pacific Coast tick, *D. occidentalis* Marx; the American dog tick, *D. variabilis* (Say); the blacklegged tick, *Ixodes scapularis* Say; and the spinose ear tick, *Otobius megnini* (Duges).

From publications on losses, it is estimated that horn flies cause a 12 lb per calf weight loss, a 10 lb per head weight loss of stocker cattle, a 4.6% weight loss of slaughter cattle, and a 1% reduction in milk flow for the 6-month horn fly season. Stable flies cause a $16.40 per head weight loss in the 12.9 million cattle in feedlots and a 5% reduction in milk flow for the 6-month stable fly season. Mosquitoes cause a $6.00 per head weight loss in cattle in feedlots for the 6-month mosquito season. The face fly, because of pinkeye transmission, causes a weight loss of 33 lb per calf in 10% of the calves. Cattle grubs cause a 12 lb per head weight loss in calves in feedlots, a 41 lb per head weight loss in cattle on finishing feed, a $1.00 per animal loss in meat trim of the 40% of the cattle slaughtered when they have cattle grubs in their backs, a $1.00 per hide loss for hides that have more than five holes, and a 10% decrease in milk flow for 3 months of the year when cattle grubs are in backs of cattle. Lice cause a 68 lb per head weight loss in 12% of the cattle slaughtered. Scabies mites cause a loss of $45.00 per head in weight gain for the 340,000 cattle found infected in feedlots (fiscal year 1977), and demodectic mange mites cause a $1.00 per hide loss due to blemishes in the 30% of the 43.3 million hides that are heavily infested. Ticks cause a 9 lb per

Table 1
ESTIMATED LOSSES IN CATTLE PRODUCTION DUE TO ARTHROPOD PESTS

Arthropod	Annual losses ($ million)
Horn fly	730.3
Stable fly	398.9
Mosquitoes	38.7
Face fly	53.2
Cattle grub	607.8
Lice	126.3
Scabies mites	15.3
Mange mites	14.4
Ticks	275.7

Table 2
ESTIMATED LOSSES IN SHEEP AND SWINE PRODUCTION DUE TO ARTHROPOD PESTS

Arthropod	Annual losses ($ million)
Sheep	
Keds	40.9
Bots	13.5
Swine	
Lice	154.4
Mange mites	77.2

head weight loss in 68 million cattle in the U.S., and Gulf Coast ticks create a $3.00 per cwt loss at marketing due to "gotch ear" and a 20 lb per head weight loss of the 2 million cattle in the tick-infested area. The total estimated losses are summarized in Table 1. The losses due to horse and deer flies, most small biting flies, most ticks, and "minor" pests are not included because they have not been determined.

Sheep and Goats

Sheep and goats are parasitized by arthropods that suck blood, live on skin scales, hair, or wool, invade tissues, inhabit nasal chambers, and cause decreased weight gain and loss in value and production of wool and mohair and in value of skin. Sheep are infested with the sheep biting louse, *Bovicola ovis* (Schrank); sucking body louse, *Linognathus ovillus* (Neumann); and the sucking foot louse, *L. pedalis* (Osborn). Goats are infested with *B. crassipes* (Rudow), *B. limbatus* (Gervais), *B. caprae* (Gurlt), *L. stenopsis* (Burmeister), and *L. africanus* (Kellogg and Paine). Sheep (and goats to a much lesser extent) are infested with the sheep ked, *Melophagus ovinus* (L.), a wingless, bloodsucking fly that is the cause of cockle, a sheepskin defect. Sheep are commonly infested with *Chorioptes bovis,* a chorioptic scab mite, and to a lesser extent with the itch mite, *Sarcoptes scabiei* (De Geer). The sheep scab mite was declared eradicated from sheep in the U.S. in 1973. Goats may be infested with a mange mite, *Chorioptes caprae* (Delafond). Sheep are infested with *Demodex ovis* Railliet and goats with *D. caprae* Railliet, which cause follicular or red mange and are usually not controlled. Both sheep and goats are attacked by larvae of the screwworm fly; by fleece worms — primarily *Cochliomyia macellaria* (F.); the secondary screwworm fly, *Phormia regina* (Meigen); the black blow fly, *Phaenicia sericata* (Meigen); and other species of lesser importance. Larvae of the sheep nose bot fly, *Oestrus ovis* L., live on mucous surfaces of the nasal passages and sinuses of sheep and occasionally goats. Sheep and goats are parasitized by a few ticks, especially the spinose ear tick. The biting gnat, *Culicoides variipennis* (Coquellet), attacks sheep and transmits bluetongue, a virus disease. Sheep are also attacked by black flies, mosquitoes, and other biting flies.

From publications on losses, it is estimated that sheep keds cause an 8% reduction in weight gain, a 15% reduction in wool production, and a 30% reduction in value of sheepskins; and sheep nose bot larvae cause a 4% decrease in weight gain (Table 2). The losses due to lice, mites, fleece worms, and other ectoparasites of sheep and goats have not been determined.

SWINE

Swine in the U.S. are parasitized by the hog louse, *Haematopinus suis* (L.), and a mange mite, *Sarcoptes scabiei suis* Gerlach, that cause irritation, itching, decreased weight gain, and stunted growth. Also, large numbers of house flies can cause serious irritation. Swine are also attacked by stable flies, mosquitoes, black flies, gnats, horse flies, and other biting flies.

Losses in production of swine (Table 2) were calculated from data on losses in Alabama presented in Cooperative Plant Pest Report;[6] hog lice cause an estimated 2% reducton in weight gain, and hog mites cause a 1% reduction. The losses due to the other arthropods have not been determined.

Poultry

Poultry are attacked by bloodsucking mites: the northern fowl mite, *Ornithonyssus sylviarum* (C. and F.); the tropical fowl mite, *O. bursa* (Berlese); and the chicken mite, *Dermanyssus gallinae* (De Geer). Poultry are also attacked by lice: the body louse, *Menacanthus stramineus* (Nitzsch) (the most common species); the shaft louse, *Menopon gallinae* (L.); the head louse, *Lipeurus heterographus* (Nitzsch); the wing louse, *L. caponis* (L.); and the fluff louse, *Goniocotes gallinae* (De Geer). Poultry are also attacked by fowl ticks, *Argas* spp.; the sticktight flea, *Echidnophaga gallinacea* (Westwood); the scalyleg mite, *Knemidokoptes mutans* (Robin and Lanquetin); and the depluming mite, *K. gallinae* (Railliet). A turkey chigger, *Neoschongastia americana* (Hirst), also infests turkeys in certain areas of the southern U.S. These pests chew skin debris and feathers, irritate the animals, suck blood and other fluids, and predispose animals to common diseases of poultry. In addition, poultry houses can be the source of nuisance flies that breed in the manure under caged layers.

From publications on losses, it is estimated that fowl mites, especially the northern fowl mite, cause a 4% reduction in weight gain and a 9% reduction in egg production for the ⅓ of the year when mites are active, and fowl lice cause a 7% reduction in weight gain and a 10% reduction in egg production. Losses in weight gain are determined for poultry other than broilers, for broilers are grown so rapidly that ectoparasite populations do not build up to loss levels. Losses caused by turkey chiggers are limited to those experienced by growers in certain southern states who market turkeys when lesions created by chiggers cause downgrading at the processing plant. Losses, summarized in Table 3, do not include losses due to the minor pests, for these values have not been determined.

Horses

Although horses are not an agricultural commodity, they are a very valuable asset in many rural and urban situations. The value of the 6 million pleasure horses in the U.S. has been estimated in excess of $4 billion. It is not possible to determine production losses, but millions of dollars are spent annually in order to protect horses from arthropod pests.

Horses are attacked by a number of bloodsucking flies — the horn fly, stable fly, a variety of horse flies and deer flies, biting gnats, black flies, biting midges, and mosquitoes — that are of particular importance because they annoy horses as well as transmit diseases. Two nonbiting flies, the face fly and often the house fly, are attracted to the moisture about the eyes of horses where they sponge liquids.

Horses are infested with the common horse bot fly, *Gasterophilus intestinalis* (De Geer); the nose bot fly, *G. haemorrhoidalis* (L.); and the throat bot fly, *G. nasalis* (L.). Horse bot larvae that hatch from eggs laid by the female flies on the front of

Table 3
ESTIMATED LOSSES IN POULTRY PRODUCTION DUE TO ARTHROPOD PESTS

Arthropod	Annual losses ($ million)
Mites	116.3
Lice	378.7
Others (ticks, chiggers)	4.0

horses irritate the tongue, lips, gums, and lining of the stomach and small intestine; and infested animals often suffer from colic or other gastric ailments. In addition, the ovipositing females cause horses to rear, run, and try to avoid the flies; such actions are often detrimental to the horse and rider.

Horses are parasitized by a variety of ticks, lice, and mites. Of special importance are the winter tick, often found in large numbers on horses in the fall and winter, and *Anocentor nitens* (Neumann), the tropical horse tick; though distributed only in Georgia, Florida, and southern parts of Texas, the tropical horse tick is significant because it is a vector of equine piroplasmosis. Horses are infested with the horse biting louse, *Bovicola equi* (Denny), and the horse sucking louse, *Haematopinus asini* (L.). Horses may be infested with the itch mite, the sheep scab mite, and the chorioptic scab mite. In addition, chigger mites and *Demodex* spp. may cause severe dermatitis.

DISEASES, NUTRITIONAL DISORDERS, AND PARASITES

Diseases, nutritional disorders, and parasites are "pests" that seriously affect the production of livestock in the U.S. Nutritional disorders, which include metabolic dysfunctions and noninfectious diseases, such as grass tetany, ketosis, and mineral and vitamin deficiencies, are included because they do affect the well-being of livestock and may create greater losses when found in combination with other "pests" than when each is found individually. Greatest losses are associated with diseases of the reproductive, respiratory, and digestive tracts.[7,8] The diseases may be further characterized by their etiology into viral, bacterial, and mycotic. About ¼ to ½ of the losses due to diseases, nutritional disorders, and parasites are directly attributable to the death of the animal. The rest of the losses are associated with morbidity and are reflected by the individual's failure to reach the production potential of the breed.[8]

Beef and Dairy Cattle

Greatest losses to the cattle industry from diseases are due to bacteria and viruses. Some diseases, such as the respiratory and enteric infections, have a combined bacterial and viral etiology. Respiratory tract infections, the number one cause of economic loss, involve bacteria, such as *Pasteurella haemolytica* and various species of mycoplasma, and viruses, such as the virus of infectious bovine rhinotracheitis (IBR), parainfluenza-3 (P*I*-3), bovine viral diarrhea (BVD), adenovirus, bovine respiratory syncytial virus, and rhinovirus.[7,9] The IBR virus is the most prevalent.[9,10] Losses are in the form of decreased milk production, weight loss, retarded growth, abortion, or death from bronchopneumonia. The major cause of enteric infections in feedlot cattle and in young calves is bovine viral diarrhea virus.[7,10] Rotavirus, coronavirus, and adenovirus, frequently involved in outbreaks of enteric infections of calves, may be the sole agent isolated, but frequently they are found combined with bacterial agents such as *Escherichia coli*, a major cause of calf scours.[7,11]

Losses from bovine leukosis virus are related to carcass condemnation and loss of export sales of purebred cattle and semen. However, the losses are difficult to evaluate because of the decreased acceptance by the consumer of beef and dairy products from leukotic cattle.

Reproductive diseases, including brucellosis, leptospirosis, vibriosis, IBR virus, and BVD virus, result in abortions, failure to conceive, death of potential seed-stock, lower milk production, and inefficient utilization of feed and facilities.[7,10,12]

Infectious keratitis (pinkeye), a bacterial disease that affects the eyes of cattle, causes a loss due to lower milk production, weight loss, and retarded growth rate.[7]

Bovine mastitis of bacterial origin (*Streptococcus agalactiae, E. coli,* and *Staphylococcus aureus*) is the most serious disease of dairy cattle in the U.S. In addition to losses in milk production, the cost of mastitis must include the cost of discarded milk and drugs, veterinary fees, increased labor, decreased sale value of cows, and increased costs of herd replacements.[7]

Mycotic diseases and mycotoxicoses cause losses associated with abortions, reduced weight gains, and lower milk production. Recently, mycotoxins in the feed have been shown to interfere with the development of acquired immunity to some infectious diseases.[7]

Major nutritional disorders of cattle include milk fever, ketosis, and grass tetany. Losses are due to decreased milk production and death of the animal.[7]

Major losses are also due to the presence of parasites such as gastrointestinal worms (including *Haemonchus, Ostertagia,* and *Trichostrongylus*), coccidia (primarily *Eimeria* species), anaplasma, and liver fluke (*Fasciola hepatica*). Helminthic parasites (worms) of cattle are widespread in the U.S. and affect both dairy and beef animals.[13] Coccidiosis occurs in two forms — clinical and subclinical. Animals with clinical coccidiosis often become moribund and die; those with subclinical coccidiosis convert feed inefficiently, lag in weight gain and growth, have low resistance to other diseases and stress, and produce less milk. Anaplasmosis, an infectious disease caused by *Anaplasma marginale* that invades red blood cells and causes their destruction, is transmitted by flies, mosquitoes, and ticks. Losses are due to anemia, poor growth, weight loss, lowered milk production, and deaths.[7]

Estimated losses (Table 4) in beef and dairy cattle production are based on numbers of cattle and production from Agricultural Statistics.[5] For example, an estimated 10% of the 50 million calves die because of calf scours; and an estimated 10% loss of milk production ($11 billion) is due to mastitis.

Sheep and Goats

Of major importance to the sheep industry are several bacterial diseases including clostridial infections (such as blackleg and malignant edema), colibacillosis (lamb dysentery), vibrionic abortion, polyarthritis, brucellosis, listeriosis, and foot rot. These diseases result in losses from poor growth rate, neonatal deaths, infertility, abortion, and inefficient feed converson.[14]

Viral diseases include bluetongue, which is transmitted by an insect, *Culicoides* spp., and chronic progressive pneumonia.[7]

In general, goats are susceptible to the same bacteria that infect sheep. Losses due to salmonellosis, brucellosis, clostridial infections, and neonatal enteric infections are economically significant. Mastitis is a major problem in dairy goats. Viral infections of the reproductive, nervous, respiratory, and digestive systems are of economic sig-

Table 4
LOSSES IN CATTLE PRODUCTION DUE TO DISEASES, NUTRITIONAL DISORDERS, AND PARASITES

Cause	Annual losses ($ million)
Diseases	
Respiratory	
Feedlots	600
Dairy	200
Enteric	
Calf scours	250
Feedlots	200
Dairy	100
Reproductive	200
Bacterial	
Pinkeye	200
Mastitis	1000
Mycotic	100
Nutritional disorders	
Milk fever	25
Ketosis	15
Grass tetany	10
Parasites	
Helminths	250
Coccidiosis	70
Anaplasmosis	200

Table 5
LOSSES IN SHEEP AND GOAT PRODUCTION DUE TO DISEASES, NUTRITIONAL DISORDERS, AND PARASITES

Cause	Annual loss ($ million)
Diseases	
Bacterial	62
Viral	
Bluetongue	10
Chronic progressive pneumonia	20
Mycotic	11
Nutritional disorders	5
Parasites	45

nificance. Mycotic infections may cause abortion in sheep and goats and are responsible also for respiratory diseases, mastitis, and skin infections.

Nutritional disorders in sheep and goats are similar to those of cattle. Mineral and vitamin deficiencies are of major concern to the sheep and goat producers. Losses from deficiencies of copper, calcium, phosphorus, cobalt, and iodine are particularly widespread.[7,15]

Sheep and goats suffer losses because of internal (helminth) parasites,[14-16] especially tapeworms, hookworms, stomach worms, lungworms, and liver flukes. The major tapeworm affecting sheep is the large tapeworm, *Moniezia expansa*, which chiefly affects lamb, and causes diarrhea, weight loss, and death. *Haemonchus contortus*, the large stomach worm, the most widespread parasite of the abomasum of sheep, causes poor growth rate, weight loss, anemia, weakness, and poor quality wool. The nodular worm, *Oesophagostomum columbianum,* causes losses in the form of poor growth, weight loss, diarrhea, emaciation, and weakness. Additionally, the nodules lower the value of intestines for use as sausage casings and surgical sutures. The large lungworm, *Dictyocaulus filaria*, causes severe losses to sheep and frequently predisposes them to bacterial and viral pneumonias and death. Losses in sheep and goat production are summarized in Table 5.

Swine

Losses from diseases of swine are generally related to infections of the respiratory, enteric, and reproductive tracts.[7,16] Major bacterial pathogens of the respiratory tract include species of *Pasteurella, Corynebacterium, Bordetella, Mycoplasma*, and *Myco-*

bacterium. Viral pathogens of the respiratory tract include the herpes virus, influenza virus, and coronavirus. Losses from respiratory pathogens are associated with neonatal deaths, poor growth, and condemnations due to lesions at the slaughter house. Enteric infections include the major bacterial diseases of colibacillosis, salmonellosis, and swine dysentery, and also transmissible gastroenteritis and other viral enteric diseases. Losses are in the form of neonatal deaths, poor growth, and inefficient utilization of feed.[7] Reproductive diseases such as parvovirus, pseudorabies, leptospirosis, and brucellosis also cause substantial annual losses to fetal death, abortion, or birth of weak or stillborn pigs. The most important of these diseases is parvovirus. Other significant diseases of swine are erysipelas, jowl abscesses, and mycotoxins. The etiology of the syndrome of Metritis-Mastitis-Agalactia (MMA), a perennial problem to the pork producer, remains obscure and apparently involves metabolic factors and infectious agents. Additional nutritional disorders of neonatal pigs include iron deficiency and acute hypoglycemia, which frequently cause poor growth or death.

The swine industry incurs heavy losses from numerous parasites.[17] Losses result from inefficient utilization of feed, morbidity, mortality, and condemnation of parts and carcasses at abbatoirs. Helminths are the most important internal parasites of swine. *Ascaris lumbricoides,* the large roundworm, is found in the small intestine, where the female produces large numbers of eggs. Eggs, when ingested, hatch into larvae that penetrate the intestinal wall and enter the portal circulation where they later damage the lungs and liver extensively. Other major internal parasites of swine include *Strongyloides, Oesophagostomum,* and *Trichuris* spp. Losses to the swine industry from internal parasites include the cost of drugs, medicated feeds, and other items to prevent infestation. The losses in swine production are summarized in Table 6.

Poultry

Major causes of losses to the poultry industry include a variety of pathogens of the respiratory, reproductive, and digestive tracts. Bacterial pathogens of the respiratory tract of etiologic significance include *Mycoplasma* species (airsacculitis), *Pasteurella multocida* (fowl cholera), and *Chlamydial* agents (ornithosis). Financial losses are related to lowered egg production, poor growth, and carcass condemnation at the processing plant from respiratory tract lesions. Other bacterial diseases of economic significance include salmonellosis, clostridial infections, and colibacillosis. Viral respiratory disease entities of economic significance to the poultry industry include Newcastle disease and infectious bronchitis. Losses are mainly related to lowered egg production and quality, poor growth, or death of broilers. Neoplasms of viral origin also create substantial annual losses. Losses from mycoses and mycotoxicoses are usually in the form of broiler deaths. Evidence indicates that low levels of mycotoxins in the feed may lower natural and acquired resistance to infectious diseases.[7]

The poultry industry also suffers losses from nutritional disorders. Of particular significance are vitamin deficiencies, inorganic mineral imbalances, and mycotoxicoses.[18] Additional losses, due to stress-related conditions associated with intensive confinement housing, are in poor growth or death of young birds. Coccidiosis is the major parasitic disease of poultry. Although several genera are involved, the most common is *Eimeria,* which causes pathologic changes in the small intestine and ceca. *Eimeria tenella* and *E. necatrix* are the most common species. Losses are attributable to severe diarrhea, dehydration, and death of young birds. An extremely large percentage of the estimated losses is related to the cost of drugs and medications.[7]

Histomoniasis, also known as blackhead or infectious enterohepatitis, another major disease of turkeys, is caused by the protozoan *Histomonas meleagridis.* Bacterial infections contribute to the pathology of the liver and ceca. Most losses are attributed to death of young birds under 3 months old.[17] Other significant losses from parasitic

Table 6
LOSSES IN SWINE PRODUCTION DUE TO DISEASES, NUTRITIONAL DISORDERS, AND PARASITES

Cause	Annual losses ($ million)
Diseases	
Respiratory	200
Enteric	200
Reproductive	60
Other	10
Nutritional disorders	
Metritis-Mastitis-Agalactia	100
Iron deficiency	10
Hypoglycemia	10
Parasites	355

Table 7
LOSSES IN POULTRY PRODUCTION DUE TO DISEASES, NUTRITIONAL DISORDERS, AND PARASITES

Cause	Annual losses ($ million)
Diseases	
Bacterial	290
Salmonellosis	128
Viral	
Newcastle	215
Infectious bronchitis	150
Neoplasms	134
Mycotic	50
Nutritional disorders	200
Parasites	225

diseases are due to toxoplasmosis and leucocytozoon infections.[17] Losses in poultry production are summarized in Table 7.

Horses

Horses, a very valuable aspect of the U.S. economy, are attacked by a number of bacterial, viral, and mycotic diseases. Major bacterial diseases include streptococcal infections of the reproductive and respiratory tract resulting in abortions, pneumonia, and neonatal septicemias.[7]

The equine encephalomyelitides are arthropod-born diseases that are leading causes of financial loss to the horse industry. Losses are in the form of death or nervous disorders that render the animal unfit for service. Equine infectious anemia (swamp fever) is another viral disease that accounts for losses in the form of debilitation and death.[7] Respiratory diseases of viral origin are also significant causes of losses.[17] Mycotic infections and mycotoxicoses affect horses in the form of abortions, respiratory diseases, and toxemias.

Nutritional disorders, hormonal imbalances, and other noninfectious diseases account for an additional annual loss.

Horses suffer from a number of internal parasites (helminths). Three species of large strongyli, *Strongylus vulgaris, S. edentatus,* and *S. equinus,* occur in American equids and are major causes of losses. Both the adult parasites, which occur in the large intestines and ingest blood and intestinal tissues, and the immature stages, which migrate to various locations in the abdominal organs and induce a variety of serious lesions, are extremely damaging to the health of horses. Particularly serious are thrombi and aneurysms induced in intestinal and other major arteries. Intestinal parasites cause particularly heavy losses in young horses. An estimate is that the 2 million foals born annually are treated on an average of four times with anthelminthics; also, the mortality in foals from helminths is about 5% or 100,000 per year.

Piroplasmosis caused by *Babesia caballi* and *B. equi,* a protozoan disease transmitted by ticks in the southeastern U.S. has a severe impact on restricting the movement of racing horses into and out of the U.S. Losses in the horse industry are summarized in Table 8.

Table 8
LOSSES IN THE HORSE INDUSTRY DUE TO DISEASES, NUTRITIONAL DISORDERS, AND PARASITES

Cause	Annual losses ($ million)
Diseases	
Bacterial	50
Viral	50
Mycotic	5
Nutritional disorders	5
Parasites	
Helminths	78
Piroplasmosis	5

WEEDS, BRUSH, AND POISONOUS PLANTS

Weeds, brush, and poisonous plants are very properly classified as pests, for their presence affects not only the quantity and quality of forage available for livestock consumption but also the life, well-being, and productivity of livestock themselves. The losses in reduced quality and yield of crops and forage due to weeds, plus cost of control in developed countries of the temperate zones, estimated at 10 to 15% of the total value of agricultural and forest products,[3] stem from a variety of single or compounded factors such as competition for nutrients, water, light, and space; reduced quantity and quality of crops and livestock forage; harborage for insects, nematodes, and disease organisms; clogging of farm water supplies; and costs of equipment, manpower, and chemicals for control measures. Among field crops, losses in yield due to weeds range from 10% in corn to 17% in oats and soybeans (average annual loss [1951 to 1960] of $1,572 million).[3]

Brushy plants such as mesquite and sagebrush are about 4 to 6 ft tall, generally larger than weeds, have no defined trunk, and cover an enormous amount of land that otherwise could be productive. The presence of brush and herbaceous weeds causes losses in forage production on an estimated billion acres of land: 230 million acres in rangeland and pastures in 31 eastern states and 769 million acres in rangeland in 17 western states.[3] These figures are estimates and may not present the extent of the problem. In the southern part of the western states, the problems are much more acute; for instance, 107 million acres of grazing land in Texas alone is infested with brush. Weeds and brush are extremely hardy, very easily propagated, and highly invasive; thus, the landowner must use continuous measures to control brush. Although cultural, ecological, mechanical, biological, and other nonchemical methods of brush and weed control are used in over ⅔ of the land, the use of chemical herbicides has increased from a total cost of $473 million in 1965 to treat over 120 million acres, to over $1 billion in 1974 to treat over 300 million acres.

Poisonous plants are one of the most persistent problems confronting the livestock producer, and economic losses are heavy. It is difficult to separate losses according to classes of livestock; therefore, losses will be presented for all of livestock production. The losses in livestock as a result of poisonous plants may be direct or indirect. Direct costs induced by poisonous plants include death, debilitation, chronic illness, failure to gain weight, photosensitization, abortions, and birth defects. Indirect costs, often less understood but possibly more costly, include loss of productive pastures or ranges, loss of animal production because of unrecognized subclinical effects, time spent by ranchers watching and handling stock away from infested range, treatment of poisoned animals, and increased cost of providing a ration to rehabilitate affected stock.

Recent estimates[19] place the annual losses due to poisonous plants at $51 million in the U.S., but other estimates place them between $50 and $100 million in Texas alone. Another U.S. estimate, based on 1976 figures, indicates the loss in cattle and sheep to be greater than $133 million annually, or 5.5% of the total value.[19] These estimates were based on numbers of livestock in which specific problems occur, economic value of each species, percentage of morbidity and average cost in productivity, percentage of mortality multiplied by the value of the individual animal, and cost of feed and labor in attempts to reduce losses. Figures vary not only with locality but also with climatic changes within a locality. For instance, estimates of deaths in cattle from ingestion of larkspur generally average between 3 and 5%; however, in some years, death losses in livestock from grazing larkspur-infested ranges are higher than 15%.[20]

Of the more than 700 plants in the U.S. listed as known or suspected to be toxic to livestock, many may not cause financial loss, may be of historic inclusion, may have been suspected to be toxic without any scientific investigation, or may be just folklore. Of these, fewer than 100 are considered to be of economic importance; again, the importance depends on the locality and prevalence of the plant within a given area. The poisonous plant problem remains more of a problem in the western U.S. than the eastern U.S.

The estimated average annual loss in livestock caused by 36 of the 700 species of plants known or suspected to be poisonous is presented in Table 9.

In addition to the specific plants listed in Table 9, losses caused by specific toxic entities, such as those listed in Table 10, may be caused by a number of plants that have the same toxic effect.

About 1.33% of all beef cattle in the U.S. are affected by some form of nutritional disease; of these, ingestion of poisonous plants accounts for 16.7% of the nutritionally ill cattle and 33.2% of the death losses in nutritionally ill cattle.[21]

The samples itemized in Tables 9 and 10 are not complete; their inclusion is based on our present knowledge, data recorded in specific instances, or information gathered in field investigations from knowledgeable people such as range specialists, veterinarians, and extension agents. Nevertheless, these incomplete data show that livestock must be protected from the effects of poisonous plants before production of livestock can be maximized in the U.S.

PREDATORS

Livestock, including poultry, are attacked by a variety of predators that include coyotes, dogs, bears, bobcats, mountain lions, eagles, and other carnivorous animals. These animals create losses because of their direct destruction of livestock for their own food. Unfortunately, with the exception of the sheep industry, there is little factual information available on the losses in livestock production in the U.S. to predators. Terrill[22] reported that total numbers of livestock (except sheep) lost in millions in the 22 western states for 1971, 1972, and 1973 were 0.17, 0.17, and 0.18, respectively, for cattle and calves; 0.23, 0.17, and 0.20 for swine; and 11.7, 10.7, and 7.7 for poultry. Estimated total monetary losses of all livestock to predators increased from $105 million in 1971 to $170 million in 1973. Monetary losses of cattle were 1.5 times sheep losses in 1973, swine 0.09 times, and poultry 0.6 times. Percentage of losses of livestock to predators for Nebraska for 1971, 1972, and 1973[23] were 0.2 each year

Table 9
LOSSES IN LIVESTOCK PRODUCTION DUE TO 36 MOST PREVALENT POISONOUS PLANTS

Plant	Common name	Annual losses ($ million)
Acacia berlandieri	Guajillo	0.5
Agave lecheguilla	Lechuguilla	0.5
Amaranthus spp.	Pigweed	0.5
Asclepias spp.	Milkweed	0.5
Astragalus spp.	Locoweed	10.0
Baileya multiradiata	Desert baileya	2.0
Brassica spp.	Mustards	2.0
Cassia occidentalis	Coffee senna	0.5
Centaurium spp.	Texas star	0.1
Crotalaria spp.	Rattlebox	1.0
Delphinium spp.	Larkspur	2.0
Euphorbia spp.	Spurge	1.0
Festuca spp.	Fescue grass	0.5
Flourensia cernua	Tarbrush, blackbrush	0.2
Gutierrezia microcephala	Broomweed, snakeweed	10.0
Helenium spp.	Sneezeweeds	2.0
Hymenoxys odorata	Bitter rubberweed	5.0
Isocoma wrightii	Rayless goldenrod	0.1
Kallstroemia spp.	Carpetweed	1.0
Lobelia inflata	Indian-tobacco	0.1
Lupinus spp.	Lupines	1.0
Nerium oleander	Oleander	0.5
Nolina texana	Texas sacahuista	0.5
Panicum spp.	Panicgrass	0.1
Peganum harmala	African-rue	0.1
Phyllanthus abnormis	Abnormal leafflower	0.1
Prosopis glandulosa	Honey mesquite	2.0
Psilostrophe spp.	Paperflower	0.5
Pteridium aquilinum	Bracken fern	1.0
Quercus spp.	Oaks	30.0
Ricinus communis	Castorbean	0.5
Senecio spp.	Groundsels	10.0
Sesbania spp.	Sesbania	0.5
Solanum spp.	Nightshades	2.0
Veratrum spp.	False hellebore	0.5
Xanthium spp.	Cocklebur	0.5
Miscellaneous plants		15.0

for cattle and calves; 0.5, 0.4, and 0.5, respectively, for swine; and 7.0, 6.7, and 6.0 for poultry. Gee[24] reported losses to predators of 11% of calves with a value of $31 million in 1975. Goats are also subject to a high level of predation, but the relatively small scale of this industry in the U.S. results in an overall monetary loss that is much less than that for the other species of livestock.[25] Most exact figures on losses to predators have been obtained for the sheep industry, and these losses will be examined in some detail.

Shelton,[26] who summarized percentage of losses of sheep to various predators as coyote (84.7), dog (3.9), bobcat (3.4), bear (2.6), mountain lion (0.7), eagle (0.5), and others (4.2), concludes that sheep and goats are the primary prey, and the coyote is the primary predator. Wolves were prominent predators of livestock when this country was first settled, but coyotes have become more important in recent times. Numerous surveys[27] show that losses of lambs to predators have been important and are increasing.

Table 10
LOSSES IN LIVESTOCK PRODUCTION DUE TO SPECIFIC TOXIC CONDITIONS CAUSED BY ANY OF SEVERAL PLANTS WITH SIMILAR ACTIONS

Condition	Plant example	Annual losses ($ million)
Ergotism	*Claviceps* spp.	1.5
Mycotoxicosis	*Fusarium* spp.	5.0
Prussic acid poisoning	Johnsongrass, sorghums	1.0
Nitrate poisoning	Many plants accumulate nitrate	10.0
Oxalate Poisoning	*Halogeton* spp.	1.0
Photosensitization	*Ammi* spp., many plants cause poisoning and/or secondary photosensitization	6.0
Pulmonary edema		0.5
Pulmonary adenomatosis		0.5
Wheat pasture poisoning		5.0
Grass tetany		0.5
Bermudagrass staggers		5.0

Terrill[28,29] found that deaths of lambs minus deaths of sheep as reported by the U.S. Department of Agriculture[30] could serve as an index to determine trends in losses of sheep and lambs over the years because this index was highly related to the trends in losses to predators in the National Forests (Rummell).[31] This study, along with the finding of Gee et al.[27] that predators kill more lambs than sheep (along with detailed survey reports on 15 western states in 1974), provides a means of estimating losses to predators by using data from Statistical Reporting Service without conducting expensive surveys.

Differences in percentage of deaths of lambs minus percentage of deaths of sheep were calculated from 1953 through 1977 for each state in which predator losses were evident. These years were used as a base period because predator losses tended to be lowest in those years. Deviations from the base period were then calculated from 1958 through 1977 and were adjusted by a correction factor from Gee et al.[27] for each state, and for other years which were corrected accordingly. Correction factors for states not surveyed by Gee et al.[27] were used from Oregon for Washington, from Nebraska[23] for Nebraska, Iowa, and Missouri, and from the average of North Dakota, Nebraska, and Kansas for all other farm states. Nebraska data from Gee et al.,[27] when applied by the above method, showed no loss in 1 year when the Nebraska survey[23] revealed substantial losses; so substitute correction factors were used to provide realistic but still conservative estimates. The corrected percentage of lamb deaths was then multiplied by the lamb crop for each state each year to obtain an estimate of total lambs killed by predators. Then the ratio of adult sheep killed to lambs killed[27] was applied to lamb losses to obtain the number of adult sheep lost to predators. Value of sheep and lambs lost was calculated from the average value per head for each state and year.[30] Many young lambs were killed, and some would have died from causes other than predators, but the cost of production was much the same as if they had all lived to market age. Therefore, the market value is a reasonable estimate of trends of loss.

Limited dynamics of annual losses from 1958 through 1977 are presented in Table 11 by ecosystem areas. Losses were greatest in the intermountain west. Losses were somewhat similar in Texas and in the west north central area. Texas is unique in that sheep are raised in a compact area, and sheep and lambs toward the center of the area tend to be better protected than those on the periphery. The pattern of loss of the west

Table 11
DYNAMICS OF ESTIMATED LOSSES OF SHEEP AND LAMBS TO PREDATORS

Year	Sheep and lambs killed (1000)	% Losses of inventory plus lamb crop	Value of loss ($ thousand)
	West Coast[a]		
1958	139	2.65	3,091
1968	198	5.17	4,887
1977	156	5.98	9,179
	Intermountain West[b]		
1958	700	4.10	16,068
1968	705	5.34	17,637
1977	581	7.30	33,124
	Texas		
1958	313	4.09	5,483
1968	325	4.93	5,460
1977	264	6.66	12,024
	West North Central[c]		
1958	385	5.15	7,904
1968	307	4.39	6,970
1977	250	6.91	12,778
	East North Central[d]		
1958	23	2.05	413
1968	22	1.51	435
1977	23	3.13	923
	Northeast[e]		
1968	1.9	1.04	40
1977	16.8	4.70	773
	Mid-East and South[f]		
1969	0.8	2.13	111
1977	4.6	2.12	187
	U.S.[g]		
1958	1561	4.04	32,960
1968	1559	4.83	35,430
1977	1295	6.66	68,997

[a] California, Oregon, and Washington.
[b] Arizona, Colorado, Idaho, Montana, Nevada, New Mexico, Utah, and Wyoming.
[c] Iowa, Kansas, Minnesota, Missouri, Nebraska, North Dakota, and South Dakota.
[d] Illinois, Indiana, Michigan, and Wisconsin.
[e] New York and West Virginia.
[f] Arkansas, Kentucky, Louisiana, and Oklahoma.
[g] Totals were added for the areas to provide a weighted average for the U.S.

coast area was similar to that of the other western areas except that percentage of loss was somewhat lower. All of the eastern states showed relatively minor losses when compared with losses in the western states. In general, losses in adult sheep equaled about ⅓ of the lamb losses in the range states and ⅔ of the lamb losses in the farm states.

A total of 29 states showed losses to predators. Maine, Ohio, Pennsylvania, New Hampshire, and Vermont were not included, even though they probably had losses, because clear trends were not shown. Indiana was included for 1974 only because both the data and independent observations showed losses in that year. All contiguous states west of the Mississippi River showed losses beginning in 1958 except Minnesota, which began in 1959, Kansas in 1968, Louisiana in 1969, Arkansas and Missouri in 1970, and Oklahoma in 1973. States east of the Mississippi showed losses beginning in 1958 for Michigan and Wisconsin, in 1968 for Illinois and New York, in 1970 for West Virginia, and in 1974 for Indiana. Thus, 16 of the 29 states showed losses for the entire 20-year period. Obviously, the area of loss to predators was enlarged by movement of the coyote both eastward and southward.

Losses for the U.S. were obtained by adding the totals for the areas to give weighted averages. The percentage of loss and the value of the loss increased fairly steadily from 1958 to 1977 even though the total number of animals lost declined to about 1.3 million in 1977. The value of individual animals increased because of inflation and the decline in sheep numbers.

Dynamics of losses in inventory and marketing of sheep and lambs, and in value of products from the sheep industry caused by predators, are presented in Table 12. In some years (6 of 20 during 1958 to 1977), the losses from predators exceeded reductions of both inventory and marketing. In general, these years of overwhelming predator losses were followed by large reductions in inventory, possibly because farmers discontinued sheep production. The large predator losses in relation to reduction of inventory and marketings are evidence that predator loss was a major factor in the decline of the sheep industry. Although the gross income from sales of sheep, lambs, and wool was somewhat stable from 1958 through 1977, the value (and percentage of loss) of products lost to predators increased steadily. The fact that losses to predators affected the income of families with high production efficiency is illustrated by the figures of Gee,[32] which showed a loss of $4.16 per ewe from predators, a $3.88 per ewe return to capital and family labor in the western states, and a loss of $2.44 return to capital after deducting the value of family labor.[18]

According to the figures of Terrill[22] for the ratio of value of losses in production of cattle, swine, and poultry to that of sheep and the estimated 1977 losses of $69 million for sheep, the losses would be $110 million in cattle, $6 million in swine, and $41 million in poultry. Thus, the total losses in the livestock industry (including poultry) would be an estimated $226 million.

From these figures, the efficiency of production of livestock in the U.S. is affected by predators. In fact, predator control may be a key factor in the maintenance of a viable sheep industry in this country.

Table 12
INVENTORY AND MARKETING AND MARKET VALUE OF SHEEP AND LAMBS AND ESTIMATED LOSSES CAUSED BY PREDATORS

Year	Total reduction in inventory and marketing from previous year (1000)[a]	Estimated total sheep and lambs killed by predators (1000)	Loss from predators as % of reduction in inventory and marketing	Gross income from sale of sheep, lambs, and wool ($1000)	Monetary loss from sheep and lambs killed by predators ($1000)	Loss from predators as % reduction of gross income from sales of sheep, lambs, and wool
1958	1729[b]	1561		449,750	32,960	7.3
1963	2961	1788	60	431,056	25,681	6.0
1968	2039	1559	76	391,606	35,430	9.0
1973	2829	1540	54	515,319	51,539	10.0
1977	1022	1295	126	476,080	68,997	14.5

[a] Includes home slaughter.
[b] An increase in inventory.

REFERENCES

1. **Hyslop, J. A.,** Losses occasioned by insects, mites, and ticks in the United States, E-444, Bureau of Entomology and Plant Quarantine, U.S. Department of Agriculture, Washington, D.C., 1938.
2. **Anon.,** Losses in agriculture: a preliminary appraisal for review, ARS-20-1, Agricultural Research Service, U.S. Department of Agriculture, Washington, D.C., 1954.
3. **Anon.,** Losses in agriculture, Agricultural Handbook 291, Agricultural Research Service, U.S. Department of Agriculture, Washington, D.C., 1965.
4. **Steelman, C. D.,** Effects of external and internal arthropod parasites on domestic livestock production, *Annu. Rev. Entomol.*, 21, 155, 1976.
5. **Anon.,** Agricultural Statistics, U.S. Department of Agriculture, Washington, D.C., 1977.
6. **Anon.,** Estimated losses and production costs attributed to insects and related arthropods — 1976, in Cooperative Plant Pest Report 3, Animal and Plant Health Inspection Service, U.S. Department of Agriculture, Washington, D.C., 1978.
7. **Anon.,** ARS National Research Programs, Agricultural Research Service, U.S. Department of Agriculture, Washington, D.C., 1976.
8. **Anon.,** *World Food and Nutrition Study: The Potential Contributions of Research,* National Academy of Sciences, Washington, D.C., 1977.
9. **Anon.,** National Cattlemen's Association Panel on Bovine Respiratory Disease Report, Denver, Colo., 1978.
10. **Kahrs, R. F.,** Infectious bovine rhinotracheitis: a review and update, *J. Am. Vet. Med. Assoc.*, 171, 1055, 1977.
11. **McClurkin, A. W.,** Probable role of viruses in calfhood diseases, *J. Dairy Sci.*, 60, 278, 1977.
12. **Faulkner, L. C.,** Ed., *Abortion Diseases of Livestock,* Charles C Thomas, Springfield, Ill., 1968.
13. **Herlich, H. and Douvres, F. W.,** Parasitism and calfhood diseases, *J. Dairy Sci.*, 60, 283, 1977.
14. **Guss, S.,** Management and diseases of dairy goats, *Dairy Goat J.*, Scottsdale, Ariz., 1978.
15. **Marsh, H.,** *Newsom's Sheep Diseases,* 2nd ed., Williams and Wilkins Co., Baltimore, Md., 1958.
16. **Dunne, H. W. and Leman, A. D.,** Eds., *Diseases of Swine,* 4th ed., Iowa State University Press, Ames, Iowa, 1975.
17. **Siegmund, O. H.,** Ed., *Merck Veterinary Manual,* 4th ed., Merck & Co., Rahway, N.J., 1973.
18. **Hofstad, M. S.,** Ed., *Diseases of Poultry,* 7th ed., Iowa State Press, Ames, Iowa, 1978.
19. **Dollahite, J. W. and Smalley, H. E.,** Estimates of animal economic losses in livestock in the United States from consuming poisonous plants, unpublished data, 1977.
20. **Ensminger, M. E., Galgan, M. W., and Slocum, W. L.,** Problems in Protecting American Cattlemen, *Wash. Agric. Exp. Stn.*, Bull. 562, Institute of Agricultural Science, State College of Washington, Pullman, Wash., 1955.
21. **Keeler, R. F., Van Kemper, K. R., and James, L. F.,** Eds., *Effects of Poisonous Plants in Livestock,* Academic Press, New York, 1978, chaps. 3, 23.
22. **Terrill, C. E.,** Livestock losses to predators, *Rangeman's J.*, 13, 1976.
23. **Anon.,** Nebraska livestock and chicken losses, Statistical Report Service, Nebraska Department of Agriculture, U.S. Department of Agriculture, Lincoln, Neb., 1971, 1972, 1973.
24. **Gee, C. K.,** Cattle and calf losses in feeder cattle production, 1975, Agricultural Economic Report 409, U.S. Department of Agriculture, Washington, D.C., 1978.
25. **Shelton, M. and Klindt, J.,** Interrelationship of coyote density and certain livestock and game species in Texas, MP-1148, *Tex. Agric. Exp. Stn.*, College Station, Tex., 1974.
26. **Shelton, M.,** Symposium on livestock predators, nature of the problem — livestock, paper presented at Ann. Meet. Am. Assoc. Adv. Sci., Denver, Colo., 1977.
27. **Gee, C. K., Magleby, R. S., Bailey, W. R., Gum., R. L., and Arthur, L. M.,** Sheep and lamb losses to predators and other causes in the western United States, Agricultural Economic Report 369, U.S. Department of Agriculture, Washington, D.C., 1977.
28. **Terrill, C. E.,** Sheep and lamb losses to predators, *J. Anim. Sci.*, 41, 281 (Abstr. 159), 1975.
29. **Terrill, C. E.,** Monitoring losses to coyotes, paper presented at Symp. Predator Control Livestock Production, Ann. Meet. Am. Soc. Anim. Sci., College Station, Tex., 1976.
30. **Anon.,** Agricultural Statistics, 1977 and previous years, and Statistical Bulletins 184, 278, 284, 389, 400, 502, 510, 587, 598, MtAn 1-1 (78), Sheep and Goats, U.S. Department of Agriculture, 1978.
31. **Rummell, R. S.,** Unpublished data from Regions 1 to 6 on losses of sheep and lambs on National Forest Land, Forest Service, U.S. Department of Agriculture, Washington, D.C., 1975.
32. **Gee, C. K.,** Enterprise budgets for western commercial sheep businesses 1974, ERS-659, Economic Research Service, U.S. Department of Agriculture, Washington, D.C., 1977.

POTENTIAL IMPACT OF ALIEN ARTHROPOD PESTS AND VECTORS OF ANIMAL DISEASES ON THE U.S. LIVESTOCK INDUSTRY

Neal O. Morgan

INTRODUCTION

Because of increased and expanded international trade, the existence of a disease in any part of the world poses a threat elsewhere. Thus, foreign animal diseases are no longer regarded as exotic curiosities of purely academic interests. In fact, such unfamiliar exotic diseases may well be both the most devastating and also the most insidious since they can gain a firm foothold before methods of control or eradication can be initiated.

The relation of arthropods to the transmission of animal diseases must therefore be recognized and considered seriously. For example, arthropod vectors of foreign animal diseases retain their infectivity for weeks or even months and some survive severe environmental changes. These may be transported, unknowingly, to disease-free areas, where others locate susceptible hosts and cause an outbreak of some fatal or debilitating disease; or, disease reservoir animals may be transported to a disease-free area where potential vectors have existed for years, but without disease vector incrimination. As a result, an outbreak could occur among susceptible hosts.

Since U.S. livestock are subject to a wide variety of arthropod pests and parasites, routine insecticidal treatment has become integrated into livestock management programs. Even so, the estimated annual losses in 1974 in livestock production due to arthropod involvement exceeded $2.8 billion.[1] In view of these losses, and the severe economic damage to livestock production that could result from foreign animal diseases transmitted by arthropods, extreme caution must be exercised to prevent introduction of the exotic diseases into the U.S. For example, within recent years, it has been necessary to eradicate tick-borne babesiosis of cattle, Venezuelan equine encephalomyelitis (VEE), viscerotropic velogenic Newcastle disease (VVND), and hog cholera from the U.S. In each case the price of eradication was minimal compared with the impact of the disease on the industry.

Before 1906 the protozoan parasite *Babesia bigemina* that causes piroplasmosis and its indigenous natural vector *Boophilus microplus* Canestrini cost the cattle industry $100 million annually.[2] Eradication of the tick vector from the southern states has eliminated the disease from the U.S. Since 1939 there has been no bovine babesiosis in the U.S., although there have been frequent reinfestations of the vectors from Mexico and other countries where the disease is prevalent. Constant vigilance has prevented the disease from becoming reestablished in the U.S. livestock industry. Much later *Rhipicephalus evertsi* Neuman was discovered when an inspector touring a Florida zoo in 1960 noticed several ticks on an eland. These were identified as the red-legged tick that is common to Africa and is also a known vector of piroplasmosis. Since this species infests all species of equines and ruminants, an eradication program was initiated November 1, 1960 to prevent the spread to other zoos (especially by transfer of livestock) or to the surrounding farming community. Animal and area treatment was terminated 14 months later after eradication was declared successful. In view of that success, a nationwide eradication program was initiated against *B. microplus,* now the only remaining vector of the disease. At a cost of just under $5 million, this tick too was eliminated.[3]

VEE was introduced into the $6 billion horse industry in the summer of 1971 by infected mosquitoes flying north from Latin America. Rapid implementation of an emergency disease eradication program by the Animal and Plant Health Inspection Service, U.S. Department of Agriculture in July limited the outbreak to Texas. The program was very successful, and eradication was announced December 31, 1971. The cost for the 6-month program was $19 million.[4]

During 1972—73, a VVND outbreak was centered in southern California and Arizona. More than 11 million birds had to be destroyed to eradicate the disease and protect the nation's $5 billion poultry industry. Although the recognized means of VVND transmission include contact with contaminated birds and equipment and inhalation of infected air-borne particles, the virus was isolated from pools of the lesser house fly, *Fannia canicularis* (L.) and *F. femoralis* (Stein), which were collected near the infected birds.[5]

The hog cholera virus is commonly transmitted by direct contact, fomites, and ingestion of uncooked, infected garbage. It has been mechanically transmitted by tabanid and muscid flies. A 15-year nationwide hog cholera eradication program was completed in 1977 at a cost of $140 million. Before the eradication campaign began, hog cholera was costing U.S. hog producers $50 million a year — $10 million in death losses and $40 million in the cost of preventive vaccination. An estimated 5000 to 6000 swine herds were affected annually. Even though direct benefits from eradication of hog cholera go to the producers, consumers also gain from the subsequent decreased production costs.[6] The pronouncement of eradication also enables the U.S. swine industry to participate in the annual multibillion dollar international market for swine products.

The following sections describe exotic arthropods, diseases they transmit, and their potential impact on the U.S. livestock industry should they become established.

TICKS

Ticks are among the most efficient of arthropod vectors of animal diseases; they can transmit viruses, rickettsia, and protozoa, and some can cause paralysis via their neurotoxic salivary secretions. The abilities of ticks to attach firmly, suck blood very slowly, and remain unnoticed for long periods explain their potential as vectors. Also, many species are resistant to environmental stresses and may live for years. Ticks have few natural enemies and a wide range of hosts. Some transmit disease organisms by transovarial and transtadial transmission.[7]

The exotic tick-borne diseases that would be especially damaging to the U.S. livestock industry include African swine fever, East Coast fever, heartwater, louping ill, Nairobi sheep disease, and sweating sickness.

African swine fever (ASF) is a highly contagious vital disease of swine that causes pronounced hemorrhages of the internal organs and is almost 100% fatal. The recognized vectors are the argasid ticks *Ornithodorus moubata porcinus* in Africa and *O. erraticus* in Spain. In Africa the wart hog serves as a reservoir of the virus, and dissemination of the virus requires either contact between susceptible swine and infected wart hogs, garbage, or other contaminated medium or transmission by ticks. In Spain *O. erraticus* demonstrated transmissibility of ASF in the laboratory. ASF invaded the western hemisphere for the first time in 1971 when it was found in the Republic of Cuba. Eradication was accomplished the same year by slaughtering almost 26% of the Cuban swine population.[8] The most recent invasion of ASF in the western hemisphere occurred in 1978. The virus was isolated from swine in the Dominican Republic and Brazil.[6]

East Coast fever (ECF) is an acute, highly pathogenic disease of cattle caused by the protozoa *Theileria parva,* which parasitizes the red blood cells and produces swelling of the external lymph nodes, emaciation, and high mortality. Tick vectors of ECF include 7 species of *Rhipicephalus* and 3 species of *Hyalomma;* all may harbor the protozoas for 6 months.[9]

Heartwater, an infectious septicemic disease of sheep, goats, and cattle is caused by *Rickettsia ruminatium,* which invades the endothelial cells of the blood vessels. In the acute form of the disease, the reflex response to stimulation resembles the response to strychnine poisoning or tetanus. The bont tick *Amblyomma hebraeum* is the most common vector, but three other species of *Amblyomma (A. variegatum, A. gemma,* and *A. pomposum)* are also known vectors.[10]

Louping ill is a tick-borne, nonfatal disease of sheep caused by a neurotropic virus. The disease has also been recognized in cattle, horses, swine, and man. The symptoms include impaired locomotion, hyperexcitability, and a peculiar leaping gait. The disease occurs only on premises infested with *Ixodes ricinus, I. persulcatus,* and *Rh. appendiculatus* Neumann.[9]

Nairobi sheep disease (NSD) is a noncontagious viral disease of sheep and goats. The notable vectors include *Rh. appendiculatus, Rh. parvus, Rh. hurti* and *Rh. variegatum.* Transovarial transmission is known to occur in *Rh. appendiculatus,* the principal vector. Unfed adult ticks retain the virus for long periods, and tick-infested pastures are considered dangerous for at least 18 months. The bont tick, *A. variegatum,* has lived without a host for 732 days.[9]

Sweating sickness (tick-borne toxicosis) is usually a disease of calves, but it may also affect sheep and swine. It is caused by a tick-borne toxin that produces a severe moist eczema that may cover the entire body surface. A fly-attractive, foul odor is emitted, the skin is very sensitive, and the animals have difficulty walking. From 30 to 70% of affected calves die, and the survivors develop immunity. The principal vector, *Hyalomma trucatum* can transmit the disease transovarially for more than 10 generations. Prevention or eradication of the disease lies in the effective control or eradication of infective ticks.[12]

BITING FLIES

The category of biting flies includes such arthropods as culicoides, mosquitoes, tsetse flies, horse flies, deer flies, and stable flies. All suck blood, are usually short lived, and reproduce rapidly. Some of the exotic livestock diseases transmitted by biting flies include African horsesickness, ephemeral fever, lumpy skin disease, Rift Valley fever, and nagana (trypanosomiasis).

African horsesickness is an acute, or subacute, usually fatal viral disease of equines that is transmitted by many species of biting gnats *(Culicoides spp.)* and by other biting flies. When infected animals are introduced into areas free of horsesickness, native gnats feeding first on infected and then on susceptible horses rapidly spread the disease. Of the 4 clinical forms of horsesickness the most fatal is the pulmonary form wherein the lungs may be filled with fluids and large quantities of yellow froth flow from the nostrils, though the horse may continue eating until shortly before death.[11] When a horse population is susceptible, at least a 90% mortality can be anticipated.[12] With the cardiac form of horsesickness, the clinical signs develop slowly and may last 12 days, but recovery is common; in fatal cases there are distinct signs of heart failure. The third form, which is mixed, affects the lungs and heart, the disease develops rapidly, and the death rate is variable. The mild form, sometimes called horsesickness fever, is frequently undetected and the animals usually recover quickly.[11] The virus is only spread by flying insects such as culicoides and mosquitoes; it is not mechanically transmitted from an infected horse to a totally susceptible horse.[12] African horsesickness virus remained infectious in *Aedes aegypti* for more than 5 weeks after engorgement.[13]

Ephemeral fever (or 3-day sickness) is a virus disease of cattle transmitted by insects, primarily the biting midges, punkies, of the family Ceratopogonidae. Fatal cases are rare, and fever, stiffness, lameness, and quick recovery are characteristic. Since affected animals often refuse or are unable to swallow, there is danger of causing mechanical pneumonia by attempting to force medication.[9]

Lumpy skin disease is an acute, highly infectious virus disease of cattle. The disease is characterized by a spectacular appearance of lumps (nodules) in the skin on all parts of the body. The death rate is low, but economic losses are often high because infected cattle rapidly lose weight and condition. Circumstantial evidence indicates that mosquitoes are vectors because major outbreaks have been associated with large mosquito populations.[12]

Rift Valley fever (RVF), an acute febrile virus disease that affects sheep and cattle primarily, was first isolated and identified from sheep in the Rift Valley of Kenya. The disease causes high rates of abortion among pregnant ewes and cows and heavy mortality among young lambs and calves. The virus has been isolated from three species of *Aedes* and six of *Eretmapodites* mosquito.[13] *A. (Ochlerotatus) caballus* (Theobald) is an experimental vector in South Africa, and the subgenus *Ochlerotatus* is represented in North America by 50 species including *A. dorsalis* (Meigen), *A. nigromaculis* (Ludlow), and *A. sollicitans* (Walker). *A. caballus* readily bites man and animals in the daytime and is easily transported by automobiles, trains, airplanes, and animals.[14] During the 1977 outbreak of RVF in Egypt, the primary method of transmission among animals and humans appeared to be by arthropod vectors, primarily the *Culex pipiens* L. mosquito.[15] Experimental transmission of RVF virus has been achieved with the brown ear tick, *Rh. appendiculatus.*[9]

Trypanosomiasis (nagana) is an acute or chronic infectious protozoan disease of domestic animals caused by one or more of three species of trypanosomes: *Trypanosoma congolense, T. brucci,* and *T. vivax.* A tsetse fly, *Glossina morsitans* Westwood, is the most efficient vector of nagana, but at least 20 other species of tsetse are known vectors of trypanosomiasis. Also species of *Tabanus* and *Stomoxys* are capable of direct mechanical transmission and immune or inapparent carrier cattle, sheep, goats, and wild animals are reservoirs of trypanosomes. The disease is perpetuated through cyclic transmission resulting from development and multiplication of trypanosomes within the tsetse fly.[9] However, mechanical transmission by *Glossina spp.* and other flies does occur. There is evidence that nagana moves into areas free of tsetse flies. Thirty species of wild animals have been found to harbor pathogenic trypanosomes of man and animals. Dogs are more often chronically infected and could easily introduce trypanosomes into clean countries. Moreover, *T. vivax* does not require tsetse flies to persist and can become established where tabanids, *Stomoxys, Simulinum, Chrysops,* and other biting flies are common. It could easily be introduced into the southern U.S. and established quickly if surveillance of foreign animal diseases and vectors becomes relaxed.[16]

ARTHROPOD PESTS

In 1970 a dipteran parasite, *Hippobosca longipennis* (louse-fly), was introduced into the U.S. on a shipment of cheetahs from East Africa. The cheetahs were then distributed to several zoological parks. Within 1 year infestations of the louse-fly had become established in California, Oregon, Texas, and Georgia. The fly is a pest of man and animals though it has been considered of low economic importance. However, people who have handled the infested cheetahs have been very painfully bitten by the aroused louse-flies. The hematophagous parasite attaches to all hairy animals including livestock, dogs, cats, and wild game animals. In the U.S. the louse-fly has adapted to the cold climate of Oregon winters and to the heat of Texas summers, and in 1973, it was found on wild and domestic animals near the infested zoological parks.[3] By 1975 the State of California Public Health Service, in cooperation with the State Department of Food and Agriculture and the U.S. Department of Agriculture eradicated the pest within the state and recommended methods for eradication in other infested areas.[17]

The face fly, *Musca autumnalis* De Greer, a pest of horses and cattle, came to North America from northern Europe in 1952. In the 25 years since then, the species has spread to 45 of 48 contiguous states. Face fly annoyance can greatly reduce livestock production; however, it is much more important that this fly is a vector of an eyeworm, *Thelazia sp.*, that can cause blindness of cattle and deer and is capable of mechanically transmitting hog cholera.[19] Presently millions of dollars are spent annually to control this pest. However, if a disease transmissible by the face fly were to become epizootic anywhere within the U.S., the cost of eradicating the disease would be very great. For example, the fly spends 75% of the time on vegetation instead of on livestock, so control measures would have to include applying insecticides to the entire epizootic area.

In Australia, the buffalo fly, *Haematobia exigua* (de Meijere), is a notorious pest of livestock but is not known to be involved in the transmission of arboviruses.[20]

The red imported fire ant, *Solenopsis invicta* Buren, a native of South Africa, entered the U.S. during World War I through the port of Mobile, Alabama. Since 1918, the ant has spread through much of the South and now occupies more than 120 million acres in nine southern states. Ant mounds are found in pastures and hay fields and there may be as many as 50 mounds per acre. When the ants are disturbed they become viciously aggressive and will attack livestock and man. The painful, burning stings may result in the formation of blisters. Had the importance of the imported fire ant to the economy of the U.S. been realized in 1918, immediate controls and eradication probably would have been attempted. The imported fire ant control program is an annual multimillion dollar effort, and the invasion is being extended to the North and West at an alarming rate.[3]

PREVENTION AND PROTECTION

On the basis of past experience, alien arthropods will continue to gain entry to the U.S. In fact, the possibility has been greatly increased by the opening of midwestern airports, as at Chicago and Kansas City, to international traffic. Obviously, improved quarantine inspection procedures will be needed to offset this increase of commerce with foreign countries.[21]

The economic importance of diseases such as African swine fever, African horsesickness, bluetongue disease, and RFV in the regions where they were first identified was evident. They have even greater potential for damage in other areas. Moreover, the presence of such diseases in different parts of the world indicates both the ease with which arthropod vectors can be introduced and the susceptibility of livestock. There is, therefore, a vital need for experienced medical entomologists, veterinarians, and/or customs inspectors at every sea and air port of entry to the U.S. and they all should be trained to recognize known and potential vectors of foreign animal diseases. The fact that many ticks are capable of harboring pathogens for several months suggests that U.S. visitors abroad could unknowingly be the physical or mechanical hosts for such vectors as they return home. Passengers and their clothing are seldom inspected for arthropods at our ports of entry. Nevertheless, in 1975, U.S. Customs cleared 252 million people, 75 million cars, 353 thousand aircraft, and 123 thousand vessels and examined over 77 million pieces of luggage. The magnitude of the threat potential to our 240 million cattle, sheep, goats, swine, etc. and millions of domestic fowl is evident.[22]

For all these reasons, the establishment and maintenance of an effective national foreign animal disease surveillance program is one of the more important responsibilities of state and federal animal health officials. It is also most important to determine the presence of a foreign animal disease or known arthropod vectors of a foreign animal disease before the disease has time to spread by movement of livestock, people or vectors.

The U.S. Department of Agriculture therefore has overseas ports of debarkation livestock quarantine facilities where cattle are inspected prior to shipment to the U.S. Department of Agriculture's Fleming Key Animal Import Center at Fleming Key, Florida. Although U.S. law prohibits the direct importation of domestic swine and ruminants from countries where foot-and-mouth disease and rinderpest are declared to exist (this includes much of Europe and all of South America, Africa, and continental Asia), valuable breeding stock will be permitted to enter after it has been carefully screened, had long-term isolation, and been subjected to stringent veterinary testing and examination. The purpose is to make possible the use of exotic breeds of cattle in livestock improvement programs. Cattle are transported by ship to Key West, Florida, where they are offloaded and transported to Fleming Key in self-contained trucks. The cattle entering the center are identified, inspected, and a precautionary dipping for ectoparasites is conducted. Testing procedures for various diseases are conducted during the quarantine. Upon completion of a 5-month quarantine, the livestock are released to their destination without further restrictions. It is estimated that the cost to the importer will be about $2500 per animal.[23]

However, the methods employed at Fleming Key to protect against invasion by foreign, infected arthropods are inapplicable to people, wild animals, and animal products. There is a distinct possibility that viral diseases such as RVF could spread in the U.S. if animals, including monkeys, in the viremic stage were imported from endemic areas or if infected vectors were introduced. Insect control procedures including disinsectization of all aircraft at airports having international air traffic will help to prevent the entrance of infected insect vectors into the country. Control of importation of animals, animal products, and wild animals from endemic areas is necessary.

REFERENCES

1. **Drummond, R. O.,** Effects of arthropod parasites on livestock production, CRC Handbook of *Pest Management,* CRC Press, Boca Raton, Fla., (in press).
2. **Hunter, W. D. and Hooker, W. A.,** Information concerning the North American fever tick, Bull. 72, Bureau of Entomology, U.S. Department of Agriculture, Washington, D.C., 1907, 1.
3. Anon., Imported Ectoparasites and Pests of Man and Animals, Foreign Animal Disease Report, U.S. Department of Agriculture, Animal and Plant Health Inspection Service, November — December 4, 1947.
4. **Omohundro, R. E.,** Venezuelan equine encephalomyelitis emergency operation, *J. Am. Vet. Med. Assoc.,* 161, 1516, 1972.
5. **Rogoff, W. M., Carbrey, E. C., Bram, R. A., Clark, T. C., and Gretz, G. H.,** Transmission of Newcastle disease virus by insects: detection in wild *Fannia* species, *J. Med. Entomol.,* 12, 225, 1975.
6. Anon., United States declared hog cholera free, Foreign Animal Disease Report, U.S. Department of Agriculture, Animal and Plant Health Inspection Service, January — February 1978, 1.

7. **James, M. T. and Harwood, R. F.,** *Hermes' Medical Entomology,* 6th ed., Collier-MacMillan Limited, London, 1969, 1.
8. **Sharman, E. C.,** African swine fever, in Proc. 75th Ann. Meet., U.S. Animal Health Association, U.S. Department of Agriculture, Washington, D.C., 176, 1971.
9. Anon., Foreign animal diseases: their prevention, diagnosis, and control, U.S. Livestock Sanitary Association, Richmond, Va., 243, 1964.
10. **Diamont, G.,** The bont tick, exotic vector of heartwater found in the United States, *Vet. Med. Small Anim. Clin.,* August, 847, 1965.
11. Anon., Emerging diseases of animals, Agricultural Studies No. 61, Food and Agriculture Organization of the United Nations, Geneva, 1963, 71.
12. **Multon, W. M.,** African horse sickness, Proc. Foreign Animal Disease Seminar, U.S. Department of Agriculture, Animal and Plant Health Inspection Service, Hyattsville, Md., January 15—16, 1976.
13. **Ozawa, Y., Nakata, G., Shad-del, F., and Navai, S.,** Transmission of African horse-sickness by a species of mosquito, *Aedes aegypti* Linnaeus, *Arch. Inst. Razi,* 18, 147, 1966.
14. **Hermes, W. B. and James, M. T.,** Mosquitoes as vectors of disease, *Medical Entomology,* MacMillan, New York, 1961, 215.
15. Anon., Rift Valley Fever, Foreign Animal Disease Report, U.S. Department of Agriculture, Animal and Plant Health Inspection Service, Washington, D.C., July—August, 1978, 4.
16. **Maré, C. J.,** African trypanosomiasis, U.S. Department of Agriculture, Animal and Plant Health Inspection Service, Proc. For. Anim. Dis. Sem., Washington, D.C., February, 1978, 8—9.
17. **Keh, B. and Hawthorne, R. M.,** The introduction and eradication of an exotic ectoparasitic fly, *Hippobosca longipennis,* in California, *J. Zoo Anim. Med.,* December, 8, 1977.
18. **Chitwood, M. B. and Stoffolano, G.,** First report of *Thelazia* sp. in the face fly, *Musca autumnalis,* in North America, *J. Parasitol.,* 57, 1363, 1971.
19. **Morgan, N. O. and Miller, L. D.,** Muscidae: experimental vectors of hog cholera virus, *J. Med. Entomol.,* 12, 657, 1976.
20. **Standfast, H. A. and Dyce, A. L.,** Potential vectors of arboviruses of cattle and buffalo in Australia, *Aust. Vet. J.,* 48, 224, 1972.
21. **Sailer, R. I.,** Our immigrant insect fauna, *Bull. Entomol. Soc. Am.,* 24, 3, 1978.
22. Anon., Awareness of emergency diseases, Foreign Animal Disease Report, U.S. Department of Agriculture, Animal and Plant Health Inspection Service, Washington, D.C., July—August, 1976, 4.
23. Anon., Foreign Animal Disease Report, Animal and Plant Health Inspection Service, U.S. Department of Agriculture, Fleming Key Animal Import Center, April—May, 1977, 2.

Estimated Losses without Pesticides and Substituting Only Readily Available Nonchemical Controls

ESTIMATED CROP LOSSES WITHOUT THE USE OF FUNGICIDES AND NEMATICIDES AND WITHOUT NONCHEMICAL CONTROLS

Einar W. Palm

According to a National Science Foundation RANN Program report in 1976,[12] the annual economic benefits to U.S. society from the use of agricultural pesticides are probably between $5.0 to $10.0 billion/year. This kind of estimate is usually developed by estimating the negative effects (i.e., the costs of restricting or eliminating pesticides).

Economic benefits from the use of a certain pesticide — whether it be fungicide, nematicide, insecticide, or herbicide — should be measurable in terms of the income from the increased yield, less the cost of the pesticide and its application. Such information is made available on selected pest-crop situations usually as a result of efficiency tests conducted by the producers of the chemicals and/or state agricultural experiment stations. Efficacy or yield benefits are often obtained when pest control methods are practiced alongside untreated controls for a number of years.

Considerable differences in these experiments result from the fact that there are many factors involved, and lack of standardization in experiments and experiments conducted in different locations often bias results. Factors affecting yield data are: rainfall patterns, unusual weather conditions, fertilizer use patterns, soil types, cropping programs prior to test, region of the country, pest infestation levels, infection patterns in a particular year, frequency, rates, and methods of pesticides applications, and many other variables. Many tests and several years of testing are often required to obtain significant results and data. The aggregate costs of these tests are often very high, but they are necessary before any pest control materials can be introduced into marketing channels, and evaluated once they are being used.

Careful testing is also essential in order to determine whether a pesticide may create certain harmful effects on man or the environment. This testing begins prior to manufacture and marketing and continues into the future of every agricultural chemical.

EXTENT OF THE USE OF FUNGICIDES AS COMPARED TO OTHER PESTICIDES

Agricultural pesticides are used primarily against weeds, insect pests, and plant pathogens. Of the pesticides used annually in agriculture in the U.S., about 51% are herbicides, 35% insecticides, and 14% fungicides (Table 1). If crop land devoted solely to pastures is removed from the 890 million acres total cropland in the U.S., 34% of the crop land is treated with herbicides, 12% with insecticides, and 20% with fungicides.

Many plant pathogens are controlled primarily by nonchemical methods, especially in the use of resistant varieties and a number of basic cultural methods such as crop rotation, sanitation, timing of operations, and fertility management. Disease resistant varieties are used extensively in the major cereal crops — corn, wheat, barley, oats, and rye.

However, the nature of pesticide use is more important than acreage figures to a proper understanding of the effects of these materials on particular crops. For instance, the use of fungicides and nematicides on agronomic crops such as corn and soybeans, except for seed treatments, is relatively low. By contrast, the use of fungicides and nematicides on fruit and vegetable crops is relatively extensive and is important to quality and quantity of these crops.

Table 1
SOME EXAMPLES OF PERCENTAGES OF CROP ACRES TREATED, OF PESTICIDE AMOUNTS USED ON CROPS, AND OF ACRES PLANTED TO SPECIFIC CROP (USBC, 1973A, B, C, D, USDA, 1975A)

Crops	Insecticides		Herbicides		Fungicides		Portion of total crop acres (%)
	Acres (%)	Amount (%)	Acres (%)	Amount (%)	Acres (%)	Amount (%)	
Nonfood	NA[a]	50	NA	NA	<0.5	NA	NA
Cotton	61	47	82	9	4	1	1.11
Tobacco	77	3	7	NA	7	NA	0.11
Field crops	NA	33	NA	NA	NA	15	NA
Corn	35	17	79	45	1	NA	7.43
Peanuts	87	4	92	2	85	11	0.16
Rice	35	1	95	3	0	NA	0.22
Wheat	7	1	41	5	0.2	NA	6.11
Soybeans	8	4	68	16	2	NA	4.19
Pasture hay and range	0.5	2	1	4	0	NA	68.40
Vegetables	NA	7	NA	2	NA	24	NA
Potatoes	77	2	51	NA	49	10	0.16
Fruit	69	9	NA	1	NA	60	NA
Apples	91	3	35	NA	61	18	0.07
Citrus	72	2	22	NA	47	24	0.08
All crops	6	35	17	51	0.9	14	100

[a] Not available.

Adapted from Pimentel, D., Krummel, J., Gallahan, D., Hough, J., Merrill, A., Schreiner, I., Vittum, P., Koziol, F., Back, E., Yen, D., and Fiance, F., *A Cost-Benefit Analysis of Pesticide Use in U.S. Food Production,* New York College of Agriculture and Life Sciences, Cornell University Agricultural Experiment Station, Ithaca, 1978. With permission.

Fungicides are used on more than half of the acreage of apples, certain other fruits, several vegetables, and peanuts (Tables 1 and 2).

ESTIMATED CROP LOSSES WITHOUT THE USE OF FUNGICIDES

Estimates of expected crop losses for major crops if they were grown without the use of fungicides are often made by evaluating data obtained from field tests using fungicides as compared to the controls. To some extent, these loss data may be somewhat higher than actual losses by growers, but they provide some guidelines. For some crops, especially certain fruits and vegetables, there are few or no experimental data that compare yields with or without fungicide use. It has been estimated that crop loss to diseases without the use of pesticides is 15%, as compared to 18% and 9% to insects and weeds, respectively.[14]

Losses to agronomic crops from plant pathogens range from 7 to 28% if no fungicides are used (Table 2). For a crop such as peanuts with a 28% loss, having 85% of the acres treated would sustain serious economic loss should fungicide use be terminated. By contrast, the 10% loss estimate means very little to corn with only 1% of the acreage treated. The peanut losses would be about $296 million, whereas losses to corn would be of little consequence.

Losses to vegetables range from 8 to 25% (Table 2). Crops such as potatoes and tomatoes, with half or more of the acres treated and a potential loss of over 20%, could be seriously affected by the loss of appropriate fungicides. The estimated total annual loss would be $702 million.

Most fruit and nut crops have from 26 to 65% of the acres treated with fungicides (Table 2). With loss estimates of from 8 to 29% if fungicides were not used, the costs due to the removal of strategic fungicides would be great, estimated at $991 million annually.

IMPACT OF THE WITHDRAWAL OF FUNGICIDES AND NEMATICIDES

Since 1977, several agricultural pesticides have been reviewed by the Environmental Protection Agency (EPA) in respect to their benefits and risks. The process is known as "Rebuttable Presumption Against Registration" (RPAR).[9,10,13,15] It is a regulatory procedure designed to evaluate the possible adverse effects of an agricultural chemical on man or his environment. In this procedure, information and data are gathered by the EPA from many sources, and the information is evaluated. Preliminary assessments of its validity are made prior to any further action. Data that appear to exceed certain hazard criteria established by the EPA act as "flags" that a compound should be reviewed.

An RPAR notice does not ban a given compound, nor is it a notice of intent to cancel the registration. It does, however, suggest that a compound poses some hazard, and it provides opportunity for the manufacturer and others acquainted with the nature of the compound to "rebut" or argue that the presumed hazard does not exist, is not unreasonable, or that the benefits of the compound outweigh the apparent risk.

A number of fungicides and nematicides have come under RPAR review. The future use of these chemicals will be determined by the finalization of RPAR proceedings. The impact on agricultural production can be estimated in part on the basis of what will occur in the event a material is withdrawn. The ultimate impact will be contingent on many factors: the prevalent use of the material, the economic impact on each crop on which the chemical is used, the number and economic importance of the crops

Table 2
LOSS ESTIMATES FROM PLANT PATHOGENS WITH CURRENT FUNGICIDE USE AND AN ESTIMATE OF LOSSES IF NO FUNGICIDES WERE USED

				Fungicide treatment				Additional crop pest pest loss without fungicide control		
Crop	Acres (× 10^3)	Crop value ($\$\times10^6$)	Crop feed value (kcal×10^{12})	Acres treated (%)	Cost ($/acre)	Control cost ($\$\times10^3$)	Current crop pest loss,(%)	%	$\$\times10^5$	kcal×10^{12}
Field Crops										
Corn	65,194	13,717	499	1	11	7,171	10	0	0	0
Cotton	12,547	2,382	32.8	4	5	2,509	12	1	24	0.33
Wheat	65,459	7,242	160	0	0	0	14	0	0	0
Soybeans	52,460	8,246	135	2	5	5,246	14	1	82	1.35
Rice	2,569	1,195	19	0	0	0	7	0	0	0
Tobacco	963	2,160	0.3	7	5	337	11	3	65	0.01
Peanuts	1,472	658	6.9	85	16	20,019	28	45	296	3.08
Sorghum	13,917	1,751	53	0	0	0	9	0	0	0
Sugar beets	1,217	1,307	10	13	6	949	16	3	39	0.30
Other grain	38,000	3,034	102	0	0	0	20	0	0	0
Alfalfa	26,642	3,376	298	0	0	0	24	0	0	0
Other hay	33,904	2,394	436	0	0	0	15	0	0	0
Other field crops	6,533	2,453	15.3	1	5	327	13	0	0	0
Pasture	563,000	20,000	5,148	0	0	0	5	0	0	0
Total	883,877	69,915	6,915.3			36,558			506	5.07
Vegetable Crop										
Lettuce	226	250	0.28	21[g]	12	570	12	5[h]	13	0.01
Cole	196	144	0.35	15[g]	12	353	9	5[h]	7	0.02
Carrots	80	76	0.36	25[g]	16	320	8	6[h]	5	0.02
Potatoes	1,380	1,121	9.50	49[g]	8	5,410	20	20[h]	224	1.9
Tomatoes	465	798	1.60	98[g]	18	8,203	23	44[h]	351	0.7
Sweet corn	628	212	0.85	9[g]	11	622	8	1[h]	2	0.01
Onions	110	158	0.52	70[g]	18	1,386	21	28[h]	44	0.15

Cucumbers	177	125	0.11	42[g]	18	1,338	15	17[h]	21	0.02
Beans	444	195	0.23	24[g]	18	1,918	25	6[h]	12	0.01
Cantaloupe	70	95	0.06	53[g]	18	668	21	9[h]	9	0.01
Peas	426	42	0.17	2[g]	18	153	23	0[h]	0	0
Peppers	48	70	0.04	48[g]	18	415	14	10[h]	7	>0.00
Sweet potatoes	119	107	0.69	1[g]	18	21	18	0[h]	0	0
Watermelons	215	88	0.27	40[g]	18	1,548	10	2[h]	2	0.01
Asparagus	112	74	0.03	7[g]	18	141	9	1[h]	1	0
Other vegetables	128	182	0.23	10[g]	18	230	10	2[h]	4	0.01
Total	4,824	3,737	15,29			23,296			702	2.87
Fruit and nut crops										
Apples	526	538	1.7	61	50	16,043	8	60	323	1.02
Cherries	129	114	0.15	51	50	3,290	24	35	40	0.05
Peaches	301	259	0.43	65	40	7,826	21	30	78	0.13
Pears	112	125	0.41	34	40	1,523	17	35	44	0.14
Prunes and plums	161	115	0.40	35	40	2,254	10	35	40	0.14
Grapes	720	610	1.76	40	40	11,520	27	15	91	0.26
Oranges	862	612	2.67	48	4	1,655	16	25	153	0.67
Grapefruit	174	153	0.48	52	4	362	2	45	69	0.22
Lemons	83	109	0.11	47	4	156	29	20	22	0.02
Other citrus	87	51	0.19	28	4	97	18	20	10	0.04
Strawberries	40	153	0.80	42	20	336	26	10	15	0.01
Other fruits	147	176	0.45	26	20	764	20	30	53	0.14
Pecans	382	65	0.43	46	22	3,866	21	45	29	0.19
Other nuts	415	240	2.11	46	22	4,200	12	10	24	0.21
Total	4,139	3,320[j]	11.37[k]			53,892			991	3.24

Adapted from Pimentel, D., Krummel, J., Gallahan, D., Hough, J., Merrill, A., Schriener, I., Vittum, D., Koziol, F., Back, E., Yen, D., and Fiance, S., A Cost-Benefit Analysis of Pesticide Use in U.S. Food Production, New York College of Agriculture and Life Sciences, Cornell University Agricultural Experiment Station, Ithaca, 1978. With permission.

involved, the potential for alternative use chemicals, and the cost of alternative non-chemical methods when used without the supplementary chemical benefits.

A purpose of this paper is to present some hypothetical information with respect to what crop losses might occur as a result of the withdrawal of certain fungicides and nematicides that are currently being used in crop production in the U.S.

THE IMPACT OF THE WITHDRAWAL OF FUNGICIDES

Among the fungicides involved in RPAR review are benomyl (Benlate®), the ethylene bisdithiocarbamates, the so-called EBDC fungicides (e.g., maneb, mancozeb, metiram, diammonium EBDC, zineb) and the pentachloronitrobenzene (PCNB) fungicides. Each material or group of materials was subjected to RPAR review on the basis of selected presumptions. In each case, manufacturers and assigned impact assessment groups made careful reviews of the presumptions and made appropriate replies. Since the RPAR process will likely be part of the future evaluation of all pesticides, the exercises of evaluating benefit to risk data and the economic impacts on crop production will also be on-going processes.

For the purpose of brevity in this paper, the economic assessment of the EBDC fungicides was selected as a hypothetical example of the impact of the potential withdrawal of a fungicide or class of fungicides that have historically been of great importance to agriculture.

Since their introduction in the mid 1940s, these fungicides have become the most widely used plant disease control chemicals in food production in the U.S. and in the world. Major agricultural uses in the U.S. are in the control of diseases of potatoes, tomatoes, other vegetables, fruit and nut crops (of which citrus crops, grapes, and apples are of great importance), and certain cereal grains. The EBDC fungicides are presently registered on 271 crops and for 1296 diseases.[4]

EBDC uses have been prevalent because the fungicides have broad-spectrum capabilities, they are protectants that do not promote pathogen resistance, they are relatively non-phytotoxic, they have been safe for humans and animals, and they have been relatively inexpensive.

IMPACT OF EBDC WITHDRAWAL ON CROP PRODUCTION

Since the dithiocarbamates are a major group in the arsenal of fungicidal chemicals, the withdrawal of such a group would create extremely serious problems in the production of many important crops, specifically on vegetable and fruit crops. Large scale removal of EBDC fungicides would also create serious problems in agricultural chemical manufacturing and marketing. Costs of alternative chemicals could possibly increase and be reflected in higher farm production costs and increased consumer food prices. There may be serious doubts as to whether the quantity of alternative materials could meet the demand.[15] For the purpose of illustrating the hypothetical impact of EBDC withdrawal, only two well-known agricultural commodities have been selected — potatoes and apples. Data compiled by a U.S. Department of Agriculture/State/EPA Assessment team report[15] have been used as the principal basis for the following information. For potatoes, data from Cetas and Corsini[6] have been used also.

Impact of EBDC on Withdrawal on Potato Production

The EBDC fungicides on potatoes for control of several major fungus diseases represents a very large use of these chemicals. For that reason, potato production has been selected as an example of the potential impact should these plant protectants be withdrawn from the agricultural scene.

Table 3
RESULTS OF TREATING POTATO SEEDPIECES FOR CONTROL OF *FUSARIUM* DECAY IN VIRGINIA, 1974

Fungicidal treatment	Stand (%)	Fusarium[a] infection (%) *Superior*	Yield (cwt/acre)	Stand (%)	Fusarium[a] infection (%) *Norchip*	Yield (cwt/acre)
None (not inoculated)	58	95 a	56 a	52 a	98	75 a
None (inoculated)	59	99 a	62 a	64 a	96	101 ab
Captan 7.5% dust	95	51 b	135 b	88 b	68	143 ab
Mancozeb 8% dust	96	30 c	152 b	92 b	48	177 b
Metiram 7% dust	94	55 c	142 b	92 b	42	176 b

Note: Small letters indicate groupings of treatments that are not significantly different at the 5% level.

[a] Percentage of seedpieces with *Fusarium* decay about 20 weeks after treating in 1974.

From Cetas, R. C. and Corsini, D. L., Evaluation of Ethylenebisdithiacarbamate (EBDC) Fungicide Use on Potatoes in the U.S., Economic Research Administration, U.S. Department of Agriculture, Washington, D.C., 1978.

Potatoes are grown in every state in the continental U.S. and Alaska, with 32% of the potato acreage grown east of the Mississippi River and 68% west of it. However, 58% of the potato acreage is contained in the top five potato producing states: Idaho, Maine, Minnesota, North Dakota, and Washington, according to the U.S. Department of Agriculture data.[2]

Fungicides are used to control the following major potato diseases in the U.S.: (1) seed piece decay, (2) late blight, and (3) early blight. Because of their prevalent use, efficacy, availability, and economy, EBDC fungicides account for approximately 65% of late blight treatment and 60% of all early blight treatment in the U.S. About 1,156,000 acres have one or more foliar applications for blight control and about 1,078,000 acres have seed piece treatment.[6]

Fungicidal dusts are applied to the seedpieces prior to planting for the control of Fusarium seed piece decay *(Fusarium* spp.), Rhizoctonia stem canker *(Rhizoctonia solani* Kuhn), common scab *(Streptomyces scabies* (Thaxt) Waks and Henrici), and other pathogens that may be carried on the seed tubers.

Before seedpiece treatment became a standard practice in the late 1940s, stand reductions of 10 to 90% due to Fusarium seed piece decay were common, according to Cunningham and Reinking.[7] Replanting was a costly operation and reduced yields from late planting were often the result.

The 8% mancozeb dust became commercially available in the late 1960s and is currently used by more growers than either 7.5% or 10% captan or 7% metiram (Polyram®) dust for the control of Fusarium seed piece decay, Rhizoctonia stem canker, and scab in the U.S.

According to experimental results with the above materials (Table 3), mancozeb performance has been somewhat, but not significantly, superior to captan or metiram for Fusarium infection. Therefore, captan could serve as an acceptable alternative chemical in seed piece protection even though it is not quite as effective. However, there are also concerns about the future of captan, which is also subject to RPAR review.

Foliar sprays are used primarily in the control of the two blights. Late blight control is more important in the eastern region of the U.S., but wherever late blight control is carried out, early blight control will also be obtained.

The use of protective fungicides for late blight and early blight control is extensive. Consistent use of fungicides such as the EBDCs (e.g., mancozeb 80W), and others such as captafol (Captafol 4F®) and chlorothalonil (Daconil® or Bravo 6F®) has kept losses to a minimum for many years — probably less than 5% for the potato industry as a whole in the affected regions. However, the potential losses, in the event that no effective fungicides were available or permitted, could range to 30 to 40% in affected areas and as high as 100% for individual growers (Table 4). Therefore, the use of foliar fungicides is essential for a stable potato industry.

The only viable alternatives for the EBDC foliar fungicides are captafol and chlorothalonil. These fungicides will control late and early blight comparably to the EBDC fungicides. They are somewhat higher in cost. The per-acre cost of controlling late blight of potatoes in 1976 was about $18.00 with an EBDC fungicide, $23.00 with captafol, and $32.00 with chlorothalonil. The production cost of potatoes would rise somewhat should the EBDCs be withdrawn. However, since over 5,000,000 pounds of foliar EBDCs are used annually on potatoes in the U.S. it would be very difficult for the manufacturers of alternative chemicals to produce the large quantity of materials needed without considerable expansion of plant capacity. Resulting shortages of appropriate foliar protectants could likely cause increased prices. The removal of the EBDCs from the market could result in increased production costs and eventual higher consumer prices because of these changes. Since the EBDCs are used extensively in vegetables, fruits, cereal grains, ornamentals, and turf, there would be a chain reaction across the entire agricultural production sector.

From the above data it may be suggested that the widespread use of fungicides for seed piece decay control and late blight and early blight control for several decades has stabilized the disease picture in potato production. Should anything happen to these control measures, these diseases could increase dramatically and have serious impact on the potato industry.

The EBDCs have been used extensively. Only one viable alternative chemical, captan, is available for seed piece decay control. This fungicide is also under potential RPAR review.

Only two alternative fungicides, captafol and chlorothalonil, can give nearly comparable control to the EBDC fungicides. The copper fungicides have generally been much less effective. Should a sudden withdrawal of the EBDCs occur, the supplies of the alternative materials would not be adequate to meet immediate needs, nor might there be adequate industry adjustment to prevent unfortunate industry and agricultural problems.

At present there are few curative fungicides (compounds that will kill the fungus after it has entered the plant). Newer materials that rely on specific enzyme inhibition if used on a large scale could be faced with fungal resistance. The presently used protectants have not demonstrated any such problems. No cultivars of potato are currently available that carry high resistance to the late blight disease. The EBDC fungicides have been a proven part of integrated pest management programs in potatoes. Along with fungicide applications to seedpieces and foliar applications for early and late blight control, other recommended measures include the use of disease-free seed stock, complete healing of cut seed, removal of diseased materials (sanitation), crop rotation, and other cultural practices which favor the potato and suppress pathogens. Where feasible, adapted and resistant varieties are used. However, these nonchemical measures cannot replace the use of fungicides.

Impact of EBDC Withdrawal on Apple Production

Apples are one of the major fruit crops grown in the U.S. Apples were produced commercially on about 526,000 acres in the U.S. in 1969, the latest year for which acreage statistics for apples have been published.[2] In 1973, the total utilized U.S. pro-

Table 4
POTENTIAL YIELD LOSSES DUE TO LATE BLIGHT[20]

State	Year	Cultivar	% Defoliation		Cwt/acre		Loss	
			Treated[a]	Check[b]	Treated[a]	Check[b]	Cwt /acre	%
Wisconsin	1972	Russet Burbank	7	48	203	180	23	11.3
	1973	Russet Burbank	22	92	344	299	45	13.1
Maine	1973	Katahdin	3	71	290	243	47	16.2
	1972	Katahdin	16	100	388	314	74	21.9
			5	100	392	293	99	25.3
	1973	Katahdin	1	84	271	199	72	26.6
Pennsylvania	1975	?	54	100	342	248	94	27.5
Florida	1972	Red La Soda	2	100	241	172	69	28.6
Pennsylvania	1974	Kennebec	74	99	335	239	96	28.7
Maine	1974	Katahdin	7	87	313	209	104	33.2
	1976	Katahdin	2	83	337	218	119	35.3
New York	1975	Hudson	5—10	80—100	321	203	118	36.8
Pennsylvania	1974	Kennebec	43	95	207	207	137	39.8
New York	1975	Katahdin	1—5	80—100	226	130	96	42.5
Maine	1974	Katahdin	3	98	339	179	160	47.2
Pennsylvania	1974	Kennebec	62	99	329	329	166	50.4
New York	1976	Hudson	13	100	334	158	176	52.7
Michigan	1973	Russet Burbank	23	71	148	37	111	75.0

[a] Plots sprayed with either an EBDC, captafol or chlorothalonil fungicide.
[b] Plots not sprayed with a fungicide. Otherwise treated the same as those sprayed with a fungicide.

From Cetas, R. C and Corsini, D. L., Evaluation of Ethylenebisdithiocarbamate (EBDC) Fungicide Use on Potatoes in the U.S., Economic Research Administration, U.S. Department of Agriculture, Washington, D. C., 1978.

duction of apples was 6.3 billion lb. The farm value of the 1973 apple crop was about $512 million.

Apple production is concentrated in several eastern, northern, mid-atlantic, and southeastern states.

Apple scab *(Venturia inaequalis)* is the major fungus disease of apples in most apple growing areas. In addition to apple scab, powdery mildew, *(Podosphaeria leucotricha),* cedar-apple rust, *(Gymnosporangium juniperi virginianae),* several fruit rots, and some minor diseases such as sooty blotch and fly speck are prevalent.

The key to successful prevention of economic loss from apple scab is to control primary infections. To be effective, fungicides must be present on susceptible apple tissues prior to fungus infection, or within a very short period (generally less than 24 hr) after infection.

The EBDC fungicides registered for control of diseases of apples are the major materials that are used for this purpose (Table 5). Use survey data such as that prepared by S. M. Ries at the University of Illinois (Table 6) indicates that there is high grower preference and need for EBDC fungicides in apple disease control programs. Data from other parts of apple growing areas are comparable.

Losses to apples without the benefit of the EBDC fungicides would be of serious significance if they should be withdrawn. Major registered uses and the potential crop losses for apples in the East are listed in Table 7. Potential losses of from 50 to 100% are possible from various apple diseases when estimated as maximum losses under conditions in which no control measures would be used. These estimates were based on observations and infection data from disease control experiments.

The major diseases on which the EBDC fungicides provide superior control are the rusts, black rot, white rot, bitter rot, sooty blotch, and fly speck.[1] There are other fungicides that are superior for control of apple scab and powdery mildew. Benomyl, captafol, dichlone, and dodine are considered as replacements for the EBDCs. There are some problems that accrue with these replacements. Resistant strains of the scab fungus have developed to benomyl and dodine. Captafol and dichlone, although excellent materials for scab control, have problems of phytotoxicity. Phytotoxicity is true also of lime sulfur and sulfur. In addition, many of the useful alternative fungicides to the EBDC materials are themselves directly or indirectly involved in RPAR review.[8,9] The EBDC fungicides are generally less expensive than the potential alternative materials. Withdrawal of these materials could cause production costs to increase, at the least, and at the worst, there could be some chaotic experiences in obtaining stable supplies of the alternative materials for the apple industry, as there could also be severe competition for the materials from other production areas — potatoes, vegetables, citrus fruits, grapes, turf, ornamentals, etc.

From the above information it may be concluded that the EBDC fungicides are vital materials in the control of the major apple diseases in the U.S.

Losses to apples without the benefit of the EBDC fungicides would present very serious consequences to the apple industry if they should be withdrawn. The potential replacement fungicides are already being used, but exhibit some possible hazards such as phytotoxicity and development of pathogen resistance.

The costs of apple production would increase and quality would likely be adversely affected.

ESTIMATED CROP LOSSES TO NEMATODES IN U.S.

A large number of species of parasitic nematodes attack several major agricultural crops in the U.S. Nematode losses on all crops were estimated at 11% in 1970 by the

Table 5
EBDC FUNGICIDES REGISTERED FOR CONTROL OF DISEASES OF FRUITS[a]

Fruit	Disease	Registered EBDC
Apples	Bitter rot	Mancozeb
		Polyram®
		Zineb
	Black rot	Maneb
		Polyram®
		Zineb
	Blotch	Zineb
	White rot	Polyram®
		Zineb
	Brooks spot	Zineb
	Brown rot	Mancozeb
		Polyram®
	Bullseye rot	Maneb
	Cedar apple rust	Mancozeb
		Maneb
		Polyram®
		Zineb
	Fire blight	Zineb
	Flyspeck	Mancozeb
		Maneb
		Polyram®
		Zineb
	Frogeye leaf spot	Zineb
	Quince rust	Zineb
	Rust	Mancozeb
	Scab	Amobam
		Mancozeb
		Maneb
		Polyram®
		Zineb
	Sooty blotch	Mancozeb
		Maneb
		Polyram®
		Zineb

[a] Data from Environmental Protection Agency.

From USDA/State/EPA Assessment Team report on "Assessment of Ethylenebisdithiocarbamate (EBDC) Fungicide Uses in Agriculture," U.S. Department of Agriculture, Washington, D.C., 1978.

Society of Nematologists.[16] Using 1967 to 1968 farm values with costs derived from factors such as yield reduction and market quality reduction resulting primarily from nematode infection, losses were estimated at $1.59 billion annually. These loss estimates were derived after consideration of increased yield and/or quality following the use of nematicides and the additional cost of the chemical applications. Adjusted for subsequent inflation in the following decade, the Agricultural Research Service of the U.S. Department of Agriculture now estimates yearly total losses due to nematodes at $4.0 billion.

NEMATODE CONTROL MEASURES

There are several nonchemical and chemical measures used in controlling nema-

Table 6
FUNGICIDES AND THE NUMBER OF APPLICATIONS EACH USED BY MEMBERS OF THE ILLINOIS QUALITY APPLE CLUB DURING THE YEARS 1973 — 1976

Fungicide	Number of fungicide spray applications[a]					
	1973 (22)[b]	1974 (19)	1975 (25)	1976 (14)	Total	% of total
Benomyl	3	45	55	29	132	8.6
Captan	39	42	29	40	150	9.7
Captafol	4	4	9	9	26	1.2
Copper chem.	1	1	0	0	2	0.1
Dikar®	76	59	104	45	245	15.9
Dinocap®	1	0	0	0	1	0.1
Dodine	108	65	69	48	290	18.8
Ferbam®	1	2	0	0	3	0.2
Phaltan®	33	33	34	22	122	8.9
Polyram®	53	68	64	22	207	13.4
Sulfur	74	38	23	28	163	10.6
Zineb	66	29	21	22	138	8.9
Others	0	5	8	11	24	1.6
Total EBDC	195	156	189	89	629	40.8

[a] The number of times the listed fungicides were applied for disease control in the years listed.
[b] Number of growers participating in the survey.

From USDA/State/EPA Assessment Team report on "Assessment of Ethylenebisdithiocarbamate (EBDC) Fungicide Uses in Agriculture," U.S. Department of Agriculture, Washington, D.C., 1978.

Table 7
POTENTIAL UNCONTROLLED LOSSES TO MAJOR EASTERN APPLE DISEASES IN THE ABSENCE OF FUNGICIDES

Major registered uses	Causal organism	Potential maximum loss without treatment (%)[a]	Estimated area affected[b]	
			Acreage (thousand)	Production (million lb)
Scab	*Venturia inaequalis*[c]	100[c]	226	3000
Rusts	*Gymnosporangium* sp.[c]	75[c]	135	1500
Black rot	*Physalospora obtusa*	100[d]	140	1600
Bitter rot	*Glomerella cingulata*	100	140	1600
Botryosphaeria rot	*Botryosphderia ribis*	80[d]	140	1600
Sooty blotch	*Gloeodes pomigena*[e]	100[e]	225	2500
Fly speck	*Microthyriella rubi*	100[e]	225	2500
Brown rot	*Monilinia* sp.	50	25	253

[a] Based on losses on untreated trees in disease control experiments or observations in local areas.
[b] Acreages and affected production estimated from prevalance of diseases throughout eastern production regions.
[c] Severe infection one season may reduce yield for two seasons.
[d] Percent fruit loss; also capable of producing cankers that kill trees.
[e] Percent fruit infected; some superficially infected fruit may be salvaged for juice.

From USDA/State/EPA Assessment Team report on "Assessment of Ethylenebisdithiocarbamate (EBDC) Fungicide Uses in Agriculture," U.S. Department of Agriculture, Washington, D.C., 1978.

todes. The most common nonchemical control method for annual crops is crop rotation. Obviously, this method is not readily applicable to long-lived perennials, such as orchard crops, vineyards, and citrus groves. The use of resistant varieties, when available, is also an important measure. However, genetic resistance is often effective against only one or two nematode genera or species, and therefore, narrow. New races that can attack resistant varieties tend to develop rather quickly in some cases (e.g., emergence of soybean cyst nematode, Race 4, which attacks Race 3 resistant varieties). Therefore, resistant varieties are most effective when used in combination with other management practices. A number of other cultural practices such as fertility management, deep-plowing, sanitation, and flooding have been used with variable results depending upon the particular situation. Biological control measures by way of natural enemies to nematodes are still very much in infant stages, and are not currently practical.

Chemical control of nematodes has been used successfully and consistently in many situations. Nematicides are highly effective and reliable in controlling and wide range of nematode species. Since most nematicides are high in cost, they are used mainly to protect high return crops in situations where the losses are great from nematode damage and alternative control measures are not feasible. Chemical control of nematodes has been proven to be effective and profitable in the U.S. and in other parts of the world.

Approximately 2 million acres are treated with nematicides each year in the U.S. They are used principally on crops such as cotton, nut crops, orchard crops, potatoes, soybeans, sugar beets, turf and ornamentals, vegetables, and vineyard crops.

Several nematicide-fumigants are broad-spectrum in their activity and persist for variable periods of time in the soil. However, they usually degrade in time, leaving no apparent lasting adverse effects. Some of the nematicides have an intermediate adverse effect on the crop. This phytotoxicity requires that a waiting period be exercised between application and planting. This is true of materials such as methyl bromide, ethylene dibromide (EDB), and chloropicrin.

IMPACT OF WITHDRAWAL OF NEMATICIDES

Less than 25 nematicides are registered for use on important food and fiber crops. Because of their potential adverse effects on the environment, or other hazards in their production and use, some nematicides are subject to RPAR review. For example, ethylene dibromide (EDB) is under RPAR review and may have an uncertain future. In the event of a withdrawal from the market, alternative chemical or nonchemical measures would have to be evaluated.

Regulatory Actions Concerning Cancellation of DBCP Nematicides

An initial EPA suspension order was issued September 15, 1977, that stopped the sale and use of 1,2-dibromo-3-chloropropane or dibromochloropropane (DBCP) sold under trade names such as Fumagon®, Fumazone®, Nemafume®, Nemagon®, etc. This action affected production of melons, peanuts and 18 vegetable crops. Conditional suspensions were made on the use of DBCP on all other labelled crops. A cancellation order for all registered users was issued in November, 1977.[11]

The withdrawal of DBCP demonstrates an example of the potential impact on the agricultural economy in the removal of a widely used and very successful material that has few alternative control measures, either chemical or nonchemical.

The importance of DBCP was carefully investigated and assessed on the following crops: peaches, citrus, grapes, pineapples, soybeans, cotton, peanuts, vegetables, mel-

Table 8
ECONOMIC IMPACTS OF CANCELLING DBCP

	Grower		Consumer
Use	**Control cost increase/year**	**Production loss/3 years**	**Retail price increases**
Citrus	0	$26,851,00	Up to 5%
Cotton	$2,600,000	Negligible	Negligible
Grapes	0	$65,000,000	Major
Peaches and nectarines	0	$80,672,000	Up to 25%
Pineapple	$200,000	$5,793,000	Up to 9%
Soybean	$23,471,000	Negligible	Up to 0.25%
Peanut	$3,515,000	$19,500,000	None
Commercial vegetables	$7,728,000	$17,736,000	Up to 0.1%
Home gardens	Negligible	Negligible	—
Strawberry nursery stock	0	Cease production in East	Up to 18%
Almonds	0	$26,502,000	Up to 20%
Apricots, cherries, figs and walnuts	Negligible due to no use	Negligible due to no use	—
Bananas	Negligible due to no use	Negligible due to no use	—
Other vine berries	Negligible	$500,000	Negligible
Plums	0	$13,754,000	Up to 20%
Home lawns	$2,750,000 to replace lawns	—	Not applicable
Commercial turf	$2,200,000 to $5,000,000	—	Unknown
Ornamentals	$100/Acre (treated acres unknown)	Significant losses to small producers	Unknown
Totals	$44,164,000	$268,253,000	—

From Kempter, J., and Bernstein, M. H., "Dibromochoropropane (DBCP): Final Position Document" Special Pesticide Review Division, Office of Pesticide Programs, Environmental Protection Agency, Washington, D.C., 1978.

ons, strawberries, home gardens and lawns, commercial turf, and ornamentals, by a DBCP Impact Assessment Team, and reported in "USDA/State and EPA Cooperative Assessment of DBCP Uses in Agriculture."[10]

In addition, a final position document prepared by the EPA provided information not only concerning the risk assessment and benefit analysis, but also a review of the economic impacts of the major crops that will be affected by the withdrawal of the nematicide.[11] Table 8 shows the economic impacts of cancelling DBCP, illustrating control cost increases per year to the growers, production losses for the following 3 years, and a projection of possible consumer retail price increases. It was estimated that the withdrawal of DBCP would result in increased control costs of over $44 million, especially to the growers of cotton, soybeans, peanuts, commercial vegetables, lawns and turf, and pineapples. The production loss estimate projected for the next 3 years was over $268 million.

Although the consumer price impact for most commodities would be negligible, the most seriously affected crops would be peaches (up to 25% increase in retail price); almonds and plums (up to 20%); pineapples (up to 9%); and citrus (up to 5%).

For purposes of brevity all of the crops will not be reviewed. Peaches were selected as an example of a fruit crop and soybeans were selected as an example of a field crop. The examples that follow were prepared to illustrate various biological and economic impacts that may occur with the withdrawal of DBCP. The withdrawal or possible withdrawal of any other widely used nematicides would need similar evaluation. The following information was adapted from the Impact Assessment Team report.[10]

Impact of DBCP Withdrawal on Peach Production

Peach orchards in the northeast, southeast, and west have been affected by nematodes causing declining vigor, shortened life expectancy, susceptibility to other diseases, and reduced production potential on both short-term and long-term bases.

Several nematodes have been implicated, such as ring *(Macroposthonia xenoplax)*, root-knot *(Meloidogyne* spp.), lesion *(Pratylenchus* spp.), and dagger *(Xiphinema* spp.).

According to Brittain and Miller,[5] the life expectancy of peach trees in nematode-infested soils in the southeast is 6 to 7 years, whereas the life expectancy of trees receiving pre- and post-plant nematicides is estimated to be 12 to 13 years, all other management practices remaining the same. In California, the expected life of untreated trees is 10 years as compared to 20 years with chemical nematode control.

Peach trees reach their greatest yield potential from the 6th to the 8th years, after which there is a decline. If nematodes are present, trees are less productive and shorter lived. Major losses occur after the 4th or 5th year, at the time when the potential should be greatest.

In the 25 major peach growing states, over 575,000 acres are grown. Over 90% of the acres are believed to be infested with nematodes, and at least 12% are in a state of decline. Of the 38,390 acres that are planted yearly, over 10,000 acres have been preplant treated with DBCP. About 36,947 acres have had post-plant treatments annually.

DBCP has been an integral and necessary component of the nematode management program in peaches. It has become popular because of its efficacy, adaptability, and economy. DBCP has been used as a pre-plant treatment as one of several alternative nematicides. Others are: ethylene dibromide, Vorlex®, D-D®, Telone®, methyl bromide, and chloropicrin. The advantages of DBCP as a pre-plant treatment are: (1) there is less waiting time prior to planting, (2) it does not adversely affect non-target organisms (e.g., mycorrhizal fungi), and (3) the cost is less than alternatives.

Post-plant treatments of DBCP at 18 to 24 month intervals have increased tree vigor, longevity, disease resistance, and productive capacity. Table 9 shows the optimal yields that can be expected from an acre of peaches as a result of periodic applications of DBCP. By contrast, the typical yield curve without treatment shows declining production after the 6th year. The cumulative loss in a 12-year-old orchard is 4827 bu. This can be the expected loss to the peach industry in the future unless a comparable substitute chemical can be found. At this time there are no alternative post-plant nematicides.

Since peaches are a perennial tree crop, decline problems will become more serious with time as a result of the withdrawal of DBCP. There are some alternative preplant nematicides, although they are not as desirable. There are no adequate alternative post-plant nematicides at present. The cost to the peach industry will be $150 to $200 million yearly because of future losses.[10]

Table 9
PEACH YIELDS PER ACRE IN BUSHELS UNDER OPTIMAL GROWING CONDITIONS AND CONDITIONS WHERE NEMATODES ARE PRESENT IN DAMAGING PROPORTIONS

Year	Optimal yields	Yields without DBCP	Cumulative yield loss
1	0	0	0
2	50	48	2
3	200	100	102
4	400	200	302
5	600	300	602
6	800	300	1102
7	800	250	1652
8	800	200	2252
9	800	175	2877
10	800	150	3527
11	800	150	4177
12	800	150	4827

From USDA/State/EPA Assessment Team report on "Assessment of Ethylenebisdithiocarbamate (EBDC) Fungicide Uses in Agriculture," U.S. Department of Agriculture, Washington, D.C., 1978.

Impact of DBCP Withdrawal on Soybean Production

The soybean-growing states in the U.S. that sustain significant losses from nematodes are Arkansas, Mississippi, Missouri, Tennessee, Georgia, North Carolina, South Carolina, Florida, Oklahoma, Maryland, Louisiana, Kentucky, and Virginia. Over 20 million acres are involved.[3] In these states nematodes are one of the major limiting factors in soybean production. Among the species known to cause damage are: the soybean cyst nematode *(Heterodera glycines)*, the root-knot nematode *(Meloidogne* species), the stubby root nematode *(Trichodorus* spp.), sting nematode *(Belonolaimus* spp.), lesion nematode (*Pratylenchus* spp.), ring nematode (*Cricorremoides* spp.), lance nematode (*Hoplolaimus* spp.), dagger nematode *(Xiphenema* spp.), stunt nematode (*Tylenchorhynchus* spp.), and reniform nematode (*Rotylenchulus* spp.). The most seriously damaging of these nematodes are the soybean cyst nematode and the root knot nematode.

According to the Soybean Disease Workers Council, the average loss to nematodes in soybeans in the southeastern U.S. was 8.5% in 1976.[3] Estimated average yield for the net loss due to nematodes in spite of treating with nematicides was estimated to be 42,283,080 bu.

In 1976 approximately 10% of the acreage in the southeastern states was treated with DBCP and certain other nematicides. There was an average yield increase of 9.3 bu/acre. At $5/bu there was a $46.50/acre yield increase, or a net profit of $36.50/acre.

Other materials that have EPA labels for nematode control on soybeans are Dasanit®, Temik®, Mocap®, Nemacur®, and Telone®. Research results have shown that these materials have been generally more costly for the optimum rate of control and have not been as consistently effective. Some of these alternative materials need additional testing to determine efficacy.

Crop rotation as a control measure in soybeans, especially in states where the ecto-parasitic nematode populations tend to be high, has many complications, and is far from an ultimate technique.

Deep chiseling as a cultural control measure has been used on soybeans in the southeastern states. A positive yield response is obtained the first year, but in succeeding years nematodes begin to build up again and yields decline. Therefore, neither available cultural practices nor crop rotation provide adequate control of nematodes in soybeans. Supplemental use of nematicides is necessary in many fields to insure adequate production.

DBCP has been generally less harmful to nontarget organisms in the soils than the several other common soil fumigants and some of the organophosphate chemicals. For instance, there is some evidence that the organophosphate insecticide-nematicides will destroy certain beneficial predator insects, thereby increasing the need for insecticide use to control insect pests.

From the above information it can be concluded that DBCP has become a prominently used nematicide in the control of nematodes on soybeans. The most destructive and serious of the soybean nematodes are the soybean cyst nematode and the root knot nematode. Estimated losses of 8.5% or more were reported by the Soybean Disease Workers Council. Nearly 24 million acres of soybeans grown in the southeastern U.S. are affected. The average yield for the net loss due to nematodes even when nematicide treatments are used is estimated to be over 42 million bu. At $5.00/bu this would involve over $210 million. If nematicide use were withdrawn, the losses would be considerably higher.

There are a number of other registered nematicides for use on soybeans, such as Nemacur®, Mocap®, Dasanit®, Temik®, and Telone®. None of these materials has performed as well as DBCP and they are generally more costly. They have less adaptability to certain cultural practices, such as the "hipper ripper" technique which is used for energy and soil conservation in the southeast U.S.

Although crop rotation and other cultural controls are used in soybean infested areas, they are inadequate for optimum controls. The development of resistant or tolerant varieties has been only partially successful. The biological variability of nematodes, as has been demonstrated by rather rapid emergence of new races of the soybean cyst nematodes, has resulted in breeding programs not keeping up with the problem. The situation will likely become more serious and economically more costly with time, unless new control measures can be generated.

Benefit Analysis of DBCP Withdrawal

According to the EPA Review Summary, the Benefit Analysis estimates that the potential short-term impact of cancellation of DBCP to users may be as high as $400 million during the first 3 years after cancellation. This is comprised of about $280 million in yield reduction, and about $115 million in increased control costs. Reductions in yield would be most severe in the perennial crops for which there are no registered post-plant alternatives such as citrus, grapes, peaches, plums, and almonds. These crops account for about 80% of the projected dollar losses from yield reductions. Increase in control costs would be severe for soybeans, accounting for about 58% of the projected dollar increase in control costs.

The Benefits Analysis also quantitatively describes the anticipated long-term impacts of a DBCP cancellation for each use or category of use. Assessments of these impacts are difficult because of the large number of variables involved. Factors such as the development, use, and availability of new alternative control methods or chemicals, and the potential for greater productivity and increased acreage in uninfested areas, are difficult to predict. Furthermore, price elasticity and demand data for certain crops are not always available. Therefore, long-term impacts of withdrawal of DBCP cannot entirely be substantiated.

SUMMARY

Economic benefits from the use of all pesticides — including fungicides and nematicides — should be measurable in terms of the income from the increased yield or quality of product, minus the cost of the pesticide and its application. Determination of efficacy or yield benefits of various pesticides are usually obtained from experiments in which pest control materials and methods are practiced alongside untreated controls for a number of years.

Considerable variability in experimental results can be seen because many factors are involved, including geographical locations, soil types, weather conditions, pest infestation levels, infection patterns, frequency, rates and methods of pesticide applications, and many other variables. In spite of these variabilities, certain conclusions have been drawn from these data as to the benefits that are derived from the use of pesticides in the control of pests. Although they are imperfect, they represent the best educated judgments that we have in evaluating benefits in pesticide use.

Fungicides and nematicides have been successfully used in the control of a wide range of fungal diseases and nematode problems for many years. Many plant pathogens are controlled primarily by nonchemical methods, especially in the use of resistant varieties and in the use of other cultural control methods such as crop rotation, sanitation, timing of operations, and fertility management. However, there are many diseases that can be controlled only by using chemical controls. There are considerable differences in these needs. The use of fungicides and nematicides on agronomic crops such as corn and soybeans are quantitatively relatively low, although the effectiveness of such control measures such as seed treatment should not be underestimated. By contrast, the quantitative use of fungicides and nematicides on fruit and vegetable crops is extensive and is vitally important for uniform harvest of high yielding and high quality crops.

Losses to agronomic crops from fungal pathogens range from 7 to 28% if no fungicides are used. Losses to vegetables range from 8 to 25% and most fruit and nut crops losses range from 8 to 29% if no fungicides are used.

The annual cost to American agriculture, if all fungicides were removed, would be about $1.7 billion. The estimated annual cost if nematicides were removed would be at least $1.59 billion, and perhaps much more.

Several commonly used fungicides and pesticides are presently under Rebuttable Presumption Against Registration (RPAR) by the EPA, a regulatory procedure designed to evaluate the possible adverse effects of agricultural chemicals on man and his environment.

Among certain widely used fungicides that have been under RPAR review, the ethylenebisdithiocarbamates (EBDCs) were selected as a hypothetical example of possible impacts should these fungicides be withdrawn from American agriculture. Data compiled by a U.S. Department of Agriculture/State/EPA Assessment Team and other sources were reviewed. For brevity, only potatoes as a vegetable crop and apples as a fruit crop were used to illustrate the problems that would accrue if the EBDCs were no longer available. Not only would production costs be increased but large-scale removal of the EBDC fungicides would also create serious supply/demand problems in the manufacturing and marketing of alternative materials. Consumer prices would likely be increased. Several diseases now relatively low in incidence because of the use of the fungicides could become more prevalent and destructive. This, in turn, could make additional pressures in production and marketing.

To illustrate the impact of withdrawal of nematicides on crop production, the cancellation of sale and use of dibromochloropropane (DBCP) was used as a strategic example. Data prepared by a U.S. Department of Agriculture/State and EPA Cooperative Assessment Team for DBCP Uses in Agriculture and a final position document by the EPA were used to illustrate the impact of withdrawal of a nematicide. Withdrawal of DBCP on several important agricultural crops suggests costly adjustments that will have to be made in the future, both short- and long-term. For brevity, peaches and soybeans were used as examples. The cost to the peach industry alone will be $150 to $200 million annually. The cost to soybean producers will also be over $200 million annually.

In light of the economic impacts that can occur from possible withdrawal of fungicides and nematicides — and other pesticides — as a result of RPAR reviews, benefit/risk as well as benefit/cost information should be considered carefully prior to any action. The agricultural industry is highly dependent on the judicious use of chemicals as part of the overall control of plant diseases and other pest problems.

REFERENCES

1. **Adwinkle, H. S.,** Ed., The bad apple, *Newsletter of the Apple and Pear Disease Workers,* Vol. 1, No. 5, N.Y. Agric. Exp. Stn., 1976.
2. **Anon.,** Agricultural Statistics, Department of Agriculture, Washington, D.C., 1976.
3. **Blackman, C. W.,** *Fungicide-Nematicide Test Results — Soybeans,* American Phytopathological Society, St. Paul, Minn., 1976, 171.
4. **Brandes, G. A.,** *Acreage for Crops for which EBDC Fungicides are Registered,* Doc. 0959C, Rohm and Haas, Philadephia, 1977.
5. **Brittain, J. A. and Miller, R. W.,** Managing peach tree short-life in South Carolina, Circular 568, Clemson University Cooperative Extension Service, Clemson, S.C., 1976.
6. **Cetas, R. C. and Corsini, D. L.,** Evaluation of Ethylenebisdithiocarbamate (EBDC) Fungicide use on potatoes in the United States, Economic Research Administration, U.S. Department of Agriculture, Washington, D.C., 1978.
7. **Cunningham, H. S. and Reinking, O. A.,** *Fusarium* seed piece decay of potato on Long Island and its control, N.Y. *Agric. Exp. Stn. Bull. (Cornell),* 721, 1, 1946.
8. Public Affairs Dept., The RPAR Against Benomyl, DuPont Agrichemical News, DuPont, Wilmington, Del., 1978.
9. EPA Pesticide Programs, Rebuttable Presumption Against Registration and Continued Registration of Pesticide Products Containing Ethylenebisdithiocarbamate, *Fed. Regist.,* 42 (154), 1977.
10. **Good, J. M. and Smart, G. C., Eds., and** DBCP Impact Assessment Team, USDA/State and EPA Cooperative Assessment of DBCP Uses in Agriculture, U.S. Department of Agriculture, Washington, D.C., 1977.
11. **Kempter, J. and Bernstein, M. H.,** Dibromochloropropane (DBCP) Final Position Document, Special Pesticide Review Division, Office of Pesticide Programs, U.S. Environmental Protection Agency, Washington, D.C., 1978.
12. **Lawless, D. W. and Von Rumker, R., et al.,** A Technology Assessment of Biological Substitutes for Chemical Pesticides, MRI Project No. 3879-2, National Science Foundation RANN Program, Washington, D.C., 1978.
13. **McKeogh, J. F.,** EPA Rebuttable Presumption Against Re-registration, Ethylenebisdithiocarbamate Fungicides, Rohm and Hass, Philadelphia, 1977.
14. **Pimentel, D., Krummel, J., Gallahan, D., Hough, J., Merrill, A., Schriener, I., Vittum, P., Koziol, F., Back, E., Yen, D., and Fiance, S.,** A Cost-Benefit Analysis of Pesticide Use in U.S. Food Production, New York College of Agriculture and Life Sciences, Cornell University Agricultural Experimental Station, Ithaca, 1978.

15. **Powell, C. C., Dowler, W. M., Ballard, G., Bissonnette, H. L., Bratland, J., Cetas, R. C., Cole, H., Corsini, D. L., Kantzes, J. G., Lightfield, J., McMillan, R. I., Paulson, G. D., Paulus, A. O., Pelletier, N., Ragsdale, J., Rowan, S. S., Stanton, O. R., Starler, N., Steiner, P. W., and Yoder, K.,** Assessment of Ethylenebisdithiocarbamate (EBDC) Fungicides Uses in Agriculture, Part II. An Analysis of Current EBDC Uses, Their Benefit, the Role of Alternatives, and Impacts to Agriculture from Changes in EBDC Use Patterns. USDA/State/EPA Assessment Team, Coordinated by the Office of Environmental Quality Activities, U.S., Department of Agriculture, Washington, D.C., 1978.
16. Estimated Crop Losses from Parasitic Nematodes in the United States, Spec. Publ. No. 1, Society of Nematologists, Ames, Iowa, 1971.
17. Vegetable-Processing, 1976 — Annual Summary Acreage, Yield, Production, and Value, Crop Reporting Board, Statistical Reporting Service, U.S. Department of Agriculture, Washington, D.C., 1977.
18. **Wehunt, E. J. and Good, J. M.,** The Peach: Varieties, Cultures, Marketing, and Pest Control, U.S. Department of Agriculture, Washington, D.C., 1974.
19. **Whitehead, C. W.,** Chevron response to staff summary — Captan Fact Sheet, Ortho Field Research and Services, Des Moines, Iowa, 1978.
20. American Phytopathological Society, *Fungicide and Nematicide Tests,* Vols. 28-32, American Phytopathological Society, St. Paul, Minn., 1973—1977.

ESTIMATED CROP LOSSES DUE TO WEEDS WITH NONCHEMICAL MANAGEMENT

John R. Abernathy

INTRODUCTION

The competitive effect that weeds have on the production of numerous crops is well-documented in scientific journals. Herbicides have become essential for economically controlling these pests. The elimination of these chemicals and use of economically feasible nonchemical means (rotation, sanitation, biocontrol, etc.) would reduce annual farm revenue 31% resulting in economic losses (based on 1976 figures) of $13.0 billion (Table 1). This staggering sum amounts to $58 for every man, woman, and child in the U.S. or a total loss of $4649 for each American farmer. Such losses in crop production and farm revenue could produce severe shortages of food and fiber in the U.S. and result in a 50% or greater increase in food prices to the American public. Exportation of agricultural products would essentially be eliminated.

Without herbicides American farmers would need an additional 128.3 million acres of productive crop land to produce the same quantities of food and fiber using available nonchemical means of weed control. This would represent a 46% increase in acres to be farmed. Obviously, this is not a feasible alternative as the additional land is not available.

For this chapter, production and revenue losses which would result if herbicides were not available are considered for seven major crops including cotton, corn, peanuts, sorghum, soybeans, rice, and wheat and other small grains. Many economically important crops produced in the U.S. are not considered in this chapter. Among these commodities, and depending heavily on herbicides for weed management, are vegetable crops. Without herbicides, production costs in many vegetable crops would be prohibitive.

All losses are based on estimates of authorities involved in weed management for each respective crop. No considerations are given for shifts into alternative crops, future farm programs or legislation. The assumption has been made that a labor pool willing to participate in hand weeding of many crops does not exist. In many cases if this hand labor were available, the production cost would be prohibitive. Within these seven crop cultures, peanut and rice production would be damaged most severely if herbicides were not available. Yield losses of 90 and 70% respectively, could be expected for these commodities. The production of wheat and other small grains, amounting to a loss of 20% or 1.0 billion bu would be least affected by the elimination of available herbicides. The losses expected from cotton, corn, sorghum, and soybeans would be 40, 36, 35, and 24%, respectively. In the following discussion, the losses incurred by weed competition are examined in detail for each of the major U.S. crops. A listing of literature and references for weed competition will not be made as other writers for this handbook will include many of these references in their respective commodity groups.

CORN

Corn production in 1976 occupied 84.9 million acres (Table 2). The total value of this crop was $13.5 billion. A total of 207 million lb of herbicide was used to produce this crop. Based on estimates of crop loss, a 25% reduction in corn yield could be

Table 1
CROP PRODUCTION IN U.S. WITH AND WITHOUT HERBICIDES (1976)[a]

				Without Herbicides	
Crop	Total acres	Total value ($)	Total herbicide used (lbs)	Net loss ($)	Additional acres to keep same level of production
Corn	84.93	13471.79	207.00	4748.44	47.77
Cotton	11.66	3301.27	18.30	1496.15	7.77
Peanuts	1.55	750.26	3.30	730.73	13.95
Sorghum	18.40	1450.09	15.70	559.80	9.91
Soybeans	50.23	8768.98	81.00	2004.50	15.86
Rice	2.49	811.36	8.50	691.71	5.81
Wheat and other small grains	108.75	13377.79	27.20	13075.55	27.19
Total	278.01	41931.54	361.00	12951.79	128.26

[a] Values are in millions.

Table 2
CORN PRODUCTION IN U.S. WITH AND WITHOUT HERBICIDES

Crop variables (1976)	Total (× 1,000,000)	Per acre	Ref.
Acres	84.93		1
Amount produced (bu)	6,266.36	73.78	1
Value ($)	13,471.79	158.62	1
Herbicide used (lbs)	207.00	2.44	2
Herbicide cost ($)	560.00	6.59	3
Crop loss w/o herbicides			3
Competition (25%) (bu)	1,566.59	18.45	
($)	3,367.95	39.66	
Indirect (10%) (bu)	626.64	7.38	
($)	1,347.18	15.86	
Harvest (1%) (bu)	62.66	0.74	
($)	134.72	1.59	
Total (36%) (bu)	2,255.89	26.56	
($)	4,849.84	57.10	
Additional tillage needed w/o herbicides	3	3	3
Cost of extra tillage ($)	458.60	5.40	4
Hand weeding costs w/o herbicides	None	None	3
Biological control possibilities	None	None	3
Net loss w/o herbicides			
(bu)	2,255.89	26.56	
($)	4,748.44	55.91	
Additional acres to maintain current production w/o herbicides	47.77		

expected from competition by weeds without the use of herbicides. The later planting date of corn would be considered an indirect loss. Without herbicides, planting dates would have to be delayed to allow for the mechanical destruction of the first germinating population of weeds. This would eliminate the use of high-yielding, full-season hybrids in Midwest corn production and the resulting shift to shorter season cultivars

Table 3
COTTON PRODUCTION IN U.S. WITH AND WITHOUT HERBICIDES

Crop variables (1976)	Total (× 1,000,000)	Per acre	Ref.
Acres	11.66		1
Amount produced (lbs)	5078.88	435.58	1
Value ($)	3301.27	283.13	1
Herbicide used (lbs)	18.30	1.57	2
Herbicide cost ($)	120.40	10.33	5
Crop loss w/o herbicides			5
Competition (32%) (lbs)	1625.24	139.39	
($)	1056.41	90.60	
Indirect (3%) (lbs)	152.37	13.07	
($)	99.04	8.49	
Harvest (5%) (lbs)	253.94	21.78	
($)	165.06	14.16	
Total (40%) (lbs)	2031.55	174.23	
($)	1320.51	113.25	
Additional tillage needed w/o herbicides	3	3	5
Cost of extra tillage ($)	62.94	5.40	4
Hand weeding costs w/o herbicides	233.10	19.99	5
Biological control possibilities	None	None	5
Net loss w/o herbicides			
(lbs)	2031.55	174.30	
($)	1496.15	128.31	
Additional acres to maintain current production w/o herbicides	7.77		

would reduce production an additional 10%. Another 1% loss can be expected at harvest due to the interference of weeds in the actual harvesting operation.

The total loss due to weeds in corn with herbicides made unavailable would be 36% or 2.3 billion bu of corn at a cost of $4.7 billion. Without herbicides, three additional cultivations per acre would be required. The cost of this extra tillage on U.S. corn acreage would be $459 million. Hand weeding costs would not be increased because, at current labor costs, this practice is not economically feasible. Even if it were, a labor force willing to participate in this type of employment would be nonexistent. Presently, there are no biological control measures that offer promise for controlling weeds in corn. Thus, the elimination of herbicides and use of only available cultural and mechanical control practices results in a loss of 36% of the total U.S. corn crop. This amounts to a loss of 27 bu/acre at a cost of $56/acre. To maintain the same level of corn production while sustaining these losses due to weeds, an additional 47.8 million acres of crop land would be needed.

COTTON

In 1976, the U.S. cotton crop was grown on 11.7 million acres that produced 5 billion lb of lint valued at $3.3 billion (Table 3). Approximately 18.3 million lb of herbicides were used in producing this crop. Without these chemicals, a 32% yield reduction could be expected due to competition of weeds. Another 3% would be lost due to indirect causes such as weed contaminants lowering the cotton grade. A 5% reduction could be attributed to harvest loss. Weeds cause severe interference with the stripping or picker harvesting operation when present within the cotton crop. Thus, the total loss of cotton could be 40% of 2 billion lb of lint worth $1.3 billion.

Table 4
PEANUT PRODUCTION IN U.S. WITH AND WITHOUT HERBICIDES

Crop variables (1976)	Total (× 1,000,000)	Per acre	Ref.
Acres	1.55		1
Amount produced (lbs)	3750.89	2419.93	1
Value ($)	750.26	484.04	1
Herbicide used (lbs)	3.30	2.13	2
Herbicide cost ($)	30.90	19.94	6
Crop loss w/o herbicides			6
Competition (60%) (lbs)	2250.53	1451.96	
($)	450.16	290.43	
Indirect (15%) (lbs)	562.63	362.99	
($)	112.54	72.61	
Harvest (15%) (lbs)	562.63	362.99	
($)	112.54	72.61	
Total (90%)(lbs)	3375.80	2177.94	
($)	675.23	435.63	
Additional tillage needed w/o herbicides	6	6	6
Cost of extra tillage	16.72	10.79	4
Hand weeding costs w/o herbicides	69.68	44.96	6
Biological control possibilities	None	None	6
Net loss w/o herbicides			
(lbs)	3375.80	2177.94	
($)	730.73	471.44	
Additional acres to maintain current production w/o herbicides	13.95		

Three additional tillage operations would be required at a cost of $62.9 million. The hand weeding costs in production of cotton would go up drastically, rising to a level where the practice would be economically unfeasible. A realistic figure could be set at $20/acre for additional hand labor costs. A net loss of $1.5 billion could be anticipated in cotton production without herbicides. A loss of $128/acre could be expected if no herbicides were available for use in cotton and the best cultural, tillage, and hand labor methods available were employed. With such losses, an additional 7.8 million acres would be required to sustain cotton production at a 1976 level.

PEANUTS

Total peanut production in 1976 was 3.8 billion lb on 1.6 million acres at a value of $750 million (Table 4). The total amount of herbicide used on U.S. peanut acreage was 3.3 million lb. In peanut production, weed competition is severe, and direct yield losses of 60% could be anticipated. Indirect losses would contribute another 15% loss in production. Many additional cultivations would be required. These mechanical operations are known to enhance the incidence of diseases such as Southern blight which can drastically increase the loss of peanuts.

Another 15% loss could be attributed to lowered harvest efficiency. Presence of weeds or weed tissue directly relate to field losses during harvest. Also, substantial losses can be incurred due to damaged harvesting equipment when weed materials are present.

Thus, a total loss in peanut production could be set at 90% without herbicides even when using all available tillage and cultural methods. This loss would amount to 3.4 billion lb of peanuts at a value of $730 million. Without herbicides, six additional

Table 5
SORGHUM PRODUCTION IN U.S. WITH AND WITHOUT HERBICIDES

Crop variables (1976)	Total (× 1,000,000)	Per acre	Ref.
Acres	18.40		1
Amount produced (bu)	719.82	39.12	1
Value ($)	1450.09	78.81	1
Herbicide used (lbs)	15.70	0.85	2
Herbicide cost ($)	47.10	2.56	7
Crop loss w/o herbicides			7
Competition (25%) (bu)	179.96	9.78	
($)	362.52	19.70	
Indirect (7%) (bu)	50.39	2.74	
($)	101.51	5.52	
Harvest (3%) (bu)	21.60	1.17	
($)	43.50	2.36	
Total (35%) (bu)	251.94	13.69	
($)	507.53	27.58	
Additional tillage needed w/o herbicides	3	3	7
Cost of extra tillage ($)	99.37	5.40	4
Hand weeding costs w/o herbicides	None	None	7
Biological control possibilities	None	None	7
Net loss w/o herbicides			
(bu)	251.94	13.69	
($)	559.80	30.42	
Additional acres to maintain current production w/o herbicides	9.91		

tillage operations would be required costing an additional $16.7 million. The cost of additional hand weeding would amount to $70 million assuming hand labor to be available. A loss of 2200 lb/acre of peanuts could be expected without herbicides with a dollar loss of $471/acre. If herbicides were not available for use in peanuts, an additional 14.0 million acres would be required to meet 1976 production levels.

SORGHUM

Sorghum valued at $1.5 billion was produced on 18.4 million acres in 1976 with a total yield of 720 million bu. (Table 5). A total of 15.7 million lb herbicide was used to produce the crop. Losses without herbicides in sorghum would amount to 25% due to weed competition. A 7% loss would occur from dockage of the harvested sorghum and the delayed planting needed in order to kill early flushes of weeds. Later plantings of sorghum would increase losses to injurious insects. Another 3% could be lost due to harvesting as the weeds present in a mature field of sorghum can interfere with efficiency of combining. A total loss of 35% of the sorghum production could be expected if herbicides were unavailable and would amount to 252 million bu. Without herbicides, three additional tillage operations would be required at a cost of $99 million for the entire U.S. sorghum crop.

Additional hand labor would not be economically feasible in sorghum. Also, there are no biological control techniques available. The total loss in sorghum would be $560 million due to weeds without the use of herbicides. This amounts to a loss of 13.7 bu or $30/acre. Without herbicides, an additional 9.9 million acres would be required to produce the same amount of sorghum as is now produced on 18.4 million acres.

Table 6
SOYBEAN PRODUCTION IN U.S. WITH AND WITHOUT HERBICIDES

Crop variables (1976)	Total (× 1,000,000)	Per acre	Ref.
Acres	50.23		1
Amount produced (bu)	1287.56	25.63	1
Value ($)	8768.98	174.58	1
Herbicide used (lbs)	81.00	1.61	2
Herbicide cost ($)	522.00	10.39	8
Crop loss w/o herbicides			8
Competition (20%) (bu)	257.51	5.13	
($)	1753.80	34.92	
Indirect (2%) (bu)	25.75	0.51	
($)	175.38	3.49	
Harvest (2%) (bu)	25.75	0.51	
($)	175.38	3.49	
Total (24%) (bu)	309.01	6.15	
($)	2104.60	41.90	
Additional tillage needed w/o herbicides	3	3	8
Cost of extra tillage ($)	271.22	5.40	4
Hand Weeding costs w/o herbicides	150.68	3.00	8
Biological control possibilities	None	None	8
Net loss w/o herbicides			
(bu)	309.01	6.15	
($)	2004.50	39.91	
Additional acres to maintain current production w/o herbicides	15.86		

SOYBEANS

In the U.S., 50 million acres of soybeans were grown during 1976 at a value of $8.8 billion (Table 6). This acreage required the use of 81 million lb of herbicide. Direct loss due to weed competition is estimated in soybeans at 20%. A 2% loss is incurred due to delayed plantings and an additional 2% loss from harvest inefficiency. This amounts to a 24% loss in yield or 309 million bu of soybeans if herbicides were not available. Three additional tillage operations would be required at an annual cost of $271 million. Hand weeding costs in soybeans would increase by $151 million and biological control methods are not presently feasible.

The total loss in soybeans would be approximately $2.0 billion without herbicides and using cultural control measures. An additional 15.9 million acres of soybeans would be required in the U.S. to maintain current production levels. The per acre loss of soybeans would be 6 bu at an average annual cost of $40/acre without herbicide use.

RICE

Rice was grown on 2.5 million acres in 1976 at a total value of $811 million (Table 7). Herbicides applied on this acreage amounted to 8.5 million lb. Losses in rice due to weeds are severe without the use of herbicides. A loss of 50% of the rice crop could be expected due to competition of weeds. Another 20% would be lost to indirect causes such as a dockage of the product. Also without herbicides, more rice land would have to be water-planted and flooded for longer periods creating exaggerated costs as well as reducing yields.

Table 7
RICE PRODUCTION IN U.S. WITH AND WITHOUT HERBICIDES

Crop variables (1976)	Total (× 1,000,000)	Per acre	Ref.
Acres	2.49		1
Amount produced (cwt)	115.65	46.45	1
Values ($)	811.36	325.85	1
Herbicide used (lbs)	8.50	3.41	2
Herbicide cost ($)	25.54	10.26	9
Crop loss w/o herbicides			9
Competition (50%)—(cwt)	57.83	23.23	
($)	405.68	162.92	
Indirect (20%)—(cwt)	23.13	9.29	
($)	162.27	65.17	
Harvest (0%)—(cwt)			
($)			
Total (70%)—(cwt)	80.96	32.52	
($)	567.95	228.09	
Additional tillage needed w/o herbicides	None	None	9
Cost of additional water for weed control	149.30	59.96	9
Hand weeding costs w/o herbicides	None	None	9
Biological control possibilities	N. jointvetch		9
Net loss w/o herbicides			
(cwt)	80.96	32.52	
($)	691.71	277.79	
Additional acres to maintain production w/o herbicides	5.81		

There probably would be no additional losses due to harvest without herbicides in rice. A total loss of 70% could be expected if herbicides were unavailable; this would amount to 81 million cwt. Flooding cost to achieve additional weed control would amount to $149.3 million. No extra tillage operations would be required without herbicides. There is one possible biological weed control measure for use in rice at the present time: a disease organism that attacks Northern jointvetch which is a weed problem in a limited area of the rice production acreage. The total loss in U.S. rice production without herbicides and employing the best known cultural methods would amount to $692 million. On a per acre basis, 3300 lb of rice would be lost at a cost of $278. An additional 5.8 million acres of rice would be required to maintain current levels of production.

WHEAT AND OTHER SMALL GRAINS

The production of wheat and other small grains in the U.S. in 1976 occupied 109 million acres (Table 8). On this acreage, 27 million lb of herbicides were used. The value of the crop produced was $13 billion. Weed competition in wheat and other small grains is not as severe as many other crops grown in the U.S. An 18% loss could be expected due to competition of weeds. A 2% loss would arise from dockage of the grain and delayed planting for a total of a 20% reduction in yield of wheat and other small grains across the U.S. This loss would amount to 1 billion bu of grain in this country. Two additional tillage operations would be required to produce the crop which would cost about $217 million. Therefore, the total loss would be $2.8 billion if herbicides were not available for use in wheat and other small grains. On a per acre

Table 8
WHEAT AND SMALL GRAIN PRODUCTION IN U.S. WITH AND WITHOUT HERBICIDES

Crop variables	Total (× 1,000,000)	Per acre	Ref.
Acres	108.75		1
Amount produced (bu)	5,203.50	47.85	1
Value $	13,377.79	123.01	1
Herbicide used (lbs)	27.20	0.25	2
Herbicide cost ($)	48.83	0.45	10
Crop loss w/o herbicides			10
Competition (18%) (bu)	936.63	8.61	
($)	2,408.00	22.14	
Indirect (2%) (bu)	104.07	0.96	
($)	267.56	2.46	
Harvest (0%) (bu)			
($)			
Total (20%) (bu)	1,040.70	9.57	
($)	2,675.56	24.60	
Additional tillage needed w/o herbicides	2	2	10
Cost of extra tillage ($)	217.49	2.00	4
Hand weeding costs w/o herbicides	None	None	10
Biological control possibilities	None	None	10
Net loss w/o herbicides			
(bu)	1,040.70	9.57	
($)	2,844.22	26.15	
Additional acres to maintain production w/o herbicides	27.19		

Table 9
1976 CROP EXPORTS VS. CROP LOSSES TO WEEDS WITHOUT HERBICIDES[a]

Crop	Exports[11]	Crop losses
Corn	1550 bu	2255 bu
Cotton	3.3 bales	4.1 bales
Sorghum	196 bu	252 bu
Soybeans	565 bu	309 bu
Rice	62 cwt	81 cwt
Wheat	1173 bu	1040 bu

[a] Values in millions.

basis, this would amount to $26. To grow the equivalent amount of wheat and other small grains, an additional 27.2 million acres would be required to produce at the same level if herbicides were not available for use in these grain crops.

SUMMARY

In summary, a 31% loss in total crop revenue from the major crops grown in the U.S. could be expected if herbicides were not available for use and production depended only on nonchemical and cultural means. This amounts to a very substantial loss to U.S. farmers, and eventually this loss would be realized by the American public in exaggerated food costs. The loss of crop production by weeds could drastically disrupt the U.S. crop export market (Table 9). Based on 1975 and 1976 data, corn, cotton, sorghum, and rice losses without herbicides would be greater than the total amount of exports. These crop losses would also seriously cut into amounts available for domestic use. The losses in wheat and soybeans could be 55 and 89%, respectively, of the total crop exported, again causing great loss of export markets and threatening U.S. domestic supplies. Without herbicides, farming in the U.S. would be severely jeopardized.

REFERENCES

1. Field Crops, Crop Reporting Board, Economics, Statistics, and Cooperatives Service, Washington, D.C., 1978.
2. **Eikers, T. R., Andrilenas, P.A., and Anderson, T. A.,** Farmer's Use of Pesticides in 1976, Econ. Rep. No. 418, Economics, Statistics, and Cooperatives Service, U.S. Department of Agriculture, Washington, D.C., 1976.
3. **Fred W. Slife,** University of Illinois, personal communication.
4. **Marvin Sartin,** Texas Agricultural Extension Service, personal communication.
5. **Gale A. Buchanan,** Auburn University, personal communication.
6. **Ellis W. Hauser,** USDA-SEA-AR, Tifton, Georgia, personal communication.
7. **Allen F. Wiese,** Texas Agricultural Experiment Station, personal communication.
8. **Loyd M. Wax,** USDA-SEA-AR, Urbana, Illinois, personal communication.
9. **E. Ford Eastin,** Texas Agricultural Experiment Station, personal communication.
10. **John D. Nalewaja,** North Dakota State University, personal communication.
11. 1976 Handbook of Agricultural Charts, Agricultural Handbook No. 509, U.S. Department of Agriculture, Washington, D.C., 1976.

ESTIMATED LOSSES WITHOUT PESTICIDES AND SUBSTITUTING ONLY READILY AVAILABLE NONCHEMICAL CONTROLS FOR LIVESTOCK PESTS

Harry E. Smalley

A world without pesticides is hard to imagine. Many people equate the word "pesticide" with the synthetic organic chemicals developed in rather recent years beginning with the use of DDT as an insecticide in the early 1940s. While I do not pretend to be an historian, I feel that we must acknowledge the fact that "pesticides" have been known and used for many years. For instance, sulfur was used in the pesticide connotation before the Roman Empire was eminent. Botanically derived insecticides (nicotine, pyrethrin) were known and used in the 18th century; rotenone has been in continuous use since 1848. The chemical insecticides, which included metals, were developed in the early 1900s, and include lead arsenate, calcium arsenate, carbon disulfide, copper sulfate, and others. Indeed, the arsenic dip for the tick-borne disease, Texas Fever, in cattle, is largely credited for saving the cattle industry in the Southwest. Perhaps I have an advantage (although it's hard to realize at times) in that I spent my early years in a rural atmosphere when there were no "pesticides" per se; we did use lube oil, Paris green, flowers of sulfur, rotenone dust, dips, tar, and asafoetida. I firmly resist calling these "the good old days." We utilized our own energy to produce much less than we do now

In developing these estimates, I have had to draw on past experience, as well as surveys and records which were estimates even then. I have also had to use a rather battered imagination in trying to envision "worst-case scenarios" as well as being realistic. I have had to extrapolate monetary losses in the 20s and 30s to percent of present production. If this sounds like an apology, it is; what I have is a forecast (guess) based on inadequate information, estimates of estimations, and imagination.

I have had to limit these losses to direct losses to livestock, realizing that the indirect losses — such as losses due to lack of adequate forage and feed — may be even more than direct losses. Nonchemical control methods are very adequately described elsewhere, and their continued and expanded use will help in reducing losses; however, even the expanded Integrated Pest Management Programs utilize the judicious use and application of chemical insecticides.

One of the most damaging livestock insects, the screw-worm, is controlled rather successfully by a nonchemical eradication program involving the sterile male release; this one control program has eliminated much of the losses and reduced the need for much of the labor in livestock production, when compared to the early days of stock raising.

Conservatively estimated losses for the years 1974 to 1976 amount to $4 billion of the total value of $35 billion, or an 11.4% loss of production. In 1955, losses in livestock and poultry from disease, parasites, and pests came to $1 billion, of a total value of $11.2 billion, or 8.9%.[1] We must realize not only the increase in numbers of livestock and changes in production (feed lots), but the tremendous inflation that has affected us since 1955. Production efficiency increased tremendously over the years since 1936, but the total land available for agricultural pursuits has decreased significantly. This increase in production can be attributed to several factors — certainly not due to the increased use of pesticides only, but also by improved management practices, improved genetic selection, improved feed and forage plants, and improved nutrition. Pesticides have significantly contributed, of course; but it is the total efforts

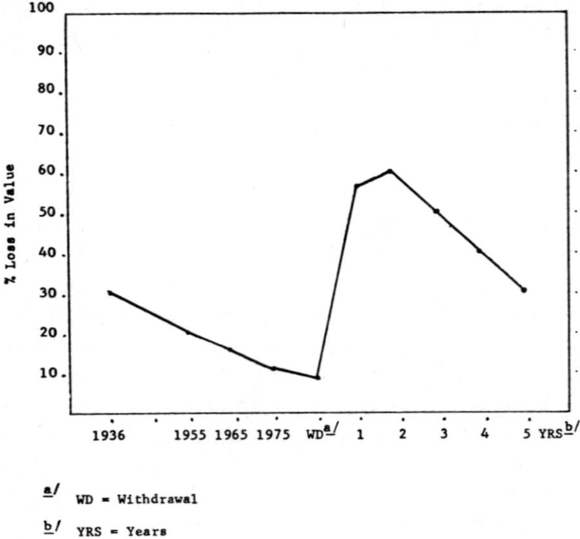

FIGURE 1. "Worst-case scenario" after immediate withdrawal of all pesticides and biocides.

of a number of disciplines that have effected this tremendous increase in efficiency. Loss of pesticides will affect livestock production unquestionably; but to what extent?

Without chemical pesticides, Borlaug, the Nobel laureate, estimates that crop losses would escalate to 50%.[2] Pimentel estimates that crop losses would increase by 7%.[3] Another source says that the output of American farms would be reduced immediately by about 25% and the price of products increased by 50%.[4] Snelson indicates that "most authorities have placed world-wide pest losses at a minimum of 35% of agricultural production." Snelson further states that these losses could exceed 50% in developing countries, primarily those in tropical climates.[5] Although it is statistically abhorrent, by taking an average of these estimates, we arrive at a mean loss of approximately 30% of production.

Looking at beneficial effects of pesticides, tests have shown that chemical control of biting flies resulted in an increase in weightgain of 15 lb/month in beef cattle and an increase in butterfat production of as much as 29.8%.[6]

If we attempt to take the average loss cited above, 30%, this would be a decrease in livestock production income of 18.6% from what we presently enjoy. This is a conservative figure, I believe, because livestock production in undeveloped countries suffers from uncontrolled parasites and diseases — estimated losses ranging from 35 to 70%, with the higher figure representing the vast areas in Africa where the tsetse fly is dominant. This figure of 30% loss of production without pesticides may not reflect the "worst-case scenario"; however, based on present losses, using the armamentarium of pest and disease control chemicals available, I believe it to be within the realm of credibility.

In the worst possible situation, if all pesticides were denied at one blow, the situation would be different. Producers and research workers would not have had opportunity to develop or expand bio-control schemes or disperse concentrations of livestock; and if we were to have a warm, wet winter, conducive to expanding populations of pests, I would expect losses in the range of 50%. After the first year or so, as we adjusted to the situation and instituted new management and bio-control practices, we could expect a decline to a level below that seen prior to the advent of the organic pesticides.

We have made great strides in the past 40 years in all areas of management, nutrition, and genetics; our studies have not been as great in nonchemical methods of pest control because we have depended, to a large extent, on chemical pesticides. I feel that were we faced with the loss of pesticides, all of our efforts would be bent on further nonchemical control measures to a much greater extent.

REFERENCES

1. **Ensminger, M. E., Galgan, N. W., and Slocum, W. L.,** Problems and practices of American cattlemen, *Wash. Agric. Exp. Stn. Bull.*, 562, 1955.
2. **Borlaug, N. E.,** Mankind and civilization at another cross-road: in balance with nature — a biological myth, *BioScience*, 22, 41, 1972.
3. **Pimentel, D.,** Extent of pesticide use, food supply, and pollution, *Proc. N.Y. Entomol. Soc.*, 81, 13, 1973.
4. **Lewert, H. V.,** *A Closer Look at the Pesticide Question for Those Who Want the Facts*, The Dow Chemical Co., 1976.
5. **Snelson, J. T.,** The importance of chlorinated hydrocarbons in world agriculture, *Ecotoxicol. Environ. Saf.*, 1, 17, 1977.
6. **Whitten, J. L.,** *That We May Live*, D Van Nostrand, Princeton, 1966, 54.

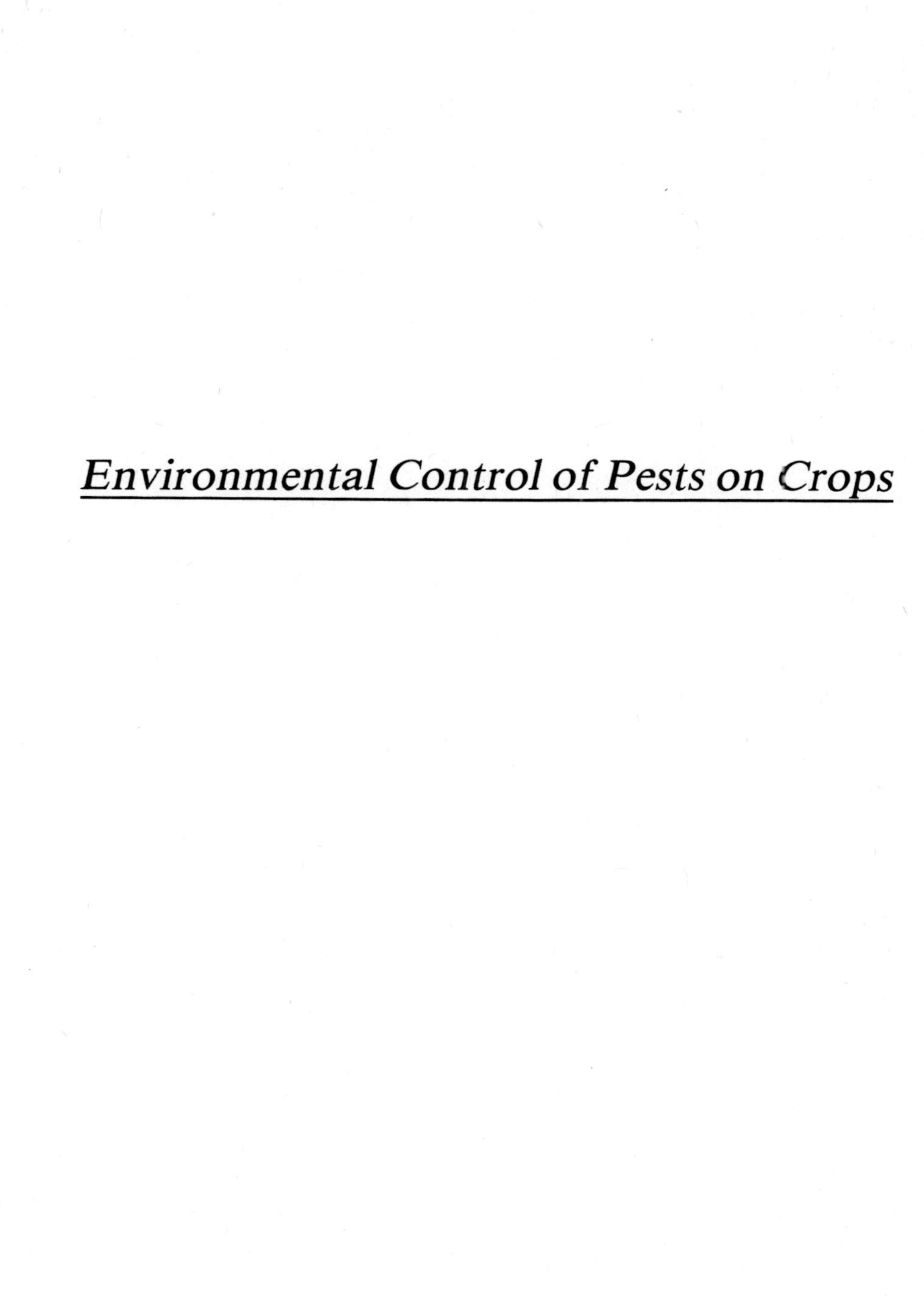

Environmental Control of Pests on Crops

THE ENVIRONMENTAL CONTROL OF INSECTS USING PLANT RESISTANCE

Ward M. Tingey

INTRODUCTION

The phenomenon of plant resistance to insects is a quality that enables a plant to avoid, tolerate, or recover from the effects of oviposition or feeding that would cause greater damage to other genotypes of the same species under similar environmental conditions. Resistance can be further described by four characteristics:

1. Resistance is heritable and controlled by one or more genes.
2. Resistance is relative and can be measured only by comparison with other plant genotypes.
3. Resistance is measurable, i.e., the mechanisms and magnitude of resistance can be qualitatively or quantitatively determined by analysis of insect behavioral and metabolic responses to the host, and by assessment of plant growth and development in response to insect feeding and oviposition.
4. Resistance is variable and can be modified by the physical and biological environment.

For additional discussion of the broad aspects of plant resistance, the reader is referred to the books by Painter,[1] those edited by Rodriguez,[2] van Emden,[3] Maxwell and Harris,[4] and other articles or chapters.[5-8]

HISTORICAL PERSPECTIVES

Long before recorded time, natural selection for environmental fitness led to the appearance of insect-resistant ancestors of modern crop plants. As man began to domesticate these plants for agriculture, unintentional selection for fitness against phytophagous insects probably aided in maintaining his primitive crop.[9] With increasing domestication and more sophisticated forms of pest control, however, much of the genetic diversity for defense against insects was lost owing to a sometimes single-minded quest for fruitfulness and quality.[10]

The first published record of varietal resistance to insect attack appeared in 1782, when an unknown writer in a farm paper described resistance to the Hessian fly, *Mayetiola destructor* (Say) in 'Underhill' wheat.[11] Lindley[12] reported resistance to the woolly apple aphid, *Eriosoma lanigerum* (Hausmann), in the apple cultivar, 'Winter Majetin' fifty years later. In both of these cases, however, resistance was a serendipitous find, as there is no firm evidence for its intentional development.

Perhaps the most striking early example of directed development of resistant cultivars is that of the grape phylloxera, *Phylloxera vitifoliae* (Fitch). This North American aphid was introduced into Europe where it threatened to devastate the French wine industry in the mid-1860s. Remarkable control was achieved in just 10 years following the grafting of susceptible European scions to resistant North American rootstocks.[13]

Despite this spectacular beginning, plant resistance to insects attracted little research interest until the pioneering and systematic programs initiated by Reginald H. Painter and co-workers of Kansas State University in the 1930s.

HOST SELECTION AND UTILIZATION

The discrimination and utilization of host plants by phytophagous insects are governed through a sequential series of responses to key plant stimuli.[14] The generalized host selection and acceptance process involves at least four essential steps: (1) selection of host habitat, (2) host finding and acceptance, (3) host acceptance, and (4) host suitability.[7,15] Plant stimuli play an active role in regulating insect responses in steps 2 to 4. The absence of positive stimuli or the presence of negative stimuli at any step in this sequence limits the probability of sustained insect-plant contact and the potential for plant damage.

INSECT-LIMITING PLANT CHARACTERISTICS

Allelochemics

The chemical profile of the host plant is extremely important in determining its acceptance and utilization by insects.[16-18] Phytochemical stimuli (allelochemic factors) can be classified into two major categories, depending on their functional role in insect behavior and physiology.[19,20] *Allomones* mediate negative responses from insects and tend to reduce the probability of host contact and utilization. These include repellants, oviposition and feeding deterrants, and toxicants. *Kairomones,* on the other hand, are disadvantageous to the plant, as they increase the likelihood of acceptance and utilization. These chemical stimuli promote host recognition, oviposition, and feeding, and include attractants, arrestants, excitants, and stimulants.

Morphological Features

Morphological characteristics are frequently associated with host suitability. Variation in foliage size, shape, color, and the presence or absence of nectar-secreting glands may regulate host acceptance and utilization, whereas pubescence and tissue hardness limit insect mobility by acting as structural barriers.[21-24]

Nutritional Factors

The nutritional composition of plant tissue also acts in regulating insect performance, although most genotypes of a host species differ little in types of essential nutrients and growth factors.[25] Genotypes may differ significantly, however, in their proportion of essential nutrients and, in general, recent workers have emphasized the quantitative aspects of food composition with regard to host selection and utilization.[26,27]

FUNCTIONAL CATEGORIES OF RESISTANCE

Although many terms have been proposed to classify mechanisms of resistance, the terms proposed by Painter,[1] i.e., *nonpreference, antibiosis, tolerance,* have had the greatest acceptance. In this system, the modality of resistance is described by the functional relationship between the plant and the insect. From a causal standpoint, these terms are not mutually exclusive, because plant factors that condition nonpreference can also mediate antibiosis.

Impaired Behavior

To be successful in evolutionary time, a phytophagous insect must discriminate among the vast array of plants in its environment, and select those most suitable as nutritional substrates for itself and its offspring. The behavioral component of host

Table 1
INDUCTION OF FEEDING BY THE SPOTTED CUCUMBER BEETLE ON COTYLEDONARY TISSUE OF THE RESISTANT CULTIVAR 'EARLY GOLDENBUSH SCALLOP' (EGBS), *CUCURBITA PEPO*, BY TOPICAL APPLICATION OF CUCURBITACINS

Treatment	Mean damage rating
Early Golden Bush Scallop (total cucurbitacins = 0.08 mg/g)	0.40
EGBS + solvent	.50
EGBS + Cucurbitacin A (0.05%)	2.35
+ Cucurbitacin A (0.10%)	2.90
EGBS + Cucurbitacin I (0.05%)	2.25
+ Cucurbitacin I (0.10%)	2.80

Adapted from Sharma, G. C. and Hall, C. V., Influence of cucurbitacins, sugars, and fatty acids on cucurbit susceptibility to spotted cucumber beetle, *J. Am. Soc. Hort. Sci.*, 96, 675, 1971. With permission.

Table 2
ATTRACTION OF THE SPOTTED CUCUMBER BEETLE TO FRUITS OF SIX SPECIES OF CUCURBITS DIFFERING IN LEVELS OF CUCURBITACINS

Species	Mean no. adults attracted in 30 min	Concentration of cucurbitacins (mg/g fresh wt)
Cucumis dipsaceus	56.4	0.51
C. longpipes	29.4	.43
C. anguria	1.0	.03
C. melo	0.8	not detectable
Cucurbita foetidissima	57.9	.56
C. pepo	3.5	.04

Adapted from Sharma, G. C. and Hall, C. V., *Environ. Entomol.*, 2, 154, 1973. With permission.

selection and utilization is the fundamental criterion in Painter's definition of nonpreference. That is, a *nonpreferred* plant is resistant based on its ability to lessen the probability of recognition and utilization by insects. This type of resistance can be conferred by either allelochemic or structural factors. The cucurbitacins, for example, a class of tetracyclic terpenes, act as specific attractants and feeding incitants for a major pest of cucurbits, the spotted cucumber beetle, *Diabrotica undecimpunctata howardi* Barber[28-30] (Table 1). Cucurbits that lack or have low levels of specific individual cucurbitacins are nonpreferred; they attract fewer beetles and suffer less feeding damage than do susceptible genotypes[31] (Table 2). For numerous additional examples of resistance factors that impair insect behavior, the reader is referred to the excellent article by Hedin et al.[32]

Nonpreference can also be conferred by structural plant factors. The cotton bollworm, *Heliothis zea* (Boddie), for example, prefers villous substrates for oviposition. Cotton genotypes lacking leaf trichomes incur much less egg-laying and subsequent damage than do normally pubescent genotypes.[33]

Table 3
EFFECT OF RESISTANT AND SUSCEPTIBLE SORGHUM SELECTIONS ON DEVELOPMENTAL BIOLOGY OF BIOTYPE C OF THE GREENBUG

Genotype	Nymphal weight (mg)	Fecundity (nymphs/♀)	Pre-reproductive period (days)	Reproductive life (days)
Shallu grain	0.08	9.8	9.2	12.2
IS 809	.09	11.5	9.2	15.0
PI 264453	.14	20.2	7.6	15.2
BOK-8	.17	44.9	6.4	20.0

Adapted from Schuster, D. J. and Starks, K. J., *J. Econ. Entomol.*, 66, 1131, 1973. With permission.

From a practical standpoint, it is important to distinguish between two types of nonpreference, i.e., nonpreference expressed in the absence of other hosts, and non-preference expressed only in the presence of other suitable hosts.[34] The limitations of the latter are obvious in view of the trends for monoculture, genetic uniformity, and freedom from weed species in modern agricultural systems. Perhaps because this modality of resistance is particularly subject to the cultural environment, nonpreference has been accorded secondary importance in breeding programs.

Finally, the term "nonpreference" has been semantically criticized, because it implies an insect response rather than an effect of the plant on an insect. Recently, Kogan and Ortman[35] have proposed term "antixenosis," which more accurately reflects the functional influence of the plant on insect behavior.

Impaired Metabolism

The growth and development of insects are strongly influenced by the quality and abundance of their host plants. Among the most important phytochemical factors influencing host suitability are the presence of essential nutrients and growth factors in the proper balance, and the absence of toxic agents, antimetabolites, and inhibitory enzymes. The disruption of normal metabolic processes is the basic criterion in Painter's definition of *antibiosis*. That is, ingestion of plants resistant by antibiosis mechanisms leads to impairment of normal metabolic physiology. Typical symptoms of antibiosis resistance include death of early instars; lengthened developmental periods; reduced growth rates and fecundity; undersized immature and adult stages; morphological malformations; shortened adult longevity and reproductive period; increased mortality during pupal and resting stages; and restlessness and other abnormal behavior (Table 3).

Antibiosis is the most frequently exploited modality of resisance and is typically assessed by analysis of growth, development and survival on the host, extraction and identification of pest-active factors, and subsequent confirming bioassays. Numerous plant chemicals resulting from secondary metabolism have been shown to condition antibiosis.[36] Some of the best known and most completely documented of these include the cyclic hydroxamic acids of maize[37] (Figure 1), gossypol and related polyphenolics of cotton,[38,39] steroidal glycosides of potato[40-42] (Figure 2), and the saponins of alfalfa.[43]

Primary plant metabolites, particularly imbalance of sugars and amino acids, have been shown to confer nutritional antibiosis. In pea, for example, resistance to the pea aphid, *Acrythosiphon pisum* (Harris), is associated with decreasing levels of amino acids and increasing sugar content,[44,45] whereas fecundity of the brown planthopper,

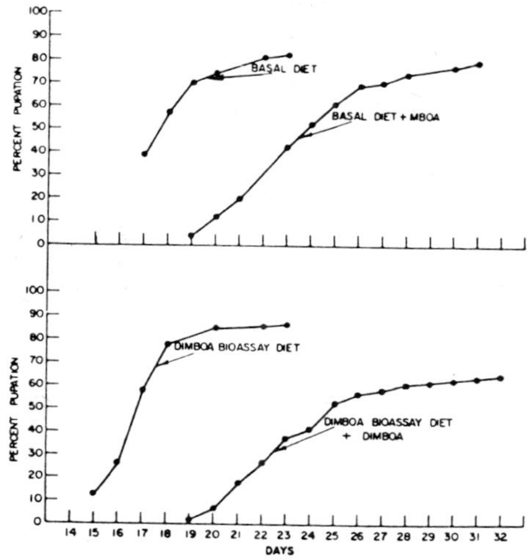

FIGURE 1. Effect of the benzoxazinones, 6-MBOA and DIMBOA, on pupation of the European corn borer reared on artificial diet. (Adapted from Klun, J. A., Tipton, C. L., and Brindley, T. A., *J. Econ. Entomol.*, 60, 1529, 1967. Courtesy of Klun, J. A., and the Entomological Society of America.)

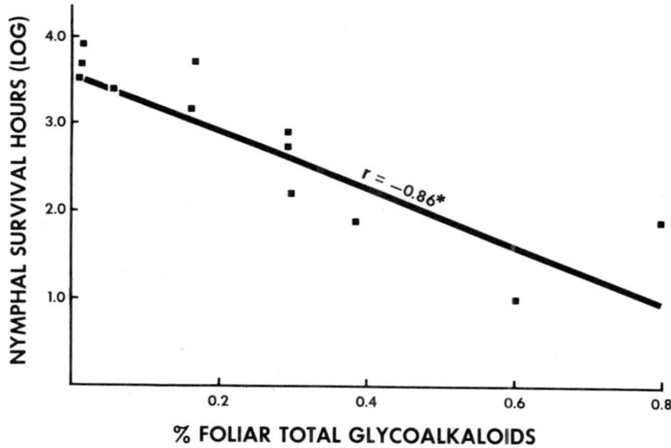

FIGURE 2. Relationship between survival of potato leafhopper nymphs and levels of potato glycoalkaloids in artificial diet. (From Raman, K. V., Tingey, W. M., and Gregory, P., *J. Econ. Entomol.*, 72, 337, 1979. Copyright of the Entomological Society of America, reproduced by permission.)

Nilaparvata lugens Stål, is reduced on rice cultivars deficient in the amino acid, asparagine.[6]

Structural Barriers

Resistance conditioned by morphological factors is sometimes difficult to categorize according to the traditional definitions of antibiosis and nonpreference, because it may not be strictly dependent on ingestion of plant tissues or on an adverse behavioral response to plant stimuli. Morphological features that physically limit oviposition, feeding, and mobility were collectively referred to as *phenetic resistance factors* by Kogan,[8] and include pubescence; tissue hardness, shape, and configuration; and

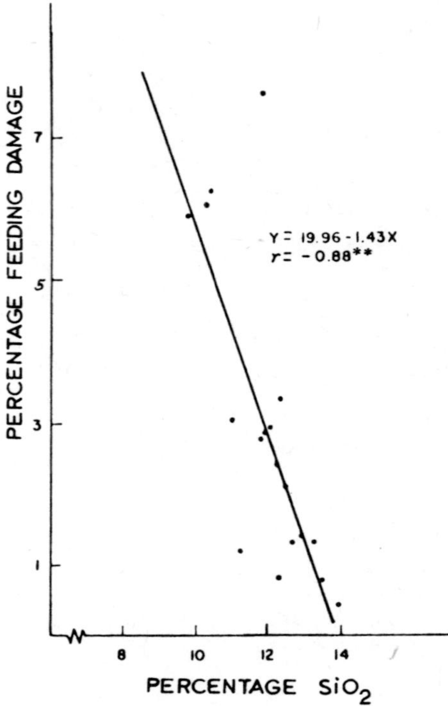

FIGURE 3. Relationship between silica content of rice stems and feeding damage of the Asiatic rice borer for 18 rice accessions. (Adapted from Djamin, A. and Pathak, M. D., *J. Econ. Entomol.*, 60, 347, 1967. Courtesy of M. D. Pathak and the Entomological Society of America.)

gummy exudates. Husk tightness and length, for example, have long been associated with resistance of maize to the corn earworm, *Heliothis zea* (Boddie);[46-48] increased hardness of stalk tissues caused by greater concentrations of silica and lignin is responsible for resistance of rice to the striped rice borer, *Chilo suppressalis* Walker[49] (Figure 3), and of sugarcane to the sugarcane borer, *Diatraeasaccharalis* (Fabricius).[50] The density of pith and node tissues in stems of *Cucurbita* species and wheat is a major factor in resistance to the squash vine borer, *Melittia cucurbitae* (Harris),[51] and wheat stem sawfly, *Cephus cinctus* Norton,[52] respectively. These morphological traits confer resistance by limiting penetration and tunneling of host tissues.

Of all structural resistance phenomena, perhaps the most common is that involving plant pubescence.[53] Length, type, and density of epidermal hairs (trichomes) have been implicated in insect resistance of at least eight crops (Tables 4, 5). Two general types of trichomes have been shown to confer resistance, secretory or glandular trichomes and nonglandular trichomes. Exudates of glandular trichomes confer resistance by physical entanglement (Figure 4), or through the action of toxicants and other allelochemic factors in the secretions. Nonglandular trichomes, on the other hand, interfere with host acceptance and utilization by spatially limiting insect access to vulnerable tissues or by capture and impalement (Figure 5).

From a practical standpoint, morphological resistance mechanisms are particularly valuable because they frequently constitute a "first line of defense" against phytophagous insects by limiting feeding and oviposition soon after host-pest contact. In addi-

Table 4
EXAMPLES OF INSECT RESISTANCE CONFERRED BY NON-GLANDULAR PUBESCENCE

Crop	Pest: Common name	Pest: Latin name	Special effect on pest	Ref.
Cotton	Leafhopper	*Empoasca fascialis*	Probably spatial impairment of feeding and oviposition	54
	Two-spotted spider mite	*Tetranychus urticae*	Reduced settling and feeding	55
Wheat	Cereal leaf beetle	*Oulema melanopus*	Reduced egg laying, increased egg and larval mortality	56, 57
Soybean	Potato leafhopper	*Empoasca fabae*	Increased egg and larval mortality	58
Common bean	Potato leafhopper	*Empoasca fabae*	Capture and impalement on hooked trichomes	59
	Cowpea aphid	*Aphis craccivora*	Capture and impalement on hooked trichomes	60
Alfalfa	Potato leafhopper	*Empoasca fabae*	Probably spatial impairment of feeding and oviposition	61

Table 5
EXAMPLES OF INSECT RESISTANCE CONFERRED BY GLANDULAR PUBESCENCE

Host species	Pest	Specific effect on pest	Ref.
Solanum berthaultii, S. polyadenium, S. tarijense	Green peach aphid *(Myzus persicae)*, Potato aphid *(Macrosiphum euphorbiae)*, Colorado potato beetle *(Leptinotarsa decemlineata)*, two-spotted spider mite *(Tetranychus urticae)*	Entrapment and impaired mobility	62
S. berthaultii, S. polyadenium	Potato leafhopper *(Empoasca fabae)*	Entrapment, impaired feeding and mobility	63
S. pennellii	Potato aphid *(Macrosiphum euphorbiae)*	Entrapment	64
Lycopersicon sp.	Potato aphid *(Macrosiphum euphorbiae)*	Entrapment	64
	Spider mites *(Tetranychus urticae, T. cinnabarinus)*	Entrapment, contact toxicity, repellency	65, 66
	Greenhouse whitefly *(Trialeurodes vaporariorum)*	Entrapment	67
Nicotiana sp.	Green peach aphid *(Myzus persicae)*, tobacco hornworm *(Manduca sexta)*	Contact toxicity	68, 69
	Two-spotted spider mite *(Tetranychus urticae)*	Entrapment, contact toxicity	70
Medicago sp.	Alfalfa weevil *(Hypera postica)*	Contact toxicity	71

tion, phenotypes bearing morphological resistance factors are relatively easily identified in screening programs. Finally, resistance factors that physically impair normal behavior or that entangle the pest can be expected to have a long period of usefulness since profound changes in pest anatomy and behavior would be necessary for the development of host-specific biotypes capable of overcoming the resistance. Leaf pubescence, for example, was utilized in the breeding of leafhopper-resistant cotton cultivars as early as 1925, and continues today as an effective control strategy.[54,55]

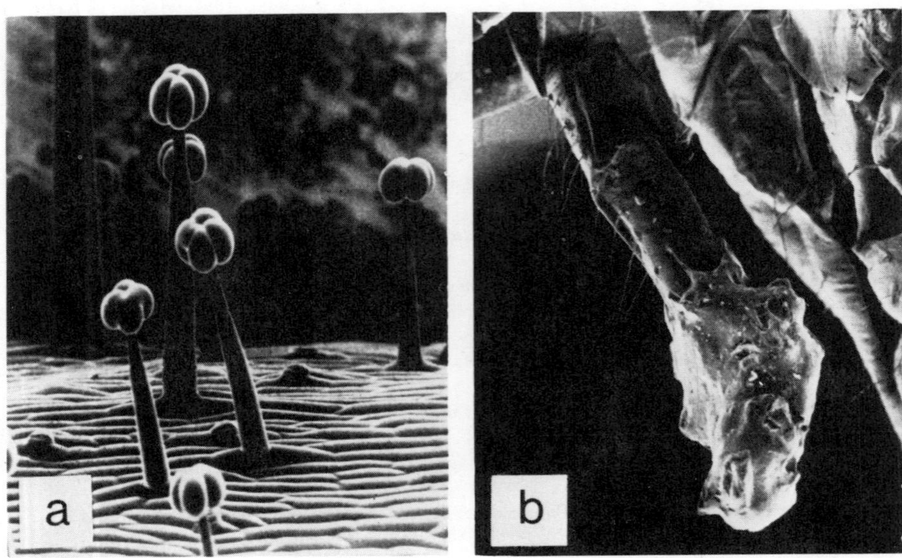

FIGURE 4. (A) Glandular trichomes on petiole of the wild potato species, *Solanum polyadenium*. (Magnification × 100.) (B) Trichome exudate on labium of potato leafhopper nymph. (Magnification × 106.) (From Tingey, W. M. and Gibson, R. W., *J. Econ. Entomol.*, 71, 856, 1978. Copyright Entomological Society of America, reproduced by permission.)

Tolerance

Unlike resistance mechanisms that interfere with insect behavior and metabolism, *tolerance* provides a plant the ability to produce satisfactory yield in the presence of a pest population that would damage a susceptible or nontolerant plant. Tolerant hosts do not depress or limit pest populations, nor do they provide selection pressure that can lead to the development of insect races or biotypes capable of overcoming resistance. The phenomenon of tolerance is generally the cumulative result of interacting but poorly understood plant growth processes, including general vigor, inter- and intraplant compensatory growth, wound compensation, mechanical strength of tissues and organs, and nutrient and growth regulator partitioning.

An excellent example of tolerance was reported by Zuber et al.,[72] who studied the injury to various maize genotypes by larval feeding of the western corn rootworm, *Diabrotica virginifera* LeConte. Larvae of this pest feed on roots of the maize plant, causing damage ranging from decreased ability to utilize soil moisture and nutrients to reduced anchorage and subsequent lodging. These workers showed that genotypes differed in their ability to repair and replace damaged root systems, as measured by comparison of infested and noninfested plots for root volume and resistance to uprooting. Tolerant genotypes actually produced greater root volume in the presence of corn rootworm feeding than they did when noninfested, whereas root volume of infested susceptible genotypes was reduced as much as 20% (Table 6). Obviously, such compensatory capacity is highly desirable when present in genotypes with acceptable agronomic qualities. Tolerance is a complex phenomenon, however, involving many interacting plant growth processes, and because of this complexity has not been exploited as frequently as other resistance modalities.

GENETIC NATURE OF RESISTANCE

Knowledge of the genetic nature of resistance can be of great value in breeding programs by providing a quantitative basis for recombination and selection, and by iden-

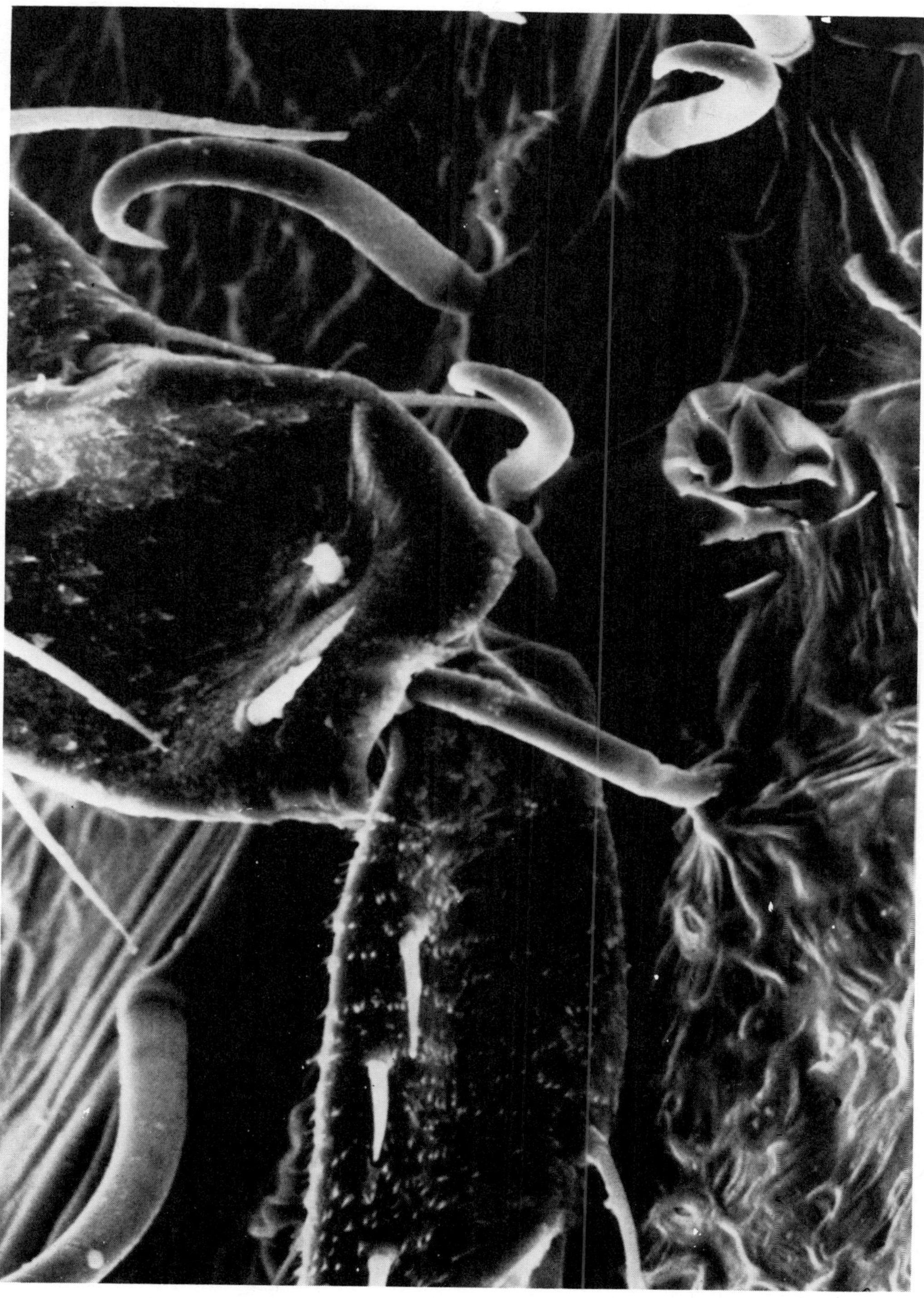

FIGURE 5. Leaf trichome of the common bean, *Phaseolus vulgaris* L., embedded in membranous tissue between leg segments of the potato leafhopper, *Empoasca fabae* (Harris) (Magnification × 970). (From Pillemer, E. A. and Tingey, W. M., *Science,* 193, 482, 1976. Copyright American Association for the Advancement of Science, reproduced by permission.)

Table 6
RELATIONSHIP BETWEEN WESTERN CORN ROOTWORM INFESTATION AND ROOT GROWTH CHARACTERISTICS OF FOUR MAIZE SINGLE CROSSES GROWN IN INFESTED AND CONTROL (UNINFESTED) PLOTS

	Root volume (mℓ)			Root pulling resistance (kg)		
Single cross	Infested	Control	Infested/Control	Infested	Control	Infested/Control
Mo 22 × T8	252	224	1.13	148	150	0.99
Mo 940 × Ky 27	256	248	1.03	152	180	.84
Mo 940 × L317	317	347	.91	167	187	.89
Mo 22 × C164	439	548	.80	170	215	.79

Adapted from Zuber, M. S., Musick, G. J., and Fairchild, M. L., *J. Econ. Entontol.*, 64, 1514, 1971. With permission.

tifying those gene products (resistance factors) most likely to be stable against the genetic plasticity of the pest. Two general classes of inheritance have been reported for insect resistance traits, *oligogenic* (major gene), and *polygenic* (minor gene).

Oligogenic resistance is controlled by one to three major nuclear genes, whose individual effects are relatively easy to detect. Oligogenic resistance is usually dominant and relatively stable from environmental influence, but it is generally thought to be less stable against the development of host-specific biotypes than polygenic resistance.[73]

Polygenic resistance is controlled by many nuclear genes, each of small effect. The inheritance of polygenic resistance is frequently complex and may be associated with such quantitative traits as plant vigor and yield. The complex nature of polygenic resistance provides considerable buffering against host-specific biotypes, although the environment may modify the expression of polygenetically controlled resistance.[74]

Cytoplasmically inherited factors are of considerable importance in resistance to some plant pathogens, but have not been reported for insect resistance.

Additional information on specific inheritance patterns for several types of insect resistance is summarized by Pathak,[6] Gallun,[75] and Russell.[76]

ANALYSIS AND MEASUREMENT OF RESISTANCE

Insect Responses

An immediate practical objective in breeding resistant crop plants is to determine the degree of control provided by a resistant host. At initial stages of assessment, most workers concentrate on identifying those plants that support the smallest populations of the target pest. Sampling techniques vary depending on the specific crop, pest, and researcher, but the excellent information provided by Southwood,[77] Ruesink and Kogan,[78] and Dahms[79] covers most situations.

Initial assessment is frequently made in free-choice field environments because of the need to screen large populations of plants. The use of natural cropping environments is also desirable because the expression and magnitude of resistance is subject to modification by environmental factors. As pointed out later in this chapter, dependence on artificial environments in the screening and measurement of resistance may lead to results inconsistent with those expressed under natural conditions. One important limitation in use of field assessment methods, however, is the unpredictive nature of insect phenology and abundance. This problem can frequently be corrected by external manipulation or confinement of the pest population to achieve the optimum timing and magnitude of infestation. External methods of pest manipulation include mass laboratory rearing or field collection followed by release; trap crops and pre-infested susceptible hosts to concentrate and magnify the pest population; use of attractants and baits; use of selective insecticides to eliminate natural enemies and competing species; and use of cultural practices favorable to the pest.

Following the identification of plant genotypes expressing reduced infestation, their specific impact on oviposition, feeding activity, growth rates, developmental time, and survival is commonly assessed using no-choice confinement techniques.

Plant Responses

A final and vital element in the selection and development of resistant genotypes is assessment of their reaction to insect feeding injury. This is commonly accomplished by direct measurement of feeding damage or defoliation, and assessment of altered morphology, growth patterns, yield, and quality.

In some cases, quantitative knowledge of plant damage and pest suppression is sufficient for identification of sources of resistance and subsequent development of resistant cultivars. Ultimately, however, knowledge of the specific plant traits conditioning resistance will be useful in understanding the genetics of resistance and the precise mode of action, and in refinement of selection and breeding procedures.

Characterization of Specific Resistance Factors

Initial leads regarding the role of specific plant factors in resistance are frequently obtained by the analysis of correlations between pest performance or plant damage and quantitative or qualitative variability among genotypes in chemical and morphological characteristics. In the case of allelochemic factors, for example, plant tissues are commonly extracted and fractions of suspected activity then tested for activity against the target pest by use of behavioral or developmental bioassays. Analysis of behavior is frequently made using choice-discrimination and activity-monitoring methods; metabolic activity can be assessed by incorporating suspect fractions in artifical or natural feeding substrates. As additional information regarding the specific physical or chemical basis for resistance becomes availablė, the input of specialists in the areas of insect and plant physiology, insect nutrition and toxicology, and natural products chemistry becomes essential.

FACTORS AFFECTING THE EXPRESSION AND PERMANENCE OF RESISTANCE

The stability and expression of plant resistance to insects is the result of interaction between the genetic characteristics of the plant and insect, and certain environmental factors. The dynamic interplay of these three systems can lead to shifts in both the magnitude and permanence of resistance.

Influence of the Physical Environment

Among the most important climatic, edaphic, and cultural factors active in modification of resistance are temperature, light intensity, soil moisture and fertility, relative humidity, and chemical pesticides. In general, the effect of these factors on expression of resistance is thought to involve change in fundamental plant physiological processes, which in turn leads to alteration in levels of allelochemic resistance factors or shifts in nutritional suitability of the host.

Temperature

Exposure to unusually high or low temperatures may lead to loss in expression of resistance, although the effect is generally reversible after conditioning at the opposing temperature extreme. The best examples of low-temperature-induced loss of resistance

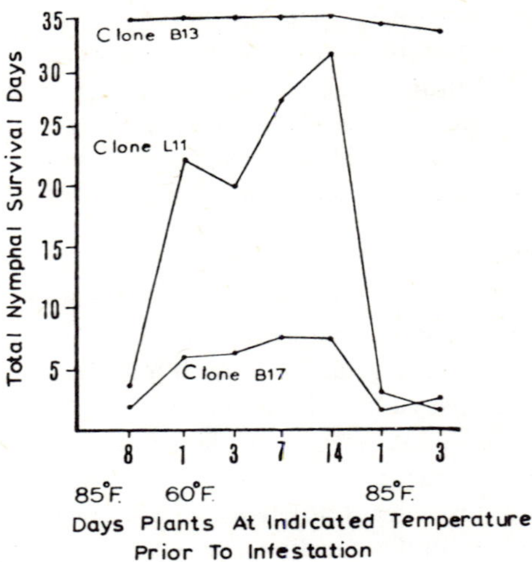

FIGURE 6. Effect of temperature on magnitude and expression of resistance to the spotted alfalfa aphid in 3 alfalfa clones. (Adapted from Isaak, A., Sorensen, E. L., and Painter, R. H., *J. Econ. Entomol.*, 58, 140, 1965. Courtesy of E. L. Sorensen and the Entomological Society of America.)

Table 7
RELATIONSHIP BETWEEN STEM SOLIDNESS, LIGHT INTENSITY, AND RESISTANCE TO WHEAT STEM SAWFLY IN THREE SELECTIONS OF SOLID-STEMMED WHEAT

	Stem solidness rating		Percentage infested stems with dead eggs		Percentage infested stems with dead larvae	
Selection	**Unshaded**	**Shaded**	**Unshaded**	**Shaded**	**Unshaded**	**Shaded**
Rescue	26	18[a]	30	11[a]	68	43[a]
H4191	27	21[a]	31	12[a]	70	34[a]
H46146	25	21	34	15[a]	70	40[a]

[a] Significantly less than that for unshaded plants at P = 0.05.

Adapted from Roberts, D. W. A. and Tyrrell, C., *Can. J. Plant Sci.*, 41, 457, 1961. With permission.

are those reported for certain alfalfa genotypes resistant to the spotted alfalfa aphid (*Therioaphis maculata* (Buckton))[80,81] (Figure 6) and pea aphid,[81] and for some sorghum genotypes resistant to the greenbug, *Schizaphis graminum* (Rondani).[82] Progressive loss of resistance to the Hessian fly in wheat, on the other hand, has been reported at temperatures in excess of 18°C.[83,84]

Light Intensity

Reduction in light intensity leads to loss in resistance of some wheat cultivars to the wheat stem sawfly by decreasing stem solidness and density[85] (Table 7). Shade-induced loss of resistance has also been reported in sugar beet and potato genotypes resistant to the green peach aphid, *Myzus persicae* (Sulzer),[86] and Colorado potato beetle, *Leptinotarsa decemlineata* (Say),[87] respectively. In the latter case, shading was associated with reduced foliar levels of steroidal glycosides, several of which are known to limit feeding and development of the Colorado potato beetle.[40]

Soil Fertility

Shifts in levels of soil nutrients also influence the expression of resistance. In alfalfa, for example, excess levels of nitrogen and magnesium, and deficient levels of potassium and calcium, have been associated with loss of resistance to the spotted alfalfa aphid.[88]

Induced Resistance

In contrast to resistance modified by the environment but primarily controlled by genetic factors, induced resistance is governed largely by external factors of the environment. Cultural practices that alter levels of soil nutrients and moisture, and the use of certain agricultural chemicals, have been implicated in many cases involving induced resistance of normally susceptible genotypes. Deficient levels of soil nitrogen or phosphorus, and excess levels of potassium, for example, have been reported to limit insect performance on a number of crop plants.[89-94]

In recent years, the induction of resistance by externally applied chemical agents has attracted considerable attention. The plant growth retardant, (2-chloroethyl) trimethylammonium chloride *(CCC),* for example, limits fecundity or survival of the cabbage aphid, *Brevicoryne brassicae* (L.), bean aphid, *Aphis fabae* (Scopoli), oleander aphid, *A. nerii* (Boyer), currant aphid, *A. varians* (Patch), and green peach aphid on treated susceptible cultivars of Brussels sprout, broad bean, oleander, black currant, and Brussels sprout, respectively.[95-99] Other agricultural chemicals implicated in induction of resistance include the plant-growth retardant, SADH,[100] and two growth stimulants, gibberellic acid and EBNU.[96,101]

The mechanism of induced resistance is not clearly understood, but most studies suggest that the nutritional quality of treated plants is affected. Reduced levels of such essential nutrients as amino acids and soluble carbohydrates have been frequently observed following plant exposure to inducing agents. Additional examples and expanded discussion of these phenomena are presented by Rodriguez,[102] El-Tigani,[103] van Emden,[104,105] and Singh.[106]

Intrinsic Biological Factors.

Host-Specific Biotypes

Biotypes are discrete populations capable of utilizing and damaging plant genotypes resistant to other populations of the same pest species. Biotypes arise as a consequence of the selection pressure imposed by a resistant host, and represent pest genotypes with superior fitness. The presence of host-specific biotypes poses one of the most serious limitations to permanence of resistance. The occurrence of biotypes has been documented in at least nine insect species, of which the Aphididae predominate (Table 8).

The most complete understanding of the relationship between insect resistance and host-specific biotypes is that involving the Hessian fly and wheat. Nine biotypes of the Hessian fly have been recognized,[116] and each gene for resistance in the wheat plant is countered by a single gene for virulence in the pest.[112] Early understanding of this gene-for-gene relationship has aided researchers in predicting and preparing for the appearance of previously unrecognized biotypes.[117]

Although the relative stability of monogenic (single gene) as opposed to polygenic (multiple gene) resistance to insects is controversial, experience with the Hessian fly and plant pathogens suggests that polygenic resistance is likely to be more effective in combating the genetic plasticity of pests than is resistance controlled by a single gene.[74] This evidence emphasizes the need to identify and characterize as many different

Table 8
INSECT PESTS FOR WHICH HOST-SPECIFIC BIOTYPES HAVE BEEN REPORTED

Species	Number of biotypes	Crop(s)	Ref.
HOMOPTERA			
Acyrthosiphon pisum (pea aphid)	9	Alfalfa, pea	108
Amphoraphora rubi (rubus aphid)	4	Raspberry	109
Dysaphis devecta (rosy apple aphid)	3	Apple	113
Nilaparvata lugens (brown planthopper)	3	Rice	114
Phylloxera vitifoliae (grape phylloxera)	2	Grapes	115
Rhopalosiphum maidis (corn leaf aphid)	5	Maize, sorghum	111
Schizaphis graminum (greenbug)	3	Sorghum, wheat	110
Therioaphis maculata (spotted alfalfa aphid)	6	Alfalfa	107
DIPTERA			
Mayetiola destructor (Hessian fly)	9	Wheat	112

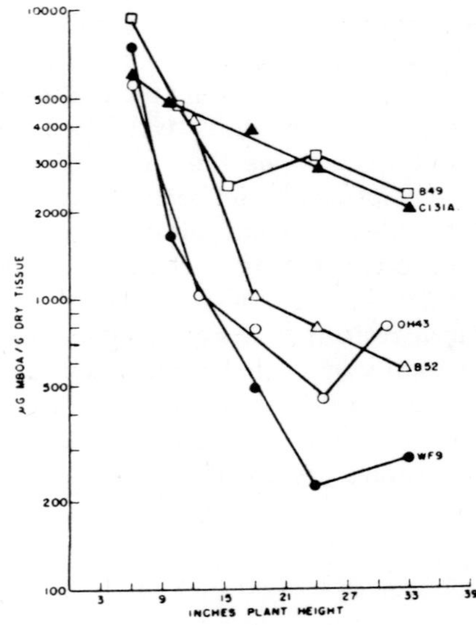

FIGURE 7. Relationship between plant growth stage and levels of 6-MBOA (a breakdown product of DIMBOA) in whorl tissues of five maize genotypes. (Redrawn from Klun, J. A. and Robinson, J. F., *J. Econ. Entomol.*, 62, 214, 1969. Courtesy of J. A. Klun and the Entomological Society of America.)

sources of resistance and their genetics as possible. It should be noted, however, that breeding for polygenic resistance factors is generally more complex and time-consuming than for resistance factors governed by single genes.

Plant Age

A plant's responses to feeding injury and its suitability as food for insects may vary with its physiological age, leading to shifts in expression of resistance. Resistance of maize to leaf-feeding of the European corn borer, *Ostrinia nubilalis* (Hübner), for example, is conferred by the cyclic hydroxamic acid, 2,4-dihydroxy-7-methoxy-1,4-

benzoxazine-3-one (DIMBOA).[37] Leaf tissues of most genotypes contain high DIMBOA levels at the seedling stage and are well defended against this pest. Later in plant development, however, DIMBOA levels decline but do so much more rapidly in susceptible than in resistant genotypes[118] (Figure 7). Thus, maximum expression of resistance is achieved during the midwhorl growth stage, which coincides with the normal phenology of first-brood infestation in the midwestern U.S.

Age-dependent expression of resistance has also been reported in alfalfa[119] and species of *Nicotiana*[120] to the spotted alfalfa aphid and green peach aphid, respectively.

CONSIDERATIONS IN IMPLEMENTING PLANT RESISTANCE PROGRAMS

Interdisciplinary Inputs

The disciplines of plant breeding and entomology form the minimum structural unit in most successful plant-resistance projects. In some instances, the interaction of specialists in these two disciplines alone has led to the development of economically competitive resistant cultivars. A comprehensive plant-resistance program, however, should include, in addition to screening and breeding objectives, (a) characterization of resistance mechanisms, (b) identification of physical and chemical resistance factors, and (c) assessment of the genetic basis for resistance. The input of specialists in plant and insect physiology, genetics, and natural-products chemistry can contribute significantly to the previous objectives and to the ultimate development of competitive cultivars possessing effective and stable resistance.

Time of Development

Although sources of resistance have been identified and cultivars placed in commercial production in as little as 3 years of development,[121] such examples are the exception rather than the rule. In many cases, resistance is identified in unadapted stocks of unacceptable agronomic quality, and considerable time is required for integration of resistance traits with those for desirable agronomic properties. Relatively long-term support by funding agencies is often a critical factor in the ultimate success of plant-resistance programs.

Host-Specific Biotypes

The appearance of host-specific biotypes parallels that of insecticide-resistant pest populations, and poses special problems in assigning priority to selection and development of specific resistance modalities. The experience to date, however, indicates that limitations posed by biotype development can be minimized by special research emphasis on identification of alternate resistance sources including tolerance, analysis of the genetic basis for resistance, and expanded use of polygenic resistance mechanisms. Additional knowledge in these crucial areas, coupled with the lack of other ecologically compatible control strategies, should guarantee increasingly greater acceptance of plant resistance in pest management.

Conflicting Resistance Factors

Differential Pest Reactions

Some plant traits confer resistance to one pest, but enhanced susceptibility to others. In cotton, for example, the frego bract trait mediates resistance to the boll weevil, *Anthonomus grandis* Boheman,[122] but magnifies susceptibility to plant bugs.[123] Pubescent cotton cultivars are well defended against leafhopper infestation,[54] but particularly attractive to the cotton bollworm for oviposition.[33]

Non-Specific Resistance

Some resistance factors are broadly toxic or interfere with processing of the crop for use by man and animals. Gossypol and related polyphenolic compounds in the

Table 9
PERCENTAGE EARWORM DAMAGE-FREE EARS OF RESISTANT AND SUSCEPTIBLE SWEET CORN HYBRIDS USING VARYING NUMBERS OF APPLICATIONS OF GARDONA® AT 0.5 lb AI/A

	Hybrid	
Number of applications	471-U6 × 81-1(R)	Stowell's EG (S)
7	74.4 a	47.3 c
5	64.5 b	35.0 d
3	60.1 bc	24.2 e
2	52.1 c	23.4 e
1	50.3 c	10.3 f
0	29.7 de	6.6 f

Adapted from Wiseman, B. R., Harrell, E. A., and McMillian, W. W., *Environ. Entomol.*, 2, 919, 1973. With permission.

cotton plant confer resistance to the cotton bollworm and other pests,[38] but are toxic to nonruminant vertebrates at pest-active levels, unless removed during processing of cottonseed oil and meal.[124] Steroidal glycosides (glycoalkaloids) of potato act in defense against the Colorado potato beetle and the potato leafhopper, *Empoasca fabae* (Harris),[40,41] but are hazardous to man at the levels needed for pest suppression.[42,125] The strong correlation between levels of these broadly toxic allelochemics in foliage and tubers limits their practical usefulness as resistance factors.

Another problem posed by nonspecific resistance mechanisms is possible interference with biological control agents. Trichome exudates from many solanaceous crop species confer resistance to a wide variety of small, soft-bodied pests, but may limit the effectiveness of small predacious and parasitic species that come into physical contact with the host plant of their prey.[126,127]

PLANT RESISTANCE AND PEST MANAGEMENT

Primary Control Strategy

Genetic resistance is an ideal protective strategy against insects for which other control measures are inadequate, and for those crops of relatively small value per unit area or with narrow limits of insecticide-residue tolerance. Outstanding successes in use of plant resistance as a principal control method have been achieved for Hessian fly and wheat stem sawfly in wheat, European corn borer in maize, greenbug in barley, and spotted alfalfa aphid in alfalfa.[121,128]

Supplement to Chemical, Cultural, and Biological Controls

Resistant cultivars offer a number of advantages for use in integrated management of arthropod pests. One of the most significant benefits is compatibility with insecticidal controls and the prospect of reduced insecticide input. McMillian et al.,[129] for example, demonstrated that the corn earworm-resistant maize hybrid, 471-U6 × 81-1, provided more than twice the level of control as that of 7 lb active ingredient of insecticide applied to a susceptible cultivar, resulting in a savings of 3 lb of insecticide per acre (Table 9).[130] In another case, Cuthbert and Jones[131] reported that resistance of sweet potato to collective damage by a soil-insect complex was at least equal to that provided by 2.24 kg/ha application of a standard soil insecticide. Even at rates of 4.48 kg/ha, damage to the susceptible cultivar, 'Goldrush', exceeded that of the untreated resistant line, "W-13".

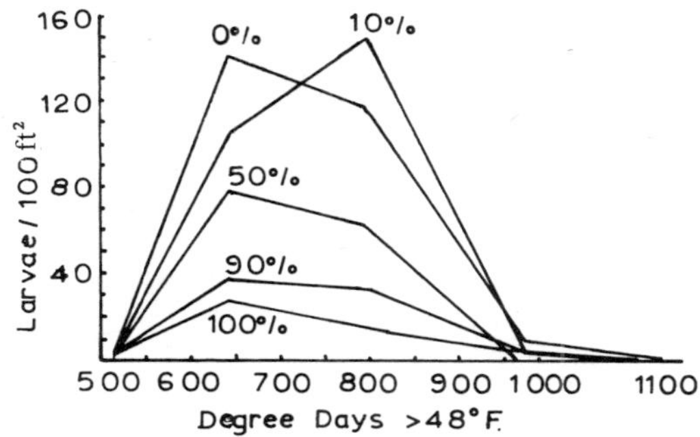

FIGURE 8. Larval densities of the cereal leaf beetle in pure and mixed plots of resistant and susceptible wheat cultivars. (Redrawn from Casagrande, R. A. and Haynes, D. L., *Environ. Entomol.*, 5, 153, 1976. Courtesy of R. A. Casagrande and the Entomological Society of America.)

An intriguing potential for integration of resistance and cultural control practices was provided by Casagrande and Haynes,[132] who compared mixed vs. pure stands of resistant and susceptible wheat cultivars for damage by the cereal leaf beetle, *Oulema melanopus* L. Their results indicate that pure stands of a resistant cultivar may not always be necessary to achieve economical control on a large planting (Figure 8). This strategy could have practical application in those crops for which resistant cultivars may be somewhat inferior to susceptible cultivars in intrinsic agronomic qualities.

An example of the complementary interaction of plant resistance and biological control was provided by Starks et al.,[133] who demonstrated that resistant cultivars of barley and sorghum enhanced the efficiency of the greenbug parasite, *Lysiphlebus testaceipes* (Cresson), thus reducing greenbug damage below that suffered in the absence of the parasite.

CURRENT RESEARCH AND IMPLEMENTATION OF RESISTANCE

Variety Development

According to Sprague and Dahms,[121] over 100 cultivars resistant to more than 25 species of arthropod pests had been developed and released to commercial production by 1972. The bulk of this effort was directed toward major forage and grain crops, i.e., alfalfa, barley, beans, maize, rice, sorghum, sugarcane, and wheat. Many other crop species investigated with varying degrees of success include cotton, crucifers (brocolli, cabbage, cauliflower), cucurbits (cucumber, muskmelon), forage and seed grasses, oats, onion, pea, raspberry, rye, potato (sweet, white), sugarbeet, tobacco, and tomato. References to research on these and many other crops are provided by Maxwell et al.[134]

Economic Impact

In 1971, 170 million lb of insecticides were applied in the U.S.[135] Schalk and Radcliffe[136] estimated that use of resistant alfalfa, barley, grain, sorghum, and maize cultivars during that year made unnecessary an additional input of 63 million lb of insecticide that would have been needed had susceptible cultivars been grown.

As for other crops, use of sawfly- and Hessian fly-resistant wheat cultivars saved growers at least $10 million annually in production losses.[137] The significance of resistance in this case is even more dramatic because alternate control measures for these

Table 10
IMPACT OF INSECT-RESISTANT CULTIVARS OF FOUR MAJOR CROPS ON CROP ACREAGE, INSECTICIDE USE, PEST POPULATIONS, AND ECONOMICS OF PRODUCTION

Crop	Pest	Total acreage (millions)	Acreage of resistant cultivars (millions)	Savings in insecticide (lb)	Reduction in pest population %	Annual value of resistance (million)
Alfalfa	Spotted alfalfa aphid	62	2	600,000	90	$60
Barley	Greenbug	10	0.5	125,000	[a]	$0.5
Maize	Chinch bug	64	10	15,000,000	[a]	[a]
	European corn borer	64	22	47,300,000	93	$150
Wheat	Hessian fly	62	8.5	[b]	95	$238
	Sawfly	62	1.5	[b]	90	$4

[a] No information available.
[b] Use of insecticides is economically impractical.

Adapted from Maxwell, F. G., *Proc. Beltwide Cotton Prod. Res. Conf.*, 83, 1972; and Schalk, J. M. and Ratcliffe, R. H., *Bull. Entomol. Soc. Am.*, 22, 7, 1976. With permission.

pests are ineffective or impractical, owing to the large acreages involved and the relatively small per-unit-area value of the crop. Yield losses from wheat stem sawfly alone can exceed 75% when resistant cultivars are not used.

As for economic value of resistance to other specific insects, Horber[128] estimated that resistant maize inbreds grown in the midwestern U.S. reduced losses by the European corn borer from $350 million in 1949 to $10 million in the 1960s. Use of alfalfa cultivars resistant to the spotted alfalfa aphid saved growers at least $35 million annually in the southwestern U.S. during the 1960s.[138] The economic and ecological benefits associated with use of insect-resistant cultivars of these crops are summarized in Table 10.

Despite the costs and time of development, the payoffs for insect-resistant crop plants are impressive. According to Luginbill,[138] costs for development of cultivars resistant to Hessian fly, wheat stem sawfly, spotted alfalfa aphid, and the European corn borer totaled about $9.3 million. Total savings to growers using these cultivars were about $308 million annually. Over 10 years of use, these cultivars provided a net savings of about $3 billion, or a 300:1 return on each research dollar invested. While these figures dramatically emphasize the economic benefits of plant resistance, they do not include the bonus effects, such as reduced pest infestations in adjacent susceptible crops, absence of secondary pest outbreaks, and minimal disturbance of the environment.

REFLECTION AND FUTURE OUTLOOK

Despite dramatic successes in insect control before the turn of this century, plant resistance did not begin to attract a favorable share of research dollars until some 25 years ago, when problems associated with dependence on chemical insecticides became apparent. The increasing appearance of insecticide-resistant pest populations, insecticide-induced outbreaks of secondary pests, and a recognition of the ecological, sociological, and economic liabilities associated with chemical insect control made plant resistance an appealing control strategy, and fueled an explosion of research and development. Today, the results of those efforts are paying off in the form of more efficient agricultural production.

For the short term, conventional methodology (i.e., identification of sources of resistance and traditional plant breeding) will continue to play the dominant role in the development of insect-resistant cultivars. But we can now envision a time when the genetic make-up of a plant will be engineered at the molecular level for optimum expression and stability of resistance to insects, as well as for the multitude of other traits conferring environmental fitness and value upon man.

REFERENCES

1. **Painter, R. H.,** *Insect Resistance in Crop Plants,* Macmillan, New York, 1951.
2. **Rodriguez, J. G., Ed.,** *Insect and Mite Nutrution — Significance and Implications in Ecology and Pest Management,* North-Holland, Amsterdam, 1972.
3. **van Emden, H. F., Ed.,** Insect/plant relationships, *Symp. Roy. Entomol. Soc.,* London, 6, 1973.
4. **Maxwell, F. G. and Harris, F. A., Eds.,** Proc. Summer Inst. Biol. Contr. Plant Insects Dis., University of Mississippi Press, Jackson, 1974.
5. **Painter, R. H.,** Resistance of plants to insects, *Annu. Rev. Entomol.,* 3, 267, 1958.
6. **Pathak, M. D.,** Genetics of plants in pest management, in *Concepts of Pest Management,* Rabb, R. L. and Guthrie, F. E., Eds., North Carolina State University, Raleigh, 1970, 138.
7. **Beck, S. D.,** Resistance of plants to insects, *Annu. Rev. Entomol.,* 10, 207, 1965.
8. **Kogan, M.,** Plant resistance in pest management, in *Introduction to Insect Pest Management,* Metcalf, R. L. and Luckmann, W., Eds., John Wiley & Sons, New York, 1975, 103.
9. **Hawkes, J. G,.** The origins of agriculture, *Econ. Bot.,* 24, 131, 1970.
10. **National Academy of Sciences,** *Genetic Vulnerability of Major Crops,* National Academy of Sciences Publishing and Printing Office, Washington, D.C., 1972.
11. **Fitch, A.,** The Hessian fly, its history, character, transformations, and habits, *Trans. N.Y. State Agric. Soc.,* 6, 12, 1847.
12. **Lindley, G.,** A Guide to the Orchard and Kitchen Garden, Longman, Rees, Orme, Brown and Green Ltd., London, 1831, cited in Painter, R. H., *Insect Resistance in Crop Plants,* MacMillan, New York, 1951.
13. **Howard, L. O.,** *A History of Applied Entomology,* Smithsonian Miscellaneous Collection 34, Washington, D.C., 1930.
14. **Dethier, V. G.,** Evolution of feeding preferences in phytophagous insects, *Evolution,* 8, 33, 1954.
15. **Kennedy, J. S.,** Mechanisms of host plant selection, *Ann. Appl. Biol.,* 56, 317, 1965.
16. **Schoonhoven, L. M.,** Chemosensory bases of host plant selection, *Annu. Rev. Entomol.,* 13, 115, 1968.
17. **Fraenkel, G.** The raison d'etre of secondary plant substances, *Science,* 129, 1466, 1959.
18. **Fraenkel, G.,** Evaluation of our thoughts on secondary plant substances, *Entomol. Exp. Appl.,* 12, 473, 1969.
19. **Whittaker, R. H. and Feeny, P. P.,** Allelochemics: chemical interactions between species, *Science,* 171, 757, 1971.
20. **Dethier, V. G., Barton-Browne, L., and Smith, C. N.,** The designation of chemicals in terms of the responses they elicit from insects, *J. Econ. Entomol.,* 53, 134, 1960.
21. **Isely, D.,** The relation of leaf color and leaf size to boll weevil infestation, *J. Econ. Entomol.,* 21, 553, 1928.
22. **Kennedy, J. S.,** Physiological condition of the host-plant and susceptibility to aphid attack, *Entomol. Exp. Appl.,* 1, 50, 1958.
23. **Schuster, M. F., Lukefahr, M. J., Maxwell, F. G.,** Impact of nectariless cotton on plant bugs and natural enemies, *J. Econ. Entomol.,* 69, 400, 1976.
24. **Maxwell, F. G.,** Morphological and chemical changes that evolve in the development of host plant resistance to insects, *J. Environ. Qual.,* 1, 265, 1972.
25. **Waldbauer, G. D.,** The consumption and utilization of food by insects, *Adv. Insect Physiol.,* 5, 229, 1968.
26. **House, H. L.,** Effects of different proportions of nutrients on insects, *Entomol. Exp. Appl.,* 12, 651, 1969.

27. **Beck, S.D. and Reese, J. C.,** Insect-plant interactions: nutrition and metabolism, *Recent Adv. Phytochem.,* 10, 41, 1976.
28. **Chambliss, O. Y. and Jones, C. M.,** Cucurbitacins: specific insect attractants in Cucurbitaceae, *Science,* 153, 1392, 1966.
29. **DaCosta, C. D. and Jones, C. M.,** Cucumber beetle resistance and mite susceptibility controlled by the bitter gene in *Cucumis sativus* L., *Science,* 172, 1145, 1971.
30. **Sharma, G. C. and Hall, C. V.,** Influence of cucurbitacins, sugars, and fatty acids on cucurbit susceptibility to spotted cucumber beetle, *J. Am. Soc. Hort. Sci.,* 96, 675, 1971.
31. **Sharma, G. C. and Hall, C. V.,** Relative atractiveness of spotted cucumber beetles to fruits of fifteen species of Cucurbitaceae, *Environ. Entomol.,* 2, 154, 1973.
32. **Hedin, P. A., Maxwell, F. G., and Jenkins, J. N.,** Insect plant attractants, feeding stimulants, repellents, deterrents, and other related factors affecting insect behavior, in *Proc. Summer Inst. Biol. Control Plant Insects Diseases,* Maxwell, F. G. and Harris, F. A., Eds., University of Mississippi Press, Jackson, 1974, 494.
33. **Lukefahr, M. J., Martin, D. F., and Meyer, J. R.,** Plant resistance to five Lepidoptera attacking cotton, *J. Econ. Entomol.,* 58, 516, 1965.
34. **Overman, J. L. and MacCarter, L. E.,** Evaluating seedlings of cantaloupe for varietal nonpreference-type resistance to *Diabrotica* spp., *J. Econ. Entomol.,* 65, 1140, 1972.
35. **Kogan, M. and Ortman, E. E.,** Antixenosis — a new term proposed to define Painter's "nonpreference" modality of resistance, *Bull. Entomol. Soc. Am.,* 24, 175, 1978.
36. **Hedin, P. A., Jenkins, J. N., and Maxwell, F. G.,** Behavioral and developmental factors affecting host plant resistance to insects, Am. Chem. Soc. Symp. Ser., 62, Washington, D. C., 1977, 231.
37. **Klun, J. A., Tipton, C. L., and Brindley, T. A.,** 2,4-dihydroxy-7-methoxy-1,4-benzoxazin-3-one (DIMBOA), an active agent in the resistance of maize to the European corn borer, *J. Econ. Entomol.,* 60, 1529, 1967.
38. **Bottger, G. T., Sheehan, E. T., and Lukefahr, M. J.,** Relation of gossypol content of cotton plants to insect resistance, *J. Econ. Entomol.,* 57, 283, 1964.
39. **Chan, B. G., Waiss, A. C., Jr., and Lukefahr, M.,** Condensed tannin, an antibiotic chemical from *Gossypium hirsutum, J. Insect Physiol.,* 24, 113, 1978.
40. **Kuhn, R. and Löw, R.,** Resistance factors against *Leptinotarsa decemlineata* (Say), isolated from the leaves of wild *Solanum* species, in *Origins of Resistance to Toxic Agents,* Sevag, M. G., Reid, R. D., and Reynolds, O. E., Eds., Academic Press, New York, 1955, 122.
41. **Dahlman, D. L. and Hibbs, E. T.,** Responses of *Empoasca fabae* Harris to tomatine, solanine, leptine I, tomatidine, solanidine, and demissidine, *Ann. Entomol. Soc. Am.,* 60, 732, 1967.
42. **Tingey, W. M., Mackenzie, J. D., and Gregory, P.,** Total foliar glycoalkaloids and resistance of wild potato species to *Empoasca fabae* (Harris), *Am. Potato J.,* 10, 577, 1978.
43. **Horber, E.,** Alfalfa saponins significant in resistance to insects, in *Insect and Mite Nutrition — Significance and Implications in Ecology and Pest Management,* Rodriguez, J. G., Ed., North-Holland, Amsterdam, 1972, 611.
44. **Maltais, J. B. and Auclair, J. L,.** Factors in resistance of peas to the pea aphid, *Acyrthosiphon pisum* (Harris) (Homoptera: Aphididae). I. The sugar-nitrogen ratio, *Can. Entomol.,* 89, 365, 1957.
45. **Auclair, J. L., Maltais, J. B., and Cartier, J. J.,** Factors in resistance of peas to the pea aphid, *Acrythosiphon pisum* (Harris) (Homoptera: Aphididae). II. Amino acids, *Can. Entomol.,* 89, 457, 1957.
46. **Douglas, W. A.,** The effect of husk extension and tightness on earworm damage to corn, *J. Econ. Entomol.,* 40, 661, 1947.
47. **Luckmann, W. H., Rhodes, A. M., and Wann, E. V.,** Silk balling and other factors associated with resistance of corn to corn earworm, *J. Econ. Entomol.,* 57, 778, 1964.
48. **Cameron, J. W. and Anderson, L. D.,** Husk tightness, corn earworm egg numbers, and starchiness of kernels in relation to resistance of corn to the corn earworm, *J. Econ. Entomol.,* 59, 556, 1966.
49. **Djamin, A. and Pathak, M. D.,** The role of silica in resistance to Asiatic rice borer, *Chilo suppressalis* (Walker) in rice varieties, *J. Econ. Entomol.,* 60, 347, 1967.
50. **Agarwal, R. A.,** Morphological characteristics of sugarcane and insect resistance, *Entomol. Exp. Appl.,* 12, 767, 1969.
51. **Howe, W. L.,** Factors affecting the resistance of certain cucurbits to the squash borer, *J. Econ. Entomol.,* 42, 321, 1949.
52. **O'Keefe, L. E., Callenbach, J. A., and Lebsock, K. L.,** Effect of culm solidness on the survival of the wheat stem sawfly, *J. Econ. Entomol.,* 53, 244, 1960.
53. **Levin, D. A.,** The role of trichomes in plant defense, *Quart. Rev. Biol.,* 48, 3, 1973.
54. **Parnell, F. R., King, H. E., Ruston, D. F.,** Jassid resistance and hairiness of the cotton plant, *Bull. Entomol. Res.,* 39, 539, 1949.

55. **Kamel, S. A. and Elkassaby, F. Y.,** Relative resistance of cotton varieties to spider mites, leafhoppers, and aphids, *J. Econ. Entomol.*, 58, 209, 1965.
56. **Schillinger, J. A., Jr. and Gallun, R. L.,** Leaf pubescence of wheat as a deterrent to the cereal leaf beetle, *Oulema melanopus, Ann. Entomol. Soc. Am.*, 61, 900, 1968.
57. **Hoxie, R. P., Wellso, S. G., and Webster, J. A.,** Cereal leaf beetle response to wheat trichome length and density, *Environ. Entomol.*, 4, 365, 1975.
58. **Broersma, D. B., Bernard, R. L., and Luckmann, W. H.,** Some effects of soybean pubescence on populations of potato leafhopper, *J. Econ. Entomol.*, 65, 78, 1972.
59. **Pillemer, E. A. and Tingey, W. M.,** Hooked trichomes: a physical plant barrier to a major agricultural pest, *Science*, 193, 482, 1976.
60. **Johnson, B.,** The injurious effects of the hooked epidermal hairs of French beans *(Phaseolus vulgaris* L.) on *Aphis craccivora* Koch., *Bull. Entomol. Res.*, 44, 779, 1953.
61. **Taylor, N. L.,** Pubescence inheritance and leafhopper resistance relationships in alfalfa, *Agron. J.*, 48, 78, 1956.
62. **Gibson, R. W. and Turner, R. H.,** Insect-trapping hairs on potato plants, *PANS (Pest Artic. News Summ.)*, 22, 272, 1977.
63. **Tingey, W. M. and Gibson, R. W.,** Feeding and mobility of the potato leafhopper impaired by glandular trichomes of *Solanum berthaultii* and *S. polyadenium, J. Econ. Entomol.*, 71, 856, 1978
64. **Gentile, A. G. and Stoner, A. K.,** Resistance in *Lycopersicon* and *Solanum* species to the potato aphid, *J. Econ. Entomol.*, 61, 1152, 1968.
65. **Stoner, A. K., Frank, J. A., and Gentile, A. G.,** The relationship of glandular hairs on tomatoes to spider mite resistance, *Proc. Am. Soc. Hort. Sci.*, 90, 324, 1968.
66. **Aina, O. J., Rodriguez, J. G., and Knavel, D. E.,** Characterizing resistance to *Tetranychus urticae* in tomato, *J. Econ. Entomol.*, 65, 641, 1972.
67. **Gentile, A. G., Webb, R. E., and Stoner, A. K.,** Resistance in *Lycopersicon* and *Solanum* to greenhouse whiteflies, *J. Econ. Entomol.*, 61, 1355, 1968.
68. **Thurston, R., Smith, W. T., and Cooper, B. P.,** Alkaloid secretion by trichomes of *Nicotiana* species and resistance to aphids, *Entomol. Exp. Appl.*, 9, 428, 1966.
69. **Thurston, R.,** Toxicity of trichome exudates of *Nicotiana* and *Petunia* species to tobacco hornworm larvae, *J. Econ. Entomol.*, 63, 272, 1970.
70. **Patterson, C. G., Thurston, R., and Rodriguez, J. G.** Two spotted spider mite resistance in *Nicotiana* species, *J. Econ. Entomol.*, 67, 341, 1974.
71. **Shade, R. E., Thompson, T. E., and Campbell, W. R.,** An alfalfa weevil larval resistance mechanism detected in *Medicago, J. Econ. Entomol.*, 68, 399, 1975.
72. **Zuber, M. S., Musick, G. J., and Fairchild, M. L.,** A method of evaluating corn strains for tolerance to the western corn rootworm, *J. Econ. Entomol.*, 64, 1514, 1971.
73. **Day, P. R.,** Crop resistance to pests and pathogens, in *Pest Control Strategies for the Future*, National Academy of Science Washington, D.C., 1972, 257.
74. **Simons, M. D.,** Polygenic resistance to plant disease and its use in breeding resistant cultivars, *J. Environ. Qual.*, 1, 232, 1972.
75. **Gallun, R. L.,** Genetic interrelationships between host plants and insects, *J. Environ. Qual.*, 1, 259, 1972.
76. **Russell, W. A.,** Breeding and genetics in the control of insect pests, *Iowa State J. Res.*, 49, 527, 1975.
77. **Southwood, T. R. E.,** *Ecological Methods*, Chapman and Hall, London, 1971.
78. **Ruesink, W. G. and Kogan, M.,** The quantitative basis of pest management: sampling and measuring, in *Introduction to Pest Management*, Metcalf, R. L. and Luckmann, W. H., Eds., John Wiley & Sons, New York, 1975, 309.
79. **Dahms, R. G.,** Techniques in the evaluation and development of host-plant resistance, *J. Environ. Qual.*, 1, 254, 1972.
80. **Hackerott, H. L. and Harvey T. L.,** Effect of temperature on spotted alfalfa aphid reaction to resistance in alfalfa, *J. Econ. Entomol.*, 52, 949, 1959.
81. **Isaak, A., Sorensen, E. L., and Painter, R. H.,** Stability of resistance to pea aphid and spotted alfalfa aphid in several alfalfa clones under various temperature regimes, *J. Econ. Entomol.*, 58, 140, 1965.
82. **Starks, K. J., Wood, E. A., Jr., and Teetes, G. L.,** Effects of temperature on the preference of two greenbug biotypes for sorghum selections, *Environ. Entomol.*, 2, 351, 1973.
83. **Cartwright, W. B., Caldwell, R. M., and Compton, L. E.** Relation of temperature to the expression of resistance in wheats to Hessian fly, *J. Am. Soc. Agron.*, 38, 259, 1946.
84. **Sosa, O., Jr. and Foster, J. E.,** Temperature and the expression of resistance in wheat to the Hessian fly, *Environ. Entomol.*, 5, 333, 1976.
85. **Roberts, D. W. A. and Tyrrell, C.,** Sawfly resistance in wheat. IV. Some effects of light intensity on resistance, *Can. J. Plant Sci.*, 41, 457, 1961.

86. **Lowe, H. J. B.,** Testing sugar beet for aphid-resistance in the glasshouse: a method and some limiting factors, *Z. Ang. Entomol.*, 76, 311, 1974.
87. **Pierzchalski, T. and Werner, E.,** Changes in glycoalkaloid contents in leaves of cultivated and wild potatoes and their hybrids during growth and their influence on the Colorado potato beetle *(Leptinotarsa decemlineata* Say), *Hodowla. Rosl. Aklim. Nasienn.*, 2, 157, 1958.
88. **Kindler, S. D. and Staples, R.** Nutrients and the reaction of two alfalfa clones to the spotted alfalfa aphid, *J. Econ. Entomol.*, 63, 938, 1970.
89. **Coon, B. F.,** Aphid populations on oats grown in various nutrient solutions, *J. Econ. Entomol.*, 52, 624, 1959.
90. **Cram, W. T.,** Fecundity of the root weevils *Brachyrhinus sulcatus* and *Sciopithes obscurus* on strawberry at different conditions of host plant nutrition, *Can. J. Plant Sci.*, 45, 219, 1965.
91. **Mistric, W. J., Jr.,** Effects of nitrogen fertilization on cotton under boll weevil attack in North Carolina, *J. Econ. Entomol.*, 61, 282, 1968.
92. **Rodriguez, J. G., Chaplin, C. E., Stoltz, L. P., and Lasheen, A. M.,** Studies on resistance of strawberries to mites. I. Effects of plant nitrogen, *J. Econ. Entomol.*, 63, 1855, 1970.
93. **Daniels, N. E., Wilson, G. C., and Clarke, C. S.,** Nitrogen and phosphorus content of greenbugs reared on fertilized wheat, *J. Econ. Entomol.*, 61, 1746, 1968.
94. **van Emden, H. F.,** Studies on the relations of insect and host plant. III. A comparison of the reproduction of *Brevicoryne brassicae* and *Myzus persicae* (Hemiptera: Aphididae) on Brussels sprout plants supplied with different rates of nitrogen and potassium, *Entomol. Exp. Appl.*, 9, 444, 1966.
95. **van Emden, H. F.,** Effect of (2-chloroethyl) trimethyl-ammonium chloride on the rate of increase of the cabbage aphid (*Brevicoryne brassicae* (L.)), *Nature (London)*, 201, 946, 1964.
96. **Honeyborne, C. H. B.,** Performance of *Aphis fabae* and *Brevicoryne brassicae* on plants treated with growth regulators, *J. Sci. Food Agric.*, 20, 388, 1969.
97. **Tahori, A. S., Halevy, A. H., and Zeidler, G.,** Effect of some plant growth retardants on the oleander aphid, *Aphis nerii* (Boyer), *J. Sci. Food Agric.*, 16, 568, 1965.
98. **Smith, B. D.,** Spectra of activity of plant growth retardants against various parasites of one host species, *J. Sci. Food Agric.*, 20, 398, 1969.
99. **van Emden, H. F.,** Plant resistance to *Myzus persicae* induced by a plant growth regulator and measured by aphid relative growth rate, *Entomol. Exp. Appl.*, 12, 125, 1969.
100. **Tauber, M. J., Salucha, B., and Langhans, R. W.,** Succinic acid-2,2-dimethylhydrazide (SADH) prevents whitefly population increase, *HortScience*, 6, 458, 1971.
101. **Rodriguez, J. G. and Campbell, J. M.,** Effects of gibberelin on nutrition of the mites, *Tetranychus telarius* and *Panonychus ulmi, J. Econ. Entomol.*, 54, 984, 1961.
102. **Rodriguez, J. G.,** Nutrition of the host and reaction to pests, *Am. Assoc. Adv. Sci.*, 61, 149, 1960.
103. **El-Tigani, M.E.-A.,** Der Finfluss der Mineraldungüng der Pflanzen auf Entwicklung und Vermehrung von Blattlaüssen, *Wiss. Z. Univ. Rostock.*, 11, 307, 1962.
104. **van Emden, H. F.,** Plant resistance to insects induced by environment, *Sci. Hort.*, 18, 91, 1966.
105. **van Emden, H. F.,** Plant resistance to aphids induced by chemicals, *J. Sci. Food Agr.*, 20, 385, 1969.
106. **Singh, P.,** Host-plant nutrition and composition: effects on agricultural pests, *Can. Dep. Agric. Res. Inst. Inf. Bull.*, 1970.
107. **Nielson, M. W., Don, H., Schonhorst, M. H., Lehman, W. L., and Marble, V. L.,** Biotypes of the spotted alfalfa aphid in western United States, *J. Econ. Entomol.*, 63, 1822, 1970.
108. **Cartier, J. J., Isaak, J. A., Painter, R. H., and Sorensen, E. L.,** Biotypes of pea aphids, *Acyrthosiphon pisum* (Harris) in relation to alfalfa clones, *Can. Entomol.*, 97, 754, 1965.
109. **Briggs, J. B.,** The distribution, abundance, and relationships of four strains of the rubus aphid *(Amphorophora rubi* (Kalt)) in relation to raspberry breeding, *J. Hort. Sci.*, 49, 109, 1965.
110. **Wood, E. A., Jr.,** Designation and reaction of three biotypes of the greenbug cultured on resistant and susceptible species of sorghum, *J. Econ. Entomol.*, 64, 183, 1971.
111. **Singh, S. R. and Painter, R. H.,** Reaction of 4 biotypes of corn leaf aphid, *Rhopalosiphum maidis* (Fitch), to differences in host plant nutrition, *Proc. 12th Int. Cong. Entomol.*, 1965, 543.
112. **Gallun, R. L. and Hatchett, J. H.,** Interrelationships between races of Hessian fly, *Mayetiola destructor* (Say) and resistance in wheat, *Proc. 3rd Int. Wheat Genetics Symp., Austral. Acad. Sci.*, 1968, 258.
113. **Alston, F. H. and Briggs, J. B.,** Resistance genes in apple and biotypes of *Dysaphis devecta, Ann. Appl. Biol.*, 87, 75, 1977.
114. **Cheng, C. H.,** New biotypes of the brown planthopper and interaction between biotypes and resistance in rice, *Bull. Taiwan Agric. Res. Inst.*, 32, 29, 1975.
115. **Stevenson, A. B.,** Strains of the grape phylloxera in Ontario with different effects on the foliage of certain grape cultivars, *J. Econ. Entomol.*, 63, 135, 1970.
116. **Sosa, O., Jr.,** Biotype L, ninth biotype of the Hessian fly, *J. Econ. Entomol.*, 71, 458, 1978.
117. **Gallun, R. L.,** Genetic basis of Hessian fly epidemics, *Ann. N.Y. Acad. Sci.*, 287, 223, 1977.

118. **Klun, J. A. and Robinson, J. F.,** Concentration of two 1,4-benzoxazinones in dent corn at various stages of development of the plant and its relation to resistance of the host plant to the European corn borer, *J. Econ. Entomol.*, 62, 214, 1969.
119. **Howe, W. L. and Pesho, G. R.,** Influence of plant age on the survival of alfalfa varieties differing in resistance to the spotted alfalfa aphid, *J. Econ. Entomol.*, 53, 142, 1960.
120. **Abernathy, C. O. and Thurston, R.,** Plant age in relation to the resistance of *Nicotiana* to the green peach aphid, *J. Econ. Entomol.*, 62, 1356, 1969.
121. **Sprague, G. F. and Dahms, R. G.,** Development of crop resistance to insects, *J. Environ. Qual.*, 1, 28, 1972.
122. **Jenkins, J. N. and Parrott, W. L.,** Effectiveness of frego bract as a boll weevil resistance character in cotton, *Crop Sci.*, 11, 739, 1971.
123. **Lincoln, C., Dean, G., Waddle, B. A., Yearian, W. C., Phillips, F. R., and Roberts, L.,** Resistance of frego-type cotton to boll weevil and bollworm, *J. Econ. Entomol.*, 64, 1326, 1971.
124. **Lambou, M. G., Shaw, R. L., Decossas, K. M., and Vix, H. L. E.,** Cottonseed's role in a hungry world, *Econ. Bot.*, 20, 256, 1966.
125. **Zitnak, A. and Johnson, G. R.,** Glycoalkaoid content of B5141-6 potatoes, *Am. Potato J.*, 47, 256, 1970.
126. **Rabb, R. L. and Bradley, J. R.,** The influence of host plants on parasitism of eggs of the tobacco hornworm, *J. Econ. Entomol.*, 61, 1249, 1968.
127. **Elsey, K. D. and Chaplin, J. F.,** Resistance of tobacco introduction 1112 to the tobacco budworm and green peach aphid, *J. Econ. Entomol.*, 71, 723, 1978.
128. **Horber, E.,** Plant resistance to insects, *Agric. Sci. Rev.*, 10, 1, 1972.
129. **McMillian, W. W., Wiseman, B. R., Widstrom, N. W., and Harrell, E. A.,** Resistant sweet corn hybrid plus insecticide to reduce losses from corn earworm, *J. Econ. Entomol.*, 65, 229, 1972.
130. **Wiseman, B. R., Harrell, E. A., and McMillian, W. W.,** Continuation of tests of resistant sweet corn hybrid plus insecticides to reduce losses from corn earworm, *Environ. Entomol.*, 2, 919, 1973.
131. **Cuthbert, F. P., Jr. and Jones, A.,** Insect resistance as an adjunct or alternative to insecticides for control of sweet potato soil insects, *J. Am. Soc. Hortic. Sci.*, 103, 443, 1978.
132. **Casagrande, R. A. and Haynes, D. L.,** The impact of pubescent wheat on the population dynamics of the cereal leaf beetle, *Environ. Entomol.*, 5, 153, 1976.
133. **Starks, K. J., Muniappan, J. R., and Eikenbary, R. D.,** Interaction between plant resistance and parasitism against greenbug on barley and sorghum, *Ann. Entomol. Soc. Am.*, 65, 650, 1972.
134. **Maxwell, F. G., Jenkins, J. N., and Parrott, W. L.,** Resistance of plants to insects, *Adv. Agron.*, 24, 187, 1972.
135. **Andrilenas, P. A.,** Farmer's use of pesticides in 1971 — Quantities, *USDA-ERS Agric. Econ. Rep.*, 252, 1974, 56.
136. **Schalk, J. M. and Ratcliffe, R. H.,** Evaluation of ARS program on alternative methods of insect control: host plant resistance to insects, *Bull. Entomol. Soc. Am.*, 22, 7, 1976.
137. **Gallun, R. L., Starks, K. J., and Guthrie, W. D.,** Plant resistance to insects attacking cereals, *Annu. Rev. Entomol.*, 20, 337, 1975.
138. **Luginbill, P., Jr.,** Developing resistant plants — an ideal method of controlling insects, *USDA-ARS Prod. Res. Report III*, 1969.

ENVIRONMENTAL CONTROL OF INSECTS USING TRAP CROPS, SANITATION, PREVENTION, AND HARVESTING

Vernon Stern

INTRODUCTION

Agroecosystems vary widely in stability, complexity, and the area they occupy. The kinds of crops, agronomic practices, changes in land use, and weather are important elements affecting the degree of stability of an agroecosystem. All of these, except perhaps weather, can be manipulated to influence pest or natural enemy populations. Quite often the agroecosystem may lack only a minor key factor or feature to adversely affect a pest or favorably modify the environment to increase the effectiveness of its natural enemies. These needs can often be met by proper use of environmental or cultural control.[1]

The two basic principles in the environmental control of insect pests are (1) manipulation of the agroecosystem to make it less favorable to the pest, and (2) manipulation to make it more favorable to their natural enemies. Both may be used together to prohibit, reduce, or delay pest population increase.

The classical use of environmental or cultural control has generally been achieved by agronomic practices. However, new techniques such as preharvest chemicals, modification of crop varieties such as short season cotton, manipulation of pest populations using pheromones, and selective employment of insecticides have also helped to suppress the pest and preserve natural enemies.

Environmental methods require a thorough knowledge of production of the crop and the biology and ecology of the pest and its natural enemies in order to integrate the techniques for pest control into proved agronomic procedures for crop production. This knowledge permits assessment of the agroecosystem that favors a particular pest and enables environmental changes that can be made to reduce pest population numbers and damage.

TRAP CROPS

Many experiments have been conducted with trap crops to attract and retain a pest species or to provide a more favorable habitat to increase natural enemies. This method of pest control has had limited success.[2] The methods are often too difficult to control most insects or growers are reluctant to rely on them in comparison to pesticides which are relatively cheap and usually give consistently good results.

Moreover, the trap crop may have little or no commercial value, or, if utilized, it is usually of less value than the main crop. In such cases the reduction of pesticide cost must be substantial to offset the return and cost of growing the trap crop.

Alfalfa Interplanting

Lygus bugs are a key pest of cotton in the San Joaquin Valley, Calif. Stern et al.[3,4] showed that *Lygus hesperus* Knight and *L. elisus* Van Duzee prefer alfalfa over cotton as long as the alfalfa remains in a lush growing condition.

Studies were initiated in 1967 to interplant 20-ft strips of alfalfa in cotton at 300 to 500 ft intervals (Figure 1).[4]

At that time cotton in the U.S. was subsidized by a parity price support program and cotton production was limited. A specific amount of cotton acreage was allocated

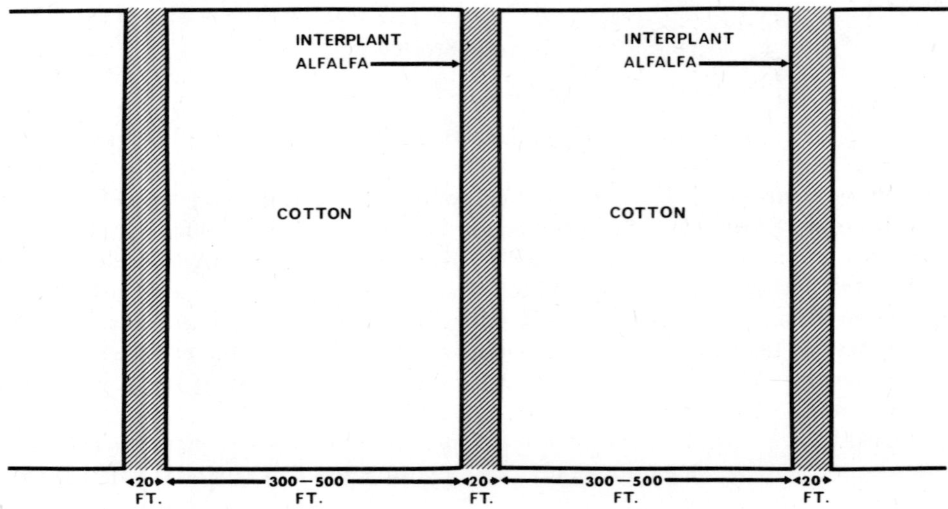

FIGURE 1. Schematic diagram of alfalfa interplanted in cotton. Most trap crops are small plantings within a main crop.

to each cotton grower. This acreage depended on the past history of cotton planting for that farm, which meant that growers had an excess of idle land.

Since there was no alternate high value summer crop, many growers in Arizona and California found a loophole in the federal law that was designed to regulate cotton production. The growers found they could increase cotton yield by alternating planting two rows of cotton and leaving the next two rows unplanted. Thus, in a 160-acre field there would be 80 acres of actual cotton and 80 acres of idle land.

There were additional labor, equipment, water, and energy costs in this farming method. The entire field had to be plowed, furrowed for irrigation, cultivated for weeds, and treated for insect pests because most pesticides used on field crops in the irrigated southwest are applied by air. Nevertheless, the profits from increased cotton yield exceeded the additional cost of this farming method.

There were about 10,000 acres of cotton interplanted with alfalfa in the San Joaquin Valley in 1968. The lygus bug populations were followed in a number of interplant fields and the cost of interplanting was obtained from two growers.

One grower planted 20-ft alfalfa strips to be harvested as a seed crop. There were 5 acres of strips in the 80-acre field of alternating two rows of cotton and two rows unplanted.[4] This field was treated one time for lygus bugs when the alfalfa strips became dry and would no longer attract and hold the bugs. The cost for this DDT-Toxaphene treatment was $7.10/acre because the cost doubles when treating "two in, two out" cotton.

The alfalfa seed production cost $226/acre and the seed crop returned $146/acre for a loss of $400 on the five interplanted alfalfa acres.

The adjacent 80-acre cotton field of "two rows in, two rows out" cotton was treated four times for lygus bugs, and the cost of three additional treatments was $21.30/acre. The $400 loss from 5 acres of alfalfa seed in the 80-acre interplanted field was spread over 75 acres of cotton at $5.33/acre. The net gain from interplanting alfalfa as a trap crop was $15.97/acre.

The second grower farmed in an area where water costs were low. His cost of interplanting 8 acres of alfalfa in a 160-acre field of solid planted cotton was about $160/

acre. Moving equipment to harvest the small acreage of alfalfa was not economical and harvesting was abandoned. This field was never treated for *Lygus*.

Most cotton fields in the area were treated 4 to 6 times for lygus bugs or worms at a treatment cost of about \$4.50/acre. This was slightly higher than the first grower because DDT-Toxaphene could not be used in alfalfa hay growing areas. Five treatments would have amounted to \$22.50/acre. A 152-acre field treated five times would have cost \$3420. The expense of interplanting 8 acres of alfalfa was \$1280 for a reduced pest control cost of \$14.08/acre. Since the grower was prohibited by law from planting more cotton, the 8 acres of alfalfa do not represent a loss in cotton profits, which at that time netted about \$50 to \$60/acre in California.

Both of these growers and others showed a profit from planting alfalfa strips in cotton. However, there were problems in using this alternate method of pest control.

Much of California's certified alfalfa seed crop (40% of U.S. total) is grown on the west side of the San Joaquin Valley. Regulations require that certified alfalfa seed cannot be grown unless the field has been free of any alfalfa crop for three years. Interplanting alfalfa in cotton disrupted the cotton-alfalfa seed rotation programs.

Alfalfa requires more frequent irrigation than cotton to keep it attractive to *Lygus*. Moving water in irrigation ditches or in pipes from one field to another to irrigate alfalfa strips presented water management problems for most growers. Consequently many alfalfa strips dried out and the lygus bugs moved into the cotton. The fields required treatment and this discouraged growers.

Federal restrictions on cotton acreage were eventually eliminated and growers believed they could make a greater profit by growing solid fields of cotton rather than diverting small acreages to alfalfa strips. The data do not show this. Nevertheless, interplanting alfalfa in cotton was soon abandoned.

Corn Interplanting

In Peru, the planting of corn either in neighboring fields or rows of corn interplanted in cotton fields with one row to every fifth or seventh row of cotton has been practiced for many years. This favors the reproduction of many natural enemies of cotton pests and they build up earlier in corn than in cotton. Stern et al.[1] comment that the benefits are real although not easy to establish. Corn is also used in Peru as a food crop and grown on many farms thus representing a trap crop that is utilized.

Moving Parasites on Host Crops

As a substitute for, or supplement to interplanting special plants within a main crop to increase natural enemies, the transfer of plants densely populated with natural enemies can be used to advantage. The introduced parasites, *Praon palitans* Mues. and *Trioxys utilis* Mues., of the spotted alfalfa aphid, *Therioaphis maculata* (Buckton), were rapidly spread throughout California by cutting and spreading infested hay.[5] This technique, along with growers planting resistant alfalfa varieties, growers using a selective insecticide (demeton) to permit maximum survival of the parasites and lady beetles, and aid of a fungus disease, reduced the cost of damage and chemical control from about 13 million in 1955 to about 1 million by 1958, and the reafter, this insect became a very minor alfalfa pest.[6] Similar parasite movements have been used with success in the USSR and in Peru.[1]

SANITATION

Field crop, glasshouse, and grain storage sanitation to reduce pest infestations by removing the breeding and hibernating sites, destroying diseased or infested plants, and fumigating soil for potted plants is a sound procedure of pest management.

Field Sanitation

The clover seed chalcid, *Bruchophagus roddi* Guss., has several generations per year and is a constant threat to alfalfa seed production in the western U.S. where most of the seed crop is grown.[7-10] The small wasp lays the egg through the seed pod into the developing green seed and the larvae devour the seed, leaving nothing but the seed coat. The adult gnaws a hole through the seed coat and pod and emerges.

This pest overwinters in larval diapause in the seed. Adults become active in early spring and infestations are usually found on bur clover, *Medicago hispida* Goertn., and on volunteer alfalfa along roads and ditch banks before the alfalfa seed fields are mature enough to infest.[7-10]

Chemical control is difficult since the eggs are laid and the larvae develop within the seeds. Continuous emergence of the adults from the infested seeds and wasps migrating from outside sources into the seed fields require frequent application. The adult chalcid wasps are most active in the fields during bloom when honey bees are pollinating the crop. Excessive treatments increase the possibility of bee kill during that period.

Cultural and sanitation practices recommended by Sorenson, Wildermuth, and Bacon et al. have markedly reduced alfalfa seed damage.[7-9] These include: destruction of volunteer alfalfa and bur clover near seed fields, burning of straw and chaff in the field after harvest, tillage and irrigation after harvest to cover the seeds left in the field, prevention of seed set on regrowth after harvest, covering of trucks hauling seed to the cleaning mills to prevent scattering of seeds along roads to give volunteer alfalfa plants, and production of the maximum seed crop by midsummer.

Surveys conducted by Bacon et al.[9] and Bacon[11] in the southern San Joaquin Valley, Calif. showed that an average of 12 to 14% of the crop was destroyed during the 1950s. A more recent survey of alfalfa seed fields in the same area showed damage from the alfalfa seed chalcid to be less than 1%. The reduction was the result of cultural and sanitation practices mentioned above and amounted to a saving of $13.5 million between 1970 and 1977.[11]

Stored Grain Sanitation

About 10% of the world's cereal storage is lost each year through insect infestation.[12] Much of this is due to poor sanitation practices. A number of publications discuss the more common insects injuring stored grain and methods of controlling them.[12-13]

In the U.S. losses from cereal pests are not as great as in some parts of the world because of climatic conditions and facilities for handling and storing cereal grains. Nevertheless, losses to grain and their products from insects including the cost of prevention and control have been estimated to be about $600 million annually.[12]

The small size of stored-product insects in storage and in processing plants makes cleanliness or hygiene so important as a control measure. Small accumulations of debris, broken grains lodged in corners, crevices, sills, and on ledges all afford food and habitat for stored-product insect pests.[12,13]

Farm storage facilities vary so greatly that no detailed rules for sanitation can be laid down, except to keep them clean. In the broadest sense, sanitary procedures blend with prevention and may include fumigation or chemical treatment of a silo to eradicate an existing stored-product pest population before filling it with grain.[12]

PREVENTION

Prevention by chemical control rather than environmental control may be the only available way to combat pests that breed to high numbers in one habitat or crop where they are not a pest, but move to another crop where they become a pest.

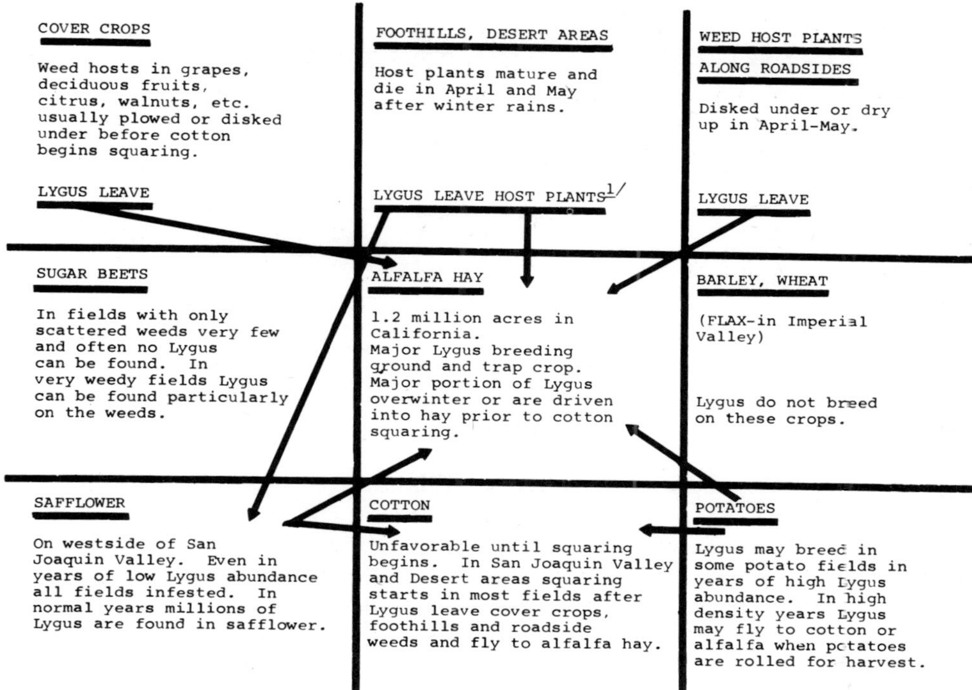

FIGURE 2. Diagram of hosts and the movement of *Lygus* into alfalfa and safflower from other crops, range lands, and weedy areas. When an entire alfalfa field is cut the lygus bugs fly to cotton and other crops.

This occurs with the beet leafhopper, *Circulifer tenellus* (Baker) in California. This species mostly overwinters in the nonirrigated range land foothills of the Central Valley. It increases to high numbers on the range land plants and picks up the curly-top virus. When the foothills begin to dry after the winter rains the leafhoppers move to the valley floor and feed on and transmit the virus to a variety of commercial crops.

The California Department of Food and Agriculture treats a large portion of the overwintering area where the leafhoppers congregate to prevent them from spreading the curly-top virus to commercial plantings. Total crop loss occurred in some tomato fields and 10 to 15% in bean fields before the spray program was initiated. Losses were reduced to less than 1% after the spray program started.

A similar situation occurs with lygus bugs which overwinter on range land plants and fly to safflower in late March and April on the west side of the San Joaquin Valley, Calif. (Figures 2, 3).[3,4,15] Millions of *Lygus* can be found in safflower fields, but they are rarely if ever a pest in this crop. When the safflower begins to mature in late May and mid-June the adults fly to adjacent cotton where they are a severe pest. Since the bugs reach the adult stage over a 3- to 4-week period, cotton fields may require 3 to 5 treatments to control the invaders.

Mueller and Stern studied *Lygus* population trends in safflower over a 3-year period, and in 1970 initiated a single areawide treatment of safflower based on the time when 70% of the nymph population were in the 3 to 5 instar.[15] Pesticide costs on cotton from 1967 to 1969 averaged $32.33/acre and were reduced to $14.43/acre in 1970. Subsequently, every year since, the success of the program has been repeated.[16]

An important feature of the program is that every safflower grower also produces cotton and there is about three times as much cotton produced in this area as safflower. Thus, when the cost of the safflower treatment is charged as a cotton expense the cost is minimal.

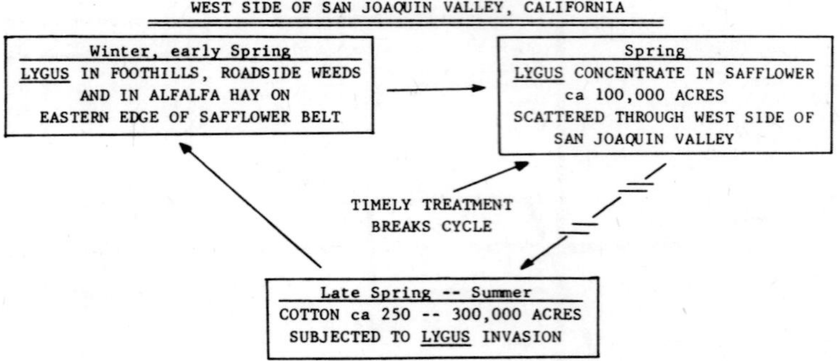

FIGURE 3. Annual movement of lygus bugs on the west side of the San Joaquin Valley, Calif.

HARVESTING

Harvesting procedures involving crop maturity, time of harvest or cutting practices, selective harvesting, and strip-harvesting can be of considerable assistance in suppressing a variety of insect pests, increasing their natural enemies, and affecting yields.[1]

Time of Harvest

Many pest species overwinter in or on the stalks, stems, and other parts of their host plant, and destruction of these parts by shredding, burning, plowing, and so forth can greatly reduce overwintering populations. A stalk destruction program followed by plowing was developed in Texas for control of the pink bollworm, *Pectinophora gossypiella* (Saunders).[17,18] This pest lays its eggs on the fruiting forms of cotton and, immediately after hatching, the larvae bore into the flower buds and bolls. This behavior makes chemical control expensive and unsatisfactory.

Early research showed the pink bollworm diapauses in the last larval instar, mainly in the seeds of bolls that remain in the field after harvest.[19] Adkisson et al.[20] and Lukefahr et al.[21] showed that diapause is controlled by photoperiod and induction begins in September when day length becomes less than 13 hr. Diapause incidence then increases rapidly and attains a maximum in mid-October and early November. The seasonal onset of diapause can be predicted at any latitude.

The place and time of diapause in the pink bollworm was then used to achieve heavy suppression of overwintering populations by modification of certain cotton cultural practices.[22] Until this time, cotton growers usually allowed the plants to grow in a way that resulted in a lengthy period of boll opening and harvesting; plants often continued to grow and bolls to open after harvest was completed; plants were left undisturbed in the fields through the winter; and the next year's cotton planting times were determined by each grower. Such practices provided excellent overwintering conditions for this pest. Anti-pink bollworm cultural controls eliminated or modified all these conventional practices. The mature cotton plants were managed by use of defoliants and desiccants so that bolls opened at nearly the same time and were promptly harvested. Soon after harvest the plants were shredded mechanically and plowed deeply into the soil. The next year's cotton crops were not allowed to be planted in the spring until after a designated time, which was set well after most adult moths from the overwintering generation emerged and died.[17]

Early defoliation provides the initial step for pink bollworm suppression, followed by the remaining cultural procedures because most of the overwintering generation comes from eggs laid after mid-September. If a preharvest defoliant or desiccant is applied in late August or early September (before days are short enough to induce diapause) the mature bolls open and immature fruiting forms either shed or dry up, and little suitable food is left for larval development. But, if the application of the desiccant or defoliant is delayed until early October, it is relatively ineffective in reducing potential overwintering larval numbers because most of the larvae have already reached the diapause stage by this time.

When cotton is mechanically stripped, virtually all pink bollworm larvae are carried to the gin since the stripper leaves almost no bolls in the field. Almost 100% of the larvae are killed by the ginning process.[23] When cotton is harvested by a spindle picker, some immature bolls are left in the field and these may constitute a source of infestation in the next season. However, a rotary stalk cutter may kill from 50 to 85% of the larvae that remain on standing stalks after harvest.[18] When the stalks are plowed under immediately after shredding, the combined mortality from stalk shredding, plowing, and winter weather may exceed 90%.[18]

These cultural procedures have been very successful in Texas and insecticides are seldom needed for the control of this pest. The effectiveness of this program has largely depended on legislation that prohibits planting before an established date in the spring and requires plowing up of the crop by another date in the fall. Growers who do not comply with these dates are subject to a fine. There is also a "social" aspect to the success of this program because few if any growers care to be accused of contaminating their neighbors' fields.

A similar type of cultural program utilizing stalk destruction and blacklight trapping of the tobacco hornworm, *Manduca sexta* (Johan.), and tomato hornworm, *M.quinquemaculata* (Haworth), attacking tobacco has been developed in North Carolina.[24,25] Insecticidal treatments on tobacco for these two pests in a 113-mi^2 study area were reduced more than 90% following a wide-scale implementation of the program, and there was a reduction by 60% in applications for all tobacco pests.[24]

Strip Harvesting

Lygus bugs, primarily *L. hesperus,* are a key pest of cotton in the San Joaquin Valley, Calif. During May, most cotton fields are still in the seedling stage and unattractive to *Lygus.* By mid-June the situation changes and often large numbers of adults can be found in the fields (Figure 2).[3,4]

The adults come from two sources, alfalfa and safflower.[26] Thousands of acres of alfalfa are grown and adjoin cotton fields through the middle part of the Valley (N—S) and across the southern part of the Valley (E—W). When an entire alfalfa field is suddenly mowed *Lygus* is drastically affected. The humidity drops and the temperature rises quickly. This sudden change destroys the lygus bug's habitat, food, shelter, and oviposition sites. During hot weather, nearly all the adults leave solid cut fields within 24 hr after cutting (Figure 4).[3] The problem then is to stabilize the alfalfa hay environment to keep the lygus bug adults in the alfalfa where they do little or no damage.

This can be done by strip cutting the alfalfa fields. Under the strip cutting technique, the alfalfa is harvested in alternate strips so that two different aged hay growths occur in a field simultaneously. When one set of strips is cut, the alternate strips are about half grown. The field becomes a rather stable habitat because the lygus bugs move from the cut strips to the half-grown strips instead of flying to adjoining crops such as cotton, as they do when the entire field is cut at one time.

Since natural enemies (predators) of the *Lygus* also move from strip to strip, there is no increase in the lygus bug population in the alfalfa. Moreover, when the *Lygus* adults move into the uncut strips they deposit eggs in the half-grown hay. However,

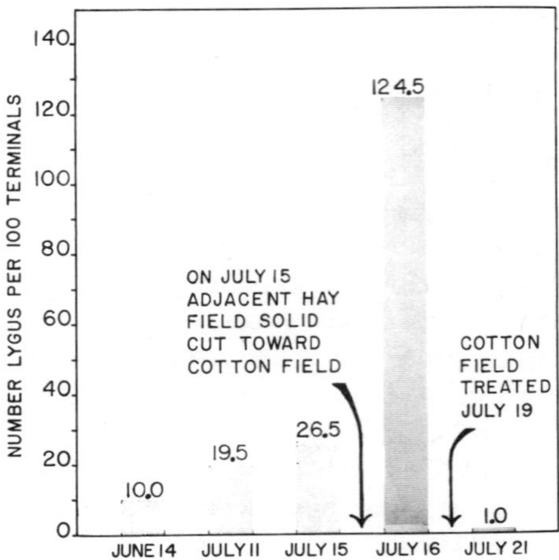

FIGURE 4. Movement of lygus bugs from alfalfa into cotton when an entire alfalfa field is suddenly cut.

these strips mature and are cut in about 2 weeks. The time required for *Lygus* development (egg to adult) is much longer and most of the nymphs and the unhatched eggs are destroyed by high temperatures and the drying of the mowed alfalfa.

Strip harvesting also reduces other pest problems in alfalfa itself, particularly as a result of more effective biological control. Van den Bosch et al.[27] found a reduction in pea aphid, *Acyrthosiphon pisum* (Harris), populations because strip harvesting favors its parasite *Aphidius smithi* Sharma and Rao. In strip-cut fields both the aphid and the parasite persist during midsummer, a time when populations of both species are disrupted in solid-cut alfalfa. In the latter fields, *A. smithi,* being host-density dependent, is particularly hard hit, not only by the adverse physical conditions but also by the protracted scarcity of its host. It is virtually eradicated from the fields and does not show vigorous activity again until autumn. By contrast, in strip-cut alfalfa the parasite remains abundant in continuous interaction with its host and quickly responds to the aphid upsurge in late summer.

Although intensive analyses have not been made of other parasite-host relationships, there are some indications that parasites of lepidopterous pests are also favored by strip cutting. This appears to be true for parasites of the alfalfa caterpillar, *Colias eurytheme* Boisduval, the beet armyworm, *Spodoptera exigua* (Hübner), and the western yellow-striped armyworm, *S. praefica* (Grote).[3]

REFERENCES

1. **Stern, V. M., Adkisson, P. L., Beingolea,G. O., and Viktorov, G. A.,** Cultural controls, in *Theory and Practice of Biological Control,* Huffaker, C. B. and Messenger, P. S., Eds., Academic Press, New York, 1976, chap. 24.
2. National Academy of Sciences, Cultural controls, in *Insect-Pest Management and Control,* Vol. III. National Academy of Science, Washington, D.C., 1969, chap. 10.
3. **Stern, V. M., van den Bosch, R., and Leigh, T. F.,** Strip cutting alfalfa for lygus bug control, *Calif. Agric.,* 18, 4, 1964.

4. **Stern, V. M., Mueller, A., Sevacherian, V., and Way, M.,** Lygus bug control in cotton through alfalfa interplanting, *Calif. Agric.,* 23, 8, 1969.
5. **van den Bosch, R., Schlinger, E. I., Dietrick, E. J., Hagen, K. S., and Holloway, J. K.,** The colonization and establishment of imported parasites of the spotted alfalfa aphid in California, *J. Econ. Entomol.,* 52, 136, 1959.
6. **DeBach, P.,** The scope of biological control, in *Biological Control of Insect Pests and Weeds,* DeBach, P., Ed., Reinhold, New York, 1964, chap. 1.
7. **Sorenson, C. J.,** The alfalfa- seed chalcid-fly in Utah, *Utah Agric. Exp. Stn. Bull.,* 218, 1930.
8. **Wildermuth, V. L.,** Chalcid control in alfalfa-seed production, *U.S.D.A. Farmers Bull.,* 1642, 1931.
9. **Bacon, O. G., Riley, W. D., Burton, V. E., and Sarquis, A. V.,** Clover seed chalcid in alfalfa, *Calif. Agric.,* 13, 7, 1959.
10. **Strong, F. E., Bacon, O. G., and Russell, J. R.,** Flight habits of the alfalfa seed chalcid *Bruchophagus roddi* Guss. (Hymenoptera: Eurytomidae), *Hilgardia,* 35, 1, 1963.
11. Anon., Alfalfa seed research pays dividends, *Western Hay and Grain Grower,* Cline, H., Ed., Munford, Fresno, Calif., 1978, 14.
12. **Munro, J. W.,** *Pests of Stored Products,* Hutchinson & Co., New York, London, 1966.
13. **Cotton, R. T.,** *Pests of Stored Grain and Grain Products,* 4th ed., Burgess, Minneapolis, 1963.
14. **Harper, R. W.,** Beet leafhopper, *Circulifer tenellus, Insect Pest Control Manual,* Bureau of Entomology, California Department of Food and Agriculture, Sacramento, Calif., 1959.
15. **Mueller, A. J. and Stern, V. M.,** Timing of pesticide treatments on safflower to prevent *Lygus* from dispersing to cotton, *J. Econ. Entomol.,* 67, 77, 1974.
16. **Sevacherian, V., Stern, V. M., and Mueller, A. J.,** Heat accumulation for timing *Lygus* control measures in a safflower-cotton complex. *J. Econ. Entomol.,* 70, 399, 1977.
17. **Adkisson, P. L. and Gaines, J. C.,** Pink bollworm control as related to the total cotton insect control program of Central Texas, *Texas Agric. Exp. Stn. Misc. Publ.,* 444, 1960.
18. **Noble, L. W.,** Fifty years of research on the pink bollworm in the United States, Agric. Handb. No. 357, U.S. Department of Agriculture, Washington, D.C., 1969.
19. **Ohlendorf, W.,** Studies of the pink bollworm in Mexico. *U.S. Dept. Agric. Bull.,* 1374, 1926.
20. **Adkisson, P. L., Bell, R. A., and Wellso, S. G.,** Environmental factors controlling the induction of diapause in the pink bollworm, *Pectinophora gossypiella* (Saunders), *J. Insect Physiol.* 9, 299, 1963.
21. **Lukefahr, M. J., Noble, L. W., and Martin, D. F.,** Factors inducing diapause in the pink bollworm, *U.S. Dep. Agric. Tech. Bull.,* 1304, 1964.
22. **Adkisson, P. L.,** Timing of defoliants and dessicants to reduce populations of the pink bollworm in diapause, *J. Econ. Entomol.,* 55, 949, 1962.
23. **Robertson, O. T., Stedronsky, V. L., and Currie, D. H.,** Kill of pink bollworms in the cotton gin and the oil mill, *U.S. Dep. Agric. Prod. Res. Rep.,* 26, 1959.
24. **Gentry, C. R., Lawson, F. R., Knott, C. M., Stanley, J. M., and Lam, J. J., Jr.,** Control of hornworms by trapping with blacklight and stalk cutting in North Carolina, *J. Econ. Entomol.,* 60, 1437, 1967.
25. **Rabb, R. L., Neunzig, H. H., and Marshall, H. V., Jr.,** Effects of certain cultural practices on the abundance of tobacco hornworms, tobacco budworms, and corn earworms on tobacco after harvest, *J. Econ. Entomol.,* 57, 791, 1964.
26. **Sevacherian, V. and Stern, V. M.,** Movement of lygus bugs between alfalfa and cotton, *Environ. Entomol.,* 4, 163, 1975.
27. **van den Bosch, R., Lagace, C. F., and Stern, V. M.,** The interrelationship of the aphid, *Acyrthosiphon pisum,* and its parasite, *Aphidius smithi,* in a stable environment, *Ecology,* 48, 993, 1967.

THE ENVIRONMENTAL CONTROL OF INSECTS USING PLANTING TIMES AND PLANT SPACING

George L. Teetes

INTRODUCTION

Insects become pests when they are present in sufficient numbers to cause economic damage and coincide temporally and spatially with host crops. Certain crop production practices can sometimes be manipulated to lessen crop loss by insect pests. When the length of the growing season exceeds the time required to produce a given crop, the time of planting may be varied. The ability to do so results in a logical management tactic which disrupts a synchronized crop-pest association. Likewise, plant spacing, by varying seeding rates or patterns, can influence crop-pest relationships and reduce insect damage. These are not new revelations, as the founders of applied entomology stressed the importance of such cultural practices to lessen crop loss by insect pests. The manipulation of crop planting time and plant spacing inevitably held a position of paramount importance in those earlier days of pest control because there were few alternatives as effective. The advent of chemical insecticides reduced the need for such cultural control practices as major pest control methods. Accumulated data on the subject collected dust on bookshelves and in files during the "chemical era." The effective insect control provided by chemical insecticides resulted in a void in the knowledge of these less dramatic forms of pest control.

The rejuvenation of an integrated pest management approach has renewed the interest in crop planting time and plant spacing in an insect control strategy. In the context of pest management, no implication is made that either of these cultural control methods can or should function alone as control tactics. Most frequently, each tactic is applicable to one or a few insect species of a usually large pest complex. Also, other factors (biotic and abiotic) influence the plant-pest relationship, which is certainly complex and a challenge for the scientist to understand. The pest manager seeks to use knowledge of this relationship to assure crop production, but in a manner which optimizes benefit-risk ratios. So, the assumption is made that crop planting time and plant spacing are cultural control methods that, in combination with other compatible direct control tactics, formulate a defense strategy against insect pests.

PLANTING TIMES

The objective of manipulating time of crop planting is largely to avoid a pest. By changing or carefully selecting the time when a crop is planted it may be possible to avoid the egg-laying period of a particular pest, get young plants well established to a tolerant stage before the attack occurs, allow a shorter period of susceptibility during which the insect will attack, or even get a crop matured before a certain pest becomes abundant.[2] Certainly, other objectives might be conceptualized, such as adjusting planting times to (1) synchronize insect pests and their natural enemies, (2) synchronize crop production with climatic conditions which adversely affect the pest, (3) synchronize crop production with available preferred alternate host plants of the pest, and (4) produce and destroy the crop before the insect can enter diapause.

For any of the objectives mentioned above, the intent of manipulating planting time is to render the desired crop less vulnerable to the insect. The nature of the insect-plant interaction is considered and used to select a planting time that results in a temporal separation of the potential pest and crop (asynchrony) or puts into force some biotic or abiotic factor that detrimentally affects the pest or leads the pest away from the crop.

Insect Pest Dynamics

The avoidance of inset pest damage by crop planting time relates to the population dynamics of the pest. Dynamics refers to the fluctuations (rise and fall) in numbers of pest individuals over periods of time. Insect numbers increase as long as intrinsic and extrinsic factors favor reproduction. But as one or more factors limit or retard the growth of a population, the rate of growth is decelerated, often to the point that the growth rate is negative, whereupon the population declines.

Insect density may fluctuate according to the rate of birth and death and according to migration. Insects increase in number according to natality and decline in number as a result of mortality for a stationary or sedentary species, but migration often governs the dynamics of an insect population.

If an insect pest population could continue to grow unchecked, it would, in theory, reach its full potential size, which is dependent on the carrying capacity of the system supporting the population. The continuing reproduction of the organism at a maximum rate under favorable conditions is called the biotic potential of that organism. In nature full biotic potential is never achieved, for conditions favorable to the reproduction of an organism never remain favorable indefinitely. Instead, factors (animate and inanimate) unfavorable to maximum reproductivity come into play, even at the outset, and limit or retard the rates of growth and result in fluctuations in population density.

Measures exercised in the attempts to control crop pests are, in essence, moving forces behind fluctuations in pest densities, and a successful control measure inhibits the rise and hastens the fall of the pest population. So, if a crop pest is to be kept below economically damaging densities, the population must be reduced rapidly and/or the rate of growth slowed down and/or the crop placed in asynchrony with the pest or favorable conditions for the pest. One way to accomplish the latter is by manipulating planting times. Concomitant with any of these approaches is an understanding of the concept of economic threshold levels.

Crop Dynamics

The avoidance of insect pest damage by crop planting time relates to the plant's biology, i.e., developmental characteristics and fruiting habits, as well as to the nature of damage that the pest under consideration causes. Like insect pest population dynamics, biotic and abiotic factors influence plant growth and development, and it is the science of agronomy that deals with these issues.

Crop plants may be classified in a number of ways, but our interest basically lies in plant growth and fruit production in periods of time and in relation to the nature of damage caused by certain insect pests. Therefore, the underlying force of crop planting time lies in the ability to adjust cropping systems in our favor and against insect pests. This ability is influenced by the inherent nature of the crop's growth characteristics. Flexibility may appear to be greater with annual crops, those that complete their life cycle in one year or less, than perennial crops which grow more or less indefinitely from year to year and may produce seed more than once. This may not always be true however, as some perennial plants are grown as annuals for agricultural production. Also, plants may senesce as a result of physiological factors, or, of course, growth and development of perennial plants may be terminated by climatic conditions.

The duration of growth and fruit development of crop plants also is an important consideration. One might expect that determinate crop plants are more amenable to planting time manipulation to escape pests than indeterminate crop plants. Determinate is used in reference to the growth of a plant with a stem that terminates in a floral bud and thus will not grow in length indefinitely, or in reference to flowering that occurs during a specific period, regardless of environmental factors. Indeterminate is used in reference to the growth exhibited by a stem that terminates in a vegetative bud and will thus continue to elongate. A plant that flowers more or less continuously until climatic conditions become unfavorable also is said to exhibit indeterminate growth.

However, the determinate-indeterminate classifications are not intended to relate solely to the time required for crop maturity and/or crop harvest, although these are certainly important characteristics. It is true that time of harvest surely is an important consideration relative to pest insects either during the production season or subsequent ones. Consequently, planting and harvest time may go hand in hand as a control tactic for some, but not all, pests.

Climate

Insect pests and crop plants are affected by climate. Climate is a regulating force underlying field crop-insect pest relationships, but is a nonmanipulatable one. This certainly does not preclude the use of climatic effects on insects and crops to some advantage to alter the crop-pest relationship. Such utilization requires a basic understanding of climatic factors, which combine in a variety of ways to engender local conditions.

For example, the importance of temperature in crop production is indicated by the fact that the major crops are classified as either "cool season" or "warm season" plants. In any biological system, the rate of growth, development, and reproduction is dependent on temperature, and, within limits, as temperature increases so do the rates until the higher developmental threshold is reached and the rates decline

Temperature affects plants and insects throughout their life cycle. Although a crop producer can do little to alter the temperature range, it is possible to adjust crop planting time to take advantage of temperature.

Notwithstanding the influence of temperature on the rate of plant growth and development, it is the influence of temperature on the insect pest itself, as expressed through its biology and habits, that makes this natural force so vital and useful in insect pest management.

In temperate zones, temperature (also photoperiod, diet, etc.) is responsible for the termination of insect activity (death, diapause, or estivation) at the end of the crop production period, and for the initiation of insect activity at the onset of the crop production period. In tropical zones, activity has evolved around and is regulated by factors associated with "wet" and "dry" seasons. In any case, there are initiating forces, then terminating forces that "break" the crop-pest association for some period of time, and which must be reestablished. Resumption of crop plant and insect pest association can be disrupted by adjusting the time of planting.

Use of Crop Planting Times

It is not the intent of the author to provide a review of the literature relating to crop planting times as an insect pest control tactic. Instead, examples of the tactic will be discussed. Appropriate examples will be used that relate to the three major or basic means to avoid insect pest damage, i.e., early planting, late (delayed) planting, and planting times that provide injury-free periods during the course of crop growth and development. One very interesting aspect of the literature relating to adjusting crop planting times to avoid insect damage is that in many cases the information on which the practice is based was empirically derived. Also, much of the data on the subject were collected prior to the "chemical era." One additional confounding problem is the regional effectiveness of the tactic. In other words, manipulating planting time in one locality may be quite effective, while in another it is not, or the planting schedule may change among regions. Consequently, the examples that are discussed are intended only to describe the principle under consideration.

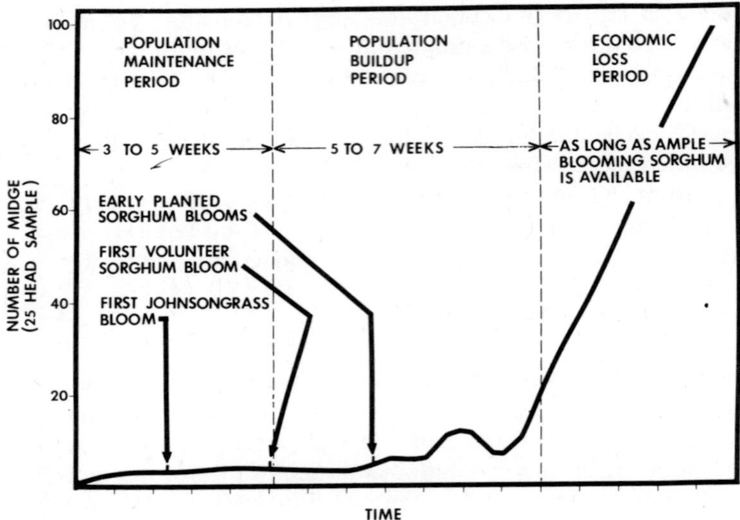

FIGURE 1. Seasonal sorghum midge build-up, illustrating the effectiveness of early sorghum planting to avoid damage.[20]

Early Planting

There are several classic examples regarding the use of early crop planting to avoid pest damage. In most cases the approach allows the production of a harvestable crop before the insect pests reach damaging numbers. The pest complex to which this practice is applicable is commonly referred to as ''late season'' pests, inferring that, inevitably, number and subsequent damage increase with lateness of crop development. As explained earlier, usually this is the consequence of the population dynamics of the pest as it reestablishes its density after a crop-free period. In many cases, alternate host plants play an important role in the pest reaching damaging proportions. The dynamics of these pests also may be an artifact of a chemically inundated system.

One of the most damaging insect pests of sorghum is the sorghum midge, *Contarinia sorghicola* (Coquillett) [Diptera:Cecidomyiidae]. This key pest deposits its eggs in floral spikelets during flowering and subsequent larval feeding prevents seed development. Overwintering occurs in the larval stage in the spikelets of host plants. Pupation and adult emergence occurs in the spring about the time johnson grass, *Sorghum halepense* (L.), begins to flower. Spring and early-summer generations are maintained on johnsongrass and other wild hosts, during which time one to three generations may occur. This is significant, since early sorghum is generally required in order for midge to reach damaging numbers. As early sorghum begins to flower, midges migrate from wild hosts to nearby early sorghum, the preferred host. Infestations in the sorghum planted early are not sufficiently damaging to warrant control. But, these early generations contribute directly to later generations which cause economic damage. The sequence of events described above is shown graphically in Figure 1.[20] Based on the population dynamics of the sorghum midge, early, uniform planting of sorghum on a regional basis is an effective means of avoiding midge damage to sorghum.[25]

Late Planting

There are several examples of delayed crop planting that could be cited to illustrate this principle. One of the more classic is provided by the use of so-called ''fly-free'' planting dates to avoid the Hessian fly, *Mayetiola destructor* (Say)[Diptera: Cecidomyiidae], in wheat. Before the development of fly-resistant varieties this was the most effective method of combating the pest.

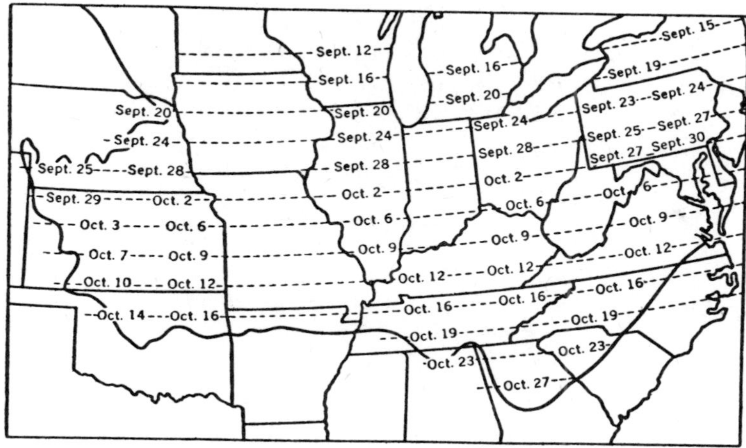

FIGURE 2. Fly-free dates for sowing winter wheat to avoid Hessian fly damage, illustrating effectiveness of delayed planting.[1]

Hessian fly eggs are laid on the upper surface of young winter wheat leaves. Larvae make their way to the leaf axis where they feed on plant juices, killing tillers and young plants in the fall. Summer and winter periods are passed inside the plant sheaths in the puparium or "flaxseed" stage. Based on these habits, the idea behind the practice is to delay fall seeding so that wheat will not be available for fly attack until most of the oversummering adult flies have emerged and died.

This cultural practice was so important that fly-free planting dates were established for infested areas of the U.S. (Figure 2).[1] The effectiveness of the practice is shown by data presented in Table 1 which also indicate that adequate plant development could be attained before winter in spite of the delayed planting.[2]

In the future, these planting date schedules will be improved by the use of thermal accretion unit models which can predict when optimal fly mortality has occurred and will predict the probability of having sufficient heat units to establish the crop. Thus, rather than using calendar dates which may be set later than necessary for some years, earlier plantings could be made without risk of fly injury. The same tactic has application to other crop-pest relationships.

Planting to Attain Injury-Free Periods

Some crop plantings may be adjusted in time to allow production during an injury-free period. This practice is amenable for pests whose numbers decline drastically for a period of time. This often occurs when the pest has discrete, nonoverlapping generations. Adjusting sunflower planting to escape damage by the sunflower moth, *Homoeosoma electellum* (Hulst) [Lepidoptera: Phycitidae], provides a good example.[3] The insect damages the seed of sunflower in the larval stage after eggs are deposited during flowering. The species has two to three generations per year. The first and second generations are very pronounced, with a decline in larval numbers occurring between the two peaks in larval density (Figure 3). Planting of the crop can consequently be adjusted to allow the crop to produce flowering heads during a period of low larval densities between generations. In a similar manner, Japanese beetle *(Popillia japonica* Newman) injury to corn silks may be avoided by planting the crop to silk either before[28] or after[29] the peak of adult beetle activity in the field.

Table 1
AVERAGE YIELDS OF WHEAT AND PERCENTAGE OF INFESTATION BY THE HESSIAN FLY FOR 8 YEARS IN FIELDS PLANTED BEFORE THE SAFE-SEEDING DATE AS CONTRASTED TO THOSE PLANTED AFTER THE SAFE-SEEDING DATE[2]

	Average yield		Average % of infestation	
Location of field	From wheat sown before the safe-seeding date, (bu)	From wheat sown after the safe-seeding date, (bu)	In wheat sown before the safe-seeding date (%)	In wheat sown after the safe-seeding date (%)
Rockford, Ill.	21.8	28.1	24.5	1.7
Bureau, Ill.	27.4	32.9	45.5	5.2
La Harpe, Ill.	30.8	36.5	38.0	1.8
Urbana, Ill.	29.5	37.1	32.6	5.4
Virden, Ill.	23.6	28.4	48.0	6.3
Centralia, Ill.	14.5	21.9	81.0	8.0
Carbondale, Ill.	21.5	23.9	16.0	1.0
Grand Chain, Ill.	15.5	21.4	32.3	1.0
Average	23.1	28.8	39.7	3.8

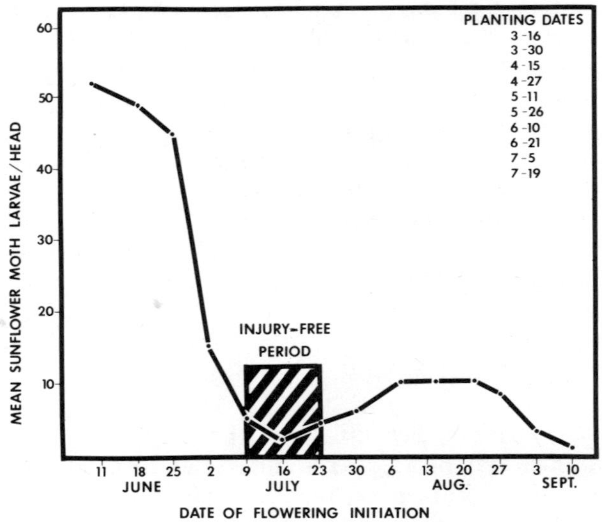

FIGURE 3. Seasonal sunflower moth larval density, illustrating injury free period to avoid damage.[3]

Factors Influencing Crop Planting Time

he time to seed or plant various field crops is governed not only by the necessity of evading the ravages of insect pests and diseases, but by the environmental requirements of the crop. One would surely recognize at this point that crop planting time might certainly result in a conflict of interest with regard to different pests and agronomic practices. In some cases, the latter may in fact override the manipulation of planting dates to avoid insect pest damage. For this reason, a basic understanding of crop production requirements is mandatory. The factors influencing crop planting time are discussed below with respect to both crop environmental requirements and the effects on insect pest population densities or damage. The major field crops, corn, sorghum, cotton, and wheat were chosen because of their importance and also because of the cross-section of crop types they represent.

1. Corn = determinate annual, warm season
2. Sorghum = determinate perennial grown as an annual, warm season
3. Cotton = indeterminate perennial grown as an annual, warm season
4. Wheat = determinate annual, cool season

Corn: A decided trend toward earlier corn planting in the central and northern states has been evident because research has shown corn grain yield advantages.[4] Earlier planting has become more practical because of effective weed control from herbicides, seed treatment, and improved seed quality.[5] Perhaps the best guide for choosing a corn planting date is to plant when the soil temperature at the 7.6-cm soil depth has reached an average of 15°C for several days. Corn planting begins about February 1 in southern Texas and progresses northward through the eastern two thirds of the country at an average rate of 13 mi/ day. General guides for optimum corn planting dates in the U.S. are available in published literature.[21] Planting dates apparently are less critical in the southern states where the growing season is longer than in the northern states.

Information is presented in Table 2 that shows the possible effect of crop planting time on insect pest density or damage. Early planting of corn would appear to be the preferred practice against most insect pests of the crop. However, early corn planting tends to increase population density or damage by seed-corn beetle, *Stenolophus lecontei* (Chaudoir), seed-corn maggot, *Hylemya platura* (Meigen), wireworms, grape colaspis, *Colaspis brunnea* (F.), armyworm, *Pseudaletia unipuncta* (Haworth), and European corn borer, *Ostrinia nubilalis* (Hübner). For the other listed pests, early planting tends to reduce population density or damage, or have no effect. Early planting would appear to be an important management tactic for corn root aphid, *Anuraphis maidiradicis* (Forbes), northern, *Diabrotica longicornis* (Say), western, *D. virgifera* Leconte, and southern corn rootworm, *D. undecimpunctata howardi* Barber, thrips, southwestern corn borer, *Diatraea grandiosella* (Dyar), corn earworm, *Heliothis zea* (Boddie), fall armyworm, *Spodoptera frugiperda* (J. E. Smith), chinch bug, *Blissus leucopterus* (Say), corn leaf aphid, *Rhopalosiphum maidis* (Fitch), mites, *Oligonychus* spp., and grasshoppers (several spp.).

Sorghum: Sorghum is usually planted after the soil becomes warm, generally 10 to 20 days after corn-planting time in a particular locality. However, a wide range of planting dates is possible, especially in states such as Texas, due to the length of the growing season and sorghum maturity groups available. Grain sorghum hybrids vary in time from 90 to 135 days from planting to maturity, and this results in a wide range of possibilities for sorghum planting dates. However, a number of factors influence dates of planting in relation to highest expected yield:[16]

1. Late maturing hybrids have higher yield potentials but must be planted earlier than medium- or early-maturing hybrids.
2. Seeds should be planted after the last spring frost when the soil is sufficiently warm to produce a strong, vigorous seedling.
3. In dryland areas, planting dates should take advantage of seasonal rainfall patterns.
4. Extreme high air temperatures result in lower yields.

Table 2
EFFECT OF CROP PLANTING TIME ON INSECT ABUNDANCE AND/OR POTENTIAL DAMAGE[a]

Corn[b]			Sorghum			Cotton			Wheat		
	Planting time			Planting time			Planting time			Planting time	
Insect	Early	Late	Insect	Early	Late	Insect	Early	Late	Insect	Early	Late
Seed-corn beetle[c]	+	−	Wireworm	+	−	Thrips	+	−	Wireworm	+	−
Seed-corn maggot	+	−	White grub	+	−	Cotton aphid	+	−	White grub	+	−
Wireworm	+	−	Flea beetles	−	+	Cotton fleahopper	+	−	Grasshoppers	+	−
White grub	?	−	Chinch bug	−	+	Boll weevil	−	+	Hessian fly	+	−
Corn root aphid	−	+	Thrips	+	−	Bollworm	−	+	Greenbug	+	−
Crape colaspis	+	−	Cutworms	+	−	Tobacco budworm	−	+	Corn leaf aphid	+	−
N. corn rootworm[c]	−	+	S. corn rootworm	−	+	Pink bollworm	−	+	Oat bird-cherry aphid	+	−
W. corn rootworm[c]	−	+	Greenbug	+	−	Cotton leafworm	−	+	Cutworms	+	−
S. corn rootworm	−	+	Corn leaf aphid	−	+	Cabbage looper	−	+	Armyworms	+	−
Cutworms	0	+	Fall armyworm	−	+	Lygus bug	−	+	Chinch bug	0	0
Billbug	0	0	Beet armyworm	−	+	Stink bug	−	+	Wheat strawworm	0	0
Thrips	−	+	Corn earworm	−	+	Spider mites	−	+	Wheat stem maggot	+	0
Mite	−	+	Sorghum midge	−	+	Cotton square borer	+	−	Common stalk borer	0	0
Eur. corn borer	+	−	Sorghum webworm	−	+	Salt marsh cat.	0	0	Lesser cornstalk borer	+	−
S. W. corn borer	−	+	False chinch bug	?	?	Garden webworm	−	+	Flea beetles	+	−
Corn earworm	−	+	Leaf-footed plant bug	−	+	Cutworms	+	−	Leafhoppers	−	+
Fall armyworm	−	+	Stink bugs	−	+	Leaf miners	−	+	Leaf-feeding sawflies	?	?
Armyworm	+	−	S. W. corn borer	−	+				Thrips	+	−
Chinch bug	−	+	Sugarcane borer	−	+				Wheat curl mite	+	−
Corn leaf aphid	−	+	Sugarcane rootstock weevil	0	0				Wheat stem sawfly	0	0
Grasshoppers	?	+	Spider mites	+	1				Wheat jointworm	0	0
			Yellow sugarcane aphid	+	−						

Note: + = Increase in population density or damage, − = reduction in population density or damage, 0 = no effect, ? = effect unknown.

[a] Compiled by the author in consultation with entomologists familiar with the subject. Most of the information presented is from published research results, but some may represent the views of entomologists.

[b] North Central States.

[c] Influence is different in Southeast.

5. Conditions likely to exist at harvest time also need to be considered, as delayed harvesting due to unfavorable drying conditions can result in losses due to lodging, weathered grain, grain sprouting in the head, and immature green heads from late tillering.

Planting sorghum in South Texas, particularly the Coastal Bend, begins in mid-February and proceeds northward at successively later dates. Most of the sorghum from North Texas to South Dakota is planted in late May and June.

For most insect pests of sorghum, early planting would appear to be an effective management tactic, since population densities and damage increase with the lateness of the crop. The exceptions are several species of soil pests, thrips, greenbug, *Schizaphis graminum* (Rondani), yellow sugarcane aphid, *Sipha flava* (Forbes), and spider mites (Table 2). Apparently, spider mites are less severe on late-planted sorghum than early-planted sorghum because mites increase during the reproductive stages of the crop. Cooler, wetter fall weather tends to discourage mite build-up.

Historically, the chinch bug and sorghum midge have greatly influenced sorghum planting dates. In many areas of the U.S. an early planting date has proved essential to be an effective tactic against these pests. In Asia and Africa early planting lessens damage by the sorghum shootfly, *Atherigona varia soccata* Rondani and various grasshoppers.[26]

Cotton: Planting dates for cotton become later from south to north and always follow the date of the last killing frost, frequently by as much as two weeks. Ideally, planting should be delayed until the soil temperature reaches at least 60°F (16°C). This occurs in early March along the Gulf Coast and in Arizona, and after mid-April in most of the cotton belt and in California.

Most cotton pests increase numerically, resulting in greater damage, during later parts of the growing season. Thus, late-planted cotton is exposed to large numbers of pests such as the boll weevil, *Anthonomus grandis* Boheman, tobacco budworm, *Heliothis virescens* (F.), bollworm, *Heliothis zea* (Boddie), and pink bollworm, *Pectinophora gossypiella* (Saunders). However, delayed planting of cotton to cause suicidal emergence of overwintering pink bollworms and boll weevils has been shown to constitute promise as a management tactic. The damage by thrips, fleahoppers and aphids is sometimes more severe in early-planted cotton as lower temperatures retard growth which results in an intensification of pest-induced plant damage.

Wheat: Medium-season seeding of winter wheat for any locality is usually most favorable. Late-sown wheat generally suffers more winter injury, tillers less, and may ripen later the next season. Wheat sown too early may use up soil moisture, joint in the fall, and suffer from winter injury and disease. In the semiarid Great Plains the optimum date for seeding winter wheat is about September 1 in Montana and progressively later in the southwest to northwestern Texas, where the best time for seeding is about October 15. Planting earlier than the optimum seeding time is a common practice in the Great Plains, either to take advantage of favorable soil moisture conditions when they occur or to provide wheat pasture.

A disadvantage of too early seeding of winter wheat is the likelihood of injury from fungus diseases that develop under warm conditions. Also, early planting of wheat generally increases the severity of insect pests (Table 2). In the central and eastern states the optimum seeding date is about the time of, or a few days earlier than, the Hessian fly-free date (Figure 2).

Wheat, being a cool season crop, is an interesting candidate for manipulating planting time, exemplified by the practice to avoid Hessian fly damage. The tactic is also effective for avoiding damage by white grubs. Wheat planting can be delayed until after white grubs have stopped feeding and have entered the diapause state.

Time of wheat planting in the fall would appear to have little effect on pest populations during the following spring.

Other crops: There are certainly crops in addition to those discussed to this point for which planting time influences pest density or damage and where selected planting times are used to lessen crop damage. For example, among vegetable crops, cucumbers may be planted early to avoid the pickleworm, *Diaphania nitidalis* (Stoll), as the pest migrates northward in the summer from overwintering in the Gulf Coast states.[17] Field observations have indicated that early-planted potatoes are more subject to attack by psyllids, *Paratrioza cockerelli* (Sulc), than later plantings.[12,13] In some localities, sugar beets are planted early to escape beet leafhopper, *Circulifer tenellus* (Baker), the vector of the virus causing curly top disease, *Ruga verrucosans* Carsner and Bennet. In other areas it is advisable to delay planting until after the spring leafhopper migration.[14]

For centuries, Chinese peasants have used the proper selection of grain-sowing dates to prevent damage to rice by the rice borer, *Schoenobius incertellus* (Walker); to wheat by the wheat stem maggot, *Meromyza americana* (Fitch); and to millet by *Chilotraea infescatellus* (Smell).[22,23]

An example involving more permanent crops involves the weevil, *Hylobius pales* (Herbst). Weevil damage to young pines planted in areas recently cleared of pine stands by harvest or fire can be minimized by delaying reforestation.[14] A 9-month delay is suggested in the southern U.S. and at least 2 years in northern areas.[24] The delay allows weevils to breed and mature in stumps and remnant escape trees, and then disperse from the area when suitable habitats are no longer available.

Early harvest: Losses due to pest damage often can be greatly reduced by early harvesting. In some cases optimum planting time influences optimum harvesting. Such early harvest practices have been useful against such pests as the sugar cane borer, *Diatraea saccharalis* (F.), sweet potato weevil, *Cylas formicarius elegantulus* (Summers), potato tuberworm, *Phthorimaea operculella* (Zeller), cabbage looper, *Trichoplusia ni* (Hübner), imported cabbage worm, *Pieris rapae* (L.), and a multitude of field-infesting stored grain pests such as the pea weevil, *Bruchus pisorum* (L.). Selective harvesting of lumber by a system of early detection of bark beetle damage and removal of infested trees has reduced economic losses considerably in many forest areas.[11]

In alfalfa, a crop which is planted only once in a 3- to 10-year period and repeatedly harvested throughout the season, early cutting of the first harvest can often provide good control of the alfalfa weevil, *Hypera postica* (Gyllenhal).[15,27]

PLANT SPACING

The major objective in spacing crop plants is to obtain the maximum yield on a unit area without sacrifice of quality. As this statement implies, most concern, even in research, has been for yield advantages and less for influences on insect pest density or damage. With regard to the effect of plant spacing on insect pests or their damage, there are at least three basic considerations: (1) plant spacing as it influences the microenvironment of the pest species, (2) plant spacing as it relates to the health, vigor, and strength of the plant, and (3) plant spacing as it influences the pattern or duration of crop growth and development. It is likely that these influences on pest density or damage often work together. In such instances, individual effects might be difficult to distinguish.

Plant Spacing and Microenvironment

The chinch bug tends to attack in the thinner or poorer stand areas of grain fields.[8] They seldom do serious injury to heavy stands. Consequently, recommendations for

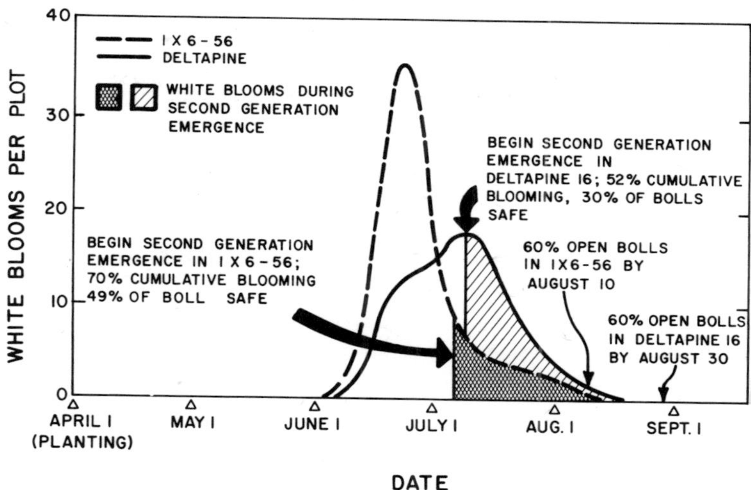

FIGURE 4. A hypothetical figure showing how the rapid fruiting sequence of double-drilled cotton might reduce boll weevil damage and allow early defoliation, illustrating the influence of plant spacing on damage.[11]

control of the pest frequently include practices that encourage thick, vigorous stands of grain crops. Such practices are intended to increase shading, which appears to be a condition not preferred by the pest. Such shading also tends to increase the incidence of fungal pathogen infection which can be an important mortality factor of the chinch bug.

Plant Spacing in Relation to Plant Health, Vigor, and Strength

There are numerous examples of control suggestions which encourage healthy and vigorous plant growth to reduce crop loss by insects.

The use of plant spacing as an insect management tactic to ensure strong plant stalks will reduce lodging due to stalk borer attack. This approach is especially effective for reducing lodging of corn due to the feeding and girdling of second generation southwestern corn borer.[6] A reasonable plant population (22,000 to 24,000 plants per acre) ensures large, healthy stalks. Proper fertilization and adequate irrigation will help prevent lodging of borer-infested plants. Densely planted corn is more likely to be girdled than less densely planted corn.[7]

Plant Spacing and Crop Growth Duration

Plant spacing which encourages rapid crop maturation is obviously a candidate as a control tactic. There are numerous examples that might be cited, but the modern cotton production systems being utilized in Texas are very appropriate at this point.

Cotton genotypes with early fruiting characteristics were recognized long ago as being valuable in producing a crop in areas where the boll weevil, *Anthonomus grandis* Boheman, is a pest.[9] Several of the modern varieties in Texas possess early fruiting characteristics, but, unfortunately, the expression of earliness is related to the environment in which the cotton is grown. If these varieties are grown in irrigated, highly fertile land, the early maturing potential often is lost. However, recently released cotton varieties retain early fruiting characteristics under diverse growing conditions. Such earliness is further enhanced by narrow-row planting.[10] This concept is being used, in conjunction with early season insecticide application, in management schemes against the boll weevil in Texas.[11,19] The system is graphically illustrated in Figure 4, and although designed for the boll weevil, it is also effective in dealing with pink bollworm.

CONCLUSION

The ecological simplicity of field crops produced in monoculture demands contemporary insect control strategies that optimize pest control in terms of the overall economic, social and environmental values. To this end, integrated pest management (IPM) schemes must take a production systems approach utilizing multiple pest suppression tactics that work together in a compatible manner to maintain pests below economically damaging levels. Although IPM by its very definition implies an applied science, the basis for this systems approach is an understanding of fundamental, ecological crop-pest relationships. In short, IPM is an ecological approach to insect control. The adversities of the use of chemical insecticides as a unilateral control tactic have largely been a result of the failure to consider ecological consequences. The lessons learned in respect to chemical control must be put to use, as IPM suggests it will.

The advantages to the use of such cultural practices as planting times and plant spacing to minimize crop loss by insects and mites result from the intimate level of crop-pest association. The demand for a comprehensive understanding of how these practices might formulate part of the IPM system is great and must begin with a fundamental knowledge of the biological and ecological relationship of the pest and its host(s).

It has been impossible for the author to document the economic returns as a result of using planting times and plant spacing as IPM components of pest suppression. In terms of implementation, there is little or no monetary increase in input (except possibly more planting seed or stock), since the cost of planting remains essentially the same. The monetary results from these practices, although not easily definable, are tremendous. The role that planting time has played in regard to avoiding wheat losses by Hessian fly, sorghum losses by sorghum midge, corn losses by corn earworm, and cotton losses by cotton bollworm and tobacco budworm constitute millions of dollars annually. The role plant spacing has played in regard to lessening corn loss by borers has resulted in a proportionate economic return.

REFERENCES

1. **Walton, W. R. and Packard, C. M.,** The Hessian fly and how losses from it can be avoided, *U.S. Dep. Agric. Farmers' Bull.*, 1627, 1936.
2. **Metcalf, C. L. and Flint, W.P.,** *Destructive and Useful Insects,* McGraw-Hill, New York, 1951.
3. **Mitchell, T. L., Ward, C. R., Teetes, G. L., Schaefer, C. A., Bynum, E. D., and Brigham, R. D.,** Sunflower pest population levels in relation to date of planting on the Texas High Plains, *Southwest. Entomol.*, 3, 279, 1978.
4. **Larson, W. E. and Hanway, J. J.,** Corn production, in *Corn and Corn Improvement,* Sprague, G. F., Ed., American Society for Agronomy, Madison, Wisc., 625, 1977.
5. **Rossman, E. C. and Cook, R. L.,** Soil preparation and date, rate, and pattern of planting, in *Advances in Corn Production: Principles and Practices,* Pierre, W. H., Ed., Iowa State University Press, Ames, 53, 1966.
6. **McWhorter, G. M.,** Insect and mite pests of field corn, *Tex. Agric. Ext. Ser. Leafl.*, 1417.
7. **Zepp, D. B. and Keaster, A. J.,** Effects of corn plant densities on the girdling behavior of the southwestern corn borer, *J. Econ. Entomol.*, 70, 678, 1977.

8. **Packard, C. M. and Benton, C.,** How to fight the chinch bug, *U.S. Dep. Agric. Farmers' Bull.,* 1780, 1937.
9. **Hunter, W. D.,** The boll weevil problem with special reference to means of reducing damage, *U.S. Dep. Agric. Farmers' Bull.,* 344, 25, 1971.
10. **Niles, G. A.,** Growth and fruiting modifications for mechanized production, in *Beltwide Cotton Production Research Conferences,* 114, 1969.
11. **Walker, J. K., Jr. and Niles, G. A.,** Population dynamics of the boll weevil and modified cotton types, *Tex. Agric. Exp. Stn. Bull.,* 1109, 1971.
12. **Wallis, R. L.,** Ecological studies on the potato payllid as a pest of potatoes, *U.S. Dep. Agric. Tech. Bull.,* 1107, 1955.
13. **Dudley, J. E., Jr., Landis, B. J., and Shandes, W. A.** Control of potato insects, *U.S. Dep. Agric. Farmers' Bull.,* 2040, 1952.
14. **Anon.,** Cultural control, in *Insect-Pest Management and Control,* National Academy of Sciences, Washington, D.C., 1969
15. **Armbrust, E. J. and Byrisco, G. G.,** Forage crops insect pest management, in *Introduction to Insect Pest Management,* Metcalf, R. L. and Luckman, W. H., Eds., John Wiley & Sons, New York, 1975.
16. **Onker, A. B.,** Cultural practices for grain sorghum production, *Tex. Agric. Exp. Stn. Consol.,* PR-2938, 5, 1971.
17. **DeBach, P.,** *Biological Control by Natural Enemies,* Cambridge University Press, Cambridge, England, 1974.
18. **Stern, V.,** Economic thresholds, *Annu. Rev. Entomol.,* 18, 259, 1973.
19. **Walker, J. K., Niles, G. A., Gannaway, J. R., Bradshas, R. D., and Glodt, R. E.,** Narrow row planting of cotton genotypes and boll weevil damage, *J. Econ. Entomol.,* 69, 249, 1976.
20. **Thomas, J. G.,** The sorghum midge and its control, *Tex. Agric. Ext. Serv. Leafl.,* 942, 1969.
21. **Sprague, G. F. and Larson, W. E.,** USDA Agricultural Handbook, U.S. Department of Agriculture, Washington, D.C., 332, 1966.
22. **Cheng, T. H.,** Insect control in mainland China, *Science,* 140, 269, 1963.
23. **Nash, R. G. and Cheng, T. H.,** Research and development of food resources in communist China. I. *BioScience,* 15, 643, 1965.
24. **Speers, C. F. and Rauschenberger, J. L., Pales weevil,** *U.S. Dep. Agric. Forest Pest Leafl.,* 104, 1967 rev. 1971.
25. **Wiseman, B. R. and McMillian, W. W.,** Relationship between planting date and damage to grain sorghum by the sorghum midge, *Contarinia sorghicola* (Diptera: Cecidiomyiidae), in 1968, *J. Ga. Entomol. Soc.,* 4, 55, 1969.
26. **Schmutterer, H.,** *Pests of crops in Northeast and Central Africa,* Gustav Fischer, Stuttgart, 21, 1969.
27. **Casagrande, R. A. and Stehr, F. W.,** Evaluating the effects of harvesting alfalfa on alfalfa weevil (Coleoptera: Curculionidae) and parasite populations in Michigan, *Can. Entomol.,* 105, 1119, 1973.
28. **Woodside, A. M.,** Japanese beetle damage to corn as influenced by silking date, *J. Econ. Entomol.,* 47, 349, 1954.
29. **Cory, E. N. and Langford, G. S.,** The Japanese beetle retardation program in Maryland, *Univ. Md. Ext. Bull.,* 156, 1955.

THE ENVIRONMENTAL CONTROL OF INSECTS USING CROP DIVERSITY

William J. Cromartie, Jr.

INTRODUCTION

A wide range of evidence exists that pest problems in crops are aggravated by the reduction in environmental diversity associated with cultivation. Ecological research has begun to reveal some of the reasons that insect attack is often more severe on plants grown in monoculture than in crop mixtures or in diverse natural vegetation. The applicability of this knowledge to problems of crop production, however, is uncertain. Practical tests of crop diversification as an economical means of controlling pests are still very few.

Monoculture and Pest Outbreaks

Agricultural crops are usually grown in monocultures (fields containing a single kind of plant) or in relatively simple mixtures that are much less diverse in number of plant species present than the native vegetation they replace. With the development and spread of modern agricultural techniques, there has been a trend to more extensive monocultures; clean cultivation, with machinery or herbicides, may permit large fields to contain only a single species — the crop being grown. There is a continuing trend towards fewer crops within a given agricultural region, so that large areas are becoming less diverse than in the past.[1] Thus, modern crops are exposed to conditions very unlike the relatively rich associations of plants in which their wild ancestors probably evolved.

Along with the plants, agriculture also subjects the whole fauna (plant-associated insects, mites, spiders, nematodes, and so on) to the novel environmental conditions it creates. As Pimentel[2] and Root[3] point out, agriculture reduces the multispecies, compound community of the forest or meadow to a single component community containing the crop plant and those herbivores, predators, and parasites adapted to living upon it, under the physical conditions created by cultivation. Not all animal species are equally well suited to exploit the new conditions. Some may be unable to survive at all and will not be found in the crop situation; others will have all their needs met within the monoculture and will find the microclimate suited to rapid population growth. These species may increase greatly in abundance, becoming dominant.[4] Furthermore, the biological situation is dynamic; species will adapt to the changed conditions. Geographic range, behavior, and tolerance of physical conditions will shift as the new crop conditions and the species' own evolutionary potential permit. Species that did not previously feed upon the plant may attack it when it is grown in cultivation.

Early in this century, foresters began to note that insect outbreaks were apparently more frequent and more severe in pure, even-aged stands of forest trees, and especially in plantations.[2,4] Gibson and Jones[5] cite numerous examples where monoculture has been blamed for forest insect outbreaks. While in many cases, other factors such as poor site or poor sanitation seem to be the real problem, they conclude that in other situations monoculture is the cause. The most important single factor seems to be the reduction in average distance between individuals of the same species and the same age in plantations compared to natural forests. For example, spruce budworm outbreaks in Canada are more likely to occur in stands with a high percentage of the main host, balsam fir.[6,7] In addition, the reduction of diversity in stands where birch and poplar died out may have triggered some outbreaks.[6,7]

Recognizing that pest problems arise as a consequence of the ecological and evolutionary conditions created by agriculture, a number of workers have shown that pest populations are higher and cause greater crop losses in monocultures than in diverse stands where weeds or other crop plant species occur.[2-4, 8-17] These same studies have begun to show how specific kinds of alterations in crop diversity can influence particular pest populations. At the same time, there have been efforts to evaluate the general usefulness of diversity in agriculture.

Diversity and Stability

The observations on forest insect problems and the work on crops mentioned above, together with certain theoretical ecological considerations,[18] have led some investigators to argue that diverse systems are inherently more stable than simple ones and that pest problems might be ameliorated by increasing either within-field or regional diversity in agriculture through intercropping, allowing weeds to grow in crops, or by having a greater mix of different crops grown in small fields separated by hedgerows, woodlots, or uncultivated meadows.[19] In this view, diversity is valuable because of the stability it is thought to confer.[19,20]

Critical examination of this concept has, however, revealed that it is unlikely to be of much help in solving pest problems. First, stability by itself is not a solution if a pest population is stable at a level that causes economic loss.[18,21-23] Second, instability is characteristic of annual cropping systems because of plowing, fertilizing, cultivating, harvesting, crop rotation, abandonment of fields, and so on,[20-24] and many successful pest control techniques, both chemical and biological, involve nonstabilizing or catastrophic impacts on the target population.[18] Third, there is little sound evidence to show that diversity is a cause of stability in ecological systems.[21,22] In some cases simply increasing species diversity may actually reduce stability.[18]

Kennedy[24] and Murdoch[22] have concluded that diversity in itself is not necessarily useful in agriculture, and that the supposed stability of natural systems is not of a sort that can or should be sought for crops. Others have come to similar conclusions.[25-27]

Importance of Population Studies

The value of crop diversification needs to be understood in a more refined way than the simple diversity/stability hypothesis. Most of the authors who reject the diversity/stability approach urge a careful, case-by-case consideration of the underlying mechanisms by which diversity can help control pest losses. It is necessary to understand the plant/pest and pest/enemy interactions in various kinds of situations so that the level and kind of diversity appropriate to a particular situation can be determined. In some cases a reduction in diversity may be needed, in others a very small addition of diversity may be effective, and other times large increases may produce the best solution.[18,22,25,27] The remainder of this chapter will treat these issues in detail, beginning with a look at the critical components of the crop/pest/enemy system, and at a number of studies of specific pest populations in simplified and diversified crops (Table 1).

INFLUENCE OF DIVERSITY ON CROP PESTS

Diversity of Crop Systems

The diversity of a crop system can be affected by a variety of factors.[25] First, one or several kinds of crops may be raised in various patterns: row intercropping, strip intercropping, random mixture, or undersown cover crops. Second, efforts to eliminate weeds influence diversity. All weeds or certain selected species may be allowed to remain, either throughout the field or only in the spaces between rows. Orchards may

Table 1
STUDIES OF THE EFFECT OF CROP DIVERSIFICATION ON POPULATION DYNAMICS OF INSECT PESTS

Locality	Crop	Principal treatments	Major pest species	Ref.
New York, U.S.	Collards (*Brassica oleracea*) and other *Brassica* crops	Hoed plot; weedy old field	Flea beetles (*Phyllotreta striolata*), aphids (*Myzus persicae, Brevicoryne brassicae*), caterpillars (*Pieris rapae*)	2
Huntingdon, England	Brussels sprouts (*B. oleracea*)	Hoed plot; weedy plot; herbicide plot	Caterpillar (*P. rapae*)	10
Berkshire, England	Brussels sprouts	Hoed plot; weedy plot (weeds cut back to 15 cm 1976)	Aphids (*M. persicae, B. brassicae*), whiteflies (*Aleyrodes brassicae*), caterpillar (*P. rapae*)	11, 15
New York, U.S.	Collards	Monoculture plot; diverse row in old field; tomato and tobacco interplanted	Flea beetle (*Phyllotreta cruciferae*)	4
New York, U.S.	Collards	Monoculture plot; diverse row in old field	Flea beetle (*P. cruciferae*), aphids (*M. persicae, B. brassicae*), caterpillar (*P. rapae*)	3
Huntingdon, England	Cauliflower and Brussels sprouts	Hoed plot; undersown white or red clover	Cabbage root fly (*Erioischia brassicae*), aphids (*B. brassicae*), caterpillar (*P. rapae*)	12
Cambridge, England	Brussels sprouts	Undersown white clover, 0, 25, 50, 75, 100% cover	Same as above	78
New York, U.S.	Collards	1; 10; 100-plant monocultures; 10-plant plot in old field (mowed)	Flea beetle (*P. cruciferae*), (*P. striolata*), aphids, caterpillar (*P. rapae*)	14
Valle De Cauca, Colombia, S.A.	Beans (*Phaseolus vulgaris*)	Hoed plots; weedy plots; plots with grass border	Leafhopper (*Empoasca kraemeri*), leaf beetle (*Diabrotica balteata*)	16
San Joaquin Valley, California, U.S.	Grapes (*Vitis sp.*)	Vines without weedy grasses; vines with weedy grasses	Mites (*Eotetranychus willamettei*)	9
Los Banos, Philippines	Cabbage (*B. oleracea*)	Monoculture; intercropped with tomato	Caterpillar (*Plutella xylostella*)	13
Costa Rica	Beans, squash, maize	Monoculture; polyculture (intercropping)	Leaf beetles (*D. balteata, D. adelpha, Acalymma themei*)	51
Valle De Cauca, Colombia, S.A.	Beans (*P. vulgaris*), corn (*Zea mays*)	Monoculture; intercropped	Leafhopper (*E. kraemeri*), leaf beetle (*D. balteata*), cutworm (*Spodoptera frugiperda*)	79

also contain mixtures of permanent crops, along with annual or perennial cover crops and various weeds. Finally, the size and arrangement of fields and field margins, orchards, woodlots, and hedgerows will produce a more or less complex mosaic of crops, meadows, fallow areas, and so on, and thus influence how far on average the crop plants are from one another and from more diverse vegetation.

Components of Pest Attack

The level of damage by an insect pest to a crop depends in part on the following factors: (1) probability of individuals of the pest species finding the crop; (2) probability that, having found the crop, they will remain in it to feed or reproduce; (3) the number of offspring, if any, they produce; (4) the rate at which they or their offspring feed; (5) the length of time (before death or emigration) that the pests or their offspring feed. The effects of the first two factors may be considered together as the level of colonization, while the effects of the last three determine the cumulative amount of feeding on the crop. All of these factors can be influenced by the diversity of the environment in which the crop is grown, and the beneficial effects of crop diversification can be attributed to one or more of these effects.

Whether the effect of diversity is ultimately beneficial or not depends on the biology of the particular pest species involved. By far the greatest amount of experimental work to date has been done on the effects of diversity manipulations within fields, since experiments on a truly regional scale are obviously difficult to manage. Small scale diversity manipulations, however, can be effective in reducing pest problems even when the species involved have very high dispersal powers, such as aphids.[27] The success of pests in exploiting crops is a complex process which is subject to disruption or enhancement at several crucial points, as described in the next five sections.

Effects of Diversity on Colonization Success of Pests

For many herbivorous insects the discovery of a suitable hostplant is an essential phase in the life cycle. For species which feed on annual crops (especially where crop rotation is practiced), overwinter away from their hosts, use alternate hosts during the year, or require resources (such as nectar) from outside the crop, the discovery of the hostplant is a recurring problem. Complex visual and olfactory methods are employed by many of these insects in their search for food.[28-30] Disruption of the searching behavior through crop diversification can be a useful tool in pest control.

Insects' difficulty in finding a host is likely to be greater in situations where the foodplant is dispersed among other vegetation.[31] When fields of the same crop are small and far apart, colonization is likely to be less successful than when fields are big and close together.[26,32] In general, it can be stated that outbreaks of insects may be less severe when the environment contains a relatively low proportion of food usable by the particular insect.[18,33] The need for insecticide treatments on cotton was much lower in Costa Rica, where the crop was grown in relatively scattered fields, than in Guatemala, where it was grown over extensive areas.[26]

It should be apparent, however, that the large scale changes necessary to reduce the overall abundance of a crop within a region are seldom possible. The degree of regional diversity needed to isolate fields of the same crop from each other, and so reduce pest movement from field to field, depends on the vagility of the pest species involved,[25] and is often on the scale of many tens or hundreds of kilometers. The more promising approaches to reducing colonization rely on reducing the attractiveness of the crop to arriving colonists.

Olfactory Inhibition

Many, if not most, crop plants are derived from plants of early successional habitats, grasslands, meadows and so on. For such plants the presence of other plants of different species may be an important component of their defense against herbivorous insects. The "associational resistance"[4] which plants in diverse vegetation possess is in large part a result of their being hard for their herbivores to find.[34-36] The difficulty of discovery results from the plants being dispersed in a community of high spatial, biotic, and microclimatic complexity. In particular, the odors which such insects use in locating their hosts will be mixed with the odors of other plants, leading to confusion, antagonism, and repellance, and a breakdown of the patterns of orientation to the host.[4] Feeny[35] has termed this masking effect a reduction in plant "apparency". Agriculture, by producing large monocultures, renders apparent and easy to find plants whose ancestors survived by being the opposite.[4,34,35]

Tahvanainen and Root[4] have demonstrated on a specialized pest of cole crops the effect of increasing crop diversity on colonization. Collards grown with tomato or tobacco interplanted had many fewer individuals of the flea beetle *Phyllotreta cruciferae* (Goeze) than did collards grown alone (Table 2). Lab experiments showed that the odor of tomato and ragweed interfered with orientation and feeding by *P. cruciferae* (Table 3). Root[3] and Cromartie[14] also found great reductions in flea beetle infestation of collards set out in diverse meadow vegetation, compared to cultivated stands (Figures 1, 2, and 3). Crushed tomato leaves repel flea beetles *(Podagrica* sp.) from cotton,[27] and diamondback moth larvae *Plutella xylostella* (L.) are less abundant on cabbages interplanted with tomatoes.[13] In a study of bean cropping systems, the leafhopper *Empoasca kraemeri* Ross and Moore was found to be repelled by the odor of weedy grasses (Figures 4, 5, and 6, Tables 4 and 5).[16] Finally, organic gardeners have reported on a wide variety of repellent effects, mostly from interplanting aromatic herbs such as basil, garlic, and marigold with vegetable crops, but also from interplanting tomatoes and asparagus, potatoes and beans, etc.[37-38]

Feeny[35,36] has suggested that interplanting two or more varieties of the same crop bred to emit different odors might interfere with colonization by pests, but no experimental evidence on this is available. Deterrence of colonization by olfactory inhibition remains, however, one of the best documented beneficial effects of increased diversity.

Visual Effects

Diversification of the crop can also inhibit visual orientation by pests. Aphids have been shown to colonize plants more readily when they stand out against a background of bare soil (Figures 3 and 7, Table 6).[11,14,15,39,40] If the contrast between the crop and the background is minimized by allowing weeds to remain in the crop,[1,15] by interplanting a cover crop,[12] or even by putting down layers of green burlap between rows,[15] far fewer aphids are captured by pan or suction traps over the crop, and fewer are found on the plants (Figure 7).

Cabbage butterflies *Pieris rapae* (L.) oviposited more heavily in plots with a brown burlap background versus a green burlap background.[15] Interplanting peanuts with maize reduced the numbers of corn borers *Ostrinia furnacalis* Guenee.[41] Green burlap sacks placed between the rows of maize to simulate peanut plants were also effective, thus showing the effect was probably visual.

In considering visual effects, Smith[15] states that a bare soil background in Brussels sprout plots provided optimal visual stimulus for colonization by aphids *Brevicoryne brassicae* (L.) (Figure 7), white flies *Aleyrodes brassicae* (Wlk) (Figure 8) and cabbage butterflies *P. rapae* (Fig. 9). With aphids, part of the effect may be due to an optomotor response; they tend to land on objects that loom up at them along their flight path.[28] Thus, interplanted crops, weeds, or even denser plantings of one crop reduced aphid colonization by reducing the contrast between plants and the ground.[15,42,43] In

Table 2
POPULATION BUILDUP OF *PHYLLOTRETA CRUCIFERAE* IN COLLARD MONOCULTURES (SUBPLOTS P1-P3) AND IN MIXTURES OF COLLARD, TOMATO, AND TOBACCO (SUBPLOTS M1-M3) IN 1970 FIELD EXPERIMENT[a]

		Sampling date				
		June 26	July 13	July 24	August 6	August 18
Number of *P. cruciferae* per collard	P1	0.42	18.80	155.60	466.00	230.80
	P2	1.42	31.60	215.20	366.60	165.40
	P3	1.00	20.20	181.80	386.00	165.60
	M1	0.14	14.40	80.80	101.80	62.40
	M2	0.28	10.40	53.60	149.80	66.80
	M3	0.14	13.20	72.40	134.80	77.80
Number of *P. cruciferae* per 10 g (dry weight) of collard	P1	1.33	10.37	22.46	40.00	24.60
	P2	8.11	11.67	30.96	32.21	19.69
	P3	7.04	9.52	48.87	36.62	21.78
	M1	0.36	4.17	8.95	9.67	5.19
	M2	0.92	5.14	7.60	11.48	5.68
	M3	0.38	4.59	12.29	16.04	8.80
Average weight of collard	P1	3.14	18.12	69.25	116.48	93.80
	P2	1.75	27.06	69.50	113.80	84.00
	P3	1.42	21.20	37.20	105.40	76.00
	M1	3.85[b]	34.50[b]	90.25	105.26	120.14
	M2	2.85	20.22	70.50	130.40	117.40
	M3	3.64	28.70	58.87	84.02	88.38

[a] On June 26 seven collards, and on all other sampling dates five collards, were sampled in each subplot.

[b] Significant (5% level) difference in collard weights between monocultures and mixed stands. Calculated by use of Wilcoxon's two-sample rank-sum statistic.

From Tahvanainen, J. O. and Root, R. B., The influence of vegetational diversity on the population ecology of a specialized herbivore *Phyllotreta cruciferae* (Coleoptera: Chrysomelidae), *Oecologia (Berlin)*, 10, 321, 1972. With permission.

other cases, the presence of a taller or denser intercrop may simply hide plants from a pest.[16]

The effectiveness of visual deterrence will depend on the situation the pest insect is adapted to exploit.[27] The wild hosts of crucifer pests such as *P. rapae* and *B. brassicae* occur mainly in open, successional habitats and so are more attracted to areas where plants stand out against bare soil. In contrast, the frit fly *Oscinella frit* (L.) and the corn borer *O. furnacalis* occur in dense grass stands and will be less attracted to open planting of cereal or maize crops.[27]

Even among herbivores on a single crop, notable differences in visual and olfactory responses may occur. Different species of the crucifer fauna responded differently to both size and background of plots of collards, with the adult flea beetle *P. cruciferae* accumulating in large, bare plots and the cabbage butterfly *P. rapae* ovipositing more heavily in small or weedy plots (Fig. 3).[14] *P. rapae* shows an inconsistent colonization responses to crop diversity (Figures 2, 3, 9, Table 7), suggesting that either an uncontrolled factor such as microclimate is important in influencing oviposition or that local behavioral adaptation to food plant distribution has occurred.[2,3,10,12,14,15,44,45] Despite these variations, visual deterrence shows real potential as a control measure for some insects, especially aphids.

Table 3
NUMBERS OF FEEDING HOLES MADE BY 15 *P. CRUCIFERAE* IN COLLARD LEAVES OVER A 12-HR PERIOD

Trial	Collard alone	With intact tomato	Collard alone	With intact ragweed	Collard alone	With crushed tomato	Collard alone	With crushed ragweed
1	43.8	13.1	51.0	18.0	93.5	13.5	92.1	7.2
2	51.1	22.8	149.6	14.4	52.0	5.3	74.6	23.8
3	35.7	46.2	101.2	3.5	51.7	22.8	118.3	11.8
4	82.5	21.1	75.2	28.1	55.0	8.0	93.6	15.6
5	44.8	4.3	71.1	5.4	63.7	15.0	116.9	5.2
6	98.8	5.8	102.2	1.6	54.7	30.2	76.2	6.0
7	62.4	0.4	78.2	45.2	76.7	9.3	65.4	9.5
8	91.2	0.0	85.6	18.8	56.5	11.1	69.1	18.4
9	99.9	11.4	89.8	49.1	82.7	15.9	90.7	18.2
10	115.9	6.4	84.6	21.0	102.4	35.4	78.4	11.8
Mean ratio	5.5:1		4.3:1		4.1:1		6.9:1	
Trial period	Aug. 26—Sept. 10		Aug. 1—13		Sept. 12—26		Aug. 13—26	

Note: The beetles were given a choice between collards alone and collards accompanied by non-host materials. Each figure represents the average of four replicates.

From Tahvanainen, J. O. and Root, R. B., The influence of vegetational diversity on the population ecology of a specialized herbivore *Phyllotreta cruciferae* (Coleoptera: Chrysomelidae), *Oecologia (Berlin)*, 10, 321, 1972. With permission.

Physical Barriers

The added complexity of an interplanted or weedy crop may impose physical obstacles to colonization by pests. Trenbath[46] describes what he terms the "fly paper effect" whereby nonsusceptible plants present in a mixture trap spores of pathogens and so reduce the inoculation of the susceptible plants. With insects such as aphids, flea beetles, and possibly some Lepidoptera and Hemiptera, landing on a nonhost may actually stimulate dispersal away from the crop.[3,4,47,48] In this situation, visual and olfactory effects are probably more important than just the simple contact with the nonhost.

It has been noted that barriers to air flow over crops, such as a tall crop grown in rows between a shorter one, can cause some pests to settle out in greater numbers than if the air flow were uninterrupted (see References 41, 49, and 50). Recent work in Costa Rica, however, has shown that tall maize plants among beans and squash interfere with flight behavior of certain herbivorous beetles,[51] so physical barriers may be valuable in deterring some invaders, though they must be used with caution.

Summary

Deterrence of colonization is probably one of the most promising means of controlling pests through intrafield diversity, because only a little additional diversity in the crop field may have a very large effect. It is important, however, to be sure that the amount and kind of diversity is appropriate to the mechanism of orientation employed by the target species. Much research and practical testing of various possible methods remains to be done.

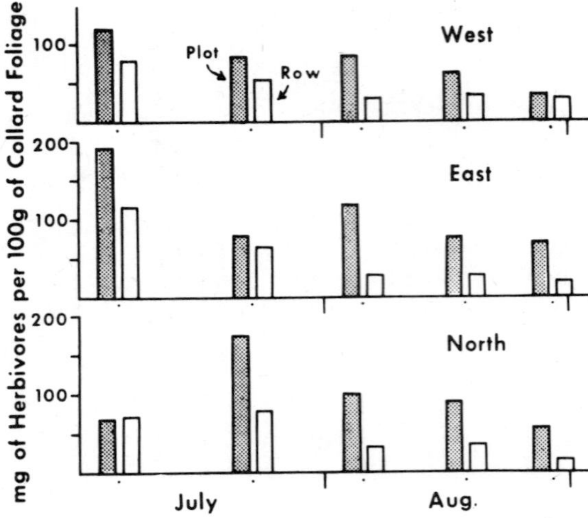

FIGURE 1. The herbivore load on collards grown in three different gardens at Savage Farm during 1967, as calculated from the dry weights of all herbivores and plant parts taken in the bag samples. Plot treatment monoculture. Row treatment diverse (plants growing among weedy old field vegetation). (Root, R. B., Organization of a plant-arthropod association in simple and diverse habitats: the fauna of collards *(Brassica oleracea)*, *Ecol. Monogr.*, 43, 95, 1973. With permission.)

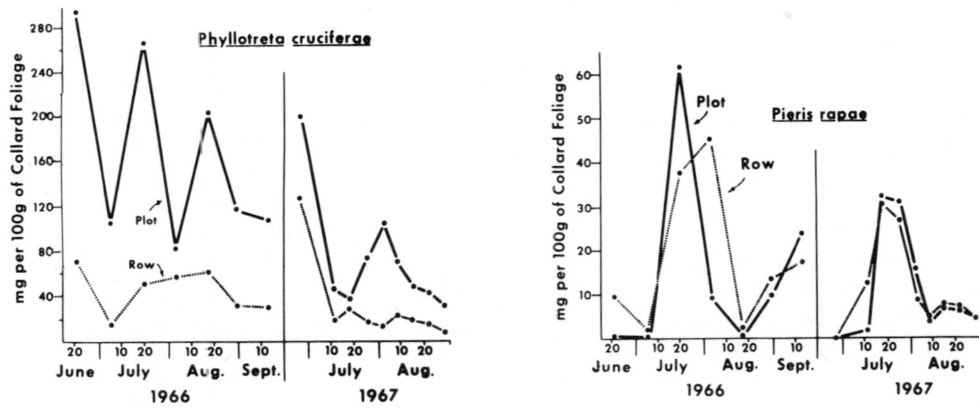

FIGURE 2. Population biomass of *Phyllotreta cruciferae* adults and *Pieris rapae* larvae in a monoculture (plot) and a diverse stand (row) of collards. (Root, R. B., Organization of a plant-arthropod association in simple and diverse habitats: the fauna of collards *(Brassica oleracea)*, *Ecol. Monogr.*, 43, 95, 1973. With permission.)

Diversionary Effects in Diverse Crops

When interplanted crops or weeds in the crop are also suitable host plants for a particular pest, they may reduce feeding damage to the main crop by diverting the pest. This diversionary effect of increased diversity can also function on a large scale. Unfortunately, both within and between fields, the effects of such diversity are likely to be extremely complex and there is apparently as much danger of increasing loss to pests as of reducing it.

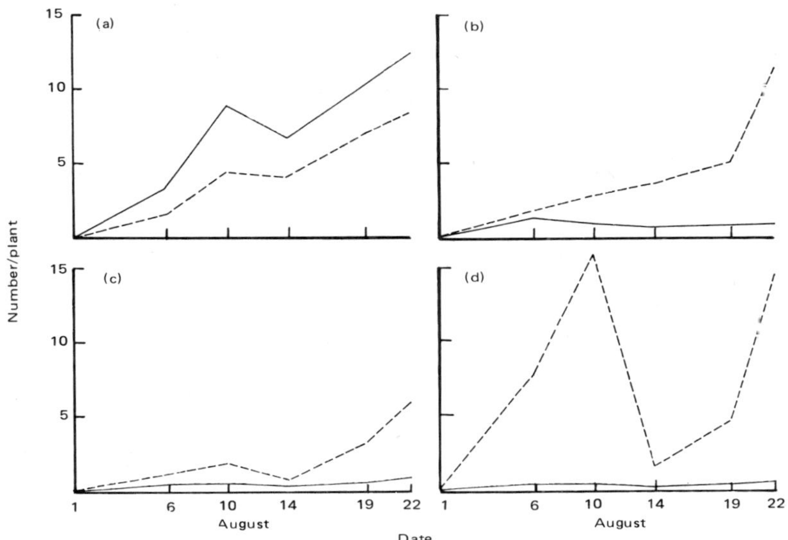

FIGURE 3. Abundance of (a) *Pieris rapae,* (b) *Phyllotreta cruciferae,* (c) *P. striolata,* and (d) alate aphids in cultivated (---) and uncultivated (—) ten-plant plots. Differences between the two treatments significant at the 0.001 level (t-test on mean abundance on each plant over entire period of experiment, log transformation of data). (Cromartie, W. J., The effect of stand size and vegetational background on the colonization of cruciferous plants by herbivorous insects, *J. Appl. Ecol.,* 12, 517, 1975. With permission.)

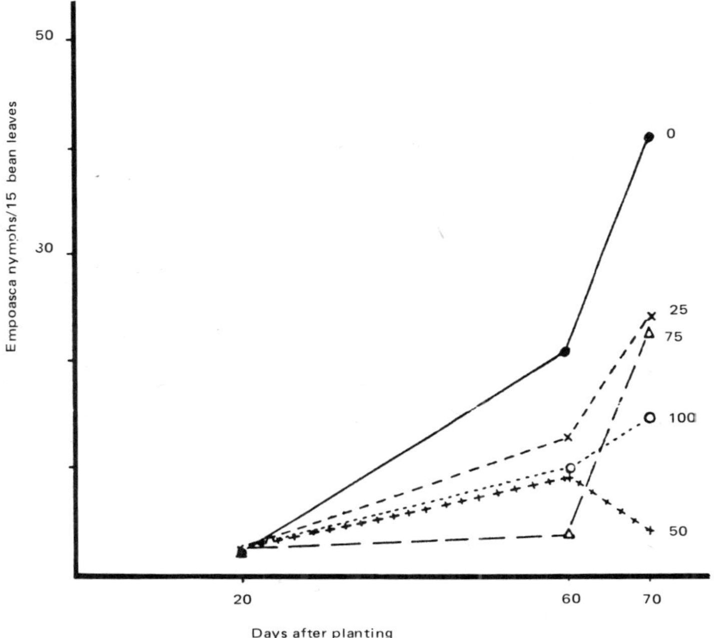

FIGURE 4. Effect of five weed densities (0—100% soil cover) on the mean nymph population of *Empoasca kraemeri* on beans. (Altieri, M. A., van Schoonhoven, A., and Doll, J., The ecological role of weeds in insect pest management systems: A review illustrated by bean *(Phaseolus vulgaris)* cropping systems, *PANS,* (Pest Artic. News Summ.), 23, 195, 1977. With permission.)

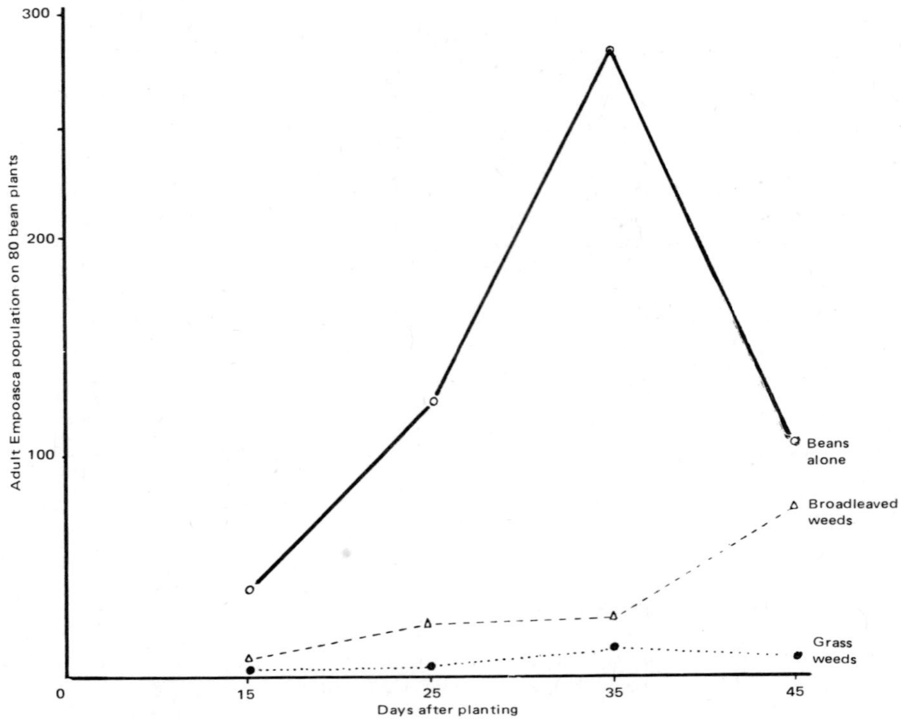

FIGURE 5. Effect of weed type (broadleaved and grass weeds) on the population of adult *Empoasca kraemeri* on beans. (Altieri, M. A., van Schoonhoven, A., and Doll, J., The ecological role of weeds in insect pest management systems: A review illustrated by bean *(Phaseolus vulgaris)* cropping systems, *PANS,* (Pest Artic. News Summ.), 23, 195, 1977. With permission.)

Diversion to Weeds

Van Emden[52] has noted numerous examples of pest problems attributable to the presence of weeds. Perennial weeds growing within an asparagus crop led to a nearly 200-fold increase in density of redbacked cutworm *Euxoa orchrogaster* (Guenee).[53] In other situations, however, removal of weeds has caused insect problems. Wolcott[54] observed that when weedy grasses were hoed out of sugar cane plantations, wingless *Aphis maidis* Fitch would crawl to the normally nonpreferred sugar cane, spreading mosaic virus to the crop. Similarly, hoeing out weeds led to cutworm damage on young sisal.[54] Damage by the springtail *Onychiurus armatus* (Tull.) to sugar beet seedlings is greater when herbicides remove the springtails' alternate hosts.[55] In these cases it is possible that the weeds might be left as valuable diversionary hosts, or weed control practices might be modified to reduce immigration to the crop. Obviously, successful use of diversion requires detailed knowledge of the biology of the particular pest to avoid the danger of making problems worse instead of better.

Diversion to Intercrops

In intercropping, the combination of plants may at times decrease economic loss by shifting insect feeding to the more tolerant or less valuable crop, as is the case with cowpea intercropped with cereals.[27] Again, the effect is not always beneficial. As with weeds, an intercrop may serve as an essential food source or shelter at some point in the life cycle or during some part of the season, enabling the pest to maintain or build up its numbers in the field and so attack the main crop more severely.[23,27]

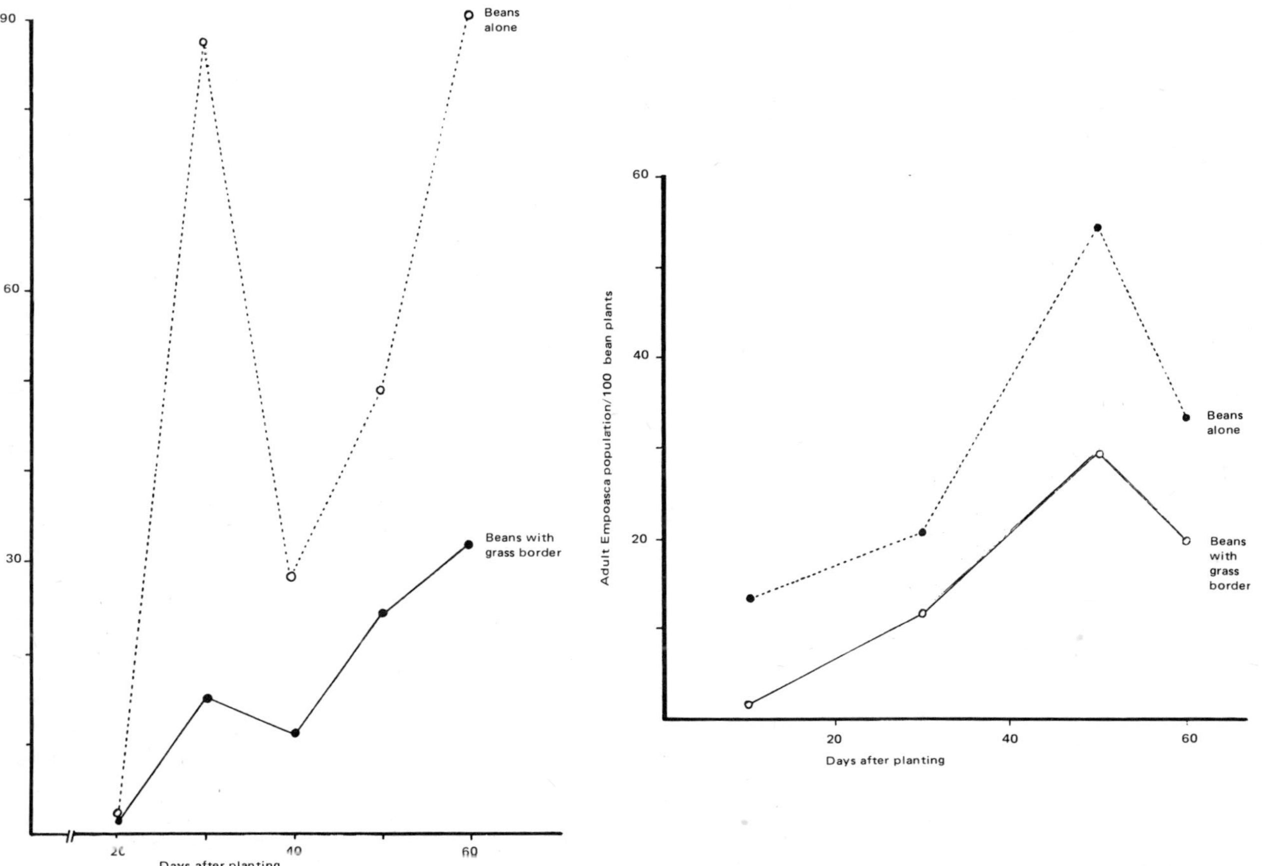

FIGURE 6. Effect of grass borders around 16m² bean plot on the population of *Empoasca kraemeri* nymphs and adults. (Altieri, M. A., van Schoonhoven, A., and Doll, J., The ecological role of weeds in insect pest management systems: A review illustrated by bean *(Phaseolus vulgaris)*, cropping systems, *PANS*, (Pest Artic. News Summ.), 23, 195, 1977. With permission.)

Table 4
INCIDENCE OF PESTS AND PREDATORS IN DIFFERENT BEAN-WEED SYSTEMS

% soil covered with weeds	*Empoasca kraemeri*				
	Adults per 80 bean plants	Nymphs per 15 bean leaves	*Diabrotica balteata* adults per 80 bean plants	Dolichopodidae adults per 80 bean plants	Reduvidae and Nabidae adults per 80 bean plants
0	52.8c[a]	22.4b	2.3	0.98	1.48
25	37.7b	13.8a	3.6	0.60	2.60
50	29.7a	10.5a	6.7	1.40	2.60
75	28.4a	11.8a	5.6	0.95	3.30
100	30.1a	6.7a	4.5	0.83	3.70

[a] Numbers followed by same letter in each column do not differ significantly (5% level).

From Altieri, M. A., van Schoonhoven, A., Doll, J., The ecological role of weeds in insect pest management systems: a review illustrated by bean *(Phaseolus vulgaris)* cropping systems, *PANS*, (Pest Artic. News Summ.), 23, 195, 1977. With permission.

Table 5
EFFECT OF CONTINUOUS WEED SOLUTION APPLICATIONS ON *EMPOASCA KRAEMERI* REPRODUCTION

Treatment	No. of nymphs per 10 leaves (60 days after planting)
Elusine indica	14.6
Leptochloa filiformis	11.3
Water	19.3
Control	23.6

From Altieri, M. A., van Schoonhoven, A., Doll, J., The ecological role of weeds in insect pest management systems: a review illustrated by bean *(Phaseolus vulgaris)* cropping systems, *PANS*, 23, 195, 1977. With permission.

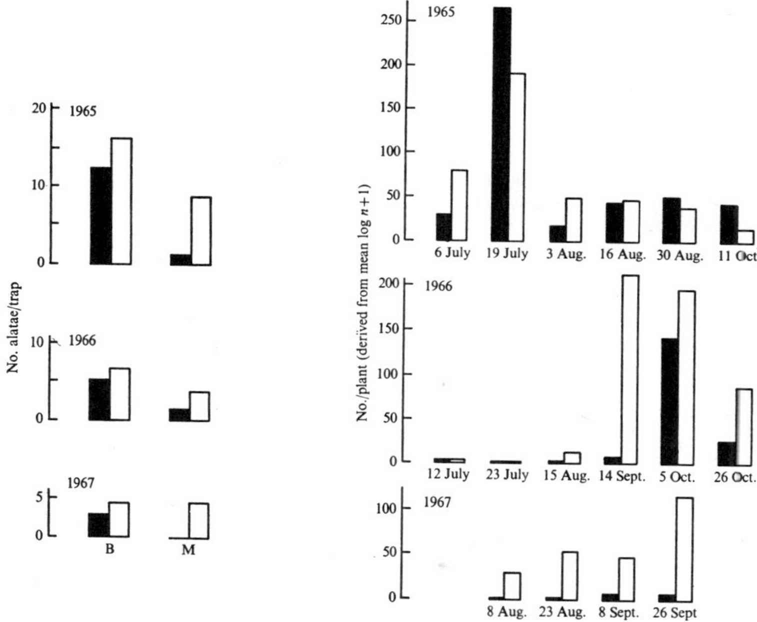

FIGURE 7. Left: The number of alate *Brevicoryne brassicae* (B) and *Myzus persicae* (M) caught in yellow pan traps over Brussels sprout plants. Right: The number of *Brevicoryne brassicae* on sprout plants. Plots surrounded by weeds (■) or bare soil (□). (Smith, J. G., Influence of crop background on aphids and other phytophagous insects on Brussels sprouts, *Ann. Appl. Biol.*, 83, 1, 1976. With permission.)

In Peru maize intercropped in cotton reduced damage by larvae of *Heliothis* sp. moths, but in Tanzania the same intercropping pattern increased loss to *Heliothis*.[26] Whether the intercrop is a useful diversionary host or a harmful alternate food supply for *Heliothis* in cotton depends on the cropping sequence, in particular on when various growth stages of the plants become available.[23] The same would apply to any other attempt to use intercrops in this way, so again detailed knowledge of the pest's biology is required.

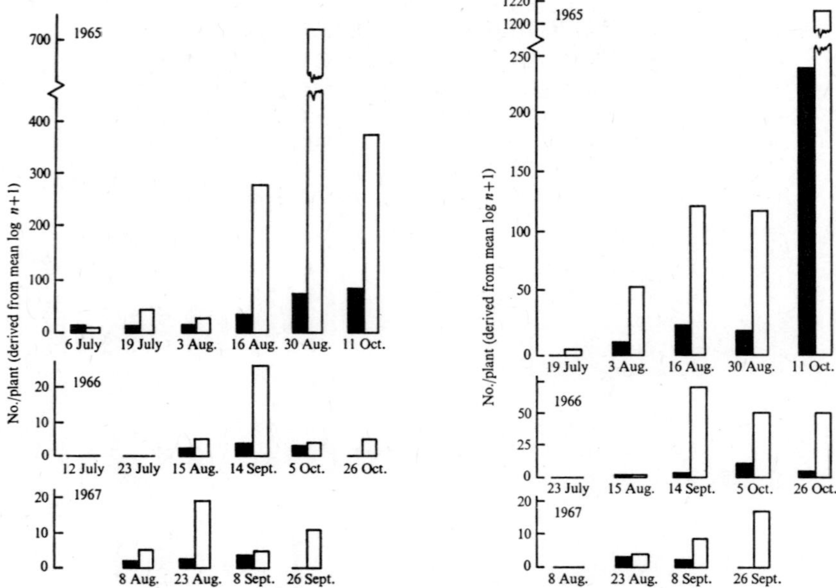

FIGURE 8. Number of *Aleyrodes brassicae* eggs (left) and nymphs (right) on Brussels sprout plants surrounded by weeds (■) or bare soil (□). (Smith, J. G., Influence of crop background on aphids and other phytophagous insects on Brussels sprouts, *Ann. Appl. Biol.*, 83, 1, 1976. With permission.)

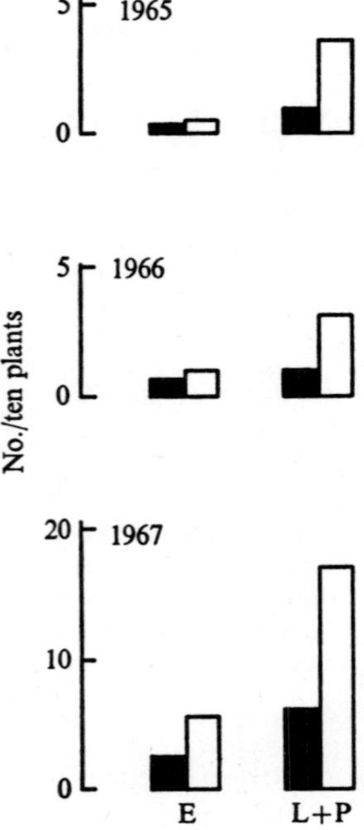

FIGURE 9. The number of *Pieris rapae* eggs (E) and larvae + pupae (L + P) on Brussels sprout plants surrounded by weeds (■) or bare soil (□) in 1965 (July-October), 1966 (July-October) and 1967 (August-September). (Smith, J. G., Influence of crop background on aphids and other phytophagous insects on Brussels sprouts, *Ann. Appl. Biol.*, 83, 1, 1976. With permission.)

Table 6
BREVICORYNE BRASSICAE CAPTURED IN PAN TRAPS IN BRUSSELS SPROUT PLOTS OR COUNTED ON SPROUT PLANTS; PLOTS A AND C UNDERSOWN WITH RED CLOVER: PLOTS B AND D HOED CLEAN

Plot	Number alatae per 7 traps, 18 June — 15 July	Mean number per plant 30 July (derived from log (n + 1) aphids per plant)
C—Clover	91	0.7 ± 0.9
A—Clover	84	0.2± 0.2
D—Hoed	349	9.4 ± 5.9
B—Hoed	344	12.0 ± 7.3

From Dempster, J. P., Coaker, T. H., Diversification of crop ecosystems as a means of controlling pests, in *Biology in Pest and Disease Control, 13th Symp. Br. Ecolog. Soc.*, Oxford, 1972, Price Jones, D. and Solomon, M. E., Eds., John Wiley & Sons, New York, 1974, 106. With permission.

Table 7
PIERIS RAPAE LARVAE ON BRUSSELS SPROUTS; PLOTS A AND C UNDERSOWN WITH WHITE CLOVER; PLOTS B AND D HOED CLEAN

	A Clover (1 m spacing)	D Hoed (1 m spacing)	C Clover (0.56 m spacing)	B Hoed (0.56 m spacing)
Eggs/plant (Mean ± SE)	4.55 ± 0.78	2.65 ± 0.56	7.65 ± 2.01	3.20 ± 0.73
Survival to instar III (%)	27.2	42.6	22.9	43.3

From Dempster, J. P. and Coaker, T. H., Diversification of crop ecosystems as a means of controlling pests, in *Biology in Pest and Disease Control, 13th Symp. Br. Ecolog. Soc.*, Oxford, 1972, Price Jones, D. and Solomon, M. E., Eds., John Wiley & Sons, New York, 1974, 106. With permission.

Another possibly useful but untested diversionary tactic is an outer encircling row of some highly preferred host to act as a trap and a barrier to dispersing pests.[27,37]

Effects of Interfield and Regional Diversity

As noted before, on a regional scale a diverse pattern of crops and noncrop habitats can divert insect pests, but it can also aggravate problems, just as within single fields. Smith and Reynolds[26] cite several cases where the presence of other crops in a region have led to more severe damage by *Heliothis* sp. in cotton, presumably by allowing the moths to continue to develop during the season when cotton is unavailable. Cotton stainers *Dysdercus* sp. in the Sudan move out of diverse bush habitats to attack cotton.[23] Thus, on a regional scale, diversity can create pest problems as well as solve them,[23,27,56] and regional diversification cannot be recommended without careful consideration and study.

Effects of Diversity on Emigration and Within-Crop Dispersal

Within a field, insects may make more or less frequent short range movement from plant to plant. Nonhosts present among the crop may considerably increase the mortality of such dispersers, as well as the chance that they will emigrate from the field. Dethier[57] and Morris and Mott[58] reported such effects for lepidopterous larvae crawling or ballooning on silk threads from plant to plant. Perrin[27] notes that the effect is particularly significant with soil inhabiting pests. The reduction in dispersal success may depend on the way mixed crops are arranged, intrarow interplantings being more effective in some cases than interrow patterns.[27] Reduction in survival of young stages during dispersal may be an especially valuable effect of crop diversity and could be especially important where such local movements spread diseases from plant to plant.

Effects on Reproduction, Growth and Survival

Olfactory

Increasing crop diversity may inhibit reproduction of insect pests, possibly because the physical and chemical complexity of a more diverse environment interferes with reproductive behavior in the same way it interferes with orientation to the host. In field and laboratory experiments, weedy grasses *Elusine indica* (L.) Gaertn. and *Leptochloa filiformis* (Lam.) Beauv. reduced reproduction of the leafhopper *Empoasca kraemeri* on beans.[16] Sprays of homogenized weeds have a similar effect, showing that olfactory inhibition may be involved (Table 5). Tahvanainen and Root[4] found that emergence of the new generation of the flea beetle, *Phyllotreta cruciferae,* was lower on collards interplanted with tomato and tobacco or grown in a weedy old field than on controls grown in monoculture (Figure 10).

These examples suggest that olfactory inhibition may work against population growth as well as colonization by pests, but more testing would be required to prove this conclusively.

Alteration of Microclimate and Nutrition in Diverse Crops

Physical and chemical complexity in diverse crops is usually accompanied by differences in microclimate as compared to monoculture. Tall intercrops or weeds will cast shade over low growing plants, while thick ground covers, as with undersown crops or weeds, will alter reflectivity, temperature, and evapotranspiration at the surface. Insects are adapted to specific microclimate regimes and may be strongly influenced by these changes.[51,52] Temperature and moisture in particular will affect growth rates, feeding rates, and survival of immature and adult stages.[27] For instance, leaf beetles do less damage to beans and squash shaded by maize plants.[51]

Nutritional status of crops will also be affected by the presence of intercrops and weeds, and this will in turn influence feeding and survival of pests.[27] Neither microclimatic nor nutrition changes have been deliberately used in pest control strategies employing diversification, and it is possible that such changes could have detrimental as well as beneficial effects, since changes which work against one pest may possibly favor another.[59]

Effects of Diversity on Natural Enemies

By far the greatest emphasis on the value of diversity in pest control has been on the enhancement of natural enemies of pests. The presence of diverse vegetation within or near the field may add essential resources for predators or parasites, and so enable them to find all their requirements near the pest population, rather than having to seek it farther away. Such resources include food (especially nectar), cover and alternate prey.[23,60] Diversity may, however, also adversely affect the orientation of predators and parasites to their prey.

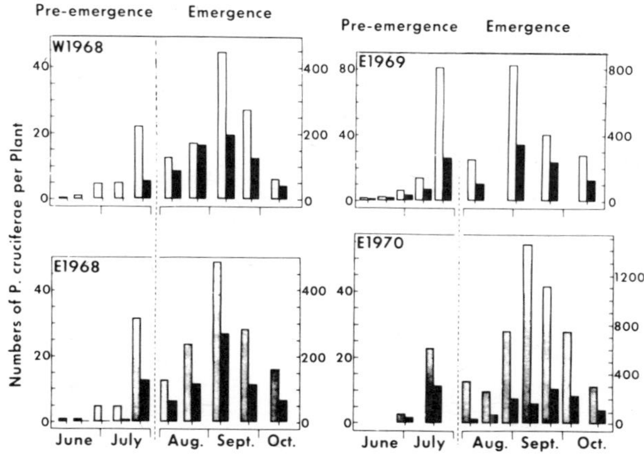

FIGURE 10. Seasonal abundance of *P. cruciferae* on collard plants in monoculture plots (gray bars) and diverse rows (black bars) in the west garden in 1967 (W1968) and in the east garden in 1968, 1969, and 1970 (E1968-1970). The scale for the pre-emergence period is on the left and for the emergence period on the right. (Tahvanainen, J. O. and Root, R. B., The influence of vegetational diversity on the population ecology of a specialized herbivore *Phyllotreta cruciferae* (Coleoptera: Chrysomelidae), *Oecologia (Berlin),* 10, 321, 1972. With permission.)

Table 8
EFFECT OF WILDFLOWER DENSITY ON PARASITISM OF APPLE PESTS IN UNSPRAYED ORCHARDS

	Mean % parasitism		
Wildflower density	**tent caterpillar papae**	**tent caterpillar eggs**	**codling moth**
High	65.2	3.7	33.6
Moderate	19.7	1.7	18.2
Low	3.7	0.9	7.1

From Leius, K., Influence of wildflower on parasitism of tent caterpillar and codling moth, *Can. Entomol.*, 99, 444, 1967. With permission.

Provision of Nectar and Other Foods

In unsprayed orchards with abundant wildflowers, parasitism on tent caterpillars and codling moths was heavier than in orchards with few or no flowers (Table 8).[61] Similarly, both fecundity and longevity of European pine shoot moth parasites are enhanced by the presence of flowers in pine plantations.[62] In both cases, nectar to provide energy for searching females was a critical resource.

Enhancement of Cover

Ground cover provided by weeds and intercrops can be valuable to predators, particularly those that feed by night and hide by day.[10] Carabid beetles *Harpalus rufipes* DeGeer were much more destructive to early instar *Pieris rapae* caterpillars on weedy vs. hoed Brussels sprouts.[10] Using white clover as a ground cover in Brussels sprouts led to lower *Pieris rapae* survival during instars I - II (22.9 - 27.2% vs. 42.6 - 43.3%, undersown clover vs. clean cultivated controls).[12] Pitfall traps in the clover plots captured more ground predators, especially *Harpalus rufipes* and harvestmen *Phalangium opilio* (L.) (Table 9). Dispersal of anthocorid bugs to aphid-infested plants is facilitated by the presence of weeds.[63] General predators may be more effective against cereal aphids in weedy fields;[17] a higher ratio of predatory insects to prey insects was found in weedy cereal fields as compared to clean fields,[1] and more artificial prey were removed by predatory beetles in an experimental test when the wheat crop was infested with the grass *Poa annua* (L.).[64] Spiders *Lycosa* spp. preyed more heavily on corn borers when maize was intercropped with peanuts.[41]

Provision of Alternate Hosts or Prey

Many predators and parasites can be more effective against pests if alternate prey or hosts are available to permit them to survive and reproduce while the pest is scarce or at an inappropriate stage in the life cycle. Crop diversification can provide plant hosts for herbivorous insects which in turn serve as alternate prey for natural enemies. Early work demonstrated the value of diversification in increasing predation.[8, 56]

The grass *Sorghum halepense* (L.) Pers. in vineyards serves as a reservoir of alternate prey for the phytoseid mites that control *Eotetranychus willamettei* Ewing.[9] In cotton, the weeds *Rumex crispus* (L.) and *Ambrosia artemisioides* (Cav.) help sustain *Rhinacloa* sp. and other enemies of *Heliothis* moths.[26] In walnut orchards, ground cover plants infested with aphids are a major source of coccinellid predators on *Chromaphis juglandicola* (Kalt.).[65] Predators and parasites build up on alfalfa grown in California orchards and then move to the trees when the alfalfa is cut.[66] In Peru, maize interplanted at the rate of one row per twelve rows of cotton increases control of *Heliothis* sp. by anthocorid bugs and other predators.[67]

Young spiders may find more food in the richer soil microfauna of intercropping mixtures, and when spider populations are high there is greater mortality of pests while dispersing from plant to plant.[27] Weeds may provide oviposition sites and food for syrphid flies, anthocorid bugs, and lacewings whose larvae later disperse to feed on aphids on the crop (Table 10).[63]

The need to maintain predator populations can also be met by diversification of variety, growth stage, or cultural practices within fields containing a single crop. Strip cutting of alfalfa prevents loss of important parasites by insuring that only a portion of the population is displaced at one time.[18] Murdoch[22] recommends interplanting small blocks of a susceptible variety of the crop (or of an alternate host plant of the pest) as a reservoir of the pest to keep the natural enemy population going.

Limitations of the Effectiveness of Natural Enemies in Diverse Crops

Within-field diversity may not always contribute to increased predation on pest insects. Rates of predation and parasitism differed little in mixed stands and monocultures of collards[3] (Table 11) or in weedy vs. weedless bean fields[16] (Table 4), and in some cases predation and parasitism on certain pests in weedless crops was higher (Tables 11 and 12).[3,4,63] Syrphids attacking *Brevicoryne brassicae* on Brussels sprouts varied in response to weedy versus cultivated fields, some responding positively, others negatively, and others showing no effect of weed cover (Tables 10 and 13).[63]

Predators and parasites, like herbivores, respond to a variety of stimuli when orienting to their prey, and crop diversification can have a similar inhibitory effect on enemies as it does on pests.[68-71] This is particularly the case if a predator or parasite orients to its host by the odor of the plant on which the host feeds.[71,72] Olfactory inhibition may, for instance, explain the lower rates of aphid parasitism sometimes observed in weedy fields (Table 11).

Table 9
GROUND-LIVING PREDATORS IN BRUSSELS SPROUT PLOTS

	A Clover 1 m spacing	D Hoed 1 m spacing	C Clover 0.56 m spacing	B Hoed 0.56 m spacing
Phalangium opilio	6.00 ± 1.1	2.6 ± 0.5	4.1 ± 0.6	0.6 ± 0.2
Harpalus rufipes	26.5 ± 2.6	9.9 ± 1.1	22.0 ± 1.7	15.8 ± 1.2
Bembidion spp.	2.2 ± 0.5	6.3 ± 0.6	2.4 ± 0.4	9.8 ± 1.0
Feronia melanaria	4.4 ± 0.6	1.9 ± 0.4	5.3 ± 1.0	1.3 ± 0.2
Staphylinidae	17.8 ± 1.2	3.1± 0.8	13.4 ± 1.5	7.5 ± 1.1

Note: Plots A and C undersown with white clover. Plots B and D hoed clean. Mean numbers in 25 pitfall traps per day ± SE.

From Dempster, J. P. and Coaker, T. H., Diversification of crop ecosystems as a means of controlling pests, in *Biology in Pest and Disease Control, 13th Symp. Br. Ecolog. Soc.*, Oxford, 1972, Price Jones, D. and Solomon, M. E., Eds., John Wiley & Sons, New York, 1974, 106. With permission.

Table 10
DENSITY OF SYRPHID EGGS, PER 10 YD² (8.36 M²) OF A WEEDY AND A WEED-FREE CROP OF BRUSSELS SPROUTS

Syrphid	Date	Weedy crop: Weeds[a]	Weedy crop: Sprouts	Weedy crop: Total	Sprouts (crop in bare soil)
Melanostoma spp.	1966 July	1430	338	1768	13
	Aug./Sept.	3520	1275	4795	513
	1967 July/Aug.	3700	63	3763	137
	Aug.	1700	712	2412	288
Syrphus balteatus	1966 July	30	2026	2056	1282
	Aug./Sept.	0	25	25	50
	1967 July/Aug.	400	588	988	500
	Aug.	140	213	353	288
Platycheirus spp.	1966 July	0	150	150	119
	Aug./Sept.	1000	226	1226	726
	1967 July/Aug.	110	126	236	26
	Aug.	140	588	728	900
Sphaerophoria spp.	1966 July	0	119	119	31
	Aug/Sept.	0	13	13	0
	1967 July/Aug.	200	0	200	113
	Aug.	0	63	63	62

[a] Standardized to the nearest unit of ten.

From Smith, J. G., Influence of crop background on natural enemies of aphids on Brussels sprouts, *Ann. Appl. Biol.*, 83, 15, 1976. With permission.

Influence of Interfield and Regional Diversity on Natural Enemies

Surrounding crops and diverse vegetation can also act as reservoirs of natural enemies and are especially important if enemies must recolonize pesticide sprayed fields.[18,19,52] Enemies find food, cover, and alternate prey in such areas. Blackberries planted near vineyards enable a parasite of grape leafhopper to survive over winter on an alternate leafhopper host, and strawberries grown near peach orchards in New Jersey similarly provided overwinter resources for a parasite of oriental fruit moth.[73,74] In Britain, stinging nettle *Urtica dioica* (L.) has been noted as a potentially valuable reservoir of predatory coccinellids; June mowing of nettles is recommended to enhance dispersal to adjacent crops.[75] Work in Poland showed that potato aphids reached lower population levels in fields near woods, where predators invaded earler and in larger numbers.[76] In one carefully studied case, however, oviposition by syrphids on aphid-infested *Brassica* crops was higher in areas with relatively little diverse vegetation, and woodland species of syrphids contributed little to oviposition on plants adjacent to the woods.[50] In balance, it has proved very difficult to determine whether uncultivated land is ultimately beneficial or harmful in controlling pests. Probably only case by case or area by area studies considering the particular crops, pests and natural enemies can provide an answer.[49,50,76,77] Meanwhile, particular plant species may be identified whose presences are beneficial or detrimental, and specific recommendations for their enhancement or control devised.[75]

Influence of Diversity on the Evolution of Pest Resistance to Control Techniques

A final potential value of crop diversification may lie in the effect diversity of the physical and biological environment has on the evolution of resistance to control measures. Planting multiple varieties of a single crop bred for different lines of resistance to pests has been suggested as a way both to reduce the risk of damage in the short

Table 11
THE NUMBERS OF PREDATORS AND ADULT PARASITOIDS TAKEN ON COLLARDS GROWN IN A MONOCULTURE (PLOT) AND A DIVERSE STAND (ROW)

	Predators				Parasitoids				Aphid parasitism[a]			
	Number of individuals		Number of species		Number of individuals		Number of species		Number of aphid individuals		Percentage mummified	
Date	Plot	Row	Plot	Row	Plot	Row	Plot	Row	Plot	Row	Plot	Row
Bag samples												
1966												
June 21	2	0	2	0	1	0	1	0	107	96	24.3	33.3
July 5	2	10	2	7	2	0	2	0	527	27	23.1	48.1
July 20	1	5	1	4	2	2	2	2	419	10	12.6	30.0
Aug. 3	5	0	5	0	4	4	3	2	56	8	10.7	12.5
Aug. 17	2	0	2	0	2	2	2	2	171	39	0	0
Aug. 30	11	11	4	4	1	5	1	4	310	186	4.8	2.7
Sept. 13	7	13	6	7	8	6	4	2	1213	513	18.8	7.0
Total	30	39	18	16	20	19	7	9	2803	879	16.1	10.2
1967												
June 27	3	0	3	0	0	0	0	0	2	2	0	0
July 11	7	1	6	1	3	1	3	1	26	16	0	0
July 18	3	4	3	2	3	1	2	1	32	27	6.3	14.8
July 26	8	8	5	5	4	3	3	1	51	22	2.0	0
Aug. 2	11	10	8	3	2	2	2	2	54	31	1.9	12.9
Aug. 8	12	6	6	2	3	2	2	2	148	25	4.7	4.0
Aug. 15	30	9	17	4	11	7	9	5	346	73	6.6	6.8
Aug. 22	35	7	15	6	21	10	11	7	1526	113	5.8	7.1
Aug. 29	30	10	17	7	26	8	10	5	1298	122	6.4	6.6
Sept. 14	22	8	12	7	16	4	10	4				
Total	161	63	45	21	89	38	25	17	3483	431	5.9	6.7
1968												
July 8	1		1		0		0		226		8.8	

Table 11 (continued)
THE NUMBERS OF PREDATORS AND ADULT PARASITOIDS TAKEN ON COLLARDS GROWN IN A MONOCULTURE (PLOT) AND A DIVERSE STAND (ROW)

	Predators				Parasitoids				Aphid parasitism[a]			
	Number of individuals		Number of species		Number of individuals		Number of species		Number of aphid individuals		Percentage mummified	
Date	Plot	Row	Plot	Row	Plot	Row	Plot	Row	Plot	Row	Plot	Row
July 22	16		9		24		5		804		26.9	
Aug. 5	29		17		41		7		625		14.2	
Aug. 19	16	17	12	12	30	18	11	9	2115	641	5.2	22.3
Sept. 3 (day)	9	10	7	5	5	5	3	4	481	448	15.0	12.3
Sept. 3 (night)	8	9	6	7	14	11	10	7				
Total	79		26		114		14		4251		11.9	
Vacuum-net samples 1967												
July 20	14	28	8	14	53	65	21	41				
Aug. 4	32	39	16	17	76	77	27	43				
Aug. 18	38	45	18	19	121	92	28	18				
Sept. 1	31	23	14	11	88	49	24	24				
Sept. 14	29	8	13	7	58	53	25	19				
Total	144	143	32	30	396	336	64	103				

[a] The percentage of the aphids parasitized was taken as the ratio of aphid mummies to the total living and mummified aphids.

From Root, R. B., Organization of a plant-arthropod association in simple and diverse habitats: the fauna of collards *(Brassica oleracea), Ecol. Monogr.*, 43, 95, 1973. With permission.

Table 12
PERCENTAGE OF *PHYLLOTRETA CRUCIFERAE* PARASITIZED BY *MICROCTONUS VITTATAE* ON COLLARD PLANTS IN MONOCULTURE PLOTS AND DIVERSE ROWS

	Number of *P. cruciferae* dissected		Percentage of *P. cruciferae* parasitized	
Date	Plot	Row	Plot	Row
1968				
June 28	250	250	18.0	12.8
Aug. 30	300	250	2.0	0.4
1969				
July 4	200	200	14.0	8.0
1970				
July 1	200	200	9.0	11.0
Aug. 25	200	200	1.0	1.5

From Tahvanainen, J. O. and Root, R. B., The influence of vegetational diversity on the population ecology of a specialized herbivore *Phyllotreta cruciferae* (Coleoptera: Chrysomelidae), *Oecologia (Berlin)*, 10, 321, 1972. With permission.

Table 13
DENSITY OF SYRPHID EGGS, PER TEN PLANTS, ON BRUSSELS SPROUTS SURROUNDED BY WEEDS OR BARE SOIL

	Melanostoma spp.		*Syrphus balteatus*		*Platycheirus* spp.	
Sampling date	Weeds	Bare soil	Weeds	Bare soil	Weeds	Bare soil
1965, Expt 1						
16 Aug.	121	150	84	130	9	54
30 Aug.	185	87	81	169	162	360
11 Oct.	38	20	7	10	10	49
1966, Expt 2						
12 July	0	0	130	64	3	3
23 July	18	1	103	65	8	6
5 Aug.	89	73	20	24	4	20
14 Sept.	128	51	8	5	24	73
5 Oct.	9	18	0	0	5	13
1967, Expt 3						
8 Aug.	29	14	59	50	6	3
23 Aug.	71	29	21	29	59	90
8 Sept.	118	33	13	46	75	154
26 Sept.	0	33	0	16	3	204

From Smith, J. G., Influence of crop background on natural enemies of aphids on Brussels sprouts, *Ann. Appl. Biol.*, 83, 15, 1976. With permission.

run and to minimize the probability that pests will evolve to overcome the plants' resistance.[1,12,35,36,60] A pest confronted with a variety of different resistance mechanisms is more likely to evolve simply to avoid the unsuitable varieties rather than to overcome their defenses.[36,60] By breeding different crops for widely different chemical defenses, one can also reduce the risk that an insect currently adapted to feed on one crop will shift over to attack another as well.[36] This may be especially important in considering plants for intercropping systems.

The above considerations probably apply most forcefully to those insects that are specialized to feed on a narrow range of host plants within a single family; more generalized feeders may present different problems.[36] Here a diverse biological environment may promote the high levels of genetic variability which is more apt to lead to rapid evolution of resistance to control measures. The question of pest evolution in simple vs. complex environments obviously requires much further study.

CROP DIVERSIFICATION IN INTEGRATED PEST CONTROL

Introduction

The preceding sections have considered the various ways in which increasing diversity can influence pest populations. Most of the quantitative experimental work cited, however, has not been concerned with producing crops economically, so there is little hard data available on the value of diversification in actual agricultural situations. Most of what is available is from tropical areas (see References 16, 27, and 41 and the references therein). Only a few studies have considered the impact of diversification on crop yields, and fewer have studied diversification in combination with other means of controlling pests.

Effect of Diversification for Pest Control on Yields of Crops

In several studies where weeds were left in the field to increase diversity, the crop plants in the weedy treatments were adversely affected by competition (Table 14).[2,10,15] Uncontrolled weed growth in crops, even when it reduces pest levels, cannot be recommended.[15] Dense cover crops may impose similar difficulties.[12,78] On the other hand, in some studies on diversification by weeds or by interplanting or undersown cover crops, yields have been increased or at least maintained compared to monocultures (Table 14).[4,12,16,79] Even if diversity does not serve to control destructive pests, yields in mixed crops may be maintained if the growth of one crop increases when the other is damaged by insects.[80] Unfortunately, no long-term studies on combined yields of intercrops in relation to pest attack seem to be available.

Further work is needed before weeds or intercrops can be recommended as a pest control strategy. The most promising examples seem to be intercropping maize with beans; maize and beans with squash; tomato and tobacco with collards;[4,13,51,79] the use of white clover as an undersown cover in cole crops;[12,78] and the use of natural or artificial weed cover in beans.[16] It appears that flea beetle damage could be substantially reduced on cole crops by diversification, eliminating the need for insecticide treatments. The question remains, however, whether the additional costs of interplanting would be greater than the savings. Interplanting clover with Brussels sprouts is only one third as expensive as pesticide treatments, but the loss of crop to competition is still a problem which needs to be evaluated more fully.[12,78] A bean/weed system showed good yields compared to beans alone, but more knowledge is needed to insure that the balance of plant competition will favor the beans under a wider range of soil and weather conditions.[16]

Timing of planting has recently been shown to exert a major influence on the degree

Table 14
EFFECT OF WITHIN-FIELD DIVERSIFICATION ON CROP YIELDS

Crop	Treatment	Yield		Ref.
		Treated	Control	
Collards	Interplanting tomato and tobacco	108.64g[a]	84.6g[a]	4
Brussels sprouts	Undersown white clover	2.19, 1.81[c]	1.26, 1.18[c]	12
Brussels sprouts	Undersown red clover; no insecticide	0.57, 0.58[c]	2.39, 2.38[c]	
Brussels sprouts	Undersown red clover with insecticide[b]	1.74, 0.96[c]	2.42, 2.59[c]	
Cauliflower	Undersown red clover; no insecticide	53, 76[d]	126, 136[d]	
Cauliflower	Undersown red clover with insecticide[b]	79, 107[d]	131, 147[d]	
Brussels sprouts	Undersown white clover	2.0[c]	1.2[c]	78
		0.6[c]	2.3[c]	
	100% cover	2.3[c]	4.4[c]	
		0.9[c]	4.8[c]	
	50% cover	2.7[c]	4.4[c]	
		2.9[c]	4.8[c]	
		0.1[c]	0.7[c]	
	100% cover with irrigation	0.3[c]	2.1[c]	
	50% cover with irrigation	0.8[c]	2.1[c]	
Brussels sprouts	Weeds (kept cut back to 15 cm high; weeds immedately beneath sprout plants hoed out)	1.7, 2.1, 1.3[e]	2.5, 2.9, 2.4[e]	11, 15
Beans	Soil covered by weeds 0%		1.70[f]	16
	Soil covered by weeds 25%	1.78[f]		
	Soil covered by weeds 50%	1.75[f]		
	Soil covered by weeds 75%	1.79[f]		
	Soil covered by weeds 100%	1.85[f]		
	Field with border of grasses *Elusine indica* and *Leptochloa filiformis* seeded artificially	0.65[f]	0.55[f]	
Beans and corn	Intercropped	7318—7631[g]	6535[g]	79
		8153—8769[g]	8205[g]	
		6926[g]	7221[g]	
		4177[g]	5600[g]	

Note: Controls are clean cultivated monocultures.

[a] Mean dry weight per collard in grams.
[b] To kill cabbage root fly *Erioischia brassicae* (Bouche).
[c] Mean fresh weight kg/plant.
[d] Curd diameter × number of curds.
[e] Mean number of "buttons" per plant.
[f] Tons/ha.
[g] Maize grain yield, kg/ha.

of pest reduction in bean/maize intercropping.[79] This work illustrates the need to continue research in order to combine modern insights into pest management with traditional intercropping systems.

Table 15
CABBAGE ROOT FLY ON BRUSSELS SPROUTS AND CAULIFLOWERS UNDER DIFFERENT TREATMENTS IN 1971; PLOTS A AND C UNDERSOWN WITH RED CLOVER; PLOTS B AND D HOED CLEAN

	C Clover	A Clover	D Hoed	B Hoed
Mean no. eggs/plant, 5 May—14 July	169.6 ± 32.1	161.6 ± 33.5	208.2 ± 33.2	257.0 ± 23.5
Root damage index on 28 June at harvest of cauliflowers without insecticide	65.7	68.2	94.0	91.2
With insecticide	9.5	10.0	36.9	36.6

From Dempster, J. P. and Coaker, T. H., Diversification of crop ecosystems as a means of controlling pests, in *Biology in Pest and Disease Control, 13th Symp. Br. Ecolog. Soc.*, Oxford, 1972, Price Jones, D., and Solomon, M. E., Eds., John Wiley & Sons, New York, 1974, 106. With permission.

Little information is available on the effect of mixed cropping on the chemical and nutritional composition of plants. The value of the crop can be altered by plant/plant interactions, as well as the yields.

Integration of Diversity with Other Control Techniques

Although crop diversity has been shown to reduce colonization by pests and to aid in biological control by natural enemies, there is still a need to evaluate crop diversification in combination with other methods of control. Little information on the effect of crop diversity on insecticide action, for example, is available. The insecticide chlorfenvinphos gave better control of cabbage root fly *Erioischia brassicae* (Bch.) on plots of Brussels sprouts and cauliflower undersown with clover than in monoculture plots, possibly because the microclimate in the clover plots enhances its action (Table 15).[12] Crop diversity, by itself, will rarely be a sufficient means of control, so it is clearly important to develop it as a component of integrated schemes and not in isolation, as has usually been done.

CONCLUSIONS

The economic gain from crop diversification depends on the balance between lowered pest control costs (including less weed control, if they are used as the source of diversity) and increased cost of maintaining an intercropped field, along with possibly decreased yield from greater plant competition.[78,81] Net profit can be increased if diversification favorably changes the balance between income and costs. At present there has simply not been detailed enough evaluation of diversity as a means of pest control to provide an answer to the question, "Is it economical?" for even a single crop. Field trials in a system such as a cole crop, where some data are already available, integrating diversification with other means of control to assess the financial returns as well as the biological effects, would be most useful.

One of the unfortunate problems in assessing the value of within-crop diversity is that the bulk of the studies outside the topics have been on cole crops (Table 1). Beans and corn seem to be a logical subject for further research in the temperate zone, and possibly also cotton.[16,26,79] The limited studies now available suggest that small amounts of added diversity could be very useful in suppressing certain key pests and maintaining or improving yields with less recourse to chemical controls. The potential of the method seems greatest in controlling specialized pests of annual row crops by inhibiting colonization, feeding and reproduction, and of orchards, vineyards, and annual crops by enhancing predation.[73]

In tropical areas, the value of diversification for pest reduction in traditional cropping systems is beginning to be more fully appreciated.[51,79,80] Where agriculture is more labor intensive the potential of intercropping may be more easily realized since the complexity of fields is less of a problem in manual than in mechanical sowing, cultivation, and harvesting.

In both the temperate and tropical zones, finding optimal combinations and spatial arrangements of plants remains a major challenge to agricultural researchers.

REFERENCES

1. **Potts, G. R. and Vickerman, G. P.,** Studies on the cereal ecosystem, *Adv. Ecol. Res.,* 8, 107, 1974.
2. **Pimentel, D.,** Species diversity and insect population outbreaks, *Ann. Entomol. Soc. Am.,* 54, 76, 1961.
3. **Root, R. B.,** Organization of a plant-arthropod association in simple and diverse habitats: the fauna of collards *(Brassica oleracea), Ecol. Monogr.,* 43, 95, 1973.
4. **Tahvanainen, J. O. and Root, R. B.,** The influence of vegetational diversity on the population ecology of a specialized herbivore, *Phyllotreta cruciferae* (Coleoptera: Chrysomelidae), *Oecologia (Berlin),* 10, 321, 1972.
5. **Gibson, I. A. S. and Jones, T.,** Monoculture as the origin of major forest pests and diseases, in *Origins of Pests, Parasite, Disease and Weed Problems,* Cherrett, J. M. and Sagar, G. R., Eds., 18th Symp. Brit. Ecol. Soc., Blackwell, Oxford, 1977, 139.
6. **Tothill, J. D.,** Some reflections on the cause of insect outbreaks, *Proc. 10th Int. Congr. Entomol.,* 4, 525, 1958.
7. **Mott, D. G.,** The forest and the spruce budworm, *Mem. Entomol. Soc. Can.,* 31, 333, 1963.
8. **Marcovitch, S.,** Experimental evidence on the value of strip farming as a method for the natural control of injurious insects with special reference to plant lice, *J. Econ. Entomol.,* 28, 62, 1935.
9. **Flaherty, D.,** Ecosystem trophic complexity and willamettei mite *Eotetranychus willamettei* Ewing (Acarina: Tetranychidae) densities, *Ecology,* 50, 911, 1969.
10. **Dempster, J. P.,** Some effects of weed control on the numbers of the small cabbage white *(Pieris rapae L.)* on Brussels sprouts, *J. Appl. Ecol.,* 6, 339, 1969.
11. **Smith, J. G.,** Some effects of crop background on populations of aphids and their natural enemies on Brussels sprouts, *Ann. Appl. Biol.,* 63, 326, 1969.
12. **Dempster, J. P. and Coaker, T. H.,** Diversification of crop ecosystems as a means of controlling pests, in *Biology in Pest and Disease Control,* Price Jones, D. and Solomon, M. E., Eds., John Wiley & Sons, New York, 1974, 106.
13. **Buranday, R. P. and Raros, R. S.,** Effects of cabbage-tomato intercropping on the incidence and oviposition of the diamondback moth, *Plutella xylostella* (L.), *Philipp. Entomol.,* 2, 369, 1975.
14. **Cromartie, W. J.,** The effect of stand size and vegetational background on the colonization of cruciferous plants by herbivorous insects, *J. Appl. Ecol.,* 12, 517, 1975.
15. **Smith, J. G.,** Influence of crop background on aphids and other phytophagous insects on Brussels sprouts, *Ann. Appl. Biol.,* 83, 1, 1976.
16. **Altieri, M. A., van Schoonhoven, A., and Doll, J.,** The ecological role of weeds in insect pest management systems: A review illustrated by bean cropping systems, *PANS (Pest Artic. News Summ.),* 23, 195, 1977.
17. **Potts, G. R.,** Some effects of increasing monoculture in cereals, in *Origins of Pest, Parasite, Disease and Weed Problems,* Cherrett, J. M. and Sagar, G. R., Eds., 18th Symp. Brit. Ecol. Soc., Blackwell, Oxford, 1977, 183.
18. **Van Emden, H. F. and Williams, G. C.,** Insect stability and diversity in agro-ecosystems, *Ann. Rev. Entomol.,* 19, 455, 1974.
19. **Elton, C. S.,** *The Ecology of Invasions by Animals and Plants,* Methuen and Co., London, 1958, Chap. 8.

20. **Graham, S. A.,** Forest insects and the law of natural compensation. *Can. Entomol.,* 88, 45, 1956.
21. **Goodman, D.,** The theory of diversity-stability relationships in ecology, *Quart. Rev. Biol.,* 50, 237, 1975.
22. **Murdoch, W. W.,** Diversity, complexity, stability, and pest control, *J. Appl. Ecol.,* 12, 795, 1975.
23. **Way, M. J.,** Pest and disease status in mixed stands vs. monocultures; the relevance of ecosystem stability, in *Origins of Pest, Parasite, Disease and Weed Problems,* Cherrett, J. M. and Sagar, G. R., Eds., 18th Symp. Brit. Ecol. Soc., Blackwell, Oxford, 1977, 127.
24. **Kennedy, J. S.,** The motivation of integrated control, *J. Appl. Ecol.,* 4, 492, 1968.
25. **Southwood, T. R. E. and Way, M. J.,** Ecological background to pest management, in *Concepts of, Pest Management,* Rabb, R. L. and Guthrie, F. E., Eds., North Carolina State University, Raleigh, 1970, 6.
26. **Smith, R. F. and Reynolds, H. T.,** Effects of manipulation of cotton agro-ecosystems on insect pest populations, in *The Careless Technology,* Farvar, M. T. and Milton, J. P., Eds., Natural History Press, N. Y., 1972, 373.
27. **Perrin, R. M.,** Pest management in multiple cropping systems, *Agro-Ecosystems,* 3, 93, 1977.
28. **Kennedy, J. S., Booth, C. O., and Kershaw, W. J. S.,** Host finding by aphids in the field. III. Visual attraction, *Ann. Appl. Biol.,* 49, 1, 1961.
29. **Dethier, V. G.,** Chemical interactions between plants and insects, in *Chemical Ecology,* Sondheimer, E. and Simeone, J. B., Eds., Academic Press, New York, 1970, 83.
30. **Schoonhoven, L. M.,** Secondary plant substances and insects, *Recent Adv. in Phytochem.,* 5, 197, 1972.
31. **Huffaker, C. B.,** Some concepts on the ecological basis of biological control of weeds, *Can. Entomol.,* 94, 507, 1962.
32. **Price, P. W. and Waldbauer, G. P.,** Ecological aspects of insect pest management, in *Introduction to Insect Pest Management,* Metcalf, R. L. and Luckmann, W. H., Eds., John Wiley & Sons, New York, 1975, 36.
33. **Watt, K. E. F.,** Community stability and the strategy of biological control, *Can. Entomol.,* 97, 887, 1965.
34. **Feeny, P. P.,** Biochemical coevolution between plants and their insect herbivores, in *Coevolution of Animals and Plants,* Gilbert, L. and Raven, P., Eds., Univ. of Texas, Austin, 1975, 3.
35. **Feeny, P. P.,** Plant apparency and chemical defense, *Recent Adv. in Phytochem.,* 10, 1, 1976.
36. **Feeny, P. P.,** Defensive ecology of the Cruciferae, *Ann. Missouri Bot. Gard.,* 64, 221, 1977.
37. **Philbrick, H. and Gregg, R. B.,** *Companion Plants and How to Use Them,* Devin-Adair Co., New York, 1966.
38. **Tirrell, R.,** Not many pests in my patch, *Org. Gard. Farming,* 17(5), 60, 1970.
39. **Gonzales, D. and Rawlins, W. A.,** Aphid sampling efficiency of Moericke traps affected by height and background, *J. Econ. Entomol.,* 61, 109, 1968.
40. **Kring, J. B.,** Flight behavior of aphids, *Annu. Rev. Entomol.,* 17, 461, 1972.
41. **Litsinger, J. A. and Moody, K.,** Integrated pest management in multiple cropping systems, in *Multiple Cropping,* Spec. Publ. 27, Papendick, R. I., Sanchez, P. A., and Triplett, G. B., Eds., American Society of Agronomy, Madison, Wis., 1976, 293.
42. **A'Brook, J.,** The effect of plant spacing on the numbers of aphids trapped over the groundnut crop, *Ann. Appl. Biol.,* 61, 289, 1968.
43. **A'Brook, J.,** The effect of plant spacing on the number of aphids trapped over cocksfoot and kale crops, *Ann. Appl. Biol.,* 74, 279, 1973.
44. **Pimentel, D.,** The influence of plant spatial patterns on insect populations, *Ann. Entomol. Soc. Am.,* 54, 61, 1961.
45. **Meyers, J. H. and Campbell, B. J.,** Distribution and dispersal in populations capable of resource depletion: a field study on cinnabar moth, *Oecologia (Berlin),* 24, 7, 1976.
46. **Trenbath, B. A.,** Plant interactions in mixed crop communities, in *Multiple Cropping,* Spec. Publ. 27, Papendick, R. I., Sanchez, P. A., and Triplett, G. B., Eds., American Society of Agronomy, Madison, Wis., 1976, 129.
47. **Kennedy, J. S., Booth, C. O., and Kershaw, W. J. S.,** Host finding by aphids in the field. I. Gynoparae of *Myzus persicae* (Sulzer), *Ann. Appl. Biol.,* 47, 410, 1959.
48. **Kennedy, J. S., Booth, C. O., and Kershaw, W. J. S.,** Host finding by aphids in the field. II. *Aphis fabae* Scop. (gynoparae) and *Brevicoryne brassicae* L.; with a re-appraisal of the role of host finding in virus spread, *Ann. Appl. Biol.,* 49, 1, 1959.
49. **Van Emden, H. F.,** The effect of uncultivated land on the distribution of the cabbage aphid *(Brevicoryne brassicae)* on an adjacent crop, *J. Appl. Ecol.,* 2, 171, 1965.
50. **Pollard, E.,** Hedges., VI. Habitat diversity and crop pests, a study of *Brevicoryne brassicae* and its syrphid predators, *J. Appl. Ecol.,* 8, 751, 1971.

51. **Risch, S.**, Effects of resource diversity on the population dynamics of several pests in a tropical agro-ecosystem: monocultures and polycultures of corn, beans and squash in Costa Rica (abstract only), *Bull. Ecol. Soc. Am.*, 59, 53, 1978.
52. **Van Emden, H. F.**, The role of uncultivated land in the biology of crop pests and beneficial insects, *Sci. Hort.*, 17, 121, 1965.
53. **Tamaki, G., Moffitt, H. R., and Turner, J. E.**, The influence of perennial weeds on the abundance of the redbacked cutworm on asparagus, *Environ. Entomol.*, 4, 274, 1975.
54. **Wolcott, G. N.**, Increase of insect transmitted plant disease and insect damage through weed destruction in tropical agriculture, *Ecology*, 9, 461, 1928.
55. **Coaker, T. H.**, Crop pest problems resulting from chemical control, in *Origins of Pest, Parasite, Disease and Weed Problems*, Cherrett, J. M. and Sagar, G. R., Eds., 18th Symp. Brit. Ecol. Soc., Blackwell, Oxford, 1977, 313.
56. **Hambelton, E. J.**, *Heliothis virescens* as a pest of cotton with notes on host plants in Peru, *J. Econ. Entomol.*, 37, 660, 1944.
57. **Dethier, V. G.**, Food plant distribution and density and larval dispersal as factors affecting insect populations, *Can. Entomol.*, 91, 581, 1959.
58. **Morris, R. F. and Mott, D. G.**, Dispersal and the spruce budworm *Mem. Entomol. Soc. Can.*, 31, 180, 1963.
59. **Batra, H. N.**, Mixed cropping and pest attack, *Indian Farming*, 11(12), 23, 1962.
60. **Atsatt, P. R. and O'Dowd, D. J.**, Plant defense guilds, *Science*, 193, 24, 1976.
61. **Leius, K.**, Influence of wild flowers on parasitism of tent caterpillar and codling moth, *Can. Entomol.*, 99, 444, 1967.
62. **Syme, P. A.**, The effect of flowers on the longevity and fecundity of two native parasites of the European pine shoot moth in Ontario, *Environ. Entomol.*, 4, 337, 1975.
63. **Smith, J. G.**, Influence of crop background on natural enemies of aphids on Brussels sprouts, *Ann. Appl. Biol.*, 83, 15, 1976.
64. **Speight, M. R. and Lawton, J. H.**, The influence of weed-cover on the mortality imposed on artifical prey by predatory ground beetles in cereal fields, *Oecologia (Berlin)*, 23, 211, 1976.
65. **Sluss, R. R.**, Population dynamics of the walnut aphid *Chromaphis juglandicola* (Kalt.) in northern California, *Ecology*, 48, 41, 1967.
66. **Stern, V. M.**, Interplanting alfalfa in cotton to control lygus bugs and other insect pests, *Proc. Tall Timbers Conf. Ecological Animal Control by Habitat Management*, 1, 55, 1969.
67. **Barducci, T. D.**, Ecological consequences of pesticides used for the control of cotton insects in Canete Valley, Peru, in *The Careless Technology*, Farvar, M. T. and Milton, J. P., Eds., Natural History Press, New York, 1972, 423.
68. **Chandler, A. E. F.**, The relationship between aphid infestations and oviposition by aphidophagous Syrphidae (Diptera), *Ann. Appl. Biol.*, 61, 425, 1968.
69. **Chandler, A. E. F.**, Some factors influencing the occurrence and site of oviposition by aphidophagous Syrphiade (Diptera), *Ann. Appl. Biol.*, 61, 435, 1968.
70. **Read, D. P., Feeny, P. P., and Root, R. B.**, Habitat selection by the aphid parasite *Diaeretiella rapae* (Hymenoptera: Braconidae) and hyperparasite *Charips brassicae* (Hymenoptera: Cynipidae), *Can. Entomol.*, 102, 1567, 1970.
71. **Shahjahan, M. and Streams, F. A.**, Plant effects on host finding by *Leiophron pseudopallipes* (Hymenoptera: Braconidae) a parasitoid of the tarnished plant bug, *Environ. Entomol.*, 2, 921, 1973.
72. **Monteith, L. G.**, Influence of plants other than the food plants of their host on host-finding by tachinid parasites, *Can. Entomol.*, 92, 641, 1960.
73. **Doutt, R. L. and Nataka, J.**, The Rubus leafhopper and its egg parasitoid: An endemic biotic system useful in grape pest management, *Environ. Entomol.*, 2, 381, 1973.
74. **Allen, H. W.**, Present status of oriental fruit moth parasite investigations, *J. Econ. Ent.*, 25, 360, 1932.
75. **Perrin, R. M.**, The role of the perennial stinging nettle, *Urtica dioica*, as a reservoir of beneficial natural enemies, *Ann. Appl. Biol.*, 81, 289, 1975.
76. **Galecka, B.**, The role of predators in the reduction of two species of potato aphids, *Aphis nausturii* Kalt. and *A. frangulae* Kalt., *Ekol. Polska Ser. A*, 14, 245, 1966.
77. **Finch, S. and Ackley, C. M.**, Cultivated and wild host plants supporting populations of the cabbage root fly, *Ann. Appl. Biol.*, 85, 13, 1977.
78. **O'Donnell, M. S. and Coaker, T. H.**, Potential of intracrop diversity for the control of *Brassica* pests, *Proc. 8th Brit. Insect. Fung. Conf.*, 1, 101, 1975.
79. **Altieri, M. A., Francis, C. A., van Schoonhoven, A., and Doll, J. D.**, A review of insect prevalence in maize *(Zea mays* L.) and bean *(Phaseolus vulgaris* L.) polycultural systems, *Field Crops Res.*, 1, 33, 1978.
80. **Norton, G. H.**, Multiple cropping and pest control — an economic perspective, *Meded. Fac. Landbouww. Rijks. Univ. Gent.*, 40, 219, 1975.
81. **Trenbath, B. R.**, Biomass productivity of mixtures, *Adv. Agron.*, 26, 177, 1974.

REGULATORY PLANT PEST MANAGEMENT

G. Gregor Rohwer

INTRODUCTION

Definition

The term pest management or, more recently, integrated pest management is used to cover the integration of all available pest control strategies. A policy statement of the U.S. Department of Agriculture (USDA) issued December 12, 1977,[1] states that pest management is a systems approach in which methods and materials are chosen to control pests while minimizing undesirable results. Among the major pest management strategies indicated in the U.S. Department of Agriculture policy statement are prevention, management of local populations, areawide pest management, and elimination of pest species from defined areas.

Regulatory plant pest management consists of all activities in multifaceted pest management programs conducted by federal and/or state agencies which are authorized by various legislative documents. Authorizing legislation defines a plant pest as any living stage of insects or other invertebrate animals, viruses, bacteria, fungi, and parasitic plants that can directly or indirectly injure or damage plants or plant products. Weeds as defined in other legislative acts are also included in regulatory plant pest management. A plant pest quarantine is but one element of the many activities in regulatory plant pest management.

Regulatory plant pest management, as with other pest control strategies, must be acceptable economically, ecologically, and sociologically. Regulatory plant pest management may be divided into three major areas, namely (1) prevention, (2) appraisal, and (3) pest population manipulation.

Authorizing Legislation

Numerous enabling federal acts provide the authority for regulatory plant pest management activities. For the purposes of this review, the following are the principal authorities:

Plant Quarantine Act[2]

The primary authority provided by the Plant Quarantine Act of 1912 is a means to control the artificial introduction of exotic plant pests associated with plants. The Act also provides authority for the invocation of domestic plant quarantines to prevent spread of pests new to and not widely disseminated within the U.S. Prior to passage of this act, there was other ineffective legislation; however, the Plant Quarantine Act of 1912 was the first major regulatory authority to provide the federal government with a sound basis for the inauguration of activities in regulatory plant protection. Prior to passage of the Plant Quarantine Act, a number of states had adopted some Acts authorizing plant pest regulatory activities.

Federal Plant Pest Act[3]

The Federal Plant Pest Act of 1957 authorizes emergency action to prevent the introduction or interstate dissemination of those plant pests which are not subject to provisions of the Plant Quarantine Act. Importantly, this act includes a broad definition of a plant pest, namely, any living stage of "any insects, mites, nematodes, slugs, snails, protozoa or other invertebrate animals, bacteria, fungi, other parasitic plants or reproductive parts thereof, viruses, or any organisms similar to or allied with any of the foregoing, or any infectious substances, which can directly or indirectly injure or cause damage in any plants or parts thereof, or any processed, manufactured, or other products of plants." The act authorizes under permit the movement of plant pests for scientific study.

Organic Act[4]

The Plant Quarantine and Plant Pest Acts' primary thrusts relate to control over the introduction and interstate movement of exotic pest organisms. The Organic Act provides basic authority for local and/or areawide pest management strategies independently or in cooperation with states, farmers' organizations, individuals, or groups when in the judgment of the Secretary of Agriculture such activities are in the public interest. The Act authorizes survey and control strategies against endemic as well as exotic pests. This Act, therefore, permits participation in cooperative survey and control activities, including areawide distribution of beneficial organisms such as parasites and predators. This Act also authorizes the issuance of phytosanitary certificates, which are used to attest that plant products for export are pest-free in accordance with the requirements of the destination states (countries).

Weed Act[5]

The previously mentioned legislation deals with plant pests. It does not include provisions for the control of the introduction, interstate spread, or management of detrimental weed species other than parasitic weeds. The Federal Noxious Weed Act of 1974 authorizes activities to prevent the introduction of unwanted exotic weeds as well as cooperative activities to control or eradicate noxious weeds of foreign origin which are new to or not widely prevalent in the U.S.

Restrictive Legislation

Federal and/or state agencies engaged in regulatory plant pest management are subject to certain legislative restrictions which also apply to pest management in the private sector. The primary legislative restrictions which must be considered in any pest management strategy are as follows:

National Environmental Policy Act[6]

The National Environmental Policy Act of 1969, as amended, as it relates to pest management, requires that an environmental analysis be made of any major action which could significantly affect the quality of the environment. If, on the basis of such an analysis, it is determined that the proposed activity may adversely affect the environment, an environmental impact statement must be completed. The public has an opportunity to comment on a draft environmental impact statement and such comments are evaluated and included in or commented on by the issuing agency in the final document. The purpose is to assure that before action is initiated all aspects of the proposal are considered and those actions most beneficial to environmental concerns are selected. If it is determined that the action has an unacceptable environmental impact the program may be canceled. From the standpoint of cooperative federal/state plant pest management programs, the primary thrust of an impact statement relates to its effect on the utilization of pesticides as one of the tools used in pest management strategies. However, provisions of the Act apply to any major activity which could adversely affect the environment.

Endangered Species Act[7]

The Endangered Species Act of 1973, as it relates to regulatory plant pest management activities, provides a mechanism to assure that biological organisms listed on the endangered species list that may be adversely affected by organized federal or state operations are provided protection so that such activities do not adversely affect or jeopardize their habitat or their survival. Accordingly, any regulatory plant pest management program must take into consideration any endangered species and their critical habitat that may be present in the area.

Cooperative Interrelationships

To be most effective, it is necessary that regulatory plant pest management activities be conducted in cooperation with international, state, and other federal agencies, and the private sector.

International

To be most effective, activities in regulatory plant pest management, particularly those relating to prevention, must involve the cooperation of foreign governments, institutions, and private concerns.

Domestic

All authorizing legislation for cooperative domestic (U.S.) plant pest management programs requires that authority for entrance upon private property be provided by cooperators, whether they be state agencies, private institutions, or individuals. Domestic programs are conducted in cooperation with other agencies, whether they be federal, state, or private institutions or groups. Participation may be on an individual, regional, or national basis. International cooperation is involved as it relates to adjoining countries.

Cooperative activities in prevention, appraisal, and pest population management involve many state, federal, and private institutions, including the military, which is usually coordinated through the Armed Forces Pest Management Board. Also, interface must be maintained with other federal agencies such as the Environmental Protection Agency, Department of Commerce, and National Cancer Institute as they relate to pesticide activities.

Liaison also must be maintained with affected agricultural industry groups to assure cooperation and the maximum effectiveness of the programs.

Research/Methods Development

Regulatory plant pest management strategies are based on research findings. For the most part, federal agencies charged wtih the conduct of regulatory plant pest management strategies are not authorized to conduct basic research. Basic research necessary for a scientific foundation of regulatory plant pest management programs is provided by other federal, state, or private agencies specifically funded for research activities. However, basic research findings are tested and adapted by regulatory plant pest management action groups in large-scale field trials. Such large-scale testing and applied research activities are referred to as methods development. Research support is required relating to an economic appraisal of the effectiveness of regulatory plant pest management programs.

Program Impact Evaluation and Analysis

Monitoring is an essential and important facet of regulatory plant pest management. These activities have many different objectives, including the accumulation of data as a basis for pesticide registration, determining effectiveness of activities under field operational conditions, and assessing the impact of operations on nontarget organisms and other aspects of the environment.

Information needed as a basis for pesticide registration is usually obtained by a pesticide formulator or manufacturer; however, many uses important for regulatory activities are not of a sufficient scale to provide an economic incentive to commercial companies. For that reason, to develop regulatory treatments, data needed as a basis for registration often must be obtained by regulatory agencies with minimal, if any, participation by companies. Also, though efficacy and residue data are obtained as a basis for registration, program operations must be systematically evaluated to assure that objectives are being met.

PREVENTION

Prevention substrategies of regulatory plant pest management may be defined as those activities having as their major goal the elimination or reduction of artificial spread of plant pests between countries. Legislation provides authority to cooperate in these activities on an international and regional basis, as well as to invoke specific quarantines or regulations to reduce the artificial introduction of named plant pest carriers. Further, such authorities provide for emergency action to be taken against any plant pest whether it is or is not included in a specific quarantine or regulation.

A fundamental concept of all activities designed to prevent or reduce artificial spread is that the most effective programs are those designed and enforced at the point of origin. This contemplates the utilization of all strategies, in the area where pests exist, to prevent or reduce artificial pest spread. Such activities include pest population suppression, since the hazard of spread is in direct proportion to pest population levels. Additionally, it is more effective to apply commodity treatments to eliminate pests at the point of origin to preclude the possibility of accidental escape enroute or at destination prior to completing treatment.

Prevention may be grouped into three main areas, namely: international cooperation, preclearance, and port of entry activities.

International Cooperation

Action taken to cooperate with regulatory agencies of other countries may be on a one-to-one basis as well as through regional and international organizations. For prevention activities to be most effective, cooperative action must involve not only governmental agencies but trade organizations and other similar groups. An understanding of prevention activities by all concerned groups is essential to maximize effectiveness.

Additionally, at the request of foreign governments, professionally trained employees are temporarily assigned in foreign countries to assist in the utilization of regulatory plant pest management strategies. Such assignments may be for the purpose of the development of prevention or appraisal strategies or participation in developing or directing pest population management activities. These assignments provide an opportunity for gathering information regarding pest conditions and strategies in foreign countries that may be utilized in detection and appraisal, as well as pest population management in event of an artificial introduction of exotics into the U.S.

International Plant Protection Convention[8]

The International Plant Protection Convention (IPPC), formed in 1951, was endorsed in principle by the U.S., but was not ratified by the U.S. Senate until 1972. The primary objective of IPPC is to prevent pest spread between countries in connection with commerce. The convention provides for the utilization of an approved model phytosanitary certificate which attests to the apparent freedom of plant pests in accordance with the entry requirements of the receiving country. Such certificates are issued by regulatory officials of the exporting countries to cover the movement of plants and plant products. Although the validity of phytosanitary certificates varies from country to country, dependent on the competence of the regulatory agency, their use has nevertheless reduced the movement of plant pests between countries. Ratification of the IPPC by member countries demands that certain activities be conducted which foster the development of more effective regulatory functions.

During the last 5 years, U.S. agricultural exports have almost tripled in value, reaching $22.15 billion in 1976. Requests for federal phytosanitary certificates have also increased from an average of 4600 a month in 1972 to 7000 per month in 1977. Such movement of agricultural products is important from the standpoint of the balance of payments. To assure acceptance of the products, it is essential that the phytosanitary aspects of the product be in accord with the entry requirements of importing countries.

The IPPC also endorses the formation of regional plant protection conventions, and numerous such regional groups are in existence. Some regional organizations are a part of the IPPC, such as the Caribbean Plant Protection Commission[9] of which the U.S. is a member. Some regional groups are not a part of the IPPC, such as the North American Plant Protection Organization (NAPPO)[10] of which the U.S., Mexico, and Canada are members. The primary purpose of NAPPO is to foster (to the extent feasible) comparable regulatory activities on a continental basis. Cooperation in the application of pest populations management strategies also is provided for when pest species involve any combination of the three countries. Under this concept, the U.S. and Mexico cooperated in eradicating the khapra beetle from Mexico and the U.S. and all three countries participate in similar activities to prevent additional artificial introductions of this exotic pest.

Through NAPPO arrangements are also made for the sharing of research, survey, and other information to more adequately function on a regional basis.

Preclearance

Preclearance denotes the removal of pest risk in connection with the movement of products, articles, or passengers at the point of origin rather than at port of entry. It involves activities by U.S. personnel at point of origin to the extent deemed necessary to eliminate or reduce pest risk. Experience has shown that the U.S. should not and, as a result, does not rely solely on the validity of phytosanitary certificates as a basis for the entry of plant products. Other entry safeguards are imposed, such as inspection and clearance at port of entry to the extent deemed necessary to provide for adequate pest prevention.

In addition to the use of phytosanitary certificates and other safeguards at ports of entry, arrangements are made with U.S. agencies and foreign governments for the preclearance of certain products and passengers at the point of origin. For example, in cooperation with the U.S. armed forces, military cargo and materials are cleaned before being returned from foreign areas to the U.S. In addition, arrangements also are made to preclear military personnel and their possessions. Similar preclearance activities are conducted regularly for civilian air passengers departing Hawaii, Puerto Rico, U.S. Virgin Islands, Bahamas, and Bermuda for the U.S. mainland.

Numerous preclearance programs are conducted which involve the clearance of certain agricultural products at point of origin before shipment to the U.S. Cost of these programs are charged to the exporters, including reimbursement for the assignment of U.S. personnel in foreign countries for the period necessary to preclear the products. Prior to the initiation of any such preclearance activities, a determination is made that the program will benefit the U.S. through the inspection, treatment, or processing of the products in a manner to handle pest risk. The volume of product movement must be sufficient to warrant the assignment of personnel, both on the part of the U.S. and the exporters. Examples of preclearance programs are: deciduous fruit from South Africa, apples from New Zealand, bulbs from Holland, Germany, France, Israel, Italy and South Africa, Unshu oranges from Japan, fern plants from Holland, grapes from Chile, and mangoes from Haiti.

Procedures used as a basis for preclearances depend on the pest problem and type of products involved. New Zealand apples, because of the presence of the light brown apple moth, are inspected prior to shipment using a statistically designed sampling system. If the light brown apple moth is detected, the entire lot is refused.

In the Unshu orange program, the movement of such oranges is allowed only from specified areas within Japan where U.S. specified safeguards are followed, including inspections cooperatively conducted by U.S. and Japanese officials, and citrus canker has not been detected. As an additional safeguard, arrangements are made for bacteriophage testing of citrus to be exported to the U.S.

In the preclearance program for fern plants from Holland, regulatory personnel of Holland supervise the production of ferns in a manner prescribed by the U.S. Monitoring by U.S. employees is conducted to the extent necessary to assure that the material is produced, packaged, and shipped in accordance with requirements. This procedure precludes the possibility of pest contamination to a much greater degree than possible through inspection and release at a U.S. port of entry.

On other programs, such as those involving fruit flies, the preclearance program provides for the use of procedures which have been developed and approved as a basis for eliminating pest spread. Such procedures include the use of chemical fumigants, such as ethylene dibromide or methyl bromide, independently or in connection with cold treatment. When cold treatment is utilized, the ships are equipped with recording devices in accordance with U.S.-approved procedures. The vessels and cold treatment equipment are checked by U.S. inspectors prior to being approved for bringing certified fruit into the U.S.

Port of Entry

Although the phytosanitary certification and preclearance programs are important in pest prevention strategies, the majority of prevention activities is still centered at ports of entry. Ports of entry include airports, seaports, and land border ports of entry, including quarantine facilities through which high-hazard products are processed before clearance.

The effective clearance of products and passengers at ports of entry requires an ongoing evaluation of modes and speed of travel as well as volume of traffic. Pest hazards vary as conditions change. For example, the transportation of products in containers not intended to be opened from point of origin to final destination has had a significant impact on pest prevention activities at ports of entry.

Port of entry clearance activities depend on trained personnel clearing products and passengers. Port of entry activities may be divided into three primary areas, namely, those involving clearance through quarantine facilities or post entry requirements, those where treatments are applied as a condition of entry, and those where entry is based on inspection and clearance.

Entry Through Quarantine Facilities

Certain products or materials, due to the potential pest risk associated with their movement, are subject to entry through quarantine facilities. For example, beneficial organisms are screened through a quarantine facility to reduce the possibility of introducing hyper-parasites or other pests that may be detrimental to American agriculture. This also provides an opportunity to determine that the beneficial organism introduced is, in fact, the species intended to be imported. Such a screening program also precludes the possibility of introducing unwanted pests in connection with the shipping media, and makes it possible to determine that the imported organism does not attack other agricultural products.

During 1977, a total of 226 shipments containing 220,000 specimens of beneficial parasites or predators was received from overseas at the quarantine facility at Newark, Delaware. This facility in turn, after quarantine clearance, shipped 82 beneficial species in 409 shipments involving 105,000 specimens to research workers in 26 states and 2 Canadian provinces. During that year, 31 species were field-released in the U.S. Additionally, the University of California operates two quarantine facilities which handle beneficial parasites and predators. Beneficial organisms for the control of weeds are cleared through the quarantine facility at Albany, California.

Due to the presence of certain plant pests in the country of origin, some plant species are forbidden entry. Other high-risk plant material and plant germplasm are cleared through government-operated plant quarantine facilities. At a plant quarantine facility, the plants are subject to tests for obscure pests and retained for a holding period necessary to ensure that no detrimental plant pests are present.

Post-Entry

Importations of some plant material must be handled under controlled growing and other conditions as a basis for entry. The growing site is preapproved and inspections made during the growing season to detect unwanted plant pests. For example, certain plants which may introduce viruses or other unwanted plant pests not known to occur in the U.S. may be introduced and grown under specialized post-entry conditions before release.

Species of plants in 42 genera are grown under post-entry requirements. In addition, 53 genera of fruits and nuts come under the post-entry program. In fiscal year 1977, there were 680 shipments of post-entry material which entered the U.S During the same year, there were 524 releases of material after having been satisfactorily cleared from a pest-risk standpoint.

Also, certain crop seeds are introduced and grown under specialized conditions to preclude the importation of unwanted pests into the U.S. Growing conditions are stipulated as a condition of entry. If no unwanted pests are found, the plant material is eligible for release and movement to other parts of the U.S.

Mandatory Treatment as a Condition of Entry

Treatment, as a condition of entry, is required for some products or articles originating in areas known to be infested with unwanted plant pests. Such treatment procedures usually are applicable to those products which cannot be satisfactorily inspected or otherwise determined to be free of plant pests. For example, most host fruits originating in areas where exotic fruit fly pests are present, such as the European cherry fruit fly and the Mediterranean fruit fly, are required to be treated as a condition of entry. Such treatments may be applied at origin under a preclearance program, treated enroute, or at the port of arrival.

In addition to treatment requirements as a condition of entry for certain host materials or articles, authority is available to require treatment whenever, upon inspection, pest conditions are found to warrant such treatment. Further, based on inspection at ports of entry, repeated findings of pest conditions often dictate that mandatory treatments be applied as a condition of entry under circumstances which normally would not apply. For example, khapra beetles have repeatedly been found in holes left by wood-boring insects in wood crating material. When such conditions are found on repeated occasions in material from specific port areas, mandatory treatment may be required as a condition of entry.

Inspection and Clearance

Inspection and clearance of passenger baggage as well as cargo and other products are essential parts of pest-entry prevention. Such inspections require the utilization of personnel highly trained in the biological sciences, who are knowledgeable of pest conditions as well as rules and regulations applicable to the introduction and movement of material between foreign countries and the U.S.

Articles and plant materials not known to routinely present a risk of pest introduction, but which may be contaminated with unwanted plant pests, are subject to inspection at ports of arrival to determine pest conditions.

If, on inspection, any regulated material is found in passengers' baggage, the material may be inspected and released, confiscated and destroyed, treated and released, or reexported. Every effort is made to work with importers, private citizens, and commercial concerns to provide for expeditious entry of requested materials without compromising pest risk.

Commercial lots of propagative plant material which may or may not require treatment as a condition of entry are processed through established plant inspection stations. These facilities provide for better inspection conditions, the holding of material found on inspection to be infested under conditions to prevent pest spread, and for treatment prior to release, if desirable.

During 1976, there were 14 established plant inspection stations in operation in the U.S. During that year, representative samples were inspected from 160 million plant units capable of propagation. Additionally, representative samples were inspected from 2 million lb of tree seeds.

A total of 16000 shipments of material was received of which 2400 shipments required treatment to eliminate pest risk prior to release. A total of 477 shipments was refused entry due to pest conditions. Plants entering the U.S. from foreign areas must be free of soil as a condition of entry, with limited exceptions, e.g., plant material from Canada.

In connection with operations at ports of entry to clear passengers, carriers, and products, many plant pests are intercepted. Records are maintained on the number of such interceptions that are of particular significance in protecting the U.S. Such interceptions are referred to as pests of quarantine significance, which does not include organisms widely distributed, but rather those which are new to or not widely prevalent within the U.S.

Table 1 indicates the number of pests of quarantine significance intercepted during a 3-year period, indicating the means of potential introduction and the type of organisms. Approximately 42,000 quarantine-significant pest interceptions were made during this period. Although it is not expected that all such pests would have become established, it can, nevertheless, be concluded that such an interception program is of significance in protecting the U.S. from additional unwanted plant pest organisms.

APPRAISAL

This pest management substrategy has two primary activities, namely, (1) early detection of new plant pests, and (2) plant pest population monitoring. Authorizing legislation is available to allow employment of personnel to conduct detection and delimiting surveys and to cooperate with individuals, groups, or other governmental agencies

Table 1
QUARANTINE SIGNIFICANT PLANT PEST INTERCEPTIONS, 1975—77

Location of interception	Type of organisms: Nematodes	Insect	Mites	Snails	Pathogens	Total
Baggage	61	18,923	71	435	2,501	21,991
Mail	1	437	1	6	194	639
Cargo	13	11,857	606	293	837	13,606
Carrier contaminant	29	4,315	66	73	2,045	6,528
Totals	104	35,532	744	807	5,577	42,764

in survey activities, as well as to accumulate and store data gathered from such surveys and to publish results thereof.

Early Detection

The early detection of exotic pest species which have gained entry in spite of prevention program activities is an extremely important part of regulatory plant pest management. The primary purpose of early detection is to locate unwanted exotic plant pests at an early stage after introduction so that they may be eradicated, contained, or suppressed to the extent that technology and resources are available.

Biometrically designed surveys are conducted for exotic species at ports of entry, including airports, seaports, and border ports. Information gathered through port of entry prevention activities, such as number of pest interceptions and volume of traffic, are utilized in evaluating the potential risk of artificial introduction at the various ports. Primary survey attention is given to ports in areas having the highest pest introduction potential. These surveys include not only the high hazard port but also an environs area within an approximate 100-mile radius. This provides an opportunity to survey in agricultural lands, home yards and gardens, idle lands, and at agricultural storage and holding facilities. All available technologies, based on information obtained from areas where the pests occur, are utilized in these detection programs, including sex, food lure, and other types of survey traps, as well as visual surveys. Specially conducted programs provide for survey personnel to be highly trained in early detection activities. To identify the volume of specimens collected, arrangements are made with other organizations under contract to screen the material and discard pests known to occur in the area. Due to the large number of ports of entry, the inspection program is rotated over a period of years to achieve coverage of all high-hazard areas on a progressive basis.

An important part of early pest detection surveys is followup inspections at destination for cargo that left the port of arrival prior to a pest infestation being found in the ship and/or cargo. For example, during ship inspections after cargo has been unladen and released, infestations of khapra beetle may be found which could have been spread with the released cargo. When such situations occur, arrangements are made to immediately determine the destination of all such cargo and followup inspections are conducted. Treatment may be required of such cargo, dependent on pest hazard. Other followup inspections are made whenever an exotic pest species is first found to be established in the U.S. Inspections are made at locations to which shipments have been made that could spread the pest. For example, should an infestation of an exotic pest be found in an ornamental plant nursery for the first time, sales records are obtained from the nurserymen and followup inspections made.

Another important part of early detection is a determination, when possible, of the source of infestation and how the pest penetrated prevention program activities. By so doing, it may be possible to strengthen the prevention program.

In addition to exotic pest-detection activities around high-hazard port areas, trained inspectors are specifically assigned to detect new pests in agricultural crops and concurrently monitor pest conditions in such areas. In the plant disease area, this, of course, would include new strains or races of plant disease organisms. The systematic monitoring of certain select crops for new plant diseases was not funded at the federal level until 1977; therefore, this is a relatively new program. Observations for endemic plant diseases also are included in this new pest detection program.

Pest Population Monitoring

If a new pest species is detected and if it is determined that neither eradication nor containment is feasible with current technologies, an appraisal program is developed to monitor pest populations. Such monitoring activities are necessary to support strategies of managing the pest on a local or areawide basis. Monitoring surveys also are conducted for specific endemic pest species to provide information for pest population management.

Pest monitoring activities are of assistance to the agricultural industry and individual growers in applying timely pest management systems. Pest monitoring also is utilized by public agencies in the conduct of areawide management programs. The extent of monitoring activities depends, to a considerable degree, on the utilization of the data which is collected. Where sterile insects are to be utilized in a pest management program, it is necessary to determine the population of the target pest species in order to estimate the overflooding ratios required to attain control.

Similarly, where other biological organisms are utilized, it is desirable to monitor pest populations in order to guide the release of parasites and predators. The assessment of beneficial insect populations in crop fields also is necessary to determine the number and proper timing of pesticide applications. This provides a basis for applying pesticides only when needed, and reduces adverse effects on beneficial insects. Data systematically collected in pest population monitoring can be utilized in developing prediction models. Pest population assessment is also necessary to determine the effectiveness of the various pest population suppression techniques, whether they be biological, cultural, or chemical.

PEST POPULATION CONTROL

Population control, one of the substrategies of regulatory plant pest management, involves activities of two different intensities, namely, management out of existence (eradication) and population manipulation. Domestic quarantines or regulations are utilized whenever their use supports the applicable fundamental approach. Quarantines are always applied in eradication programs to prevent or retard pest spread. Quarantines may or may not be utilized on programs where the objective is population manipulation rather than eradication. Legislation is available which provides necessary support for areawide management activities, including requirements to follow certain cultural practices or other population manipulation systems.

Prevention and appraisal substrategies, though necessary portions of regulatory plant pest management systems, usually do not adversely affect environmental considerations. Rather, they provide a basis for population control and have a favorable cost/benefit ratio with the least adverse affect on environmental considerations. Population control systems, especially in the short-term time frame, may have an unfavorable cost/benefit ratio and, for a limited period, could have an adverse affect on non-target organisms. Conversely, on a long-term basis, the same pest management system could have a high cost/benefit ratio and cause the least adverse affect on environmental considerations. Integrated pest population management systems, therefore, must be individually analyzed for each pest species on the basis of both their short- and long-term effects.

Regulatory plant pest management systems are designed to systematically incorporate all available management strategies. Action plans are developed for selected plant pests of known economic importance prior to their establishment in this country. The selection of such pests for which plans are developed is predicated not only on their economic potential but also on the likelihood of artificial introduction through commerce, interception information gained from the prevention system, and natural spread from adjacent land areas. Prior development of such plans makes it possible to promptly inaugurate an applicable management system if the pest is found.

Eradication

It must be emphasized that the goal of eradication with available technologies is not total elimination of a species worldwide. Rather, it is the deliberate eradication of a pest species, usually an introduced species, from a section of the geographic range of the pest. When eradication is accomplished, it prevents continued annual crop losses and eliminates, as well, costs which would be associated with managing the eradicated pest species. Additionally, if pesticides are required in the management of the pest species, it precludes the utilization of these materials on a continuing basis with possible adverse environmental impact.

Tools used in eradication strategies are dependent on the pest species involved, as well as the state of the art in the development of eradication systems by research agencies. Some introduced pest species have been eradicated through the utilization of only cultural practices, others only with pesticides; whereas, with some, several population manipulation tools have been employed. The time frame required to attain eradication is dependent not only on the life cycle of the past organism and the relative effectiveness of technologies available in manipulating the life stages of the pest, but also on appropriated funds.

It required 29 years to eradicate the citrus canker organism from the U.S. This pest had been found in citrus groves and in wild hosts from Texas to Florida. Quarantines were applied to prevent artificial spread of the organism. Eradication was attained by conducting intensive surveys and removing all host material found to be infected. Conversely, for the oriental fruit fly, a highly specific and effective lure is available which makes early and prompt detection possible. Also, the lure, when mixed with a pesticide and applied as a spot treatment, is highly effective and can be used to eliminate the pest from defined geographic areas. For this unwanted plant pest, therefore, eradication of incipient outbreaks can be accomplished in a very short time frame. Eleven separate introductions of this species in California were eliminated over a 17-year period.

Because of the many avenues of artificial introduction, prevention strategies cannot stop all repeated invasions. For certain pest species, eradication programs are conducted as each new invasion occurs. Fortunately, research may develop new population management techniques which are not only effective but also are more environmentally acceptable.

The Mediterranean fruit fly was first introduced into the U.S. in 1929. Following the eradication of that outbreak, this pest has reinvaded the U.S. on four additional occasions (years), each time being eliminated. However quite different eradication technologies have been utilized. The principal population management system in 1929 was the destruction of infected host material and the use of a relatively ineffective insecticidal spray. Detection techniques available were far less effective than the synthetic lures now available. By the time of the 1956 invasion, research had developed not only effective lures and traps, but also highly effective bait sprays which were applied by aircraft. Additionally, fumigation procedures had been developed so that host material could be marketed from regulated areas rather than destroyed. Based on continued research, recent introductions of this pest have been eradicated through the use of highly effective detection devices and the application of bait sprays, only to the extent necessary to reduce pest populations to a level to enable eradication through the rearing and release of sterile insects. The majority of host fruits can be safely moved from infested areas, primarily through the application of fumigation procedures.

Some exotic plant pests are of such concern to the U.S. that, when they are first detected, programs are undertaken to contain the infestation to the extent possible. Research and methods programs are conducted in an attempt to develop new technologies to adequately cope with the pest. Some such programs have resulted in the development of technologies that make eradication a feasible goal. Witchweed, a parasitic plant pest of corn and related species, was found in North Carolina and South Carolina in 1956. Intensive survey, quarantine, and population suppression activities confined the infestation to 37 counties from 1956 through 1977. New technologies from research and methods development now make eradication of witchweed feasible, thus providing ultimate protection from this plant pest.

Since the inauguration of organized regulatory plant protection activities, a number of programs have been conducted with the objective of eradicating introduced plant pests when feasible. Table 2 summarizes successful eradication programs including a listing of those pests where more than one invasion occurred. The pest population management strategies utilized are indicated.

Population Manipulation

For many pest species, though the initial approach was containment/eradication, the program objective necessarily was changed to one of pest population manipulation. However, containment/elimination activities usually provide protection to American agriculture while research developed technologies to make it possible to more effectively manage pest populations. The cereal leaf beetle, detected in Michigan in 1962, initially was attacked through the application of pesticides and the use of quarantine in an attempt to contain or eradicate the organism. Concurrently, research scientists determined parasites which could be used in managing this pest and, just as important, how to rear and distribute these biological organisms. When it became evident that activities to eliminate or prevent spread were not feasible, the program direction was changed to one of population management primarily through the utilization of parasites and predators, and the quarantine was revoked.

Some plant pests, especially those that are nonmigratory, may be effectively managed on a local or field-by-field basis. Through the utilization of some control strategies, migratory species may likewise be managed on a field-by-field basis. However, for pest management strategies to reach the ultimate in effectiveness, they usually must be applied on an areawide basis. The size of the geographic area varies, dependent not only on the pest species but also on the pest management strategies which are available. Federal and state authorizing legislation provides for the utilization of an areawide approach when, in the public interest, funds as well as technologies are available.

Varying strategies are utilized on organized federal/state pest management pro-

Table 2
CONTINENTAL U.S. ERADICATION PROGRAMS

			Population management technique		
Year(s)	Plant pest	State or location	Cultural	Biological	Chemical
1913—36	Parlatoria date scale	Arizona and California	Destruction of infested fronds and worthless tree; flame heating tree trunk		
1914—43	Citrus canker	All gulf states	Host destruction		
1918—74	Potato wart	Maryland, Pennsylvania, and West Virginia	Host destruction		Fungicide
1929—30	Mediterranean fruit fly	Florida	Fruit destruction		Foliar sprays
1956—57		Florida			Bait sprays
1963		Florida			Bait sprays
1966		Texas			Low volume foliar spray and bait spray
1975		California		Sterile release	Bait spray
1934—38	Citrus blackfly[a]	Florida			Oil foliar sprays
1941—58	Hall scale	California			Cultivated hosts — fumigation, wild hosts — herbicides
1953—66	Khapra beetle	Arizona, California, New Mexico, Texas, and Mexico			Fumigation and residual sprays
1957—65	Hoja blanca	Florida, Louisiana, and Mississippi	Plowing under crops		Foliar insecticides sprays and herbicides
1960—77	Oriental fruit fly[b]	California		Lure/insecticide; spot treatment	
1969—75	Giant African snail	Florida	Hand collection and destruction		Insecticide/bait; Sevin foliar spray
1970	Corn rust	Florida	Host destruction		Fumigation of soil

[a] Became established in Texas in 1971 and reinvaded Florida in 1976. [b] 11 Invasions.

Table 3
EXAMPLES OF REGULATORY POPULATION MANIPULATION PROGRAMS

Plant pest	Quarantine[a]	Population management technology: Cultural	Biological	Chemical
Black stem rust	F	Alternate host destruction	Resistant varieties	
Cereal leaf beetle	F[b]		Rearing/managing parasites	Foliar sprays[c]
Citrus blackfly	F		Rearing/managing parasites	Foliar sprays
Golden nematode	F	Crop rotation	Resistant varieties	Nematicides
Grasshopper	None	Livestock density management	Pathogen production and application	Treatment of economic population
Gypsy moth	F		Rearing/release parasites; virus	Foliar sprays of isolated, peripheral, and defoliating populations
Phony peach	F[d] S	Infected tree removal	Destruction of weeds attractive to insect vector	Herbicide symptomless noncommercial host trees
Pink bollworm	F	Planting/harvesting dates; destruction crop residues	Sterile releases	Foliar sprays
Soybean cyst nematode	F[d] S	Crop rotation	Resistant varieties	
Witchweed	F	Mechanical host destruction	Artificial seed stimulated to promote suicidal germination	Herbicides

[a] F = Federal and state quarantine; S = state quarantine only
[b] State and federal quarantine revoked.
[c] Primary technology in early stages of program, now in isolated pockets of nonparasitized populations.
[d] Federal quarantine revoked, state quarantine continued.

grams. Not only are they dependent upon technologies developed by research but also, even for same pest species, on the area where infestations are known to occur. The citrus blackfly in Mexico is managed through the utilization of parasites. Parasite populations rarely are adversely affected by the application of pesticides since the Mexican Government requires a permit for the use of pesticides in the principal citrus-producing area. It is, however, necessary to manipulate the parasites by moving them into areas where pest populations are heavy due to low parasite populations. For the same pest species in the Rio Grande Valley of Texas, again, the primary management technology is the managed utilization of parasites. Pesticides are utilized only to knock down heavy pest populations. An infestation of citrus blackfly found in the coastal area of Florida in 1976 is being managed with eradication as the ultimate objective through the application of pesticides on the outer edges of the infestation pushing inward. Meanwhile, parasites are reared and released in the main or core area of infestation to reduce populations. This makes chemical/biological elimination feasible and reduces the risk of artificial and natural spread.

Ecological conditions in the area of the same pest species also influence the management techniques utilized. Populations of the pink bollworm in Texas are adequately managed through the utilization of cultural procedures including planting and harvesting dates, as well as the destruction of crop residues. These cultural procedures are not adequate for the management of this pest in the drier regions of Arizona and the Imperial Valley of California. In these areas pesticides also must be utilized, and their use is creating problems with other pest species. On the other hand, the primary management procedure utilized in the San Joaquin Valley of California is monitoring traps to determine pest populations as a basis for overflooding with laboratory-reared, sterilized pink bollworm moths. These activities are supported by the enforcement of quarantines to prevent artificial spread into the San Joaquin Valley. In the meantime, research is continuing and it appears that new technologies for cultural control will be available to assist in reducing pest populations in the drier regions of the U.S.

Cooperative regulatory plant pest management programs are conducted against certain endemic species, especially in situations where actions by an individual landowner are not effective. Pest populations of rangeland grasshoppers are monitored to determine when pesticide applications are necessary on an areawide basis to protect range grasses. Such programs are financed on a cooperative basis usually involving the grower, as well as state and federal governments. Landowners on an individual basis are able to protect their croplands through the utilization of pesticides. Table 3 illustrates cooperative state/federal pest-population manipulation programs showing the organism and the pest management technologies which are utilized.

REFERENCES

1. U.S. Department of Agriculture Secretary's Memorandum No. 1929, "U.S.D.A. Policy on Management of Pest Problems," December 12, 1977.
2. Plant Quarantine Act of 1912, as amended (7 U.S.C. 151-165, and 167).
3. Federal Plant Pest Act, approved May 23, 1957 (7 U.S.C. 150aa-150jj).
4. Organic Act of 1944, as amended (7 U.S.C. 147a).
5. Federal Noxious Weed Act of 1974 (7 U.S.C. 2801-2813).
6. The National Environmental Policy Act of 1969, as amended (42 U.S.C. 4321 et seq.)
7. Endangered Species Act of 1973 (16 U.S.C. 1531-1543).
8. International Plant Protection Convention of 1972, Food and Agriculture Organization of the United Nations, Rome.
9. Caribbean Plant Protection Commission of 1967, Santo Domingo, Dominican Republic.
10. North American Plant Protection Organization of 1976.

CONTROL OF PLANT DISEASES BY EXCLUSION: QUARANTINES AND DISEASE-FREE STOCKS

Howard E. Waterworth

INTRODUCTION

Definitions

Control of plant diseases by prophylaxis, as opposed to disease resistance, includes such procedures as direct protection, eradication and exclusion.[43] Direct protection refers to measures used when it is assumed that the host will be exposed and that it will be infected unless preventive measures are taken. Most spraying and dusting of foliage is direct protection. Eradication consists of measures taken to eliminate the pathogen after it has become established in the sphere in which the host is growing. Exclusion includes measures designed to keep pathogens from entering the sphere in which the host is growing. With respect to quarantines, entry of the pathogen is prevented by inspection, or the host plant or its parts are banned or restricted from movement across political boundaries. This chapter deals primarily with exclusion as affected by federal and state quarantine procedures.

Terms and phrases used frequently in this chapter are definedas follows:

Country of origin: the country from which the plant material is exported to the U.S. (not the country or countries where that species evolved).

Germplasm: any propagative plant parts, vegetative or seed, brought to the U.S. for research or educational purposes. The term ''plant'' is used in reference to entire plants imported by private citizens, which do not become part of a research program.

Virus-free: plants or vegetative parts from sources that were tested and found free of known viruses and mycoplasmas. The term does not imply absolute freedom from all infectious entities known or not yet discovered.

Certified plants: plants that are apparently free from insects and diseases, as determined by visual inspection only. The quality of this inspection varies with the expertise of the inspectors, and latent virus infections are usually not detected.

Some frequently used acronyms are:

SEA: Science and Education Administration of USDA (includes Agricultural Research (AR), formerly ARS)
BARC-E: Beltsville Agricultural Research Center, east side
BARC-W: Beltsville Agricultural Research Center, west side (formerly Plant Industry Station)
GRL: Germplasm Resources Laboratory of SEA
NPGS: National Plant Germplasm System (federal, state, and private)
RPIS: Regional Plant Introduction Station (state-federal)
P.I.: Plant Introduction; usually a small lot of a specific species or cultivar
APHIS: Animal and Plant Health Inspection Service of USDA
PPQ: Plant Protection and Quarantine Programs of APHIS
NPPS: National Program Planning Staff of PPQ
PGQC: Plant Germplasm Quarantine Center of NPPS

Scope of Chapter

This chapter is limited to consideration of the following:

1. U.S. federal and state quarantine regulations and procedures
2. disease control, especially of viral diseases, although many of the principles also apply to insects

3. imported plants or their parts brought in as germplasm (usually in small quantities) and plants for private residents, as opposed to shipload quantities of plant parts destined for consumption or processing
4. sources, particularly foreign ones, of certified virus-free vegetatively propagated planting stock

History of Quarantine and Plant Introduction

It is not generally known that extremely few of the major U.S. agricultural crops are native to this country. Except for the trees of our forests and the grasses of our plains, nearly all of our crops — the cereals, vegetables, forages, fruits, legumes, cotton, tobacco, and many ornamentals — came from somewhere else[24,40] (Figure 1). Only sunflowers, cranberries, blueberries, pecans, hops, some grapes and brambles, and certain ornamental plants like laurel and dogwood are native. Even the ingredients of such well-known American foods as apple pie and pumpkin pie came from elsewhere centuries ago. We think of corn as native because the Indians taught its culture to the arriving colonists, yet corn originated in Mexico. Tobacco, the crop that the colonists grew and exported so much of, is not native either. It came from South America. In terms of native edible plant species, the U.S. was a poor country.

We know that the prehistoric inhabitants of what is now the U.S. brought many plants into this country from neighboring territories, and that this practice was thriving in pre-Columbian times.[40] When the Europeans began settling the U.S., they were advised to bring seeds of crops with them. This form of plant introduction continued well into the 19th century. Plant introduction became a function of the Federal Government in 1827 with President John Quincy Adams' order to American consuls to send home rare plants and seeds, and in 1829 with Congress's first agricultural appropriation of $1000. From 1836 to 1862 the U.S. Patent Commissioner introduced new species.[40] When the U.S. Department of Agriculture was established in 1862, plant exploration was accelerated. When the Section of Seed and Plant Introduction was established in 1898, the now well-known system of assigning a number to each lot of seeds, scions, bulbs, tubers, and whole plants was established. Thus, cabbage seed from the U.S.S.R. was assigned Plant Introduction (P.I.) No. 1. A permanent and detailed record of most items of germplasm that have subsequently entered the U.S. has been published by the USDA and is available in the libraries of most land grant agricultural colleges. This multivolume published inventory includes information on where, when, and why most of the now 450,000 plant introductions were collected.

Little attention was paid to the inadvertent importation of pests until 1912 when the first federal plant quarantine regulations were enacted by Congress. Unfortunately, by then many of the world's most destructive plant disease agents were already in the U.S. or later became established in spite of the new quarantines.[43] Among the more notable pathogens thus introduced were the white pine blister rust fungus, first found in the U.S. in 1906; the chestnut blight fungus in 1904 in New York City; the citrus canker bacterium in 1912 in Florida (subsequently eliminated at a cost of $53,000,000); the potato black wart fungus, in Pennsylvania in 1918 (since eradicated); the wheat flag smut fungus in 1919 near St. Louis; and the Dutch elm fungus in 1930 in Ohio. Hundreds of lesser known examples of pathogen introductions could be cited. Unfortunately, most of the pathogens that have been responsible for catastrophic plant diseases,[21] such as pear decline, peach yellows, potato late blight, the cereal rusts, brown spot of rice, and root rot of sugarcane, are now established in the U.S. Thousands of other pathogens, including some reported to cause immense losses, such as the soybean rust fungus, the sugarcane smut fungus, and the plum pox virus, are not yet known to occur in the 50 states.[51] Certain foreign diseases have been categorized as being of high, intermediate, or limited threat to U.S. agriculture.[37]

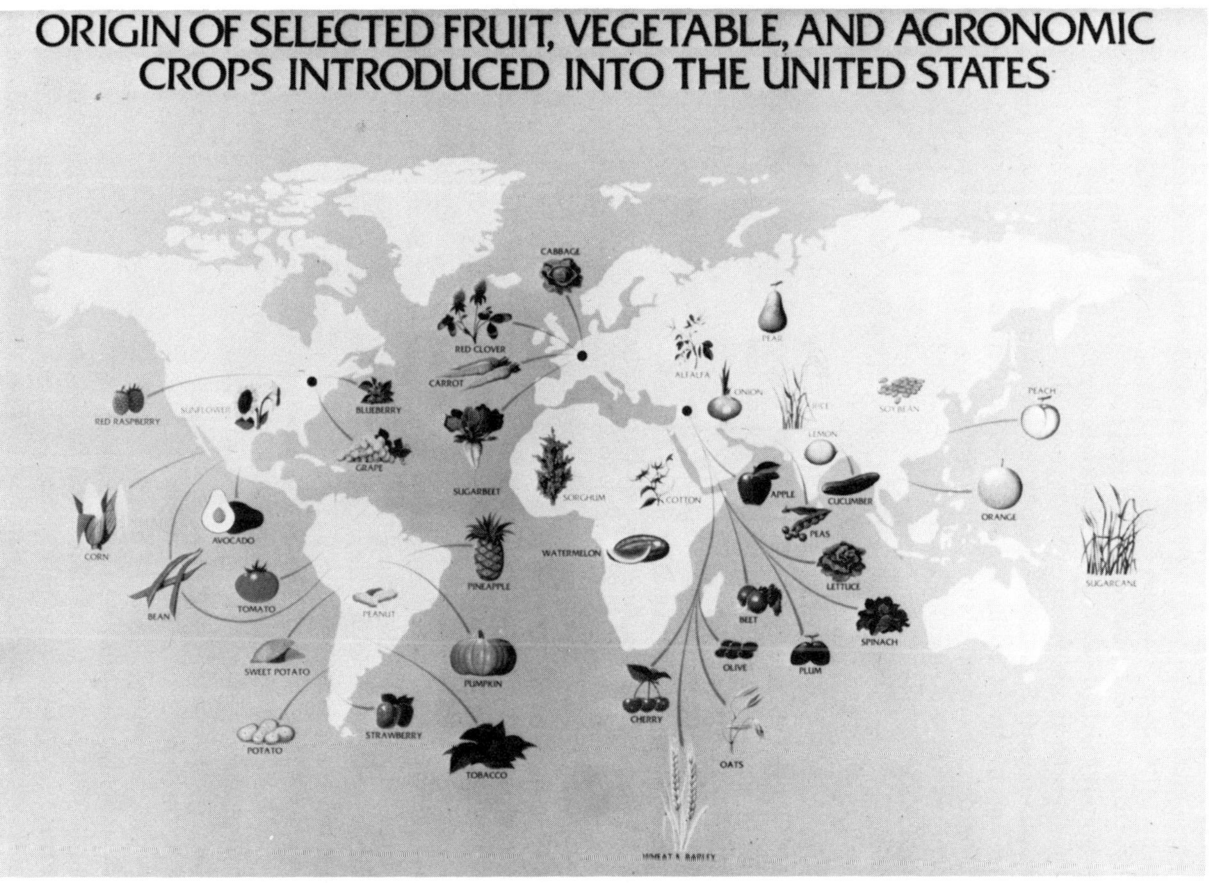

FIGURE 1. Origin of selected fruit, vegetable, and agronomic crops introduced into the U.S.

Even though 450,000 P.I.s have been brought into the U.S. since 1898, the need for foreign germplasm is still as great as ever because: (1) there are large gaps in the genetic diversity of some crops, and (2) sources of resistance to many diseases and insects are not yet available to breeders. The value of foreign germplasm as a source of disease resistance, in the form of wild species, has been summarized[22] and the native gene centers of these crops are known.[24] The gene centers of most genera are outside the U.S. (Figure 1). Some of these sources of genetic diversity are being depleted, displaced, or abandoned. Once gone, they will never again be available to mankind.[40]

Hence, plant germplasm continues to enter the U.S. at an average rate of 7500 items per year. The plant introduction and quarantine system of the USDA, unlike the systems in many countries, allows for the importation of any species of plant from any country of the world at almost any time of the year, with minimum threat to U.S. crops. Germplasm enters the U.S. today primarily as a result of USDA-sponsored collections in specific areas of the world, in efforts to broaden the genetic base of a particular crop or category of plants. The other main reason for new accessions is in response to requests by U.S. plant scientists.

FEDERAL QUARANTINE

Introduction

At the federal level two distinct but integral operations are involved with bringing plants into the U.S. One is the plant introduction scheme, which includes foreign explorations, receiving, establishment, propagation, preliminary evaluation, and distribution of new germplasm within the U.S. This activity is executed by personnel of the Science and Education Administration (SEA). The other activity is quarantine, which includes inspection of the germplasm upon its arrival in the U.S., observation of the plants during the quarantine period, indexing for viruses and mycoplasmas, identification of insect and disease agents, and release from quarantine according to regulations.[41] These procedures are carried out simultaneously with the plant introduction activities by personnel of the Plant Protection and Quarantine (PPQ) Programs of the Animal and Plant Health Inspection Service (APHIS). They will be discussed in detail later.

There are a number of excellent reviews on plant quarantine. Some deal with principles and methodology,[18] the legal basis,[41] theoretical aspects on benefits vs. risks to U.S. agriculture,[17,25] and multiple aspects of quarantine.[26] Others describe quarantine in other countries or regions of the world.[29,34] The discussion here is intended to acquaint the reader with several of the diverse programs that are administered by the PPQ section of APHIS as a means of controlling plant diseases. These programs encompass attempted prevention of entry of pests into the U.S., inspection of established P.I.s to reduce the risk of pest outbreak, and eradication of a pest in the unfortunate event that it too becomes established.

Economic Basis for Quarantine

Other reports deal with the economic basis for protection against plant pathogens by quarantine.[27,21] McGregor[27] reported that there are 550 foreign plant pathogens that, if established here, would be a significant threat to U.S. agriculture and would produce a wide range of economic impacts. The ten thought to be the most damaging if established in the U.S. are listed in Table 1. Pathogens of forest trees are at the top of the list because of the nature of the crop. The PPQ has an in-house list of the 100 most damaging foreign plant pests that includes insects, mites, and nematodes, as well as microorganisms. Soybean rust fungus heads this list. Although dollar-value loss figures are assigned to each pathogen, they undoubtedly are rough estimates at best.

Table 1
THE TEN MOST DANGEROUS EXOTIC (FOREIGN) PLANT PATHOGENS FOR THE U.S.[27]

Organism	Rank	Host(s)	Probability of establishment	EEI[a] ($, millions)
Rosellinia radiciperda (fungus)	1	Nine forest tree species	High	3126
Helicobasidium mompa (fungus)	2	Fruit trees, crucifers, peanuts	Medium	2703
Cronartium himalayense (fungus)	3	Pine	Medium	1992
Poria rhizomorpha (fungus)	4	Oak	Medium	1915
Cronartium quercuum (fungus)	5	Oak	Medium	1406
Xanthomonas acernea (bacterium)	6	Maple	Medium	1118
Melampsora pinitorqua (fungus)	7	Pine	Medium	662
Rosellinia quercina (fungus)	8	Oak and pine	Medium	579
Phakopsora pachyrhizi (fungus)	9	Soybean	Medium	551
Rhizoctonia lamellifera (fungus)	10	Pine	Medium	496

[a] Estimated economic impact once established. The dollar values are not highly accurate.

Nevertheless, the inadvertent introduction of the top ten pathogens could mean losses of billions of dollars.

The APHIS publishes periodically a list of the thousands of organisms found on plants that were intercepted upon arrival at U.S. ports and plant inspection stations.[42] Pests are listed in the several tables by Latin binomial, by country of origin, and by host. This book, of more than 100 pages, certainly presents convincing evidence that costs of quarantine programs are justifiable.

Organization

National Plant Germplasm System (SEA)

Plant introduction and plant quarantine are so closely intertwined at all steps that they will be considered together. The plant introduction operation, known as the National Plant Germplasm System (NPGS), is carried out by personnel of the SEA-AR, with lesser involvement of state experiment station personnel. The key elements of the system are (1) the SEA-AR Germplasm Resources Laboratory (GRL) at Beltsville, Md.; (2) two SEA-AR Plant Introduction Stations, at Glenn Dale, Md. and Miami, Fla.; (3) four state-federal regional plant introduction stations, located at Pullman, Wash., Ames, Iowa, Geneva, N.Y., and Experiment, Ga.; (4) the State-Federal Potato Introduction Station at Sturgeon Bay, Wis.; (5) the SEA-AR National Seed Storage Laboratory at Fort Collins, Colo.; and (6) a group of federal and state germplasm curators located throughout the U.S. The structure of the NPGS is shown in Figure 2. The flow of germplasm immediately upon arrival in the U.S. is diagrammed in Figure 3.

National Program Planning Staff (APHIS)

More or less superimposed upon the GRL (SEA-AR) germplasm-flow scheme are the activities of the APHIS-PPQ National Program Planning Staff (NPPS) and the state quarantine personnel who inspect, treat, index, and release P.I.s at the various locations.

Figure 4 will familiarize the reader with the various federal quarantine operations that may impinge upon his program. The NPPS is composed of seven staffs, as listed below.

POLICY

The National Plant Genetic Resources Board
(Office of the Secretary of Agriculture)

COORDINATION AND ADVICE

The National Plant Germplasm Committee

Representation: State Agricultural Experiment Stations; National Council of Commercial Plant Breeders; U.S. Department of Agriculture

OPERATIONS

NATIONAL	REGIONAL	STATE
Principal Plant Introduction Office, Beltsville, Md. National Seed Storage Laboratory, Fort Collins, Colo. Institute for Tropical Agriculture, Mayaguez, P.R. Interregional Projects: IR-1 Potatoes, Sturgeon Bay, Wis. IR-2 Virus-free Tree Fruits, Prosser, Wash.	Regional Plant Introduction Stations: Geneva, N.Y. Ames, Iowa Experiment, Ga. Pullman, Wash. Regional Technical Committees Regional Projects Curators of collections not located at Regional Plant Introduction Stations: Many locations	State and Private Curators of Collections
Participation: State & Federal Scientists; State & Federal Advisers	Participation: SAES — Administrative[a] Adviser, Scientists; SEA, CR, SCS, FS, BLM — Scientists, Coordinators	

[a] SAES, State Agricultural Experiment Station; SEA, Science and Education Administration; CR, Cooperative Research; SCS, Soil Conservation Service; FS, Forest Service; BLM, Bureau of Land Management (Department of the Interior).

FIGURE 2. The structure of the National Plant Germplasm System.

1. The Port Operations Development Staff plans, develops, and coordinates programs that concern importation and clearance of baggage, cargo, mail, and carriers entering U.S. ports.
2. The Pest Program Development Staff plans and guides specific programs for control of plant pests. These programs constitute the federal-state cooperative program aimed at restriction or elimination of particularly serious pests that become established in the U.S., such as the golden nematode, the gypsy moth, and witchweed.
3. The Environmental Evaluation Staff develops and evaluates biological information relating to monitoring of pesticides and efficacy testing in the pest control program.
4. The Methods Development Staff seeks to improve the effectiveness of PPQ program operations by reviewing the existing operations and developing better, cheaper techniques.

National Plant Germplasm Quarantine Center[a]
Beltsville, Maryland
(to one or more of the following:)

Federal Introduction Stations	State-Federal Regional Plant Introduction Stations	Specialized Stations	Individuals
Glenn Dale, Md. Savannah, Ga.	Ames, Ia. Experiment, Ga. Geneva, N.Y. Pullman, Wash.	Subtropical Horticulture Research Unit, Miami, Fla. National Seed Storage Laboratory, Ft. Collins, Colo. Institute for Tropical Agriculture, Mayaguez, P.R.	Regular curators Requesting plant scientists

[a] Jointly operated by Plant Protection Quarantine of APHIS and the Germplasm Resources Laboratory of SEA-AR.

FIGURE 3. Flow of germplasm immediately upon arrival in the U.S.

Director

Port Operations Development	Pest Programs Development	Environmental Evaluation	Methods Development	New Pests Detection and Survey	Regulatory Support	Plant Importation Technical Support

Plant Importation			Technical Support	
Foreign Site Inspection (The Netherlands)	Plant Quarantine Facility at the P.I.S., Glenn Dale, Md.	1. Plant Germplasm Quarantine Center 2. Post Entry Quarantine Unit, Beltsville, Md.	Biological Evaluation Unit, Hyattsville, Md.	Diagnostic Unit (Identification of organisms, nematodes, slugs, weeds, arthropods, etc.) Beltsville, Md.

FIGURE 4. Structure of a portion of the National Program Planning Staff of the Plant Protection and the Quarantine Programs, Animal and Plant Health Inspection Service.

5. The New Pests Detection and Survey Staff seeks to detect incipient colonies of new pests in the U.S., evaluate the destructive potential, and respond by eradicating, suppressing, or delimiting newly discovered pests of quarantine significance.

In 1977 a new phase of this objective was begun when APHIS hired nine plant pathologists and placed one in each department of plant pathology in the north-central states (Table 2). These members are on continual alert for unexpected diseases in commercial crops in a ten-state area. Much of their time is spent surveying fields throughout their respective states. They work in cooperation with state university pathologists, state department of agriculture personnel, agribusiness representatives, and county agricultural agents. Facilities are available for identification of suspected new disease agents and for dealing with them appropriately. Early detection of outbreaks of new diseases can mean their control and

Table 2
LOCATION OF ANIMAL AND PLANT HEALTH INSPECTION SERVICE (APHIS) NEW PESTS DETECTION AND SURVEY PATHOLOGISTS

State	University Address[a]	Phone
Dakota(s)	N. D. State Univ., Fargo, 58102	701-237-8362
Illinois	Univ. of Illinois, Urbana 61801	217-333-3170
Indiana	Purdue Univ., W. Lafayette 47907	317-749-6435
Iowa	Iowa State Univ., Ames 50010	515-281-3861
Michigan	Michigan State Univ., E. Lansing 48823	517-335-4529
Minnesota	Univ. of Minn., St. Paul 55101	612-373-0582
Missouri	Univ. of Mo., Columbia 65201	314-882-2418
Nebraska	Univ. of Nebr., Lincoln, 68503	402-472-3169
Ohio	Ohio State Univ. Columbus 43210	614-422-6397

[a] Write to: APHIS Plant Pathologist, Department of Plant Pathology, at the above university, city, state, and ZIP code.

may mean reduced food loss to society. Plans are to expand this program to other states. Persons who encounter or suspect outbreaks of new, exotic diseases are encouraged to contact the nearest APHIS pathologist.

6. The Regulatory Support Staff writes and publishes standards, regulations, and procedural guides for the importation, interstate movement, and exportation of plant materials and soil. The staff conducts public hearings on proposed quarantine actions. Its Permit Section issues permits to scientists who wish to import germplasm or soil, and provides information on the plant import requirements of foreign countries. This staff prepares summaries of state quarantine regulations and maintains liaison with other federal, state, and foreign regulatory officials.
7. The Plant Importation and Permit Staff evalutes pest risk in the importation of germplasm and importation and interstate movement of plant pests and pathogens, and advises the Permit Section (above) accordingly. This staff plans and directs the activities relating to enforcement of quarantine of nursery stock, plants, and seeds. These functions include (1) inspection and control of plants imported under post-entry quarantine and growing at several hundred locations within the U.S., (2) inspection of plants in foreign countries before their shipment to the U.S., (3) services relating to quarantine and indexing for viruses of prohibited-category germplasm, and (4) preparation for the export of germplasm by the USDA. The staff identifies pests and pathogens found on P.I.s and maintains a computerized data file on intercepted organisms and their hosts. In carrying out these duties, this staff is in charge of inspections at the Plant Germplasm Quarantine Center (PGQC) at Beltsville East, Md., and has personnel located at the Plant Introduction (Quarantine) Station at Glenn Dale, Md., and at Beltsville West, Md.

Categories of Quarantine on Plants or Non-seed Parts

Restricted

Nearly all germplasm that enters the U.S. for scientific or educational purposes falls into one of three categories of quarantine.[41] They are known as (1) "restricted," or "Regulation 6", (2) "post-entry," and (3) "prohibited." Well over 90% of the plant species of the world are in the first, or least restrictive, category. Genera in this category can be readily imported by scientists and others, with few, if any, restrictions on

Table 3
EXAMPLES OF PLANTS WHOSE NONSEED PARTS ARE IN THE INTERMEDIATE, OR "POST-ENTRY", QUARANTINE CATEGORY[41]

Plant name	"Post-entry" when imported from:
Maple	All countries except Canada, Japan, and part of Europe
Bromeliads	All countries, when destined for Hawaii
Hops	All countries
Apples, pears, peaches, cherries, plums, and apricots, when not prohibited[a]	All countries except Canada
Raspberries	All countries
Roses	All countries except 5
Elm (including seed)	All countries except Canada[b]
Pine	All countries except Canada, Europe, and Japan[c]
Oak	All Countries except Canada and Japan[d]
36 other genera	Usually, all except 1 — 5 countries

[a] See Table 10.
[b] Prohibited from Europe.
[c] Prohibited from Europe and Japan.
[d] Prohibited from Japan.

how or where the material can be used or grown. Usually an import permit is required (procedure to be described later), and all such material is subject to inspection and possible treatment upon arrival in the U.S. Many of the restricted category items are seeds, or plant parts of exotic genera, or of common crops that originate from countries with no known serious pest problems. After an item is inspected, it is usually assigned a P.I. number and immediately forwarded to the importer or appropriate curator. If the item was not specifically requested, it is sent to one or more of the Regional Plant Introduction Stations (Figure 2) or curators, where it is established and subjected to preliminary evaluation.

Species are placed into one of the other categories only when APHIS determines, from the literature or from knowledge of the disease or pest situation in the country of origin, that additional precautions should be taken as safeguards for U.S. crops. Ordinarily a more restrictive classification is made only after a public hearing, during which any affected persons or businesses may be heard.

Post-entry

Examples of postentry genera and their countries of origin are listed in Table 3. These plants, when imported from the countries listed, would present a relatively high risk of damage to U.S. crops if they were processed as those in the restricted category are. Consequently, after these plants arrive and are inspected and treated, they must be grown under close observation. They may be grown at one of the seven plant introduction stations, at a state agricultural experiment station, at a private research facility, or in one's private yard. In all cases permission must have been obtained before receiving the material from the federal quarantine office. The permit will have specified restrictions on use and propagation of the germplasm or plants. Furthermore, such material must be available for inspection for 2 years. Such inspections are made by persons in the respective state quarantine offices (Table 4) under the guidance of the federal "post-entry" quarantine unit at Beltsville. If no pest is found within 2 years, or if all pests found have been destroyed, the germplasm is released from quar-

Table 4
ADDRESSES OF THE PLANT REGULATORY OFFICES OF THE STATES, THE DISTRICT OF COLUMBIA, CANADA, AND MEXICO

State	Address[a]	Phone
Alabama	Div. of Plant Industry, Dept. of Agric. & Industries, Box 3336, Montgomery 36109	205-832-3753
Alaska	Div. of Agric., Dept. of Natural Resources, Box 1088, Palmer 99645	907-745-3236
Arizona	Comm. of Agric. and Hort. 1688 West Adams, Phoenix 85007	602-255-4373
Arkansas	Div. of Plant Industry, State Plant Board, Box 1069, Little Rock 72203	501-371-1021
California	Div. of Plant Industry, California Dept. of Food and Agric., 1220 N Street, Sacramento 95814	916-445-5537
Canada	Plant Quarantine Div., Dept. of Agric., Sir John Carling Building, Ottawa, Ontario K1A OC5.	613-995-5880
Colorado	Div. of Plant Industry, State Services Building, 1525 Sherman Street, Denver 80203	303-839-2838
Connecticut	Agric. Exp. Stn., Box 1106, New Haven 06504	203-789-7421
Delaware	State Dept. of Agric., Drawer D, Dover, 19901.	302-736-4813
Dist. of Columbia	Plant Protection and Quarantine Programs, APHIS, USDA, 103 S. Gay Street, Room 515, Baltimore 21202	301-962-4939
Florida	Div. of Plant Industry, Dept. of Agric. and Consumer Services, Box 1269, Gainesville 32602	904-372-3505
Georgia	Div. of Entomology, Dept. of Agric., Agric. Building, Capitol Square, Atlanta 30334	404-656-3641
Guam	Dept. of Agric., Gov. of Guam, Agana, Guam 96910	671-734-3941
Hawaii	Plant Quarantine Branch, State Dept. of Agric., Box 2520, Honolulu 96804	808-548-7175
Idaho	Div. of Plant Industries, State Dept. of Agric., Box 790, Boise 83701	208-334-3240
Illinois	Dept. of Agric., Div. of Agricultural Industry Regulation, Bureau of Plant and Apiary Protection, 1010 Jorie Boulevard, Oakbrook, 60520	312-920-9256
Indiana	Div. of Entomology, State Dept. of Natural Resources, 613 Indiana State Office Bldg., Indianapolis 46204	317-232-4120
Iowa	Dept. of Agric., East 7th and Court, Des Moines 50319	515-281-5861
Kansas	Div. of Entomology, State Board of Agric., 1720 S. Topeka Ave., Topeka 66612	913-296-3016
Kentucky	Dept. of Entomology, Agric. Exp. Stn., S225 Agric. Science Center North, Lexington 40506	606-258-5638
Louisiana	Bureau of Entomology and Plant Industry, Dept. of Agric., Box 44153, Capitol Station, Baton Rouge 70804	504-925-3761
Maine	Div. of Plant Industry, State Dept. of Agric., Augusta 04333	207-289-3891

Table 4 (continued)
ADDRESSES OF THE PLANT REGULATORY OFFICES OF THE STATES, THE DISTRICT OF COLUMBIA, CANADA, AND MEXICO

State	Address[a]	Phone
Maryland	Div. of Plant Industries, Maryland Dept. of Agric., Parole Plaza Office Bldg., Annapolis, 21401	301-269-2325
Massachusetts	Div. of Plant Pest Control, Dept. of Food & Agric., Leverett Saltonstall Bldg., 100 Cambridge Street, Boston 02202	617-727-3031
Mexico	Direccion General de Sanidad Vegetal, Secretaria de Agricultura y Ganaderia, Galle Guillermo Perez Valenzuela No. 127, Coyoacan 21, D.F., Mexico	554-0529
Michigan	Plant Industry Div., State Dept. of Agric., Box 30017, Lansing 48909	517-373-1087
Midway Island	Code 11433 Pacific Division, Naval Facilities Engineering Command, FPO San Francisco, CA 96610	808-471-3948
Minnesota	Div. of Plant Industry, Dept. of Agric., 670 State Office Bldg., St. Paul 55155	612-296-8448
Mississippi	Div. of Plant Industry, Dept. of Agric. and Commerce, Box 5207, Mississippi State, 39762	601-325-5713
Missouri	Plant Industrial Div., Dept. of Agric., Box 630, Jefferson City, 65101	314-751-2462
Montana	Hort. Div., Dept. of Agric., Airport Way Bldg. West, 1300 Cedar Street, Helena, 59601	406-449-3730
Nebraska	Bureau of Plant Industry, Dept. of Agric., Box 94756, Lincoln 68509	402-471-2341
Nevada	Div. of Plant Industry, Dept. of Agric., Box 11100, Reno 89510	702-784-6401
New Hampshire	State Dept. of Agric., c/o Entomology Dept., Nesmith Hall, Durham 03824	603-862-2300
New Jersey	Div. of Plant Industry, Dept. of Agric., Box 1888, Trenton 03625	609-292-5440
New Mexico	Div. of Plant Industry, Dept. of Agric., Box 3189, Las Cruces 88001	505-646-3207
New York	Div. of Plant Industry, Dept. of Agric. & Markets, Bldg. 8, State Campus, Albany 12235	518-457-2087
North Carolina	Pest Control Div., Dept. of Agric., Box 27647, Raleigh 27611	919-733-3556
North Dakota	Dept. of Agric., State Capitol, Bismarck 58505	701-224-2232
Ohio	Div. of Plant Industry, Dept. of Agric., Reynoldsburg 43068	614-866-6361
Oklahoma	Plant Industry Div., Dept. of Agric., 122 State Capitol Bldg., Oklahoma City, 73105	405-521-3883
Oregon	Plant Div., Dept. of Agric., Agricultural Bldg., Salem 97310	503-378-3776
Pennsylvania	Bureau of Plant Industry, Dept. of Agric., 2301 North Cameron St., Harrisburg 17120	717-787-4843
Puerto Rico	Plant Quarantine Section, Dept. of Agric., Box 10163, Santurce 00908	809-724-0422 & 4627

Table 4 (continued)
ADDRESSES OF THE PLANT REGULATORY OFFICES OF THE STATES, THE DISTRICT OF COLUMBIA, CANADA, AND MEXICO

State	Address[a]	Phone
Rhode Island	Div. of Agric., Dept. of Natural Resources, Veterans Memorial Bldg., Providence 02903	401-277-2781
South Carolina	Plant Pest Regulatory Service, Room 212, Barre Hall, Clemson Univ., Clemson 29631	803-656-3006
South Dakota	Div. of Agric. Regulations & Inspection, Dept. of Agric., Anderson Bldg., Pierre 57501	605-773-3724
Tennessee	Dept. of Agric., Div. of Plant Industries, Ellington Agric. Center, Box 40627, Melrose Station, Nashville 37204	615-741-1551
Texas	Dept. of Agric., Box 12847, Capitol Station, Austin 78711	512-475-4457
Trust Territories	Agric. Div., Dept. of Resources and Development, Trust Territory Pacific Islands, Saipan, Mariana Island 96950	
Utah	Div. of Plant Industry, Dept. of Agric., 147 North 200 West, Salt Lake City 84103	801-533-5421
Vermont	Div. of Plant Pest Control, Dept. of Agric., 116 State St., State Office Bldg., Montpelier 05602	802-828-2431
Virgin Islands	Dept. of Agric., P.O. Box 8, Kingshill, St. Croix 00850	809-772-0990
Virginia	Div. of Product and Industry Regulation, Dept. of Agric. and Commerce, 203 North Governor St., Richmond 23219	804-786-3515
Washington	Dept. of Agric., Plant Industry Div., 406 Gen. Admin. Bldg., Olympia 98504	206-753-5052
West Virginia	Plant Pest Control Div., Dept. of Agric., Charleston, 25305	304-348-2212
Wisconsin	Plant Industry Div., Dept. of Agric., Trade & Consumer Protection, Box 8911, Madison 53708	608-266-7130
Wyoming	Div. of Plant Industry, Dept. of Agric., 2219 Carey Ave., Cheyenne 82002	307-777-7321

[a] Write to "Plant Regulatory Office" at the address given.

antine by the Head of the "post-entry" unit. This procedure allows the importer to obtain foreign germplasm rather quickly and to begin certain experiments under phytosanitary conditions without the long wait that he must endure for prohibited genera. In recent years some 600 U.S. scientists and private citizens have held "post-entry" permits on one or more genera.

Prohibited

Genera in the "prohibited" category can be imported only by the Secretary of Agriculture.[41] Plants are in this category because their introduction from certain countries would present a serious risk to U.S. agriculture if maximum precautions were not taken. Some of the "prohibited" genera are listed in Table 5. Except for seed of corn, sorghum, millet, cotton, rice, and wheat, most of these crops are vegetatively propa-

Table 5
EXAMPLES OF PLANTS THAT ARE IN THE "PROHIBITED" QUARANTINE CATEGORY[41]

Name	Prohibited (when imported from:)	Injurious pest or disease
Fir	All countries except Canada	Rusts, *Phacidiopycnis* fungi
Barberry	All countries	Black-stem rust
Cedar	Europe	*Fusarium*, *Phacidiopycnis* fungi
Ash	Europe	*Pseudomonas*
Maple	Japan, parts of Europe	*Xanthomonas* bacterium and virus
Mahonia (seed)	All countries	*Puccinia graminis*
Tree fruits (apples, pears, peaches, plums, cherries, apricots)	All countries (except as described later)	Diverse fungi and viruses
Hydrangea spp.	Japan	*Aecidium* spp.
Grapes	All except Canada	Viruses
Willow	England, the Netherlands, and Germany	*Erwinia* bacteria
Cotton	All countries	Leaf curl virus
85 other genera	For many, a few specific countries only	A diversity of diseases

gated. These genera tend to either have a high incidence of latent viral infections, carry new races of fungi or bacteria, be alternate hosts for disease agents already established here, or originate from an area that has a widespread, serious disease in that genus which is not known to occur in the U.S.

Upon arrival in the U.S. these plants are inspected and, with a few exceptions, established and quarantined at the Plant Introduction Station at Glenn Dale, Md. Here they are observed for all pests and pathogens, but especially for those of quarantine significance — i.e., those that are listed in the regulations and those that are not known to occur in the U.S. Most "prohibited" plants remain in quarantine for at least 2 years. Others, such as the tree fruits and grapes, are in quarantine for as long as 6 years. If it is released from quarantine, the germplasm is distributed to plant breeders, nursery companies, and certain other specialists upon request.

Processing Germplasm at the Plant Introduction (Quarantine) Station

General Procedures

Most of the "post-entry" and "prohibited" introductions are grown in insect-proof greenhouses or screenhouses, none of which can be entered directly from the outdoors. A PGQC inspector makes weekly inspections of plants in all greenhouses. Any pests or pathogens found are identified at the station or by specialists in the Technical Support Staff (Figure 4). When possible, affected plants are treated to eliminate the pest.

The frequency of finding non-viral pests on introductions is low because of many factors. Among these are (1) the foreign collector of the germplasm usually selects healthy-looking plant parts, (2) the shipment is small — usually two to four fruit-tree scions, two to three small potato tubers, or a few small cuttings of ornamental plants, and (3) upon arrival in the U.S. each item is inspected and often is dipped or fumigated to kill surface-infesting organisms. However, these procedures are not effective for internal, latent pathogens such as viruses, viroids, some mycoplasmas, and spiroplasmas; a much greater effort is required to detect these agents.

Depending on the genus, one or more of five procedures are used to detect and usually to identify such agents. They are:

Table 6
PLANTS FOUND TO BE SUSCEPTIBLE TO MANY PLANT VIRUSES

Plant	No. of viruses tested	% of viruses tested that produced infection
Chenopodium amaranticolor (a species related to common lambsquarters)	53	89
Gomphrena	42	62
Cowpea	82	59
Potato	105	57
Tobacco	210	55
Zinnia	84	54
Pea	81	52
Jimson weed	137	51
Eggplant	63	49
Cucumber	144	47
Bean	131	46
Tomato	136	44
Beet	95	34
Cabbage	69	33

Electron microscopy.[7]

Serology — Sap is expressed from certain parts of the established plant and drops of the sap are allowed to react with drops of several antisera to different known viruses.[1] If a reaction occurs, virus in the expressed sap is not only detected but also identified. For such testing, an extensive collection of antisera is maintained; it includes some for viruses that are not known to occur in the U.S.

Mechanical transmission[14] — Flower petals or immature leaves of the introduction are triturated in a buffer and the liquid is rubbed onto the leaves of various herbaceous species that are known to be susceptible to viruses that may be present in the introduction.[3] Most viruses that can be mechanically transmitted incite disease symptoms in one or more of the test species within 4 to 25 days. Some test species react in a wide variety of ways to different viruses, so that the experienced person can identify a virus by the symptoms it incites on such a species.[20,47] Table 6 is a list of species that are susceptible to an unusually large number of viruses;[36] most of these species are used regularly at Glenn Dale. Mechanical transmission is the procedure most widely used on introductions other than seed and fruit, and has resulted in the detection of many viruses from a broad spectrum of species, some of which had no disease symptoms.[45,46,48-50]

Graft transmission — This is done to varieties known to be particularly sensitive to one or more viruses that could be present in the introduction.[4] This procedure, to be described below, is used on all "prohibited" woody fruit introductions.

Graft transmission — This may be done to other plants of the same species to determine whether a "virus-like" symptom is transmissible. This procedure is used with woody ornamental introductions that exhibit an "abnormal" condition that could be due to a genetic disorder, a nutritional problem, an infectious agent, air pollution, a natural mosaic, or a variety of other causes.[18] If the condition cannot be transmitted to other plants of the same species, it is assumed not to be caused by an infectious agent, and the introduction is processed as disease-free.

Indexing Fruit Introductions for Viruses

Grapes, cacao, apple, pear, peach, cherry, plum, apricot, almond, orange, grapefruit, lemon, and lime are indexed for a number of viruses. The procedure described here is general, because procedures vary somewhat with the species being indexed and the country of origin of the introduction.

Most of the virus-sensitive indicator varieties of pome and stone fruits are established by budding onto seedling rootstocks in the field. Apple, pear, and quince indicators are used 3 to 4 weeks later, whereas stone fruit indicators are used after a year.

Buds from each fruit introduction are first established singly onto seedling rootstocks in the greenhouse or screenhouse. When a given bud has developed into a big enough whip, which usually takes 12 to 18 months, enough scionwood is collected from a single tree to conduct the indexing. Two scionwood buds are placed on the stock below the virus-sensitive indicator buds. Two trees of each indicator are thus "inoculated." Check trees in the same row as each inoculated indicator are left unbudded for comparative observation during the following years. Under suitable conditions about 95% of the buds heal in, so that any virus in the P.I. test buds below passes into the rootstock and the virus-indicator buds above. The trees are read for symptoms of virus or mycoplasma infection for the rest of that year and for two more growing seasons.

Experience has shown that most strains of most viruses will cause symptoms in the leaves or wood of the respective indicators within this time. Viruses that cause symptoms only in the fruit, and they have been rare in introductions, will not yet be detectable. However, in the interest of the importer, the introduction is released from quarantine after the third season if it appears to be free of other viruses. The appropriate indicators are observed for the fruit-deforming agents when they fruit.

There are seven varieties of apples which are used to detect some 15 apple pathogens; five pear and quince varieties are used to detect eight pathogens that commonly infect these; and seven to nine varieties of stone fruits are used, depending on whether the introduction is peach, apricot, cherry, almond, or plum. Since the indicators used and the pathogens sought can change as better indicators are found or new viruses are discovered, space will not be taken here to list either indicator varieties or pathogens. For the most part the indicators used are those recommended by the Interregional Project-2[12] and the International Committee for Cooperation in Fruit Tree Virus Research.[9]

Citrus and grape introductions are also indexed for viruses by grafting them to several known virus-sensitive indicator cultivars.[6,10,15,53] These genera are indexed by special arrangement with the University of California — the grapes at Davis and the citrus at Riverside.

The percentage of P.I.s that are infected with one or more viruses or mycoplasmas upon arrival in the U.S. is shown in Table 7 for various crops. It ranges from a low of 2%, for cacao, to 68%, for apples. Certain genera, such as apples, potatoes, and grapes, are often infected with two or three pathogens.[19] Less than 10% of all viruses detected are not known to occur in the U.S. Details of some of these indexing procedures, and the identity of the pathogens by country or origin, have been published.[19,44]

Books on citrus,[6] apples, pears, and quince,[33] stone fruits,[38] and small fruits and grapes,[10] offer detailed descriptions and photographs of most of the virus and virus-like diseases of these crops. Procedures for testing for viruses and control measures are also described for some of the diseases.

Disposition of Plant Introductions

Introductions are categorized by pathogen status as follows: (1) no infectious agents detected, (2) only viruses already widespread in the U.S. detected, (3) disease agent identified and not known to occur in the U.S. or not widespread, and (4) disease agent not identified.

Table 7
PERCENTAGE OF VEGETATIVE PLANT INTRODUCTIONS THAT WERE INFECTED WITH ONE OR MORE VIRUSES OR MYCOPLASMAS UPON ARRIVAL IN THE U.S. 1968—1977, AND PROCEDURES USED FOR DETECTION

Crop	Items infected		Detection procedures usually used[a]
	No.	%	
Apples	155	68	3, 4
Pears	160	40	3, 4
Potatoes	1191	53	1, 3
Sweet potatoes	52	15	3, 4
Cacao	167	2	4
Woody ornamentals	35	14	2, 3, 5
Herbaceous ornamentals	121	57	3
Prunus (stone fruits)	199	46	3, 4
Grapes	200	50	4
Citrus	138	65	4

[a] 1 = serology, 2 = electron microscopy, 3 = mechanical transmission to herbaceous species, 4 = graft transmission to known virus-sensitive varieties, 5 = graft transmission to other plants of same species.

Generally P.I.s in category 1 are released from quarantine and made available to scientists in all states and in other countries, whereas those in categories 3 and 4 are not released for distribution. Some are treated in the effort to obtain virus free germplasm, though most are destroyed. Items in category 2 may be released from quarantine with restrictions that vary according to (1) how and where the P.I. will be used, (2) the species, (3) the parts to be distributed, (e.g., cuttings, seed, or pollen), and (4) individual state or foreign-country plant quarantine regulations. Usually P.I.s in category 2 are available to U.S. scientists by special arrangement and only with the approval of the quarantine office of the state concerned.

Each year lists of P.I.s that have been released from quarantine are distributed to plant breeders throughout the U.S. Fruits are included in two of these lists; another list contains ornamental introductions and is distributed to another group of plant breeders, in cooperation with the National Arboretum in Washington, D.C. Recipients merely indicate which new introductions they wish to include in their research program and return the form. The germplasm, as scionwood for ''prohibited'' category items and as small potted plants for most other introductions, is mailed within weeks without obligation.

How to Obtain Foreign Plants or Germplasm

Any U.S. resident wanting foreign plants or germplasm may pursue one of two courses of action, depending on the quarantine category of the item (determined by the genus and country of origin).

By Contacting the Permit Section of PPQ

For all but the ''prohibited'' genera (Table 5), plant breeders and private citizens should write to: The Permit Section, Plant Protection and Quarantine, APHIS, FCB,

Hyattsville, Md., 20782. If the item falls in the least restrictive quarantine category, a permit will be issued to the requestor who may then arrange to import it. It is subject to inspection upon arrival and may then be grown and distributed as the importer desires. If the item falls in the intermediate, or "post-entry", category, the importer will be sent appropriate forms to be completed. The importer then returns the forms through the State quarantine office (Table 4) to the federal office in Hyattsville, Md. If the state has approved the request, and the Permit Section is satisfied that the importer is taking the necessary precautions to prevent the escape of foreign pests, he will be issued a green and yellow shipping/pre-addressed label and an import permit. Copies of the permit are sent to the APHIS port of entry so APHIS personnel there will know to whom to forward the item after it is inspected; to the state quarantine office, whose personnel will inspect the item during the first 2 years after arrival; and to the Federal Post-entry Quarantine Unit, the head of which will release the item from quarantine if it is pest-free after the 2-year period. The importer then makes his own arrangements to import the plants, or germplasm if for official use. Items that enter the U.S. via the PGQC at Beltsville are reviewed for possible inclusion into the national germplasm system. Hence the requestor may be asked to share them. Some of these items are assigned P.I. numbers.

By Contacting the Germplasm Resources Laboratory of SEA-AR

Persons who desire to import plants or germplasm for educational or research purposes, usually in the public interest, may pursue a second course of action and also have access to genera in the most restrictive, or "prohibited," category. Another advantage of this route is that most of the effort is performed by federal officials. Scientists may write to: the Germplasm Resources Laboratory, USDA, BARC-W, Bldg. 001, Beltsville, Md., 20705, naming the cultivars desired, and if possible, the name of the source person or institution. The GRL will first ascertain that the item has not been previously imported, and then obtain its quarantine category by contacting the Regulatory Support Staff of APHIS (Figure 4). If the germplasm is not already in the U.S., the GRL informs the requestor of its quarantine category and asks the foreign source person for a small quantity of the item. In the meantime, the requestor may apply for a permit from the Permit Section of PPQ.

Upon arrival at the PGQC in Beltsville, the material is inspected and, if it is neither "post-entry" nor "prohibited", sent to the scientist. If it is a "post-entry" item, it will be sent to the scientist only if he has obtained the permit and completed a growing agreement. If it is a "prohibited" item, it is sent to the Glenn Dale Plant Introduction Station. If the required virus indexing tests, which take 3 to 4 years, are negative, it is released to the importer, plant breeders, nursery companies, arboretums, and the appropriate germplasm repositories.

Quarantine on Seed Introductions

About 90% of all food crops grown on earth are propagated by true seed.[28] There are nine crops — wheat, rice, corn, barley, sorghum, sugarbeet, beans, soybeans, and peanuts — which represent by far the greater part of world food production. All of these crops are attacked by devastating seedborne diseases. Seeds of all species can harbor any of hundreds of surface-contaminating microorganisms.[28] Many seedborne diseases cannot be recognized by examining the seed; only by examining the plants they produce can one be certain that seeds are free of bacteria and fungi.[23] Seeds of at least 200 genera of economic crops can become infected by 100 or so viruses.[2,5,28,32] In view of this, the importation of seed is also controlled under paragraph 319 of federal regulations.[41] Generally, seeds of field crops, vegetables, and flowers, which are essentially herbaceous, may be imported without a permit from APHIS. Seeds of some herbaceous genera or from certain countries, and generally seeds of woody species, may also be imported, but a written permit from APHIS is required. The importer should then ask the overseas source person to send seeds via the PGQC, Bldg. 320, BARC-East, Beltsville, Md., 20705. Upon arrival they will be inspected and, if necessary, fumigated or otherwise treated, and immediately forwarded to the requesting scientist.

Seeds of about 20 genera are "prohibited"; however, they may still be imported by special arrangement with APHIS. Some of the major crops whose seeds are in this category, and the reasons for the restrictions, are: corn (downy mildew); sorghum and millet (smuts); wheat (flag smut); rice (smuts); and cotton (various fungi). Other genera are barberry, lentils, avocado, coconut, and bamboo. Special precautions are required of breeders who arrange to import seeds of these genera; these are spelled out in the permit issued by APHIS. The GRL will assist the importer in obtaining seeds of these genera.

Regional Plant Introduction Stations

By far the majority of seed introductions (other than those requested by scientists) arrive as a result of contacts initiated by the GRL and USDA-sponsored collection trips made in the effort to broaden the base of genetic diversity of our crops. It is not unusual for 50 to 200 accessions of a given species to arrive in a single shipment. Table 8 is a list of germplasm accessions that entered the U.S. during October 1977 via the PGQC. It shows that a diversity of items entered the U.S. during that month, that a high percentage of items enter as seed, and that not all items become part of the P.I. system and, hence, not all are assigned a P.I. number.

Most of the seed accessions are sent to one of the Regional Plant Introduction Stations (Figure 3). Here they are processed variously according to the quarantine, if any, on that species and the station's research program. Some are sown and evaluated for specific traits; some are propagated to desired volume, inspected to insure freedom from diseases, and distributed, upon request, to plant breeders. Most items are placed in germplasm collections, whereas others are discarded as non-viable, or occasionally as duplicates, or without value. For information about research programs and germplasm collections of the RPIS, breeders may write to the individual stations.

STATE QUARANTINE

Variation Among Regulations

As one might expect, state regulations vary considerably.[39] The crops listed, the procedures followed and the definition of nursery stock are examples of the items that vary among the state regulations. Scarcely a single statement can be made about quarantine that applies to all states, except that they all have a plant regulatory office and produce certified planting stock (Table 4). Space does not permit even a general discussion of quarantine regulations and programs for production of disease-free nursery stock for each state individually. Such information soon becomes outdated anyway, because regulations are revised or amended regularly and because new programs are added and old ones abandoned among the various states. This discussion will, therefore, be general, with specific examples given to demonstrate certain points.

States publish quarantine laws and regulations, most of which define nursery stock; describe shipping requirements, field inspections, and certificates required for interstate movement of plant materials; and list regulated plant species. The regulations of about half of the states include detailed instructions for the import of certain species

Table 8
GERMPLASM ENTERING THE U.S. VIA THE PLANT GERMPLASM QUARANTINE CENTER, BELTSVILLE DURING OCTOBER 1977

Plant quarantine number[a]	P.I. number (if assigned)[b]	Item	Origin	Disposition[c]	Quarantine status upon arrival
70359		5 lots triticale seed	India	GRL	Prohibited
70361		34 lots winged bean seed	Colombia	IL	Restricted
70366		1 var. *Malus* budwood	Israel	FL	Prohibited
70367		1 var. each *Malus* and *Pyrus* budwood	Israel	P.I.S. — MD	Prohibited
70373		2 lots *Lagerstroemia* seed	India	Wash., DC	Restricted
70387		209 accs. of *Triticum, Hordeum, Avena,* and *Secale* seed	Yugoslavia	GRL	Restricted
70396	420027—050	23 lines *Zea* seed	Yugoslavia	MO	Restricted
70420	418288—563	275 ornamental items, cuttings	Japan	P.I.S. — MD	Restricted, Post-entry, and prohibited
70466		16 lots *Glycine* seed	Surinam	IL	Restricted
70473		4 lots *Oryza* seed	Venezuela	GRL	Prohibited
70496		1 lot *Zea* seed	Kenya	MD	Prohibited
70534		5 vars. *Lolium*	Japan	P.I.S. — WA	Restricted
70540	420138	1 lot *Helianthus* seed	Australia	CA and P.I.S. — IA	Restricted
70541	420139	1 lot *Cartharnus* seed	Italy	CA and P.I.S. — IA	Restricted
70544		30 lines *Triticum* and triticale	Mexico	MD	Prohibited
70548		8 lots cotton cuttings	Syria	CA	Prohibited
70554		5 lots citrus cuttings	China	P.I.S. — MD	Prohibited
70576		1 lot *Molinis* seed	France	MD	Restricted
70600		102 accs. *Setaria* seed	Taiwan	GA	Prohibited
70601		1 lot of *Panicum* seed	Taiwan	P.I.S. — IA	Restricted
70612		1 lot bamboo plants	England	P.I.S. — MD	Prohibited
70615		6 lots *Alnus* seed	W. Germany	P.I.S. — IA	Restricted

[a] = Missing numbers in the sequence were assigned to exported germplasm.
[b] = P.I. number to be assigned if the item is incorporated into the germplasm system, or after the item is established. About 15% of the imported germplasm does not become part of the P.I. system.
[c] = P.I.S. = Plant Introduction Station, MD, IA, WA, GA, or NY; GRL = Germplasm Resources Laboratory, Beltsville, Md. Others are state abbreviations.

because of an insect or disease problem. For example, Florida regulations describe requirements regarding aquatic plants, camellia flower blight, oak wilt, St. Augustine grass decline, the sweetpotato weevil, and other pests and diseases; whereas Nevada regulations deal in detail with potato bacterial ring rot, elm tree diseases, the European corn borer, mint wilt, and others. The regulations of some states, such as New York and Washington, describe the requirements for importation of fruit tree rootstocks; New Jersey and Georgia regulations are concerned specifically with the vegetable transplant industry; and Hawaii regulations single out diseases of bananas, coffee, orchids, and passion fruit for detailed discussion.

The regulations of about one third of the states contain at least one list of quarantine states, and even specific counties of certain other states, from which a specific genus may not be imported; or, if importation is allowed, a list of the requirements that must be fulfilled. Such is the case with sugarcane, peaches, and sweetpotatoes entering Louisiana from other southern areas; with peanuts entering Oklahoma from some eastern states; and with cotton entering California from specific states and counties in the Cotton Belt.

Regulations of some states describe the requirements that specific imported species must meet to become eligible for entry into that state. For example, Michigan regulations are detailed with regard to importation of strawberry plants — how they must be grown, the frequency and time of field inspections, and tolerance levels for several diseases, nematodes, and insects. Geranium cuttings from California are acceptable to Michigan only when California certifies that, after removal of the leaves, the cuttings were dipped for 90 sec in water containing a bactericide at 50 to 52°C. Similarly, Ohio regulations specify the requirements for certification and acceptability of imported raspberry plants. Many other states describe various standards for specific species which must be fulfilled before certification is given. The above is an attempt to show the scope and diversity of state regulations with regard to movement of plants, plant parts, and seeds across boundaries. The requirements of any state can be learned from the appropriate office (Table 4).

Wall Chart of Regulations

Some of the requirements of all states can be obtained from the American Association of Nurserymen, 230 Southern Building, Washington, D.C., 20005. The Association sells, for $2.50, a 21 × 33-in. (53 × 74-cm) wall chart entitled "Regulations Governing Shipments of Nursery Plants." Across the top are listed the 50 states, Puerto Rico, and the District of Columbia, and the phone number of the quarantine office. Down the left side are listed 50 diseases, plant pests, and affected plant genera, such as citrus nursery stock, strawberry diseases, San Jose scale, peach mosaic, and golden nematodes. At the intersection of the "disease" line and the "state" line is one of 30 symbols, each of which is defined in the right border. For example, from this chart we can see that Arizona, California, Florida, Hawaii, Idaho, Nevada, New Mexico, Oregon, Texas, Utah, and Washington require a special certification tag for interstate transport of corn, because of the European corn borer; and that orchids entering Florida from the Virgin Islands or Puerto Rico must meet special requirements.

Also on this chart are seven federal domestic quarantines on cereal rust, the golden nematode, the gypsy and browntail moths, the fire ant, the Japanese beetle, and witchweed. Symbols indicate whether all or part of each state is regulated and whether it is by the state or federal government, or both.

Parenthetically, this chart also has a section on state licenses and shipping requirements, which include nursery license fees, dealer's fees, agent's fees, any state requirement for a copy of certification on shipments, registration fees, filing or license fees, reciprocal arrangements between states, any requirement for bonds, any requirement for inspection at the Postal Service terminal, and other like information. The wall chart does not contain information on seeds.

National Plant Board

The reader should also be aware of the existence of the National Plant Board (NPB), its organization, and its purposes. It is composed of the chief plant regulatory official of each of the 50 states and of Puerto Rico (Table 4), and has been in existence since 1925. Its executive committee is composed of two persons from each of four regional plant boards — Western, Central, Eastern, and Southern. Its purposes are to coordinate state plant regulatory activities and to serve as an advisory body in formulation of state policies. The NPB also maintains contact with the USDA and other related federal and state agencies and confers with them about quarantine policy, which affects the U.S. at large and the individual states or groups of states. The NPB meets once a year to conduct business and elect officers. The views of any plant scientist may be expressed to a regional board or the NPB on any state or federal regulatory matter.

Seed Improvement Associations

Each of the 50 states also has an agency that is responsible for seed certification programs within the state. Most are members of an international organization, the Association of Official Seed Certifying Agencies (AOSCA), with headquarters in the P & AS Building, Clemson, S.C., 29631. Most of the agencies are known as the Crop Improvement Association; and most write and publish, usually annually, the standards for growing, inspecting, and labeling seed for all of the crops grown for certification.

Unfortunately most of the states' standards for most crops say little about disease control. Many certification standards deal with field growing conditions, germination, inert matter, labeling, and other nondisease considerations. Some of the states' standards include only a general comment, such as "shall be free of diseases." With only four or five exceptions, such as lettuce mosaic and bean common mosaic, nothing is stated about ascertaining freedom from any of the hundreds of viruses carried in seed. Seed improvement associations would do well to include in their programs procedures for detection of viruses, many of which are latent in the inspected plants from which the seeds originate. Various procedures are available and have been described.[28,32] A copy of state certification standards and other publications can be obtained by writing to the appropriate association (Table 9). Also available is a Certification Handbook which contains the genetic and seed standards for all crops under certification, where standards have been developed. It contains names and addresses of all certification agencies, foundation seed stock organizations, members of all general and commodity committees, the Plant Variety Protection Board, inspection suggestions and the like. This handbook is available for $8 from AOSCA. More on the subject of disease-free seed stocks is given in the chapter by G. Mink.

USE OF CERTIFIED VIRUS-FREE PLANTING STOCK

Virus Infections

Once a virus infects a plant it remains in the plant for the balance of that plant's life and cannot be eliminated from the entire plant. Vegetative propagations, such as cuttings or buds, carry the virus infection into the new plant, where it continues to cause adverse effects.

There are no generally available and effective viricides with which to spray a virus-

Table 9
LIST OF CROP IMPROVEMENT ASSOCIATIONS OR OTHER STATE AGENCY IN CHARGE OF CERTIFICATION OF PLANTING MATERIAL IN THAT STATE OR PROVINCE

Alabama Crop Imp. Assoc., South Donahue Dr., Auburn Univ., Auburn, 36830
Alaska Crop Imp. Assoc., Inc., Box 895, Palmer, 99645
Arizona Crop Imp. Assoc., College of Agric., Room 140, Univ. of Arizona, Tucson, 85721
Arkansas State Plant Board, 421½ W. Capitol, Little Rock, 72203
California Crop Imp. Assoc., 231 Hunt Hall, Univ. of California, Davis, 95616
Canada, Plant Products Division, Agriculture Canada, Sir John Carling Building, 930 Carling Ave., Ottawa, Ontario K1A 0C5
Canadian Seed Growers Assoc., Box 8455, Ottawa, Ontario K1G 3T1
Colorado Seed Growers Assoc., Plant Science Bldg., Colorado State Univ., Fort Collins, 80521
Delaware Crop Imp. Assoc., Dept. of Plant Science, Univ. of Delaware, Newark, 19711
Florida Dept. of Agric. and Consumer Services, Certification Section, Mayo Bldg., Tallahassee, 32304
Georgia Crop Imp. Assoc., Inc., 925 West Whitehall Road, Athens, 30605
Hawaii, Weed Branch, Division of Plant Industry, Hawaii Dept. of Agric., Honolulu, 96814
Idaho Crop Imp. Assoc.,Inc., 5284 Overland Road, Boise, 83705
Illinois Crop Imp. Assoc., 508 South Broadway, Urbana, 61801
Indiana Crop Imp. Assoc., 3510 U. S. 52 South, Lafayette, 47905
Iowa Crop Imp. Assoc., 112 Agronomy Bldg., Ames, 50010
Kansas Crop Imp. Assoc., 205 Call Hall, Kansas St. Univ., Manhattan, 66505
Kentucky Seed Imp. Assoc., Inc., Box 12008, Lexington, 40579
Louisiana Dept. of Agric., Box 44153, Capitol Station, Baton Rouge, 70804
Maine Seed Potato Board, State Office Bldg., Augusta, 04330
Maryland Dept. of Agr., Turf & Seed Certification Sect., Division of Plant Industries, 1103 H. J. Patterson Hall, Univ. of Md., College Park, 20742
Massachusetts State Seed Laboratory, West Experiment Station, Univ. of Mass., Amherst, 01002
Michigan Crop Imp. Assoc., Box 21008, Lansing, 48909
Minnesota Crop Imp. Assoc., Univ. of Minnesota, 1900 Hendon Ave., St. Paul, 55108
Mississippi Seed Imp. Assoc., P. O. Drawer MS, Mississippi State, 39762
Missouri Seed Imp. Assoc., Box 852, Univ. South Farm, Columbia, 65211
Montana Seed Growers Assoc., Leon Johnson Hall, Montana State Univ., Bozeman, 59717
Nebraska Crop Imp Assoc., College of Agr., Keim Hall, East Campus, Univ. of Nebraska, Lincoln, 68533
Nevada State Dept. of Agric., Div. of Plant Industry, 350 Capitol Hill Ave., Box 11100, Reno, 89510
New Jersey State Dept. of Agric., P.O. Box 1888, John Fitch Plaza, Trenton, 08625
New Mexico Crop Imp. Assoc., Box 3CI-NMSU, Las Cruces, 88003
New York Seed Imp. Coop., Inc., Box 474, Ithaca, 14850
North Carolina Crop Imp. Assoc., Inc., Box 5155, N. C. State Univ., Raleigh, 27607
North Dakota State Seed Dept., State Univ. Station, Fargo, 58102
Ohio Seed Imp. Assoc., Box 477, 6150 Avery Road, Dublin, 43017
Oklahoma Crop Imp. Assoc., 369 Ag. Hall, Oklahoma State Univ., Stillwater, 74074
Oregon Seed Certification Service, 102 Crop Science, Oregon State Univ., Corvallis, 97331
Pennsylvania Dept. of Agric., 2301 North Cameron Street, Harrisburg, 17120
South Carolina, Dept. of Seed Certification, Room 265 P & AS Bldg., Clemson Univ., Clemson, 29631
South Dakota Seed Certification Service, Plant Science Dept., S.D. State Univ., Brookings, 57006
Tennessee Crop Imp. Assoc., Box 11019, 5201 Marchant Drive, Nashville, 37211
Texas Dept. of Agriculture, P.O. Drawer 12847, Austin, 78711
Utah Crop Imp. Assoc., Utah Ag. Experiment Station, Utah State Univ., UMC 48, Logan, 84322
Vermont Dept. of Agric., State Office Bldg., 116 State Street, Montpelier, 05602
Virginia Crop Imp. Assoc., 10 Sandy Hall, VPI&SU, Blacksburg, 24061
Washington State Crop Imp. Assoc., Inc., 513 North Front St., Yakima, 98901
West Virginia Associated Crop Growers, 1090 Agricultural Sciences Bldg., W.V. Univ., Morgantown, 26506
Wisconsin Crop Imp. Assoc., 349 Moore Hall, Univ. of Wisconsin, Madison, 53706
Wyoming Seed Certification, Plant Science Div., Box 3354, Univ. of Wyoming, Laramie, 82070

diseased crop. Hence, one of the most effective means of controlling virus diseases, especially in long-term perennial crops, is to begin with virus-free trees or vines. Unless the supplier states that his merchandise has been indexed for virus infections, the grower should not assume that the nursery stock is free of damaging viruses.

The damage caused by viruses or mycoplasmas is well known in diseases such as pear decline, peach stem pitting, apple mosaic, raspberry bushy dwarf, plum pox, sour cherry yellow, citrus tristeza, grape fanleaf, and hundreds of others.[8]

However, the latent or chronic virus infections that occur throughout the U.S. in pome and stone fruits should be of even greater concern to the fruit grower.[8] These infections cause inconspicuous symptoms in trees, and are sometimes present in a large percentage of trees sold by nurseries. And although their effects on growth appear to be slight, some of these "mild" viruses are known to appreciably reduce long-term yield and income, sometimes making the difference between a loss and a profit.

Since most fruits and nuts, certain woody ornamental crops, and some vegetables must be propagated vegetatively to maintain trueness-to-type features, virus infections have plagued growers for decades.

Sources of Virus-Free Planting Stocks

Procedures are now available whereby some virus-free growing points or buds can be obtained from infected mother plants after growing the plants at 95 to 100°F (35 to 38°C) for several weeks.[16,30] These healthy buds are grafted onto disease-free rootstocks and, after they have grown into small trees, are tested to assure that the viruses are indeed absent. There are basically three sources of virus-free crops. One source includes the federal- and state-funded special projects, such as IR-1 (potatoes) and IR-2 (fruits) in the U.S., and the projects of other countries (Table 10). A second source is several state departments of agriculture. A third source, primarily for fruits, is specifically listed foreign nursery companies and experiment stations.[41]

Special Projects

Potatoes

One large vegetatively propagated crop is potato. Most of the major potato-producing states and Canada have had well-known, successful certification programs, some for decades. The tolerances for many bacterial and fungal diseases are clearly spelled out in the regulations of each of the state programs. However, few, if any, of the state programs appear to offer germplasm certified disease-free with regard to latent virus infections because certification has been based upon visual inspections of field plants rather than on indexing.[35] Consequently, a buyer of certified seed potatoes obtains virus-free stock only by chance.

However, modifications of traditional production and inspection procedures are being developed for the control of two latent viruses, S and X, and for the bacterium which causes black leg disease. These viruses, which used to be universal in most North American varieties, have been eliminated from over 100 varieties and seedlings during the past decade at the Canada Agriculture Research Station in Vancouver. Tubers from this collection have been used as basic seed stock on official seed farms in Canada and in several states. Tolerance levels for virus infections have now been proposed.[52] For current information on available virus tested varieties, write to the Canada Agricultural Research Station, 6660 N.W. Marine Drive, Vancouver, B.C. V6T 1X2. Other sources of known virus-free varieties are given in Table 10.

In the U.S., Interregional Project No. 1, headquartered at the Potato Breeding Station, Sturgeon Bay, Wis., 54235, has begun an effort to produce certified virus-free potatoes. Most of the items in this program are foreign varieties and breeding lines. For current information on varieties available, one should write to these sources.

Fruit Crops

One of the best known sources of tested disease-free tree fruits in the U.S. is the Interregional Project 2, at Prosser, Wash.[11-13] During the past decade several hundred

Table 10
SPECIAL PROJECTS THAT PRODUCE CERTIFIED VIRUS-FREE, VEGETATIVELY PROPAGATED CROPS THAT ARE AVAILABLE FOR RESEARCH PURPOSES FROM OTHER COUNTRIES

Country	Crops	Address for details
Australia	Grapes, pome and stone fruits, citrus, avocados	Fruit Variety Foundation, Dept. of Health, Box 100, Woden A.C.T. 2606
Canada	Pome and stone fruits	Plant Protection Division, 8801 East Saanich Road, Sidney, B. C. V8L 1H3
	Potatoes	Canada Agriculture, Research Station, 6660 N. W. Marine Drive, Vancouver, B. C. V6T 1X2
Denmark	Pome and stone fruits, potatoes, strawberries, raspberries, chrysanthemums	The State Pathology Institute, Lottenborgvej 2, IK-2800, Lyngby
France	Pome and stone fruits, miscellaneous	Centre Technique, Des Fruits et Legumes, 22 Rue Bergere, 75009 Paris
West Germany	Pome, stone, and other fruits	Institut fur Obstbau u. Gemusebau, Friedrich Wilhelm University, Auf dem Hagel 6, Bonn 5300
	Grapes, potatoes	Biol. Bundesanst. Land.U. Forstwirtschaf. Inst. Virusforschung, Messeweg 11/12, 3300 Braunschweig
Great Britain	Pome and stone fruits, ornamentals	EMLA, East Malling Research Station, Maidstone, Kent
Israel	Citrus, pome and stone fruits, ornamentals	The Volcani Institute of Agricultural Research, Bet Dagen
South Africa	Pome and stone fruits, potatoes, ornamentals	Plant Protection Research Institute, Plant Quarantine Station, Stellenbosch
The Netherlands	Pome, stone, and small fruits, potatoes, flower bulbs, cutflowers	Algemene Keuringsdienst voor Boomkwekerijgewassen, Laan van Meerdevoort 144, 2517 Be's-Gravenhage, The Hague

virus-free varieties of pome and stone fruits have been produced and distributed to state departments of agriculture and nursery companies in the U.S. and abroad.[12] More on this program is given in the chapter by Mink elsewhere in this volume.

Several other countries also have programs for the production, maintenance, and distribution of known virus-free nuclear stock (Table 10). Most include deciduous fruits; some also have varieties of raspberries, potatoes, chrysanthemums, carnations, citrus, avocados, grapes, and strawberries. These programs continue to expand in terms of the number of varieties of vegetatively propagated crops included in them.

One of the best-known European collections of certified virus-free pome and stone fruits is that of the East Malling — Long Ashton (EMLA) research stations.[8] This collection also includes rootstocks for apples, pears, plums, and cherries, as well as for many ornamental crabapples and cherries. Lesser known in the U.S. but equally successful are the projects of Australia, Canada, Israel, South Africa, and several European countries. Most of these projects were initiated within the decade and all are expanding. Ordinarily, most of these genera would be in quarantine upon arrival in the U.S. but, when the country of origin certifies that an item has been indexed and determined to be free of viruses, the quarantine is reduced or eliminated.[41]

It should be emphasized that most of these projects can supply only limited amounts of germplasm, usually at a nominal cost, from which the purchaser can make propagations. Again, to know what is available, and its cost in a given year, one should write to the project headquarters.

State Departments of Agriculture

A number of states have initiated programs for the production of indexed virus-free plants. These programs are operated cooperatively, in some instances, by the state government and the growers of the planting stock. These programs vary in scope; some include only a single crop and some include pome and stone fruits, citrus, grapevines, strawberries, brambles, roses, avocados, nut crops, seed potatoes, and certain vegetable crops. California probably has the most extensive program in that all of these crops are produced and in the number of nursery companies that participate in the programs. California even requires fumigation of the soil in which certified fruit tree nursery stock is to be grown to eliminate disease-spreading nematodes. The programs in Washington, Oregon, Pennsylvania, New York, Michigan, and Maryland also include many crops and are developing innovative approaches to virus-free crop registration and certification. Certain other states have major programs on fewer crops, such as those of North Carolina and Delaware with strawberries, New Jersey with orchids and blueberries, Arizona with citrus, and South Carolina with stone fruit.

The state regulations include dozens of defined terms that are used in connection with certified virus-free plant programs. The producer or purchaser of such plants should know the meanings of these terms as defined in his state; definitions of some terms differ among states and with the crop under consideration. Among the terms in state regulations are nuclear block, registered block, certified block, foundation block, increase block, mother block, indexed, off-type, virus-free, stocks, candidate tree, progeny plant, selected plant, and others. Some of these terms are synonymous with others.

State regulations also describe, for producers of virus-indexed planting stock, the inspection and testing procedures; fees; location and maintenance requirements for various blocks; responsibilities of the producers; harvesting, grading, and labeling procedures; container requirements; refusal, cancellation, and rejection procedures; and other information relating to production and sale of disease-free plants.

Some states offer certified planting stock of some vegetatively propagated crops, such as sweet potatoes, mint, rhubarb, and certain root crops including Irish potatoes, but again the certification of "freedom from viruses" is based only upon visual inspection rather than on virus indexing of the mother plants. Visual inspection does not necessarily detect viruses that are latent in some cultivars or under some growing conditions.

Foreign Nurseries and Research Centers

About 140 commercial nurseries and research stations in Europe (15 in Belgium, 12 in England, 10 in Germany, and 103 in the Netherlands) and 22 in Canada produce one or more cultivars of certified virus-free pome or stone fruits. Any variety from these nurseries which is virus-indexed and certified free of known viruses by the government of the country of origin can be imported. Unlike most fruit germplasm introductions, material from these sources is not in the "prohibited" quarantine category and hence is immediately available to the importer. For a current list and details on how to import material from these sources, the importer should write to: Permit Section, Plant Protection and Quarantine, APHIS, FCB 1, Hyattsville, Md., 20782. Prospective importers may also write to the foreign nurseries for details and fees, if any.

CONCLUSION

The value of excluding disease agents by quarantine cannot be overemphasized. One has only to look at what happened when the chestnut blight fungus or the Dutch elm fungus were carelessly brought to the U.S. several decades ago outside the realm of quarantine. These two tree species have vanished from the U.S. in the wake of these fungi except where protected at great expense.

The gypsy moth and Japanese beetle are examples of insect pests that cause inestimable losses to U.S. crops every year. Fortunately the U.S. plant introduction and quarantine system works in that 325,000 kinds of germplasm have been brought into the U.S. since 1898 without a single major disease outbreak attributable to careless quarantine. It is recognized that existing regulations on some crops are excessively strict and that others are far too lax. Regulations are continuously updated as we become enlightened by research and as the world disease and pest situation changes. In view of modern-day international travel by people and the increased activity in germplasm exchange among more of the developing countries, inspection and quarantine are more important than ever.

The value of excluding disease agents beginning with certified disease-free plants has also been demonstrated. Healthy planting stock, to include freedom from viruses, is generally available. There are also a few sources of certified virus-free planting material of less common woody crops, small fruits, potatoes, and other vegetatively propagated vegetable crops. More and more commercial nursery companies in the U.S. and abroad are offering these virus-free varieties. Not all nurseries indicate in their catalogues that some of their cultivars have been grown from virus-free source plants. Usually these varieties cost more than varieties not certified virus-free, but research has shown that these initial costs are quickly offset by earlier production and higher yields.

The situation with regard to disease-free seed is very different. The "certified" tag does not mean that the seed has been tested for viruses. Usually such seed originates from crops that were inspected for visible signs of disease and in which none or only a low incidence was found.

Since both state and federal governments are encouraging an increase in the availability of tested, disease-free seeds and plants, the programs that produce such stock are continuously expanding and being improved upon. The best way to know what is available, and from which sources, is to contact the appropriate state or federal offices.

REFERENCES

1. **Ball, E. M.,** *Serological Tests for the Identification of Plant Viruses,* The American Phytopathology Society, St. Paul, Minn., 1974.
2. **Bennett, C. W.,** Seed transmission of plant viruses, *Adv. Virus Res.,* 14, 221, 1969.
3. **Bos, L.,** Methods of studying plants as virus hosts, in *Methods in Virology,* Vol. 1, Maramorosch, K. and Koprowski, H., Eds., Academic Press, New York, 1967, 129.
4. **Bos, L.,** Graft transmission of plant viruses, in *Methods in Virology,* Vol. 1, Maramorosch, K. and Koprowski, H., Eds., Academic Press, New York, 1967, 403.
5. **Bos, L.,** Seed-borne viruses, in *Plant Health and Quarantine in International Transfer of Genetic Resources,* Hewitt, W. B. and Chiarappa, L., Eds., CRC Press, Cleveland, 1977, chap. 5.
6. **Childs, J. F.,** Indexing procedures for 15 virus diseases of citrus trees, U.S. Department of Agriculture Handbook No. 333, U.S. Department of Agriculture, Washington, D.C., 1968.
7. **Corbett, M. K.,** Detection of viruses and diagnosis of plant viral disease by electron microscopy, in Third International Symp. on virus diseases of ornamental plants. Lawson, R. H. and Corbett, M. K., Eds., *Acta Hortic.,* 36, 141, 1974.
8. **Cutting, C. V. and Montgomery, H. B. S., Eds.,** *More and Better Fruit With EMLA, a Practical Guide to Virus-free Planting Material,* East Malling Research Station, Maidstone, Kent, England, 1973.

9. **Dunez, J.,** Tenth Int. Symp. Fruit Tree Virus Diseases, *Mitt. Biol. Bundesanst. Land-forstwirtsch. Berlin-Dahlem,* 170, 1977.
10. **Frazier, N. W.,** *Virus Diseases of Small Fruits and Grapevines,* University of California, Berkeley, 1970.
11. **Fridlund, P. R.,** IR-2, A project with a "blood bank" of virus disease-free fruit trees, *Wash. Agric. Exp. Stn. Circ.,* 401, Washington State University, Prosser, 1962.
12. **Fridlund, P. R.,** The objectives and methods of the IR-2 virus-free deciduous fruit tree repository, *HortScience,* 12, 487, 1977.
13. **Fridlund, P. R.,** Fruit Tree Scions Available From the IR-2 Repository, Fall 1977 Through Spring 1978, Irrigated Agric. Res. Ext. Center, Prosser, mimeo, 1977.
14. **Fulton, R. W.,** Mechanical transmission of viruses of woody plants, *Annu. Rev. Phytopathol.,* 4, 79, 1966.
15. **Goheen, A. C.,** Virus and virus-like diseases of grapes, *HortScience,* 12, 465, 1977.
16. **Bollings, M.,** Disease control through virus-free stock, *Annu. Rev. Phytopathol.,* 3, 367, 1965.
17. **Kahn, R. P.,** International plant quarantine, in *Genetic Resources in Plants: Their Exploration and Conservation,* Frankel, O. H. and Bennett, E., Eds., Blackwell Scientific, Oxford, 1970, 403.
18. **Kahn, R. P.,** Plant quarantine: Principles, methodology, and suggested approaches, in *Plant Health and Quarantine in International Transfer of Genetic Resources,* Hewitt, W. B. and Chiarappa, L., Eds., CRC Press, Cleveland, 1977.
19. **Kahn, R. P., Monroe, R. L., Hewitt, W. B., Goheen, A. C., Wallace, J. M., Roistacher, C. N., Nauer, E. M., Ackerman, W. L., Winters, H. F.,, Seaton, G. A., and Pifer, W. A.,** The incidence of virus detection in vegetatively propagated plant introductions under quarantine in the U.S., 1957 to 1967, *Plant Dis. Rep.,* 51(9), 715, 1967.
20. **Kahn, R. P. and Monroe, R. L.,** *Datura metel L.* as a virus-indicator plant, *Phytopathology,* 60, 1183, 1970.
21. **Klinkowski, M.,** Catastrophic plant diseases, *Annu. Rev. Phytopathol.,* 8, 37, 1970.
22. **Knott, D. R. and Dvorak, J.,** Alien germplasm as a source of resistance to disease, *Annu. Rev. Phytopathol.,* 14, 211, 1976.
23. **Kreitlow, K. W., LeFebvre, C. L., Presley, J. T., and Zaumeyer, W. J.,** Diseases that seeds can spread, in Seeds, the Yearbook of Agriculture, U.S. Department of Agriculture, Washington, D.C., 265, 1966.
24. **Leppik, E. E.,** Gene centers of plants as sources of disease resistance, *Annu. Rev. Phytopathol.,* 8, 323, 1970.
25. **Mathys, C.,** Thoughts on quarantine problems, *EPPO Bull.,* 5, 55, 1975.
26. **McCubbin, W. A.,** *The Plant Quarantine Problem,* Ejnar Munksgaard, Copenhagen, 1954.
27. **McGregor, R. C.,** The Emmigrant Pests, report Adm. Animal Plant Health Inspection Service, U.S. Department of Agriculture, Hyattsville, Md., 1973.
28. **Neergaard, P.,** *Seed Pathology,* Vols. I and II, Halsted Press, New York, 1977.
29. **Nel, A. C.,** Plant quarantine in the Republic of East Africa, *E. Afr. Agric. J.,* 21, 10, 1955.
30. **Nyland, G. and Goheen, A. C.,** Heat therapy of virus diseases of perennial plants, *Annu. Rev. Phytopathol.,* 7, 331, 1969.
31. **Ordish, G. and DuFour, D.,** Economic bases for protection against plant diseases, *Annu. Rev. Phytopathol.,* 7, 31, 1969.
32. **Phatak, H. C.,** Seed-borne plant viruses — identification and diagnosis in seed health testing, *Seed Sci. Techn.,* 2, 3, 1974.
33. **Posnette, A. F.,** *Virus Diseases of Apples and Pears,* Tech. Comm. No. 30, Commonwealth Agricultural Bureaux, Maidstone, Kent, England, 1963.
34. **Renjehn, S.,** Plant quarantine in India, *Sci. Cult.,* 28, 215, 1962.
35. **Shepard, J. F. and Claflin, L. E.,** Critical analysis of the principles of seed potato certification, *Annu. Rev. Phytopathol.,* 13, 271, 1975.
36. **Thornberry, H. H.,** Suggested procedures and differential hosts for identifying viruses, in *Proc. 2nd Conf. Org. Citrus Virologists,* University of Florida Press, Gainesville, 1961, 256.
37. **Thurston, H. D.,** Threatening plant diseases, *Annu. Rev. Phytopathol.,* 11, 27, 1973.
38. U.S. Department of Agriculture, Virus Diseases and Non-infectious Disorders of Stone Fruits in North America, Agric. Handbook No. 437, U.S. Department of Agriculture, Washington, D.C., 1976.
39. Summary of State Regulations Governing the Interstate Movement of Plants and Plant Pest Carriers, Number 805.21A, Animal and Plant Health Inspection Service, PPQ, U.S. Department of Agriculture, Washington, D.C., 1976.
40. The National Plant Germplasm System, Prog. Aid #1188, Agricultural Research Service, U.S. Dept. of Agriculture, Washington, D.C., 1977.

41. U.S. Department of Agriculture, Code of Federal Regulations, Title 7, Agriculture, Animal and Plant Health Inspection Service, U.S. Department of Agriculture, Hyattsville, Md., 1977, chap. III.
42. List of intercepted plant pests 1972, Animal and Plant Health Inspection Service, U.S. Department of Agriculture, Hyattsville, Md., 1974.
43. **Walker, J. C.,** *Plant Pathology,* McGraw-Hill, 2nd ed., 1957.
44. **Waterworth, H. E.,** Virginia crab apple as an indicator for a widespread latent virus of pear, *Plant Dis. Rep.,* 55, 983, 1971.
45. **Waterworth, H. E.,** Purification, serology and properties of a virus from lilac, *Syringa oblata affinis, Plant Dis. Rep.,* 56, 923, 1972.
46. **Waterworth, H. E.,** Purification of arabis mosaic virus isolated from a jasmine plant introduction, *Phytopathology,* 65, 927, 1975.
47. **Waterworth, H. E.,** *Chenopodium quinoa* as a host species for plant virus research, *Ind. Phytopath.,* 29, 155, 1976.
48. **Waterworth, H. E., Lawson, R. H., and Kahn, R. P.,** Purification, electron microscopy and serology of dioscorea latent virus, *J. Agr. Univ. P.R.,* 58, 351, 1974.
49. **Waterworth, H. E., Lawson, R. H., and Monroe, R. L.,** Purification and properties of hibiscus chlorotic ringspot virus, *Phytopathology,* 66, 570, 1976.
50. **Waterworth, H. E. and Povish, W. R.,** A virus related to cucumber mosaic virus isolated from imported Ixora plants, *Phytopathology,* 65, 728, 1975.
51. **Watson, A. J.,** Foreign bacterial and fungus diseases of food, forage, and fiber crops: An annotated list, U.S. Department of Agriculture Handbook No. 418, U.S. Department of Agriculture, Washington, D.C., 1971.
52. **Wright, N. S., Cochran, J. E., Manzer, F. E., and Munro, J.,** Report of committee on PVX testing, *Am. Potato J.,* 53, 333, 1976.
53. **Wutscher, H. K.,** Citrus tree virus and viruslike diseases, *Hort Science,* 12, 478, 1977.

THE ENVIRONMENTAL CONTROL OF PLANT PATHOGENS USING ERADICATION

Thor Kommedahl

INTRODUCTION

To eradicate means to uproot, to exterminate, to extirpate. To control disease by eradication means to remove physically the pathogen from a plant, a crop, or an area, after the disease has become established. The discussion that follows deals with eradication by changing the biotic environment and, to a limited extent, the edaphic environment, and not by the use of chemicals.

The biotic environment can be altered directly by collection and destruction of plant refuse, by burning of infected plants, or by flooding fields of infected plants or plant parts when the pathogens are soil-borne. Also, whole plants can be eradicated (rogued) in fields, orchards, and, to a limited extent, in wooded areas; or, portions of plants can be eradicated (pruned). Plants eradicated, or rogued, may be primary hosts, alternate hosts, or alternative hosts. Indirectly, the pathogens may be controlled, if not eradicated, by altering the biotic and edaphic environment through crop rotation and by management of crop residues. Controlling a disease may mean reduction of losses from it to economically acceptable levels.

This topic is treated under sanitation, crop rotation, and crop residue management.

SANITATION

Although sanitation usually includes all activities designed to destroy or reduce amounts of infected materials from a field or greenhouse, only two aspects of sanitation (based on environmental control procedures) are being treated: (1) removal or destruction (in situ) of infected residues after harvest, and (2) removal and disposal of infected plant parts (pruning) or roguing of infected plants or alternate hosts from the field or its vicinity.

Frequently no single method of disease control is used exclusively. Effective control or management may depend on the application of several control methods simultaneously or in succession. Illustration of control by sanitation does not preclude control by other methods either as a substitute for, or a supplement to, sanitation. The methods of destroying infected plant materials by tillage operations, burning, or flooding applies generally to field and garden crops whereas the destruction of plant parts by pruning applies mainly to woody plants. Roguing of infected plants applies to annual crops, weed hosts, and perennial woody plants that are economically important.

Removal of Crop Refuse

The removal of infected plant parts, fallen infected leaves, or branches may be an effective control measure but may not be justified economically, except in selected instances. For example, Meredith[1] has shown that the removal of trash (badly spotted leaves) in banana plantations reduced the incidence of the Sigatoka disease caused by *Mycosphaerella musicola*. This is recommended only because it reduces the sources of inoculum so that subsequent chemical spray applications will be more effective.

The destruction of cull piles of potato tubers and vines has been cited as an example of control of blackleg *(Erwinia atroseptica)* and late blight *(Phytophthora infestans)*. Bonde and Schultz[2] found that cull piles 150 to 180 m upwind from potato fields must be destroyed to prevent or reduce severity in epidemics of late blight. On the other hand, Van Der Plank[3] estimated that the likely effect of destroying cull piles is to delay a blight epidemic by 5 to 10 days. Moreover seed transmission is now considered to be the main source of inoculum for blackleg. Thus destruction of cull piles appears to have limited value as a control measure.

Another special situation was reported where greenhouses were located near corn fields. Carnations growing in the greenhouses became infected with *Fusarium roseum* 'Graminearum' which causes stub dieback of carnation and stalk rot of corn. Disease in this instance was attributed to inoculum surviving in perithecia on corn stubble and wind-blown into greenhouses through the ventilation system.[4] Removal of such corn stubble eliminated inoculum.

In general, removal of crop refuse is not an economically feasible control measure and other methods of eradication are usually more promising.

Burning

Fire effectively destroys inoculum and has been used since ancient times for control of pests and diseases. In modern times, fire has been effective in selected instances, e.g., it is the principal field method for controlling brown spot needle disease caused by *Scirrhia acicola* on longleaf pine *(Pinus palustris).*[5] Longleaf pines have to grow to a stem diameter at the ground line of 2 to 3 cm before seedlings begin their "height" growth. During this period, fascicled needles appear and these are subject to infection by the fungus. Brown spot on needles at this time will arrest growth and confine seedlings to a height of about 15 cm. Controlled burning in winter should not begin before the second growing season, just as seedlings start their "height" growth. The best time to burn is in late winter just before spring growth begins. To avoid heavy mortality from fire, the burning should be started in midwinter, and before 35% of the foliage is infected. Foliage killed but not burned by fire is free from inoculum because temperatures high enough to kill leaf tissue will also kill the fungus. In general, the sanitary effect of fire varies directly with the size of the burned-over tract and availability of inoculum near the area burned.

Fire can be used to eradicate dwarf mistletoe *(Arceuthobium pusillum)* on black spruce *(Picea mariana),* by removing harvestable timber and leaving slash for the fire.[6] Burning eliminates infected trees not removed by logging, including those with latent infections. Complete mortality of spruce occurred upon allowing slash to dry for 50 days and then burning when weather conditions were favorable for fire (20°C and 50% relative humidity). However, burning is effective in control only if (1) the fire spreads sufficiently and is intense enough to kill all living spruce in the treated area, (2) weather conditions before, during, and after the fire permit reasonably safe operations, and (3) the total cost of eradication is less than the value of losses prevented by eliminating the dwarf mistletoe.

Citrus canker caused by *Xanthomonas citri* was the first eradication program to be successful in the U.S. and the first eradication program to be sponsored by the federal government (1914). This disease was eradicated mainly in Florida, but also in Louisiana and Texas by the burning of infected trees, and this is the only successful control method in the Gulf Coast states.[7] More than three million trees were burned in Florida and adjacent states in the first few years of eradication. If one tree was infected, the entire grove was burned. By 1934, nearly 20 million trees had been destroyed. The host range includes grapefruit, *Poncirus trifoliata,* Key lime, navel orange, sweet orange, lemon, Satsuma orange, tangerine, mandarin, king orange, and kumquat, in decreasing order of susceptibility. The disease is dangerous because it is spread so easily by mechanical means, and by insects, birds, animals, and humans. Long-distance spread occurs by transport of nursery stock.

Burning is also effective in eradication of diseases of herbaceous plants. The single most important cultural practice in grass seed production in the Pacific Northwest is the burning of grass fields after harvest.[8] *Gloeotinia temulenta,* which causes blind seed disease of perennial ryegrass *(Lolium temulentum),* is carried on straw and stubble, and this material has been burned in most fields in Oregon since 1949. Such burning also destroys ergot *(Claviceps purpurea)* in the same fields.

Instead of open burning of crop residues or of slash, flame has been used. Horner[9] obtained excellent control of mint rust *(Puccinia menthae)* by directing a propane flame at the soil surface after mint shoots emerged from soil. Flaming kills teliospores and the early emerging infected shoots, and removes foliage at a time when aeciospores must infect leaves or perish.

In Hardison's review[8] are described many additional examples where fire or flame has been tried for disease control. Some of these uses of fire are not usually applied or necessarily recommended. In special situations it may be feasible to collect and burn dead vines and pods of snap beans to eradicate inoculum of *Diaporthe phaseolorum,*[10] or to burn crop residues of alfalfa and sweet clover to destroy inoculum of fungi causing leaf spots,[11] or to collect and burn smut galls from corn and sugarcane.[12] However, better control procedures are available in the latter examples and the use of fire would serve as a control measure only in exceptional circumstances.

Flooding

Flooding has had varying success as a means to eradicate pathogens. Flood fallow programs were developed in banana plantations in Honduras and Panama from 1945 to 1955, to control the Panama disease caused by *Fusarium oxysporum.*[13] By 1956, about 6100 ha were flooded in Honduras and about 4000 in Panama. A continuous program was developed based on four criteria: (1) selection of areas with a history of slow disease development, (2) elimination of banana culture at least 4 to 5 years before flooding to lower inoculum concentration, (3) alternately flooding and plowing land, and (4) providing good drainage with a low constant water table. Using a moldboard plow between periods of flooding consistently reduced disease incidence in both Honduras and Panama which was attributed to burial of surviving spores that were destroyed by subsequent flooding. The recommended procedure is to flood land for 3 months, drain and plow it, then flood it for 2 months, drain and plow, and finally flood land for 2 months. In addition to reducing incidence of Fusarium wilt, the bacterial wilt pathogen *(Pseudomonas solanacearum)* and the burrowing nematode *(Radopholus similis)* were destroyed. Despite these benefits, however, the costs of labor, engineering, and equipment have increased so much that today this flood-fallow program is not economical.

Klisiewicz[14] found that post-harvest flooding at 39°C or above for 7 days gave effective control of seedling rust of safflower caused by *Puccinia carthami.* Jenkins and Taylor[15] reported that 1 to 2 years of total submergence is required to eliminate root knot nematodes from most plants.

Flooding is seldom recommended as a control measure to eradicate pathogens because it is expensive, it depends upon a source of water, it affects soil structure adversely, and the effect is temporary.

Pruning

Pruning as a disease control or eradicative measure has value mainly for specific woody ornamentals or shade trees in urban settings or in orchards, but not in forests or forest nurseries. Generally, trees with cankers, galls, and twig blights are pruned back to sound, uninfected wood, to rid the plant of infection and enable callus to form. Although in cities and towns pruning is practicable and recommended for eradicating cankers of ash, basswood, oak, and sumac, it is not recommended for other cankers such as Cytospora canker of poplar; in fact, pruning spreads this disease.[16] Similarly, pruning is often recommended for control of fire blight of apple, pear, hawthorn, or mountain ash, as well as for control of some other bacterial diseases of woody plants; however, this is not always successful. Unless pruning tools are disinfected between cuts, and wood is pruned to uninfected tissue when dry, this method of eradication may not only be ineffective but disease severity may be enhanced. If pruned early, new growth of susceptible tissue becomes available for new infection. Pruning for fire blight is effective only when combined with other methods.

In greenhouses, or with house plants, any plant part that has leaf spot or blight can be pruned, e.g., the prompt and persistent picking of flowers infected with *Ovulinia azaleae* reduces incidence and spread in azalea and rhododendron.[16] Jenkins and Taylor[15] report that populations of foliar nematodes in greenhouse-grown plants can be reduced by removal and destruction of infected plant parts. They also reported that root-propagated plants such as peony, when infected with the root knot nematode and roots become galled, that such roots should be pruned and destroyed when plants are divided. In a similar example, lesions on banana rhizomes caused by the nematode *Radopholus similis* are pruned from propagating material at planting.[15]

Roguing

Roguing is not always a satisfactory control measure. By the time symptoms appear and the plant is rogued from the field, the disease has probably already spread to nearby plants.[17] Also, because it is a hand operation, labor costs are high. Sometimes it serves as an adjunct to other control practices.

Roguing is probably more feasible with perennial than annual plants and probably more effective against virus than bacterial or fungus diseases. Labor costs limit use of this practice to high-value cash crops. Five crops where roguing of the primary host is practiced are banana, cacao, peach, potato, and sugarcane. Roguing of plantains near banana plantations is used in control of banana mosaic virus.[18]

Swollen shoot of cacao is economically one of the most important plant diseases of the world, accounting for a 10% loss in the world's cocoa crop.[19] It is one of the most important factors that limit cocoa production in West Africa, Sri Lanka, and Trinidad. The virus causes swelling of stems because of increases in amounts of phloem, xylem, and, to some extent, cambium. The disease has been eradicated in Trinidad but not in West Africa, partly because growers are reluctant to eradicate trees that lack symptoms. In Nigeria, a *cordon sanitaire* was established around cocoa groves in which both infected and symptomless trees were rogued: all trees in a 5-m belt were rogued if 1 to 5 trees were infected; all trees in a 9-m belt were rogued if 6 to 50 trees were infected; and all trees in a 14-m belt were eradicated if 51 to 200 trees were infected. This method proved effective, especially for small outbreaks of swollen shoot disease (<50 infected trees), which emphasizes the importance of eradication in the early stages of disease outbreaks.

A 20-year program in 26 states in which peach and plum trees were rogued to control phony peach disease was evaluated.[20] It was found that roguing had eliminated the disease in six states and in 175 of 295 counties in eight other states. Roguing proved effective if done annually on an area-wide basis, often in combination with insecticides for vector control. Similar results were reported in Colorado for peach mosaic.[21]

Several potato viruses can be controlled by roguing: calico caused by the alfalfa mosaic virus, haywire caused by sugar beet curly top virus, leafroll, mild mosaic caused by potato virus A, rugose mosaic caused by potato virus X and Y (plants 12 to 15 cm tall rogued every 7 to 10 days to midseason), and witches' broom.[22] Early roguing of tobacco plants infected with tobacco mosaic virus[23] retards spread in the field. However, in some crop cultivars, virus symptoms may be so mild as to make roguing impractical, e.g., spindle tuber; and in other cultivars, a virus may spread so rapidly as to make roguing ineffective. Even if effective, roguing is usually impractical in field crops unless seed certification is sought, e.g., roguing of soybean plants infected with soybean mosaic virus in seed-production fields.[24]

The sugarcane downy mildew *(Sclerospora sacchari)* has been one of the most destructive diseases of sugarcane in Australia, the Republic of China, and the Fiji Islands. It was nearly eradicated in these countries by thorough roguing, as well as by use of disease-free seeds and sets, and the use of resistant cultivars.[25]

Many of the rust fungi have alternate hosts and their eradication has been done either with individual plants or with extensive acreages. The alternate hosts to rust diseases of crops are usually woody plants, and costs of eradication usually prohibit all but those that affect the most extensively grown food crops. The classic example has been the eradication of barberry and mahonia as a control measure for stem rust of cereals *(Puccinia graminis)*.

The campaign to eradicate the barberry was probably the largest plant disease control program ever attempted. The barberry plant was first outlawed in Rouen, France, in about 1660. Late in the 19th century, laws were passed for eradication of barberry in Austria and Denmark, and, in the following century, in six additional European nations as well as in New South Wales, Australia. The barberry was banned in three New England states before 1766. However, it was the great epidemic of 1916, in which 200 million bu (70 million hℓ) of wheat were lost to stem rust in the U.S. and 100 million (35 million hℓ) in the prairie provinces of Canada, that led to 19 states passing laws to control the barberry and related species *(Berberis* and *Mahonia* spp.). Barberry bushes were destroyed in about 1 million miles^2 (2.6 million km^2) of the best prairie land of the northerly wheat-growing region of the U.S.[26]

Nearly a half-billion barberry bushes were destroyed and it was calculated that this eradication campaign reduced the annual losses from stem rust from 40 million to less than 15 million bu (14 to 5 million hℓ) of wheat.[26] The total cost of barberry eradication, since its start in 1918, has been less than 1% of the value of the wheat, oats, barley, and rye crop harvested in the 19 states.

Teliospores of the stem rust fungus *(Puccinia graminis)* overwinter on straw of grains and grasses. These spores cannot reinfect grain but must go to the barberry. On the barberry leaf, the sexual stage occurs which provides for recombination of genes and the production of new races.[27,28] Spores from infected barberry are carried to wheat, where the recurring stage is initiated, and these spores spread from wheat to wheat and from field to field. As wheat ripens, the overwintering black teliospore stage again occurs on wheat straw. Thus, the barberry not only serves as a source of rust inoculum, but as a source of new stem rust races.

For several decades following the start of the barberry eradication campaign, it was generally concluded that losses from stem rust could be appreciably reduced by eradication of barberries; however, the devastating epidemics of 1935, 1953, and 1954 forced a reappraisal of this conclusion. Presumably enough barberries had been destroyed to make epidemics less likely. In Van Der Plank's analysis,[3] the early outbreaks of rust had originated locally from barberries. However, massive numbers of spores, blown from southern states to northern wheat fields, produced abundant inoculum that brought about widespread rust epidemics. Barberry eradication campaigns could not guarantee that wheat would escape rust epidemics. Nevertheless, eradication undoubtedly reduced the number and size of epidemics over the years. The federal campaign, started in 1918 in the U.S., was scheduled to end in 1979.

Cronartium ribicola causes blister rust on eastern *(Pinus strobus)* and western *(P. monticola)* white pine, and on sugar pine *(P. lambertiana)*. This rust was introduced to the U.S. on nursery trees from Europe and occurred first in the eastern states. By 1921, it had spread to all important white pine regions of the nation.[26] Because currants and gooseberries *(Ribes* spp.) are alternate hosts for this rust, a campaign to eradicate these plants was started in 1909 in the U.S., and it became the largest tree disease control program ever undertaken. In the period from the 1920s to the 1950s, at which time Ribes eradication was abandoned as a federal program, more than 1.5 billion bushes had been destroyed.

The damaging stage of blister rust occurs on white pine, but the recurring stage is on *Ribes* spp. During cool, wet weather in July and August, basidiospores are wind-borne to white pines. These basidiospores are short-lived and it was assumed that if *Ribes* ssp. were eradicated within a 300-m radius of pines, blister rust would not occur. However, subsequent analysis of the eradication campaign indicated that eradication did not reduce incidence of blister rust to acceptable levels, especially in western white pine.[29] Forest Service surveys of both the Rocky Mountain and Great Lakes regions indicated no significant correlation between numbers of infected pines and Ribes plants per acre, unless all but a few bushes were eradicated per acre. In the high-hazard areas in eastern white pine forests, there is strong circumstantial evidence for long-distance dissemination (8 km if near large lakes) of rust basidiospores, making Ribes eradication ineffective as a control measure. However, in the intermediate- and low-hazard areas, Ribes eradication is feasible.

Cedar-apple rust *(Gymnosporangium juniperi-virginianae)* occurs on apples and on red cedar *(Juniperus virginiana)*.[26] Basidiospores of this rust travel 3 to 5 km. Thus, red cedars are eradicated (in some states by law) in or near apple orchards, or apples from red cedar stands, whichever is the preferred tree. The lack of a repeating stage, the long incubation time on red cedar, and the short time that the galls on red cedar are active makes eradication successful. Hawthorn rust *(G. globosum)* and quince rust *(G. clavipes)* are also controlled by eradication of red cedar, hawthorn, or quince, whichever is the preferred plant species. On poplars, several leaf rusts *(Melampsora* spp.) occur which complete their life cycles on hemlock and larch, provided that such trees are within 900 m of poplars. In nurseries or special habitats it may be profitable to eradicate the host of least preference.

The role of the alternate host in disease of the primary host is much reduced when (1) there is no repeating stage, (2) the repeating stage is on the primary host, or (3) the infective propagule from the alternate host is a basidiospore. These points are illustrated with the three rusts. Wheat stem rust uredospores of the repeating stage are blown long distances from field to field of wheat, the primary host. Spore numbers are high and, given favorable environmental conditions and susceptible wheats, epidemics occur despite eradication of the alternate host. Uredospores are durable and can withstand the hazards of the atmosphere.

With white pine blister rust, the spores that travel great distances are usually the aeciospores that go from pine (the primary host) to currant, and the uredospores that go only from currant to currant (the alternate host). In cedar-apple rust there is no repeating stage at all.

Because the blister rust spore that infects the pine, the primary host, is the basidiospore, and this spore is more sensitive to changes in temperature and moisture, and because its success in long-distance transport is less assured than that of uredospores or aeciospores, the alternate host need only be eradicated in the immediate vicinity of the primary host, at least in the intermediate-hazard areas.

Inoculum sources other than that from the alternate host need to be considered.

Wheat rust multiplying on wheats and grasses from fields in Mexico to those in Canada, can potentially produce large amounts of inoculum over a wide geographic area, and thereby negate the local influence of rust-infected barberries. Other rusts may produce a limited supply of inoculum because the primary hosts are not as numerous over a sufficiently wide area.

Weeds or wild plants can serve as alternative hosts for pathogens of crops, especially for virus diseases. Sometimes a given pathogen has a wide host range among weeds, e.g., aster yellows occurs on 41 species of weeds in 31 genera of 14 plant families under natural conditions in California. Eradication of these weeds by roguing would be impracticable under most circumstances. Again, the roguing of weeds in the Chenopodiaceae found in beet fields aids in control of curly top virus in beets. Roguing cruciferous weeds in cruciferous crops protect them from Alternaria leaf spot. Roguing solanaceous weeds in pepper can protect pepper from anthracnose and bacterial spot, or tomato from several bacterial diseases. However, though effective as a disease control measure, these plants can be controlled more economically by means other than roguing.

Obviously, roguing as a method of eradication can be used only in extraordinary circumstances. Its use of hand labor militates against it in most agricultural situations. It is often of value in gardens, in specialty crops of high cash or ornamental value, or in seed-production fields.

CROP ROTATION

The origin of crop rotation practices is probably lost in antiquity. Certainly agriculturists followed some form of rotation in biblical and Roman times even when reasons for the rotation were not known. Rotation practices developed at one time and place may not be applicable to other times and other places. Much of the history and development of crop rotation in recent times has been reviewed by Curl,[30] Berkeley,[31] and Glynne,[32] and some of the principles have been discussed by Stevens.[17] Leighty[33] has tabulated 24 diseases on 17 crops in which crop rotation controlled that disease either wholly or in part. Although crop rotation is treated here for its benefits in disease control, other benefits accrue from rotation. Any rotation developed to reduce populations of pathogens and keep them low must also be a rotation that aids in productivity and microbial balance in soil, helps control insects and weeds as well as pathogens, preserves the physical condition of soil, aids in reducing erosion by wind and water, and be practical agriculturally and economically.

Crop rotations are applicable more to root than to foliage or stem pathogens and are generally thought to be more effective against fungi than other pathogens. However, bacterial diseases in general are reduced in severity by crop rotation, except in crown gall and bacterial wilt (but not the banana race of *Pseudomonas solanacearum*). In nematode control, the potential benefits from rotation have not been fully achieved in farming practice, according to Nusbaum and Ferris,[34] and rotation is usually complementary to chemical treatment and host resistance.

As Stevens[17] pointed out, the chief obstacles to success of crop rotation are the pronounced longevity of inoculum in soil and the frequently wide host ranges of pathogens. Rotations are more effective against root inhabitants than soil inhabitants because root inhabitants are specialized, have a more limited host range, and have a lower competitive saprophytic ability than soil inhabitants.[35] Root inhabitants survive in soil as long as crop residues remain and survive longer than that only if they produce a survival structure in host tissue. Thus, rotations can hasten the demise of root inhabitants by host deprivation. Both short and long rotations are used, but farmers are often reluctant to use rotations longer than 1 to 4 years.

To be effective, both the duration and the kinds of crops that make up the rotation should be varied from time to time to avoid buildup of pathogens of the crops in the rotation. Also, crops in the rotation should be of different plant families, e.g., potatoes, tomatoes, pepper, and even sweet potato, should not be used in the same rotation with tobacco, because all these crops are susceptible to the same pathogens that cause severe losses in tobacco.[24] The crop immediately preceding the crop for which the rotation is planned should be selected with care.

Some of the effects of rotation are illustrated in work by Williams and Schmitthenner[36] where they studied the effects of 3 years of monocropping and of all possible 2-year sequences of corn, oats, wheat, soybeans, and alfalfa for number and kinds of soil fungi. They found a richer and more variable soil mycoflora in crops in the rotation (66 fungus groups affected) than in crops not in the rotation (32 fungus groups affected). Ullstrup[37] stated that for corn, rotation was most effective (1) where the crop was grown in a limited area, (2) the disease-producing agents were strictly soil borne, and (3) the disease-causing organism survives in old corn debris on the soil surface. The influence of a given crop may not show in the field until 2 years later.[36]

For diseases caused by nematodes, the length of the rotation may be dictated by the nematode population in soil and the rate of decline during the rotation period.[15] For nematodes lacking a resistant stage, rotations of 1 or 2 years may be sufficient; however, with species having a resistant stage, e.g., *Heterodera schachtii,* a 3- to 7-year rotation may be necessary. Some nematodes decline more rapidly than others in the absence of a suitable host.

Care must be selected in choice of alternate crops because even if they are not hosts they may enable the pathogen to multiply. For example, barley (a nonhost for *Fusarium oxysporum* causing cotton wilt), planted between crops of cotton, supported on its roots substantial populations of the wilt fungus, so that the inoculum concentration after barley was greater than that after cotton.[38]

Sometimes a rotation used to control one disease can favor another. For example, Cook[39] found that the root rot pathogen of wheat, *Fusarium roseum* 'Culmorum', multiplied more rapidly in fields cropped to oats than to wheat, yet oats are useful in a wheat rotation to reduce take-all of wheat. Similarly, the presence of infected corn stalks in soil increases the likelihood of subsequent seedling blight of wheat caused by *Fusarium roseum* even though it decreases the chances of infection by *Rhizoctonia solani.*[40]

Where fallow is part of the rotation, either as an ecofallow approach or as a substitute for a crop, the intent is to starve the pathogen by denying it sufficient substrate for multiplication and survival. This is more effective if the pathogen lacks a survival structure in soil. The advantage of fallow is that many unrelated pathogens are affected equally.

There are nine major diseases in which crop rotation is usually not effective unless rotations are long (5 to 10 years).[41] These are as follows:

1. Bacterial wilt caused by *Pseudomonas solanacearum* — There are 60 hosts, especially potato, pepper, tomato, and eggplant; immune plants are used in rotations.
2. Crown gall caused by *Agrobacterium tumefaciens* — Plants in 40 plant families are susceptible. Long rotations with corn, oats, grasses, sorghum, asparagus, and onion or other immune crops are recommended.
3. Common scab caused by *Streptomyces scabies* — This occurs on beet, cabbage, carrot, eggplant, mangel, onion, parsnip, potato, radish, rutabaga, salsify, spinach, and turnip. Long rotations with small grains, grasses, or corn are recommended.

4. Sclerotinia diseases caused by *S. sclerotiorum* and other species — Nearly all succulent plants are susceptible (flowers, shrubs, weeds, and nearly all vegetables). Long rotations with immune crops are recommended, e.g., beets, onions, spinach, corn, small grains, and grasses.
5. Rhizoctonia diseases, caused by *R. solani* — Nearly all crop species are susceptible. Long rotations with unrelated crops offer some control.
6. Verticillium wilt caused by *V. albo-atrum* — This fungus occurs on at least 200 host species (flowers, fruits, trees, weeds, vegetables, and field crops). Long rotations are used with unrelated crops, especially corn, small grains, and grasses.
7. Black root rot caused by *Thielaviopsis basicola* — This has a wide host range. Rotation is not effective and outbreaks of disease are more severe after alfalfa, clover, beans, lupines, and tobacco; beets and tomatoes are resistant.
8. Sclerotium wilt caused by *S. rolfsii* — More than 200 host species are susceptible. Long rotations with corn, sorghum, and small grains can reduce disease.
9. Texas root rot caused by *Phymatotrichum omnivorum* — There are about 2,000 hosts in dicotyledonous crops. Long rotations with corn, small grains, sorghum, and grasses help reduce disease.

Nearly 100 diseases on 36 crops are listed in Table 1, in which crop rotation has been documented for some location at some time. This list is not complete for there are many diseases in which crop rotation is used and is helpful in reducing the effect of diseases, but which is not by itself a major factor in disease control. Perhaps all crops would benefit to some degree from crop rotation for many reasons. The list includes examples in which rotation appears to have a major effect on disease control at least for one location. One would have to use judgment as to whether a rotation used in Idaho, for example, would be applicable in New York. Moreover, crop rotation generally is used in conjuction with other disease control measures.

CROP RESIDUE MANAGEMENT

Crop residue management represents a type of crop husbandry designed to hasten the natural disappearance of a pathogen from crop residues in soil. The pathogens in residues may have colonized them after their burial in soil or they may have invaded them while they were in live plants before harvest. *Cephalosporium gramineum*, for example, invades live wheat plants; at harvest and subsequent burial by tillage operations, the fungus continues to grow until stopped by soil inhabitants that colonize such plant materials.[42] When this straw has eventually decayed, the pathogen no longer survives in soil and dies from lack of substrate. Other fungi may produce survival structures and persist in soil until new substrates or susceptible plants appear. The residues themselves may be toxic to other plants;[43-45] their effects are known as allelopathic.[46] The destruction of such residues would be beneficial to succeeding crops.

Deep plowing of residues hastens the disappearance of root-inhabiting types of pathogens, and is especially destructive to foliage pathogens. In fact, plowing aerial plant parts into soil shortly after harvest destroys most, if not all, pathogens causing leaf spots, anthracnose, rusts, smuts, mildews, and leaf blights. This can be done by chopping, harrowing, or plowing material into soil prior to the subsequent growing season. Some examples are listed in Table 2.

In plowing down residues, there are many variations of the theme. In banana wilt, it was important to plow in residues of a previous crop of sugarcane in amounts of at least 29 metric tons/ha to reduce Fusarium wilt of banana.[47] Amendment of residues to soil prior to planting a given crop can reduce bean root rot caused by *Thielaviopsis*

Table 1
CROPS AND DISEASES IN WHICH ROTATION HAS HAD A MAJOR EFFECT IN DISEASE CONTROL

Crop	Disease	Pathogen	Crop interval (year)	Commentary	Ref.
Alfalfa	Root rot	*Cylindrocladium ehrenbergii*	1—2	Rotate with corn, small grains, or forage grasses	50
	Stem nematode	*Ditylenchus dipsaci*	2	Use any cultivated crop	33,50
	Violet root rot	*Helicobasidium purpureum*	1—2	Rotate with cereal crops	50
Barley	Leaf blotch	*Septoria passerinii*	2	Rotate when barley fields not contiguous or close	51
Bean, dry	Anthracnose	*Colletotrichum lindemuthianum*	2	Use any crop but beans	33
	Ashy stem blight	*Macrophomina phaseoli*	2—3	Avoid beans or sweet potato	10
	Common blight	*Xanthomonas phaseoli*	2	Avoid bean crops	10
	Dry root rot	*Fusarium solani*	5	Avoid bean crops	41
	Halo blight	*Pseudomonas phaseolicola*	2	Avoid bean crops	10
	Leaf spot	*Cercospora cruenta*	5 +	Rotate with nonlegumes	41
	Root rot	*Pythium butleri, Rhizoctonia solani, Sclerotium rolfsii*	4—5	Rotate with cereals, clover, or alfalfa	10
	Watery soft rot	*Sclerotinia sclerotiorum*	2	Use small grains, corn, or hay crop	10
	Wilt	*Corynebacterium flaccumfaciens*	2	Avoid bean crops	10
Bean, lima	Pod blight	*Diaporthe phaseolorum*	1	Avoid beans in rotation	41
	Scab	*Elsinoë phaseoli*	1	Avoid beans in rotation	41
Bean, red kidney	Root rot	*Fusarium solani*	3	Wheat before bean	52
Bean, snap	Hypocotyl rot	*Rhizoctonia solani*	1	After corn; plow down residues	53
Beets, sugar	Black root	*Aphanomyces cochlioides*	2—5	Beets after corn, soybeans, and cereals, but not legumes Montana: beets, barley, alfalfa (3 years); oats, beans, alfalfa (2 years); corn, beets; beets, fallow, potato Iowa: 3—4 years between beets; >6 with 3—4 years alfalfa where disease severe Ohio: corn better than oats before beets. Legumes effective if fall but not spring plowed	54—57
	Cyst nematode	*Heterodera schactii*	2—5	If spotty, 2 years legumes, grains, or solanaceous crops between beet crops; if >25% of beet crop lost, plant 3—5 years between beet crops	58

Beet, table	Leaf spot	*Cercospora beticola*	4—5	Rotate with corn, potato, other crops	33,41
		Xanthomonas beticola	2+	Rotate with corn, potato, other crops	41
		Phoma betae	5+	Rotate with corn, potato, other crops	41
Bentgrass	Nematode	*Anguina agrostis*	1	Seed fields: permit no *Agrostis* spp. to go to seed for 1 year	59
Broccoli	Anthracnose	*Gloeosporium concentricum*	1	Rotate with noncrucifers	41
Cabbage	Black leg	*Phoma lingam*	3+	Rotate with noncrucifers	41
Carrot	Root knot	*Meloidogyne hapla*	1	Plant corn or weed-free sod (*Poa pratensis*) 1 year between root crops	60
	Root scab and blight	*Xanthomonas carotae*	2	Long rotations preferable	41
Cantaloupe	Root knot	*Meloidogyne incognita*	2	Rotate with soybean or soybean and crotalaria	61
Cauliflower	Anthracnose	*Gloeosporium concentricum*	1	Fungus on cabbage, broccoli, and Brussels sprouts	41
	Peppery leaf spot	*Pseudomonas maculicola*	1	Rotation with noncrucifers and destruction of residues	41
Celery	Brown spot	*Cephalosporium apii*	1	Rotate with lettuce, onions or potato (N.Y.)	62
	Pin nematode	*Paratylenchus hamatus*	2	Lettuce or spinach between celery crops	63
Cereals and grasses	Cyst nematode	*Heterodera avenae*	1—2	Use legumes in rotation	64
	Root knot	*Meloidogyne naasi*	1	Use root crops in rotation	28
Corn	Downy mildew	*Sclerospora sorghi*	1	Rotate with nongrass crop	37
	Leaf blight	*Helminthosporium carbonum, H. maydis, H. turcicum*	1—3	Rotate with nongrass crop	37
	Meadow nematode	*Pratylenchus leiocephalus*	1—2	Alternate with peanut crop	65
	Stalk rot	*Fusarium moniliforme, F. roseum*	1	After soybeans	36
Cotton	Root rot	*Macrophomina phaseoli*	1	After alfalfa or barley	66
		Phymatotrichum omnivorum	4	After cereals or grasses; plow residues deeply	67
	Seedling blight	*Ascochyta gossypii*	1	After any other crop; plow residues after harvest	68
Crucifers	Leaf spot	*Alternaria brassicae*	1	Rotate with noncrucifers	41
Cucumber	Anthracnose	*Colletotrichum lagenarium*	2	Any crop but muskmelon, watermelon, gourd	33
Hops	Wilt	*Verticillium albo-atrum*	1	Rotate with cereals, root crops, vegetables, not potato or raspberry	69
Lettuce	Leaf spot	*Cercospora lactucae*	4+		41
		Septoria lactucae	4+		41

Table 1 (continued)
CROPS AND DISEASES IN WHICH CROP ROTATION HAS HAD A MAJOR EFFECT IN DISEASE CONTROL

Crop	Disease	Pathogen	Crop interval (year)	Commentary	Ref.
Onion	Bloat	*Ditylenchus dipsaci*	2	Rotate with beets, carrots, crucifers, lettuce, spinach (N.Y.)	41
	Nematode	*Balonolaimus gracilis*	5—10	Use watermelon or tobacco and eradicate weeds	41
	White rot	*Sclerotinia cepivorum*	8—10		41
Parsnip	Canker and leaf spot	*Itersonilia perplexans*	5—10		41
Pea	Anthracnose	*Colletotrichum pisi*	2+	Rotate with non-host crops	41
	Bacterial blight	*Pseudomonas pisi*	2+		
	Leaf spot	*Septoria pisi*	4+		
	Root rot	*Aphanomyces euteiches*	6—10	Use more than half crop sequence in corn, grains, vegetables; not forage	54
	Seedling blight	*Rhizoctonia solani*	1+	Rotate with crops such as cereals or corn	70
Pepper	Anthracnose	*Colletotrichum piperatum, C. capsici*	1	Avoid solanaceous crops in rotation	41
	Bacterial spot	*Xanthomonas vesicatoria*	1	Control solanaceous weeds	41
Peppermint	Wilt	*Verticillium albo-atrum*	5	Peppermint for 2 years; corn or corn and onions 5 years (midwest U.S.)	71
Potato	Corky ring spot (Spraing)	Tobacco rattle virus	1	Barley better than corn, potato, or sugar beet in rotation	72
	Golden nematode	*Heterodera rostochiensis*	8	Exclude potato or tomato in rotation; in short rotations, use beans, corn, red clover, rye-grass	22
	Root knot	*Meloidogyne hapla*	1	Use corn or weed-free sod *(Poa pratensis)* 1 year between crops	60
	Root rot	*Rhizoctonia solani*	3—6	Use alfalfa before potato (Neb. irrigated field)	73
	Scab	*Streptomyces scabies*	1+	Soybean as green manure before potato (Calif.)	74
	Wilt	*Fusarium oxysporum*	3—6	Alfalfa before potato (Neb. irrigated fields)	73
		Verticillium albo-atrum	2—3	Grain-clover (1—2 years) potato (in some locations, red clover is a symptomless carrier and increases wilt of potato in a rotation)	22

Sorghum	Root knot	*Meloidogyne naasi*	1	Rotate with root crops	28
	Stalk rot	*Fusarium moniliforme*	2	Winter wheat-grain sorghum-fallow-with no tillage	75
Soybean	Brown spot	*Septoria glycinea*	1	Use any nonsusceptible crop	24
	Brown stem rot	*Cephalosporium gregatum*	5	5 years corn before soybean or corn-soybean-oats-clover	76
	Downy mildew	*Peronospora manshurica*	1	Use nonhost crops	24
	Cyst nematode	*Heterodera glycines*	1—5	Use nonhost crops	24,77
	Leaf spot	*Cercospora sojina*	1	Use nonhost crops	24
		Phyllosticta sojaeicola	1	Rotation and plow residues	24
		Pseudomonas glycinea	1	Rotate and plow residues	24
Spinach	Downy mildew	*Peronospora spinaciae*	3		41
Squash	Root and stem rot	*Fusarium solani*	2—3		78
Sweet potato	Black rot	*Ceratocystis fimbriata*	2+	Use any other crop	41
Tobacco	Black shank	*Phytophthora parasitica*	4	Rotate with cotton, lespedeza, peanut, soybean, oats, rye, or wheat, but not legumes before tobacco	23
	Black root rot	*Thielaviopsis basicola*	5	Many crops in rotation but use small grain just before tobacco	23
	Granville wilt	*Pseudomonas solanacearum*	3—5	Rotate with corn, cotton, cowpea, soybean, velvet bean, red top, or small grains	23
	Root knot	*Heterodera marioni*	2—3	Corn-oats-tobacco; cotton-peanut-tobacco; peanut-oats-tobacco	23
Tomato	Anthracnose	*Colletotrichum phomoides*	2	Rotate only if susceptible crops not nearby	41
	Bacterial canker	*Corynebacterium michiganense*	1	Rotate with other crops and control solanaceous weeds	41
	Bacterial speck	*Pseudomonas tomato*	1		
	Bacterial spot	*Xanthomonas vesicatoria*	1		
	Blight and fruit rot	*Helminthosporium lycopersici*	1	Rotate and plow diseased residues	41
	Septoria blight	*Septoria lycopersici*	4	Rotate other crops and control horse nettle	41
Tomato	Soil rot	*Rhizoctonia solani*	1	Rotate with pangola grass (Florida)	79
Vegetables	Nematodes	*Pratylenchus* spp.	1	Vegetable crops after oats and peanuts then after corn or lupines	65 41

Table 1 (continued)
CROPS AND DISEASES IN WHICH CROP ROTATION HAS HAD A MAJOR EFFECT IN DISEASE CONTROL

Crop	Disease	Pathogen	Crop interval (year)	Commentary	Ref.
Wheat, spring	Common root rot	*Helminthosporium sativum* *Fusarium roseum*	4	Use noncereal crops (Sask.)	80
	Eyespot	*Cercosporella herpotrichoides*	2—3	Rotate other crops but not barley or wheat (England)	81
	Flag smut	*Urocystis tritici*	1	Any crop but susceptible wheat	82
	Foot rot	*Helminthosporium sativum*	1	Use nongrass crops (e.g., legumes)	82
	Root gall nematode	*Subanguina radicicola*	1	Use legumes and root crops	28
	Seed gall	*Anguina tritici*	1—2	Rotate with noncereals	15
	Scab	*Fusarium roseum*	1	Where scab severe, 1 year between cereals and plow in residues	28
	Septoria leaf and glume blotch	*Septoria avenae, S. nodorum, S. tritici*	2	Rotate with noncereals	28
Wheat, winter	Cephalosporium stripe	*Cephalosporium gramineum*	1—2	Rotate with corn or legumes	83,84
	Snow mold	*Fusarium nivale, Sclerotinia borealis, Typhula* spp.	1	Rotate with spring cereals or legumes	28
	Take-all	*Gaeumannomyces graminis*	2—3	Rotate with noncereals (Kansas)	83

Table 2
CROPS AND DISEASES IN WHICH CROP RESIDUE MANAGEMENT HAS HAD A MAJOR EFFECT IN DISEASE CONTROL

Crop	Disease	Pathogen	Recommendation	Ref.
Banana	Wilt	*Fusarium oxysporum*	Plant prior crop sugarcane and plow in residues	47
Bean, snap	Root rot	*Thielaviopsis basicola*	Plow in alfalfa and cabbage residues, or castor, peanut, safflower, or soybean meals (5000 ppm)	85
Beet, table	Powdery mildew	*Erysiphe polygoni*	Plow under bean refuse	41
	Bacterial blights and leaf spots	*Xanthomonas beticola* and other bacteria	Plow in or destroy beet refuse	41
	Leaf spot	*Cercospora beticola*	Plow in refuse	41
Bentgrass	Nematode	*Anguina agrostis*	Plow under bentgrass sod at least 15 cm with complete sod turnover	59
Cotton	Root rot	*Phymatotrichum omnivorum*	Green manure before cotton; spring: cereals; winter: cowpea, guar, sesbania, sweet clover with other control methods (e.g., 4-year rotation)	67
	Seedling blight	*Ascochyta gossypii*	Plow down residues after harvest	68
Corn	Anthracnose	*Colletotrichum graminicola*	Plow down corn refuse to prevent early infection	86
	Gray leaf spot	*Cercospora zeae-maydis*	Plow under infected leaves to delay early onset of disease	37
	Southern corn leaf blight	*Helminthosporium maydis*	Inoculum on crop refuse reduced by chopping residues, harrow and plow	87
	Stalk rot	*Fusarium moniliforme, F. roseum*	Leave corn residues on soil surface	88,89
	Yellow leaf blight	*Phyllosticta maydis*	Clean plow to bury debris to delay early onset of disease	37
Peanut	Stem rot	*Sclerotium rolfsii*	Deep plow crop refuse with "nondirting" procedure	48
Sorghum	Downy mildew	*Sclerospora sorghi*	Bury surface soil and debris to 25—30 cm	90
Soybean	Anthracnose	*Colletotrichum dematium Glomerella glycines*	Plow down residues after harvest	24
Wheat	Seedling blight	*Rhizoctonia solani*	If wheat after corn, bury corn residue by plowing	90
Wheat, winter	Cephalosporium stripe	*Cephalosporium gramineum*	Deep plow stubble, bury below 8 cm	28,91
	Take-all	*Gaeumannomyces graminis*	Drill wheat directly in stubble and leave residues above ground; early rotavation between successive wheat crops	48,92

basicola (Table 2). The amendment of live plants in the form of green manure is effective with certain diseases, such as cotton root rot; here, it makes a difference if the green manure crop is plowed down in the spring or winter (Table 2). Even during spring, leaf diseases of corn that appear early may be eliminated by plowing under residues, but inoculum from other sources may cause infection later in the season; thus, plowing residues only protects young plants.

Depth of burial sometimes is important in control. Burial to at least 8 cm is necessary to control *Cephalosporium* stripe of wheat, to at least 15 cm for the bentgrass nematode, and to at least 25 to 30 cm for sorghum downy mildew. In contrast, allowing residues to deteriorate at the soil surface may be better than burial, e.g., in stalk rot of corn, caused by *Fusarium moniliforme,* there is less disease by leaving the chopped residues to decay at the soil surface. Moreover, the infected residues at the soil surface are out of reach of roots of the succeeding crop of corn. Similarly, the direct drilling of winter wheat into the stubble and residues of the previous crop of wheat, in which tillers were infected with the take-all fungus, meant that residues were out of reach of wheat roots of the next wheat crop.[48]

Conservation tillage, or ecofallow, decreased incidence of stalk rot in grain sorghum grown in a winter wheat-grain sorghum-fallow rotation. The decrease ranged from 39% in conventionally tilled plots, to 23% under minimum tillage, and to 11% under no tillage (Table 1). By planting one crop directly into the residue of another type of crop, the residues left on the soil surface reduce stress on sorghum, by increasing water conservation, reducing fluctuations in soil temperature, and lowering mean temperature, thereby reducing incidence of stalk rot. Thus, the rotation and the management of residues play important roles in reducing disease incidence of sorghum.

Earlier, it had been common practice to use a "dirting" procedure for peanut crops, in which soil and debris would be heaped around peanut plants in the row to control weeds. This had the effect of heaping inoculum of *Sclerotium rolfsii* onto the stem bases of peanuts and providing conditions favorable for infection. By using a "nondirting" method and plowing down crop refuse to at least 10 cm, the inoculum is destroyed or will be out of reach of peanut stems.[49] Weed control by herbicides may be necessary with this "nondirting" procedure.

In conclusion, almost none of the disease control procedures of sanitation, crop rotation, or crop residue management are used exclusively as control measures. Seldom is a pathogen eradicated completely. Control is attained by a combination of control measures that often include breeding for resistance or application of fungicides, and is aimed at reducing pathogen populations to acceptable levels.

REFERENCES

1. **Meredith, D.S.,** Banana leaf spot disease (Sigatoka) caused by *Mycosphaerella musicola* Leach, *Phytopathol. Pap.,* No. 11, Commonwealth Mycological Institute, Kew, 1970.
2. **Bonde, R. and Schultz, E. S.,** Potato refuse piles as a factor in the dissemination of late blight, *Maine Agric. Exp. Stn. Bull.,* No. 416, 1943.
3. **Van Der Plank, J. E.,** *Plant Diseases: Epidemics and Control,* Academic Press, New York, 1963, 128.
4. **Nelson, P. E., Pennypacker, B. W., Toussoun, T.A., and Horst, R.K.,** Fusarium stub dieback of carnation, *Phytopathology,* 65, 575, 1975.
5. **Wahlenberg, W. G.,** *Longleaf Pine,* C. Lathrop Pack Forestry Foundation, Washington, D.C., 1946, 175.
6. **Irving, F. D. and French, D. W.,** Control by fire of dwarf mistletoe in black spruce, *J. For.,* 69, 28, 1971.

7. **Dopson, R. N., Jr.,** The eradication of citrus canker, *Plant Dis. Rep.,* 48, 30, 1964.
8. **Hardison, J. R.,** Fire and flame for plant disease control, *Annu. Rev. Phytopathol.,* 14, 355, 1976.
9. **Horner, C. E.,** Control of mint rust by propane gas flaming and contact herbicide, *Plant Dis. Rep.,* 49, 393, 1965.
10. **Zaumeyer, W. J. and Thomas, H. R.,** Bean diseases — how to control them, *U.S. Dep. Agric. Agric. Handb.,* 225, Washington, D.C., 1962.
11. **Dixon, J. G.,** *Diseases of Field Crops,* 2nd ed., McGraw-Hill, New York, 1956, chaps. 13, 14.
12. **Fischer, G. W. and Holton, C. S.,** *Biology and Control of the Smut Fungi,* Ronald, New York, 1957, 439.
13. **Stover, R. H.,** Fusarial wilt (Panama disease) of banana and other Musa species, *Phytopathol. Pap.* No. 4, Commonwealth Mycological Institute, Kew, 96, 1962.
14. **Klisiewicz, J. M.,** Effect of flooding and temperature on incidence and severity of safflower seedling rust and viability of *Puccinia carthami* teliospores, *Phytopathology,* 67, 787, 1977.
15. **Jenkins, W. R. and Taylor, D. P.,** Seed and leaf gall nematodes: Anguina, in *Plant Nematology,* Reinhold, New York, 1967, chap. 12.
16. **Pirone, P. P.,** *Diseases and Pests of Ornamental Plants,* 5th ed., John Wiley & Sons, New York, 1978.
17. **Stevens, R. B.,** Cultural practices in disease control, in *Plant Pathology: an Advanced Treatise,* Vol. 3, Horsfall, J. G. and Dimond, A. E., Eds., Academic Press, New York, 1960, chap. 10.
18. **Adam, A. V.,** An effective program for the control of banana mosaic, *Plant Dis. Rep.,* 46, 366, 1962.
19. **Thorold, C. A.,** *Diseases of Cocoa,* Clarendon Press, Oxford, 1975, chap. 7.
20. **Persons, T. D.,** Phony peach disease — a review of organized control from 1929 to 1951 and the effect of recent developments on future control programs, *Phytopathology,* 42, 286, 1952.
21. **List, G. M., Landblom, N., and Sisson, M. A.,** A study of records from the Colorado peach mosaic suppression program, *Colo. Agric. Exp. Stn. Bull.,* No. 59, 1956.
22. **O'Brien, M. J. and Rich, A. E.,** Potato diseases, Agric. Handbook, No. 474, U.S. Department of Agriculture, Washington, D.C., 1976.
23. **Lucas, G. B.,** *Diseases of Tobacco,* 3rd ed., Biological Consultants Association, Raleigh, N.C., 1975.
24. **Sinclair, J. B. and Shurtleff, M. C., Ed.,** *Compendium of Soybean Diseases,* American Phytopathological Society, St. Paul, 1975.
25. **Frederiksen, R. A. and Renfro, B. L.,** Global status of maize downy mildew, *Annu. Rev. Phytopathol.,* 5, 249, 1977.
26. **Stakman, E. C. and Harrar, J. G.,** *Principles of Plant Pathology,* Ronald, New York, 1957, 412.
27. **Stakman, E. C., Levine, M. N., Cotter, R. U., and Hines, L.,** Relation of barberry to the origin and persistence of physiologic forms of *Puccinia graminis,* J. Agric. Res., 48, 953, 1934.
28. **Wiese, M. N.,** *Compendium on Wheat Diseases,* American *Phytopathological* Society, St. Paul, 1977.
29. **Anderson, R. L.,** A summary of white pine blister rust research in the lake states, U.S. Dep. Agric. For. Serv. Gen. Tech. Rep. NC6, North Central Forestry Experiment Station, University of Minnesota, Minneapolis, 1973, 9.
30. **Curl, E. A.,** Control of plant diseases by crop rotation, *Bot. Rev.,* 29, 413, 1963.
31. **Berkeley, G. H.,** Root rots of certain non-cereal crops, *Bot. Rev.,* 10, 67, 1944.
32. **Glynne, M. D.,** Crop sequence in relation to soil-borne pathogens, in *Ecology of Soil-borne Plant Pathogens,* Baker, K. F. and Snyder, W. C., Eds., University of California Press, Berkeley, 1965, 423.
33. **Leighty, C. E.,** Crop rotation, *U.S. Dep. Agric. Agric. Yearb.,* 1938, 406.
34. **Nusbaum, C. J. and Ferris, H.,** The role of cropping systems in nematode population management, *Annu. Rev. Phytopathol.,* 11, 423, 1973.
35. **Garrett, S. D.,** *Pathogenic Root-infecting Fungi,* University Press, Cambridge, 1970
36. **Williams, L. E. and Schmitthenner, A. F.,** Effect of crop rotation on soil fungus populations, *Phytopathology,* 52, 241, 1962.
37. **Ullstrup, A. J.,** Corn diseases in the U.S. and their control, *USDA Agric. Handbook,* No. 199, 1978.
38. **Smith, S. N., Snyder, W. C., and Moynihan, F.,** Populations of *Fusarium oxysporum f. vasinfectum* in field soil in relation to cotton wilt, in *Proc. Conf. Beltwide Cotton Production Research,* National Cotton Council, Dallas, 1970, p. 69.
39. **Cook, R. J.,** Influence of oats on soil-borne populations of *Fusarium roseum* f. sp. *cerealis* 'Culmorum', *Phytopathology,* 58, 957, 1968.
40. **Kommedahl, T. and Young, H. C.,** Effect of host and soil substrate on the persistence of Fusarium and Rhizoctonia in soil, *Plant Dis. Rep.,* 40, 28, 1956.
41. **Chupp, C. and Sherf, A. F.,** *Vegetable Diseases and Their Control,* Ronald, New York, 1960.
42. **Bruehl, G. W. and Lai, P.,** Prior-colonization as a factor in the saprophytic survival of several fungi in wheat straw, *Phytopathology,* 56, 766, 1966.

43. **Cochrane, V. W.,** The role of plant residues in the etiology of root rot, *Phytopathology,* 38, 185, 1948.
44. **Patrick, Z. A., Toussoun, T. A., and Snyder, W. C.,** Phytotoxic substances in arable soils associated with decomposition of plant residues, *Phytopathology,* 53, 152, 1963.
45. **Kommedahl, T., Old, K. M., Ohman, J. H., and Ryan, E. W.,** Quackgrass and nitrogen effects on succeeding crops in the field, *Weed Sci.,* 18, 29, 1970.
46. **Rice, E. L.,** *Allelopathy,* Academic Press, New York, 1974.
47. **Sequeira, L.,** Influence of organic amendments on survival of *Fusarium oxysporum* f. *cubense* in the soil, *Phytopathology,* 52, 976, 1962.
48. **Brooks, D. H. and Dawson, M. G.,** Influence of direct-drilling of winter wheat on incidence of take-all and eyespot, *Ann. Appl. Biol.,* 61, 57, 1968.
49. **Garren, K. H.,** Control of *Sclerotium rolfsii* through cultural practices, *Phytopathology,* 51, 120, 1961.
50. **Graham, J. H., Frosheiser, F. I., Stuteville, D. L., and Erwin, D. C.,** *A Compendium of Alfalfa Diseases,* American Phytopathological Society, St. Paul, 1979.
51. **Lutey, R. W. and Fezer, K. D.,** The role of infected straw in the epiphytology of Septoria leaf blotch of barley, *Phytopathology,* 50, 910, 1960.
52. **Maloy, O. C. and Burkholder, W. H.,** Some effects of crop rotation on the Fusarium root rot of bean, *Phytopathology,* 49, 583, 1959.
53. **Manning, W. J. and Crossan, D. F.,** Field and greenhouse studies on the effects of plant amendments on Rhizoctonia hypocotyl rot of snapbean, *Plant Dis. Rep.,* 53, 227, 1969.
54. **Papavizas, G. C. and Ayers, W. A.,** Aphanomyces species and their root diseases in peas and sugarbeet, U.S. Dep. Agric. Tech. Bull., 1485, 1974.
55. **Buchholtz, W. F.,** Crop rotation and soil drainage effects on sugar beet tip rot and susceptibility of other crops to *Aphanomyces cochlioides, Phytopathology,* 34, 805, 1944.
56. **Deems, R. E. and Young, H. C.,** Black root of sugar beets as influenced by various cropping sequences and their associated mycofloras, *J. Am. Soc. Sugar Beet Technol.,* 9, 32, 1956.
57. **Coons, G. H.,** Some problems in growing sugar beets, *U.S. Dep. Agric. Agric. Yearb.,* 1953, 509.
58. **Thorne, G.,** Control of the sugar beet nematode, *U.S. Dep. Agric. Farmers' Bull.,* 2054, 1952.
59. **Courtney, W. D. and Howell, H. B.,** Investigations on the bentgrass nematode, *Anguina agrostis* (Steinbuch) 1799) Filipjev 1936, *Plant Dis. Rep.,* 36, 75, 1952.
60. **MacDonald, D. M.,** personal communication, 1975.
61. **Crittenden, H. W.,** Control of *Meloidogyne incognita acrita* by crop rotation, *Plant Dis. Rep.,* 40, 977, 1956.
62. **Lewis, G. D.,** Effects of temperature and crop rotation on the occurrence of brown spot of celery in southern New York, *Plant Dis. Rep.,* 43, 1079, 1959.
63. **Rich, A. E.,** The occurrence and control of *Paratylenchus hamatus* on celery in New Hampshire, *Plant Dis. Rep.,* 39, 307, 1955.
64. **Meagher, J. W. and Brown, R. H.,** Microplot experiments on the effect of plant hosts on populations of the cereal cyst nematode *(Heterodera avenae)* and on the subsequent yield of wheat, *Nematologica,* 20, 337, 1974.
65. **Good, J. M., Robertson, W. K., and Thompson, L. G., Jr.,** Effect of crop rotation on the populations of meadow nematode, *Pratylenchus leiocephalus,* in Norfold loamy fine sand, *Plant Dis. Rep.,* 38, 178, 1954.
66. **Ghaffar, A., Zentmyer, G. A., and Erwin, D. C.,** Effect of organic amendments on severity of Macrophomina root rot of cotton, *Phytopathology,* 59, 1267, 1969.
67. **Streets, R. B. and Bloss, H. E.,** *Phymatotrichum Root Rot,* Monogr. No. 8, American Phytopathological Society, St. Paul, 1973.
68. **Smith, A. L.,** Ascochyta seedling blight of cotton in Alabama in 1950, *Plant Dis. Rep.,* 34, 233, 1950.
69. **Keyworth, W. G.,** Verticillium wilt of the hop (*Humulus lupulus*), *Ann. Appl. Biol.,* 29, 346, 1942.
70. **Hagedorn, D. J.,** *Handbook of Pea Diseases,* Wisc. Agric. Exp. Stn., University of Wisconsin, Madison, 1976, A1167.
71. **Green, R. J., Jr.,** Control of Verticillium wilt of peppermint by crop rotation sequences, *Plant Dis. Rep.,* 51, 449, 1967.
72. **Maas, P. W. T.,** Soil fumigation and crop rotation to control spraing disease in potato, *Neth. J. Plant Pathol.,* 81, 183, 1975.
73. **Jensen, J. H. and Livingston, J. E.,** Potato diseases in Nebraska, *Neb. Agric. Exp. Stn. Bull.,* No 378, 1945.
74. **Weinhold, A. R., Oswald, J. W., Bowman, T., Bishop J., and Wright, D.,** Influence of green manures and crop rotation on common scab of potato, *Am. Potato J.,* 41, 265, 1964.

75. **Doupnik, B., Jr., Boosalis, M. G., Wicks, G., and Smika, D.,** Ecofallow reduces stalk rot in grain sorghum, *Phytopathology,* 65, 1021, 1975.
76. **Dunleavy, J. M. and Weber, C. R.,** Control of brown stem rot of soybeans with corn-soybean rotations, *Phytopathology,* 57, 114, 1967.
77. **Ross, J. P.,** Crop rotation effects on the soybean cyst nematode population and soybean yields, *Phytopathology,* 52, 815, 1962.
78. **Sumner, D. R.,** Etiology and control of root rot of summer squash in Georgia, *Plant Dis. Rep.,* 60, 923, 1976.
79. **Hayslip, N. C. and Stall, R. E.,** Severity of tomato soil-rot caused by *Rhizoctonia solani* as influenced by former pasture crops, *Plant Dis. Rep.,* 43, 818, 1959.
80. **Ledingham, R. J.,** Crop rotations and common root rot in wheat, *Can. J. Plant Sci.,* 41, 479, 1961.
81. **Macer, R. C. F.,** The survival of *Cercosporella herpotrichoides* Fron in wheat straw, *Ann. Appl. Biol.,* 49, 165, 1961.
82. **Atkins, I. M. and Futrell, M. C.,** Diseases of small grains in Texas, *Texas Agric. Exp. Stn. Bull.,* No. 921, 1958.
83. **Willis, W.,** Disease control, in Wheat Production Handbook, Kansas State University Cooperative Extension Service, Manhattan, C-529, 1976.
84. **Wiese, M. V. and Ravenscroft, A. V.,** *Cephalosporium gramineum* populations in soil under winter wheat cultivation, *Phytopathology,* 65, 1129, 1975.
85. **Papavizas, G. C., Lewis, J. A., and Adams, P. B.,** Survival of root-infecting fungi in soil. XIV. Effect of amendments and fungicides on bean root rot caused by *Thielaviopsis basicola, Plant Dis. Rep.,* 54, 114, 1970.
86. **Hooker, A. L.,** Corn anthracnose leaf blight and stalk rot, *Proc. Annu. Corn Sorghum Res. Conf.,* 31, 167, 1976.
87. **Sumner, D. R. and Littrell, R. H.,** Influence of tillage, planting date, inoculum survival, and mixed populations on epidemiology of Southern corn leaf blight, *Phytopathology,* 64, 168, 1974.
88. **Parker, D. T. and Burrows, W. C.,** Root and stalk rot in corn as affected by fertilizer and tillage treatment, *Agron. J.,* 51, 414, 1959.
89. **Simpson, W. R.,** Influence of cultural practices on the incidence of stalk rot of corn, *Plant Dis. Rep.,* 51, 540, 1967.
90. **Frederiksen, R. A., McComb, D. B., Tuleen, D., and Reyes, L.,** Cultural control of sorghum downy mildew. I. The effect of deep-plowing on the distribution of soil-borne inoculum and the incidence of downy mildew, *Texas Agric. Exp. Stn.,* PR-2950, 1971.
91. **Pool, R. A. F. and Sharp, E. L.,** Some environmental and cultural factors affecting Cephalosporium stripe of winter wheat, *Plant Dis. Rep.,* 53, 898, 1969.
92. **Scott, P. R.,** Control of survival of *Ophiobolus graminis* between consecutive crops of winter wheat, *Ann. Appl. Biol.,* 63, 37, 1969.

CONTROL OF PLANT DISEASES USING DISEASE-FREE STOCKS

Gaylord I. Mink

INTRODUCTION

Principle Involved

The principle is simple. Provide growers with a continuous supply of plant materials which are free from important plant pathogens. Most reputable seed companies, nurseries, etc. attempt this as a matter of course. Some have even developed extensive testing programs at their own expense to reduce or eliminate serious disease problems in specific crops. However, many disease problems are of such magnitude or complexity that the efforts of a few are insufficient to provide adequate control. In such cases various state and federal regulatory agencies, in cooperation with commodity interest groups and researchers, have developed formal or informal programs to certify plant materials to be free from specific pathogens, providing the materials are produced and tested in prescribed ways. Under most certification schemes growers support the activities through increased costs of materials even though participation may be entirely voluntary. The basic premise, of course, is that net returns from the use of certified materials will more than offset the initial cost increases.

Strategy of Control

The strategy of control through use of disease-free stocks is to dilute the effects of diseases through the supply of large quantities of healthy planting material. In many instances these materials will eventually become infected and can be regarded as expendable.[1] The essence of this approach to control lies in maintaining disease-free foundation materials which are protected from reinfection, leavig only the materials propagated from it exposed to the hazards of reinfection.

It should be pointed out here that use of disease-free materials to control plant diseases is not a new concept. Nor is it a concept suddenly made popular by the recent concerns about the effect of chemical control methods on the environment. The concept has been traced back at least to 1723[2] with active programs developed for potatoes during the early 1900s.[3] Through the years as more has been learned about various disease agents, their modes of spread and their ability to thrive despite various control efforts, the use of disease-free materials has become increasingly attractive to many growers.

Elements of Disease-Free Stock Programs

In order to develop successful certification programs, it is necessary to resolve technical, financial, and political problems on a crop-by-crop, area-by-area basis. Consequently, no two certification programs are alike. Yet all programs have in common variations of the following:

1. A set of published standards defining the purpose, specific terminology, and the protocol necessary for certification
2. Sources of plant material considered to be free from specified pathogens and true to horticultural or genetic type
3. A scheme for preserving the above materials or for developing new ones as the old become infected or obsolete
4. Procedures for increasing materials free from specific pathogens and distributing them on a commercial scale

5. A system to monitor the established standards
6. A method for financing all operations

Available Information

Only a limited number of crops may be involved in certification programs in any one area. However, there are a wide variety of programs in the U.S. and elsewhere which together account for most of the cultivated crops. Information about individual programs is usually available from the regulatory agency in charge. Occasionally programs are described in articles which appear in specialized journals.[4-8] However, summary information about certification programs is limited.

In 1960 a symposium, held at the joint meeting of the American Phytopathological Society and the American Society for Horticultural Science in Eugene, Oregon, on the subject of pathogen-free stock was thought to be the first ever on this subject.[2] Here various authors presented concepts and progress in obtaining, maintaining, and distributing pathogen-free stock of numerous crops.[9-13] Subsequently, Hollings[1] reviewed disease control through use of virus-free stocks. More recently, schemes involved with various fruit tree programs have been reviewed.[14-16]

Because of the worldwide diversity of programs which are designed to provide stocks free from various pathogens and the difficulties involved in assembling the relevant information, this presentation is limited to programs in the U.S. In March 1978, letters of inquiry were sent to each state department of agriculture and crop improvement association requesting current information on certification programs active at that time. The information below is based in part on the replies.

USAGE OF TERMS

Some general terms when used in reference to plant certification have become working jargon with implications not necessarily expressed in their dictionary definitions. It is necessary to define some of these here.

"Disease-free" or similar terms such as "virus-free", "bacteria-free", etc. are used as oral shorthand to mean that the designated material has been tested by means other than visual inspection and found to be free from specified pathogens within the limitations of the detection procedures. "Disease-free" is frequently used to mean pathogen-free even though these terms are not necessarily equivalent.[2] Since many certification programs are supported largely or entirely by industry participation, the use of practical terms whose meanings are generally understood is of more than just academic interest. Hollings,[1] in discussing disease control through use of virus-free stock, pointed out that alternatives to "virus-free" such as "virus-tested" or "free from the following viruses..." are cumbersome and lack the appeal of the magic word "free". Similar concessions to practicality can also be found written into various certification schemes where "clean" and "dirty" are used to identify "virus-free" and "virus-infected" plants, respectively. Here the term "pathogen-free" will be used in referring to concepts, and implies freedom from one or more specific pathogens rather than freedom from all organisms capable of producing disease in a particular plant.

"Stock" is most frequently used to designate vegetatively propagated plant materials, particularly perennial nursery plants. However, it is also used in reference to commercial seed lots. In this review, "stock" will be interpreted in its broadest sense in order to discuss certification programs for seed stocks as well as those for vegetatively propagated plants.

CONTROLLING DISEASES BY USE OF PATHOGEN-FREE MATERIALS

Purpose

Not all diseases can be controlled, or controlled to the same degree, by the use of pathogen-free materials. Whether or not a disease can be controlled through certification depends to a great extent upon how the pathogen is spread naturally. The idealized concept of disease control implies that efforts are made to eliminate pathogens completely from any given situation.[1] Practically, however, pathogen testing and certification accomplishes any one or more of the following:

1. Establish and keep individual plants or entire fields free from specific pathogens
2. Establish healthy plants before infection occurs and thereby reduce effects of the disease
3. Reduce the incidence of diseased plants to a level where it is no longer economically significant

Diseases Controllable

In Perennial Fruit Crops

In virtually all perennial fruit crops diseases occur for which the only known means of spread is either graft transmission or by perpetuation of diseased plant parts. Table 1 lists 111 such diseases affecting major fruit crops grown in the U.S. All the transmissible diseases are known or suspected to be caused by viruses or mycoplasma-like organisms. The nongraft-transmissible, graft-perpetuated diseases are thought to result from genetic abnormalities. Although lacking natural vectors, many of the diseases listed in Table 1 occur worldwide and some have caused serious economic problems. The fact that these diseases are entirely dependent on propagation techniques for their dispersal makes use of disease-free propagation materials an effective means of control.

In recent years certification programs have been established in many states (see a following section) which utilize inspection and virus testing procedures to develop sources of important varieties free from the diseases listed in Table 1. Since most fruit plantings are expected to remain in place for several years it is important to begin plantings with materials free from as many diseases as possible. Consequently, all programs designed to certify perennial fruit crops also develop varieties free from diseases caused by viruses or other organisms which are transmitted by vectors (Table 2). Even though reinfection can and often does occur after a few years, in many instances the effects of the disease are much less severe than if diseased plants were initially used.

In Vegetatively Propagated Ornamentals

Several fungal and bacterial diseases have been controlled in vegetatively propagated ornamental crops by commercial companies operating their own pathogen-testing programs.[10,11] Most of these diseases are caused by pathogens which invade their hosts systemically and consequently are not amenable to most chemical control methods.

For several years a "cultured-cutting" program was used with success to reduce the effects of systemic pathogens in the propagation of chrysanthemums, carnations and geraniums.[10] Basal segments of candidate cuttings were surface sterilized and slices of basal tissue plated on potato dextrose agar to detect fungi. Other slices were placed in nutrient broth for detecting bacteria. Still other slices were indexed for viruses. The tip of the cutting was stored or rooted in sterilized medium until test results were available. If all indexing results were negative, the rooted plant was retained as part of an indexed mother-block from which further propagations were made.

Table 1
DISEASES OF PERENNIAL FRUIT PLANTS WHICH ARE ONLY KNOWN TO BE SPREAD BY GRAFT TRANSMISSION (T) OR PERPETUATION OF DISEASED PLANT PARTS (P)

Crop	Disease	Graft	Distribution	Ref.
Strawberry	Leafroll	T	Eastern Canada	17
	Feather-leaf	T	U.S.	18
	Band mosaic	T	Hungary	19
	Necrotic shock	T	California, Oregon, B.C.	20
	Lethal decline	T	Washington, Oregon, B.C.	21
	Pallidosis	T	U.S.	22
	June yellows	P	Europe, North America	23
Black Currant	Yellows	T	England	24
	Infectious variegation	T	Europe	25
Red Currant	Yellow leaf spot	T	Europe	26
	Leaf malformation	T	Europe	27
	Budlessness	P	Europe	28
Raspberry	Yellows	T	Europe	29
	Streak	T	Eastern U.S.	30
	Necrotic shock	T	California	31
	Sterility	T	U.S.	32
	Fern leaf	T	Canada	33
	Olallic disease	P	California	34
Loganberry	Calico	T	California	35
	Degeneration	T	England	33
Blueberry	Red ringspot	T	Eastern U.S.	36
	Mosaic	T	Eastern U.S.	37
	Shoestring	T	Eastern U.S.	38
	Vein mosaic	T	Czechoslovakia	39
	Heather dwarf	T	Czechoslovakia	39
Grapevine	Yellow speckle	T	Worldwide	40
	Corky bark	T	Worldwide	41
	Leafroll	T	Worldwide	42
	Fleck	T	Worldwide	43
	Asteroid mosaic	T	California	44
	Legno riccio	T	Europe	45
	Shoot necrosis	T	Italy	46
	Spindle shoot	P	California	47
Almond	Bud failure	T	California	48
	Corky spot	P	California	49
	Noninfectious bud failure	P	California	50
Apricot	Pucker leaf	T	Utah	51
	Ring pox	T	Western U.S.	52
	Gum boil	P	California	53
Peach	Asteroid spot	T	Western U.S.	54
	Blotch	T	Worldwide	55
	Calico	T	Worldwide	56
	Mottle	T	Idaho	57
	Rosette	T	Southern U.S.	58
	Wart	T	Western U.S.	59
	Red suture	T	Eastern U.S.	60
	Weak peach	T	Georgia	61
	Variegation	P	Idaho	62
Plum	Crinkle leaf	P	Western U.S.	63
	Shot hole	P	Western U.S.	64
	Rusty blotch	P	California	65
Prune	Diamond canker	T	Western U.S.	66
	Constriction mosaic	P	California	67
	Sparse leaf	P	Washington	68

Table 1 (continued)
DISEASES OF PERENNIAL FRUIT PLANTS WHICH ARE ONLY KNOWN TO BE SPREAD BY GRAFT TRANSMISSION (T) OR PERPETUATION OF DISEASED PLANT PARTS (P)

Crop	Disease	Graft	Distribution	Ref.
	Crinkle leaf	P	Washington	69
	Leaf casting mottle	P	Western U.S.	70
	Pustule canker	P	California	71
Sour cherry	Green ring mottle	T	Worldwide	72
	Pink fruit	T	Washington	73
	Gummosis	T	Washington	74
Sweet cherry	Black canker	T	Washington, B.C.	75
	Freckle fruit	T	Oregon	76
	Rough bark	T	California	77
	Crinkle leaf	P	Worldwide	78
	Deep suture	P	Worldwide	78
	Variegation	P	Worldwide	78
Apples	Chlorotic leafspot	T	Worldwide	79
	Leaf pucker	T	Worldwide	80
	Rosette	T	Netherlands	81
	Chat fruit	T	Europe	82
	Dapple apple	T	North America, Japan	83,84
	Green crinkle	T	Worldwide	85
	Green dimple	T	Italy	86
	Ring spot	T	Australia, N.Z., Italy	87
	Rough skin	T	Worldwide	88
	Russet wart	T	England	89
	Star crack	T	Worldwide	90
	Scar skin	T	North America	91
	Flat limb	T	Worldwide	92
	Platycarpa dwarf	T	England	93
	Platycarpa scaly bark	T	Worldwide	94
	Stem pitting	T	Worldwide	95
	Stem grooving	T	Worldwide	96
	Leaf curl	P	Worldwide	97
	Narrow leaf	P	Worldwide	97
	Variegation	P	Worldwide	97
	Fruit cracking	P	England	97
Pear	Ring pattern	T	Worldwide	98
	Vein yellows	T	Worldwide	99
	Stony pit	T	Worldwide	100
	Blister canker	T	England	101
	Rough bark	T	England, Denmark	102
	Rosette	P	Europe	103
Quince	Sooty ringspot	T	Worldwide	104
	Stunt	T	Worldwide	105
	Yellow blotch	T	England	106
Citrus	Psorosis	T	Worldwide	107
	Blind pocket	T	Worldwide	107
	Crinkley leaf	T	Worldwide	107
	Infectious variegation	T	Worldwide	107
	Satsuma dwarf	T	Japan	107
	Concava gum	T	Worldwide	107
	Sweet mottle	T	California	107
	Cachexia-xyloporosis	T	Worldwide	107
	Cristacortio	T	Mediterranean countries	107
	Impietrature	T	Mediterranean countries	107
	Yellow vein	T	California	107
	Leaf curl	T	Brazil	107

Table 1 (continued)
DISEASES OF PERENNIAL FRUIT PLANTS WHICH ARE ONLY KNOWN TO BE SPREAD BY GRAFT TRANSMISSION (T) OR PERPETUATION OF DISEASED PLANT PARTS (P)

Crop	Disease	Graft	Distribution	Ref.
	Tatter leaf	T	U.S.	107
	Citrange stunt	T	U.S.	107
	Ringspot	T	Worldwide	107

Table 2
VIRUSES AND VIRUS-LIKE DISEASES OF PERENNIAL FRUIT CROPS TRANSMITTED BY VECTORS

Crop	Disease or virus	Vector	Distribution	Ref.
Strawberry	Mottle	Aphid	Worldwide	110
	Veinbanding	Aphid	North America	111
	Necrosis	Aphid	Germany	112
	Chlorotic fleck	Aphid	Louisiana	113
	Pseudo mild yellow edge	Aphid	California	114
	Mild yellow edge	Aphid	Worldwide	115
	Latent C	Aphid	North America	116
	Crinkle	Aphid	Worldwide	117
	Stunt	Aphid	Western U.S.	118
	Witches' broom	Aphid	North America	119
	Multiplier disease	Aphid	North America	119
	Aster yellows	Leafhopper	North America	120
	Bronze leaf wilt	Leafhopper	England	120
	Green petal	Leafhopper	Europe, North America	120
	Delphinium yellows	Leafhopper	England	120
	Raspberry ringspot virus	Nematode	Britain	121
	Tomato black-ring virus	Nematode	Britain	121
	Arabis mosaic virus	Nematode	Europe	121
	Strawberry latent ringspot virus	Nematode	Britain	121
	Tomato ringspot virus	Nematode	California	121
Gooseberry	Vein banding	Aphid	Europe	122
	Spoon leaf	Nematode	Netherlands	123
Black currant	Reversion	Mite	Worldwide	124
	Vein clearing	Aphid	Britain	125
	Green mottle	Aphid	Britain	126
	Yellow mottle	Nematode	England	127
Red currant	Vein banding	Aphid	Europe	128
	Green mottle	Aphid	Europe	129
	Spoon leaf	Nematode	Europe	130
	American currant mosaic	Nematode	U.S.	131
	Arabis mosaic virus	Nematode	Britain	132
	Strawberry ringspot virus	Nematode	Britain	132
Raspberry	Mosaic	Aphid	Worldwide	133
	Vein chlorosis	Aphid	Britain, Canada, N.Z.	134
	Leaf curl	Aphid	Eastern U.S.	135
	Cucumber mosaic virus	Aphid	Scotland	136
	Thimbleberry ringspot	Aphid	British Columbia	137
	Dwarf	Aphid	Western U.S.	138
	Stunt	Leafhopper	Europe	139
	Raspberry ringspot	Nematode	Britain, Netherlands	140
	Tomato black-ring virus	Nematode	Scotland	140

Table 2 (continued)
VIRUSES AND VIRUS-LIKE DISEASES OF PERENNIAL FRUIT CROPS TRANSMITTED BY VECTORS

Crop	Disease or virus	Vector	Distribution	Ref.
	Yellow dwarf	Nematode	Europe	140
	Strawberry latent ringspot virus	Nematode	Europe	140
	American red raspberry ringspot	Nematode	North America	140
	Apple chlorotic leafspot virus	Pollen	Europe	141
	Black raspberry latent virus	Pollen	Maryland	142
Blueberry	Stunt	Leafhopper	Eastern U.S.	143
	Witches' broom	Leafhopper	Europe	144
	Necrotic ringspot	Nematode	Eastern U.S.	145
Cranberry	False blossom	Leafhopper	North America	146
Grapevine	Pierce's disease	Leafhopper	U.S.	147
	Flavescence doree	Leafhopper	Europe, Eastern U.S.	148
	Fan leaf	Nematode	Worldwide	149
	Yellow mosaic	Nematode	Worldwide	150
	Vein banding	Nematode	Worldwide	151
	Arabis mosaic	Nematode	Europe	152
	Tomato black-ring virus	Nematode	Germany	153
	Hungarian chrome mosaic	Nematode	Hungary	154
	Yellow vein	Nematode	California	155
Peach	Mosaic	Mite	U.S., Mexico	156
	Rosette mosaic	Nematode	Michigan	157
	Stem pitting	Nematode	U.S.	158
	Yellow bud mosaic	Nematode	California, Oregon	159
	Yellows	Leafhopper	Eastern U.S.	160
	X disease	Leafhopper	U.S.	161
	Phony	Leafhopper	Eastern U.S.	162
	Prunus ringspot virus	Pollen	Worldwide	163
Plum	Sharka	Aphid	Europe	164
Sour cherry	Prunus ringspot virus	Pollen	Worldwide	163
Sweet cherry	Little cherry	Mites	Worldwide	165
	Mottle leaf	Mites	Western U.S.	166
	Stem pitting	Nematode	California	167
	Rasp leaf	Nematode	Europe, Western U.S.	168
	Twisted leaf	Unknown	Western U.S.	169
	Rusty mottle	Unknown	Western U.S., England	170
Apple	Flat apple	Nematode	Washington	171
	Union necrosis	Nematode	U.S.	172
	Proliferation	Leafhopper	Europe	173
Pear	Decline	Psyllid	Western U.S.	174
Citrus	Tristeza	Aphid	Worldwide	107
	Seedling yellows	Aphid	Worldwide	107
	Stem pitting	Aphid	Worldwide	107
	Hassaku dwarf	Aphid	Japan	107
	Vein enation	Aphid	Worldwide	107
	Exocortis	Tools	Worldwide	107
	Greening	Psyllids	Asia, Africa	107
	Stubborn	Leafhopper	Worldwide (except Australia)	107

Although this practice was efficient it raised some new problems. The mother-block type operation occasionally encouraged spread and development of organisms hitherto relatively unimportant.[11] One such disease was Fusarium stem rot of carnations incited by *F. roseum f. cerealis.* This fungus was often carried externally on cuttings which indexed negative by cultured-cutting technique, and developed explosively under conditions used to grow mother-block plants. This problem was controlled by use of "shoot tip" culture techniques in which shoot tips (0.2 to 1.0 mm long for carnations) surrounded by a few upper leaves were excised and transferred aseptically to a suitable nutrient medium for rooting. Eventually they were transplanted into sand or gravel watered with nutrient solution. Although this technique reduced problems incited by external pathogens neither it nor the cultured-cutting technique effectively reduced problems caused by many viruses.

In recent years, tissue culture has received considerable attention as a means of propagating pathogen-free ornamentals. Even though many viruses invade plants systemically, most can be eliminated by excising very small portions from the apical meristems.[108] However, in commercial operations such small tissues are difficult to keep alive. Therefore, large tips are often used, which results in a high incidence of some viruses.[109] Consequently it is necessary to index all tissue culture propagants for viruses before selecting individuals for vegetative propagation.

In Seed

Many fungal and bacterial pathogens, as well as a few viruses, are carried on or in seeds in such a way as to be vulnerable to various heat and chemical treatments.[175,176] For these it may be superfluous to attempt to produce pathogen-free seed, especially if the pathogens are likely to recontaminate seeds during the many handling operations between threshing and planting. Seed treatment prior to planting is usually sufficient. On the other hand, most viruses are carried in seeds in such a way as to be highly tolerant or immune to various seed treatment. Phatak[177] lists 85 viruses which are transmitted through seed of various cultivated plant species. For any of these, certification programs designed to produce virus-free seed could effectively reduce or eliminate seed stocks as an initial inoculum source.

METHODS OF OBTAINING PATHOGEN-FREE STOCKS

Seed

Although it is by no means certain that pathogens carried on or in seeds will always produce infected seedlings, they can and frequently do.[178] Seeds usually remain viable for much longer periods than do vegetative propagules. This prolongs the possibility of pathogen dissemination and facilitates spread over long distances. Consequently, efforts to reduce or eliminate pathogens from seed stocks are often more effective in controlling disease development than efforts made to control spread after planting. Elimination of pathogens from seeds can be accomplished by two approaches: exclusion or eradication.

Exclusion

An effective method for obtaining seed free from specific pathogens is to produce the seed crop in areas where the pathogen rarely, if ever, occurs. Commercial bean seed production has been concentrated in the arid areas of western U.S., mainly Idaho, for several years because low humidities in surface irrigated fields throughout the growing season present an unfavorable environment for development of seed-borne bacterial pathogens. In Idaho, the state assists nature through a mandatory certification program involving full field inspection as well as plant and seed testing procedures. By directive, the Director of the Idaho Department of Agriculture or his deputies are authorized to enter and inspect any and all bean fields in the state.[179] Verification of the presence of any one of five plant pathogenic bacteria: (Halo blight,

Pseudomonas medicaginis var. *phaseolicola;* common bean blight, *Xanthomonas phaseoli;* fuscous blight, *Xanthomonas phaseoli* var. *fuscans;* brown spot, strains of *Pseudomonas syringae* pathogenic to *Phaseolus* sp.; and bean bacterial wilt, *Corynebacterium flaceumfaciens)* requires that the entire field be destroyed. A similar program is now being considered for parts of eastern Oregon.[180]

For pathogens which are limited to soils, or those which depend upon soil inhabitants as vectors, it is possible to avoid infection of seed crops by planting in noninfested land or by fumigating prior to planting. These procedures are especially effective against seedborne, nematode-transmitted viruses.[181]

Eradication

The success of eradicative procedures depends largely upon the relationship between the seed and the pathogen. Pathogens can be transmitted by seed in three general ways.[178]

1. The pathogen remains independent of the seed but accompanies it. Examples are seeds of parasitic plants such as dodder[182] *(Cuscuta* spp.) or broomrape[183] *(Orobanche* spp.) mixed with seed crops. For these, various types of separatory procedures have been developed for removing the pathogenic plant seeds.[184]
2. The pathogen is carried on the exterior of the seed or in the seed coat. Examples are bacterial canker *(Corynebacterium michiganense)* of tomato,[185] safflower rust *(Puccinia carthami),*[186] and tobacco mosaic virus (TMV) on tomato seed.[176,187] Pathogens in this group can usually be eradicated by chemical (Table 3) or hot water treatment.[184] Specific treatments vary with seed species and the pathogen involved.
3. The pathogen is contained within the seed. Examples are *Rhizoctonia solani* in endosperm of pepper seed,[193] and lettuce mosaic virus in lettuce seed embryos.[194] Pathogens in this group are difficult, if not impossible, to eradicate from infected seed.

Vegetable Plants Grown from Seed

The principle of reducing initial inoculum levels by planting pathogen-free stocks is fundamental to commercial production of vegetable seedlings. States such as Georgia and North Carolina which certify vegetable seedlings for sale as transplants require that certified plants be produced from certified seed which has been hot water treated or produced in specified areas free from specific pathogens.[195-197] Seedling beds are inspected at least three times before transplants are taken and must be found to be free from a variety of diseases. In addition, workers in tomato plant beds are cautioned to wash their hands with soap and water prior to handling plants and to avoid the use of tobacco products while pulling plants to avoid introduction and spread of "mosaic".[195]

Vegetatively Propagated Plants or Plant Parts

Rapid vegetative progagation of plants invites epidemic spread of many pathogens, especially those which systemically invade the tissues used for increase. Fundamental to certification programs involving vegetatively propagated plants is the selection or development of pathogen-free materials which form the nucleus from which all subsequent propagants are made. Pathogen-free materials are derived by many of the same principles of escape or eradication that were discussed above, even though specific methods may vary. Furthermore, after pathogen-free propagation materials have been obtained they must be maintained under conditions which prevent or greatly reduce chances of reinfection.

Table 3
CHEMICALS EFFECTIVE IN ERADICATING PATHOGENS FROM SEEDS

Compound	Pathogen	Seed crop	Ref.
Benomyl	*Ascochyta pisi*	Pea	188
	Erisyphe graminis	Barley	189
	Aphelenchoides leesseyi	Rice	190
Carboxin	*Ustilago madis*	Wheat	191
Thiabendazole	*Tillitia controversa*	Wheat	192

Escape and Exclusion

Avoiding the Pathogen

Selection of plants grown in areas free from pathogen or its vectors — Although obviously worth initial consideration, it is seldom possible to select vegetatively propagated plants which are pathogen-free simply by collecting them from an area free of one disease. After decades or even centuries of subdivision and distribution most vegetatively propagated plants have accumulated one or more systemic pathogens which can affect horticultural performance even though the plants may not express overt disease symptoms. Among the most important of these are viruses.[1] Consequently, most certification programs for vegetatively propagated plants concentrate on developing virus-free plant materials by one or more of the procedures discussed later in this section.

Once pathogen-free plant materials have been obtained for use in private pathogen-free stock programs or in state regulated certification programs it is essential that these source materials be maintained pathogen-free. For vegetatively propagated ornamentals the nuclear stocks are usually maintained in fumigated greenhouses. All potato "seed" certification programs establish foundation farms isolated from important pathogens and vectors where foundation stocks are produced.[3] This concept also resulted in the establishment of the IR-2 Interregional Deciduous Fruit Tree Repository where virus-free deciduous fruit trees are assembled, maintained, tested and distributed on a national basis.[15]

Management practices to avoid pathogens — Management practices such as timing plantings to avoid inoculum or vectors, use of barrier crops, or use of crop-free periods to break a disease cycle can effectively reduce inoculum or vector density. However, in practice these rarely result in complete pathogen control. As a result they are rarely relied upon for maintaining pathogen-free source material.

Selecting Pathogen-Free Plant Parts for Propagation

Naturally occurring pathogen-free propagules — Even though they may systemically invade plants, some pathogens occur erratically in at least some tissues. For example, apple buds from mosaic virus-[198] or chlorotic leafspot virus-[199] infected budwood can often produce healthy trees. Gladiolus cormels from mother cors infected with cucumber mosaic virus are usually free from that pathogen.[200] Cuttings of vegetatively propagated ornamentals that are free from several fungal and bacterial diseases have been developed and used by commercial flower growers.[10]

Meristem or tip propagation — Shoot tips from rapidly growing stems may escape infection by pathogens which are otherwise systemic.[1,10,11,201] With some viruses and viroids it is necessary to excise small portions of the apical meristem to insure elimination of the pathogen.[108,201,202] With other viruses, virus-free cuttings have been obtained after infected plants were exposed for prolonged periods to high tempera-

Table 4
CROP PLANTS FREED OF ONE OR MORE PATHOGENS BY MERISTEM OR TIP PROPAGATION WITH OR WITHOUT HEAT TREATMENT

Plant	Pathogen	Ref.
Alfalfa	Alfalfa mosaic virus	204
Apple	Chlorotic leafspot	205
	Rubbery wood mycoplasma-like agent	205
Asparagus	Several asparagus viruses	202
Carnation	Several viruses	206,207
	Fungi and bacteria	10, 11
Chrysanthemum	Chrysanthemum stunt viroid	208
	Several viruses	207
	Fungi and bacteria	10,11
Citrus	Exocortis viroid	209
	Citrus viruses	209
	Citrus stubborn spiroplasma	209
Cymbidium	Cymbidium mosaic virus	210
Dahlia	Dahlia mosaic virus	207
Dutch iris	Dutch iris mosaic virus	211
Fig	Fig mosaic virus	212
Freesia	Unknown virus	213
Garlic	Garlic mosaic virus	207
Geranium	Tomato ringspot virus	214
	Fungi and bacteria	10,11
Gooseberry	Vein banding virus	215
Hop	Prunus necrotic ringspot virus	215
Horseradish	Turnip mosaic virus	207
Iris	Iris mosaic virus	207
Lily	Cucumber mosaic virus	207
Narcissus	Arabis mosaic virus	1
Pear	Pear vein yellows	216
Petunia	Tobacco mosaic virus	207
Poplar	Poplar mosaic virus	217
Potato	Spindle tuber viroid	218
	Several viruses	207
Rhubarb	Several viruses	219
Rose	Rose mosaic virus	220
Rubus	Raspberry mosaic virus (heat-labile components)	221
Strawberry	Several viruses	222
Sugarcane	Sugarcane mosaic virus	207
Sweet clover	Bean yellow moasic virus	223
Sweet potato	Several viruses	207
Sweet william	Several viruses	206
Tomato	Tomato spotted wilt virus	224

tures.[1,203] Table 4 lists numerous crop plants which have been successfully freed from a variety of pathogens by various tip propagation techniques.

Nucellar embryony — Varieties of *Citrus, Fortunella,* and *Poncirus* spp. produce two distinct types of embryos: (1) gametic embryos which are produced sexually from egg cells fertilized by sperm cells, and (2) nucellar embryos which are produced asexually from somatic cells of the nucellus of the seed parent. Most citrus species produce nucellar seedlings which, because of their asexual origin, conform closely to the parent tree but are generally more vigorous than the parent tree.[225] With the possible exception of xyloporosis in sweet lime, most citrus viruses are not transmitted through the seed in significant amounts. Use of nucellar seedlings has effectively freed some citrus varieties of important virus problems. However, some objections have been raised to the use of nucellar citrus because of their many juvenile characteristics. In most cases these objections have not been considered to be too serious.[225]

Eradication

Chemotherapy

There are two approaches to chemotherapy: protective or curative.[226] Only the latter is considered here. Although benomyl has been used successfully to eradicate several bulb rot fungi[227] and Triarimol to eraicate stripe smut from bluegrass sod,[228] eradicative chemotherapy is not widely used to develop pathogen-free stocks.

Hollings[1] reviewed numerous reports of substances that presumably cure virus-infected plants. However, none of the reports have been confirmed and no such treatments are currently used.

Heat Therapy

Heat treatment has been the most successful and widely used therapeutic method for developing pathogen-free, vegetatively propagated plants. In practice, these procedures have been employed most widely for developing virus-free nuclear plants for the production of foundation stocks in virus certification programs. There are two distinct types of heat treatment based on physiological interactions of host and pathogen:

Treatment of dormant plant materials — Dormant plant tissues are exposed to hot water (30 to 54°C) or hot air (35 to 40°C) treatment for relatively short periods (several minutes to a few hours). Cures appear to depend upon heat destruction of the pathogen and require that the plant material withstand higher temperatures for longer periods than the pathogen. Such treatments have effectively eliminated internal fungal and bacterial pathogens as well as a few viruses. Nyland and Goheen[203] separated over 100 "viruses" into four groups based on ease of heat treatment. Most of the "viruses" with short inactivation times are now known to be mycoplasma-like organisms. However, at least one virus (potato leaf roll virus) was reported to be inactivated by treating potato tubers 17 min at 50°C.[229]

Treatment of actively growing plants — Plants are exposed to constant 37°C or to temperatures fluctuating between 35 and 43°C for weeks or even months[203] with or without 1 to 2 weeks preconditioning at somewhat lower temperatures. In almost all cases meristem culture or the grafting of tips or axillary buds to pathogen-free rootstocks was used to produce pathogen-free plants. Although such "cures" are frequently reported as due to virus inactivation in the treated tissue it is likely that a majority of these result from failure of the virus to reproduce in or to invade the meristem tissues produced during exposure to high temperatures. Three of the "viruses" listed by Nyland and Goheen[203] as not inactivated by heat treatment in vivo are now known to be viroids.

Cold Therapy

White clover plants have been free of bean yellow mosaic virus[223] after exposure to 10°C for 6 weeks and sweet clover freed of clover wound tuber virus[230] after 8 weeks exposure to 14°C. Work in progress suggests that low winter temperatures of the Pacific Northwest are sufficient to eradicate the causal bacterial from grapevines infected with Pierce's disease.[231]

TYPES OF CERTIFICATION PROGRAMS IN THE U.S.

Official and unofficial certification programs in the U.S. can be grouped into five broad categories: seed certification, certification of vegetatively propagated ornamentals and grasses, certification of vegetable plants and plant parts, certification of small fruits and grapevines, certification of fruit trees. Although regulations and procedures vary greatly from crop to crop and from state to state, all certification programs are designed to produce and distribute superior plant materials while distributing the fewest possible disease problems. Most states include statements in their regulations acknowledging the need to distribute materials free from the major pathogens. However, only a few programs include pathogen testing as an integral part of the requirements for certification. Some reasons for this will be discussed later.

Seed Certification

State seed certification programs may be conducted by one or more of the following agencies: a branch of the state department of agriculture or its equivalent, a crop or seed improvement association, and the agriculture research branch of the state land grant university (Table 5). Most of the state agencies belong to the Association of Official Seed Certifying Agencies (AOSCA) which attempts to coordinate certification efforts by establishing minimum requirements for the member agencies. The principle objective of each state program is to make available to the public high quality seed which is grown and distributed under conditions to insure genetic identity and purity with a minimum of seed-borne diseases.[232] While the primary consideration of AOSCA is genetic purity and identity, this organization does provide some standards involving diseases for the guidance of member agencies.

Each member state of AOSCA requires evidence that reasonable precautions have been taken to control seed-borne diseases in fields for which seed certification will be requested. To meet certification requirements fields must not contain more than the established tolerances of seed-borne diseases when visually inspected at the appropriate times during the growing season. Tolerance levels depend upon the seriousness of the disease.

For all certified seed crops four classes of seed are recognized: breeder seed, foundation seed, registered seed and certified seed. Classes are defined on the basis of origin and purpose as follows:[195]

1. Breeder seed — Seed directly controlled by the originating, or in certain cases the sponsoring, plant breeder or institution. This seed provides the source for foundation seed.
2. Foundation seed — The progeny of breeder or foundation seed so handled as to most nearly maintain specific genetic identity and purity. Production must be carefully supervised and approved by a certifying agency.
3. Registered seed — The progeny of breeder or foundation seed so handled as to maintain satisfactory genetic identity and purity and that has been approved and certified by a certifying agency.
4. Certified seed — The progeny of breeder, foundation or registered seed so handled as to maintain satisfactory genetic identity and purity and that has been approved and certified by a certifying agency.

Although numerous seed crops are certified by many states (Table 6), none of the existing seed certification programs utilize pathogen testing as a requirement for certification. On the other hand, the California citrus certification program[233] and the deciduous fruit tree certification programs of California,[234] Washington,[235] and New York[236] include testing of rootstock propagants for seed-borne viruses as an integral part of their requirements. In addition there are industry-developed programs for indexing lettuce seed for lettuce mosaic virus,[237] commercial pea seed lots for pea seed-borne mosaic virus,[238] and barley seed for barley strip mosaic virus.[239]

Table 5
AGENCIES RESPONSIBLE FOR SEED CERTIFICATION PROGRAMS IN THE VARIOUS STATES

State	Certification agency
AL	Crop Improvement Association
AK[a]	Crop Improvement Association
AZ	Crop Improvement Association
AR	State Plant Board
CA	Crop Improvement Association
CO	Seed Grower's Association
CN[a]	None
DE	Crop Improvement Association
FL[a]	Department of Agriculture and Consumer Service
GA	Crop Improvement Association
HA[a]	Department of Agriculture
ID	Crop Improvement Association
IL	Crop Improvement Association
IN	Crop Improvement Association
IA	Crop Improvement Association
KS	Crop Improvement Association
KY	Seed Improvement Association
LA	Department of Agriculture
ME[a]	Seed Potato Board
MD[a]	State Board of Agriculture
MA[a]	State Seed Laboratory and Agriculture Experiment Station, University of Massachusetts
MI	Crop Improvement Association
MN	Crop Improvement Association
MS	Seed Improvement Association
MO	Seed Improvement Association
MT	Seed Grower's Association
NE	Crop Improvement Association
NV	Department of Agriculture
NH[a]	None
NJ	Department of Agriculture
NM	Crop Improvement Association
NY	Seed Improvement Coop., Inc.
NC	Crop Improvement Association
ND	State Seed Department
OH	Seed Improvement Association
OK	Crop Improvement Association
OR	Seed Certification Service, Oregon State University
PA	Department of Agriculture
RI[a]	None
SD	Seed Certification Service, South Dakota State University
TN	Crop Improvement Association
VT	Department of Agriculture
WA	Department of Agriculture and Crop Improvement Association
WV[a]	Associated Crop Growers
WI	Crop Improvement Association
WY	Seed Certification, University of Wyoming

[a] Not a member of AOSCA.

Certification of Vegetatively Propagated Ornamentals and Grasses

Vegetatively Propagated Ornamentals

Most vegetatively propagated ornamentals (carnations, chrysanthemums, geraniums, gladiolus, lilies, poinsettias, etc.) are produced by a relatively few large-scale

Table 6
LIST OF SEED CROPS CERTIFIED IN VARIOUS STATES

Crop	States certifying
Asparagus	NJ
Avocado	CA
Beans	CA, CO, ID, MI, MN, NE, NM, NY, ND, VT, WA, WY
Buckwheat	TX
Corn	AL, CA, CO, GA, IL, IN, IA, KY, LA, MI, MN, MO, NE, NJ, MN, NY, NC, SC, SD, TN, TX, VA, WI
Cotton	AL, AZ, AR, CA, GA, LA, MA, MO, NM, NC, OK, SC, TN, TX
Cowpeas	AL, CA, GA, LA, MS, SC, TX, VA
Flax	CA, IA, MN, MT, ND, SD, TX
Forbs (wildflowers)	KS, NE, NM
Grasses	AK, AZ, CA, CO, GA, ID, IA, KS, LA, MN, MS, MO, MT, NE, NJ, ND, OK, OR, PA, SC, SD, TN, TX, VT, VA, WA, WY
Guar	TX
Lentils	WA
Millet	AL, AZ, CO, KS, LA, MN, MO, NE, NM, ND, OK, SD, TX
Mungbean	OK
Okra	AL, GA, LA, NM, SC, TX
Onion	LA, NM
Peanuts	AL, CA, GA, NM, NC, OK, SC, TX, VA
Peas	CO, ID, MN, NJ, NM, ND, WA
Pepper	NM
Rice	AR, CA, LA, MS, MO
Safflower	CA, MT
Sesame	CA
Shallot	LA
Small grains (wheat, oats, barley, rye)	AK, AZ, CA, CO, GA, IL, ID, IN, IA, KS, KY, LA, MI, MN, MS, MO, MT, NE, NV, NJ, NM, NY, NC, ND, OH, OK, OR, PA, SD, TX, UT, VT, VA, WA, WI, WY
Small seeded legumes (alfalfa, clovers, lespedeza, etc.)	AL, AK, CA, CO, GA, IL, ID, IN, IA, KS, KY, LA, MI, MN, MS, MO, MT, NE, NV, NJ, NM, NY, NC, ND, OK, OR, SD, TN, TX, UT, VT, VA, WA, WI, WY
Sorghum (broomcorn, sudangrass)	AL, CA, CO, GA, IL, IA, KS, KY, LA, MS, MO, NE, NM, OK, SD, TX, WA
Soybeans	AL, AR, GA, IL, IN, KS, KY, LA, MI, MN, MS, MO, NE, NJ, NY, NC, ND, OH, OK, SC, SD, TN, TX, VT, VA, WA, WI
Squash	NY
Sunflower	CA, LA, MN, ND
Tobacco	GA, KY, NC, SC, TN, VA, WI
Tomatoes	NJ, OK
Trees and shrubs	GA, LA, MS, NM, SD, WA
Triticale	MT, NE, SD
Watermelon	FL, OK, TX

propagators, most of whom maintain their own pathogen control programs.[10] Consequently, there has been little need or desire for state-regulated certification programs.[109] However, Pennsylvania implemented a program in 1978 to certify geraniums for freedom from virus diseases.[240]

Vegetatively Propagated Grasses

Several states including Louisiana, Mississippi, Nebraska, New Jersey, North Carolina, Oklahoma, Pennsylvania, South Carolina, Texas, Vermont, and Virginia certify vegetatively propagated grasses or turfgrass sod. In each case certification is based on visual observation that the material is reasonably free of diseases, nematodes and other pests.

Certification of Vegetable Plants and Plant Parts

Vegetable Plants

Vegetable plants for use as transplants are eligible for certification in Georgia and North Carolina providing the following conditions are met.

1. In Georgia all cabbage seed lots must be assayed for black rot *(Xanthomonas campestris)* and black leg *(Phoma lingum)* by both the Georgia Department of Agriculture and the Plant Pathology Section, Georgia Agricultural Experiment Station, Experiment, GA, and meet tolerance standards established for both pathogens.[241] Pepper and tomato seed must be disinfected and tested for fungi and bacteria by the Georgia Department of Agriculture[196,197] In North Carolina all seed must be hot water treated or be produced in pathogen-free areas of the western U.S.[195]
2. All fields and plant beds must be inspected at regular intervals for diseases, insect infestations, nematode infestations, and general physical condition of plants.
3. Plants must meet variety and grade requirements.

Seed Potatoes

The earliest and the most extensive efforts to control plant diseases through certification have been with potatoes.[3] Seed potato certification began in Europe in the early 1900s but did not become established in the U.S. until around 1914. Currently 20 states have seed potato certification programs (Table 7).

In the case of potatoes most certification standards relate to disease, and with the exception of giant hill, they do not include such factors as genetic or horticultural differences within a given variety.[3] Although specific procedures vary among states (Table 7) all states require at least two visual field inspections during the growing season; the first as early as possible to recognize diseases, the second between bloom and vine maturity. During either inspection there is zero allowable tolerance for bacterial ring rot *(Corynebacterium sepedonicum).* Tolerance levels for other pathogens vary among states and tolerance differentials between the first and last inspection permit growers to rogue out infected plants between inspections. Following harvest an additional inspection is made of the tubers to detect bacterial ring rot, tuber net necrosis, certain tuber rots, physiological diseases, and occasionally nematodes.

Bacterial, fungal, mycoplasmal, and many viral diseases are nearly always diagnosed on the basis of visible field symptoms. Because symptom expression can be markedly influenced by environment or variety, certification personnel need extensive field experience before they can become familiar with wide variations in symptoms induced by various pathogens. At times laboratory tests can be used to aid uncertain field diagnosis. For example, three types of tests can be made to verify the presence of the ring rot organism.[3]

Some viruses common in potatoes produce few if any foliar or tuber symptoms. They are referred to as the latent virus group (potato virus X, potato virus S, and potato virus M). Diagnosis of latent virus infections requires laboratory procedures. Test-plant inoculation or electron microscope techniques may be useful for initial screening of plants. However, serological techniques must be used eventually for more accurate identification.[3]

At least eight states have complex systems for producing tubers as free from pathogens as is possible and for increasing this material through a system of foundation block and registered plantings (Table 7). Pathogen-free tubers are increased initially in isolated foundation blocks which are rogued intensively. Subsequent increases are made in registered plantings with stringent disease tolerances. However, some infec-

Table 7
STATES HAVING SEED POTATO CERTIFICATION PROGRAMS

State	Certifying agency[a]	Foundation plots	Pathogen testing	Greenhouse or winter test	Visual inspection
Alabama	SDA	No	No	No	Yes
Alaska	CIA	No	No	No	Yes
California	SDA	Yes	Yes	Yes	Yes
Idaho	CIA	Yes	Yes	Yes	Yes
Louisiana	SDA	Yes	Yes	No	Yes
Maine	SDA	Yes	Yes	Yes	Yes
Mississippi	SDA	No	No	No	Yes
Montana	CIA	Yes	Yes	Yes	Yes
Nebraska	SDA	—[b]	—[b]	—[b]	Yes
Nevada	SDA	No	No	No	Yes
New Jersey	SDA	No	No	Yes	Yes
New Mexico	CIA	No	No	Yes	Yes
New York	CIA	Yes	Yes	Yes	Yes
North Carolina	CIA	No	No	No	Yes
North Dakota	SDA	No	No	Yes	Yes
Oregon	SAES	Yes	Yes	Yes	Yes
Pennsylvania	SDA	No	No	Yes	Yes
Washington	SDA	No	No	Yes	Yes
Wisconsin	SAES	Yes	Yes	Yes	Yes
Wyoming	SDA	No	No	No	Yes

[a] SDA = State Department of Agriculture, CIA = Crop Improvement Association, SAES = State Agriculture Experiment Station.

[b] Information not available.

tions may occur too late in the growing season to be detectable. Consequently all such programs require that a "winter index" be performed. Sample tubers are shipped to southern states such as Florida or California, planted in test plots around November and observed in January or February for disease incidence. In some cases similar evaluations may also be made in greenhouses during the off season.

Sweet Potatoes

For certification purposes "sweet potato" refers to any plant of the species *Ipomoea batatas* grown for its root tubers. Nine states have active certification programs on this crop with four of the nine supplying foundation seed stock for use in planting fields for which certification will be requested (Table 8). In all states certification is based entirely on visual inspection of plants growing in seed beds or production fields and of tubers during storage (Table 8).

Vegetatively Propagated, Nonwoody Specialty Crops

A few states certify vegetatively propagated, nonwoody crops which are either grown in limited geographic areas or are produced for special uses such as hops and mint (Table 9). In the case of asparagus crowns (New Jersey), Easter lily bulbs (Louisiana), mint rootstocks (Utah), and onion bulbs (Louisiana) pathogen testing is not a published requirement for certification. Plantings are certified on the basis of visual inspections only. On the other hand, state department of agriculture regulations require that hop rootstocks (Washington), mint rootstocks (Oregon, Washington), and seed garlic (California) be tested for important pathogens and insects before materials can be included in foundation plantings. Once plant materials are released for registered or certified plantings, only visual inspections are required for certification.

Table 8
STATES WITH CERTIFICATION PROGRAMS FOR SWEET POTATOES

State	Certifying agency[a]	Foundation program	Visual Inspection: Seed bed	Field	Storage
AL	SDA	No	Yes	Yes	Yes
GA	SDA	Yes	Yes	Yes	Yes
MS	SDA	No	Yes	Yes	Yes
MO	SDA	No	Yes	Yes	Yes
NM	CIA	Yes	Yes	Yes	Yes
NC	CIA	Yes	Yes	Yes	Yes
OK	SDA	No	Yes	Yes	Yes
SC	SAES	—[b]	—[b]	—[b]	—[b]
VA	CIA	Yes	Yes	Yes	Yes

[a] SDA = State Department of Agriculture, CIA = Crop Improvement Association, SAES = State Agriculture Experiment Station.
[b] No information.

Table 9
VEGETATIVELY PROPAGATED SPECIALTY CROPS WHICH ARE CERTIFIED IN VARIOUS STATES

Crop	State	Certifying agency[a]	Foundation program	Pathogen testing required	Visual inspection
Asparagus crowns	NJ	SDA	No	No	Yes
Easter lily bulbs	LA	SDA	No	No	Yes
Hop rootstocks	WA	SDA	Yes	Yes	Yes
Mint rootstocks	OR	SAES	Yes	Yes	Yes
	UT	SDA	No	No	Yes
	WA	SDA	Yes	Yes	Yes
Onion bulbs	LA	SDA	Yes	No	Yes
Seed garlic	CA	SDA	Yes	Yes	Yes

[a] SDA = State Department of Agriculture, SAES = State Agriculture Experiment Station.

Certification of Small Fruits and Grapevines

Strawberry

Twelve states have formal programs to certify strawberry plants (Table 10). Eleven of these provide virus-tested nuclear stock to establish foundation plantings. Although specific regulations and definitions may vary among states, the following scheme is representative of the certification program in many states.

Nuclear stocks of important varieties are produced by the state land grant university or state department of agriculture from plants which are indexed annually to verify freedom from viruses. These stocks are grown in screenhouses under conditions which will maintain the stocks at maximum freedom from diseases and pests.

Foundation stock is produced from nuclear stock only by growers who are approved by the certifying agency, for the grower's own use. The foundation blocks must be located at least ½ mi from any uncertified strawberry plantings and at least 100 ft from registered or certified plants on ground that has not been used to grow strawber-

Table 10
STATES CERTIFYING STRAWBERRY PLANTS

State	Certifying agency[a]	Nuclear stock virus-tested
AR	SDA	Yes
CA	SDA	Yes
DE	SDA	Yes
MD	SDA	Yes
MI	SDA	Yes
MO	SDA	Yes
NJ	SDA	Yes
NC	CIA	Yes
OH	SDA	No
OR	SDA	Yes
VT	SDA	Yes
WA	SDA	Yes

[a] SDA = State Department of Agriculture, CIA = Crop Improvement Association.

Table 11
STATES CERTIFYING CULTIVATED *RUBUS* SPECIES

State	Certifying agency[a]	Certified	Nuclear stock virus-tested	Foundation block: Under screen	Foundation block: Virus tested
CA	SDA	*Rubus* ssp.	Yes	Yes	Yes
MD	SDA	*Rubus* spp.	Yes	Yes	Yes
MI	SDA	*Rubus* spp.	Yes	Yes	Yes
MO	SDA	*Rubus* spp.	No	No	No
NJ	SDA	Blueberry	No	No	No
NY	SDA	*Rubus* spp.	Yes	No	No
OH	SDA	Raspberry	No	No	No
OR	SDA	*Rubus* spp.	Yes	Yes	Yes
VT	SDA	Raspberry	Yes	Yes	Yes
WA	SDA	*Rubus* spp.	Yes	No	No

[a] SDA = State Department of Agriculture.

ries within the past 5 years. Foundation blocks must be fumigated with methyl bromide before planting and sprayed for leaf spot diseases after planting. Visual inspections are made periodically for virus diseases, nematodes, and red stele disease. Foundation stocks are increased in registered plantings under regulations similar to those of foundation plantings except that registered plantings may be adjacent to certified plantings.

Certified plantings are made from registered stock and must be located at least 300 ft from uncertified strawberry plantings on ground where strawberries have not grown within the past 3 years. Visual inspections are similar to those given foundation plantings.

Caneberries

Cultivated *Rubus* sp. known collectively as caneberries are certified in ten states (Table 11). In seven of the ten states nuclear stock demonstrated by index testing to

Table 12
STATES CERTIFYING GRAPEVINES

State	Certifying agency[a]	Indexed for viruses: Parent or nuclear plants	Foundation plants	Mother plants
AR[b]	SDA	—	—	—
CA	SDA	Yes	No	No
MD	SDA	Yes	Yes	No
MI	SDA	Yes	No	No
NY	SDA	Yes	Yes	Maybe[c]
OR	SDA	Yes	No	No
WA	SDA	Yes	Yes	No

[a] SDA = State Department of Agriculture.
[b] Certification requirements still in draft form.
[c] At the discretion of the State Department of Agriculture.

be free from virus or virus-like diseases are grown in screenhouses and used to provide material for foundation plantings. In five of the ten states foundation plants are also grown in screenhouses and a percentage of the plants indexed each year for viruses. In the remainder foundation plants or the equivalent are examined visually.

Foundation stock is increased in registered blocks which provide materials for planting fields for which certification will be requested. Registered blocks and certified blocks are inspected visually for disease and pests.

Grapevine

Six states now have active programs to certify grapevines through their respective departments of agriculture (Table 12). In each of these states parental vines of important scion and rootstock varieties are indexed for known viruses, examined for trueness to horticultural type, and propagants are established in a foundation planting which is maintained by the certifying agency or a cooperator such as the state agricultural experiment station of the land grant university. In all states foundation plants are inspected two to three times each year. In Maryland each foundation vine is indexed for viruses each year while in New York and Washington foundation vines are reindexed every four to five years.

Propagation materials from the state foundation planting are used to establish mother blocks at each cooperating nursery. Propagants from these mother blocks which are collected, grown and inspected in a manner prescribed by each state are eligible for certification.

Certification of Fruit Trees

At the present time there are certification programs for avocado, citrus, and a variety of deciduous fruit crops (Table 13). In all of these programs parental trees are indexed for important viruses, mycoplasmas, spiroplasmas, and viroids, and are examined for graft-perpetuated genetic disorders before selection as foundation materials. Registered foundation trees provide propagation materials for mother blocks established in cooperating nurseries which in turn provide propagants for which certification is requested. In crops where seed-transmitted viruses occur, blocks of certified seed trees are established to provide virus-free seed for producing certified rootstocks.

Table 13
STATES CERTIFYING FRUIT TREES

Crop	State	Certifying agency[a]	Source trees virus indexed
Avocado	CA	SDA	Yes
Citrus	AZ	SDA	Yes
	CA	SDA	Yes
	FL	SDA	Yes
Deciduous fruit trees[b]	CA	SDA	Yes
	MD	SDA	Yes
	MI	SDA	Yes
	MN[c]	SDA	Yes
	MO	SDA	Yes
	NY	SDA	Yes
	OR	SDA	Yes
	PA	SDA	Yes
	TN	SDA	Yes
	WA	SDA	Yes

[a] SDA = State Department of Agriculture.
[b] *Malus, Prunus, Pyrus* species.
[c] Certification of *Prunus* species only.

POLITICS AND PROBLEMS INVOLVED WITH CERTIFICATION PROGRAMS

In many states the department of agriculture or its equivalent is empowered with the legal authority to develop and promulgate certification programs, but only after public hearings have been held to explain alternatives and to determine the needs and desires of interested growers and commodity groups. Once standards for certification are promulgated the department or its designated affiliate usually assumes total responsibility for administering the program and for whatever testing and inspection may be required. In recent years, however, a majority of states have found themselves financially unable to support pathogen testing even though this may be considered desirable by all interested parties. Consequently, it has become necessary for participants in nearly all certification programs to generate most, if not all, of the necessary funds by assessments, fees, or by taxing sales of the final product.

In some states a particular crop may have sufficient turnover of certified materials or may have strong financial support from a commodity group, or both, so as to generate sufficient revenue to support pathogen testing as part of the requirements for certification. The citrus registration and certification program in California is one such example. Even though citrus trees would not normally be considered a rapid turnover crop, diseases and other factors result in a continuing need for large numbers of certified citrus nursery trees.[16] After an initial investment by the University of California State Agriculture Experiment Station (SAES) and the U.S. Department of Agriculture in research and development together with interim support from various grower groups, the current program supports an extensive indexing program for important virus and virus-like diseases. With the widespread use of certified trees, diseases such as psorosis, severe exocortis, concave gum, crinkly leaf, and cochexia have practically disappeared from young citrus orchards.[241]

On a more limited scale the Washington fruit tree certification program receives a 1% annual assessment on the gross sale price of all fruit trees, fruit tree seedlings and fruit tree rootstocks sold within the state or shipped from the state by any licensed nursery dealer.[235] These funds cover administrative costs, maintenance of the foundation planting, and whatever virus testing is deemed appropriate.

By contrast, many certified crops are grown by a relatively few individuals with turnover volumes insufficient to generate enough money to equip and staff facilities needed for pathogen testing. Such programs are forced to rely heavily on the financial resources and expertise of research personnel from the SAES of the land grant university for whatever pathogen tests are necessary or to rely entirely upon visual inspection regardless of its real value.

The long tradition of reliance on SAES or USDA specialists to develop test procedures and then to do service testing for certification programs continues to aggravate the overall problem of developing adequate pathogen testing facilities in many states. The use of state and/or federal research funds to support routine pathogen testing for some state certification programs has discouraged or delayed development of private organizations capable of doing service testing on a broad range of pathogens. Local or regional service testing laboratories recognized by state regulatory agencies could play a useful role in upgrading and standardizing certification efforts on crops which are unable to support these efforts on present funds.

Finally, many crops are grown in one or a few states and then distributed nationwide. In some cases states with strict certification standards which include pathogen testing find it necessary to impose specific plant quarantines as an adjunct to their certification efforts. As a result, plants or seeds which may be certified in states without pathogen-testing requirements may not be eligible for sale in states with such programs. However, the use of state plant quarantines often raises a highly emotional issue among industry personnel. Occasionally, individuals or groups fear that imposing a quarantine will trigger retaliatory quarantines by other states. This issue has delayed or even prevented the development of effective certification programs in some states.

In at least one instance, establishment of a state plant quarantine resulted in the development of more uniform certification standards among several other states. Prior to 1970, only California used virus testing procedures in the certification of grapevines. Because of the apparent absence of several virus diseases within the state, Washington, in 1970, adopted a quarantine banning importation of all grapevines and grapevine propagants except those produced under certification schemes which included virus testing as a requirement for certification.[242] Within 5 years most states producing grapevine nursery stock established certification programs (Table 12) which could provide stocks eligible for sale in Washington. A contributing factor in the development of uniform testing procedures was the fact that each state (except Maryland) had at least one representative on the Western Regional Coordinating Committee for Grapevine Diseases and Pests (WRCC-24). This group meets annually to exchange information on research and regulatory efforts affecting grapevines in the various states.

Because states are highly individualistic and each state's certification programs are dependent upon industry for their support, it is not surprising that certification efforts in most states are far less inclusive than many which have been developed in other countries where certification is done at the federal level. With the increasing exchange of plant materials both interstate and internationally more uniform procedures for pathogen testing and certification could result in greater control of many plant diseases. However, such uniformity does not appear likely unless ways are found to finance local certification programs adequately.

REFERENCES

1. **Hollings, M.**, Disease control through virus-free stock, *Annu. Rev. Phytopathol.*, 3, 367, 1965.
2. **Wilhelm, S.**, Symposium on pathogen-free stock: Introduction, *Phytopathology*, 52, 1234, 1962.
3. **Shepard, J. F. and Claflin, L. E.**, Critical analysis of the principles of seed potato certification, *Annu. Rev. Phytopathol.*, 13, 271, 1975.
4. **Dimock, A. W.**, Suggestions for the control of carnation diseases, *Carnation Craft*, 1, 5, 1948.
5. **Wagnon, H. K., Williams, H. E., and Millican, A. A.**, Virus free nursery stock, *Annu. Rep. Dept. Agric. Calif.*, 46, 177, 1957.
6. **Gaseorkiewicz, E. C.**, Development and maintenance of pathogen-free geraniums, *Flor. Exch. Hort. Trade World*, 131, 14, 1958.
7. **Mather, S. M.**, Nursery stock registration and certification in California, Bull. Dept. Agric. Calif., 50, 173, 1961.
8. **Posnette, A. F.**, The mother tree scheme, *East Malling Res. St. Maidstone, Engl. Rep.*, 1961, 125, 1962.
9. **Nyland, G. and Milbrath, J. A.**, Obtaining virus-free stock by indexing techniques, *Phytopathology*, 52, 1235, 1962.
10. **Dimock, A.**, Obtaining pathogen-free stock by cultured-cutting technique, *Phytopathology*, 52, 1239, 1962.
11. **Baker, R. and Phillips, D. J.**, Obtaining pathogen-free stock by shoot tip culture, *Phytopathology*, 52, 1242, 1962.
12. **Baker, K. F.**, Thermotherapy of planting material, *Phytopathology*, 52, 1244, 1962.
13. **Stout, G. L.**, Maintenance of "pathogen-free" planting stock, *Phytopathology*, 52, 1255, 1962.
14. **Welsh, M. F.**, Control of stone fruit virus diseases, *U.S. Dep. Agric. Agric. Handb.*, 437, 10, 1974.
15. **Fridlund, P. R.**, IR-2, the interregional deciduous tree fruit repository, *U.S. Dep. Agric. Agric. Handb.*, 437, 16, 1974.
16. **Calavan, E. C., Mather, S. M., and McEachern, E. H.**, Registration, certification, and indexing of citrus trees, in *The Citrus Industry*, Reuther, W., Calavan, E. C., and Carman, G. E., Eds., Division of Agricultural Science, University of California, Richmond, 4, 185, 1978.
17. **Bolton, A. T.**, Strawberry leafroll, in *Virus Diseases of Small Fruits and Grapevines*, Division of Agricultural Science, University of California, Berkeley, 1970, 46.
18. **McGrew, J. R.**, Strawberry feather-leaf, in *Virus Diseases of Small Fruits and Grapevines*, Division of Agricultural Science, University of California, Berkeley, 1970, 46.
19. **Maassen, H.**, Strawberry band mosaic, in *Virus Diseases of Small Fruits and Grapevines*, University of California, Division of Agricultural Science, Berkeley, 1970, 48.
20. **Frazier, N. W. and Stace-Smith, R.**, Necrotic shock in strawberry, in *Virus Diseases of Small Fruits and Grapevines*, Division of Agricultural Science, University of California, Berkeley, 1970, 48.
21. **Schwartze, C. D. and Frazier, N. W.**, Lethal decline in strawberry, in *Virus Diseases of Small Fruits and Grapevines*, Division of Agricultural Science, University of California, Berkeley, 1970, 50.
22. **Stubbs, L. L. and Frazier, N. W.**, Strawberry pallidosis, in *Virus Diseases of Small Fruits and Grapevines*, Division of Agricultural Science, University of California, Berkeley, 1970, 51.
23. **Wills, A. B.**, June yellows of strawberry, in *Virus Diseases of Small Fruits and Grapevines*, Division of Agricultural Science, University of California, Berkeley, 1970, 55.
24. **Thresh, J. M.**, Black currant yellows, in *Virus Diseases of Small Fruits and Grapevines*, Division of Agricultural Science, University of California, Berkeley, 1970, 88.
25. **Thresh, J. M.**, Infectious variegation of black currant, in *Virus Diseases of Small Fruits and Grapevines*, Division of Agricultural Science, University of California, Berkeley, 1970, 88.
26. **van der Meer, F. A.**, Yellow leafspot of red currant, in *Virus Diseases of Small Fruits and Grapevines*, Division of Agricultural Science, University of California, Berkeley, 1970, 97.
27. **van der Meer, F. A.**, Leaf malformation of red currant, in *Virus Diseases of Small Fruits and Grapevines*, Division of Agricultural Science, University of California, 1970, 100.
28. **van der Meer, F. A.**, Budlessness of red currant, in *Virus Diseases of Small Fruits and Grapevines*, Division of Agricultural Science, University of California, Berkeley, 1970, 102.
29. **Cadman, C. H.**, Yellow disease of raspberry, in *Virus Diseases of Small Fruits and Grapevines*, Division of Agricultural Science, University of California, Berkeley, 1970, 154.
30. **Converse, R. H.**, Black raspberry streak, in *Virus Diseases of Small Fruits and Grapevines*, Division of Agricultural Science, University of California, Berkeley, 1970, 155.
31. **Frazier, N. W.**, Necrotic shock in *Rubus*, in *Virus Diseases of Small Fruits and Grapevines*, Division of Agricultural Science, University of California, Berkeley, 1970, 158.
32. **Hemphill, D. D.**, Sterility in *Rubus*, in *Virus Diseases of Small Fruits and Grapevines*, Division of Agricultural Science, University of California, Berkeley, 1970, 159.

33. **Converse, R. H.,** Minor viruses and virus diseases associated with *Rubus,* in *Virus Diseases of Small Fruits and Grapevines,* Division of Agricultural Science, University of California, Berkeley, 1970, 161.
34. **Rosenberg, D. Y., Nichols, C. W., and Wilhelm, S.,** Olallic disease, in *Virus Diseases of Small Fruits and Grapevines,* Division of Agricultural Science, University of California, Berkeley, 1970, 162.
35. **Williams, H. E. and Wagnon, H. K.,** Loganberry calico, in *Virus Diseases of Small Fruits and Grapevines,* Division of Agricultural Science, University of California, Berkeley, 1970, 160.
36. **Stretch, A. W. and Varney, E. H.,** Red ringspot of blueberry, in *Virus Diseases of Small Fruits and Grapevines,* Division of Agricultural Science, University of California, Berkeley, 1970, 184.
37. **Varney, E. H.,** Blueberry mosaic, in *Virus Diseases of Small Fruits and Grapevines,* Division of Agricultural Science, University of California, Berkeley, 1970, 185.
38. **Stretch, A. W. and Hilborn, M. T.,** Blueberry shoestring, in *Virus Diseases of Small Fruits and Grapevines,* Division of Agricultural Science, University of California, Berkeley, 1970, 186.
39. **Blattny, C., Sr. and Blattny, C., Jr.,** Minor virus diseases of *Ericaceae* in Europe, in *Virus Diseases of Small Fruits and Grapevines,* Division of Agricultural Science, University of California, Berkeley, 1970, 189.
40. **Taylor, R. H. and Woodham, R. C.,** Grapevine yellow speckle — a newly recognized graft-transmissible disease of *Vitis, Aust. J. Agric. Res.,* 23, 447, 1972.
41. **Beukman, E. F. and Goheen, A. C.,** Grape corky bark, in *Virus Diseases of Small Fruits and Grapevines,* Division of Agricultural Science, University of California, Berkeley, 1970, 207.
42. **Goheen, A. C.,** Grape leafroll, in *Virus Diseases of Small Fruits and Grapevines,* Division of Agricultural Science, University of California, Berkeley, 1970, 209.
43. **Hewitt, W. B., Goheen, A. C., Raski, D. J., and Gooding, G. V., Jr.,** Studies on virus diseases of grapevine in California, *Vitis,* 3, 57, 1962.
44. **Refatti, E.,** Asteroid mosaic of grapevine, in *Virus Diseases of Small Fruits and Grapevines,* Division of Agricultural Science, University of California, Berkeley, 1970, 212.
45. **Graniti, A. and Martelli, G. P.,** Legno riccio, in *Virus Diseases of Small Fruits and Grapevines,* Division of Agricultural Science, University of California, Berkeley, 1970, 243.
46. **Martelli, G. P.,** Shoot necrosis, in *Virus Diseases of Small Fruits and Grapevines,* Division of Agricultural Science, University of California, Berkeley, 1970, 245.
47. **Hewitt, W. B.,** Spindle shoot of Colombard grapevine, in *Virus Diseases of Small Fruits and Grapevines,* Division of Agricultural Science, University of California, Berkeley, 1970, 246.
48. **Nyland, G.,** Almond virus bud failure, in *U.S. Dep. Agric. Agric. Handb.,* 437, 33, 1974.
49. **Wagnon, H. K., Williams, H. E., and Traylor, J. A.,** Almond corky spot, in *U.S. Dep. Agric. Agric. Handb.,* 437, 259, 1974.
50. **Kester, D. E.,** Noninfectious bud failure in almond, in *U.S. Dep. Agric. Agric. Handb.,* 437, 278, 1974.
51. **Wadley, B. N.,** Apricot pucker leaf, in *U.S. Dep. Agric. Agric. Handb.,* 437, 43, 1974.
52. **Hansen, A. J., Parish, C. L., and Pine, T. S.,** Apricot ring pox, in *U.S. Dep. Agric. Agric. Handb.,* 437, 45, 1974.
53. **Nyland, G.,** Apricot gumboil, in *U.S. Dep. Agric. Agric. Handb.,* 437, 261, 1974.
54. **Williams, H. E., Wadley, B. N., and Wagnon, H. K.,** Asteroid spot, in *U.S. Dep. Agric. Agric. Handb.,* 437, 50, 1974.
55. **Nichols, C. W.,** Peach blotch, *U.S. Dep. Agric. Agric. Handb.,* 437, 56, 1974.
56. **Wagnon, H. K., Traylor, J., and Williams, H. E.,** Peach calico, *U.S. Dep. Agric. Agric. Handb.,* 437, 58, 1974.
57. **Helton, A. W.,** Peach mottle, *U.S. Dep. Agric. Agric. Handb.,* 437, 71, 1974.
58. **KenKnight, G.,** Peach rosette, *U.S. Dep. Agric. Agric. Handb.,* 437, 73, 1974.
59. **Blodgett, E. C.,** Peach wart, *U.S. Dep. Agric. Agric. Handb.,* 437, 88, 1974.
60. **Klos, E. J.,** Red suture, *U.S. Dep. Agric. Agric. Handb.,* 437, 133, 1974.
61. **KenKnight, G.,** Weak peach, *U.S. Dep. Agric. Agric. Handb.,* 437, 144, 1974.
62. **Helton, A. W.,** Peach variegation, *U.S. Dep. Agric. Agric. Handb.,* 437, 290, 1974.
63. **Pine, T. S.,** Noninfectious plum crinkle leaf, *U.S. Dep. Agric. Agric. Handb.,* 437, 283, 1974.
64. **Pine, T. S. and Welsh, M. F.,** Noninfectious plum shot hole, *U.S. Dep. Agric. Agric. Handb.,* 437, 285, 1974.
65. **Pine, T. S.,** Plum rusty blotch, *U.S. Dep. Agric. Agric. Handb.,* 437, 294, 1974
66. **Wagnon, H. K. and Williams, H. E.,** Prune diamond canker, *U.S. Dep. Agric. Agric. Handb.,* 437, 176, 1974.
67. **Wagnon, H. K.,** Standard prune constricting mosaic, *U.S. Dep. Agric. Agric. Handb.,* 437, 304, 1974.
68. **Blodgett, E. C.,** Italian prune sparse leaf, *U.S. Dep. Agric. Agric. Handb.,* 437, 277, 1974.
69. **Pine, T. S.,** Noninfectious plum crinkle leaf, *U.S. Dep. Agric. Agric. Handb.,* 437, 283, 1974.

70. **Parker, K. G. and Gilmer, R. M.,** Prune leaf casting mottle, *U.S. Dep. Agric. Agric. Handb.,* 437, 299, 1974.
71. **Pine, T. S.,** Pustule canker of Tragedy prune, *U.S. Dept. Agric. Agric. Handb.,* 437, 302, 1974.
72. **Parker, K. G., Fridlund, P. R., and Gilmer, R. M.,** Green ring mottle, *U.S. Dep. Agric. Agric. Handb.,* 437, 193, 1974.
73. **Cheney, P. W., Parish, C. L., and Johnson, F.,** Pink fruit, *U.S. Dep. Agric. Agric. Handb.,* 437, 200, 1974.
74. **Blodgett, E. C. and Aichele, M. D.,** Virus gummosis of Montmorency, *U.S. Dep. Agric. Agric. Handb.,* 437, 202, 1974.
75. **Cheney, P. W., Blodgett, E. C., and Parish, C. L.,** Black canker, *U.S. Dep. Agric. Agric. Handb.,* 437, 209, 1974.
76. **Williams, H. E. and Cameron, H. R.,** Freckle fruit, *U.S. Dep. Agric. Agric. Handb.,* 437, 226, 1974.
77. **Nichols, C. W.,** Rough bark of sweet cherry, *U.S. Dep. Agric. Agric. Handb.,* 437, 240, 1974.
78. **Blodgett, E. C. and Nyland, G.,** Sweet cherry crinkle leaf, deep suture and variegation, *U.S. Dep. Agric. Agric. Handb.,* 437, 306, 1974.
79. **Luckwill, L. C. and Campbell, A. I.,** Apple chlorotic leafspot, *Tech. Commun. Bur. Hort. East Malling,* 30, 7, 1963.
80. **Welsh, M. F. and Keane, F. W. L.,** Apple leaf pucker and associated fruit disorders, *Tech. Commun. Bur. Hort. East Malling,* 30, 7, 1963.
81. **van Katwijk, W. and Meijneke, C. A. R.,** Apple rosette, *Tech. Commun. Bur. Hort. East Malling,* 30, 23, 1963.
82. **Luckwill, L. C.,** Apple chat fruit, *Tech. Commun. Bur. Hort. East Malling,* 30, 27, 1963.
83. **McCrum, R. C., Barrat, J. G., Hilborn, M. T. and Rich, A. I.,** Dapple apple, *Tech. Commun. Bur. Hort. East Malling,* 30, 29, 1963.
84. **Yamaguchi, A., Mink, G. I., Yanase, H., and Hiraragi, K.,** Dapple apple — a virus disease, *Ann. Phytopath. Soc. Japan,* 38, 214, 1972.
85. **Kristensen, H. R.,** Apple green crinkle, *Tech. Commun. Bur. Hort. East Malling,* 30, 33, 1963.
86. **Mezzetti, A.,** Apple green dimple and ring blotch, *Tech. Commun. Bur. Hort. East Malling,* 30, 35, 1963.
87. **Canova, A.,** Apple ring spot, *Tech. Commun. Bur. Hort. East Malling,* 30, 39, 1963.
88. **van Katwijk, W. and Meijneke, C. A. R.,** Apple rough skin, *Tech. Commun. Bur. Hort. East Malling,* 30, 43, 1963.
89. **Posnette, A. F. and Cropley, R.,** Apple russet wart, *Tech. Commun. Bur. Hort. East Malling,* 30, 44a, 1963.
90. **Cropley, R.,** Apple star crack, *Tech. Commun. Bur. Hort. East Malling,* 30, 47, 1963.
91. **Millikan, D. F.,** Apple scar skin, *Tech. Commun. Bur. Hort. East Malling,* 30, 51, 1963.
92. **Kristensen, H. R.,** Flat limb of apple, *Tech. Commun. Bur. Hort. E. Malling,* 30, 55, 1963.
93. **Luckwill, L. C. and Campbell, A. I.,** Platycarpa dwarf, *Tech. Commun. Bur. Hort. East Malling,* 30, 59, 1963.
94. **Luckwill, L. C. and Campbell, A. I.,** Platycarpa scaly bark, *Tech. Commun. Bur. Hort. East Malling,* 30, 61, 1963.
95. **Posnette, A. F. and Cropley, R.** Apple stem pitting, *Tech. Commun. Bur. Hort. East Malling,* 30, 77, 1963.
96. **DeSequeira, O. A. and Posnette, A. F.,** Apple stem grooving, *Tech. Commun. Bur. Hort. East Malling,* 30, 76a, 1963.
97. **Posnette, A. F. and Cropley, R.,** Genetical disorders of apple with virus-like effects, *Tech. Commun. Bur. Hort. East Malling,* 30, 81, 1963.
98. **van Katwijk, W. and Meijneke, C. A. R.,** Ring pattern of pear, *Tech. Commun. Bur. Hort. East Malling,* 30, 89, 1963.
99. **Posnette, A. F.,** Vein yellows and red mottle of pear, *Tech. Commun. Bur. Hort. East Malling,* 30 81, 1963.
100. **Kristensen, H. R.,** Stony pit of pear, *Tech. Commun. Bur. Hort. East Malling,* 30, 99, 1963.
101. **Cropley, R.,** Blister canker of pear, in *Tech. Commun. Bur. Hort. East Malling,* 30, 103. 1963.
102. **Kristensen, H. R.,** Rough bark of pear, *Tech. Commun. Bur. Hort. East Malling,* 30, 107, 1963.
103. **Luckwill, L. C.,** Rosette of pear, *Tech. Commun. Bur. Hort. East Malling,* 30, 119, 1963.
104. **Posnette, A. F. and Cropley, R.,** Quince sooty ring spot, *Tech. Commun. Bur. Hort. East Malling,* 30, 111, 1963.
105. **Posnette, A. F. and Cropley, R.,** Quince stunt, *Tech. Commun. Bur. Hort. East Malling,* 30, 115, 1963.
106. **Posnette, A. F. and Cropley, R.,** Quince yellow blotch, *Tech. Commun. Bur. Hort. East Malling,* 30, 117, 1963.
107. **Klotz, L. J., Calavan, E. C., and Weathers, L. G.,** Virus and virus-like diseases of citrus, *Calif. Agric. Exp. St. Circ.,* 1972, 559.

108. **Hollings, M. and Stone, O. M.,** Investigations of carnation viruses. I. Carnation mottle, *Ann. App. Biol.,* 53, 103, 1964.
109. **Langhans, R. W., Horst, R. K., and Earle, E. D.,** Disease-free plants via tissue culture propagation, *HortScience,* 12, 149, 1977.
110. **Mellor, F. C. and Frazier, N. W.,** Strawberry mottle, in *Virus Diseases of Small Fruits and Grapevines,* Division of Agricultural Science, University of California, Berkeley, 1970, 4.
111. **Miller, P. W. and Frazier, N. W.,** Strawberry vein banding, in *Virus Diseases of Small Fruits and Grapevines,* University of California, Division of Agricultural Science, Berkeley, 1970, 8.
112. **Maassen, H.,** Strawberry necrosis, in *Virus Diseases of Small Fruits and Grapevines,* Division of Agricultural Science, University of California, Berkeley, 1970, 11.
113. **Horn, N. L. and Carver, R. G.,** Strawberry chlorotic fleck, in *Virus Diseases of Small Fruits and Grapevines,* Division of Agricultural Science, University of California, Berkeley, 1970, 11.
114. **Frazier, N. W.,** Strawberry pseudo mild yellow edge, in *Virus Diseases of Small Fruits and Grapevines,* Division of Agricultural Science, University of California, Berkeley, 1970, 12.
115. **Mellor, F. C. and Frazier, N. W.,** Strawberry mild yellow edge, in *Virus Diseases of Small Fruits and Grapevines,* Division of Agricultural Science, University of California, Berkeley, 1970, 14.
116. **McGrew, J. R.,** Strawberry latent C, in *Virus Diseases of Small Fruits and Grapevines,* Division of Agricultural Science, University of California, Berkeley, 1970, 16.
117. **Frazier, N. W. and Mellor, F. C.,** Strawberry crinkle, in *Virus Diseases of Small Fruits and Grapevines,* Division of Agricultural Science, University of California, Berkeley, 1970, 18.
118. **Frazier, N. W.,** Strawberry stunt, in *Virus Diseases of Small Fruits and Grapevines,* Division of Agricultural Science, University of California, Berkeley, 1970, 23.
119. **Boone, D. W.,** Witches' broom and multiplier diseases of strawberry, in *Virus Diseases of Small Fruits and Grapevines,* Division of Agricultural Science, University of California, Berkeley, 1970, 25.
120. **McGrew, J. R. and Posnette, A. F.,** Leafhopper-borne virus diseases of strawberry, in *Virus Diseases of Small Fruits and Grapevines,* Division of Agricultural Science, University of California, Berkeley, 1970, 26.
121. **Lister, R. M.,** Nematode-borne viruses as pathogens in strawberry, in *Virus Diseases of Small Fruits and Grapevines,* Division of Agricultural Science, University of California, Berkeley, 1970, 32.
122. **Posnette, A. F.,** Gooseberry vein banding, in *Virus Diseases of Small Fruits and Grapevines,* Division of Agricultural Science, University of California, Berkeley, 1970, 79.
123. **van der Meer, F. A.,** Spoon leaf of gooseberry, in *Virus Diseases of Small Fruits and Grapevines,* Division of Agricultural Science, University of California, Berkeley, 1970, 81.
124. **Thresh, J. M.,** Reversion of black currant, in *Virus Diseases of Small Fruits and Grapevines,* Division of Agricultural Science, University of California, Berkeley, 1970, 82.
125. **Thresh, J. M.,** Vein clearing and vein net of black currant, in *Virus Diseases of Small Fruits and Grapevines,* Division of Agricultural Science, University of California, Berkeley, 1970, 84.
126. **Thresh, J. M.,** Green mottle of black currant, in *Virus Diseases of Small Fruits and Grapevines,* Division of Agricultural Science, University of California, Berkeley, 1970, 85.
127. **Thresh, J. M.,** Yellow mottle of black currant, in *Virus Diseases of Small Fruits and Grapevines,* Division of Agricultural Science, University of California, Berkeley, 1970, 86.
128. **van der Meer, F. A.,** Red currant vein banding, in *Virus Diseases of Small Fruits and Grapevines,* Division of Agricultural Science, University of California, Berkeley, 1970, 91.
129. **van der Meer, F. A.,** Green mottle of red currant, in *Virus Diseases of Small Fruits and Grapevines,* Division of Agricultural Science, University of California, Berkeley, 1970, 92.
130. **van der Meer, F. A.,** Spoon leaf of red currant, in *Virus Diseases of Small Fruits and Grapevines,* Division of Agricultural Science, University of California, Berkeley, 1970, 93.
131. **Williams, H. E. and Holdeman, Q. L.,** American currant mosaic, in *Virus Diseases of Small Fruits and Grapevines,* Division of Agricultural Science, University of California, Berkeley, 1970, 95.
132. **van der Meer, F. A.,** Latent infection of red currant with arabis mosaic and strawberry latent ringspot viruses, in *Virus Diseases of Small Fruits and Grapevines,* Division of Agricultural Science, University of California, Berkeley, 1970, 95.
133. **Converse, R. H., Stace-Smith, R., and Cadman, C. H.,** Raspberry mosaic, in *Virus Diseases of Small Fruits and Grapevines,* Division of Agricultural Science, University of California, Berkeley, 1970, 111.
134. **Cadman, C. H. and Stace-Smith, R.,** Raspberry vein chlorosis, in *Virus Diseases of Small Fruits and Grapevines,* Division of Agricultural Science, University of California, Berkeley, 1970, 119.
135. **Stace-Smith, R. and Converse, R. H.,** Raspberry leaf curl, in *Virus Diseases of Small Fruits and Grapevines,* Division of Agricultural Science, University of California, Berkeley, 1970, 120.
136. **Harrison, B. D.,** Cucumber mosaic virus in raspberry, in *Virus Diseases of Small Fruits and Grapevines,* Division of Agricultural Science, University of California, Berkeley, 1970, 123.

137. **Stace-Smith, R.**, Thimbleberry ringspot, in *Virus Diseases of Small Fruits and Grapevines,* Division of Agricultural Science, University of California, Berkeley, 1970, 124.
138. **Wagnon, H. K. and Williams, H. E.**, Raspberry dwarf, in *Vir us Diseases of Small Fruits and Grapevines,* Division of Agricultural Science, University of California, Berkeley, 1970, 126.
139. **van der Meer, F. A. and de Flinter, H. J.**, *Rubus* stun t, in *Virus Diseases of Small Fruits and Grapevines,* Division of Agricultural Science, University of California, Berkeley, 1970, 128.
140. **Murant, A. F.**, Soil-borne viruses and diseases in *Rubus*, in *Virus Diseases of Small Fruits and Grapevines,* Division of Agricultural Science, University of California, Berkeley, 1970, 132.
141. **Cadman, C. H.**, Apple chlorotic leaf-spot in raspberry, in *Virus Diseases of Small Fruits and Grapevines,* Division of Agricultural Science, University of California, Berkeley, 1970, 149.
142. **Converse, R. H. and Lister, R. M.**, Black raspberry latent virus, in *Virus Diseases of Small Fruits and Grapevines,* Division of Agricultural Science, University of California, Berkeley, 1970, 151.
143. **Stretch, A. W. and Varney, E. H.**, Blueberry stunt, in *Virus Diseases of Small Fruits and Grapevines,* Division of Agricultural Science, University of California, Berkeley, 1970, 175.
144. **Blattný, C., Sr. and Blattný, C., Jr.**, Blueberry witches' broom, in *Virus Diseases of Small Fruits and Grapevines,* Division of Agricultural Science, University of California, Berkeley, 1970, 177.
145. **Varney, E. H.** Necrotic ringspot of blueberry, in *Virus Diseases of Small Fruits and Grapevines,* Division of Agricultural Science, University of California, Berkeley, 1970, 181.
146. **Stretch, A. W.**, Cranberry false blossom, in *Virus Diseases of Small Fruits and Grapevines,* Division of Agricultural Science, University of California, Berkeley, 1970, 180.
147. **Hewitt, W. B.**, Pierce's disease of *Vitis* species, in *Virus Diseases of Small Fruits and Grapevines,* Division of Agricultural Science, University of California, Berkeley 1970, 196.
148. **Caudwell, A. and Schuester, D.**, Flavescence Dorée, in *Virus Diseases of Small Fruits and Grapevines,* Division of Agricultural Science, University of California, Berkeley, 1970, 201.
149. **Viuttenez, A.**, Fanleaf of grapevine, in *Virus Diseases of Small Fruits and Grapevines,* Division of Agricultural Science, University of California, Berkeley, 1970, 217.
150. **Dias, H. F.**, Grapevine yellow mosaic, in *Virus Diseases of Small Fruits and Grapevines,* Division of Agricultural Science, University of California, Berkeley, 1970, 228.
151. **Taylor, R. H.**, Vein banding of *Vitis,* in *Virus Diseases of Small Fruits and Grapevines,* University of California, Division of Agricultural Science, Berkeley, 1970, 230.
152. **Stellmach, G.**, Arabis mosaic in *Vitis,* in *Virus Diseases of Small Fruits and Grapevines,* Division of Agricultural Science, University of California, Berkeley, 1970, 233.
153. **Stellmach, G.**, Tomato black ring in *Vitis,* in *Virus Diseases of Small Fruits and Grapevines,* Division of Agricultural Science, University of California, Berkeley, 1970, 234.
154. **Martelli, G. P., Lehoczky, J., and Quacquarelli, A.**, Hungarian chrome mosaic, in *Virus Diseases of Small Fruits and Grapevines,* Division of Agricultural Science, University of California, Berkeley, 1970, 236.
155. **Gooding, G. V., Jr. and Téliz, D.**, Grapevine yellow vein, in *Virus Diseases of Small Fruits and Grapevines,* Division of Agricultural Science, University of California, Berkeley, 1970, 238.
156. **Pine, T. S.**, Peach mosaic, *U.S. Dep. Agric. Agric. Handb.,* 437, 61, 1974.
157. **Klos, E. J.**, Rosette mosaic, *U.S. Dep. Agric. Agric. Handb.,* 437, 135, 1974.
158. **Mircetich, S. M. and Fogle, H. W.**, Peach stem pitting, *U.S. Dep. Agric. Agric. Handb.,* 437, 77, 1974.
159. **Schlocker, A. and Traylor, J. A.**, Yellow bud mosaic, *U.S. Dep. Agric. Agric. Handb.,* 437, 156, 1974.
160. **Pine, T. S. and Gilmer, R. M.**, Peach yellows, *U.S. Dep. Agric. Agric. Handb.,* 437, 91, 1974.
161. **Gilmer, R. M. and Blodgett, E. C.**, X-disease, *U.S. Dep. Agric. Agric. Handb.,* 437, 145, 1974.
162. **Cochran, L. C. and Hutchins, L. M.**, Phony, *U.S. Dep. Agric. Agric. Handb.,* 437, 96, 1974.
163. **Nyland, G., Gilmer, R. M., and Moore, J. D.**, "Prunus" ringspot group, *U.S. Dep. Agric. Agric. Handb.,* 437, 104, 1974.
164. **Kassanis, B. and Sutic, D.**, Some results of recent investigations on Sharka (plum pox) virus disease, *Zast. Bilja,* 15, 335, 1965.
165. **Welsh, M. F. and Cheney, P. W.**, Little cherry, *U.S. Dep. Agric. Agric. Handb.,* 437, 231, 1974.
166. **Cheney, P. W. and Parish, C. L.**, Cherry mottle leaf, *U.S. Dep. Agric. Agric. Handb.,* 437, 216, 1974.
167. **Mircetich, S. W.**, Personal communication. 1978.
168. **Nyland, G.**, Cherry rasp leaf, *U.S. Dep. Agric. Agric. Handb.,* 437, 219, 1974.
169. **Hansen, A. J. and Cheney, P. W.**, Cherry twisted leaf, *U.S. Dep. Agric. Agric. Handb.,* 437, 222, 1974.
170. **Wadley, B. N. and Nyland, G.**, Rusty mottle group, *U.S. Dep. Agric. Agric. Handb.,* 437, 242, 1974.
171. **Hansen, A. J., Nyland, G., McElroy, F. D., and Stace-Smith, R.**, Origin, cause, host range and spread of cherry rasp leaf disease in North America, *Phytopathology,* 64, 721, 1974.

172. **Stouffer, R. F., Hickey, K. D., and Welsh, M. F.,** Apple union necrosis and decline, *Plant Dis. Rep.,* 61, 20, 1977.
173. **Bovey, R.,** Apple proliferation, *Tech. Commun. Bur. Hort. East Malling,* 30, 62, 1963.
174. **Refatti, E.,** Pear decline and moria, *Tech. Commun. Bur. Hort. East Malling,* 30, 108c, 1963.
175. **Malone, J. P. and Muskett, A. E.,** Seed-borne fungi, *Proc. Int. Seed Test. Assoc.,* 29, 1977, 1964.
176. **Crowley, N. C.,** Studies on seed transmission of plant virus diseases, *Aust. J. Biol. Sci.,* 10, 449, 1957.
177. **Phatak, H. C.,** Seed-borne plant viruses — identification and diagnosis in seed health testing, *Seed Sci. Technol.,* 2, 3, 1974.
178. **Baker, K. F. and Smith, S. H.,** Dynamics of seed transmission of plant pathogens, *Annu. Rev. Phytopath.,* 4, 311, 1966.
179. Anonymous, Rules and regulations concerning bacterial diseases of beans, *Phaseolus* species, Idaho Department of Agriculture, Boise, Idaho, 1975.
180. **Wright, B. D.,** Personal communication. 1978.
181. **Seinhorst, J. W.,** Dynamics of the plant nematode system, in *Nematode Vectors of Plant Viruses,* Vol. 2, Lamberti, F., Taylor, C. E., and Seinhorst, J. W., Eds., Plenum Press, London, 1974, 409.
182. **Dawson, J. H., Lee, W. O., and Timmons, F. L.,** Controlling dodder in alfalfa, *U.S. Dep. Agric. Farmers' Bull.,* 2211, 1, 1965.
183. **Wilhelm, S., Benson, L. C., and Sagen, J. E.,** Studies on the control of broomrape on tomatoes. Soil fumigation by methyl bromide is a promising control, *Plant Dis. Rep.,* 45, 645, 1958.
184. **Martin, H.,** *The Scientific Principles of Plant Protection with Special Reference to Chemical Control,* 2nd ed., Edward Arnold & Co., London, 1936.
185. **Grogan, R. G. and Kendrick, J. B.,** Seed transmission, mode of overwintering and spread of bacterial canker of tomato caused by *Cornybacterium michiganense, Phytopathology,* 43, 473, 1953.
186. **Zimmer, D. E.,** Spore stages and life cycle of *Puccinia carthami, Phytopathology,* 53, 316, 1963.
187. **Taylor, R. H., Grogan, R. G., and Kimble, K. A.,** Transmission of tobacco mosaic virus in tomato seed, *Phytopathology,* 51, 837, 1961.
188. **Maude, R. B. and Lyle, A. M.,** Seed treatments with benomyl and other fungicides for the control of *Ascochyta pisi* on peas, *Ann. Appl. Biol.,* 66, 37, 1970.
189. **Evans, E., Richard, M., and Whitehead, R.,** Effect of benomyl on some diseases of spring barley, *Proc. 5th Br. Insecticide Fungicide Conf.,* 3, 610, 1969.
190. **Templeton, G. E., Johnston, T. H., and Daniel, J. T.,** Benomyl controls rice white top disease, *Phytopathology,* 61, 1522, 1971.
191. **Maude, R. B. and Shuring, C. G.,** Seed treatments with carboxin for control of loose smut of wheat and barley, *Proc. Br. 5th Insecticide Fungicide Conf.,* 2, 328, 1969.
192. **Hoffman, J. A.,** Control of common and dwarf bunt of wheat by seed treatment with thiabendazole, *Phytopathology,* 61, 1071, 1971.
193. **Baker, K. F.,** Seed transmission of *Rhizoctonia solani* in relation to control of seedling damping off, *Phytopathology,* 37, 912, 1947.
194. **Couch, H. B.,** Studies on seed transmission of lettuce mosaic virus, *Phytopathology,* 45, 63, 1955.
195. **Anon.,** *North Carolina Certification Handbook,* North Carolina Crop Improvement Association, Raleigh, 1975.
196. **Anon.,** Regulations for the production of Georgia certified pepper plants, Rules of Georgia Department of Agriculture Entomology and Plant Industry, Georgia Department of Agriculture, Atlanta, 1977, 40-4-5.
197. **Anon.,** Regulations for the production of Georgia certified tomato plants, Rules of Georgia Department of Agriculture Entomology and Plant Industry, Georgia Department of Agriculture, Atlanta, 1977, 40-4-6.
198. **Luckwill, L. C.,** Virus diseases of fruit trees. IV. Further observations on rubbery wood, chat fruit, and mosaic in apples, *Res. Sta. Long Ashton Rep.,* 1953, 40, 1954.
199. **Fridlund, P. R.,** Distribution of chlorotic leafspot virus in apple budsticks, *Plant Dis. Rep.,* 57, 865, 1973.
200. **Brierley, P.,** Gladiolus cormels free from cucumber mosaic from infected parent corms, *Plant Dis. Rep.,* 47, 863, 1963.
201. **Murashige, T. and Jones, J. B.,** Cell and organ culture methods in virus disease therapy, in *Proceedings, 3rd International Symposium on Ornamental Plant Viruses,* Lawson, R. H. and Corbett, M. K., Eds., ISHS, The Hague, Netherlands, 1974, 207.
202. **Yang, H. J. and Clore, W. C.,** Obtaining virus-free plants of *Asparagus officinalis* L. by culturing shoot tips and apical meristems, *HortScience,* 11, 474, 1976.
203. **Nyland, G. and Goheen, A. C.,** Heat therapy of virus diseases of perennial plants, *Annu. Rev. Phytopathol.,* 7, 331, 1969.
204. **Frosheiser, F. I.,** Freeing alfalfa clones from alfalfa mosaic virus by heat treatment, *Phytopathology,* 59, 391, 1969.

205. **Welsh, M. F. and Nyland, G.,** Elimination and separation of viruses in apple clones by exposure to dry heat, *Can. J. Plant Sci.,* 45, 443, 1965.
206. **Stone, O. M.** Elimination of four viruses from carnation and sweet william by meristem culture, *Ann. Appl. Biol.,* 62, 119, 1968.
207. **Mori, K.,** Production of virus-free plants by means of meristem culture, *Jpn. Agric. Res. Quart.,* 6, 1, 1971.
208. **Hollings, M. and Stone, O. M.,** Attempts to eliminate chrysanthemum stunt from chrysanthemum by meristem-tip culture after heat-treatment, *Ann. Appl. Biol.,* 65, 311, 1970.
209. **Murashige, T., Bitters, W. P., Rangan, T. S., Naver, E. M.,** Roistacher, C. N., and Holliday, P. B., A technique of shoot apex grafting and its utilization towards recovering virus-free *Citrus* clones, *HortScience,* 7, 188, 1972.
210. **Morel, G. M.,** Producing virus-free cymbidiums, *Am. Orchid Soc. Bull.,* 29, 495, 1960.
211. **Baruch, E. R. and Quak, F.,** Virus-free plants of iris "Wedgewood" obtained by meristem culture, *Neth. J. Plant Pathol.,* 72, 270, 1966.
212. **Casalicchio, G.,** Applicazione della termoterapia nel risanaments della piante di fico affette de mosaico, *Phytopathol. Medit.,* 3, 184, 1964.
213. **Brants, D. H. and Vermeulen, H.,** Production of virus-free freesias by means of meristem culture, *Neth. J. Plant Pathol.,* 71, 25, 1965.
214. **Pillai, S. K. and Hilderbrandt, A. C.,** Geranium plants differentiated in vitro from stem tip and callus cultures, *Plant Dis. Rep.,* 52, 600, 1968.
215. **Jones, O. P. and Vive, S. J.,** The culture of gooseberry shoot tips for eliminating virus, *J. Hort. Sci.,* 43, 289, 1968.
216. **Posnette, A. F., Cropley, R., and Wolfswinkle, L. D.,** Heat inactivation of some apple and pear viruses, *Ann. Rep. East Malling Res. Stn.,* 1961, 94, 1962.
217. **Berg, T. M.,** Studies on popular mosaic virus and its relation to the host, *Meded. Landbouwhogesch. Wageniningen,* 64, 1, 1964.
218. **Stace-Smith, R. and Mellor, F. C.,** Eradication of potato spindle tuber virus by thermotherapy and axillary bud culture, *J. Hort. Sci.,* 43, 283, 1968.
219. **Walkey, D. G. A.,** The production of virus-free rhubarb by apical tip-culture, *J. Hort. Sci.,* 43, 283, 1968.
220. **Holmes, F. O.,** Rose mosaic cured by heat treatment, *Plant Dis. Rep.,* 44, 46, 1960.
221. **Converse, R. H.,** Effect of heat treatment on the raspberry mosaic virus complex in Latham red raspberry, *Phytopathology,* 56, 556, 1966.
222. **Vive, S. J.,** Improved culture of apical tissues for production of virus-free strawberries, *J. Hort. Sci.,* 43, 293, 1968.
223. **Baxter, L. W., Jr. and McGlohon, N. E.,** A method of freeing white clover plants of bean yellow mosaic virus, *Phytopathology,* 49, 810, 1959.
224. **Hollings, M.,** Heat treatment in the production of virus-free ornamental plants, *G. B. Nat. Agric. Advis. Serv. Quart. Rev.,* 57, 31, 1962.
225. **Weathers, L. G. and Calavan, E. C.,** Nucellar embryony — a means of freeing citrus clones of viruses, in Citrus Virus Diseases, University of California, Division of Agricultural Science, July 1959, 197.
226. **Sbragia, R. J.,** Chemical control of plant diseases. An exciting future, *Annu. Rev. Phytopathol.,* 13, 257, 1975.
227. **Erwin, D. C.,** Systemic fungicides: disease control, translocation and mode of action, *Annu. Rev. Phytopathol.,* 11, 389, 1973.
228. **Hardison, J. R.,** Chemotherapeutic erradication of *Ustilago striiformis* and *Urocystis agropyri* in *Poa pratensis* 'Merion' by root uptake of α-(2,4,-dichlorophenyl)-α-phenyl-5-pyrimidinemthanol (EL-273), *Crop Sci.,* 11, 345, 1971.
229. **Nagaich, B. B. and Upreti, G. C.,** Heat inactivation of potato leafroll virus, *Indian Potato J.,* 6, 96, 1964.
230. **Selsky, M. I. and Black, L. M.,** Effect of high and low temperatures on survival of wound tumor virus in sweet clover, *Virology,* 16, 190, 1962.
231. **Purcell, A. H.,** Personal communication.
232. *AOSCA Certification Handbook,* Publication 23, Association of Official Seed Certifying Agencies, 1971.
233. **Anon.,** Regulation for registration and certification of citrus trees, Department of Food and Agriculture, State of California, 1974.
234. **Anon.,** Regulations for registration and certification of deciduous fruit and nut trees, Department of Food and Agriculture, State of California, 1968.
235. **Anon.,** Fruit trees — registration and certification, Order. No. 1331, Washington Department of Agriculture, Olympia, 1974.

236. **Anon.,** Voluntary program for the production of virus-tested plant materials, *Secs. 18, 164 and 167,*
237. **Kimble, K. A., Grogan, R. G., Greathead, A. S., Paulus, A. O., and House, J. K.,** Development, application and comparison of methods for indexing lettuce seed for mosaic virus in California, *Plant Dis. Rep.,* 59, 461, 1975.
238. **Mink, G. I. and Parsons, J. L.,** Detection of pea seedborne mosaic virus in pea seed by direct-seed assay, *Plant Dis. Rep.,* 62, 249, 1978.
239. **Davis, D. J. and Rubye Wallace,** Sero-diagnosis for barley stripe mosaic virus in Montana, *Proc. Assoc. Official Seed Analysts,* 57, 132, 1967.
240. **Forer, L. B.,** Division of Plant Pathology, Pennsylvania Department of Agriculture, Harrisburg, Pa., Personal communication.
241. **Anon.,** Regulations for production of Georgia-certified cabbage plants, Rules of Georgia Department of Agriculture Entomology and Plant Industry, Georgia Department of Agriculture, Atlanta, 1976, 40-4-3.

ENVIRONMENTAL CONTROL OF PLANT PATHOGENS USING AVOIDANCE

Simeon Leach

INTRODUCTION

Avoidance, when used in a pest management program, is where a pathogen and a susceptible host are prevented from coming in contact with each other, either at all or to the extent that inoculum is at a low level so that a crop can be produced with minimum loss. Complete avoidance of pathogens is difficult under most conditions; therefore, something less than complete is generally most commonly attempted. Before using avoidance as a pest management tool, a thorough knowledge of pathogen, host, host-pathogen interactions, and optimum environment for both is necessary. Also, a complete understanding of natural defense mechanisms a host possesses against a pathogen, or pathogens, is necessary in preparing a pest management program where avoidance is to be incorporated.

The most effective method of preventing plant pathogens from causing losses to economic crops, then, would be to plant the crop in areas where the pathogen is not known to occur; or, if present, is ineffective in causing a serious disease epidemic, allowing a successful crop to be produced. The first criterion, therefore, is to identify a location or locations where the pathogen does not exist or has difficulty surviving under the cultural practices utilized, yet where a crop can be produced. An example would be where crops which are highly susceptible to specific fungal or bacterial foliar diseases, when grown in humid climates, are grown relatively disease-free in arid regions where moisture is supplied from surface irrigation. When overhead sprinkler systems are used to apply moisture in these arid regions, the environment around the plants can become such that the diseases could cause losses equal to or greater than if the crop was grown in a humid climate. The environment can be used when selecting an area to grow a crop, and the producer has determined that control of a specific disease will insure prosperity, in comparison to selecting a planting site in a local area, e.g., a certain field on a certain farm, where the history of previous crops grown and diseases encountered on that site will determine what crop or crops will be grown, rather than vice versa.

Site selection is very important in avoiding many of the soil-borne organisms which cause losses to crops. Included in these are numerous fungi, bacteria, and nematodes. Crop residues should be checked for sources of primary inoculum for airborne foliar disease organisms and soil-borne disease organisms. If the pathogen can be carried by insects and transmitted to the plant, e.g., viruses by aphids, care must be exercised not to locate a crop where the flying forms of the aphids may be deposited because of wind currents.

As in all methods of pest management, for avoidance to be successful, strict sanitation measures should be carried out to insure that cultural implements and equipment used to prepare the soil for planting are free from disease organisms. Also, it is necessary to use only disease-free seed, transplants, and plant parts, and growing media adhering to plants or plant parts must be pathogen-free, as well as the containers in which they are transported. Otherwise, a source of inoculum has been put in direct contact with the host, a prerequisite for initiation of a disease epidemic. In conjunction with site selection, soil drainage is very important to insure optimum conditions for growth of the crop. With poor soil drainage, many soil-borne pathogens cause more damage, than with well-drained soil; and large amounts of water in the soil reduce root growth and oxygen content of soil, supplies a mode of transport for some pathogens, and predisposes plant parts to infection.

With this brief introduction to what avoidance is and how it is used in disease control, the specific topics to be discussed, as they relate to avoidance, will be site selection, timing of crop planting, water management, trap crops, plant spacing, harvesting, and storage management.

Site Selection

The selection of an area to grow a crop, whether it be a field on a farm or an area of the country, is very important in reducing losses caused by plant pathogens. On-farm sites are generally chosen because of soil texture, soil moisture and internal drainage, air drainage, soil-borne pathogens content, etc., whereas regional areas are chosen because of climatic condition which allows the production of a disease-free crop. The latter areas are often selected to produce disease-free certified seed.

On-farm sites are often chosen to produce a certain crop because it is free of root-infecting organisms, such as *Verticillium* or *Fusarium,* an example of which would be potatoes. Potatoes are susceptible to both *Verticillium* and *Fusarium,* and when planted in soils contaminated with these fungi, yields are reduced by the *Verticillium* and storage losses increased because the *Fusarium* causes tuber rotting. The red stele disease of strawberry caused by *Phytophthora fragariae* can be controlled by selecting well-drained soil for planting which has been fallow the previous year to reduce inoculum.[1]

A classic example of avoidance of disease by site selection is reported by Van Arsdel et al.[2] They studied the effects of given sites on the incidence of white pine blister rust (*Cronartium ribicola* Fischer) when Ribes, the alternate host, was present at the same site. In a second paper, Van Arsdel[3] describes methods of reducing damage by silviculture which, in turn, modifies the environment such that infection is reduced even when inoculum is present.

The choice of a specific geographic area for crop production is generally based on the temperature and humidity of that area and the major disease problems associated with the crop to be grown. Texas root rot of cotton (*Phymatotrichum omnivorum*) can survive freezing temperatures for only 1 or 2 days, thus, cotton grown north of the latitude of the southern Oklahoma border is not affected by this pathogen. Apple scab (*Venturia inaequalis*), a severe disease of apple in the eastern U.S., is rare in California because of lack of sufficient moisture during the growing season to disseminate and germinate conidia. Bean anthracnose [*Colletotrichum lindemuthianum* (Sacc. and Magn)] Briosi and Car. does not occur in the western U.S. because little or no rain occurs, and rain is necessary for spore dissemination.[4]

Certified seed production often occurs in specific areas of the country or would occur because the climate is such that the major diseases of a crop do not occur in the chosen area. For example, cabbage and cauliflower seed grown in California is free of black leg rot (*Phoma lingam*) even when grown from diseased plants.[4] Cucumber seed free from angular leaf spot (*Pseudomonas lachrymans*) can be produced in the arid conditions of California.

Potato is another crop where location of certified seed production is important in producing seed with low aphid-borne virus disease content. Location is important because of its effect on the overwintering, multiplication, and spread of the aphid vectors, which cannot survive extreme cold and cannot fly except during warm dry periods. Therefore, a good site for potato seed production is an area with cold winters, cool humid growing season, and hilly areas; such an area in the U.S. is Aroostook County in northern Maine.

Timing of Planting

Timing of planting a crop to avoid a disease is based on the hypothesis that if a crop is not subjected to high amounts of inoculum when the crop is in a susceptible stage of growth, disease incidence and severity will be low. Thus, by knowing the life cycle of a pathogen and the natural resistance of a crop, one can determine a planting date which will allow the production of a crop that is almost disease-free,with little or no use of pesticides. This method of disease control is good but can be used only in areas where the climate and length of growing season are suitable and the crop adaptable. Short growing seasons do not permit a delay of 1 or 2 weeks for planting, which can be crucial for disease avoidance, especially in a crop planted in the fall. However, in areas like Florida or California, or areas with comparable climate, time of planting can be used readily as a method of reducing incidence of disease in crops.

In Australia, black leg (*Phoma lingam*) of rapeseed caused severe losses until it was discovered that by delaying planting until after August, the disease caused only minor losses.[5] Bunt (*Tilletia caries*) of winter wheat can be controlled by seeding wheat when the environment is unfavorable for infection of the wheat, but favorable for wheat growth, thus allowing the wheat to grow beyond the susceptible stage before the bunt organism become active.[6,7] This system is used in the Pacific Northwest where wheat planted before mid-September and after mid-October is almost disease-free, compared to that planted during the September-October period when a high disease incidence is evident. Bunt on fall-sown wheat is reduced by early planting because the soil temperature is higher than if planted later, and conditions which favor the host are suboptimal for the pathogen.[8]

Cercosporella foot rot (*Cercosporella herpotrichoides*) is an important disease of winter wheat in Washington when the wheat is planted by September 15 on summer fallow. By contrast, if planting is delayed for 1 month, almost no foot rot is observed.[9] How this 1-month delay in planting reduces the incidence of foot rot is attributed to the differences in the density of the leaf canopy from late fall through spring, when infection takes place. A dense leaf canopy helps maintain a moist environment favorable for the sporulation of the pathogen, whereas a less dense canopy will not provide the environment necessary for sporulation. The later planting date produces small plants with a less dense canopy, thereby not providing the optimum conditions necessary for *Cercosporella* to sporulate, and little or no disease is observed.

Another disease which may be controlled by the environment created by plants is curly top of tomato and beets where, when planted early, the crop produces a canopy such that the micro-climate around the plants is unfavorable to the leaf hoppers which transmit the virus.[10]

Timing of planting can also be used as a control method in areas where the growing season is short by planting cool-season crops, such as peas and spinach, so they emerge before high temperatures prevail and increase the chance of attack by the damping-off fungi, *Pythium*, and yet will produce good stands and yields because of optimum conditions for the crops. The opposite of this is with warm weather crops, such as melons and beans, which require high temperatures for optimum germination and growth if they are to avoid severe disease loss caused by *Pythium* and *Rhizoctonia*.[6]

Timing of planting is a very useful method of controlling diseases which are transmitted by insects. In California, common or western celery mosaic is controlled if celery is not planted in the field from September to December, and in the greenhouse from September to October, not because the disease may not be present but because of a lack of aphids to transmit the virus. This shows how time of planting indirectly controls a disease by avoiding the vector and not the causal agent.[11] Another example is in the North, where planting peas as early as possible insures that the plants will be almost mature before the aphids carrying the pea mosaic virus can transmit the virus and cause much damage.

Losses caused by another plant pest, the nematode, can also be reduced by timing of planting, because many of the plant pathogenic nematodes are inactive during the periods of low temperatures. The sugar beet nematode (*Heterodera schachtii*) can cause high losses in California when beets are planted in March or April; but, if the beets are planted in January and February, good yields are obtained because of inactivity of the nematode during the colder months. The root-knot nematode (*Meloidogyne* sp.) does not damage the spring crop of potatoes in North Carolina because the crop grows at temperatures below which the nematodes are active; but, in potatoes planted later and grown at higher temperatures, a high degree of nematode damage, as well as a reduction in yield, is observed. In Britain, potato growers can grow potatoes with little damage from the golden nematode (*Heterodera rostochiensis*) by planting early varieties early, so that the plants develop roots and grow before the soil temperature is favorable for the hatching and development of the nematode.[12]

Timing of planting to avoid plant pathogens is a tool which can be used to protect certain crops in specific areas of the world, but it is not a satisfactory method of control for all diseases on all crops. If, after careful study of pathogen and host, timing of planting seems to be a good control method, then one must determine if it will also allow production of a good, high-quality crop. Timing of planting will generally allow the production of a crop so that it will not be subjected to the plant pathogen in question in such a quantity as to cause an epidemic, but may attack the crop to some extent. Often an integrated control system using some pesticide may be necessary.

In conjunction with time of planting, variety selection is important in that early-maturing varieties can complete their growth before the disease intensity is a limiting factor. Early harvesting can have a similar effect. Harvesting strawberries in early morning reduces losses from rot when compared to those harvested in the afternoon.[4]

Water Management

The area of water management and its relationship to disease incidence and severity is one of the most intensively studied areas in determining methods to avoid losses due to disease. Research in this area was begun many years ago and is still flourishing today. A large degree of the early work conducted in biological control of plant pathogens was devoted to moisture and how it affects plant diseases.[10] Some plant pathogens cause more losses during wet-soil periods and others during dry-soil periods. Water management as a control for avoiding diseases, therefore, is generally most effective when used in areas where water is a limiting factor, both in amounts available and time of availability, and can easily be regulated through irrigation. In humid and temperate regions where rainfall and temperature are optimum for both plants and pathogens, so that plants are in a susceptible stage and sufficient inoculum is present, control of disease by water management is very difficult, if not impossible, for most crops. Since moisture plays a very important role in plant disease incidence and severity, all forms of moisture must be considered, including heavy dews.[13] Therefore, when studying the effects of moisture on plant diseases and how they may be controlled through use of water management, strict attention must be paid to soil moisture, rainfall, atmospheric humidity, plant microenvironment, and irrigation. Water acts as a medium for spore germination, source of transport of motile spores, and movement of pathogens from one area to another through runoff and irrigation water.

As one would expect, not all plant pathogens are enhanced by moisture or by dry conditions. This is easily seen with some of the soil-borne disease organisms. Cook et al.[14] listed some diseases which are favored by dry soil and wet soil and how water potentials affected disease incidence. Water management is generally considered in terms of satisfying the crop's water requirement and not the effect it can have on diseases. Any stress applied to a plant will often reduce the effectiveness of its natural defenses against pathogens, and water, whether too much or too little, plays an important role in a plant's natural defenses.

Soil moisture is associated with many diseases. Wet soils enhance the severity of club root (*Plasmodiophora brassicae*) of crucifers, silver scurf (*Helminthosporium solani*) of potatoes, and *Cercosporella* on wheat, to mention a few. Diseases most severe in dry soils include those which cause white mold (*Whetzelinia sclerotiorum*) on onion, common scab (*Streptomyces scabies*) on potato, and *Fusarium* diseases of cereals.[13] With this in mind, it should be fairly easy to reduce the incidence and severity of these diseases by strict water management. This is not necessarily true, however, as a thorough knowledge of the effects of water on a particular disease complex and its host must be known before a satisfactory and sound water management program can be devised to protect a crop from a disease. Included in any study of water management should be atmospheric moisture, which plays a very important role in some disease complexes.

The control of plant pathogens by water management is most easily accomplished in the arid and semi-arid areas, or areas where, during certain parts of the year, water is the limiting factor for crop growth, and conditions for optimum crop growth with little stress can be easily maintained through the use of irrigation.[15] Irrigation often makes it possible to avoid diseases completely by growing crops out of season. For example, potatoes and wheat are often grown in Mexico and other areas under irrigation during the dry season, to avoid attacks of potato late blight (*Phytophthora infestans*) and wheat stem rust (*Puccinia graminis*). Irrigation is not a cure-all, however, and care must be taken when using irrigation not to change moisture and temperature conditions around plants to encourage development of diseases. In many areas where irrigation is now practiced on a regular basis, diseases originally absent from crops have become common, especially where overhead sprinkler systems are used.[15] There have been excellent reviews written both on water and its effects on root diseases, and on irrigation and plant diseases. These reviews show how water affects pathogens, both in increasing and decreasing losses caused by their respective diseases.[14,15]

Since most diseases are affected in some manner by moisture, or lack of moisture, some examples of various diseases will be discussed in relation to water and water requirements. Damping off of seedlings caused by *Pythium* and *Rhizoctonia* can be reduced by maintaining a dry soil surface. The degree of bacterial wilt of many species of plants is in direct proportion to the soil moisture when the temperature is optimum for bacterial growth, and control is accomplished by planting on soils with good drainage, to insure good root growth and no stress on the root system. Common scab of potato, on the other hand, can be controlled if high soil moisture is maintained during tuber formation. Sclerotinia diseases, Verticillium wilts, and Fusarium wilts require good drainage with no excess water present; and if irrigation is used, only enough water is applied to keep crop vigorous and healthy. Blossom end rot of tomato is greatly reduced if grown in moist soil which will retain moisture during drought, and irrigation is applied only when a hot, drying wind occurs to keep fruit and leaves dry. *Alternaria* leaf spot and blight of cruciferous crops grown in the field can be controlled by using overhead irrigation only in the morning so that plants can dry rapidly. Hard dry rot (*Rhizoctonia solani*) of strawberry is controlled by applying a mulch which keeps the fruit high above the soil surface and exposes it to the air for rapid drying, thus reducing infection.[1]

Flooding of soil has been used, to some extent, to control plant pests such as nematodes and some fungal diseases, e.g., those caused by *Sclerotinia* sp. The flooding reduces the number of nematodes present in the soil and promotes the decay of *Sclerotinia* sclerotia, thereby reducing a source of inoculum for future crops.[6]

Many diseases of greenhouse-grown plants are controlled by water management. Watering seedlings only in the morning to allow the soil surface to dry prevents damping off and root rots. Watering all greenhouse plants so the foliage can dry before night decreases the incidence of *Botrytis* blights and other foliar diseases which are increased when the plants are moist for long periods. Almost all soil moisture related diseases are easily controlled in greenhouses, as many methods of water application are available as well as measuring devices to prevent over- or under-watering. A greenhouse is the one location where all aspects of avoidance can be used completely and produce the results expected.

Controlling plant pathogens by using water management is a very precise method, and often the time that the water is applied is more important than the amount applied. Water management and timing of planting are often used together to produce crops which are high-yielding and disease-free. In areas where rainfall and dew are common and growing seasons are short, water management is a less-acceptable method of disease control. The only way to insure good crop production and low disease incidence in these areas is to plant crops on well-drained soil and supply water by irrigation when needed to maintain healthy, vigorous growing plants.

Trap Crops

Trap crops, or catch crops, are species of plants which are highly susceptible to a pathogen of a primary crop and are planted so as to be attacked by the pathogen at an appropriate time, and both the trap crop and pathogen are then destroyed in a single operation, leaving the primary crop relatively free from attack by the pathogen.[16] The major use of trap crops has been restricted to control of nematodes.

The planting of one or more trap crops (such as tomato in fields infested with the pineapple root knot nematode) and destroying the plants before the nematode eggs are produced reduces damage to the following pineapple crops. If the tomato plants are not destroyed before the eggs are produced, the net result is to increase the population of nematodes, thereby increasing damage to the subsequent pineapple crops. A more effective use of a trap crop is to select a crop that is highly susceptible to invasion by the nematode but is resistant to the development of larvae into adults. The crop then does not have to be destroyed to control the nematode and may be harvested or used as a green manure. *Crotalaria* has been successfully used to reduce the populations of certain species of root-knot nematodes.[10,12]

Other than for nematode control, trap crops have been used to help control bacterial wilt (*Erwinia tracheiphila*) of cucumber. In this system, muskmelons are used as the trap crop. Seeds of muskmelon are thickly planted a few days before the main crop of cucumber is planted. The melons attract the cucumber beetle which carries the disease. The beetles feed on the melons, thus leaving the cucumbers alone to produce a disease-free crop. Sometimes, however, the beetles are not content to stay on the melons, and they migrate to the cucumbers. In this case, all the catch crop has done is increase the number of insects which can transmit the disease.

Neither trap nor catch crops provide a sufficient degree of control to offset the expense and uncertainty of producing a disease-free crop.[17] If a trap or catch crop is used in a pest management program, it must be closely controlled so as not to increase rather than decrease the losses incurred by the plant pathogen one is trying to control.

Plant Spacing

Plant spacing is important in many ways when looking to control certain plant pathogens. Close plantings tend to produce microenvironments ideal for some fungal diseases, while with others, the same environment prevents disease by discouraging an insect vector from feeding. One of the main attributes of plant spacing is to provide good air drainage between plants to reduce the length of time the foliage of the plants is moist. *Botrytis* in greenhouses is controlled by insuring adequate space between plants for air drainage. Curly top of sugar beets is reduced by planting two plants in the same hill or by reducing the distance between plants by one half the normal distance to provide shade, which in turn reduces the feeding of the leafhopper vector.

As important as spacing within rows is the distance at which crops are planted from sources of inoculum. Yellow mosaic of bean and lima bean mosaic are controlled by planting bean crops at least 200 m from any possible weed or crop host. Other diseases, e.g., potato mosaic, can be controlled in this same manner.[6]

Increased seeding rates (closer plantings) are often used to offset losses by diseases that may be introduced but do not spread from plant to plant. In comparison, Berger[18] found that wide plant-spacing of celery reduced the amount of celery blight (*Cercospora apii*) when compared to close plant-spacing. Steadman et al.[19] also found that wider row spacing decreases the incidence and disease development of white mold (*Whetzelinia sclerotiorum*) on dry beans, as does the type of growth habit of variety grown. Bacterial wilt of cucumber can be controlled by planting 10 to 20% more seed than required to produce a crop, and pulling the diseased plants as they appear. Tomato transplants raised in the western coastal area of the U.S. may carry spotted wilt, a virus, but when planted inland where vectors are scarce and spread of inoculum is small, the original infected plants die. Therefore, by planting at a high density, a good stand which is almost disease-free can be obtained. Dense seeding of cotton reduces loss in yield per acre from Verticillium (*Verticillium albo-atrum*) wilt by providing more plants per acre, which produces less cotton per plant, but total yield is still more than when planted at normal spacing.

Plant spacing and water management are often factors which must be considered together when trying to determine whether either or both of them can be used to control a specific disease. Plant spacing is very important when the microenvironment may be an important part of a disease complex, and it is often difficult to ascertain its exact role in the disease.

Harvesting and Storage Management

The handling procedures used when harvesting a crop is very important to the subsequent condition of the product. Avoiding wounding at harvest is an important practice to prevent storage rots from developing, as most storage rots are caused by wound pathogens. Because wounds are a site of entry for pathogens, white potatoes and sweet potatoes are subjected to a curing period during which time the optimum environment for wound periderm formation is maintained, thus providing a barrier against pathogen entrance, infection, and rot. Other storage diseases of potato that are controlled by harvest practices are water rot (*Pythium* sp.) and late blight tuber rot (*Phytophthora infestans*). The incidence and severity of water rot is greatly reduced when tubers are allowed to mature before harvest and the harvest is carried on at cool temperatures.[20] Late blight tuber rot incidence can be reduced by not harvesting the crop until infected vines are completely dead prior to harvest, thus preventing direct contact of tubers and pathogen.[4]

The methods of handling and storing fruits greatly affect the losses encountered from decay-causing organisms. Rhizopus fruit rot (*Rhizopus nigricans*) of strawberry can be greatly reduced by preventing wounding and by pre-cooling the berries to 5 to 7°C soon after harvest and maintaining that temperature until they are consumed. The cooling of the berries to 5 to 7°C produces an environment in which the fungus is unable to grow while being stored, but will grow when the environment is modified for sale and consumption.[1,20,21] Blue mold rot (*Penicillium expansium*) of apple and pears is controlled by avoiding injuries during harvest and packing, sanitation in orchard and packing house, and prompt cooling to low storage temperature of 3 to 5°C. Gray mold rot (*Botrytis ceneria*) of apple, the second most important disease of eastern-grown apples, can be controlled by keeping the cover crops which are in the orchard under the trees closely mowed, and by prompt cooling and storage to 1°C.[21]

Low temperature storage reduces decay of tropical fruits such as bananas, pineapple, mango, and papaya, mainly by delaying the ripening process, since the chilling temperature for the fruits does not prevent growth of all pathogenic fungi which attack them. Storage atmospheres of 1% oxygen and 99% nitrogen also delay ripening of peaches, bananas, and tomatoes and reduces total decay caused by pathogens.[20] While low oxygen storage prevents post-harvest decay of cabbage by *Botrytis, Fusarium* is still pathogenic under 1.4% oxygen atmosphere. These are only a few examples of ways in which losses to fruits and vegetables may be reduced by reducing injuries, wound healing, and providing a storage environment or atmosphere which retards or prevents the growth of the pathogens but maintains the crop at a high-quality state.[20]

Summary

Avoidance of plant pathogens is one of the best methods of disease control, but it is very difficult to realize for all diseases of a crop in all locations. Under some conditions, such as growing crops for seed, avoidance is the best way to insure that seed is disease-free. For this reason, much of the seed in the U.S. is produced in areas where avoidance can be used to prevent disease incidence. Avoidance can often be used to reduce incidence of a serious plant pathogen attacking a crop while using other control methods to control minor pests. Therefore, if avoidance is to be used as a method of controlling diseases, it must be determined what pests are to be avoided and how a crop management program can be altered to include the various techniques of avoidance while still producing a profitable crop. Avoidance as a control of plant diseases cannot stand alone, but can often be used to some extent in all pest management programs. The little use now made of avoidance in general pest management programs is mainly because of a lack of understanding and feeling that unless the principles are followed in total then the desired control will not be observed. This is incorrect, as many people practice avoidance but do not realize it, e.g., planting on well-drained soil for early crops, utilizing certified seed, and many of the harvesting and storage practices now used.

The place where avoidance is followed most closely is in greenhouses, where total control of soil conditions, temperature, humidity, air flow, water application, and cultural practices are completely in the hands of the producer, and most greenhouse crops can be produced with little or no losses caused by plant pathogens.

As more research on epidemiology and ecology provide information on pathogens and host response, more effective avoidance measures should be forthcoming, allowing a reduction in the use of pesticides.

REFERENCES

1. **Plakidas, A. G.,** *Strawberry Diseases,* Louisiana State University Press, 1964, chap. 2.
2. **Van Arsdel, E. P., Riker, A. J., Kouba, T. F., Suomi, V. E., and Bryson, R. A.,** *The Climatic Distribution of Blister Rust on White Pine in Wisconsin,* Lake States Forest Exp. Stn., Paper No. 87, 1961, 1.
3. **Van Arsdel, E. P.,** Growing White Pines to Avoid Blister Rust, Research Note LS-42, U.S. Forest Service, St. Paul, Minn., 1964.
4. **Zentmyer, G. A. and Bald, J. G.,** Management of the environment, in *Plant Disease: An Advanced Treatise,* Vol. 1, Horsfall, J. G. and Cowling, E. B., Eds., Academic Press, New York, 1977, chap. 7.
5. **McGee, D. C. and Emmett, R. W.,** Black leg (*Leptosphaeria maculans* (Desm.) Ces. et de Not.) of rapeseed in Victoria: Crop losses and factors which affect disease severity, *Aust. J. Agric. Res.,* 28, 47, 1977.
6. Plant-disease development and control, in *Principles of Plant and Animal Pest Control,* Vol. I, National Academy of Sciences, Washington, D.C., 1968.
7. **Fischer, G. W. and Holton, C. S.,** *Biology and Control of the Smut Fungi,* Roland Press, New York, 1957.
8. **Walker, John C., Ed.,** *Plant Pathology,* 2nd ed., McGraw-Hill, New York, 1957.
9. **Bruehl, G. W., Nelson, W. L., Koehler, F., and Vogel, O. A.,** Experiments with Cercosporella foot rot (straw breaker) disease of winter wheat, *Wash. Agric. Exp. Stn. Bull.,* 694, 1968.
10. **Baker, K. F. and Cook, R. J.,** *Biological Control of Plant Pathogens,* W. H. Freeman and Co., San Francisco, 1974.
11. **Chupp, C. and Sherf, A. F.,** *Vegetable Diseases and Their Control,* The Roland Press, New York, 1960.
12. Control of plant-parasitic nematodes, in *Principles of Plant and Animal Pest Control,* Vol. 4, National Academy of Sciences, Washington, D.C., 1968.
13. **Calhoun, J.,** Effects of environmental factors on plant disease, in *Annu. Rev. Phytopathol.,* 11, 343, 1973.
14. **Cook, R. J. and Papendick, R. I.,** Influence of water potential of soils and plants on root diseases, in *Annu. Rev. Phytopathol.,* 10, 349, 1972.
15. **Rotem, J. and Palti, J.,** Irrigation and plant diseases, in *Annu. Rev. Phytopathol.,* 7, 267, 1969.
16. *Plant Pathology: An Advanced Treatise,* Vol. 3, Horsfall, J. G. and Dimond, A. E., Eds., Academic Press, New York, 1960.
17. **Agrios, George N.,** Ed., *Plant Pathology,* Academic Press, New York, 1969.
18. **Berger, R. D.,** Disease incidence and infection rates of *Cercospora apii* in plant spacing plots, *Phytopathology,* 65, 485, 1975.
19. **Steadman, J. R., Coyne, D. P., and Cool, G. E.,** Reduction of severity of white mold disease on great northern beans by wider row spacing and determinate plant growth habit, *Plant Dis. Rep.,* 57, 1070, 1973.
20. **Eckert, J. W.,** Post harvest diseases of fresh fruits and vegetables — the etiology and control, in *Symp. Postharvest Biol. Handling Fruits Vegetables,* Haard, N. F. and Salunkhe, D. K., Eds., AVI Publishing Co., Westport, Conn., 1975, 81.
21. **Ryall, A. L. and Pentzer, W. T.,** *Handling, Transportation and Storage of Fruits and Vegetables,* AVI Publishing Co., Westport, Conn., 1974.

THE USE OF FERTILIZERS AND ORGANIC AMENDMENTS IN THE CONTROL OF PLANT DISEASE

Don M. Huber

INTRODUCTION

Fertilizers are applied to supplement nutrients available in the soil or air in order to maintain the yield, nutritional quality, or aesthetic value of crops produced. Since diseases are a major cause of reduced yield, quality, or aesthetic value, the interaction of nutrition and disease assumes considerable importance in crop production.

The value of plant nutrition in reducing the incidence and severity of plant pathogens has been recognized for many years. Although most metabolic or physiological mechanisms involved in host-pathogen interactions are not clearly understood, specific nutrients are known to reduce disease severity by affecting virulence of the pathogen, enhancing resistance of the plant, compensating for pathogenic damage, or activating indigenous biological control mechanisms. The source of nutrients may be either inorganic or organic and nutrients from either source generally have comparable effects. However, organic sources, as complex mixtures of nutrients, may have a much more complex relationship to disease incidence and severity than many inorganic sources.

There is a large volume of literature available on the interactions of nitrogen (N), phosphorus (P), and potassium (K) with disease.[1-5] These macronutrients are required in large quantity and are frequently limited in soil. Micronutrients have not been studied in relation to disease in as much detail as the macronutrients, yet they have a critical function in metabolism and disease resistance.

Since a primary effect of pathogens is to disrupt the mineral nutrition of the plant,[2] the integration of a well balanced nutrition program with other practices generally provides the best opportunity for disease control. This necessitates a general understanding of the source and function of mineral elements before they can be effectively utilized for disease control.

THE SOURCE AND FUNCTION OF MINERAL NUTRIENTS

Green, autotrophic plants assimilate the substances necessary for growth and development directly in the form of inorganic elements. These elements are carried from the roots to the leaves in a continuous flow of water through the vascular bundles to be intimately involved in all aspects of plant growth. Nutrient elements function as substrates, components, activators, or regulators of metabolism. In addition to specific primary effects of a mineral shortage, far-reaching changes in metabolism and growth occur when tissue is deficient in an essential element as a result of secondary and tertiary interactions of the mineral which eventually reverberate throughout the total metabolic network of the plant.

Thirteen mineral nutrients in addition to carbon, hydrogen, and oxygen are generally essential for the growth and functioning of plants.[6,7] Plants frequently contain other elements, i.e., silicon, sodium, selenium, cobalt, aluminum, etc., which may partially replace an essential nutrient or be "required" under specific environmental conditions. Of the generally essential elements, only potassium, boron, and chlorine are not components of plant constituents. These three elements, along with all the others, function in inducing, regulating, or inhibiting various metabolic processes. Many of the elements are integral components of enzymes and electron carriers. There are certain other functions such as maintaining internal osmotic pressure or electrical neutrality which are not dependent on a specific element, but toward which inorganic and organic ions solubilized in the cell solution contribute. Electrical neutrality is maintained through the counter-ion function of anions and cations. This electrical neutrality, or buffering, influences the physical state and conformation of colloids, polymers, proteins, nucleic acids, etc. as well as the activity and rate of various physiological processes. Thus, each element is part of an intricate system of delicately interdependent reactions. The deficiency or excess of one element greatly influences the activity of others and has catastrophic effects on functioning of the whole plant as evidenced by the symptoms manifested.

The presence of an element in soil does not necessarily imply its availability for plant growth. Nutrient availability is dependent on the abundance of an element, its form and solubility, presence of competing or toxic entities, microbial associations, assimilative capacity of the plant, and environmental factors such as pH, moisture, and temperature. For example, low soil pH may cause direct plant injury from hydrogen ions; physiologically impaired absorption of calcium, magnesium, and phosphorus; toxic levels of solubilized aluminum, manganese, and iron; reduced availability of molybdenum and phosphorus; or a low level of basic salts causing deficiency of calcium, magnesium, and potassium. In contrast, calcareous soils are high in calcium, pH, carbonate, and microbial activity; rich in nutrients; and low in solubility of potentially toxic heavy metals. Many nutrients in soil become available for plant growth through microbial activity which is influenced greatly by the soil environment. Soil nitrogen, sulfur, phosphorus, and various other elements are made available to plants primarily through microbial action that changes their form or solubility, or the absorptive capacity of the root. Microbial activity may also render some nutrients less available, i.e., iron and manganese, through oxidation or immobilization. Microbial activity in the rhizosphere becomes especially important because this is the region of greatest nutrient uptake. Nitrogen, phosphorus, potassium, and calcium are the more generally limiting nutrients; however, chlorine is the only element that has not been found limiting somewhere under natural conditions.

Nitrogen

Nitrogen, as the fourth most abundant element in plants, is a component of numerous compounds as well as involved in many metabolic processes. Nitrogen is available to plants through biological mineralization of complex soil organic matter, microbial fixation of atmospheric nitrogen, or as organic and inorganic fertilizer amendments. Mineralization and symbiotic fixation are dynamic processes providing nitrogen initially in the ammoniacal or reduced form. The subsequent biological oxidation of ammonium nitrogen (NH_4-N) to nitrate nitrogen (NO_3-N) results in the availability of several forms of nitrogen throughout plant growth. Since nitrification may be rapid in many cultivated soils, plants will assimilate primarily NO_3-N which is then internally reduced to NH_2^+ prior to utilization. Most plants can use either form of nitrogen as well as some organic nitrogen, although some plants may be better adapted to one form or the other. Plant amine, amide, and protein composition are higher with NH_4-N than with NO_3-N.[1]

Microbial activities may also reduce the availability of nitrogen for plant use through denitrification (atmospheric) and leaching losses of NO_3-N or immobilization during decomposition of low nitrogen, carbonaceous materials. These activities are all influenced by the soil environment. Nitrification is greatly reduced in acidic, poorly aerated soils; by various sulfur metabolites formed in soil; and by various crop residues and soil amendments. Denitrification is rapid but an anaerobic process. Since the negatively charged NO_3-ion is freely mobile in the soil solution and susceptible to leaching and denitrification losses, the inhibition of nitrification in certain climax ecological systems has been considered a natural mechanism for nitrogen conservation. This has prompted commercial development of nitrification inhibitors to "stabilize" applied NH_4-N fertilizers.

Interactions of nitrogen with other nutrients is common. Potassium increases NO_3-N uptake and promotes the synthesis of organic nitrogen substances, while phosphate and chloride decrease uptake of NO_3-N and enhance uptake of NH_4-N. Chloride also reduces amino acid and protein synthesis and promotes protein degradation. Manganese is required for NO_3-N assimilation and its synthesis into proteins. Molybdenum, magnesium, ferrous ion, ascorbic acid, and energy are also required for the reduction of NO_3^- to NH_2^+. Bacteria symbiotically fixing nitrogen require cobalt, calcium, and energy (in the form of carbohydrates).

Phosphorus

Phosphorus is a part of many plant components and is essential for carbohydrate utilization, nucleic acid synthesis, and energy relationships in the plant. Root growth, cell division, seed development, and a shortened vegetative period are promoted by adequate supplies of phosphorus. Phosphorus and potassium together promote strong mechanical tissues; in contrast, excessive levels of nitrogen produce succulent tissues. Like nitrogen (NH_4^+ or NO_3^-) and sulfur (SO_4^-), phosphorus is absorbed as a complex ion (primarily as the dihydrogen phosphate ion, $H_2PO_4^-$). However, unlike nitrate (NO_3^-) and sulfate (SO_4^-), the phosphate ion is not reduced to a lower oxidation state in the cell. Phosphorus is obtained from mineral or organic phosphate in soil or from applied fertilizers such as rock phosphate, superphosphate, ammonium phosphate, basic slag, or bone meal. Inorganic phosphorus in soil occurs in many forms depending on pH, but it must be in the soluble phosphate form to serve as a plant nutrient. In acidic soils, phosphorus is readily available but physiological absorption is impaired. Plant roots may exert a solvent action on soil particles through root exudates to bring more phosphorus into solution. Although chemical considerations would suggest competition between phosphate and molybdenum, phosphate spectacularly promotes the absorption of molybdate by plants.

The availability of insoluble soil phosphorus, like nitrogen, is primarily dependent on microbial activity in the rhizosphere[8] or through mycorrhiza.[9] Mycorrhizae of herbaceous as well as woody plants appear as important to phosphorus nutrition under the limiting conditions found in many soils as symbiotic nitrogen fixation is for nitrogen.

Potassium

Potassium occurs as the primary weathered mineral, or in exchangeable and water-soluble forms. The exchangeable and water-soluble forms are most readily available for plant nutrition. Potassium is exchanged from soil particles or supplied in fertilizers as potassium chloride (muriate), potassium sulfate, potassium nitrate, or wood ashes. The source of potassium may influence uptake of other nutrients through its accompanying anion. It is not available in many textured soils because its salts are readily leached from the root profile or soil.

Potassium, unlike most other essential nutrients, does not become a structural component of the plant and is considered a mobile regulator of cellular activity. It cannot be completely replaced by the chemically similar elements sodium or lithium, and occurs principally as soluble inorganic or organic salts. As a mobile regulator of enzyme activity, it is involved in essentially all cellular functions including photosynthesis, phosphorylation, protein synthesis, translocation, water maintenance, reduction of nitrates, and reproduction. A balanced level of potassium induces thicker cell walls, accumulation of amino acids (arginine), and production of new tissues.

The potassium content of a plant depends on the level of magnesium and calcium which are influenced by soil reaction. Potassium availability in soil is enhanced by calcium in neutral, but not acidic soils. The potassium to calcium balance also affects the differential permeability of membranes.

Calcium

The most important primary calcium minerals in soil are anarthite ($CaAl_2Sl_2O_8$) and pyroxenes ($CaMg[SiO_3]_2$); however, calcite or dolomite may be the dominant source in some soils. Calcium is supplied in commercial fertilizer as calcium carbonate (calcite), calcium magnesium carbonate (dolomite), calcium hydroxide (hydrated lime), calcium sulfate (gypsum), and calcium phosphate. It is a structural component of the middle lamellae of plants and is involved in cell division, cell development, carbohydrate movement, and the neutralization of metabolic acids. Calcium, unlike many nutrients, is relatively immobile and is not readily redistributed within the plant. Mature plant parts may have large calcium reserves while younger tissues are deficient. Its functions are probably complementary to those of potassium in maintaining cell organization, hydration, and permeability. In these capacities it is involved in mitosis, enzyme activation and regulation, and membrane function. Calcium deficiency is characterized by weak stems and limited root development. Calcium interacts with phosphorus to reduce adverse effects of excessive phosphorus and other elements on the growth of some plants. Reduced nodulation of legumes by *Rhizobium* caused by high phosphorus levels may actually result from an induced calcium deficiency. Acidic soils tend to be deficient in calcium and one important function of calcium fertilizers is the neutralization of soil acidity. By decreasing soil acidity, calcium decreases toxicity of such heavy metals as aluminum, boron, manganese, and others which are soluble at an acidic pH. Phosphorus and molybdenum solubility is enhanced with ''liming'' and may be reflected in increased nitrogen efficiency via nitrate reductase. Excess calcium, generally associated with alkaline soils, may induce deficiencies of iron, manganese, copper, boron, and zinc.

Sulfur

Sulfur is present in soil in organic and inorganic forms such as pyrite (FeS_2O), saphalerite (ZnS), chalcopyrite ($CuFeS_2$), cobaltite (CoAsS); or soluble sulfate of calcium (gypsum), magnesium, potassium, and sodium. Sulfur is usually absorbed by roots as the sulfate ion (SO_4) or enters the leaves as sulfur dioxide (SO_2) from gas present in the atmosphere. Sulfur deficiency is rare in industrial countries because of adequate atmospheric levels of various sulfur oxides. It is reduced in the plant (similar to NO_3^-) and is incorporated into amino acids, proteins, enzymes, vitamins, aromatic oils, and ferredoxins. Root growth and nodulation of legumes is promoted by sulfur fertilization. Sulfur, like calcium, is relatively immobile in plants.

Magnesium

Magnesium exists in soil in water-soluble, exchangeable, fixed, and primary mineral forms. The content and forms available vary, depending on geological origin, rainfall, and the presence of potassium or other exchangeable cations. Acid soils tend to be deficient in magnesium because of impaired absorption of this ion as well as calcium, molybdenum, and phosphorus. High levels of potassium or calcium relative to magnesium may inhibit uptake of magnesium and vice versa.

As a constituent of chlorophyll, magnesium functions in photosynthesis. It is also associated with rapid growth, active mitosis, high protein levels, carbohydrate metabolism, and oxidative phosphorylation in physiologically young cells. Unlike calcium, magnesium is translocated from mature to actively growing regions of the plant.

Iron

Iron is present in soil in mineral, oxide, sulfide, and organic complexes. The ferrous forms of iron are the most available for plant nutrition. As soils become more alkaline, ferrous iron is oxidized to the ferric form which is not available for growth. "Lime-induced" chlorosis from iron deficiency is accompanied by an increase in the potassium to calcium ratio in chlorotic as compared with green foliage. Heavy metal excesses such as manganese and copper may also interfere with iron absorption. Deficiencies of potassium, phosphorus, or calcium may induce or enhance iron deficiency by interfering with its translocation. A change in the potassium to calcium ratio greatly alters potassium and iron nutrition. Phosphorus may immobilize iron as ferriphosphate in marginal soils. Thus, the apparent deficiency of iron may not be due to a lack of iron, but to its unavailability because of high pH or a mineral imbalance.

Iron in the plant occurs primarily in the form of porphyrins (hemes). These are critical elements of terminal oxidation systems (electron transport) and other oxidative enzymes. Iron is also essential for chlorophyll synthesis and intimate with the reactions of photosynthesis. Balanced potassium increases the efficiency of iron utilization in chlorophyl production.

Zinc

Since most zinc-bearing minerals in soil are readily weathered, the released zinc is generally absorbed onto colloids or complexed by organic matter. Zinc is the metal component of several dehydrogenases, peptides, and other metalloenzymes. It, like iron, manganese, copper, and molybdenum, has intermediate mobility in the plant which accounts for the intensity of deficiency symptoms being more pronounced in young tissues. This is contrasted to the highly mobile relationship of potassium, rubidium, sodium, magnesium, phosphorus, sulfur, and chlorine, but intermediate relative to lithium, calcium, barium, boron, and strontium. The most prominant physiological role known for zinc is its interrelationship with auxin. Addition of zinc to deficient plants greatly stimulates auxin synthesis. It is thereby essential for cell elongation and growth, as well as being functional in respiration and enzyme regulation. Large increases in amino acids are noted with Zn deficiency.

Copper

Copper sulfide derived from chalcopyrite is the most important primary source of this element in soil. Atmospheric sources may provide significant amounts of copper, as well as iron, zinc, manganese, boron, and molybdenum, as volatile compounds or in precipitation. Like zinc, copper is a component of several enzymes and is involved in protein and carbohydrate synthesis, and nitrogen fixation.

Boron

Various complexes containing boron are present in soil. Boron is easily leached from light, acidic soils and plant uptake is reduced by increasing soil pH. There is a relatively narrow range of concentration between deficiency and toxicity of boron. Toxicity may result in arid areas where sodium and calcium borates accumulate in surface soils, or in acidic soils where it is more soluble. It is functional in translocation, cellular differentiation and development, carbohydrate metabolism, pollen germination, and the uptake or translocation of calcium.

Manganese

The availability of soil manganese is determined largely by pH and oxidation-reduction conditions present. pH values below 6 favor reduction and values above 6.5 favor oxidation to non-available states. Biological oxidation in the rhizosphere is generally responsible for immobilization at pHs of 6 to 7.9. Manganese, as well as aluminum, may be toxic in acidic soils because of increased solubility. Manganese is a constituent of only one known plant component, manganin, but it activates various enzymes involved in nitrate reduction, carbohydrate metabolism, and respiration. It plays a direct and primary role in photosynthesis. At high concentration, manganese is a competitive inhibitor of iron absorption, translocation, and binding.

Chlorine

Chlorine is the only "essential" element for which a deficiency has not been observed in nature. This is probably due to the prevalence of cyclic salts as well as its gaseous abundance in the atmosphere relative to requirements for growth. Excess chloride is very much a matter of concern in many areas. Chlorine functions in oxygen evolution by photosytem II; however, wilting and amino acid accumulation in deficient plants may indicate an additional role in transpiration and amino-acid metabolism.

Sodium

Sodium has been demonstrated to improve the growth of celery, mangold, sugar beets, turnips, and several other plants, but has not been determined "essential" for most of them. It may partially replace potassium, and functions to lower osmotic pressure of certain halophytes. High concentrations of sodium found in "saline" soils are toxic to many plants.

Cobalt

Although not known to be essential for higher plants, cobalt is required for some symbiotic nitrogen fixing relationships and is a constituent of vitamin B-12. In soil, cobalt occurs in an available or extractable form as phosphate or sulfate salts. It is deficient in some light, calcareous, sandy soils.

USE OF NUTRIENTS FOR DISEASE CONTROL

Nonparasitic Diseases

The deficiency or excess of essential nutrients cause disease which can generally be corrected by supplying the essential nutrient or reducing its concentration.

Sufficiently severe deficiencies manifest themselves in more or less distinct symptoms. However, it should be remembered that visible symptoms are late manifestations of metabolic derangements which occurred long before the effects became apparent. Diagnosis of mineral deficiencies is complicated because: (1) reduced growth and quality may occur without being obvious; (2) the same element may induce different symptoms in different plants or under varying environmental conditions; and (3) similar or identical symptoms may result from deficiencies of different elements. For example, both manganese and iron deficiency result in a similar chlorosis of foliar tissues in moist, alkaline soils. It is important to determine whether iron or manganese deficiency is the problem in each specific situation since treatment of manganese-deficient plants with iron further suppresses manganese uptake, and symptoms become even more severe. Both nitrogen and sulfur deficiencies cause a similar chlorosis, and symptoms are relieved only after both elements are available in sufficient quantity. Multiple deficiencies magnify the problem of diagnosis, and parasites or toxicants may cause similar symptoms.

When an element is deficient, its content in the plant is reduced and tissue testing may be used as a general guide to the nutritional status of the plant.

Resistance to this type of disease is manifested as tolerance to excess levels or greater efficiency in nutrient absorption or utilization. Tolerance to various toxicants may be influenced by the nutrient status of the plant, and "balanced" nutrition becomes as important as the presence of one or another element.

Excess of a specific nutrient may be as damaging as a deficiency. Toxicity may result from increased availability of a nutrient in soil, i.e., manganese or aluminum, as soil conditions become more acid, or from foliar uptake of atmospheric sources. Drastically altered ratios of some nutrients may also induce excess uptake of toxic quantities of another element. Many plants have the capacity to absorb "luxury" amounts of readily available nutrients whereby toxicity may result from excess ion accumulation. An example is the excess accumulation of NH_4-N by tobacco plants growing in fumigated soils under environmental conditions restricting photosynthesis. The amination of organic acids may induce an energy deficiency relative to other metabolic needs. This "ammonia" toxicity is avoided if some NO_3-N is also available to the plant. Another possible explanation of NH_4-N toxicity after soil fumigation is the differential effects of some fumigants on *Nitrobacter* compared with *Nitrosomonas*. In these situations toxic levels of nitrite (NO_2-N) accumulate because the population of *Nitrosomonas* that converts NH_4-N to NO_2-N recovers more rapidly than *Nitrobacter* which oxidizes NO_2-N to NO_3-N. Use of *Nitrosomonas* specific nitrification inhibitors such as nitrapyrin (2-chloro-6-(trichloromethyl)-pyridine) prevent NO_2-N toxicity which was earlier attributed to excess NH_4-N.[10]

In addition to deficiencies caused by diseases, nutrient deficiency may be imposed by biological agents through the immobilization or competitive utilization of nutrients in the rhizosphere. An example of microbially induced deficiency is the gray-speck disease of oats which is caused by manganese-oxidizing bacteria growing in the rhizosphere of gray-speck susceptible, but not resistant, oats.[11] Nitrogen deficiency may be induced during the microbial decomposition of carbonaceous residues in soil. In contrast to these examples of microbial induced deficiencies, microorganisms frequently prevent many deficiency diseases. The symbiotic relationship involving nitrogen fixation by root nodule bacteria (*Rhizobium*) on the roots of many leguminous plants provides nitrogen for the plant in return for plant-produced carbohydrate for the bacteria. Microorganisms forming mycorrhizae may increase absorption of phosphorus, potassium, and nitrogen by plants. Factors which influence either the mycorrhizae or nitrogen-fixing ability of *Rhizobium* also influence the nutritional health of plants.[8,9]

It is not always possible to clearly distinguish parasitic (infectious) from nonparasitic diseases since most parasitic diseases impair nutrient uptake, translocation, or utilization. Thus, many of the primary and secondary symptoms associated with pathological problems are similar to those expressed by mineral deficiencies.[2] Stunting, chlorosis, wilting, mottle, rosette, dieback, leaf spot, abnormal growth, etc. may be induced by specific mineral deficiencies. Increased amino acids, auxin, and other materials commonly associated with pathogenesis are also manifestations of specific mineral deficiencies. Potassium deficiency causes soluble nitrogen compounds to accumulate resulting in necrotic leaf spots similar to symptoms induced by several foliar pathogens. The marginal chlorosis and leaf blotch induced by severe infections of wheat by *Ophiobolus graminis* on sandy soils is sometimes incorrectly diagnosed as magnesium deficiency. Treatment with dolamitic lime, however, fails to correct the pathological problem. The dwarfing, rosetting, and bronzing effects of zinc deficiency are common symptoms of virus infection.

Parasitic Diseases

Nutrition, as one of the environmental factors influencing parasitic disease, affects survival and virulence of pathogens, vigor and resistance of plants, or mechanisms of biological control. The process of parasitism is defined as a nutritional relationship where one organism obtains all or part of its food from another living organism. This concept of parasitism is compatible with the hypothesis that a parasite grows in or on a particular plant because the kinds and quantities of foods required for its survival are found only there, or that resistant plants contain inhibitors as well as nutrients.[12-14] In this relationship the nutrient status of the environment may have a direct effect on disease through food requirements of a pathogen, or an indirect effect through resistance of the plant.

Resistance to parasitic diseases is generally a dynamic process involving the production of various metabolites which may be formed prior to, or in response to, penetration of a pathogen. Resistance may also be expressed as the lack of response to a pathogen (especially obligate) if modification of the host is required for disease development. The production of mechanical barriers and inhibitory compounds, or the continued availability of nutrients essential for parasitic growth, requires energy and chemical skeletons from various metabolic pathways mediated by mineral nutrients. Thus, although resistance is genetically controlled, it is intimately interrelated with the nutritional status of the plant or pathogen through physiological processes.

It is not possible to generalize the effects of any particular nutrient for all host/pathogen combinations, and the influence of nutrition on infectious plant diseases must be considered on an individual disease basis. Variables such as host, environment, and pathogen must be taken into consideration. It is not often that a disease can be eliminated by a given type of fertilizer applied to the crop, but the severity of many diseases is reduced by specific nutrients. In some cases fertilizing may not increase the resistance of the plant as much as stimulate growth of the crop to minimize disease damage.

A severely nutrient-stressed plant is frequently more vulnerable to disease than one at a nutritional optimum. A few plants receiving large excesses of a mineral element, especially nitrogen which promotes succulence and excessive vegetative growth, also may be severely diseased. An example of this type of nutrient interaction is *Pythium* root rot of wheat and sugar cane which increases following nitrogen fertilization of low fertility soils. However, this disease is reduced by simultaneous applications of nitrogen with phosphorus that stimulates rapid development of new roots when the two nutrients are added together.

Differences in disease severity may be observed where the nutritional level is adequate but where the "balance" or form of nutrients is modified. These effects are sometimes as great as observed with nutrient deficiencies or excesses. This is exemplified by the contradictory effects of specific forms of nitrogen on many diseases.[1,15] Reduced disease through altered host resistance generally results from an alteration of metabolic pathways affecting growth, plant constituents, or exudates rather than a direct effect of the nutrient per se. Nutritionally induced changes in host composition are manifest in plant exudates (leaf and roots) which influence the survival, germination, growth, and virulence of many plant pathogens. Mineral nutrients appear to influence disease potential more than inoculum potential, and some nutrients may decrease disease even though the population of a pathogen is increased. Thus, the intricate relationship of plant pathogens with other microorganisms, environmental factors, and the host is dynamic as well as extremely complex. Nevertheless, knowledge of host nutrition in relation to disease development provides a basis for modifying current agricultural practices to reduce disease severity.

Although a deficiency or excess of any plant nutrient may influence disease severity, nitrogen, potassium, and phosphorus have been most extensively investigated.

Nitrogen

Nitrogen has been intensively studied in relation to host nutrition and disease severity for many years because of its essential requirement for plant growth, its limited availability in soil, and its effect on cell size and wall thickness. It promotes vigorous growth, delays maturity, and is essential for the production of amino acids, proteins, growth hormones, and new protoplasm. Many plant constituents altered by nitrogen are correlated with resistance or susceptibility to disease, i.e., amino acids, total nitrogen, carbohydrate: nitrogen ratios, etc. A common generality has been expressed that nitrogen tends to increase disease while potassium tends to increase disease resistance, and phosphorus may elicit either response. However, there are many exceptions to such generalizations which fail to account for effects of the form of nitrogen, rate or time of application, or soil conditions. Some foliar pathogens are able to penetrate, multiply, and develop more rapidly in succulent tissues promoted by nitrogen, while soil-borne, root disease pathogens may be reduced as the more vigorus seedlings and plants escape severe disease damage. Nitrogen is the main mineral factor changing the amount of cellulose and thereby affecting the mechanical strength of cell walls. Reduced cellulose content of cell walls after high nitrogen applications (unbalanced) has been associated with increased susceptibility to some root diseases.

The effect of specific forms of nitrogen on disease severity depends on many factors and is not the same for all host-parasite associations.[1] Assimilation of nitrogen is more complicated than other essential elements because it is assimilated both as a cation (NH_4^+) and as an anion (NO_3^-). Water, pH, other elements, and temperature influence the uptake and utilization of specific forms of nitrogen by plants, pathogens, and interacting microorganisms. Although a wide range of interactions of pathogens and their hosts are influenced by nitrogen (Table 1), it is frequently the form of nitrogen available to the host or pathogen that affects disease severity or resistance rather than the amount of nitrogen (Table 2).

Disease control achieved with a specific form of nitrogen may also depend on host response or preference, previous cropping, nitrogen rate and stability, residual nitrogen, time of application, soil microflora, or the disease complex present. Inhibition of nitrification can reduce nitrogen losses, increase utilization efficiency, and establish a predominantly ammoniacal form of nitrogen available for plant uptake. This provides an effective tool to reduce the severity of diseases that are affected by NH_4-N.

The time of application of nitrogenous fertilizers also may have a pronounced effect on disease expression. ''Side-dressing'' nutrients to plants after emergence avoids predisposition to seedling disease such as damping-off caused by *Rhizoctonia* or *Pythium*, but may increase *Fusarium (Gibberella)* and other root rots due to mechanical damage to root systems or other plant parts. Delayed application of nitrogen to winter wheat until spring frequently results in an early nitrogen deficiency and predisposition to take-all and *Cercosporella* foot rot (Figure 1).

Nitrogen influences host plant resistance by reducing the frequency of successful pathogen penetration or by retarding pathogenesis after penetration. Increased resistance of wheat to take-all is indicated by reduced infection and smaller lesions with NH_4-N than NO_3-N.[1] Increased root growth permitting wheat to escape take-all also has been reported as a benefit of nitrogen fertilization in a balanced fertilizer program.[16] Neither form of nitrogen prevents infection of stone fruits by *Xanthomonas pruni*, although NH_4-N reduced bacterial canker of prunes by hastening periderm development so that cankers healed more promptly. In contrast, NO_2^{-N} reduced defoliation of peaches and plums.

Table 1
DISEASES INFLUENCED BY NITROGEN WITHOUT REFERENCE TO THE FORM OF NITROGEN

Disease	Pathogen	Host plant	Effect of nitrogen
Bacterial diseases			
Wildfire	*Pseudomonas tabaci*	Tobacco	Decrease
Leaf spot	*P. pruni*	Peach	Decrease
Bacterial spot	*Xanthomonas pruni*	Peach/plum	Decrease
Shot-hole	*Bacterium pruni*	Peach	Decrease
Bacterial wilt	*Erwinia tracheiphila*	Cucumber	Decrease
Blight	*P. solanacearum*	Tomato	Decrease
	Pseudomonas	Tobacco	Decrease
Frenching	*Bacillus subtilis*	Tobacco	Decrease
Stem rot	*X. pelargonii*	Pelargonium	Increase
Bacterial streak	*X. translucens*	Rice	Increase
Leaf blight	*X. oryzae*	Rice	Increase
Common blight	*X. phaseoli*	Bean	Increase
Bacterial blight	*Xanthomonas* spp.	Many	Increase
	Pseudomonas spp.	Many	Increase
Brown spot	*Pseudomonas syringae*	Lima bean	Increase
Fire blight	*E. amylovora*	Pear/apple	Increase
Blackfire	*P. angulata*	Tobacco	Increase
Wildfire	*P. tabaci*	Tobacco	Increase
Angular leaf spot	*P. tabaci*	Tobacco	Increase
Stewart's wilt	*X. stewartii*	Maize (corn)	Increase
Southern wilt	*P. solanacearum*	Tobacco	Increase
Citrus canker	*Xanthomonas citri*	Citrus	Increase
Crown gall	*Agrobacterium tumefaciens*	Beet	Increase
Saprophyte	*A. radiobacter*	Legume	Decrease
Fungus diseases			
Club root	*Plasmodiophora brassicae*	Cabbage	Decrease
Take-all	*Ophiobolus graminis*	Wheat/barley	Decrease
Red thread	*Corticium fusiforme*	Turf grass	Decrease
Dollarspot	*Sclerotinia homeocarpa*	Blue grass	Decrease
Root rot	*Fusarium* spp.	*Trifolium pratense*	Decrease
	Aphanomyces euteiches	Pea	Decrease
	Phymatotrichum omnivorum	Cotton	Decrease
	Rhizoctonia solani	Many	Decrease
Stem canker	*R. solani*	Soybean	Decrease
Damping off	*R. solani*	Many	Decrease
Root rot	*Helminthosporium sativum*	Wheat	Decrease
	Pythium arrhenomanes	Cereal	Decrease
Foot rot	*Cercosporella herpotrichoides*	Cereal	Decrease
Sclerotium rot	*Sclerotium rolfsii*	Sugar beet	Decrease
Black spot	*Diplocarpon rosae*	Rose	Decrease
Wilt	*Fusarium oxysporum* f. *lycopersici*	Tomato	Decrease
	Fusarium spp.	Tomato	Decrease
	F. oxysporum f. *niveum*	Watermelon	Decrease
	F. oxysporum f. *vasinfectum*	Cotton	Decrease
	Fusarium spp.	Red clover	Decrease
	Verticillium spp.	Various	Decrease
	Verticillium albo-atrum	Cotton	Decrease
		Tomato	Decrease

Table 1 (continued) DISEASES INFLUENCED BY NITROGEN WITHOUT REFERENCE TO THE FORM OF NITROGEN

Disease	Pathogen	Host plant	Effect of nitrogen
Yellows	*F. oxysporum* f. *conglutinans*	Cabbage	Decrease
Snowmold	*Typhula* spp.	Wheat	Decrease
	F. nivale	Bentgrass	Decrease
Late blight	*Phytophthora infestans*	Potato	Decrease
Seedling smuts	*Tilletia* spp.	Cereals	Decrease
Bunt	*Tilletia* spp.	Wheat	Decrease
Rust	*Puccinia* spp.	Blue grass	Decrease
Stem rust	*Puccinia graminis*	Wheat	Decrease
Root rot	*Phymatotrichum omnivorum*	Cotton	Increase
	Pythium spp.	Sugarcane	Increase
Browning root rot	*Pythium* spp.	Wheat	Increase
Root rot	*R. solani*	Bean	Increase
	F. solani f. *phaseoli*	Bean	Increase
Damping off	*F. oxysporum*	Conifer	Increase
	Pythium debaryanum	Tomato	Increase
Brown patch	*R. solani*	Blue grass	Increase
Patch	*Ophiobolus graminis*	Turf grass	Increase
Brown patch	*R. solani*	Bent grass	Increase
Basal rot	*F. oxysporum* f. *narcissi*	Narcissus	Increase
Black root	*Phoma betae*	Sugar beet	Increase
Dollarspot	*Sclerotinia homeocarpa*	Blue grass	Increase
Stem rot	*H. sigmoidum*	Rice	Increase
Stem rot	*Leptosphaeria saloinii*	Rice	Increase
Sheath rot	*Acrocylinderium oryzae*	Rice	Increase
Stalk rot	*Diplodia zeae*	Maize	Increase
Fruit rot	*Gleosporium* spp.	Apple	Increase
Rot	*Botrytis cinerea*	Apple	Increase
Stalk rot	*Gibberella zeae*	Maize	Increase
Brown spot	*O. miyabeanus*	Rice	Increase
Sheath blight	*Corticium sasaki*	Rice	Increase
Black stem	*Phoma herbarium* var. *medicaginis*	Alfalfa	Increase
Sesame leaf spot	*Cochliobolus miyabeanus*	Rice	Increase
Brown leaf spot	*Cercospora oryzae*	Rice	Increase
Leaf scald	*Rhynchosporium oryzae*	Rice	Increase
Leaf spot	*H. hevea*	Hevea	Increase
	Helminthosporium spp.	Rice	Increase
Leaf blight	Unlisted	Tomato	Increase
Blast	*Piricularia oryzae*	Rice	Increase
Eye spot	*H. sacchari*	Sugarcane	Increase
Scab	*Venturia inaequalis*	Apple	Increase
Wet rot	Unlisted	Potato	Increase
Late blight	*Phytophthora infestans*	Potato	Increase
Dutch elm disease	*Ceratocystis ulmi*	Elm	Increase
Canker	*Fusicoccum amygdali*	Peach	Increase
Gray rot	*B. cinerea*	Grape	Increase
Early blight	*Alternaria solani*	Tomato	Increase
Blister blight	*Exobasidium vexens*	Tea	Increase

Table 1 (continued)
DISEASES INFLUENCED BY NITROGEN WITHOUT REFERENCE TO THE FORM OF NITROGEN

Disease	Pathogen	Host plant	Effect of nitrogen
Wilt	*Verticillium albo-atrum*	Hop	Increase
		Cotton	Increase
	Fusarium oxysporum f. *lycopersici*	Tomato	Increase
	F. oxysporum f. *vasinfectum*	Cotton	Increase
	F. oxysporum f. *dianthi*	Carnation	Increase
	F. oxysporum f. *melonis*	Melon	Increase
Yellows	*F. oxysporum* f. *conglutinans*	Cabbage	Increase
	F. oxysporum f. *gladioli*	Gladiolus	Increase
Basal rot	*F. oxysporum* f. *narcissi*	Narcissus	Increase
Stem rust	*Puccinia graminis*	Wheat	Increase
Leaf rust	*P. recondita*	Wheat	Increase
Stripe rust	*P. striiformis*	Wheat	Increase
Rust	*Puccinia* sp.	Wheat	Increase
Crown rust	*P. coronata*	Oat	Increase
Rust	*P. secalina*	Rye	Increase
	P. sorghi	Sorghum	Increase
	Puccinia spp.	Barley/maize	Increase
	Melampsora lini	Flax	Increase
Powdery mildew	*Erysiphe graminis*	Cereals	Increase
Fusiform rust	*Cronartium fusiforme*	Loblolly pine	Increase
Mildew	*Bremia lactucae*	Lettuce	Increase
Rust	*Sphaerotheca mors-uvae*	Gooseberry	Increase
Downy mildew	*Peronospora parasitica*	Cabbage	Increase
Flag smut	*Urocystis tritici*	Wheat	Increase
Smut	*Ustilago zeae*	Maize	Increase
Stalk smut	*Urocystis occulata*	Rye	Increase
Virus diseases			
Mosaic	Tobacco mosaic virus	Bean	Decrease
Ring spot	Cherry ringspot virus	Cherry	Decrease
Virus	Cucumber virus-1	Spinach	Increase
	Turnip virus-1	*Nicotiana glutinosa*	Increase
		Nicotiana multivalvis	Increase
Mosaic	Lettuce mosaic virus	Lettuce	Increase
	Common bean mosaic virus	Bean	Increase
	Barley stripe mosaic virus	Wheat	Increase
Wilt	Tomato spotted wilt virus	Tomato	Increase
Virus	Yellow dwarf virus	Onion	Increase
Mosaic	Tobacco mosaic virus	Tobacco	Increase
		Tomato	Increase
Insect diseases			
Aphid	*Myzus persicae*	Tobacco	Decrease
Cabbage fly	*Phorbia brassicae*	Cabbage	Decrease
Chinch bug	*Blissus leucopterus*	Corn	Decrease
Frit fly	*Oscinella frit*	Barley	Decrease
Mite		Bermuda grass	Decrease
Rice weevil	*Sitophilus oryzae*	Wheat	Decrease
Wilt	Mealy bug	Pineapple	Decrease
Shot-hole borer	*Xylebrous fornicatus*	Tea	Decrease
White fly	*Aleurolobus barodensis*	Sugarcane	Decrease
Thrips	*Heliothrips haemorrhoidalis*	Spinach	Decrease

Table 1 (continued)
DISEASES INFLUENCED BY NITROGEN WITHOUT REFERENCE TO THE FORM OF NITROGEN

Disease	Pathogen	Host plant	Effect of nitrogen
Chinch bug	*B. leucopterus*	St. Augustine grass	Increase
		Sorghum	Increase
Armyworm	*Pseudolitia unipuncta*	Turf grass	Increase
Sod webworm	*Crmbus nutabilis*	Turf grass	Increase
Aphid	*Aphis gossypii*	Cotton	Increase
Leafhopper	*Empoasca terra-reginae*	Cotton	Increase
	Cactoblastis spp.	Prickly pea	Increase
Silkworm	*Bombyx mori*	Mulberry	Increase
Mealy bug	*Pseudococcus comstocki*	Apple	Increase
Gout fly	*Chlorops taeniopus*	Barley	Increase
Greenbug	*Toxoptera graminum*	Wheat	Increase
Nematode diseases			
Nematode	*Pratylenchus penetrans*	Pea	Decrease
Root knot	*Heterodera marioni*	Cotton	Decrease
Nematode	*Meloidogyne javanica*	Tomato	Increase
	Aphelenchoides oryzae	Rice	Increase

From Huber, D. M., in *CRC Handbook Series in Nutrition and Food,* Rechcigl, M., Ed., CRC Press, Boca Raton, Fla., accepted for publication, 1980. With permission.

Increased infection of broad bean by *Botrytis fabae* was correlated with higher levels of sugars and amino acids in leaf exudates when roots were supplied NH_4-N as opposed to NO_3-N. NH_4-N increased host-cell permeability and fungal spore germination, and increased infection 2.5-fold compared with NO_3-N.[17]

The effect of specific forms of nitrogen on soil-borne diseases has been reviewed by Huber and Watson.[1] More severe damping-off of lettuce, beans, and sugar beets (black root rot) caused by *Rhizoctonia solani* with NH_4-N compared with NO_3-N is similar to effects observed with increasing levels of asparagine. NH_4-N increases the level of glutamine and asparagine in plants compared with NO_3-N and growth of *R. solani* on stems, production of infection cushions, penetration, and lesion enlargement increase as asparagine levels increase. Levels of carbon and nitrogen that reduce damping-off do not reduce the dry weight or colony diameter of *Rhizoctonia* in culture. Increased severity of diseases caused by *R. solani* with NH_4-N also has been correlated with the production of extra-cellular macerating enzymes (pectinases, cellulase) with NH_4-N or organic sources of nitrogen and the inhibition of these enzymes by NO_3-N.[18] These factors may account for the reduction in root rots caused by *R. solani* and *Fusarium solani* after application of NO_3-N. Weinke[19] found that *Fusarium* root rot of bean was increased only when NH_4-N was placed in the hypocotyl zone where the pathogen mainly invaded. Nitrogen applied to the root zone or to foliage to supply adequate nitrogen for growth of bean plants did not result in increased disease. In his extensive studies, increased disease was associated with more rapid development of a larger hyphal thallus, increased pathogenicity, and more rapid lesion coalescence with NH_4-N than NO_3-N.

Similar observations were made with *Aphanomyces* root rot of peas. This disease increased when NH_4-N was applied to soil but not when amounts adequate for plant growth were applied to foliage. Carley[20] demonstrated an indirect effect of nitrogen

Table 2
EFFECT OF INORGANIC FORMS OF NITROGEN ON PLANT DISEASE

Disease	Pathogen	Host plant	Nitrogen form: Nitrate	Nitrogen form: Ammonium
Bacterial diseases				
Angular leaf spot	*Xanthomonas malvacearum*	Cotton	Decrease	
Canker	*X. pruni*	Peach/prune	Decrease	
	Corynebacterium michiganense	Tomato	Increase	
Ring spot	*C. sepedonicum*	Potato	Increase	
Southern wilt	*Pseudomonas solanacearum*	Tobacco/tomato	Increase	Decrease
		Tomato		Increase
Stewart's wilt	*X. stewartii*	Maize	Increase	
Crown gall	*Agrobacterium tumefaciens*	Tomato	Increase	
		Beet		Increase
Fungus diseases				
Club root	*Plasmodiophora brassicae*	Cabbage	Decrease	
Scab	*Streptomyces scabies*	Potato	Increase	Decrease
Damping-off	*Pythium ultimum*	Beet	Decrease	
Seedling blight	*Rhizoctonia solani*	Sugar beet	Decrease	Increase
Black root	*Aphanomyces cochlioides*	Sugar beet	Increase	Decrease
Root rot	*A. euteiches*	Pea	Increase	Decrease
		Pea/soybean	Decrease	Increase
		Maize/tomato	Decrease	Increase
Black root	*R. solani*	Sugar beet	Decrease	Increase
Root rot	*Pythium* spp.	Pea	Increase	Decrease
		Maize	Increase	Decrease
		Cereal	Increase	
	P. debaryanum	Oat	Decrease	
	Phytophthora citrophthora	Citrus	Increase	Decrease
			Decrease	Increase
	R. solani	Cereal		Decrease
Root and stem rot	*R. solani*	Potato	Decrease	Increase
Root rot	*R. solani*	Bean	Decrease	Increase
	Phymatotrichum omnivorum	Cotton	Increase	Decrease
	Thielaviopsis basicola	Bean	Increase	Decrease
		Tobacco	Decrease	Increase
	Fusarium solani f. *phaseoli*	Bean	Decrease	Increase
	F. roseum f. *cerealis*	Wheat	Decrease	Increase
	Helminthosporium sativum	Wheat	Decrease	
	Poria weirii	Pine	Decrease	Increase
	Armillaria mellea	Pine	Decrease	Increase
	Unknown	Tobacco	Decrease	
	Fusarium spp.	Citrus		Increase

Table 2 (continued)
EFFECT OF INORGANIC FORMS OF NITROGEN ON PLANT DISEASE

Disease	Pathogen	Host plant	Nitrogen form	
			Nitrate	Ammonium
Take-all	*Ophiobolus graminis*	Cereals	Increase	Decrease
Patch	*O. graminis*	Turf grass		Decrease
Stem rot	*Sclerotium rolfsii*	Peanuts	Decrease	
Southern rot	*S. rolfsii*	Sugar beet	Decrease	Increase
Southern blight	*S. rolfsii*	Tomato	Decrease	Increase
Sharp eye spot	*R. solani*	Wheat	Decrease	Increase
Eye spot	*Cercosporella herpotrichoides*	Wheat	Decrease	Increase
Stalk rot	*Diplodia zeae*	Maize	Increase	Decrease
	Gibberella zeae	Maize	Increase	Decrease
	F. moniliforme	Maize	Decrease	Increase
Leaf spot	*Helminthosporium* spp.	Blue grass		Decrease
	Curvularia spp.	Blue grass		Decrease
	Septoria apii	Celery	Increase	
Northern leaf blight	*Helminthosporium*	Maize	Decrease	Increase
Blast	*Piricularia oryzae*	Rice	Decrease	Increase
Chocolate spot	*Botrytis cinerea*	Broad bean	Decrease	Increase
Anthracnose	*Colletotrichum phomoides*	Tomato	Increase	Decrease
Wilt	*Phialophora* spp.	Carnation	Increase	Decrease
	F. oxysporum f. *vasinfectum*	Cotton	Decrease	Increase
	F. cubense	Banana	Decrease	Increase
	F. oxysporum f. *lycopersici*	Tomato	Decrease	Increase
	F. oxysporum f. *chrysanthemi*	Chrysanthemum	Decrease	
	F. oxysporum f. *niveum*	Muskmelon	Increase	
	Verticillium albo-atrum	Potato/tomato	Increase	Decrease
	Verticillium spp.	Antirrhinum	Increase	Decrease
Virus diseases				
Mosaic	Tobacco mosaic virus	Tobacco		Increase
Virus	Potato virus X	Potato		Decrease
Nematode diseases				
Cyst nematode	*Heterodera tabacum*	Tobacco		Decrease
Cyst nematode	*Heterodera*	Soybean	Increase	Decrease
Root nematode	*Heterodera*	Tobacco	Increase	Decrease
Root knot	*Meloidogyne incognita*	Lima bean	Increase	Decrease
Other disease				
Cadang-cadang	Unknown	Coconut		Decrease

From Huber, D. M., in *CRC Handbook Series in Nutrition and Food*, Rechcigl, M., Ed., CRC Press, Boca Raton, Fla., accepted for publication, 1980. With permission.

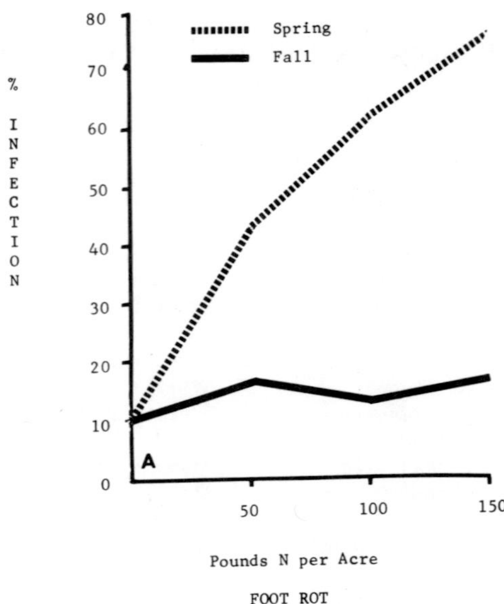

FIGURE 1. Influence of nitrogen rate and time of application on *Cercosporella* foot rot of winter wheat.

forms on *Aphanomyces* root rot by dividing the root system so the same plant could be grown with different treatments . Only those roots exposed to NH_4-N developed severe root rot. Interplanting red alder with conifers has been proposed to control *Poria* and *Armillaria* root rots. Red alder stimulates nitrification and reduces the levels of NH_4-N below that required for growth of these two pathogens.[21]

Take-all of cereals, caused by *Ophiobolus graminis,* has been associated primarily with intensive wheat culture on low fertility alkaline soils.[22] However, take-all has also been a serious problem following some legume crops which leave the soil at a high level of fertility and enhance nitrification. Application of NH_4-N to field soils reduces take-all while NO_3-N may increase disease severity (Figure 2). Increased root infection with NO_3-N may be partially offset by the improved nutritional status of the plant so that grain yield is increased.

Much of the inconsistency related to the effect of nitrogen and potassium on stalk rot of maize probably results from a failure to consider the differential effects of these nutrients on the various pathogens involved. Applications of potassium chloride, ammonium sulfate, or ammonium chloride decrease stalk rot caused by *Diplodia zeae* whereas similar rates of potassium nitrate increase disease.[23] Disease reduction with KCl may be due to the competitive inhibition of NO_3-N uptake by the chloride ion and is not dependent on potassium. Application of NO_3-N without chloride increased stalk rot, while application of chloride without nitrogen had little effect on disease. In contrast, maize stalk rot caused by *Fusarium moniliforme* is reduced by NO_3-N and increased by NH_4-N.[24]

Fungal and bacterial wilts also are influenced by forms of nitrogen. Nitrogen-deficient maize seedlings supported poor growth of *Erwinia stewartii* which is dependent on nitrogen in the tracheal sap for growth. Increasing nitrogen levels in soil, which increases nitrogen in the tracheal sap, results in better growth of the pathogen and greater wilting.[25] Potassium deficiency, which also results in a high nitrogen concentration in the sap, also increases disease severity. Unlike *E. stewartii, Corynebacterium*

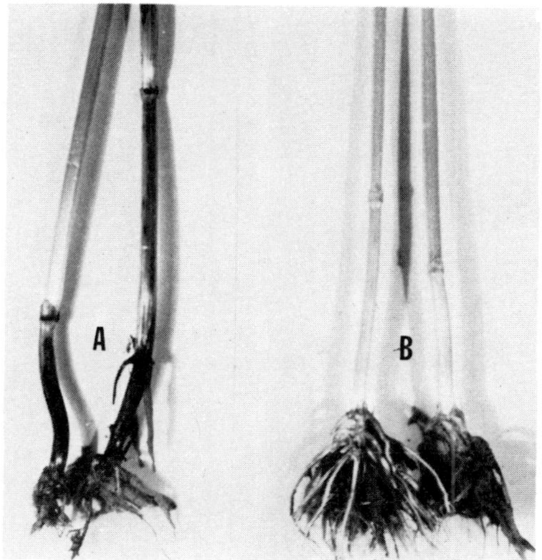

FIGURE 2. Increased take-all *(Gaeumannomyces graminis)* root and crown rot on winter wheat fertilized with nitrate (A) compared with ammonium (B) fertilizers. Dark, discolored, dead area at the base of stems is a symptom of severe disease. Inhibiting nitrification to stabilize applied nitrogen in the ammoniacal form enhances control of this disease.

michiganese (bacterial canker of tomato) is a phloem invader and growth is increased by NO_3-N.[26]

Infection by *Verticillium* and its distribution throughout the potato plant takes place early in the season, yet visible symptoms generally do not appear until early tuber set or later. By delaying plant maturity, NH_4-N reduces *Verticillium* wilt while NO_3-N hastens maturity and increases disease severity (Figure 1).[1] Compounds which inhibit nitrification and stabilize nitrogen in the ammonical form improve the control of *Verticillium* wilt with nitrogen. Inhibition of nitrification also enhances the control of potato scab caused by *Streptomyces scabies* and take-all of wheat (Figure 2). *Verticillium* wilt of hop, cotton, and tomato increases with increasing levels of NO_3-N. Nitrogen-deficient plants generally fail to show symptoms even though the pathogen can be isolated from lower plant parts. In contrast, *Fusarium* wilt of cotton, cabbage yellows, tomato wilt, and pea wilt decrease as the concentration of NO_3-N increases. The development and severity of *Fusarium* wilt of pea depend on temperature and may be related to the temperature sensitivity of nitrate reductase which is required for utilization of NO_3-N.

Temperature modifies the effect of nutrients on other diseases (Figure 3). Nutrients affect development of cabbage yellows at 28°C but not at 19°C. Thatcher wheat and stem rust race 56 (*Puccinia graminis* f. *tritici*) are in delicate balance (mesothetic) at 16 to 25°C. Nonfertilized Thatcher plants and those supplied NH_4-N were more resistant to rust, while plants supplied NO_3-N exhibited a mesothetic (X) reaction which approached complete susceptibility. At low temperatures, Thatcher and the susceptible variety Marquis receiving NO_3-N were moderately susceptible while those receiving NH_4-N were completely resistant.[27] This effect of temperature and nitrogen on cereal rusts is thought to reflect changes in protein metabolism of the host plant.[28]

Induced susceptibility to powdery mildew by NO_3-N also suggested that physiological changes within the leaves took place rather than a change in the external environment. The ability of NO_3-N to reduce northern corn leaf blight was nullified if potas-

FIGURE 3. Severe *Verticillium* wilt of potato plants receiving nitrate nitrogen (foreground) compared with the 2 to 3 week delay in symptom expression of plants receiving ammoniacal nitrogen (background).

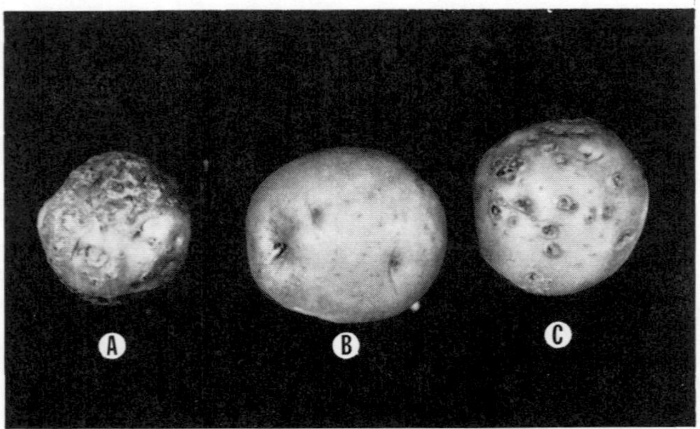

FIGURE 4. *Streptomyces* scab of potato after fertilization with (A) nitrate nitrogen; (B) ammoniacal nitrogen with a nitrification inhibitor; (C) ammoniacal nitrogen without the nitrification inhibitor. (Courtesy of R. D. Watson, University of Idaho, Moscow.)

FIGURE 5. Increased snowmold of winter wheat (outlined area) by applying 220 kg/ha ground wheat straw before snowfall in November compared with adjacent nonamended wheat.

sium chloride, which inhibits nitrate uptake, was applied with the NO_3-N rather than potassium sulfate which has little effect on nitrate uptake.[23] Increasing severity of rice blast (*Piricularia orvzae*) with NH_4-N is correlated with increasing amide nitrogen levels in plant tissues which may also be increased by low (20°C) night temperature in conjunction with NH_4-N.[29] Normally, susceptible plants grown above 26°C at night were resistant, probably because of impaired amide synthesis from decreased nitrate reduction. Ammonium sulfate, asparagine, or low temperature (which increases asparagine) induce complete susceptibility of resistant cotton seedlings to anthracnose caused by *Colletotrichum gossypii.*[30]

Much of the observed effect of nitrogen on virus infection or multiplication may be related to effects on plant growth.[31] However, the intrinsic susceptibility of tobacco and potato plants to potato virus Y (PVY) is only slightly changed by differences in nutrition that have large effects on plant growth. NH_4-N increases the number of tobacco plants infected with PVY but reduces the number of infected potato plants. Because no one form of nitrogen controls all diseases or favors disease control on any one group of plants, each disease must be considered individually and control practices integrated for optimum plant growth and production.

Phosphorus

Both phosphorus and potassium are reported to increase disease resistance; however, this effect is best established for potassium. Application of phosphorus is most beneficial in reducing seedling diseases and other fungal diseases where vigorous root development permits plants to escape disease (Table 3).[4,32]

Phosphorus has been effective in reducing *Thielevia* root rot of tobacco and ginseng, *Septoria* leaf spot of tomato, eye spot of sugar cane, dodder on clover and vetch, downy mildew of cabbage, and corn root rot caused by *Gibberella saubinetti.*[3] Phosphate fertilization of wheat has almost eliminated economic losses from *Pythium* root rot in the central wheat growing area of the U.S. Corn root rot is most severe where either potassium or phosphorus is deficient, and this disease can generally be controlled by application of potassium, phosphorus, and lime. In contrast, phosphorus may increase the severity of *Sclerotinia libertiana* on many garden plants because it restores the acidity of plant tissues and enables the fungus to survive in soil. Take-all of cereals may be severe in soils low in available phosphorus, and phosphorus applications reduce losses from take-all by stimulating root development. *Pythium* root rot of wheat in the central U.S. is seldom a problem since adequate levels of phosphorus fertilization has been practiced.

By shortening the vegetative period, phosphorus reduces the infective period for rusts and other foliar pathogens and is especially beneficial in counteracting effects of high levels of nitrogen.[33] Phosphorus and potassium together result in strong mechanical tissues in contrast to the succulent tissue produced by high levels of nitrogen. Since wheat stem rust (*Puccinia graminis*) lives only in chlorenchyma tissues, the increased relative proportion of sclerenchyma to collenchyma tissue by fertilization with phosphates and potassium may limit the area in which the rust mycelium can grow. Balanced levels of phosphorus and potassium may thus offset the effect of high nitrogen on this disease. In contrast, phosphorus reduces resistance of lettuce to mildew, *Bremia lactuca.* Phosphorus deficient wheat plants were most resistant to powdery mildew but more susceptible to flag smut.[5]

Phosphorus, or nitrogen and phosphorus amendments increased rice yields while at the same time increasing severity of stem rot caused by *Sclerotium oryzae.*[34] Potassium, although only slightly beneficial in increasing yields, reduced stem rot when applied with nitrogen and phosphorus. Since phosphorus is essential for virus multiplication, excesses may increase plant susceptibility to virus disease.

Potassium

The relationship of available potassium in soil to plant disease severity has been observed for many years (Table 4). It is a common practice to fertilize with potassium to reduce the severity of many tomato diseases. The accumulation of inorganic nitrogen in tracheal sap of potassium-deficient maize plants is associated with increased susceptibility to Stewart's wilt caused by *Erwinia stewartii.*[4] Potassium deficiency may predispose maize to root rot and lodging because of an accumulation of iron in nodal tissues of the stalk that interferes with translocation of nutrients to the roots.

Potato tubers from plants supplied nitrogen and phosphorus but deficient in potassium were most susceptible to *Phytophthora infestans.* Susceptibility was associated with high total nitrogen, protein nitrogen, non-protein nitrogen, and alpha-amino nitrogen in the non-protein fraction. These values were highly dependent on the nitrogen:potassium ratio. Arginine inhibited germination of *Phytophthora* sporangia and was lowest in the absence of potassium when nitrogen and phosphorus were high. Generally, arginine increased as potassium increased.[35]

Club root of cabbage is most severe when there is a deficiency or excess of nitrogen or an excess of potassium. The ratio of calcium to potassium is also important, and club root diminishes as the level of calcium increases in the soil.[36] Potato scab, in contrast, is increased as available potash or calcium in the soil increases. Brown rot

Table 3
PLANT DISEASES INFLUENCED BY PHOSPHORUS

Disease	Pathogen	Host plant	Effect of phosphorus
Bacterial diseases			
Blight	*Pseudomonas syringae*	Lima bean	Decrease
Blackfire	*P. angulata*	Tobacco	Decrease
Wildfire	*P. tabaci*	Tobacco	Decrease
Bacterial wilt	*P. caryophylli*	Carnation	Increase
Fireblight	*Erwinia amylovora*	Apple/pear	Increase
Stewart's wilt	*Xanthomonas stewartii*	Maize	Increase
Stem rot	*X. pelargonii*	Pelargonium	Increase
Fungus diseases			
Damping-off	*Fusarium oxysporum*	Conifer	Decrease
Browing root rot	*Pythium arrhenomanes*	Cereal	Decrease
Take-all	*Ophiobolus graminis*	Cereal	Decrease
Damping-off	*Rhizoctonia solani*	Pea/cucumber	Decrease
Patch	*O. graminis*	Turf	Decrease
Root rot	*P. arrhenomanes*	Cereal	Decrease
	Gibberella saubinetti	Corn	Decrease
	Fusarium spp.	*Trifolium pratense*	Decrease
	Thielaviopsis basicola	Tobacco	Decrease
		Ginseng	Decrease
	Polyporus schweinitzii	White pine	Decrease
	R. solani	Soybean	Decrease
	Helminthosporium sativum	Wheat	Decrease
Rot	*Phoma* spp.	Beets	Decrease
Stalk rot	*Gibberella zeae*	Maize	Decrease
Snowmold	*Sclerotinia borealis*	Timothy	Decrease
Leaf blight		Tomato	Decrease
Leaf spot	*Septoria* spp.	Tomato	Decrease
Eye spot	*H. sacchari*	Sugarcane	Decrease
Early blight	*Alternaria solani*	Tomato	Decrease
Yellows	*F. oxysporum* f. *gladioli*	Gladiolus	Decrease
Wilt	*F. lycopersici*	Tomato	Decrease
	Fusarium spp.	Red clover	Decrease
	F. vasinfectum	Cotton	Decrease
Late blight	*Phytophthora infestans*	Potato	Decrease
Stalk smut	*Urocystis occulata*	Rye	Decrease
Bunt	*Tilletia* spp.	Wheat	Decrease
Covered smut	*Sphacelotheca sorghi*	Sorghum	Decrease
Flag smut	*U. tritici*	Wheat	Decrease
Rust	*Puccinia* spp.	Cereal	Decrease
	Melampsora lini	Flax	Decrease
Powdery mildew	*Erysiphe graminis*	Cereal	Decrease
Downy mildew	*Peronospora parasitica*	Cabbage	Decrease
	Plasmopara viticola	Grape	Decrease
Club root	*Plasmodiophora brassicae*	Cabbage	Increase
Root rot	*Phymatotrichum omnivorum*	Cotton	Increase
	Thielaviopsis basicola	Citrus	Increase
Stalk rot	*Diplodia zeae*	Maize	Increase
Stem rot		Rice	Increase
Sclerotinia	*Sclerotinia libertiana*	Many	Increase
Scab	*G. zeae*	Wheat	Increase
Yellows	*F. oxysporium* f. *conglutinans*	Cabbage	Increase

Table 3 (continued)
PLANT DISEASES INFLUENCED BY PHOSPHORUS

Disease	Pathogen	Host plant	Effect of phosphorus
Wilt	*F. oxysporum* f. *vasinfectum*	Cotton	Increase
	F. oxysporum f. *lycopersici*	Tomato	Increase
Late blight	*Phytophthora infestans*	Potato	Increase
Rust	*Melampsora lini*	Flax	Increase
Stem rust	*Puccinia graminis*	Cereal	Increase
Bunt	*Tilletia* spp.	Wheat	Increase
Downy mildew	*Plasmopora viticola*	Grape	Increase
Powdery mildew	*E. graminis*	Wheat	Increase
Mildew	*Bremia lactucae*	Lettuce	Increase
Fusiform rust	*Cronartium fusiforme*	Loblolly pine	Increase
Virus diseases			
Mosaic	Tobacco mosaic virus	Bean	Decrease
TMV	Tobacco mosaic virus	Tomato	Increase
		Tobacco	Increase
		Nicotiana glutinosa	Increase
Virus	Cucumber virus-1	Spinach	Increase
	Turnip virus-1	*Nicotiana glutinosa*	Increase
		Nicotiana multivalvis	Increase
Yellow mosaic	Yellow tobacco mosaic virus	Tobacco	Increase
Nematode diseases			
	Pratylenchus penetrans	Pea	Increase
Root-knot	*Meloidogyne incognita*	Pea	Increase
Insect diseases			
Aphid	*Myzus persicae*	Tobacco	Increase
Frit fly	*Oscinella* frit	Barley	Decrease
Chinch bugs	*Blissus leucopterus*	Sorghum	Decrease
Gout fly	*Chlorops taeniopus*	Barley	Decrease
Leafhopper	*Empoasca terra-reginae*	Cotton	Decrease
Other diseases			
Dodder	*Cuscuta* spp.	Clover/vetch	Decrease

From Huber, D. M., in *CRC Handbook Series in Nutrition and Food*, Rechcigl, M., Ed., CRC Press, Boca Raton, Fla., accepted for publication, 1980. With permission.

gumosis, *Phytophthora parasitica*, is also more severe when high potassium is associated with low calcium, but is generally not a serious problem when the potassium to calcium ratio is close to one.[37] This effect on disease may be related to the differential permeability of cell membranes as affected by the potassium to calcium balance.

Barnyard manure is thought to reduce *Fusarium* wilt of cotton by correcting potassium deficiency and providing a more balanced nutrition. Late blight of potato, *Phytophthora infestans*, is severe on plants supplied only nitrogen and phosphorus, but decreases as the level of potassium increases.[35] Resistance to *Phytophthora* was associated with the potassium induced accumulation of fungistatic levels of arginine in leaves. A similar effect was reported for cereal rusts where potassium partially offset the increased disease with high nitrogen rates by reducing the soluble nitrogen compo-

Table 4
PLANT DISEASES INFLUENCED BY POTASSIUM NUTRITION

Disease	Pathogen	Host plant	Effect of potassium
Bacterial diseases			
Bacterial blight	*Pseudomonas syringae*	Lima bean	Decrease
Angular leaf spot	*P. tabaci*	Tobacco	Decrease
Wildfire	*P. angulata*	Tobacco	Decrease
Blackfire	*P. angulata*	Tobacco	Decrease
Wilt	*P. solanacearum*	Tobacco	Decrease
Angular leaf spot	*P. lachrymans*	Cucumber	Decrease
Bacterial wilt	*P. caryophylli*	Carnation	Decrease
Angular leaf spot	*Xanthomonas malvacearum*	Cotton	Decrease
Bacterial		Tomato	Decrease
Stem rot	*X. pelargoni*	Pelargonium	Decrease
Stewart's wilt	*X. stewartii*	Maize	Decrease
Bacterial blight	*X. oryzae*	Rice	Decrease
Soft rot	*Erwinia carotovora*	Cabbage	Decrease
Fire blight	*E. amylovora*	Pear	Decrease
Wildfire	*Pseudomonas tabaci*	Tobacco	Increase
Bacterial wilt	*E. tracheophila*	Cucumber	Increase
Fire blight	*E. amylovora*	Apple/pear	Increase
Canker	*Corynebacterium michagense*	Tomato	Increase
Fungal diseases			
Damping-off	*Pythium ultimum*	Beet	Decrease
Root rot	*Aphanomyces euteiches*	Pea	Decrease
	Phythophthora cinnamomi	Avocado	Decrease
		Pineapple	Decrease
	Fusarium spp.	*Trifolium pratense*	Decrease
	Gibberella saubinetti	Maize	Decrease
	Phymatotrichum omnivorum	Cotton	Decrease
	Rhizoctonia solani	Jute	Decrease
Canker	*Rhizoctonia solani*	Potato	Decrease
Basal rot	*Fusarium oxysporum* f. *narcissi*	Narcissus	Decrease
Stalk rot	*Fusarium moniliforme*	Maize	Decrease
	Gibberella zeae	Maize	Decrease
	Diplodia zeae	Maize	Decrease
Stem rot	*Leptosphaeria salvinii*	Rice	Decrease
Patch	*Fusarium nivale*	Turf	Decrease
	Ophiobolus graminis	Turf	Decrease
Stem end rot	*Fusarium* spp.	Potato	Decrease
Net blotch	*Helminthosporium teres*	Barley	Decrease
Leaf blight	*Helminthosporium turcicum*	Maize	Decrease
Leaf spot	*Helminthosporium* spp.	Turf	Decrease
Northern leaf blight	*Helminthosporium turcicum*	Maize	Decrease
Leaf spot	*Heterosporium phlei*	Timothy	Decrease
	Pleospora herbarum	Mangolds	Decrease
Brown spot	*Helminthosporium* spp.	Rice	Decrease
Sheath blight	*Corticium sasaki*	Rice	Decrease
Eye spot	*Helminthosporium sacchari*	Sugarcane	Decrease
Stem rot	*Helminthosporium sigmoidum*	Rice	Decrease

Table 4 (continued)
PLANT DISEASES INFLUENCED BY POTASSIUM NUTRITION

Disease	Pathogen	Host plant	Effect of potassium
Brown spot	*Ophiobolus miyabeanus*	Rice	Decrease
Blast	*Piricularia oryzae*	Rice	Decrease
Sclerotial disease	*Sclerotium oryzae*	Rice	Decrease
Leaf spot	*Cercospora oryzae*	Rice	Decrease
Late blight	*Phytophthora infestans*	Potato	Decrease
Yellows	*Fusarium oxysporum* f. *conglutinans*	Cabbage	Decrease
Wilt	*Fusarium oxysporum* f. *lycopersici*	Tomato	Decrease
	Fusarium spp.	Red clover	Decrease
	Fusarium oxysporum f. *vasinfectum*	Cotton	Decrease
	Fusarium spp.	Carnation	Decrease
Wilt	*F. oxysporum* f. *melonis*	Melon	Decrease
	F. lini	Flax	Decrease
Bunt	*Tilletia* spp.	Wheat	Decrease
Mildew	*Phyllactinia guttata*	Rose	Decrease
	Bremia lactucae	Lettuce	Decrease
Powdery mildew	*Erysiphe graminis*	Cereal	Decrease
Rust	*Puccinia* spp.	Cereal	Decrease
Virus diseases			
Stem rust	*Puccinia graminis*	Wheat	Decrease
Leaf rust	*Puccinia recondita*	Wheat	Decrease
Stripe rust	*Puccinia striiformis*	Wheat	Decrease
Rust	*Melampsora lini*	Flax	Decrease
Stalk smut	*Urocystis occulata*	Rye	Decrease
Rust	*Peridermium* spp.	White pine	Decrease
Club root	*Plasmodiophora brassicae*	Cabbage	Increase
Scab	*Streptomyces scabies*	Potato	Increase
Root rot	*Rhizoctonia solani*	Bean	Increase
Canker	*Rhizoctonia solani*	Potato	Increase
Brown rot gumosis	*Phytophthora parasitica*	Citrus	Increase
Pod rot	*Rhizoctonia solani*	Peanut	Increase
Leaf blight	Fungal	Tomato	Increase
Brown leaf blight	Unknown	Rice	Increase
Wilt	*Fusarium oxysporum* f. *niveum*	Watermelon	Increase
	Fusarium oxysporum f. *lycopersici*	Tomato	Increase
Downy mildew	*Peronospora parasitica*	Cabbage	Increase
Bunt	*Telletia* spp.	Wheat	Increase
Rot	*Botrytis* spp.	Casterbean	Increase
Flag smut	*Urocystis tritici*	Wheat	Increase
Virus diseases			
Mosaic	Tobacco mosaic virus	Tobacco	Decrease
Blotchy ripening	Tobacco mosaic virus	Tomato	Decrease
Mosaic	Tobacco mosaic virus	Bean	Decrease
Yellow mosaic	Yellow tobacco mosaic virus	Tobacco	Decrease
Mosaic	Potato mosaic virus	Potato	Decrease

Table 4 (continued)
PLANT DISEASES INFLUENCED BY POTASSIUM NUTRITION

Disease	Pathogen	Host plant	Effect of potassium
TMV	Tobacco mosaic virus	Tomato	Increase
Virus	Cucumber virus-1	Spinach	Increase
Leaf-roll	Potato leaf-roll virus	Potato	Increase
Nematode diseases			
Root-knot	*Meloidogyne incognita*	Lima bean	Decrease
Nematode	*Heterodera schachtii*	Sugar beet	Decrease
White tip	*Aphelenchoides oryzae*	Rice	Increase
Nematode	*Pratylenchus penetrans*	Pea	Increase
Root-knot	*M. incognita*	Cucumber	Increase
Insect diseases			
Mosquito	*Heliopeltis theiovora*	Tea	Decrease
Gout fly	*Chlorops taeniopus*	Barley	Decrease
Aphid	*Aphis rumicis*	Beans	Increase
Leafhopper	*Empoasca terra-reginae*	Cotton	Decrease

From Huber, D. M., in *CRC Handbook Series in Nutrition and Food,* Rechcigl, M., Ed., CRC Press, Boca Raton, Fla., accepted for publication, 1980. With permission.

nents in plant tissue.[23] Low levels of potassium, relative to other nutrients, increase the severity of cabbage yellows, *Fusarium* wilt of tomato, Stewart's wilt of maize, and downy mildew of tobacco. The application of potassium fertilizers has the greatest value in reducing these diseases when used in conjunction with the more resistant varieties. *Fusarium* wilt of tomato is enhanced by a low level of potassium but reduced by a low level of nitrogen.[38] *Fusarium* wilt of pea decreased with an increase in nutrient concentration when plants were growing in long summer days with high light intensity, but was not influenced in short mid-winter days with low light intensity. High rates of potassium reduce the severity of wild-fire of tobacco, *Pseudomonas tabaci,* and may offset the effect of high nitrogen rates.[39] Resistance of wheat and flax to rust may be lost under potassium-deficient conditions.[40,41] In contrast, *Verticillium* wilt of cotton may be increased by potassium fertilization.[4]

Potassium fertilization has been generally recommended to reduce the severity of maize stalk rot. However, inconsistent results are reported because this disease may be caused by several pathogens, i.e., *Diplodia zeae, Giberella zeae, Fusarium moniliforme, Pythium* spp., and *Colletotrichum graminicola,* which may respond differently to potassium fertilization. Particular attention has been devoted to the nutrient balance and the nitrogen to potassium ratio. Stalk rot increases with increasing levels of nitrogen only when potassium levels are low.[23] The effect attributed to potassium in reducing *Diplodia* stalk rot may actually be due to the anion associated with it. Thus, stalk rot is reduced with applications of potassium chloride, ammonium sulfate, or ammonium chloride, while similar rates of potassium nitrate increase severity of this disease, and potassium sulfate or potassium metaphosphate have little effect. The reduction in stalk rot with potassium chloride appears to result from competitive inhibition of NO_3-N uptake by the chloride ion and is not dependent on potassium. High rates of nitrate nitrogen may overcome the competitive inhibition imposed by chloride and explain why application of nitrate nitrogen without chloride increases stalk rot, while application of chloride without nitrogen has little effect on disease.[23] Other stalk rot pathogens are not affected the same as *Diplodia* and *Gibberella* and much of the confusion generated in the literature could be avoided if the pathogen, rather than the symptom, was identified.

Calcium

Calcium affects plant diseases directly by protecting pectate from maceration by extracellular enzymes; and indirectly by reducing soil acidity, by "neutralizing" effects of some toxins, and by affecting cell division where abnormal growth is involved (Table 5). The calcium and potassium balance is important for development of disease galls because both materials contribute to cell growth and division. Excess calcium reduces resistance of wheat to flag smut but reduces susceptibility of wheat to rust and materially lowers the water requirement in rusted as well as rust-free plants.[42]

Resistance of tomato to *"Erwinia phytophthora"* was increased by calcium fertilization. Resistance to this pathogen was correlated with resistance to cell-free macerating enzymes produced in culture filtrates.[43] Plants deficient in calcium or magnesium are also more susceptible to colonization by *Rhizoctonia solani.*[44] Tissues of deficient plants are poorly organized and have thin cell walls, large intercellular spaces, and poorly defined middle lamella. Fertilization of tomatoes with calcium also increases resistance to *Sclerotium rolfsii.* Resistant varieties of castor bean contain less potassium, but their higher levels of calcium and magnesium are associated with resistance of their pectates to the macerating enzymes of *Botrytis.*[45] Boron affects the accumulation of calcium and interacts with calcium to influence resistance of tomatoes to *Fusarium* wilt. Both elements have an effect on the development and structure of cell walls.

Streptomyces scab of potatoes becomes more severe as calcium is increased. Although limeing is frequently thought to influence scab directly by increasing the soil pH, it also increases the calcium content of tubers correlated with susceptibility to scab. Since similar effects are obtained when calcium is applied as calcium carbonate or calcium sulfate, its effect on scab appears to be independent of a direct effect of soil pH. The correlation of scab severity with the calcium:potassium ratio of tissue and the effect of pH on this ratio by altering the calcium ion concentration in soil indicates a possible indirect effect of pH on disease severity.[36]

Decreased peanut pod breakdown after gypsum applications are correlated with increased calcium content of pods and increased yields. Magnesium and potassium decrease the calcium content of pods and also increase pod breakdown.[46]

Other Elements

Silicon affects the availability of potassium and may be combined with other materials to give cell walls greater strength. The silicon content of rice and wheat plants is thought to increase resistance to blast (*Piricularia oryzae*) and powdery mildew (*Erysiphe graminis*) by strengthening cell walls and impeding penetration of those pathogens.

Minor elements, notably iron, manganese, and zinc, can affect pathogenicity by their influence on the synthesis of pectolytic enzymes. Reduction of *Fusarium* wilt has been attributed to enzyme inhibition while control of snowmold by boron is thought to be a direct effect on the pathogen.[47] Diseases influenced by these micronutrients are listed in Tables 6 to 15. Post-inoculation application of nickel is thought to reduce wheat rust by favoring the oxidation of phenols to compounds more inhibitory to the pathogen.[48]

Conclusion

The nutritional condition of a plant determines in large measure the degree of resist-

Table 5
PLANT DISEASES INFLUENCED BY CALCIUM

Disease	Pathogen	Host plant	Effect of calcium
Bacterial diseases			
Wilt	*Pseudomonas caryophylli*	Carnation	Decrease
Frenching	*Bacillus subtilis*	Tobacco	Decrease
Stem rot	*Xanthomonas pelargonii*	Pelargonium	Decrease
Black rot	*Xanthomonas campestris*	Crucifer	Decrease
Soft rot	*Erwinia carotovora*	Cabbage	Decrease
Rot	*Erwinia aroideae*	Solanaceae	Decrease
Bacterial blight	*Xanthomonas malvacearum*	Cotton	Increase
Fungal disease			
Club root	*Plasmodiophora brassicae*	Cabbage	Decrease
Damping-off	*Rhizoctonia solani*	Soybean	Decrease
		Bean	Decrease
	Fusarium oxysporum	Conifer	Decrease
Root rot	*Aphanomyces cochlioides*	Pea	Decrease
Pod rot	*Pythium myriotylum*	Peanut	Decrease
Brown rot gumosis	*Phytophthora parasitica*	Citrus	Decrease
Pod rot	*Rhizoctonia solani*	Peanut	Decrease
Root rot	*Rhizoctonia solani*	Bean	Decrease
		Cereal	Decrease
		Soybean	Decrease
	Fusarium solani	Citrus	Decrease
	Rhizoctonia solani	Many	Decrease
	Aphanomyces euteiches	Pea	Decrease
Root rot	*Polyporus schweinitzii*	White pine	Decrease
Sclerotinia disease	*Sclerotinia libertiana*	Many	Decrease
Southern blight	*Sclerotium rolfsii*	Tomato	Decrease
		Pepper	Decrease
		Sweet potato	Decrease
Leaf spot	*Septoria* spp.	Celery	Decrease
Inflorescence blight	*Botrytis* spp.	Castor bean	Decrease
Chocolate spot	*Botrytis fabae*	Broad bean	Decrease
Blight	*Phytophthora* spp.	Solanaceae	Decrease
Leaf blotch	*Heterosporium gracile*	Iris	Decrease
Dutch elm disease	*Ceratocystis ulmi*	Elm	Decrease
Wilt	*Fusarium oxysporum* f. *vasinfectum*	Cotton	Decrease
	Fusarium oxysporum f. *lycopersici*	Tomato	Decrease
	Fusarium oxysporum f. *niveum*	Muskmelon	Decrease
Covered smut	*Sphacelotheca sorghi*	Sorghum	Decrease
Stripe rust	*Puccinia striiformis*	Wheat	Decrease
Leaf rust	*Puccinia recondita*	Wheat	Decrease
Stem rust	*Puccinia graminis*	Wheat	Decrease
Scab	*Streptomyces scabies*	Potato	Increase
Damping-off	*Rhizoctonia solani*	Cotton	Increase
Root rot	*Phymatotrichum omnivorum*	Cotton	Increase
Canker	*Dasyscypha* spp.	Larch	Increase
Flag smut	*Urocystis tritici*	Wheat	Increase
Stalk smut	*Urocystis occulata*	Rye	Increase

Table 5 (continued)
PLANT DISEASES INFLUENCED BY CALCIUM

Disease	Pathogen	Host plant	Effect of calcium
Wilt	*Fusarium* spp.	Cotton	Increase
	Verticillium albo-atrum	Cotton	Increase
Mildew	*Uncinula necator (Oidium)*	Grape	Increase
Stripe smut	*Puccinia striiformis*	Cereals	Increase
Stem rust	*Puccinia graminis*	Cereals	Increase
Virus disease			
Mosaic	Tomato mosaic virus	Tomato	Decrease
Nematode disease			
Stem nematode	*Ditylenchus dipsaci*	Alfalfa	Decrease
Insect disease			
Aphid	*Myzus persicae*	Tobacco	Increase

From Huber, D. M., in *CRC Handbook Series in Nutrition and Food,* Rechcigl, M., Ed., CRC Press, Boca Raton, Fla., accepted for publication, 1980. With permission.

Table 6
PLANT DISEASES INFLUENCED BY MAGNESIUM

Disease	Pathogen	Host plant	Effect of magnesium
Bacterial diseases			
Blight	*Xanthomonas malvacearum*	Cotton	Decrease
Fungal disease			
Club root	*Plasmodiophora brassicae*	Cabbage	Decrease
Damping-off	*Rhizoctonia solani*	Cotton	Decrease
Root rot	*Rhizoctonia solani*	Bean	Decrease
		Soybean	Decrease
Leaf spot	*Botrytis* spp.	Castor bean	Decrease
	Helminthosporium spp.	Rice	Decrease
Wilt	*Verticillium albo-atrum*	Cotton	Decrease
	Fusarium oxysporium f. *conglutinans*	Cotton	Decrease
Wet smut		Wheat	Decrease
Root rot	*Phymatotrichum omnivorum*	Cotton	Increase
Pod rot	*Rhizoctonia solani*	Peanut	Increase
	Pythium myriotylum	Peanut	Increase
Southern leaf blight	*Helminthosporium maydis*	Maize	Increase
Stalk smut	*Urocystis occulata*	Rye	Increase
Flag smut	*Urocystis tritici*	Wheat	Increase
Stem rust	*Puccinia graminis*	Cereals	Increase
Stripe rust	*Puccinia striiformis*	Cereals	Increase
Insect diseases			
Purple scale	Insect	Citrus	Increase
Aphid	*Myzus persicae*	Tobacco	Increase

Table 7
PLANT DISEASES INFLUENCED BY SULFUR

Disease	Pathogen	Host plant	Effect of sulfur
Bacterial diseases			
Stewart's wilt	*Erwinia stewartii*	Maize	Decrease
Fungus diseases			
Club root	*Plasmodiophora brassicae*	Crucifer	Decrease
Scab	*Streptomyces scabies*	Potato	Decrease
Patch	*Fusarium nivale*	Turf grass	Decrease
Root rot	*Rhizoctonia solani*	Soybean	Decrease
Leaf spot	*Ramularia beticola*	Sugar beet	Decrease
Stem rust	*Puccinia graminis*	Cereal	Increase
Stripe rust	*Puccinia striiformis*	Cereal	Increase
Virus diseases			
Mosaic	Tobacco mosaic virus	*Nicotiana glutinosa*	Decrease

From Huber, D. M., in *CRC Handbook Series in Nutrition and Food,* Rechcigl, M., Ed., CRC Press, Boca Raton, Fla., accepted for publication, 1980. With permission.

Table 8
PLANT DISEASES INFLUENCED BY SODIUM

Disease	Pathogen	Host plant	Effect of sodium
Fungus diseases			
Club root	*Plasmodiophora brassicae*	Cabbage	Decrease
Root rot	*Phymatotrichum omnivorum*	Cotton	Decrease
Leaf spot	*Helminthosporium* spp.	Barley	Decrease
Loose smut	*Ustilago tritici*	Wheat	Decrease
Leaf rust	*Puccinia recondita*	Wheat	Decrease
Root rot	*Rhizoctonia solani*	Bean	Increase
Dutch elm disease	*Ceratocystis ulmi*	Elm	Increase
Wilt	*Fusarium oxysporum* f. *lycopersici*	Tomato	Increase

From Huber, D. M., in *CRC Handbook Series in Nutrition and Food,* Rechcigl, M., Ed., CRC Press, Boca Raton, Fla., accepted for publication, 1980. With permission.

ance or susceptibility to disease. The plant's histological or morphological structure or properties and the function of cells or tissues may be modified in such a way as to retard or hasten pathogenesis.

The use of fertilizers in conjunction with other practices such as disease resistance, crop rotation, weed control, and insect management, are necessary to promote maximum productivity. Manipulation of host nutrition provides an effective tool to enhance the chemical, genetic, or biological control of many plant pathogens. None of the nutrient elements completely alters the inherent reaction of a plant to pathogens. The greatest benefits from nutrition have been observed with partially resistant varieties. Since most horticultural varieties are in this latter group, proper fertility improves their resistance or opportunity to escape the consequences of serious disease.

Table 9
PLANT DISEASES INFLUENCED BY MANGANESE

Disease	Pathogen	Host plant	Effect of manganese
Bacterial diseases			
Bacterial blight	*Pseudomonas* spp.	Oats	Decrease
Fungus diseases			
Root rot	*Curvularia ramosa*	Cereal	Decrease
	Fusarium culmorum	Cereal	Decrease
	Helminthosporium sativum	Cereal	Decrease
Stem canker	*Rhizoctonia solani*	Potato	Decrease
Late blight	*Phytophthora infestans*	Potato	Decrease
Leaf spot	*Helminthosporium* spp.	Rice	Decrease
	Cercospora spp.	Sugar beet	Decrease
Wilt	*F. udum*	Pigeon pea	Decrease
	F. oxysporum f. *vasinfectum*	Cotton	Decrease
Powdery mildew	*Erysiphe graminis*	Cereals	Decrease
Rust	*Puccinia* spp.	Wheat	Decrease
Wet smut		Wheat	Decrease
Leaf spot	*Helminthosporium* spp.	Barley	Increase
Virus diseases			
Mosaic	Tobacco mosaic virus	Tobacco	Decrease
Insect diseases			
Root borer	Insect	Sugar beet	Decrease

From Huber, D. M., in *CRC Handbook Series in Nutrition and Food,* Rechcigl, M., Ed., CRC Press, Boca Raton, Fla., accepted for publication, 1980. With permission.

Table 10
PLANT DISEASES INFLUENCED BY IRON

Disease	Pathogen	Host plant	Effect of iron
Fungus diseases			
Root rot	*Curvularia ramosa*	Cereals	Decrease
	Fusarium culmorum	Cereals	Decrease
	Helminthosporium sativum	Cereals	Decrease
	Rhizoctonia solani	Soybeans	Decrease
Stem rust	*Puccinia graminis*	Wheat	Decrease
Leaf spot	*Cercospora beticola*	Sugar beets	Increase
Wilt	*F. oxysporum* f. *vasinfectum*	Cotton	Increase
Stem rust	*P. graminis*	Cereals	Increase
Stripe rust	*P. striiformis*	Cereals	Increase
Virus diseases			
Mosaic	Tobacco mosaic virus	Bean	Decrease
		Nicotiana glutinosa	Decrease
Insect diseases			
Root borer		Sugar beets	Decrease

From Huber, D. M., in *CRC Handbook Series in Nutrition and Food,* Rechcigl, M., Ed., CRC Press, Boca Raton, Fla., accepted for publication, 1980. With permission.

Table 11
PLANT DISEASES INFLUENCED BY ZINC

Disease	Pathogen	Host plant	Effect of zinc
Fungus disease			
Root rot	*Aphanomyces coclioides*	Pea	Decrease
	Helminthosporium sativum	Cereals	Decrease
	Fusarium culmorum	Cereals	Decrease
Leaf spot	*Helminthosporium* spp.	Barley	Decrease
Late blight	*Phytophthora infestans*	Potato	Decrease
		Tomato	Decrease
Wilt	*F. udum*	Pigeon pea	Decrease
Covered smut	*Sphacelotheca sorghi*	Sorghum	Decrease
Rust	*Puccinia* spp.	Wheat	Decrease
Leaf spot	*Cercospora* spp.	Sugar beets	Increase
Loose smut	*Ustilago tritici*	Wheat	Increase
Leaf rust	*Puccinia recondita*	Wheat	Increase
Rust	*Puccinia* spp.	Wheat	Increase
Virus disease			
Mosaic	Tobacco mosaic virus	Tobacco	Decrease
TMV	Tobacco mosaic virus	*Nicotiana glutinosa*	Decrease
		Tobacco	Increase
		Bean	Increase
Insect diseases			
Root borer		Sugar beet	Decrease

From Huber, D. M., in *CRC Handbook Series in Nutrition and Food,* Rechcigl M., Ed., CRC Press, Boca Raton, Fla., accepted for publication, 1980. With permission.

No general rules about fertilizing plants to increase resistance to disease are available. Each disease must be considered by itself in relation to the environment in which the host and pathogen are growing, the availability of essential nutrients, and the complex of pathogens potentially present. Obvious nutrient deficiencies which limit yield or quality should be corrected and nutrient levels adjusted to avoid or reduce predisposition to disease where possible. There is no need to starve the plant into an unproductive state in order to escape disease. Instead, other measures should be utilized for disease control.

USE OF ORGANIC AMENDMENTS FOR DISEASE CONTROL

Organic amendments, either as green manure crops, residue from crop rotation, or direct residue application, have been incorporated with soil to prevent the increase of disease in newly cultivated soils and to produce conditions less conducive for disease development in established crop land. Their effects on the physical, chemical, and biological properties of soil are so varied that it has been a challenge to determine the specific effects responsible for disease control. Contributions made by these amendments to the nutritional status of crop plants and the soil microflora may be sizeable and interact with disease as previously discussed.

Disease Prevention

Disease prevention is essential on the increasing acreage of virgin arid land brought under cultivation each year to meet the food needs of an expanding world population.

Table 12
DISEASES INFLUENCED BY BORON

Disease	Pathogen	Host plant	Effect of boron
Bacterial diseases			
Bacterial	Unknown	Linseed	Decrease
Fungus diseases			
Spot blotch	*Helminthosporium sativum*	Barley	Decrease
Root rot	*Phoma betae*	Sugar beet	Decrease
Black heart	*Ceratostomella fimbriata*	Sweet potato	Decrease
Wilt	*Fusarium oxysporium* f. *lycopersici*	Tomato	Decrease
	F. udum	Pigeon pea	Decrease
Snowmold	Low temperature Basidiomycete	Legumes	Decrease
Powdery mildew	*Erysiphe graminis*	Cereals	Decrease
	E. cichoracearum	Sunflower	Decrease
Wet smut		Wheat	Decrease
Loose smut	*Ustilago tritici*	Wheat	Decrease
Club root	*Plasmodiophora brassicae*	Turnip	Decrease
Rust	*Melampsora lini*	Flax	Decrease
Stem rust	*Puccinia graminis*	Wheat	Decrease
Leaf rust	*P. recondita*	Wheat	Decrease
Rust	*Puccinia* spp.	Wheat	Decrease
Stripe rust	*P. striiformis*	Wheat	Decrease
Spot blotch	*H. sativum*	Barley	Increase
Virus diseases			
Virus	Potato virus X	Tobacco	Decrease

Table 13
PLANT DISEASES INFLUENCED BY COPPER

Disease	Pathogen	Host plant	Effect of copper
Fungus diseases			
Root rot	*Aphanomyces cochlioides*	Pea	Decrease
	Fusarium culmorum	Cereal	Decrease
	Helminthosporium sativum	Cereal	Decrease
Stem canker	*Rhizoctonia solani*	Potato	Decrease
Late blight	*Phytophthora infestans*	Potato	Decrease
Wilt	*Verticillium albo-atrum*	Cotton	Decrease
Rust	*Puccinia* spp.	Wheat	Decrease
Covered smut	*Sphacelotheca sorghi*	Sorghum	Decrease
Wet smut		Wheat	Decrease
Leaf rust	*Puccinia recondita*	Wheat	Increase

From Huber, D. M., in *CRC Handbook Series in Nutrition and Food,* Rechcigl, M., Ed., CRC Press, Boca Raton, Fla., accepted for publication, 1980. With permission.

The limited soil microflora and low ''biological buffering'' capacity of arid lands permit newly introduced pathogens to become established and multiply rapidly without competition.[49] The pathogen-suppressing property of cultivated soil is of microbial origin and influenced more by cropping and management practices than soil type. Scab (*Streptomyces scabies*) and *Verticillium* wilt may render a potato crop unmarketable

Table 14
PLANT DISEASES INFLUENCED BY SILICON[4]

Disease	Pathogen	Host plant	Effect of silicon
Bacterial diseases			
Bacterial	Unknown	Lettuce	Decrease
Fungus diseases			
Leaf spot	*Helminthosporium* spp.	Rice	Decrease
Blast	*Piricularia oryzae*	Rice	Decrease
Powdery mildew	*Erysiphe cichoracearum*	Cucumber	Decrease
	E. graminis	Cereals	Decrease

Table 15
OTHER ELEMENTS REPORTED TO INFLUENCE PLANT DISEASE[5]

Element	Disease	Pathogen	Host plant	Effect of element
Bacterial diseases				
Molybdenum	Bacterial	Unknown	Lucerne	Decrease
Fungus diseases				
Aluminum	Root rot	*Aphanomyces cochlioides*	Pea	Decrease
Cadmium	Powdery mildew	*Erysiphe graminis*	Wheat	Decrease
Cobalt	Stem rust	*Puccinia graminis*	Wheat	Decrease
Chloride	Stalk rot	*Diplodia zeae*	Maize	Decrease
Carbonate	Club root	*Plasmodiophora brassicae*	Cabbage	Decrease
	Root rot	*Phymatotrichum omnivorum*	Cotton	Decrease
Lithium	Powdery mildew	*E. cichoracearum*	Cucumber	Decrease
		E. graminis	Wheat	Decrease
	Stem rust	*Puccinia graminis*	Wheat	Decrease
	Club root	*Plasmodiophora brassicae*	Turnip	Decrease
Nickel	Stem rust	*Puccinia graminis*	Wheat	Decrease
Trace elements	Bakannae disease	*Fusarium moniliforme*	Rice	Decrease
Virus diseases				
Molybdenum	Mosaic	Tobacco mosaic virus	Tobacco	Decrease
Silver	Mosaic	Tobacco mosaic virus	Bean	Decrease

after only 3 years of monocropping newly cultivated soils.[50] Soybean and several other crop residues prevent the increase of potato scab in newly cultivated soils but fail to provide control once the pathogen is established. Dry beans (*Phaseolus vulgaris*) are more susceptible to root rot caused by *Fusarium solani* f. sp. *phaseoli* in virgin soil than in soils previously cultivated to other crops.[51] Evidence of "biological buffering" in older cultivated soils is often seen after sterilization. *Rhizoctonia solani* kills entire flats of pepper seedlings in sterilized soil, but remains localized in nonsterile soil.[52] *Verticillium albo-atrum* readily colonizes sterilized soil in contrast to its limited rhizosphere activity in natural soil.

Disease Control

Disease control with organic amendments (crop rotation, etc.) in established, cultivated soils was initially postulated as a "starving-out" process in the absence of susceptible host plants. However, under standard rotations, *F. solani* f. *phaseoli* survives for long periods without host contact and little correlation is found between the length of time out of beans and the severity of root rot. The immediately preceding crop in the rotation, or organic amendment, has much more effect on severity of diseases caused by *F. solani* f. *phaseoli, Phytophthora parasitica* var. *nicotianae, Rhizoctonia solani, Aphanomyces cochlioides, Gibberella zeae, Streptomyces scabies, Pythium aphanidermatum, Verticillium albo-atrum,* and *Phymatotrichum omnivorum* than preceding crops of several years. The effect of a specific crop sequence or organic amendment on disease may, or may not, be correlated with the pathogen's population.[50,53]

Disease Control with a Corresponding Reduction in Pathogen Numbers

Adequate control of some pathogens can be accomplished by reduction of pathogen propagules to a low level rather than by their complete elimination. Partial eradication of several pathogens from field soils has been accomplished during the anaerobic decomposition of organic amendments under heavy irrigation. Resting structures of *Sclerotinia sclerotiorum, Fusarium oxysporum* f. *cubense, Verticillium albo-atrum,* and *Phymatotrichum omnivorum* have been destroyed under anaerobic soil conditions resulting in the control of diseases caused by these pathogens.[50] Anaerobic decomposition of residues results in a low population of other soil microorganisms which may affect the "biological buffering" capacity of soil similar to sterilization.[54] Heavy irrigation of soybean-amended field soil has resulted in increased root rot of the subsequent bean crop because of the rapid recolonization of these partially sterile soils by *F. solani* f. *phaseoli.* Similar disease increases have been observed following soil fumigation.

Disease Control without a Reduction in Pathogen Numbers

Disease control is not always related to the population of the pathogen.[19,55-58] Nonhost crops may actually stimulate germination and recycling of chlamydospores of *F. solani* f. *phaseoli,* resulting in an increased population but reduction in disease severity. Organic amendments influence disease through their effect on the availability of nutrients in soil. Nitrogen has a greater effect than any other single element on individual soil fungi.[59] The carbon to nitrogen (C:N) ratio of organic amendments was correlated initially with their influence on disease.[60,61] Since an exogenous source of nitrogen is needed for germination of *F. solani* f. *phaseoli* chlamydospores, incorporation of residues with a high C:N ratio was postulated to prevent chlamydospore germination or result in nitrogen starvation of the pathogen. The addition of nitrogen nullifies the beneficial effects of carbonaceous material such as barley, wheat, or oat straw; however, sugar amendments also increased disease severity. Thus, the general chemical nature of organic amendments is more important than the C:N ratio.[62] Disease severity has been correlated with the effect of specific residues on nitrification in soil.[63] Residues that stimulate the biological oxidation of ammonium to nitrate nitrogen reduce the severity of bean root rot under field conditions, whereas amendments that inhibit or decrease nitrification generally result in more severe root rot. The effect of organic amendments on nitrification accounts for some of the inconsistencies regarding the effects of specific nutrients and residues on disease. For example, barley amendments reduce bean root rot in the absence of added nitrogen but greatly increase this disease when nitrogen fertilizers are applied. Similar results were obtained with ammonium and uramite (a urea-formaldehyde product). These nitrogen sources reduced the severity of *Rhizoctonia* root rot when added with sawdust, but ammonium nitrate resulted in the greatest disease reduction in the absence of sawdust.[64]

The correlation of nitrification with disease severity indicates the importance of a specific form of nitrogen available to the plant or pathogen after residue incorporation. There is a definite correlation of disease severity with a specific form of nitrogen as previously discussed. Different disease reactions occur with different forms of nitrogen. Potato scab, *Verticillium wilt, Gibberella* stalk rot of corn, and take-all of wheat are decreased by ammoniacal nitrogen, but *Fusarium* stalk rot of corn and bean root rot are increased.[1] An effect of the form of nitrogen was not observed with *Verticillium* wilt of potatoes when high levels of residual nitrate nitrogen were present; however, tie-up of residual nitrate nitrogen with barley straw resulted in the anticipated reduction of *Verticillium* wilt with ammoniacal nitrogen. Use of materials which inhibit nitrification such as the non-fungicidal nematocide, 1,3-dichloropropene (''Telone''), or the *Nitrosomonas* spp. specific compound, 2-chloro-6-(trichloromethyl) pyridine, (nitrapyrin) intensify the disease response when applied with nitrogen fertilizers without reducing pathogen populations in soil. Practical field control of *Verticillium* wilt and scab of potatoes has been achieved with stabilized ammoniacal nitrogen as anticipated from the effect of specific organic amendments on nitrification and these diseases.

Mechanism of Biological Control

The mechanism of biological control enhanced by organic amendment or a specific form of nitrogen involves specific microbial interactions and host physiology. Competition, antibiosis, hyperparasitism, and lysis are probably responsible where pathogen populations are reduced; i.e., ''flood-fallowing'' after residue plow-down for control of *Sclerotinia sclerotiorum, Fusarium oxysporum* f. *cubense*, and *Verticillium albo-atrum.*

Population of pathogens such as *Fusarium* or *Rhizoctonia* which possess a high ''competitive saprophytic ability'' are not necessarily related to disease severity in the field, especially when the minimal number of units needed for maximum infection are considered. Specific bacterial-fungal associations have been correlated with disease control without a corresponding reduction in pathogen numbers in the soil. Specific soil-borne bacteria are reported to reduce pathogenicity of *F. solani* f. *phaseoli* when growing along the mycelium without greatly affecting the overall growth or population of this pathogen in soil.[65] Pathogenicity was restored in the absence of the bacterium. Isolation frequency of this association from rotation plots was directly correlated with the incidence and severity of bean root rot as influenced by specific crop rotations. Disease control with the bacterium was dependent on intimate contact with the pathogen rather than its general growth in the soil. Thus, enrichment of soil with ammoniacal nitrogen to make conditions more favorable for the bacterium increased disease because association of the bacterium with *Fusarium* was not maintained.

Another microbial interaction which may be involved without a reduction in pathogen numbers is fungistasis. Although the pathogen is present in the soil, it is ''dormant'' or inactive. Soil bacteria stimulating fungal dormancy directly or through culture filtrates have been described and associated with disease control.[57,66-67] Altered nutritional status of the host following organic amendment may also alter a disease response without affecting numbers of pathogen propagules in soil. Disease control as a result of modified host physiology may result in decreased infection or delayed pathogenesis after penetration.

In summary, organic amendments and crop rotation probably influence the severity of diseases by: (1) altering the nutritional environment of the plant and pathogen; (2) increasing the biological buffering capacity of soil; (3) hastening the reduction in pathogen population during anaerobic decomposition of organic matter; (4) affecting nitrification which determines the form of nitrogen predominating in soil; and (5) denying pathogens a host during the interim of unsuitable species. Integration of these cultural practices with other pest control programs can greatly enhance the efficiency and reliability of crop production.

REFERENCES

1. **Huber, D. M. and Watson, R. D.,** Nitrogen form and plant disease, *Annu. Rev. Phytopathol.,* 12, 139, 1974.
2. **Huber, D. M.,** Disturbed mineral nutrition, in *Plant Pathology — An Advanced Treatise,* Horsfall, J. G. and Cowling, E. B., Eds., Academic Press, New York, 1978, chap. 12.
3. **Huber, D. M.,** The role of nutrients in resistance of plants to disease, in *CRC Series in Nutrition and Food,* Rechcigl, M., Ed., CRC Press, Boca Raton, Fla., accepted for publication, 1980.
4. **McNew, G. L.,** The effects of soil fertility in plant diseases, in *Plant Diseases,* U.S. Department of Agriculture Yearbook, Washington, D.C., 1953, 100.
5. **Wingard, S. A.,** The nature of disease resistance in plants. I. *Bot. Rev.,* 7, 59, 1941.
6. **Epstein, E.,** *Mineral Nutrition of Plants; Principles and Perspectives,* John Wiley & Sons, New York, 1972.
7. **Steward, F. C.,** *Plant Physiology: A Treatise,* Vol. 3, Academic Press, New York, 1963.
8. **Gerretsen, F. C.,** The influence of microorganisms on the phosphate intake by the plant, *Plant Soil,* 1, 51, 1948.
9. **Moose, B.,** Advances in the study of vesicular-arbuscular mycorrhiza, *Annu. Rev. Phytopathol.,* 11, 171, 1973.
10. **Huber, D. M., Nelson, D. W., Warren, H. L., and Tsai, C. Y.,** Nitrification inhibitors — new tools for food production, *BioScience,* 27, 523, 1977.
11. **Timonin, M. E.,** Interaction of higher plants and soil microorganisms, in *Microbiology and Soil Fertility,* Gilmore, C. M. and Allen, O. N., Eds., Oregon State University Press, Corvallis, 1965, 135.
12. **Gaumann, E.,** *Principles of Plant Infection* (trans.), Crosby Lockwood, London, 1950, 543.
13. **Stakman, E. C. and Harrar, J. G.,** *Principles of Plant Pathology,* Ronald Press, New York, 1957, 581.
14. **Lewis, R. W.,** A graphic presentation of the balance hypothesis of parasitism, *Acta Bot.,* 3, 27, 1957.
15. **Warren, H. L., Huber, D. M., Nelson, D. W., and Mann, O. W.,** Stalk rot incidence and yield of corn as affected by inhibiting nitrification of fall-applied ammonium, *Agron. J.,* 67, 655, 1975.
16. **Garrett, S. D.,** Soil conditions and the take-all disease of wheat. XI. Interaction between host plant nutrition, disease escape, and disease resistance, *Ann. Appl. Biol.,* 35(1), 14, 1948.
17. **Blakeman, J. P.,** The chemical environment of the leaf surface in relation to growth of pathogenic fungi, in *Ecology of Leaf Surface Microorganisms,* Preece, T. F. and Dickenson, C. A., Eds., Academic Press, New York, 1971, 255.
18. **Knivehaven, C. M., Huber, D. M., and Warren, H. L.,** Comparison of growth rate and mycelial density of *Rhizoctonia solani* on an ammoniacal and nitrate source of inorganic nitrogen, *Indiana Acad. Sci.,* 85, 1976.
19. **Weinke, K. E.,** Influence of Nitrogen on the Root Disease of Bean Caused by *Fusarium solani* f. *phaseoli,* Ph.D. thesis, University of California, Berkeley, 1962, 150.
20. **Carley, H. E.,** Factors Affecting the Epidemiology of Pea (*Pisum sativum* L.) Root Rot Caused by *Aphanomyces eutriches* Drechs, Ph.D. thesis, University of Minnesota, St. Paul, 1969, 120.
21. **Li, C. Y., Lu, K. C., Trappe, J. M., and Bollen, W. B.,** Selective nitrogen assimilation by *Poria weirii, Nature (London),* 213, 814, 1967.
22. **Butler, F. C.,** Root and Foot Rot Disease of Wheat, Sci. Bull. 77, Agricultural Research Institute, Wagga Wagga, NSW, Australia, 1961, 98.
23. **Nelson, D. W.,** The Relationship Between Soil Fertility and the Incidence of *Diplodia* Stalk Rot and Northern Leaf Blight in *Zea mays,* M.S. thesis, University of Illinois, Urbana, 1963, 56.
24. **Painter, C. G. and Simpson, W. R.,** Fertilizing sweet corn for seed production, *Idaho Agric. Exp. Stn. Res. Bull.,* 501, 12, 1969.
25. **McNew, G. L. and Spencer, E. L.,** Effect of nitrogen supply of sweet corn on the wilt bacterium, *Phythopathology,* 29, 1051, 1939.

26. **Gallegley, M. E., Jr. and Walker, J. C.,** Plant nutrition in relation to disease development. V. Bacterial wilt of tomato, *Am. J. Bot.,* 36, 613, 1949.
27. **Daly, J. M.,** The influence of nitrogen source on the development of stem rust of wheat, *Phytopathology,* 39, 386, 1949.
28. **Gassner, G. and Frank, W.,** Uber den Einfluss der Temeratur auf Stickstoffgeholt and Rostresistenz junger Getreidep flanzen, *Phytopathol. Z.,* 7(3), 315, 1934.
29. **Suryanarayanan, S.,** Role of nitrogen in host susceptibility to *Piricularia oryzae, Cav. Curr. Sci.,* 27, 447, 1958.
30. **Bollenbacher, K. and Fulton, N. D.,** Effects of nitrogen compounds on resistance of *Gossypium arboreum* seedlings to *Colletotrichum gossypii, Phytopathology,* 61, 1394, 1971.
31. **Bawden, F. C. and Kassanis, B.,** Some effects of host nutrition on the susceptibility of plants to infection by certain viruses, *Ann. Appl. Biol.,* 37, 46, 1950.
32. **Clark, F. E.,** Experiments toward the control of the take-all disease of wheat and the *Phymatotrichum* root rot of cotton, *U.S. Dep. Agric. Tech. Bull.,* 835, 1942.
33. **Hursh, C. R.,** Morphological and physiological studies on the resistance of wheat to *Puccinia graminis tritici* Erikss. Henn., *J. Agric. Res.,* 27, 381, 1924; Exp. Stn. Bull., 329, 1936.
34. **Cralley, E. M.,** Effects of fertilizer on stem rot of rice, *Arkansas Agric. Exp. Stn. Bull.,* 383, 1, 1939.
35. **Allen, F. and Orth, H.,** Untersuchungen uber den Aminosaurengehalt und die Anfalligkeit der Kartoffelgegen die Kraut- und Knollenfaule, *Phytophthora infestans* de By, *Phytopathol. Z.,* 13, 243, 1941.
36. **Gries, G. A., Horsfall, J. G., and Jacobson, H. G. M.,** The balance of calcium and potassium in relation to club root of cabbage and potato scab, *Phytopathology,* 34, 1001, 1944.
37. **Chapman, H. D. and Brown, S. M.,** Potash in relation to citrus nutrition, *Soil Sci.,* 55, 87, 1943.
38. **Walker, J. C. and Foster, R. E.,** Plant nutrition in relation to disease development. III. *Fusarium* wilt of tomato, *Am. J. Bot.,* 33, 259, 1946.
39. **Moss, E. G., McMurtrey, J. E., Lunn, W. M., and Carr, J. M.,** Fertilizer tests with flue-cured tobacco, *U.S. Dep. Agric. Tech. Bull.,* 12, 1927.
40. **Hassebrauk, K.,** Untersuchungen uber den Einfluss einger Aussenfaktoren aufdas Anfalligkeitsverhatten der Standardsorten gegenube verschiedenen physiologischen Rassen des Weizenbraunrostes, *Phytopathol. Z.,* 12, 233, 1940.
41. **Sharvelle, E. G.,** The nature of resistance of flax to *Melamposora lini, J. Agric. Res.,* 53, 81, 1936.
42. **Weiss, F.,** The effect of rust infection upon the water requirement of wheat, *J. Agric. Res.,* 27, 107, 1924.
43. **Stapp, C. and Hartwich, W.,** *Zentralbl. Bacteriol.,* 2(110), 449, 1957.
44. **Bateman, D. F. and Lumsden, R. D.,** Relation of calcium content and pectic substances in bean hypocotyls of different ages to susceptibility to an isolate of *Rhizoctonia solani, Phytopathology,* 55, 734, 1965.
45. **Thomas, C. A. and Orellana, R. G.,** Phenols and pectin in relation to browning and maceration of castor bean capsules by *Botrytis, Phytopathol. Z.,* 50, 359, 1964.
46. **Halleck, D. L. and Garren, K. H.,** Pod breakdown, yield, and grade of Virginia type peanuts as affected by Ca, Mg, and K sulfates, *Agron. J.,* 60, 253, 1968.
47. **Lebeau, J. B.,** Pathology of winter injury of grasses and legumes in western Canada, *Crop Sci.,* 6(1), 23, 1966.
48. **Forsyth, F. R.,** Inhibition by nickel of the respiration and development of established infections on Thatcher wheat caused by *Puccinia recondita* Rob ex Desm, *Can. J. Bot.,* 40, 415, 1962.
49. **Baker, K. F.,** Principles of heat treatment of soil and planting material, *J. Aust. Inst. Agric. Sci.,* 28, 118, 1962.
50. **Huber, D. M. and Watson, R. P.,** Effect of organic amendment on soil-borne plant pathogens, *Phytopathology,* 60, 22, 1970.
51. **Burke, D. W.,** Soil microflora relationships in the development of bean root in Columbia basin soils, *Diss. Abstr.,* 15, 2390, 1955.
52. **Ferguson, J.,** Beneficial soil microorganisms, *Calif. Agric. Exp. Stn. Ext. Serv. Man.,* 23, 332, 1957.
53. **Nash, S. M. and Snyder, W. C.,** Quantitative estimations by plate counts of propagules of the bean root rot *Fusarium* in field soils, *Phytopathology,* 52, 567, 1962.
54. **Watson, R. D.,** Eradication of soil fungi by a combination of crop residue, flooding and anaerobic fermentation, *Phytopathology,* 54 (abstr.), 1437, 1964.
55. **Baker, R. and Nash, S. M.,** Ecology of plant pathogens in soil. VI. Inoculum density of *Fusarium solani* f. *phaseoli* in bean rhizophere as affected by cellulose and supplemental nitrogen, *Phytopathology,* 55, 1381, 1965.

56. **Easton, G. D.,** The results of fumigating *Verticillium*- and *Rhizoctonia*-infested potato soils in Washington, *Am. Potato J.,* 41 (Abstr.), 296, 1964.
57. **Huber, D. M. and McKay, H. C.,** Effect of temperature, crop, and depth of burial on the survival of *Typhula idahoensis* sclerotia, *Phytopathology,* 58, 961, 1968.
58. **Maloy, O. C.,** Physiology of *Fusarium solani* f. *phaseoli* in relation to saprophytic survival in soil, *Phytopathology,* 50, 56, 1960.
59. **Kaufman, D. D. and Williams, L. E.,** Effect of mineral fertilization and soil reaction on soil fungi, *Phytopathology,* 54, 134, 1964.
60. **Matthews, E. D.,** A biochemical study of soil organic matter as related to brown root rot of tobacco, *J. Agr. Res.,* 71, 315, 1945.
61. **Snyder, W. C., Schroth, M. N., and Christou, T.,** Effect of plant residues on root rot of bean, *Phytopathology,* 49, 755, 1959.
62. **Tyner, L. E.,** The effect of crop debris on the pathogenicity of cereal root-rotting fungi, *Can. J. Res. Sect. C,* 18, 289, 1940.
63. **Huber, D. M., Watson, R. D., and Steiner, G. W.,** Crop residues, nitrogen and plant disease, *Soil Sci.,* 100, 302, 1965.
64. **Davey, C. B. and Papavizas, G. C.,** Effect of dry mature plant materials and nitrogen on *Rhizoctonia solani* in soil, *Phytopathology,* 50, 522, 1960.
65. **Huber, D. M., Andersen, A. L., and Finley, A. M.,** Mechanisms of biological control in a bean root rot soil, *Phytopathology,* 56, 953, 1966.
66. **Curl, E. A. and Rodriquez-Kabana, R.,** Soil fertility and root-infecting fungi, *South. Coop. Ser. Bull.,* 183, 47, 1974.
67. **Venkata Ram, C. S.,** Soil bacteria and chlamydospore formation in *Fusarium solani, Nature,* 170, 889, 1952.

GENETIC RESISTANCE FOR CONTROL OF PLANT DISEASE

Gregory Shaner

EARLY WORK

Genetic resistance is one of the oldest methods of plant disease control. Theophrastus recognized differences in disease susceptibility among crop cultivars in the 3rd century B.C.[3] Throughout the centuries farmers improved, probably inadvertently, the resistance of their crops as they selected superior looking plants for the next year's seed. Not knowing that many diseases were caused by parasitic microorganisms hindered their progress in improving resistance. Disease-escaping plants were perhaps saved more often than were truly resistant plants.

In the latter part of the 19th century, when the infectious nature of many plant diseases was finally accepted by the scientific community, it became possible to systematically select resistant plants by exposing diverse collections of varieties of a crop to high levels of inoculum of various pathogens. Orton grew cottons, both domestic and exotic, on soils infested with *Fusarium oxysporum* f. *vasinfectum* and selected types that were truly resistant to wilt.[141] Similar techniques were applied to flax, tomatoes, and cabbage to find wilt resistant plants in these crops.[3]

When resistant plants were selected from existing collections of varieties, they were not always suitable in other ways. For example, the highly wilt-resistant Egyptian cottons that Orton selected did not produce high quality cotton.[141] A significant step in overcoming this problem was Biffin's demonstration that yellow rust resistance in wheat was inherited as a simple Mendelian trait.[18] He demonstrated that disease resistance could be handled in segregating populations just as easily as could various morphologic traits. This provided impetus for genetically transferring disease resistance from agronomically less desirable cultivars into susceptible but otherwise desirable cultivars.

Since this early work on disease resistance, much has been accomplished in the control of plant disease by resistance. Resistant cultivars of many crops are now widely grown and contribute significantly to alleviation of loss from disease. Concurrently there have been many basic studies concerning disease resistance — on its various expressions, its inheritance, and its stability. Likewise, the genetics of pathogens have been studied, including the ways these organisms respond to selection pressure in the form of resistant host cultivars.

EXPRESSION OF RESISTANCE

Disease resistance is a relative trait in plants rather than an absolute quality. It can only be recognized in a genotype if there is more susceptible genotype for comparison. Susceptible "standards" are by definition available for the diseases plant breeders seek to control through genetic manipulation; if they were not the need for control would not exist. Within the plant kingdom, resistance is the rule and susceptibility is the exception. That is, any plant species is resistant to most of the microorganisms known to cause disease on plants. Even if one looks at a list of pathogens of a given host species, many of these are only incidental and rarely cause serious disease.[266,280] However, all crop species are attacked by certain pathogenic microorganisms that incite severe disease. It is against this comparatively small group of pathogens of a crop that the plant breeder directs his attention.

From the biological point of view, any genotype species that restricts the pathogen significantly more than a "susceptible" genotype does is resistant. This may range from restriction that is barely measurable to complete restriction, i.e., immunity. It is not uncommon to find this full spectrum of resistance in a single crop species against a single pathogen species. Resistance has a more limited meaning when one is speaking about a method of control. For resistance to have value as a control measure it must confer substantial protection against damage from disease. Complete control of a disease is the goal. A resistant cultivar should perform as well in the presence of the disease as in its absence. The obvious way to achieve this level of control is through a high level of resistance, i.e., resistance that approaches immunity. There are many examples of this kind of resistance in the hypersensitive response plants show toward invading pathogens. Infection triggers a rapid response by host cells near the point of invasion and the infection is contained. The visible evidence of this resistance is a small chlorotic or necrotic fleck, sometimes too small to be seen with the unaided eye.

Against some diseases, high levels of resistance have never been found, forcing breeders to work with intermediate levels of resistance. This is often the only level of resistance available against root rots and some virus diseases. When a breeder is working with a high level of resistance that keeps a plant essentially free of disease, he can assume with some justification that resistant plants will sustain no yield reduction under conditions that would favor an epidemic on susceptible cultivars. This assumption is not necessarily valid for intermediate levels of resistance. Some disease develops on plants with this resistance; the amount of disease may be a function not only of host genotype, but environment and abundance of inoculum. The breeder must conduct yield loss studies with intermediate resistant genotypes to determine if the level of resistance is adequate to warrant the effort necesssary to transfer it to commercial cultivars.

TOLERANCE

The term "tolerance" frequently appears in the literature of plant disease resistance but its meaning is not always clear. Schafer confines the term to cases in which a plant endures severe disease without severe losses in yield or quality.[215] A tolerant cultivar does not inhibit the development of the pathogen, but is less damaged by any given level of infection than susceptible cultivars. "Tolerance" is also used to describe what may be a moderate level of resistance, especially in diseases such as root rots, vascular wilts, and virus mosaics, in which quantification of the pathogen in the host is difficult. Crill et al. used "resistant" to describe alfalfa clones in which alfalfa mosaic virus failed to establish itself after inoculation, and "tolerant" to describe clones that yielded as much dry matter when infected as when healthy.[50] True tolerance is difficult to demonstrate experimentally. Although the tolerant alfalfa clones had a "relatively high concentration" of virus, it was not shown that the virus titer was equivalent to that in fully susceptible clones and the apparent tolerance may have been the result of partial resistance.

From the point of view of one who is interested in growing a cultivar that is protected by its genetic makeup from the ravages of disease, distinction between tolerance and intermediate resistance is not important. In either case, the yield loss resulting from infection is lower than if a susceptible cultivar were grown. From an epidemiological point of view, the distinction between tolerance and resistance may be quite important, because a tolerant cultivar can be a reservoir of inoculum for other susceptible cultivars or species.

PATHOGEN VARIABILITY

Following some early successes in developing disease-resistant cultivars, either by direct selection within old land varieties or by hybridization and selection, plant breeders were optimistic about the ease of achieving permanent control of many plant diseases. However, it was not long after these early breeding efforts began that evidence appeared which suggested that the control of plant diseases by resistance was not as simple as it might have first seemed. It was discovered that plants resistant in one region might be susceptible in another. Furthermore, parasites in some cases seemed to be able to adapt themselves to resistant cultivars. In this connection, H. M. Ward proposed his "bridge theory" saying that a parasite could adapt itself to a resistant cultivar by growing for a time on a cultivar with intermediate resistance.[141]

In 1911 Barrus reported that some students in his plant pathology class isolated cultures of *Colletotrichum lindemuthianum* that were capable of causing anthracnose on bean cultivars he had previously found to be resistant to the disease.[11] Moreover, these cultures could not induce disease on some other cultivars that had been susceptible to the original cultures. Barrus recognized that this difference was not in overall disease-producing capacity of some cultures of *C. lindemuthianum,* but was a differential interaction between host cultivar and pathogen culture. A few years later Stakman and Piemeisel reported finding a culture of *Puccinia graminis tritici* in the Pacific Northwest that infected club wheats *(Triticum compactum = Triticum aestivum)* and several grasses in other genera, just as the midwestern cultures did, but failed to infect several common wheats *(Triticum vulgare = Triticum aestivum)* that were susceptible to these midwestern cultures.[247] Over the next few years Stakman and his colleagues found many cultures of *P. graminis tritici* that differed according to the cultivars of wheat they could infect.

Since Barrus's report, pathogen variability, often termed physiological specilization, has been found in a wide array of pathogens. These pathogenic variants are called physiologic races. Physiologic races are now so common among pathogen species against which some cultivars within a host species show a high level of resistance, that failure to find them is the exception. The extent of pathogenic variability in a pathogen is revealed only gradually as breeders continue to manipulate the host genetically. For example, highly specific races of *Phytophthora infestans* were not detected until "R"-genes that conferred immunity to blight were bred into the cultivated potato. When a new resistant cultivar of a crop is released, there is a period of freedom from disease, but then often a new race appears that can infect it. Initially the race is rare, but it becomes more prevalent with time and may eventually nullify the resistance of the cultivar.

GENERAL VS. RACE-SPECIFIC RESISTANCE

The existence of physiologic races of pathogens that infect their hosts differentially leads to the concept of race-specific resistance. This is resistance that is effective against some races of the pathogen but not against others. The antithesis of race-specific resistance is resistance that does not interact differentially with pathogen races. This is termed race-nonspecific resistance or general resistance. Van der Plank suggested the terms 'vertical' and 'horizontal' to describe race-specific and race-nonspecific resistance, respectively, believing that they would be less ambiguous than the numerous other terms various workers had used to describe these phenomena.[267] However, Caldwell pointed out that 'horizontal' and 'vertical' have clear meanings in English that have nothing to do with disease resistance, and therefore require redefinition before they become clear.[36]

There are difficulties with the concepts of race-specific vs. general resistance beyond choosing names for these phenomena. Race-specific resistance is rather clear-cut. If a cultivar is resistant to one culture of a pathogen and susceptible to another culture, its resistance is race-specific. However, failure to find a culture of a pathogen that is virulent to a resistant cultivar is not proof that its resistance is general. In many cases, breeders have found a gene for resistance, incorporated it into a commercial cultivar, and seen the cultivar attain wide use, only then to discover a race virulent to it. A plant breeder can only expose his experimental lines to a small part of a pathogen population and may easily miss a rare virulent race. Moreover, the race that eventually overcomes the resistant cultivar may not even have existed during the time the cultivar was being developed. Because no investigator can ever claim to have tested a resistant host genotype to all races of the pathogen, he can never prove that a cultivar has race-nonspecific resistance. Such claims are always tentative and must rest on empirical observation.[36,221] A cultivar that has been widely grown for many years and presumably exposed to much of the variability in a pathogen species, and that still retains its resistance everywhere, may be strongly suspected of having a general resistance. This assertion carries more weight if the cultivar in question has remained resistant over a period of time, during which other resistant cultivars have succumbed to new virulent races of the pathogen. The resistance of Knox wheat to powdery mildew is an example.[221] Suspected general resistance may also be tested by evaluating it against many pathogen cultures that differ in genes for virulence.[136]

Most cases of suspected general resistance are of the quantitative type.[36,233] Quantitative resistance is that kind in which there is no sharp distinction between resistant and susceptible genotypes. Rather, among a large collection of genetically distinct lines there is continuous variation in some metric character associated with pathogen reproduction or symptom expression, such as amount of sporulation, latent period, or lesion size. Quantitative resistance retards but does not completely inhibit the spread of disease. Quantitative resistance, as seen in the field, may be the product of several components of resistance. For a polycyclic disease caused by a fungus, a slower rate of spread could result from reduced rates of spore germination, appresorium formation, penetration, restricted spread of infection hyphae, a longer latent period, and a shorter infectious period. The "field resistance" of potatoes to late blight and the slow-rusting and slow-mildewing resistances in cereals are examples that have been studied extensively.[186,221,224,239,261]

The term virulence is often used in relation to pathogen variation of the race-specific kind.[110] Virulence is regarded as a qualitative trait, in that a specific pathogen culture may be described as virulent toward one host genotype and avirulent toward another. Virulence is the counterpart of race-specific resistance in the host. Just as there are different levels of race-specific resistance in the host, there are degrees of virulence within a pathogen, as indicated, for example, by the cereal rust infection type codes.[247] Thus, in certain host/parasite combinations the host may be described as moderately resistant or the pathogen described as weakly virulent.

Aggressiveness is defined as nonspecific variation in the pathogen's ability to cause an epidemic, i.e., a change in aggressiveness affects all host gentypes equally.[110] It is the counterpart of general resistance. Just as quantitative resistance seen in the field can be shown to consist of components in the host, so can aggressiveness be viewed as consisting of corresponding components in the pathogen.

For control of disease, partial, race-specific resistance has the same short-term effect as non-specific resistance in the host. Although the pathogen's ability to infect is not completely inhibited, its rate of spread is retarded, and this may give adequate control of a polycyclic disease. Very modest levels of race-specific resistance may greatly reduce the pathogen's reproductive potential.[110] The important distinction between partial, race-specific resistance and general resistance lies in the durability of the latter.

GENETICS OF RESISTANCE

The genetics of resistance have been the subject of numerous studies on many crops.[3,100,107,183,236,275,284] In some cases resistance is inherited as a monogenic trait. These genes may be partially or completely dominant, or recessive. Complementary gene action is known in some cases.[3,93,100] The preponderance of cases of simply inherited resistance may reflect research emphasis rather than the true frequency of this condition in higher plants. The clear-cut hypersensitive types of resistance are those that are simply inherited. Resistance that is clearly expressed is easier to follow in segregating populations, and consequently is more attractive for scientific investigation than is quantitative resistance that may have a large environmental variance associated with it.

Although in many cases resistance is governed by a single major gene, minor or modifier genes are often suspected to act together with the major gene to further enhance resistance.[100,284] Minor genes may act independently of major genes to confer a high level of resistance if they can be accumulated into a single genotype.[147,148]

Work on inheritance of resistance in hosts and on inheritance of virulence in pathogens that display pathogenic variability has done much to explain why such variability is so common. In the 1940s Flor showed that the genes for rust resistance in flax *(Linum usitatissimum)* were matched by genes for virulence in *Melampsora lini.*[77] There are 26 genes for rust resistance in flax, occurring as multiple alleles at five loci. Any one of these genes can confer a high level of resistance to certain races of *M. lini.* A flax cultivar is resistant to a culture of *M. lini* only if it carries a dominant gene for resistance, such as N^1, and the fungus carries a complementary gene for avirulence ($A_N{}^1$). No matter how many other genes for virulence the culture carries, if it has a gene for avirulence corresponding to any one gene for resistance in the host, the culture will be avirulent. These observations led Flor to his hypothesis of complementary genetic systems in the host and its obligate parasite.[77] If a flax cultivar is resistant to rust because of a single gene, say N^2, then it follows that all of the cultures to which it was tested are of geneotype $A_N{}^2A_N{}^2$ or $A_N{}^2a_N{}^2$ (the infecting urediniospores of *M. lini* are dikaryotic and function essentially as diploids). Should any urediniospore of *M. lini* have the genotype $a_N{}^2a_N{}^2$ it would be able to infect the resistant cultivar. Such a genotype could arise from mutation and sexual or parasexual recombination. Since the work with flax rust, complementary genetic systems have been demonstrated or strongly implicated for several other diseases, including some not caused by obligate parasites.[78] However, these cases represent only a small proportion of diseases for which resistance is known.

There are far fewer papers that document polygenic resistance compared to monogenic resistance,[174,233] but inheritance of this type is suspected in many cases.[236,266,275,284] Because general resistance is so often a quantitative trait it is assumed to be polygenically inherited.[233] In some cases genetic studies have confirmed this.[174,233] The belief that general resistance is polygenically controlled rests partly on negative evidence. The majority of monogenic resistances have proven to be race-specific. In several cases a gene-for-gene relationship between host and pathogen has been demonstrated.[78] If virulence in the pathogen is monogenically controlled, there is a high probability that a new strain capable of overcoming a gene for resistance in the host will evolve through mutation or recombination, given the reproductive potential of most plant pathogens. Resistance that is suspected of being general is usually an intermediate type, as described above. A pathogen strain against which this resistance is elicited is, by definition, virulent. The pathogen can infect and cause symptoms and reproduce, but it does so less frequently or more slowly than it does on susceptible host genotypes. If such resistance is governed by several genes, one would suspect that no single allele substitution in the pathogen would condition a major shift in aggressiveness.

Quantitative "field" resistance, i.e., resistance that is expressed as a slow rate of disease progression in the field, may be the cumulative effect of several simply-inherited forms of resistance. Whether polygenic resistance bears a gene-for-gene relationship with the pathogen, as Parlevliet and Zadoks hypothesize, is debatable.[187] Nonetheless, if several genes in the pathogen govern the rate at which it spreads on a host, then there would perhaps need to be allele substitutions at several of these loci to produce a much more aggressive strain. If the strain is already quite aggressive on susceptible host genotypes, then to overcome the nonspecific resistance of some other genotype, it must either become more aggressive toward all host genotypes or somehow circumvent the restrictions placed on it by the nonspecifically resistant host. If several genes in the host confer this resistance, then presumably the pathogen is restricted in several ways and it would have to overcome all of these restrictions to become fully aggressive.

If each of several resistance mechanisms were controlled by a separate, single gene, their combined effect on an epidemic in the field would be seen as a polygenic trait. Further research is needed to determine whether such mechanisms of resistance are indeed governed by distinct genes. In wheat leaf rust, a long latent period and smaller uredinia may be two manifestations of the same underlying mechanism, a retardation of the growth of infection hyphae, and be under fairly simple genetic control despite continuous variation in segregating populations.[110,180,224] Nelson argues quite persuasively that there is no sharp genetic distinction between race-specific and general resistance and that general resistance arises from the interactions of several "major", race-specific genes that no longer confer immunity.[174]

A difficulty in working with quantitative resistance is that within a homozygous line there is often considerable variation for this trait, indicating that environment contributes significantly to its expression. When the environmental component of phenotypic variation is large, it is possible for a population segregating for a trait that is conditioned by genes at only two loci to show continuous variation with no clear segregation. It is likely that some cases of quantitative resistance, which are assumed to be examples of polygenic traits, are actually under simpler but only moderately heritable genetic control.

GENE-ENVIRONMENT INTERACTIONS

The expression of genes that confer high levels of resistance is usually not significantly influenced by the environment in which the plant is growing. These genes are effective under all conditions under which the crop would be grown. They can be relied upon to protect a crop year after year, regardless of planting date, rate of crop development, and harvest date, providing no pathogen races capable of overcoming them evolve. However, temperature-sensitive genes for disease resistance are known,[100,274] Gene *Sr6* for stem rust resistance is highly effective at 20°C but completely ineffective at 25°C.[238] Genes that confer intermediate infection types are more likely to be influenced by temperature than are genes that confer complete resistance.[100] Expression of the monogenic recessive resistance in onion to pink root depends on temperature, and also on the quality of the seed.[274]

If a gene is known to be sensitive to temperature, it should only be used in cultivars intended for an area where the critical temperature is not common. If this is not possible, then the resistance gene should be used in conjunction with other genes. For example, *Sr6* can protect young wheat plants from rust because sustained temperatures of 25°C are not common during that stage of plant development. Later in the season this gene would not be effective and other genes would be needed.

Minor genes that modify a single dominant gene for stripe rust resistance in PI 178383 wheat at low temperatures can confer a high level of resistance when combined into a single line.[147] These genes are only effective in a cool temperature regime, however. Minor genes in Rego wheat are only effective in a warm regime.[148] When combined, these genes confer resistance under both cool and warm regimes, and, because several genes confer this resistance, the chance of a virulent race of *Puccinia striiformis* evolving is lower.[148]

Intermediate, quantitative resistance is more sensitive to environment than the high-level, monogenic kind. Disease development depends on the host, the pathogen, and the environment. A change in any one factor can alter the course of an epidemic. When a host has only intermediate resistance, one would expect disease to be more severe as environment becomes more favorable to disease development. A classical example of this is cabbage yellows caused by *Fusarium oxysporum* f. *conglutinans.* Some plants that were selected for resistance in the field proved to be susceptible in greenhouse tests. It was shown later that soil temperature is critical to the expression of this resistance. At 24°C or above the resistance is ineffective, and soil temperatures were often higher than this in greenhouse tests.[274] There is also a monogenic resistance in cabbage to yellows that is effective at high as well as low temperatures. Closely controlled rooting-medium temperatures are used to separate the two kinds of resistance.[274] Tomatoes likewise have two types of resistance to Fusarium wilt: the temperature-sensitive polygenic type and the temperature-insensitive monogenic type.[274,275] Although polygenic resistance to Fusarium wilts breaks down in warm soils, it has the advantage of stability toward the pathogen and it is widely used. Soil temperatures in the field may exceed the cutoff point for expression of this resistance but they generally do not remain there for extended periods. For example, the muskmelon cultivar Saticoy has an intermediate polygenic resistance to Fusarium wilt. On the sandy soils of southern Indiana Saticoy will show wilt symptoms under hot, dry conditions, and the fungus can be isolated from plants, but with the coming of rain the plants recover and produce a crop of muskmelons.[251]

In some areas polygenic wilt resistance may not be adequate. Walter reports that only two successive tomato crops would be possible on Florida sandland for a cultivar like Rutgers, which has polygenic resistance.[275] This indicates that a combination of favorable environment (warm, droughty soil) and abundant inoculum of the pathogen will render intermediate, polygenic resistance ineffective.

High levels of nitrogen fertilizer exacerbate wheat powdery mildew. This intermediate, race-nonspecific type of resistance to the disease is characterized by a slower rate of disease spread.[221] Because nitrogen fertilizer level likewise influences the rate of disease spread, one would expect the disease to progress more rapidly on a slow-mildewing wheat cultivar just as it does on a susceptible wheat cultivar at higher N levels. When this was investigated, high N was found to increase mildew less on the slow-mildewing cultivar Knox than on the susceptible cultivar Vermillion.[223] Nonetheless, if a farmer were to grow a slow-mildewing wheat he would be advised to use less than the highest recommended rate of fertilizer. Presumably there would be a point at which the yield gain from adding more N would be offset by loss from more severe powdery mildew. This would be considered along with the straw strength of the cultivar in deciding how much N to apply.

Date of planting can affect expression of intermediate resistance. Wheat spindle streak, yellow dwarf, Septoria leaf blotch, and take-all root rot are all diseases of wheat that are more severe on plants that are sown early.[222] Wheat genotypes that show partial resistance to any of these diseases show more resistance when they are sown later in the fall.

The so-called field resistance to late blight in potato is an intermediate race-nonspecific, polygenic resistance.[261] This resistance, like the slow-mildewing in wheat described above, slows but does not completely halt the spread of *Phytophthora infestans* infections. When weather is especially favorable for the disease, field resistance does not provide adequate control.[184] In "blight years" nearly all the potato hectarage in the northeastern U.S. is sprayed with fungicide, regardless of cultivar. In spray trials using Russet Rural, a fully susceptible cultivar, and Sebago, the best commercially-adapted, field-resistant cultivar, Fry showed that Sebago's resistance was equivalent to 0.42 to 0.67 kg/ha (active ingredient) of mancozeb per week.[83] This was equivalent to 25% of the total fungicide requirement for blight control. In a subsequent study, disease control on Sebago that received 75% of the normal dose of fungicide was better than disease control on Russet Rural that received the full dose, when both cultivars were sprayed according to recommendations from a disease forecast system.[84] Results of these studies suggest that even if intermediate resistance cannot always give adequate control alone, it may enhance the effectiveness of or reduce the need for other controls. The fact that intermediate resistance which is race-nonspecific may need to be supplemented with chemical or cultural control measures when conditions are especially favorable for disease may still outweigh the risk associated with using a higher level, race-specific resistance.

ADVANTAGES AND DISADVANTAGES OF RESISTANCE AS A CONTROL MEASURE

Disease resistance is sometimes referred to as the ideal method of disease control because it comes "prepackaged" in the seed or planting stock and the farmer does not have the worry and expense of fungicide application, uneconomical crop rotations, or other cultural practices that constrain his planting options. If a well-adapted, disease resistant cultivar is available to the farmer, then it will indeed reduce his burden of disease control. However, as a control method, there are many complications associated with disease resistance. Aside from the environmental sensitivity of some resistances discussed above, it is the plant breeder more than the farmer who must struggle with these complications.

One of the problems of using resistance as a control method is to find a source of resistance. If a disease is serious enough to warrant control measures, then there is evidently insufficient resistance in the locally grown cultivars. A resistant cultivar from some other area could be used, but this is often ineffective. Cultivars transported far from their region of development usually do not perform well and these deficiencies outweigh the benefits from their disease resistance. Exotic cultivars, although useless to the farmer, may form the starting point for a disease-resistance breeding program.

In many cases, high levels of resistance are found in wild relatives of cultivated crops. These sources of resistance are valuable because they may provide genes for resistance to a pathogen in addition to those in the cultivated species, and serve as a reserve against virulence changes in the pathogen. Wild relatives may also have a level of disease resistance superior to that in the cultivated species. Interspecific crosses have been used many times to transfer resistance from wild species to cultivated species.[126] The difficulty of making such crosses varies greatly. Some species are so closely allied that genetic or cytoplasmic isolating mechanisms are not strong, and fertile types are easily recovered. More often, interspecific hybrids are sterile. If sterility is mainly in the pollen, several cycles of backcrossing, using the cultivated species as the pollen source, coupled with selection for disease resistance at each step, will permit recovery of a fertile line with resistance. Embryo culture may be used to circumvent some sterility mechanisms.[126]

Recovery of a suitable plant type may be difficult owing to undesirable linkages.[126] For example, undesirable fruiting characteristics and resistance to *Heterodera marioni* are linked in *Lycopersicon esculentum-L. peruvianum* or *-L. chilense* hybrid derivatives. Resistance to *Puccinia recondita* in *Triticum aestivum-Agropyron elongatum* derivatives is linked with an undesirable yellow flour color. If translocation lines can be obtained from interspecific hybrid derivatives, undesirable linkages can sometimes be broken.[126]

Regardless of whether resistance comes from within a cultivated species or from wild relatives, it is sometimes difficult to transfer all the resistance from the donor parent into a modern cultivar.[45] It may be that the genes are linked to traits that are incompatible with the requirements for a modern cultivar, or that the resistance gene may not express itself fully in certain genotypes.[174] For example, many older wheat cultivars have moderate resistance to *Septoria tritici*. They are also late-maturing. Transferring this resistance to early-maturing cultivars is difficult.[222]

Assuming resistance can be found, it takes several years to incorporate it into a commercially valuable cultivar. Depending on how unadapted the source of resistance is, this may require only one or as many as nine or ten crosses before selfing and selection can begin. If a disease suddenly becomes a serious problem in an area, because of a change in climate, the accidental introduction of the pathogen, or because breeders inadvertently incorporated extreme susceptibility into the widely grown cultivars, it takes several years to develop resistant cultivars. In the meantime, growers must use other control measures, if available, or sustain losses.

The plant breeder needs a severe epidemic every year in his nurseries in order to effectively select lines that are resistant. However, plant diseases normally vary in prevalence and severity from year to year and in some years a disease may be almost nonexistent. Quantity of primary inoculum is one of the variables that influences severity of disease, so breeders often inoculate their nurseries to increase the probability of severe disease. Inoculation also reduces spatial variation in disease, which is also characteristic of most natural epidemics. Spatial variation means that a breeder may select plants that have escaped disease, mistakenly believing they are resistant. Spatial variation is more of a problem when working with monocyclic diseases such as root rots, vascular wilts, and those caused by root-infecting viruses. Growing standard check cultivars frequently throughout the nursery can reveal this variation but does not reduce it.

Weather is another critical element in disease development and the breeder has little control over this. If weather is extremely critical for effective selection the breeder may need to use controlled-environment facilities for his selection. This is costly and limits the amount of material he can examine.

In many breeding programs, resistance to more than one disease is sought and in practice a severe epidemic of all of these diseases does not occur each year. If one or two diseases are severe each year the breeder can make adequate progress. However, if several years elapse without a good test for resistance to disease, the nurseries become cluttered with unselected material that greatly reduces the efficiency of a program.

Another problem inherent in resistance as a disease control measure is its impermanence. Two things account for this. In breeding for resistance to infectious plant disease, the genotype of one organism (the host) is altered to the detriment of another organism (the pathogen). The pathogen also carries genes that adapt it to its environment and a resistant cultivar becomes a powerful selection force. With the tremendous reproductive rates of many pathogens, the appearance of mutants or recombinants with greater virulence becomes almost a certainty, especially when virulence is under simple genetic control. In time, more virulent strains may predominate in the pathogen population and cause significant yield loss in the previously resistant cultivar. The cultivar's resistance is said to have broken down, and new genes for resistance must be found and incorporated into useful cultivars.

Resistance cannot be automatically applied to all cultivars. Once a fungicide is proven effective for control of a disease and is registered for use, it can be applied to any cultivar of that crop the farmer grows, except in special cases where some cultivars are more sensitive to a chemical. However, if resistance is used to control a disease and a farmer switches to another cultivar because it has greater yield potential, has better quality, etc., he will lose his disease control unless the new cultivar has also been bred for resistance to the disease. Thus, as breeders continue to improve adaptability of a crop, by selecting for superior agronomic traits and quality, tolerance to abiotic stress, and resistance to other diseases, they must continue to select for disease resistance they have already incorporated into existing cultivars. Ironically, the more improved a cultivar is, including resistance to several diseases, the more difficult it is to add resistance to yet another disease and still maintain all that was in the original cultivar.

Another obstacle in the way of successfully controlling disease by resistance is the need to incorporate resistance to all diseases that fall into a certain control class. For example, early sowing of winter wheat is desirable in many areas where wheat is to be grazed and because larger plants survive the winter better. However, Hessian fly is more likely to infest early-sown wheat. Resistance to Hessian fly in soft red winter wheats permitted farmers to plant wheat much earlier than they could when only fly-susceptible cultivars were available.[37] As Hessian fly-resistant cultivars became available and the crop was planted earlier, other diseases became more severe. Date-of-planting experiments showed that Septoria leaf blotch, take-all, yellow dwarf, and wheat spindle streak were all worse on wheat sown early.[222] Even though resistance to Hessian fly, Septoria leaf blotch and take-all is available, farmers are still advised to avoid planting wheat before the fly-free date. This minimizes the chance of yellow-dwarf virus and wheat spindle streak mosaic virus infection. The intermediate resistance to take-all is more effective when wheat is sown later. Thus, the higher level of resistance to Hessian fly and Septoria leaf blotch does not eliminate the need for a cultural control method, because other diseases in that same control class are not yet completely controlled by resistance. Early planting does not eliminate the risk of these diseases so the resistance available now is nonetheless quite valuable.

Walter points out that resistance to either bacterial or *Fusarium* wilt alone in tomato would be of no value to a grower whose land was infested with both *Pseudomonas solanacearum* and *Fusarium oxysporum* f. *lycopersici.*[275] Also, gray leaf spot, late blight, black spot, and leaf mold may all be hazards to tomatoes in one area. Resistance to less than all four fungal diseases would not eliminate the need for the farmer to spray with fungicide. The breeder should try to incorporate resistances that allow the farmer to eliminate classes of control measures.

One problem associated with resistant cultivars is an ironical consequence of the success this control measure often achieves. Often, when a breeder begins to incorporate resistance to a disease into a crop, his efforts culminate in the release of a single cultivar. Then he may turn his attention to improving that resistant cultivar for other traits. In the meantime, the resistant cultivar, if it is a good one and has wide adaptation, may rapidly come to occupy a large share of the hectarage of the crop in a state or region. The wide use of a single genotype leads to what is termed genetic vulnerability. If a new virulent strain of the pathogen arises, the extensive hectarage of the resistant cultivar is an excellent "selective medium" on which it can multiply. This kind of vulnerability exists even if several cultivars are released over several years and none of them occupies a large share of the hectarage of the crop. If they collectively occupy a large hectarage and all owe their resistance to a common gene, the effect is the same. The widespread use of a single cultivar or group of related cultivars may also create vulnerability to previously minor diseases. Examples are the use of Texas male sterile cytoplasm in maize, which led to the southern corn leaf blight epidemic, and the use of the Victoria gene for crown rust resistance in oats, which led to the Victoria blight epidemic.[265, 60]

At the outset of this section, it was pointed out that one of the advantages of resistance as a disease control measure was that it shifted the burden of control away from the farmer. Resistant cultivars can generally be planted and raised like any other. There are some other features of resistance that make it more desirable than other control measures. It is perhaps the cheapest form of disease control. Much of the effort to develop resistant cultivars has historically been done at state agricultural experiment stations or by the Agricultural Research Service (now Scientific and Education Administration) of the U.S. Department of Agriculture, and much of the funding comes from general state and federal tax revenues. Often, commodity grower groups will contribute money to aid these breeding programs but the cost per hectare of harvested crop is still low compared to chemical control. Private breeding companies often rely on these public agencies as sources of adapted, disease-resistant germplasm. Whether a resistant cultivar is developed by a public agency or private company, price for its seed is never much greater than what is charged for seed of susceptible cultivars.

Disease resistance is a non-polluting form of control. Although fungicides as a group are less toxic to man and other animals than are insecticides, some of the materials used to kill fungi are highly toxic to non-target organisms.

It is often a nuisance that the genotype of a cultivar is greatly disrupted during the process of adding a gene for disease resistance to it, but this is not always a disadvantage. Breeding in many crops began in an effort to improve the disease resistance of existing cultivars. As breeders crossed diverse types they often found segregants that were not only resistant, but superior in yield or other traits, as well. The soft red winter wheat cultivar Arthur is just one example. It arose from a complex cross intended to combine resistance to powdery mildew, stem rust, and leaf rust.[190] However, it yielded much more than any of its parents or any other cultivars being grown in the soft wheat region at the time of its release. This yield advantage seems to be due to more than just disease resistance, because it is seen under essentially disease-free conditions. This is illustrated by wheat yields in replicated drill-strip plots in Tippecanoe County, Indiana in 1977. All cultivars survived the winter well and were free of disease, facts reflected by the overall high yields in the trial. Arthur yielded 5.97 metric ton/ha, which was 7, 44, and 43% more, respectively, than the yield of Monon, Vigo, and Trumbull, popular cultivars that preceded it. As crop cultivars become more highly adapted to certain regions, through long term efforts of plant breeders, the chance of a higher-yielding line fortuitously arising from a cross between any adapted cultivar and any exotic resistant cultivar diminishes. Nonetheless, many of the notable gains in genetic potential for yield and quality realized to date arose from efforts to improve disease resistance.

PLANT-BREEDING ACCOMPLISHMENTS

A complete account of what has been accomplished so far in the control of disease through genetic resistance is not possible in a brief chapter. Table 1 summarizes the disease resistance that is reported for most of the crops grown in the U.S. One difficulty with compiling such a table is that a report of resistance in a crop species does not necessarily mean that the resistance is being used on a commercial scale. Unfortunately, cultivar-use surveys are rare in the U.S., so it is difficult or impossible to determine on how many hectares a specific cultivar is grown, which would allow calculation of the percentage of the crop resistant to any given disease. Nonetheless, the author has attempted to estimate, with the aid of specialists in the various crops, the importance of genetic resistance to specific diseases.

Table 1 reveals that disease resistance is known in many of the crops used in the U.S., as well as many other countries. Most effort has gone into annual crops. Among the fruits and vegetables, resistance has been found to many diseases of beets, lettuce, crucifers, cucurbits, peas, beans, onions, peppers, tomatoes, and potatoes. Resistance is a major control method in small grains, maize, soybeans, alfalfa, and clovers. Resistance protects tobacco from many of its diseases.

The resistance categories in Table 1 are an attempt to indicate whether disease resistance in a crop is providing economical control. Because of the difficulties referred to above, these resistance categories necessarily involve judgment. When possible, specialists in each crop were consulted before assigning resistance categories in order to make this part of the table as realistic as possible.

Examination of the table reveals that there are not many diseases against which essentially all cultivars of a crop are resistant. This is not surprising. Because choice of cultivar is usually left to the farmer and is not dictated by government regulation or industry, there are often many cultivars of each crop being grown in the U.S. Some of these may have lost the resistance they once had owing to race changes in pathogens; others may never have had resistance to certain diseases but have other attributes which keep them in production. For perennial crops the transition to resistant cultivars may take several years.

The number of diseases for which there is at least one resistant, commercially acceptable cultivar is large. This gives reason for optimism. It means that resistance has been combined with other desirable traits in a crop and that this resistance can be incorporated into many more cultivars intended for use wherever the disease is a problem.

Among the various classes of disease, resistance to foliar diseases is perhaps most common. This is understandable, in that these diseases were the earliest to draw the attention of plant pathologists. Moreover, breeding for resistance requires that the breeder be able to clearly detect superior lines, and this is easier for diseases where symptoms are clearly expressed on the foliage, in comparison to diseases of belowground organs.

Resistance to damping off and seeding blights is rare. Two things account for this. These diseases are among the most easily controlled by fungicides. Seed dressings or in-furrow applications are effective and rather cheap. The fungi that cause these diseases are ubiquitous and many have a broad host range. These are low-grade pathogens that attack plants at an especially vulnerable stage in their development, and it is quite likely that a useful level of resistance in seedlings to these diseases does not exist.

Root diseases are another class of disease against which resistance is not common. These are likewise caused by comparatively unspecialized pathogens often having a broad host range. Resistance to these diseases is known in beets and carrots, crops in

Table 1
RESISTANCE OF MAJOR CROPS TO INFECTIOUS DISEASES

Crop	Disease	Pathogen	Resistance category[a]	Ref.
Betulaceae				
Filbert				
Corylus spp. and hybrids	Bacterial blight	*Xanthomonas corylina*	BP	138
	Labrella leaf spot	*Labrella coryli*		249
	Eastern filbert blight	*Crystosporella anomala*		138
Chenopodiaceae				
Beets				
Beta vulgaris	Bacterial vascular necrosis and rot	*Erwinia* sp.	B	279
	Scab	*Streptomyces scabies*	BP	140
	Downy mildew	*Peronospora schachtii*	A	154
	Cercospora leaf spot	*C. beticola*	BP	14
	Powdery mildew	*Erysiphe betae* or *E. polygoni*	BP	205, 206
	Black root	*Aphanomyces cochlioides*	BP	14
	Pythium rot	*Pythium* spp.		127
	Black root	*Pleospora bjoerlingii*	BP	230
	Yellows and stalk blight	*Fusarium oxysporum f. betae*	BP	153
	Rust	*Uromyces betae*	B	14
	Root rot	*Rhizoctonia solani*	BP	97
	Storage rot	*Botrytis cinerea*		33, 229
		Phoma betae		
		Penecillium claviformi		
	Sugar beet cyst	*Heterodera schachtii*	P	98, 126, 211, 212
	Root knot	*Meloidogyne incognita*		172
	Cyst	*Heterodera schachtii*		211, 212
	Curly top	Beet curly top virus	AP	66
	Virus-yellows	Beet yellows virus, beet western yellows virus complex	AP	155

Table 1 (continued)
RESISTANCE OF MAJOR CROPS TO INFECTIOUS DISEASES

Crop	Disease	Pathogen	Resistance category[a]	Ref.
Spinach	Blue mold	*Peronospora elfusa*		273
Spinacia oleracea				
	White rust	*Albugo occidentalis*		22
	Fusarium wilt	*F. oxysporum f. spinaciae*		273
	Blight	Cucumber virus 1		273
	Mosaic	Virus		273
	Curly top	Virus		210
Compositae				
Lettuce				
Lactuca sativa	Powdery mildew	*Erysiphe cichoracearum*		273
	Downy mildew	*Bremia lactucae*	BP	273
	Anthracnose	*Marssonina panattoniana*		273
	Gray mold	*Botrytis cinera*		271
	Sclerotinia disease	*Sclerotinia sclerotiorum*		271
	Rhizoctonia bottom rot	*Rhizoctonia solani*	B	273
	Root rot	Non-pathogenic	A	220
	Mosaic	Virus	BP	273
	Spotted wilt	Virus		273
	Big vein	Virus		273
	Beet western yellow	Virus		273
	Brown blight	Etiology unknown	A	273
	Corky smut rot	Non-pathogenic?		273
	Aster yellows	Virus		273
Safflower				
Carthamus tinctorius	Rust	*Puccinia carthami*	B	249
		Erysiphe cichoracearum		5
		Ramularia carthami		5
		Cercospora carthami		5
	Wilt	*Verticillium*	B	5
	Wilt	*Fusarium oxysporum*	B	20
	Root rot	*Phytophthora drechsleri*	B	249
	Leaf spot	*Alternaria* sp.	B	286

Sunflower				
Helianthus annuus				
Oilseed sunflower	Rust	*Puccinia helianthii*	A	286
	Downy mildew	*Plasmopora halstedii*	A	286
	Wilt	*Verticillium dahliae*	A	286
Non-oilseed sunflower	Rust	*P. helianthii*	A	286
	Downy mildew	*P. halstedii*	B	286
	Wilt	*Verticillium dahliae*	B	286
Convolvulaceae				
Sweet potato				
Ipomoea batatas	Soil rot	*Streptomyces ipomoea*		273
	White rust	*Albugo ipomoeae panduratae*		273
	Stem rot	*Fusarium hyperoxysporum* and *F. oxysporum f. batatas*	B	273
	Black rot	*Ceratocystis fimbriata*		273
	Root knot	*Meloidogyne incognita*	BP	255
	Internal cork	Virus	B	273
	Russet crack	Virus		158
Cruciferae				
Broccoli and Cauliflower				
Brassica oleraceae var. botyrtis	Clubroot	*Plasmodiophora brassicae*		273
	Yellows	*Fusarium oxysporum f. conglutinans*		273
	Downy mildew	*Peronospora parasitica*		273
	Rhizoctonia disease	*Rhizoctonia solani*		58
	Mosaic	Virus		273
Brussels sprouts				
B. oleraceae var. *gemnifera*	Powdery mildew	*Erysiphe cruciferarum*		61
		Brevicoryne brassicae		68
Cabbage				
B. oleracea var. *capitata*	Black rot	*Xanthomonas campestris*	B	273
	Club root	*Plasmodiophora brassicae*	B	273

Table 1 (continued)
RESISTANCE OF MAJOR CROPS TO INFECTIOUS DISEASES

Crop	Disease	Pathogen	Resistance category[a]	Ref.
	Alternaria leaf spot	*Alternaria* spp.		273
	Downy mildew	*Peronospora parasitica*	P	249
	Yellows	*Fusarium oxysporum f. conglutinans*	A	273
	Mosaic	Virus	B	249
Turnips *Brassica campestis*	Club root	*Plasmodiophora brassicae*	B	126
Swede turnip *Brassica napus*	Club root	*Plasmodiophora brassicae*	B	126
	Mildew	*Erysiphe cruciferarum*	B	111
	Turnip crinkle	Virus		62
Radish *Raphanus sativus*	Club root	*Plasmodiophora brassicae*	BP	176
	White rust	*Albugo candida*	BP	273
	Fusarium wilt	*F. oxysporum f. conglutinans*	A	273
	Scurfy root	*Rhizoctonia solani*	B	103
	Black root	*Aphanomyces raphani*	B	103
Oil seed rape *Brassica campestris*	White rust (staghead)	*Albugo candida*, races 2, 7	A	192
	Alternaria black spot	*Alternaria brassicae* and *A. raphani*	AP	193
	Blackleg	*Leptosphaeria maculans*	A (France) B (Australia)	32
	Brown girdling root rot	Unknown	AP	194
Mustard (Oriental, brown) *B. juncea*	Blackleg	*L. maculans*	AP	194
	White rust (staghead)	*A. candida*, race 2		188, 194

Mustard (Yellow)				
B. hirta	Blackleg	*L. maculans*	AP	194
	Alternaria black spot	*A. brassicae, A. raphani*	AP	16
	White rust (staghead)	*A. candida*, races 2, 7	A	194
Cucurbitaceae				
Cucumber				
Cucumis sativus	Bacterial wilt	*Erwinia tracheiphila*		273
	Angular leaf spot	*Pseudomonas lachrymans*	BP	236
	Downy mildew	*Pseudoperonospora cubensis*	AP	236
	Anthracnose	*Colletotrichum lagenarium*	AP	236
	Powdery mildew	*Erysiphe cichoracearum*	A	236
	Scab	*Cladosporium cucumerinum*	BP	236
	Cucumber mosaic	Cucumber mosaic virus	A	236
	Gummy stem blight	*Mycosphaerella melonis*		236
Muskmelon				
Cucumis melo	Downy mildew	*Pseudoperonospora cubensis*	AP	236
	Alternaria leaf spot	*A. cucumeria*	BP	236
	Powdery mildew	*Erysiphe cichoracearum*	AP	273
	Marsonia blight	*M. melonis*		249
	Gummy stem blight	*Mycosphaerella melonis*		236
	Fusarium wilt	*F. oxysporum f. melonis*	BP	273
	Cucumber mosaic	Cucumber mosaic virus		273
Watermelon				
Citrullus lanatus	Downy mildew	*Pseudoperonospora cubensis*		273
	Anthracnose	*Colletotrichum lagenarium*	AP	236
	Gummy stem blight	*Mycosphaerella melonis*		236
	Fusarium wilt	*F. oxysporum f. niveum*	A	236
Squashes and Pumpkin				
Cucurbita spp.	Powdery mildew	*Erysiphe cichoracearum*		236
	Scab	*Cladosporium cucumerium*		236
	Cucumber mosaic	Virus		273
Ericaceae				
Blueberry				
Vaccinium spp.				
Subgenus				
Cyanococcus	Bacterial canker	*Pseudomonas syringae*	B	85
	Powdery mildew	*Microsphaera penicillata* var. *vacinii*	B	85

Table 1 (continued) RESISTANCE OF MAJOR CROPS TO INFECTIOUS DISEASES

Crop	Disease	Pathogen	Resistance category[a]	Ref.
	Leaf rust	*Pucciniastrum myrtilli*	B	85
	Double spot	*Dothichiza caroliniana*		
	Cane canker	*Botryosphaeria corticis*	B	226
	Redleaf	*Exobasidium vaccinii*	B	1
	Mummy berry	*Monilinia vacinii-corymbosi*	B	1
	Twig and cane blight	*Botrytis cinerea*	B	1
	Fusicoccum canker	*Fusicoccum putrefaciens*	B	1
	Root rot	*Phytophthora cinnamomi*	B	85
	Necrotic ringspot	Tobacco ringspot virus	BP	85
	Red ringspot	Blueberry red ringspot virus	BP	85
	Shoestring	Blueberry shoestring virus	B	85
Fagaceae				
Chestnut				
Castanea mollissima				
C. crenata	Chestnut blight	*Endothia parasitica*	BP	108
	Twig canker	*Cryptodiaporthe castanea*		249
	Twig canker	*Botryosphaeria ribis chromogena*		249
	Blossom end rot	*Glomerella cingulata*	B	108
	Phytophthora root disease	*P. cinnamomi*	BP	108
Gramineae				
Barley				
Hordeum vulgare	Bacterial blight	*Pseudomonas atrofaciens*	B	227
	Bacterial blight	*Xanthomonas* sp.	BP	95
	Covered smut	*Ustilago hordei*		72
	Loose smut	*Ustilago nuda*	B	168
	Bunt	*Tilletia controversa*	A	59
	Leaf rust	*Puccinia hordei*	B	168
	Stem rust	*Puccinia graminis*	B	168
	Yellow rust	*Puccinia striiformis*	B	168
	Powdery mildew	*Erysiphe graminis*	B	168

Crop	Disease	Pathogen		
	Net blotch	*Pyrenophora teres*	B	167
	Scald	*Rhynchosporium secalis*	B	168
	Spot blotch	*Helminthosporium sorokinianum*	B	168
	Stripe	*Pyrenophora graminea*	B	201
	Leaf spot	*Leptosphaeria nodorum*	B	157
	Septoria leaf blotch	*Septoria passerinii*	B	168
	Ergot	*Claviceps purpurea*	BP	53
	Cyst	*Heterdoera avenae*	B	213
	Barley yellow dwarf	Barley yellow dwarf virus	B	169
	Barley yellow mosaic	Virus		170
	Barley stripe mosaic	Virus		168, 270
Oats *Avena sativa*	Bacterial stripe blight	*Pseudomonas striafaciens*	B	234
	Halo blight	*Pseudomonas coronafaciens*	B	234
	Black loose smut	*Ustilago avenae*	B	234
	Covered smut	*Ustilago kolleri*	B	234
	Crown rust	*Puccinia coronata*	B	234
	Stem rust	*Puccinia graminis*	B	234
	Powdery mildew	*Erysiphe graminis*	B	234
	Anthracnose	*Colletotrichum graminicolum*	B	234
	Helminthosporium leaf spot	*H. avenae*	BP	234
	Septoria leaf spot	*Leptosphaeria avenaria*	BP	234
	Victoria blight	*Helminthosporium victoriae*	A	234
	Helminthosporium leaf blotch	*Pyrenophora avenae*	BP	234
	Pythium root necrosis	*P. debaryanum*	BP	234
	Stem nematode	*Ditylenchus dipsaci*		43
	Cyst nematode	*Heterodera avenae*	B	234
	Mosaic	Virus	B	234
	Yellow dwarf	Virus	BP	26, 27
Rice *Oryza sativa*	Bacterial blight	*Xanthomonas oryzae*	BP	183
	Bacterial streak	*Xanthomonas transluscens f. sp. oryzicola*	B	183
	Blast	*Pyricularia oryzae*	BP	183

Table 1 (continued)
RESISTANCE OF MAJOR CROPS TO INFECTIOUS DISEASES

Crop	Disease	Pathogen	Resistance category[a]	Ref.
	Brown spot	*Helminthosporium oryzae*	B	183
	Cercospora leaf spot	*C. oryzae*	B	183
	Leaf scald	*Rhynchosporim oryzae*	B	235
	Kernel smut	*Tilletia sp.*	P	24
	Sheath blight	*Corticum sasakii*	BP	183
	Stem rot	*Leptosphaeria salvinii*		184
	Sheath rot	*Acrocylindrium oryzae*	B	112
	Bakanae	*Gibberella fujikuroi*	BP	183
	White tip	*Aphelenchoides besseyi*	B	183
	Root knot	*Meloidogyne graminicola*	B	183
	Stripe	Virus	B	150
	Hoja blanca	Virus	BP	150
	Tungro	Virus	BP	150
	Rice dwarf	Virus		150
	Yellow dwarf	Virus		150
	Blackstreaked dwarf	Virus	BP	150
	Transitory yellowing	Virus		150
Wheat *Triticum aestivum* and *Triticum durum*	Basal glume rot	*Pseudomonas atrofaciens*		231
	Bacterial blight	*Pseudomonas syringae*	B	218
	Black chaff	*Xanthomonas transluscens*	B	95
	Stem rust	*Puccinia graminis*	B	280
	Leaf rust	*Puccinia recondita*	B	280
	Stripe rust	*Puccinia striiformis*	B	280
	Powdery mildew	*Erysiphe graminis*	B	280
	Leaf blotch	*Septoria tritici*	B	191
	Glume blotch	*Septoria nodorum*	B	8
	Blight	*Alternaria triticina*	BP	137
	Leaf blight	*Pyrenophora trichostoma*	BP	225

	Flag smut	*Urocystis tritici*	BP	280
	Bunt	*Tilletia* spp.	B	280
	Loose smut	*Ustilago nuda*	B	280
	Ergot	*Claviceps purpurea*	BP	276
	Black point	*Helminthosporium sativum*	BP	248
	Take-all	*Gaeumannomyces graminis*	BP	149, 175
	Eyespot	*Pseudocercosporella herpotrichoides*	BP	280
	Snow molds	*Micronectriella nivalis*	BP	31
	Snow molds	*Typhula spp.*	BP	31
	Cereal cyst	*Heterodera avenae*	BP	28
	Soil-borne mosaic	Wheat soil-borne mosaic virus	B	280
	Wheat spindle streak mosaic	Wheat spindle streak mosaic virus	BP	106
	Yellow dwarf	Barley yellow dwarf virus	BP	280
Gramineae				
Maize				
Zea mays	Stewart's wilt	*Erwinia stewartii*	B	266
	Goss's bacterial wilt	*Corynebacterium nebraskense*	B	266
	Common rust	*Puccinia sorghi*	B	266
	Southern rust	*Puccinia polysora*	B	266
	Tropical rust	*Physopella zeae*	BP	266
	Helminthosporium leaf spot	*Helminthosporium carbonum*	B	266
	Northern corn leaf blight	*Helminthosporium turcicum*	B	266
	Southern corn leaf blight	*Helminthosporium maydis*	B	266
	Northern leaf spot	*Helminthosporium* sp.	B	266
	Yellow leaf blight	*Phyllosticta maydis*	B	266
	Brown spot	*Physoderma zeae-maydis*	B	266
	Anthracnose	*Colletotrichum graminicola*	BP	266
	Smut	*Ustilago maydis*	B	266
	Head smut	*Sphacelotheca reiliana*	B	266
	Diplodia ear rot	*Diplodia maydis*	B	266
	Gibberella ear rot	*Gibberella zeae*	BP	266
	Fusarium ear rot	*Fusarium moniliforme*	BP	266

Table 1 (continued)
RESISTANCE OF MAJOR CROPS TO INFECTIOUS DISEASES

Crop	Disease	Pathogen	Resistance category[a]	Ref.
	Diplodia stalk rot	*Diplodia maydis*	B	266
	Gibberella stalk rot	*Gibberella zeae*	B	266
	Sorghum downy mildew	*Sclerospora sorghi*	B	266
	Seedling blight	*Pythium* spp.	P	266
	Seedling blight	*Penicillium oxalicum*	BP	266
	Nematode injury	Several species	BP	266
	Sugarcane mosaic	Virus	BP	266
	Corn mosaic	Virus	B	266
	Wheat streak mosaic	Virus	BP	266
	Maize dwarf mosaic	Virus	B	266
	Corn stunt	Mycoplasma-like organism	B	266
Sorghum *Sorghum vulgare*	Rust	*Puccinia purpurea*	B	47, 189
	Rough leaf spot	*Ascochyta sorghina*	B	199
	Red lead spot	*Curvularia lunata*	B	199
	Leaf blight	*Drechslera turcica*	B	199
	Zonate leaf spot	*Gloecercospora sorghi*	B	199
	Downy mildew	*Sclerospora sorghi*		200
	Anthracnose	*Colletotrichum graminicola*	B	199
	Sugarcane mosaic	Sugarcane mosaic virus	B	47, 48
Sugarcane *Saccharum officinarum*	Leaf scald	*Xanthomonas albilineans*	B	2
	Eyespot	*Bipolaris sacchari*		56
	Yellow spot	*Cercospora koepkei*	B	204
	Leaf blight	*Leptosphaeria taiwanensis*		145
	Downy mildew	*Sclerospora sacchari*	B	2
	Smut	*Ustilago scitaminea*	B	2
	Pokkah boeng	*Gibberella fujikuroi*	B	151
	Red rot	*Physalospora tucumanensis*	B	2
	Root rot	*Pythium arrhenomanes*	BP	2

	Mosaic	Virus	B	2
	Fiji disease	Virus		250
	Ratoon stunt	Virus	BP	2
Bahiagrass				
Paspalum notatum	Leaf spot	*Helminthosporium sativum*		249
Bermuda-grass				
Cynodon dactylon	Leaf spot	*Helminthosporium cynodontis*		249
Brome				
Bromus inermis	Brown spot	*Pyrenophora bromi*		249
Bromus marginatus	Head smut	*Ustilago bullata*		249
Orchardgrass				
Dactylis glomerata	Leaf streak	*Scolecotrichum graminis*		249
Slender wheatgrass				
Agropyron trachycaulum	Leaf rust	*Puccinia recondita*		249
	Stripe rust	*P. striiformis*		249
	Stem rust	*P. graminis*		249
	Head smut	*Ustilago bullata*		249
Sudangrass				
Sorghum vulgare var. *sudanese*	Bacterial stripe	*Pseudomonas andropogonii*		249
	Leaf blight	*Helminthosporium turcicum*		249
	Anthracnose	*Colletotrichum graminicola*		249
	Zonate leaf spot	*Gloecercospora sorghi*		249
Tall fescue				
Festuca arundinacea	Crown rust	*Puccinia coronata*		249
Meadow fescue				
Festuca elatior	Crown rust	*Puccinia coronata*		249
Tall oatgrass				
Arrhenatherum elatius	Leaf rust	*Puccinia recondita*		249
Timothy				
Phleum pratense	Stem rust	*Puccinia graminis*		249
Western wheatgrass				
Agropyron smithii	Leaf rust	*Puccinia recondita*		249
Sand bluestem				
Andropogon hallii	Rust	*Puccinia* sp.		249

Table 1 (continued)
RESISTANCE OF MAJOR CROPS TO INFECTIOUS DISEASES

Crop	Disease	Pathogen	Resistance category[a]	Ref.
Side-oats grama *Bouteloua curtipendula*	Rust	*Puccinia vexans*		249
Blue grama *Bouteloua gracilis*	Rust	*Puccinia vexans*		249
Buffalograss *Buchloë dactyloides*	Rust	*Puccinia kansansis*		249
	Leaf spot	*Helminthosporium inconspicuum*		249
Juglandaceae				
Pecan *Carya illinoensis*	Downy spot	*Mycosphaerella caryigena*		249
	Scab	*Fusicladium effusum*	B	156
	Nut rot	*Botyrosphaeria ribis*	BP	208
Persian (English) walnut *Juglans regia*	Bacterial blight	*Xanthomonas juglandis*	B	80
	Phloem canker	*Erwinia rubifaciens*	B	80
	Branch wilt	*Hendersonula toruloidea*		249
	Crown rot	*Phytophthora cactorum*		249
	Root lesion	*Paratylenchus vulvus*		249
Black walnut *Juglans nigra*	Anthracnose	*Gnomonia leptostyla*	BP	38
	Leaf spot	*Cylindrosporium juglandis*		38
Labiatae				
Peppermint *Mentha piperita*	Rust	*Puccinia menthae*		282
	Powdery mildew	*Erysiphe cichoracearum*		282
	Verticillium wilt	*V. albo-atrum*	B	263
Lauraceae				
Avocado *Persea americana*	Phytophthora root rot	*P. cinnamoni*	BP	285

Leguminosae				
Alfalfa				
Medicago sativa	Wilt	*Corynebacterium insidiosum*	B	122
	Downy mildew	*Peronospora trifoliorum*	B	122
	Leaf spot	*Pseudopeziza medicaginis*	B	122
	Rust	*Uromyces striatus*	B	122
	Leaf spot	*Leptosphaerulina briosiana*		122
	Stemphylium leaf spot	*Stemphylium botryosum*		21, 122
	Summer black stem and leaf rot	*Cercospora medicaginis*		243
	Spring black stem	*Phoma medicaginis*		122
	Wilt	*Verticillium albo-atrum*	BP	122, 258
	Crown rot	*Fusarium* spp.		122
	Anthracnose	*Colletotrichum trifolii*	B	122
	Crown and stem rot	*Sclerotinia trifoliorum*		122
	Damping off and root rot	*Phytophthora megasperma*	BP	92
	Alfalfa mosaic	Alfalfa mosaic virus	BP	50
Beans				
Phaseolus vulgaris	Halo blight	*Pseudomonas phaseolicola*	B	284
	Common blight	*Xanthomonas phaseoli*	B	284
	Bacterial wilt	*Corynebacterium flaccumfaciens*		284
	Brown spot	*Pseudomonas syringae*		284
	Rust	*Uromyces phaseoli*	B	284
	Powdery mildew	*Erysiphe polygoni*	B	284
	Anthracnose	*Colletotrichum lindemuthianum*	B	284
	Angular leaf spot	*Isariopsis griseola*		284
	White mold	*Sclerotinia sclerotium*		284
	Fusarium root rot	*F. solani f. phaseoli*		284
	Rhizoctonia root rot	*R. solani*		284
	Black root rot	*Thielaviopsis basicola*		284
	Phythium blight	*P. aphanidermatum* and *P. ultimum*		284
	Root knot	*Meloidogyne* spp.		284

Table 1 (continued)
RESISTANCE OF MAJOR CROPS TO INFECTIOUS DISEASES

Crop	Disease	Pathogen	Resistance category[a]	Ref.
	Bean common mosaic	Virus	A	284
	Bean yellow mosaic	Virus		284
	Curly top	Virus	B	284
	Southern bean mosaic	Virus		284
	Pod mottle	Virus		284
	Blossom drop	Etiology unknown		251
Lima beans				
Phaseolus				
limensis	Downy mildew	*Phytophthora phaseoli*	B	249, 251
	Root knot	*Meloidogyne* spp.		273
	Lima bean mosaic	Virus		249, 251
Clover				
Trifolium spp.				
Hop clover				249
T. agrarium	Powdery mildew	*Erysiphe polygoni*		
Red clover	Leaf rust	*Uromyces trifolii*		75
T. pratense				
	Powdery mildew	*Erysiphe polygoni*		257
	Northern anthracnose	*Kabatiella caulivora*	B	257
	Southern anthracnose	*Colletotrichum trifolii*	B	257
	Stemphylium leaf spot	*Stemphylium sarcinaeforme*		171
	Snow mold	Basidiomycete fungus		249
	Crown rot	*Sclerotinia trifoliorum*	B	253
	Root rots	*Fusarium oxysporum*		253
	Root knot	*Meloidogyne incognita*		257
	Yellow mosaic	Bean yellow mosaic virus		257
Sub clover				
T. subterraneum	Powdery mildew	*Erysiphe polygoni*		257
	Northern anthracnose	*Kabatiella caulivora*	B	39
White clover				
T. repens	Curvularia leaf spot	*Curvularia* sp.	BP	118

	Crown rot	*Sclerotinia trifoliorum*	B	62
	White clover mosaic	Virus	BP	118
Lespedeza				
L. stipulacea	Powdery mildew	*Microsphaera diffusa*		249
Blue lupine				
Lupinus				
angustifolium	Anthracnose	*Glomerella cingulata*		249
Mungbean				
Vigna radiata	Bacterial leaf spot	*Xanthomonas phaseoli*	B	259
	Cercospora leaf spot	*Cercospora canescens*	B	259
	Yellow mosaic	Mungbean yellow mosaic virus	BP	259
Cowpea				
Vigna sinensis	Bacterial canker	*Xanthomonas vignicola*		249
	Powdery mildew	*Erysiphe polygoni*		249
	Fusarium wilt	*F. oxysporum*		249
		F. tracheiphilum		
	Root knot	*Meloidogyne* sp.		249
Peanut				
Arachis hypogaea				
and wild species	Leaf spot	*Cercospora arachidicola*		244
	Rust	*Puccinia arachidis*		131
	Black rot	*Cylindrocladium crotolariae*		87
	Southern stem rot	*Sclerotium rolfsii*		15
	Seedling blight, root rot	*Rhizoctonia bataticola*		202
	Pythium pod rot	*Pythium myriotylum*		202
	Pod rot	*Fusarium oxysporum*		163
		and *F. solani*		
	Stem rot	*Diplodia gossypina*		196
	Collar rot	*Aspergillus niger*		40
	Pod mold	*Aspergillus flavus*		283
	Stunt	Virus		283
	Rosette	Virus		89
	Mottle	Peanut mottle virus		135
	Lesion nematode disease	*Pratylenchus brachyurus*		240
	Northern root knot	*Meloidogyne hapla*		9
White sweetclover				
Melilotus albo	Leaf spot	*Mycosphaerella davisii*		249

Table 1 (continued)
RESISTANCE OF MAJOR CROPS TO INFECTIOUS DISEASES

Crop	Disease	Pathogen	Resistance category[a]	Ref.
	Southern anthracnose	*Colletotrichum trifolii*		249
	Stem canker	*Ascochyta caulicola*		249
	Crown rot	*Sclerotinia trifoliorum*		249
	Black stem	*Mycosphaerella lethalis*		249
	Root rot	*Phytophthora cactorum*		249
Soybeans				
Glycine max	Wildfire	*Psuedomonas tabaci*	A	123
	Bacterial pustule	*Xanthomonas phaseoli* var. *sojinsis*	A	123
	Bacterial blight	*Pseudomonas glycines*	B	123
	Downy mildew	*Peronspora manshurica*	B	6
	Frogeye leafspot	*Cercospora sojina*	A	6
	Pod and stem blight	*Diaporthe phaseolorum*	BP	6
	Stem canker	*Diaporthe phaseolorum*	B	6
	Target spot	*Corynospora cassiicola*	A	6
	Brown stem rot	*Cephalasporium gregatum*		6
	Phytophthora root rot	*P. megasperma* var. *sojae*	B	6
	Purple seed stain	*Cercospora kikuchii*	B	6
	Root knot	*Meloidogyne* sp.	B	249
	Cyst	*Heterodera glycines*	B	74
	Soybean mosaic	Soybean mosaic virus	B	67
Peas				
Pisum sativum	Powdery mildew	*Erysiphe polygoni*	B	249
	Septoria blotch	*S. pisi*		249
	Downy mildew	*Peronospora pisi*	A	249
	Root rot	*Fusarium solani f. pisi*		273
	Thielaviopsis root rot	*T. basicola*		273
	Foot rot	*Mycosphaerella pinodes* and *Ascochyta pinodella*		273
	Damping off	*Pythium ultimum*		273
	Fusarium wilt	*F. oxysporum f. pisi*	A	249

Host	Disease	Pathogen		References
	Root rot	*Aphanomyces euteiches*		249, 273
	Yellow bean mosaic	Virus	A	273
	Pea enation	Virus	B	273
	Top yellows	Pea leaf roll-virus	A	273
Winter field pea				
Pisum arvense	Root rot	*Aphanomyces euteiches*		249
Liliaceae				
Asparagus				
Asparagus officianalis	Rust	*Puccinia asparagi*	B	249
	Crown rot	*Fusarium* sp.	BP	34
Onion				
Allium cepa	Smut	*Urocystis cepulae*		249
	Downy mildew	*Peronospora destructor*		249
	Brown stain	*Botrytis cinerea*	B	44
	Smudge	*Colletotrichum circinans*	B	249
	Purple blotch	*Alternaria porri*		249
	Black mold	*Aspergillus niger*		249
	Bulb rot	Etiology unknown		249, 251
	Fusarium rot	*F. oxysporum*		251
	Pink root	*Pyrenochaeta terrestris*		249, 251
	Yellow dwarf	Virus		249
Linaceae				
Flax				
Linum usitatissimum	Powdery mildew	*Oidium lini*	B	71
	Rust	*Melampsora lini*	B	249
	Anthracnose	*Colletotrichum lini*		197
	Pasmo	*Mycosphaerella linorum*	BP	132
	Wilt	*Gibberella avenacea*	BP	181
	Wilt	*Phoma exigua*	BP	182
	Cyst disease	*Heterodera schachtii*	B	237
	Nematode disease	*Ditylenchus dipsaci* and *D. allii*	B	237
Malvaceae				
Cotton				
Gossypium spp.	Bacterial blight	*Xanthomonas malvacearum*	B	19
	Grey mildew	*Ramularia areola*	B	139

Table 1 (continued)
RESISTANCE OF MAJOR CROPS TO INFECTIOUS DISEASES

Crop	Disease	Pathogen	Resistance category[a]	Ref.
	Southwestern cotton rust	*Puccinia stakmanii*	B	19
	Blight	*Helminthosporium speciferum*	B	70
	Leaf spot	*Myrothecium roridum*	BP	245
	Leaf blight	*Alternaria macrospora*	BP	17
	Root rot	*Phymatotrichum omnivorum*	BP	17
	Seedling disease	Several fungi	BP	17
	Wilt	*Verticillium dahliae*	BP	19
	Root rot	*Thielaviopsis basicola*	B	134
	Root knot	*Meloidogyne incognita*		116
	Fusarium wilt-root knot complex	*Fusarium oxysporum* f. *vasinfectum* and *M. incognita*	B	19
	Boll rot	Several fungi	B	126
Moraceae				
Hop				
Humulus lupulus	Downy mildew	*Pseudoperonospora humuli*	B	142
	Powdery mildew	*Sphaerotheca macularis*	B	142
	Wilt	*Verticillium*	B	142
Rosaceae				
Apple				
Malus sylvestris	Bacterial blight	*Pseudomonas syringae*	BP	269
	Fire blight	*Erwinia amylovora*	BP	281
	Cedar-apple rust	*Gymnosporangium juniperae-virginianae*	B	281
	Powdery mildew	*Podosphaera leucotricha*	BP	281
	Scab	*Venturia inaequalis*	B	281
	Alternaria rot	*Alternaria mali*	AP	209
	European canker	*Nectria galligena*		25
	Collar rot	*Phytophthora cactorum*	B	25

	Blotch	*Phyllosticta solitaria*	BP	25
	Bullseye rot	*Neofabraea perennans*	B	25
	Leaf spot	*Diaporthe pomi*		207
	Brown rot	*Monilinia fructigena*	AP	90
	Bitter rot	*Glomerella cingulata*	BP	260
	Black rot	*Physalospora obtusa*		12
	Black root rot	*Enosoma lanigerum*		52
	Nematode	*Pratylenchus vulinus*		217
	Apple chat fruit	Virus		88
Apricot				
Prunus armeniaca	Bacterial leaf spot	*Xanthomonas pruni*	BP	144
	Gummosis	*Cytospora* sp.	B	144
	Anthracnose	*Gnomonia erythrostoma*	B	42
	Brown rot	*Sclerotinia laxa*		7
Blackberry				
Rubus spp.	Orange rust	*Gymnoconia interstitialis*	B	185
	Anthracnose	*Elsinoe veneta*	BP	166
	Leaf spot	*Mycosphaerella rubi*	B	185
	Double blossom	*Cercosporella rubi*	B	185
	Verticillium wilt	*V. albo-atrum*	B	185
Raspberry				
Rubus spp.	Late raspberry rust	*Pucciniastrum americanum*	B	185
	Leaf spot	*Septoria rubi*	B	185
	Anthracnose	*Elsinoe veneta*	B	185
	Spur blight	*Didymella applanata*	D	185
	Fruit rot	*Botrytis cinerea*	BP	54, 129
	Fruit rot	*Rhizopus* sp.	BP	54
	Powdery mildew	*Sphaerotheca macularis*	BP	119
		Leptosphaeria coniothyrium	B	129
		Thomasiniana theobaldi		129
	Raspberry mosaic	Virus		249
Cherry				
Prunus ovium				
and *P. cerasus*	Bacterial canker	*Pseudomonas mors-prunorum*		152
	Brown rot	*Monilinia laxa* and *M. fructicola*		79
	Leaf spot	*Coccomyces hiemalis*		79
	Verticillium wilt	*Verticillium albo-atrum*		79

Table 1 (continued)
RESISTANCE OF MAJOR CROPS TO INFECTIOUS DISEASES

Crop	Disease	Pathogen	Resistance category[a]	Ref.
	X-disease	Virus		79
Peach				
Prunus persica	Bacterial leaf spot	*Xanthomonas pruni*		99
	Crown gall	*Agrobacterium tumafaciens*		99
	Bacterial leaf spot	*Bacillus mesentericus*	B	113
	Bacterial canker	*Pseudomonas mors-prunorum f.* sp. *persicae*	BP	86
	Powdery mildew	*Sphaerotheca pannosa*		99
	Leaf curl	*Taphrina deformans*		99, 173
		Clasterosporium carpophilum	B	219
	Powdery mildew	*Podosphaera tridactyla*	B	203
	Peach canker	*Valsa incita* and *V. leucostoma*	B	99
	Twig canker	*Fusicoccum amygdali*	B	7
	Crown rot	*Phytophthora cactorum*		99
	Peach twig blight	*Glomerella cingulata*		99
	Brown rot	*Monilinia laxa* and *M. fructicola*	B	99, 256
	Root rot	*Meloidogyne incognita*		99
		Pratylenchus vulnus		217
	Peach mosaic	Virus		249
	Dark green sunken mottle	Virus		241
	Prune dwarf	Virus		241
Pear				
Pyrus spp.	Fire blight	*Erwinia amylovora*	B	143, 249
	Bacterial blight	*Pseudomonas syringae*		269
	Leaf spot	*Mycosphaerella sentina*	B	69
	Fruit rot	*Monilia* sp.		143
	Scab	*Venturia pyrina*	B	143
	Powdery mildew	*Podosphaera leucotricha*		143
	Black spot	*Alternaria kikuchiana*		133

	Leaf spot	*Phyllosticta pirina*		12
		Physalospora pinicola		124
	Leaf blight and fruit spot	*Fabraea maculata*	B	143
	Wilt	*Cytospora ambiens*		198
	Cyst	*Meloidogyne* sp.	B	143
	Pear bud drop	Virus		143
	Stoney pit	Virus		249
	Ring spot mosaic	Virus		143
	Pear decline	Mycoplasma		278
	Pear vein yellows	Virus		82
	Necrotic spot	Virus		124
	Rubbery wood	Rubbery wood virus		264
Strawberry *Fragaria ananassa* *F. chiloensis*	Powdery mildew	*Sphaerotheca macularis*	B	216
	Downy mildew	*Peronospora fragariae*		49
	Leaf scorch	*Diplocarpon earliana*	B	216
	Leaf spot	*Mycosphaerella fragariae*	B	216
	Anthracnose	*Colletotrichum fragariae*	B	216
	Crown rot	*Phytophthora cactorum*		216
	Verticillium wilt	*V. albo-atrum* and *V. dahliae*	B	216
	Red stele	*Phytophthora fragariae*	B	216
	Cactorum crown rot	*Phytophthora cactorum*	B	216
	Nematode	*Aphelenchoides fragariae*		117
	Yellows disease	*Fusarium oxysporum*		128
	Virus infections	Viruses	BP	216
Rutaceae *Citrus* spp.	Bacterial canker	*Xanthomonas citri*		161, 165
	Anthracnose	*Colletotrichum gloesporioides*		161, 268
	Mal Secco	*Deuterophoma tracheiphila*		57
	Scab	*Elsinoe fawcettii*		161
	Foot rot	*Phytophthora parasitica*		104
	Root rot	*Phytophthora citropthora*	BP	23
	Foot rot	*Diplodia natalensis*		91
	Greening virus	Virus		81
	Tristeza	Virus	BP	242
	Browning nematode	*Radopholus similis*	B	96, 242

Table 1 (continued)
RESISTANCE OF MAJOR CROPS TO INFECTIOUS DISEASES

Crop	Disease	Pathogen	Resistance category[a]	Ref.
	Citrus nematode	*Tylenchulus semipenetrans*	B	242
	Nematode	*Pratylenchus coffeae*		178
Saxifragaceae				
Black currant				
Ribus nigrum	Downy mildew	*Plasmopora ribicola*		126
	Powdery mildew	*Sphaerotheca mors-uvae*		121
Currants and gooseberries				
Ribus spp.	Septoria leaf spot	*Septoria ribis*		105
	Leaf spot	*Pseudopeziza ribis*		120
	White pine blister rust	*Cronartium ribicola*		120
	Gooseberry mildew	*Sphaerotheca pannosa*	B	120
	Cane blight	*Botryosphaeria ribis*		120
	Fruit rot	*Botrytis cinerea*		120
	Vein banding	Virus		120
	Reversion	Virus		120
Solanaceae				
Eggplant				
Solanum melongena	Bacterial wilt	*Pseudomonas solanacearum*		273
	Fusarium wilt	*Fusarium* sp.		251
	Verticillium wilt	*V. dahliae*		273
	Phomopsis blight and fruit rot	*Phomopsis vexans*		41
	Root knot	*Meloidogyne incognita*		109
	Little leaf	Virus		160

Peppers				
Capsicum annuum	Bacterial spot	*Xanthomonas vesicatoria*		273
	Bacterial wilt	*Pseudomonas solanacearum*		273
	Southern blight	*Sclerotium rolfsii*		273
	Cercospora leaf spot	*C. capsici*		273
	Verticillium wilt	*V. albo-atrum*		273
	Wilt	*Fusarium annuum*		273
	Root rot	*Phytophthora capsici*		273
	Root knot	*Meloidogyne* spp.		273
	Tobacco etch	Virus	B	273
	Hawaiian pepper	Virus		273
	Puerto Rico mosaic	Virus		273
	Tobacco mosaic	Virus	B	273
	Potato virus	Virus		273
	Potato virus Y	Virus	BP	4
	Pepper mottle	Virus	BP	4
Potato				
Solanum tuberosum	Black leg	*Erwinia phytophthora*	BP	179
	Brown rot	*Pseudomonas solanacearum*	BP	179
	Ring rot	*Corynebacterium sepedonicum*	BP	179
	Common scab	*Streptomyces scabies*	B	179
	Wart	*Synchytrium endobioticum*	B	179
	Powdery scab	*Spongospora subterranea*		102
	Leafspot	*Cercospora solani-tuberosi*		65
	Early blight	*Alternaria solani*	BP	262
	Late blight	*Phytophthora infestans*	BP	179
	Silver scurf	*Helminthosporium solani*		114
	Rhizoctonia canker	*Rhizoctonia solani*	BP	164
	Verticillium wilt	*V. albo-atrum*	BP	179
	Storage rot	*Phoma exigua* var. *faveata*		254
	Fusarium storage decay	*F. sambucinum* and *F. coerulum*	BP	273

Table 1 (continued)
RESISTANCE OF MAJOR CROPS TO INFECTIOUS DISEASES

Crop	Disease	Pathogen	Resistance category[a]	Ref.
	Golden nematode disease	*Heterodera rostochiensis*	BP	179
	Tuber-rot-nematode disease	*Ditylenchus destructor*		115
	Latent mosaic	Potato virus X	BP	179
	Mild mosaic	Potato virus A	BP	179
	Rugose mosaic	Potato virus Y	BP	179
	Leafroll	Leafroll virus	BP	179
	Net necrosis	Current season leaf roll virus infection	BP	179
	Yellow dwarf	Potato yellow dwarf virus	BP	179
	Interveinal mosaic	Potato virus S	BP	179
	Corky ringspot	Suspected virus		273
Tobacco *Nicotiana tabacum*				
	Wildfire	*Pseudomonas tabaci*	B	126
	Blackfire	*Pseudomonas angulata*	B	126
	Bacterial wilt	*P. solanacearum*	B	46
	Anthracnose	*Colletotrichum destructivum*	P	126
	Powdery mildew	*E. cichoracearum*	B	46
	Blue mold	*Peronospora tabacini*	P	46
	Brown spot	*Alternaria alternata*	P	45
	Frogeye leaf spot	*Cercospora nicotianae*	P	195
	Wilt	*Verticillium dahliae*		228
	Fusarium wilt	*F. oxysporum* var. *nicotianae*	B	46
	Black shank	*Phytophthora parasitica*	B	126
	Black root rot	*Thielaviopsis basicola*	B	46, 126
	Root knot	*Meloidogyne javanica*	B	46
	Tobacco mosaic	Tobacco mosaic virus	B	126
	Vein banding	Potato virus Y	P	35, 232
	Tomato spotted wilt	Virus	P	252

Tomato				
Lycopersicon				
esculentum	Bacterial wilt	*Pseudomonas solanacearum*	B	10
	Bacterial canker	*Corynebacterium michiganense*		275
	Leaf mold	*Cladosporium fulvum*	B	275
	Early blight	*Alternaria solani*	B	275
	Late blight	*Phytophthora infestans*	B	10
	Gray leafspot	*Stemphylium solani*	B	275
	Septoria leafspot	*S. lycopersici*		275
	Anthracnose	*Colletotrichum phomoides*		275
		C. atramenthrum		
	Black spot	*Phoma destructiva*		275
	Fusarium wilt	*Fusarium oxysporum*	B	275
		f. lycopersici		
	Verticillium wilt	*V. albo-atrum*	B	275
	Collar rot	*Alternaria solani*		249
	Root knot	*Meloidogyne incognita*	B	275
	Tobacco mosaic	Tobacco mosaic virus	B	275
	Tobacco spotted wilt	Spotted wilt virus	B	10
	Tobacco etch	Tobacco etch virus		275
	Veinbanding	Potato virus Y		275
	Curly top	Beet curly top virus	BP	10
Umbelliferae				
Carrot				
Daucus carota				
var. *sativa*	Leaf blight	*Alternaria dauci*		251
	Sclerotina disease	*Sclerotinia sclerotiorum*	BP	146
	Phoma root and crown rot	*Phoma rostrupii*	BP	146
	Root knot	*M. hapla*		76
Celery				
Apium gravcolens				
var. *dulce*	Early blight	*Cercospora apii*		273
	Late blight	*Septoria apii-graveolentis*		273
	Yellows	*F. apii*		249, 251

Table 1 (continued)
RESISTANCE OF MAJOR CROPS TO INFECTIOUS DISEASES

Crop	Disease	Pathogen	Resistance category[a]	Ref.
Vitaceae				
Grape				
Vitis vinifera				
and *Vitus* spp.	Downy mildew	*Plasmopora viticola*	B	73, 256
	Powdery mildew	*Uncinula necator*	B	256
	Anthracnose	*Sphaceloma ampelinum*		73
		Botrytis spp.	BP	177
	Anthracnose	*Elsinoe ampelina*	B	73, 159
	Excoriosis	*Phomopsis viticola*	B	64
	Black rot	*Guignardia bidwellii*	B	73
	Root knot	*M. incognita*	BP	94
	Root disease	Nematodes		73
	Grapevine fan leaf	Virus		162
	Pierce's disease	Bacterium-like organism		73

[a] "A" indicates that essentially all currently-grown cultivars of a crop are resistant to the disease. "B" indicates that there is at least one commerically useful cultivar that is resistant to the disease. "P" indicates that the resistance is inadequate to provide complete control of the disease. Other control measures must be used in conjunction with resistance.

which the root is the part of the plant that is eaten. This suggests that the absence of resistance to root disease in many crops in which the roots are not eaten is more a consequence of lack of effort rather than of its nonexistence. Nilsson reported that although he could find no total resistance in wheat against the fungus *Gaeumannomyces graminis (= Ophiobolus graminis)*, there were great differences among cultivars in levels of intermediate resistance, indicating that improvement in take-all resistance through breeding would be possible.[175]

Resistance to storage rots is also not often reported although these diseases cause significant losses in some crops. Floramerica tomato has some resistance to bacterial soft rot.[51] High levels of resistance to various storage rot organisms have likewise been found in sugarbeets.[33] There are difficulties in breeding for resistance to storage rots, in that selection often may not be possible until long after harvest and it may not be possible to make selections and crosses on an annual basis.

The development of disease-resistant cultivars is a cumulative process. Early breeding programs usually had the objective of developing a cultivar with resistance to some major disease that was of overriding importance. Resistance to yellows was a major objective of the cabbage breeding program in Wisconsin.[272] Resistance to Verticillium wilt was the goal of a large cooperative mutation-breeding scheme in peppermint.[263] Likewise, scab resistance in apple and leaf rust resistance in a soft red winter wheat were the original objectives of breeding programs for these crops. Numerous other examples could be mentioned for other crops. The early breeders were doubtless aware that many diseases afflicted these crops, but they narrowed their first efforts to what seemed the most serious one. Once a breeding program was under way, perceptive breeders would note variations in reaction to many diseases, indicating the potential for controlling disease by genetic means. As resistance to the "major" disease was incorporated into adapted types, it was possible to add resistance to other diseases. For example, once rust-resistant wheats of good agronomic type were available, these were used as parents in crosses to add resistance to Hessian fly, powdery mildew, soil-borne mosaic, etc. Using this technique, soft red winter wheat cultivars are now available that are resistant to leaf- and stem rusts, powdery mildew, Septoria leaf blotch, soil-borne mosaic, take-all root rot, and Hessian fly.[191] The original scab-resistant apples have been used to develop cultivars also resistant to cedar-apple rust, fireblight, and powdery mildew.[281] Mutation breeding in peppermint is being used to add rust resistance to the *Verticillium* wilt resistance already available. Floramerica F_1 hybrid tomato combines a high level of resistance to five infectious diseases with tolerance to four more infectious diseases and one nutritional disease, plus intermediate resistance to bacterial soft rot of the fruit.[51]

Were it not for variability in pathogenic organisms, breeders could make unimpeded progress in adding resistance to one disease after another. Unfortunately, a shift in pathogen races can eliminate effective resistance to a disease, forcing the breeder to repeat work already done. He must then find another gene or genes for resistance to the disease and add it to his breeding stocks.

FUTURE FOR DISEASE RESISTANCE

Although much has been accomplished toward the control of disease through resistance in the last 70 years, there is still much that can be done. In the major crops, resistance to many of the serious diseases is known and frequently available in useful cultivars. On some of the "minor" crops, i.e., those grown on comparatively few hectares, the list of diseases for which resistance is known is shorter. This is not necessarily because resistance is not available, but may often reflect the lack of effort on these crops. Although resistance is cheap for the individual farmer, the plant-breeding agency or company must commit money and manpower to find and utilize effective resistance. It takes just as much work to develop a disease-resistant radish that may be grown on only a few hundred hectares as it takes to develop a disease-resistant wheat that may be grown on more than a million hectares.

The future may require more effective use of the resistance that is available rather than the finding of "new" resistance. Bringing together resistance to all important diseases in an area into a single cultivar should have high priority. As improved breeding stocks become available with resistance to various maladies, combining these without sacrificing yield or quality becomes easier.

Much of the early optimism plant breeders felt about achieving permanent control of disease through resistance faded as they realized how adaptive many pathogens are. In many crops monogenic resistance is so ephemeral that its value for control of disease is seriously questioned. Leaf-infecting fungi are the most notorious pathogens for breaking resistance. Although much basic work remains to be done on the epidemiological and genetic aspects of this phenomenon, enough is known to guide plant breeders in surmounting this problem. The demonstration that monogenic resistance in the host can be circumvented by a single allele substitution in the pathogen indicates why monogenic resistance is so often ephemeral. From this it follows that genetically more complex resistance should last longer. This complexity can be achieved by combining genes for resistance into a single cultivar (gene stacking), by bulking seed of several near isogenic lines, each carrying a different gene for resistance, into a multiline cultivar, or by using several different genes for resistance in the various cultivars grown over a wide area, restricting certain genes to certain areas.[214,29,30] General resistance, such as slow rusting in cereals or field resistance to late blight in potatoes, should also provide long-term protection. The longevity of this type of resistance may be due both to reduced selection pressure on the pathogen and genetic complexity of resistance in the host.

Among these alternatives for long-lasting control, regional deployment of genes is difficult to employ. It requires cooperation from all breeders of a crop in an epidemiological area (the area over which pathogen propagules regularly spread). Gene stacking, multiline cultivars, and polygenic (general) resistance are more feasible approaches to obtain long-lasting resistance because they can be used by each breeder independently.

In recent years there has been much enthusiasm for genetic engineering as a plant breeding technique. Although this area of endeavor has many interesting possibilities, it is far from being a practical plant-breeding tool. Moreover, there is no reason to suppose that genetic engineering, in contrast to conventional plant breeding that utilizes the sexual mechanisms of plants, is essential to continued progress. Nearly all crops represent only a small portion of a complex of species, in many cases consisting of several ploidy levels. Thus, the cultivated varieties represent only a small portion of the potential variation in the species.[101] Much remains to be done, using conventional breeding techniques, in exploiting this variability. The difficulty lies in handling large gene pools in such a way that many new gene combinations are allowed to occur and then developing techniques to screen these tremendous populations.

ACKNOWLEDGMENTS

I would like to thank Mr. Darrell Cox for assisting in the literature search for this paper. Also, the following people provided valuable advice in the preparation of the table: Drs. K. L. Athow (soybeans), L. S. Bird (cotton), R. J. Green, Jr. (peppermint and spearmint), D. J. Hagedorn (peas and edible beans), R. T. Lewellen (beets), L. J. Littlefield (flax), C. C. Litton (tobacco), W. O. McIlrath and M. C. Rush (rice), A. Petrie (oil seed rape), L. Sequeira (lettuce), D. H. Thurston (potatoes), H. L. Warren (maize), R. D. Wilcoxson (alfalfa), E. B. Williams (apple and pear), P. H. Williams (crucifers), and D. E. Zimmer (sunflower and safflower). Special thanks are due to Dr. W. R. Stevenson for his considerable help with all vegetable crops throughout the preparation of this manuscript.

REFERENCES

1. **Aalders, L. E., Ismail, A. A., Hall, V. V., and Hepler, P. R.,** Augusta lowbush blueberry, *Can. J. Plant Sci.,* 55, 1079, 1975.
2. **Abbott, E. A.,** Sugarcane, in *Breeding Plants for Disease Resistance Concepts and Applications,* Nelson, R. R., Ed., Pennsylvania State University Press, Univeristy Park, Pa., 1973, 220.
3. **Allard, R. W.,** *Principles of Plant breeding,* John Wiley & Sons, New York, 1960.
4. Anonymous, Breeding a solution to pepper viruses, *Amer. Vegetable Grower,* May, 1978.
5. **Ashri, A.,** Safflower germplasm evaluation, *Plant Genetic Res. News.,* 31, 1975, 29.
6. **Athow, K. L.,** Fungal diseases, in *Soybeans: Improvement, Production, and Uses,* Caldwell, B. E., Ed., American Society of Agronomy, Madison, Wis., 1973, 459.
7. **Bailey, C. H. and Hough, L. F.,** Apricots, in *Advances in Fruit Breeding,* Janick, J. and Moore, J. N., Eds., Purdue University Press, West Lafayette, Ind., 1975, 367.
8. **Baker, C. J.,** Varietal reaction of wheat to leaf infection by *Leptosphaeria nodorum* and *Septoria tritici, Trans. Br. Mycol. Soc.,* 54, 500, 1970.
9. **Banks, D. J.,** Peanuts: germplasm resources, *Crop Sci.,* 16, 499, 1976.
10. **Barksdale, T. H., Good, J. M., and Davidson, L. L.,** Tomato diseases and their control, *U.S. Dep. Agric. Agric. Handb.,* 203, 1972.
11. **Barrus, M. F.,** Variations of varieties of beans in their susceptibility to anthracnose, *Phytopathology,* 1, 190, 1911.
12. **Barsukova, O. N.,** A study of the resistance of varieties of apple to the main diseases at the Maikop Experimental Station of the All-Union Institute of Plant Industry (trans.), *Ref. Zh.,* 10:55. 791, 1971, abstr. in *Plant Breed. Abstr.,* 44, 1031.
13. **Barsukova, O. N.,** Infection of pear varieties with *Mycosphaerella sentina* and *Phyllosticta pirina,* (trans.), *Ref. Zh.,* 6:55.809, abstr. in *Plant Breed. Abstr.,* 44, 8951, 1973.
14. **Bennett, C. W. and Leach, L. D.,** Diseases and their control, in *Advances in Sugarbeet Production,* Johnson, R. T., Ed., Iowa State University Press, 1971, 225.
15. **Beute, M. K., Wynne, J. C., and Emery, D. A.,** Registration of NC3033 peanut germplasm (Reg. No. GP9), *Crop Sci.,* 16, 887, 1976.
16. **Bhander, D. S. and Maini, N. S.,** Studies on the resistance of oleiferous Brassicas to alternaria blight, *Indian Oilseeds J.,* 9, 58, 1965.
17. **Bhaskaran, R., Natarajan, C., and Mohanvaj, D.,** Biochemistry of resistance and susceptibility in cotton to *Alternaria macrospora, Acta Phytopathol. Acad. Sci. Hung.,* 10, 33, 1975.
18. **Biffen, R. H.,** Mendel's laws of inheritance and wheat breeding, *J. Agric. Res.,* 1, 4, 1905.
19. **Bird, L. S.,** Cotton, in *Breeding Plants for Disease Resistance,* Nelson, R. R., Ed., Pennsylvania State University Press, University Park, 1973, chap. 12.
20. **Bockelman, H. E.,** The inheritance of resistance to Fusarium wilt in cultivated safflower *(Carthamus tinctorius L.), Diss. Abstr. Int. B,* 34, 1727B, 1974.
21. **Borges, O. L., Stanford, E. H., and Webster, R. K.,** Selection for resistance to *Stemphylium botryosum* in alfalfa, *Crop Sci.,* 16, 456, 1976.
22. **Bradley, G. A.,** The Department of Horticulture and Forestry, *Arkansas Farm Res.* 23, 8, 1974.
23. **Broadbent, P., Fraser, L. R., and Waterworth, Y.,** The reaction of seedlings of citrus spp. and related genera to *Phytophthora citropthora, Proc. Linnean Soc., N.S.W.,* 96, 119, 1971.
24. **Brod, C. G.,** Study of smut resistance, in *Annals, 5th General Meeting on Rice Cultivation,* 11, 1975, abstr. in *Plant Breed. Abstr.,* 46, 53, 1975.
25. **Brown, A. G.,** Apples, in *Advances in Fruit Breeding,* Janick, J. and Moore, J. N., Eds., Purdue University Press, West Lafayette, Ind., 1975, 3.

26. **Brown, C. M. and Jedlinski, H.** Registration of Lang oats (Reg. No. 288), *Crop Sci.*, 18, 525, 1978.
27. **Brown, C. M. and Jedlinski, H.**, Registration of 13 germplasm lines of oats (Reg. No. GP10 to GP22), *Crop Sci.*, 1098, 1978.
28. **Brown, J. A. M., and Ellis, S. E.**, Breeding for resistance to cereal cyst nematode on wheat, *Euphytica*, 25, 73, 1976.
29. **Browning, J. A. and Frey, K. J.**, Multiline cultivars as a means of disease control, *Annu. Rev. Phytopathol.*, 7, 355, 1969.
30. **Browning, J. A., Simons, M. D., Frey, K. J., and Murphy, H. C.**, Regional deployment for conservation of oat crown rust resistance genes, *Iowa Agric. Exp. Stn. Spec. Rep.* No. 64, Browning, J. A., Ed., 1969, 49.
31. **Bruehl, G. W., Kiyomoto, R., Peterson, C., and Nagamitsu, M.**, Testing winter wheats for snow mold resistance in Washington, *Plant Dis. Rep.*, 59, 566, 1975.
32. **Brunin, B.**, Quelques aspects anatomiques de varietes sensibles et resistantes de colze au cours de l'evolution de la necrose du collet; leur utilization possible dans les recherches de slection genetique, *Journ. Int. Colza*, 289, 1970.
33. **Bugbee, W. M.**, Registration of F1001 and F1002 sugarbeet germplasm, *Crop Sci.*, 18, 358, 1978.
34. **Bussell, W. T. and Ryan, C. L. J.**, Disease tolerance of asparagus varieties, *N. Z. J. Agric.*, 128, 27, 1974.
35. **Buzancic, A.**, The investigation of the tolerance of the Virginia and Burley type of tobacco to the YR strain potato virus Y (trans.), *Agron. Glas.*, 36, 471, 1974, abstr. in *Biol. Abstr.*, 61, 51411.
36. **Caldwell, R. M.**, Breeding for general and/or specific plant disease resistance, in *Proc. 3rd. Int. Wheat Gent. Symp.*, Finley, K. W. and Shepherd, D. W., Eds., Australian Academy of Science, Canberra, Australia, 1968, 263.
37. **Caldwell, R. M., Cartwright, W. B., Compton, L. E., Schafer, J. F., and Patterson, F. L.**, Dual: A Hessian fly-resistant soft red winter wheat, *Purdue Univ. Agric. Exp. Stn. Res. Bull.*, No. 681, 1959.
38. **Carnevali, A.**, The black walnut as a fruit crop, *Frutticoltura*, 37, 31, 1975.
39. **Chatel, D. L. and Francis, C. M.**, Sources of variation in the resistance of subterranean clover to clover scorch *(Kabatiella caulivora)* in the field, *Aust. J. Exp. Agric. Anim. Husb.*, 15, 541, 1975.
40. **Chahal, A. S., Dutt, S., and Chotran, J. S.**, Varietal resistance in groundnut *(Arachis hypogea L.)* to collar rot *(Aspergillus niger)* in the Punjab State, *J. Res. Punjab Agric. Univ.*, 11, 200, 1974.
41. **Choudhury, B., Tewari, R., Methai, P. J., Joshi, G. C., Kesavan, V., Chakravarti, A. K., Kalda, T. S., and Sivakami, N.**, Breeding for disease resistance in some tropical and subtropical vegetable crops, in *Breeding Res. Asia Oceania. Proc. 2nd Gen. Cong. Soc. Adv. Breeding Res. Asia Oceania*, Session XVIII, Indian Society of Genetics and Plant Breeding, New Delhi, India, 1974, 1222.
42. **Cifranic, P.**, Up-to-date results with the new apricot varieties, 337 (trans.), Res. Inst. Pl. Prod. Piest' any Czech., abstr. in *Plant Breed. Abstr.*, 46, 9402.
43. **Clamot, G.**, *Rev. de l'Agriculture*, 28, 637, 1975, abstr. in *Plant Breed. Abstr.*, 46, 315.
44. **Clark, C. A. and Lorbeer, J. W.**, The role of phenols in Botrytis brown stain of onion, *Phytopathology*, 65, 338, 1975.
45. **Clayton, E. E.**, Control of tobacco diseases through resistance, *Phytopathology*, 43, 239, 1953.
46. **Clayton, E. E.**, The genetics and breeding progress in tobacco during the last 50 years, *Agron. J.*, 50, 352, 1958.
47. **Coleman, O. H. and Dean, J. L.**, Inheritance of resistance to rust in sorgo, *Crop Sci.*, 1, 152, 1961.
48. **Condi, B. D., Moore, R. F., Fletcher, D. S., and Teakle, D. S.**, Inheritance of the resistance of Krish sorghum to sugarcane mosaic virus, *Aust. J. Agric. Res.* 27, 45, 1976.
49. **Coulombe, L. J.**, Repression of mildew in strawberry, *Phytoprotection*, 55, 96, 1974.
50. **Crill, P., Hanson, E. W., and Hagedorn, D. J.**, Resistance and tolerance to Alfalfa mosaic virus in alfalfa, *Phytopathology*, 61, 369, 1971.
51. **Crill, P., Bryan, H. H., Everett, P. H., Bartz, J. A., Jones, J. P., and Matthews, R. F.**, Floramerica, *Fla. Agric. Exp. Stn. Circular*, S-248, 1977.
52. **Cummins, J. N. and Aldwinkle, H. S.**, Breeding apple rootstocks, *HortScience*, 9, 367, 1974.
53. **Cunfer, B., Mathre, D. E., and Hockett, E. A.**, Diversity of reaction to ergot among male-sterile barleys, *Plant Dis. Rep.*, 58, 679, 1974.
54. **Daubeny, H. A. and Pepin, H. S.**, Fruit rot resistance in red raspberry, *Fruit Var. J.*, 30, 21, 1976.
55. **Dean, J. L. and Coleman, O. H.**, Necrotic and resistant reactions to the sugarcane mosaic virus in sorghum, *Plant Dis. Rep.*, 43, 522, 1959.
56. **Dean, J. L. and Miller, J. D.**, Field screening of sugarcane for eye spot resistance, *Phytopathology*, 65, 955, 1975.
57. **Demetradze, T. Ya., Bakanidze, M. Sh., Taoodumadze, E. I.**, Carbohydrate-nitrogen exchange in the leaves of various species and varieties of citrus plants in relation to mal secco and frost resistance (trans.), *Ref. Zh.*, 10.55.635, abstr. in *Plant Breed. Abstr.*, 45, 3910.

58. **D'Ercole, N.**, Prove di suscettibilita varietale del cavolfiore all'infezione di *Rhizoctonia solani* Kuhn in serra (trans.), *Inf. Fitopatol,* 24, 11, 1974, abstr. in *Rev. Plant Pathol.,* 54, 1486.
59. **Dewey, W. G. and Hoffman, J. A.**, Susceptibility of barley to *Tilletia controversa, Phytopathology,* 65, 654, 1975.
60. **Dickson, J. G.** *Diseases of Field Crops,* 2nd ed., McGraw-Hill, New York, 1956.
61. **Dixon, G. R.**, Field studies of powdery mildew *(Erysiphe cruciferarum)* on Brussels sprouts, *Plant Pathol.,* 23, 105, 1974.
62. **Dixon, G. R.**, Resistance of red and white clover cultivars to clover rot *(Sclerotinia trifoliorum), Ann. Appl. Biol.,* 81, 276, 1975.
63. **Dixon, G. R., Hague, J. M., and Braybrooks, J.**, Field reaction of swede cultivars to turnip crinkle virus, *Plant Pathol.,* 24, 31, 1975.
64. **Doazan, J. P.**, Sensibilite de varietes de vigne *(Vitis vinifera L.)* a l'excoriose *(Phomopsis viticola Sacc.);* distribution du charactere dans quelque descendances, *Vitis,* 13, 206, 1974.
65. **Dubey, L. N. and Joshi, R. D.**, Varietal reaction of potato to leafspot disease, *Science and Culture,* 42, 231, 1976.
66. **Duffus, J. K. and Skoyan, I. O.**, Relationship of age of plants and resistance to a severe isolate of the beet curly top virus, *Phytopathology,* 67, 151, 1977.
67. **Dunleavy, J. M.**, Viral diseases, in *Soybeans: Improvement, Production, and Uses,* Caldwell, B. E., Ed., American Society of Agronomy, Madison, Wis., 1973, 505.
68. **Dunn, J. A. and Wheatley, G. A.**, Brussels sprouts, in *Nat. Veg. Res. Stn. UK. Ann. Rep. 1974,* 1975.
69. **Dushutina, K. K. and Mitsek, N. F.**, Susceptibility of varieties and hybrid seedlings of pear to leaf spot (trans.), *Ref. Zh.,* 12.55.339, 1973, abstr. in *Plant Breed. Abstr.,* 46, 4664.
70. **Dutt, S., Bedi, P. S., and Singh, T. H.**, Evaluation of the genetic stock of cotton vis-a-vis the attack of blight in the Punjab State, *J. Res. Punjab Agric. Univ.,* 10, 401, 1973.
71. **Dutt, S., Chahal, A. S., Lobana, K. S., and Singh, M.**, Evaluation of genetic stock of linseed *(Linum usitatissimum L.)* against powdery mildew, *J. Res. Punjab Agric. Univ.,* 12, 6, 1975.
72. **Ebba, T.**, Genetic studies of the host-parasite relationship between *Ustilago hordei* and *Hordeum vulgare, Diss. Abstr. Int.* B, 35, 2812B, 1974.
73. **Einset, J. and Pratt, C.**, Grape in *Advances in Fruit Breeding,* Janick, J. and Moore, J. N., Eds., Purdue University Press, West Lafayette, Ind., 1975, 130.
74. **Endo, B. Y.**, Histological responses of resistant and susceptible soybean varieties, and backcross progeny to entry and development of *Heterodera glycines, Phytopathology,* 55, 375, 1965.
75. **Engelke, M. C., Smith, R. R., Maxwell, D. P.**, Monogenic resistance to red clover leaf rust associated with seedling lethality, *Crop Sci.,* 17, 465, 1977.
76. **Estey, R. H.**, A study of nematode problems in organic soil (trans.), *Rech. Agron.,* 19, 64, 1973/74, abstr. in *Plant Breed. Abstr.,* 45, 10553.
77. **Flor, H. H.**, Host-parasite interaction in flax rust — its genetics and other implications, *Phytopathology,* 45, 680, 1955.
78. **Flor, H. H.**, Current status of the gene-for-gene concept, *Annu. Rev. Phytopathol.,* 9, 275, 1971.
79. **Fogle, H. W.**, Cherries, in *Advances in Fruit Breeding,* Janick, J. and Moore, J. N., Eds., Purdue University Press, West Lafayette, Ind., 1975, 348.
80. **Forde, H. I.**, Walnuts, in *Advances in Fruit Breeding,* Janick, J. and Moore, J. N., Eds., Purdue University Press, West Lafayette, Ind., 1975, 439.
81. **Fraser, L. R. and Singh, D.**, Reaction of Indian citrus varieties to greening virus, 1st Int. Citrus Symp., Riverside, Calif., 1968, 365, abstr. in *Hortic. Abstr.,* no. 7188, 1978.
82. **Fridlund, R. R.**, Pear vein yellows virus symptoms in greenhouse - grown pear cultivars, *Plant Dis. Rep.,* 60, 891, 1976.
83. **Fry, W. E.**, Integrated effects of polygenic resistance and a protective fungicide on development of potato late blight, *Phytopathology,* 65, 908, 1975.
84. **Fry, W. E.**, Integrated control of potato late blight — effects of polygenic resistance and techniques of timing fungicide applications, *Phytopathology,* 67, 415, 1977.
85. **Galleta, G. J.**, Blueberries and cranberries, in *Advances in Fruit Breeding,* Janick, J. and Moore, J. N., Eds., Purdue University Press, West Lafayette, Ind., 1975, 154.
86. **Gardan, L., Luisetti, J. W., and Pranier, J. P.**, Premiers resultats sur la sensibilite varietale du pecher au deperissement bacterien *(Pseudomonas mors-prunorum* f. sp. *persicae)* (Trans.), *C. R. Seances Acad. Agric. Fr.,* 57, 1090, 1971, abstr. in *Plant Breed. Abstr.,* 43, 628.
87. **Garren, K. H. and Coffelt, T. A.**, Reaction to cylindrocladium black rot in Virginia-type peanut cultivars, *Plant Dis. Rep.,* 60, 175, 1976.
88. **Gerginova, T. S.**, Apple chat fruit virus. I. Identification and investigation on the incidence on some cultivars and clonal rootstocks (trans.), *Virusni Bolyni po Rasteniyata, Bulg. Acad. Sci.,* 165, 1974, abstr. in *Rev. Plant Pathol.,* 55, 4761.

89. **Gillien, P.,** New horizons for growing groundnuts resistant to drought and rosette (trans.), *Oleagineux,* 33, 25, 1978, abstr. in *Rev. Plant Path.,* 57, 3711.
90. **Glodeanu, C., Mariniscu, G., and Glodeanu, E.,** Biochemical differences in apple fruits with different resistance to *Monilinea fructigena* attack (trans.), *An. Univ. Craiovax III,* 5, 280, 1973, abstr. in *Plant Breed. Abstr.,* 45, 7665.
91. **Granada, C. G. A. and Sanchez, P. A.,** The etiology of citrus foot rot and a rootstock resistance test in the Canca Valley, Colombia, *Proc. Trop. Reg. Am. Soc. Hortic. Sci.,* 13, 132, 1969.
92. **Gray, F. A., Hine, R. B., Schonhorst, M. K., and Naik, J. D.,** A screening technique useful in selecting for resistance in alfalfa to *Phytophthora megasperma, Phytopathology,* 63, 1185, 1973.
93. **Habgood, R. M. and Hayes, J. D.,** The inheritance of resistance to *Rhynchosporium secalis* in barley, *Heredity,* 27, 25, 1971.
94. **Hackney, R. W. and Ferris, H.,** Infection development, and reproduction of *Meloidogyne incognita* in eight grapevine cultivars, *J. Nematol.,* 7, 323, 1975.
95. **Hagborg, W. A. F.,** Notes on bacterial diseases of cereals and some other crop plants, *Can. Plant Dis. Surv.,* 54, 129, 1974.
96. **Hearn, C. J., Hutchinson, D. J., and Barrett, H. C.,** Breeding citrus rootstocks, *HortScience,* 9, 357, 1974.
97. **Hecker, R. J. and Ruppel, E. G.,** Inheritance of resistance to Rhizoctonia root rot in sugarbeet, *Crop Sci.,* 15, 487, 1975.
98. **Heijbroek, J., McFarlane, J. S., and Doney, D. L.,** Breeding for tolerance to beet-cyst eelworm *Heterodera schachtii* Schm. in sugarbeet, *Euphytica,* 26, 557, 1977.
99. **Hesse, C. O.,** Peaches, in *Advances in Fruit Breeding,* Janick, J. and Moore, J. N., Eds., Purdue University Press, West Lafayette, Ind., 1975, 285.
100. **Hooker, A. L.,** The genetics and expression of resistance in plants to rusts of the genus Puccinia, *Ann. Rev. Phytopathol.,* 5, 163, 1967.
101. **Horsfall, J. G. et al. Eds.,** Genetic Vulnerability of Major Crops, Committee on Vulnerability of Major Crops, Agriculture Board, Division of Biology and Agriculture, National Research Council, National Academy of Sciences, 1972.
102. **Howard, H. W., Cole, C. S., Fuller, J. M., and Jellis, G. J.,** Seed potato, in *Plant Breed. Inst. Cambridge Rep.,* 15, 77, 1975, abstr. in *Plant Breed. Abstr.* 46, 7232.
103. **Humaydan, H. S., Williams, P. H., Jacobsen, B. J., and Bissonnette, H. L.,** Resistance in radish to *Aphanomyces raphani* and *Rhizoctonia solani, Plant Dis. Rep.,* 60, 156, 1976.
104. **Hutchinson, D. J. and Grim, G. R.,** Citrus clones resistant to *Phytophthora parasitica,* 1973 screening results, *Proc. Fla. State Hortic. Soc.,* 86, 88, 1974.
105. **Ishchenko, L. A. and Kuminov, E. P.,** Assessment of breeding material of black currant for resistance to septoriosis (trans.), *Mikol. Fitopatol.,* 10, 28, 1976, abstr. in *Rev. Plant Pathol.,* 55, 2806.
106. **Jackson, A. O., Bracker, C. E., and Shaner, G.,** Mechanical transmission, varietal reactions and electron microscopy of a disease in Indiana with properties of wheat sprindle streak mosaic, *Plant Dis. Rep.,* 60, 202, 1976.
107. **Janick, J. and Moore, J. N., Eds.,** *Advances in Fruit Breeding,* Purdue University Press, West Lafayette, Ind., 1975, 623.
108. **Jaynes, R. A.,** Chestnuts, in *Advances in Fruit Breeding,* Janick, J. and Moore, J. N., Eds., Purdue University Press, West Lafayette, Ind., 1975, 490.
109. **Jenkins, S. F., Jr.,** Interaction of *Pseudomonas solanacearum* and *Meloidogyne incognita* on bacterial wilt resistant and susceptible cultivars of eggplant, *Proc. Am. Phytopathol. Soc.,* 1, 69, 1974.
110. **Johnson, R. and Taylor, A. J.,** Spore yields of pathogens in investigations of the race-specificity of host resistance, *Annu. Rev. Phytopathol.,* 14, 97, 1976.
111. **Johnston, T. D.,** Mildew and club root in swedes and rape: a breeder's comments, *Ann. Appl. Biol.,* 81, 278, 1975.
112. **Jyung, H. S.,** Studies on *Acrocylindrium oryzae* on rice; testing varietal resistance, and culture filtrates from the fungal pathogen, *Korean J. Plant Prot.,* 14, 23, 1975.
113. **Kalinchenko, R. L.,** Bacterial leaf spot of stone fruits in the Crimea, *Byull. Vses. Ordena Lenina Inst. Rastenievodstva Imeni: N.I. Vavilova,* 27, 63, 1972, abstr. in *Plant Breed. Abstr.,* 43, 9967.
114. **Kamara, A.-El-M. M.,** Role of phenolics and phenolic oxidizing enzymes in the development of potato silver scurf, *Diss. Abstr. Int. B,* 34, 3033B, 1974.
115. **Kameraz, A. Ya. and Ponin, I. Ya.,** Initial material and proposals of using it in breeding potatoes for resistance to the potato nematode *(H. rostochensis* Woll.) (trans.), *Tr. Prikl. Bot. Genet. Sel.,* 53, 199, 1974, abstr. in *Plant Breed. Abstr.,* 46, 2578, 1976.
116. **Kappelman, A. J., Jr. and Sappenfield, W. P.,** Development of a set of upland cotton lines for comparisons of Fusarium wilt-root knot nematode data from several tests, *Crop Sci.,* 13, 280, 1973.
117. **Katinskaya, Yu. K.,** Comparative study of the resistance of strawberry varieties to strawberry nematodes, *Ref. Zh.,* 3.55.787, 1974, abstr. in *Plant Breed. Abstr.,* 46, 1817.

118. **Kawahara, E., Wakamatsu, T., Sasaki, T., Murasato, S., Ota, K., Sato, T., and Meguro, R.,** Characteristics and breeding history of the new white clover variety Kitaoha (trans.), *Bull. Tohoku Nat. Agric. Exp. Stn.,* 48, 31, 54, 1974, abstr. in *Plant Breed. Abstr.,* 45, 5522.
119. **Kazakov, L. V.,** Cultivating pure-strain planting material of raspberry, Bryansk province (trans.), *Sb. Nauch. Rabot. N. -i. Zonal'n. in-t Sadovodstva Nechernozem. Polosy,* 5, 152, 1972. From *Ref. Zh.,* (1974) 4.55. 719, 1974, abstr. in *Plant Breed. Abstr.,* 46, 1794.
120. **Keep, E.,** Currants and gooseberries, in *Advances in Fruit Breeding,* Janick, J. and Moore, J. N., Eds., Purdue University Press, West Lafayette, Ind., 1975, 197.
121. **Keep, E.,** North European cultivars as donors of resistance to American gooseberry mildew in black currant breeding, *Euphytica,* 26, 817, 1977.
122. **Kehr, W. R., Frosheiser, F. J., Wilcoxson, R. D., and Barnes, D. K.,** Breeding for disease resistance in alfalfa, in *Alfalfa Science and Technology,* Hanson, C. H., Ed., American Society of Agronomy, Madison, Wis., 1972, 335.
123. **Kennedy, B. W. and Tachibana, H.,** Bacterial diseases, in *Soybeans: Improvement Production and Uses,* Caldwell, B. E., Ed., American Society of Agronomy, Madison, Wis., 1973, 491.
124. **Kitajima, H.,** Disease damage in fruit trees XI, Physolospora piricola (trans.), *Agric. Hortic.,* 52, 337, 1977, abstr. in *Plant Breed. Abstr.,* 47, 11916.
125. **Kitajima, H.,** Disease damage in fruit trees. XII. Pear necrotic spot, *Agric. Hortic.,* 52, 459, 1977, abstr. in *Plant Breed. Abstr.,* 48, 687.
126. **Knott, D. R. and Dvorak, J.,** Alien germ plasm as a source of resistance to disease *Annu. Rev. Phytopathol.,* 14, 211, 1976.
127. **Koch, F.,** Studies on the prevention and control of seedling diseases in sugar beet (trans.), *Sementi Elette,* 20, 5, 1974, abstr. in *Plant Breed. Abstr.,* 45, 6576.
128. **Kodama, T.,** Characteristics of strawberry yellows caused by Fusarium and difference of its effect on the grown varieties, *Bull. Nara Agric. Exp. Stn.,* 6, 68, 1974, abstr. in *Rev. Plant Pathol.,* 55, 3667.
129. **Kollanyi, L.,** Use of some Rubus species in raspberry breeding (trans.), *Proc. XIX Int. Hortic. Congr. 1A.* Sec. VII, 1974, 340, abstr. in *Plant Breed. Abstr.,* 46, 8291.
130. **Komarova, R. and Vlasova, E.,** Diseases of lettuce under cover, *Kartofel Ovoshchi,* 9, 40, 1974.
131. **Kono, A.,** Varietal susceptibility to rust in peanut (trans.), *Bull. Agric. Miyazaki Univ.,* 24, 217, 1977, abstr. in *Rev. Plant Pathol.,* 57, 4714.
132. **Korneeva, E. M. and Loshakova, N. I.,** The flax disease pasmo (trans.), *Len Konoplya* 6, 17, 1976, abstr. in *Plant Breed. Abstr.,* 46, 10316.
133. **Kozaki, I.,** Studies on breeding pear for resistance to *Alternaria kikuchiana;* the genetics of resistance to *A. kikuchiana* (trans.), *Bull. Hortic. Res. Stn.,* No. 12, 17, 1973, abstr. in *Plant Breed. Abstr.,* 44, 4017.
134. **Krasichkov, V. P. and Mansurov, N. M.,** Breeding Soviet fine-fibred cotton in Tajikistran (trans.), *Ref. Zh.* 9.55.161, 1973, abstr. in *Plant Breed. Abstr.,* 46, 2604.
135. **Kuhn, C. W., Paguio, O. R., and Adams, D. B.,** Tolerance in peanuts to peanut mottle virus, *Plant Dis. Rep.,* 62, 365, 1978.
136. **Kuhn, R. C., Ohm, H. W., and Shaner, G. E.,** Slow leaf-rusting resistance in wheat against 22 isolates of *Puccinia recondita, Phytopathology,* 68, 651, 1978.
137. **Kumar, S.,** Quantitative and qualitative variations in leaf-surface wax in three varieties of wheat varying in susceptibility to blight caused by *Alternaria triticina, Indian Phytopathol.,* 27, 508, 1974.
138. **Lagerstedt, H. B.,** in *Advances in Fruit Breeding,* Janick, J. and Moore, J. N., Eds., Purdue University Press, West Lafayette, Ind., 1975, 456.
139. **Lanjewar, R. D., Sapkal, K. N., and Shukla, V. N.,** Reaction of some cotton varieties against grey mildew disease, *Punjabrao Krishi Vidyapeeth Res. J.,* 3, 93, 1975, abstr. in *Rev. Plant Pathol.,* 55, 5765.
140. **Lapwood, D. H., Adams, M. J., and Crisp, A. F.,** The susceptibility of red beet cultivars to Streptomyces scab, *Plant Pathol.,* 25, 31, 1976.
141. **Large, E. C.,** *The Advance of the Fungi,* J. Cape, London, 1940, 488.
142. **Laws, R. J.,** Hops and hop products — the present position, *Brewer,* 61, 12, 1975, abstr. in *Plant Breed. Abstr.,* 45, 9220.
143. **Layne, R. E. C. and Quamme, H. A.,** Pears, in *Advances in Fruit Breeding,* Janick, J. and Moore, J. N., Eds., Purdue University Press, West Lafayette, Ind., 1975, 38.
144. **Layne, R. E. C. and Harrison, T. B.,** "Haggith" apricot: rootstock seed source, *HortScience,* 10, 428, 1975.
145. **Leu, L. S., Wang, Z. N., Lin, K. H., Kwo, C. Y., and Shang, K. C.,** Leaf blight of sugarcane in Taiwan. VI. Preliminary studies on the mode of inheritance of resistance in sugarcane against the disease (trans.), *Rep. Taiwan Sugar Res. Inst.,* 71, 45, 1976, abstr. in *Plant Breed. Abstr.,* 47, 4459.

146. **Levandovskaya, L. I.**, Resistance of carrot varieties to *Phoma rostrupii* and Sclerotina diseases, *Byull. Vses. Ordena Lenina Ordean Druzhby Narodov Inst. Rastenievo-dstva Imeni N. I. Vavilova*, 50, 48, 1975, abstr. in *Plant Breed. Abstr.* 46, 7582.
147. **Lewellen, R. T., Sharp, E. L., and Hehn, E. R.**, Major and minor genes in wheat for resistance to *Puccinia striiformis* and their responses to temperature changes, *Can. J. Bot.*, 45, 2155, 1967.
148. **Lewellen, R. T. and Sharp, E. L.** Inheritance of minor reaction gene combinations in wheat to *Puccinia striiformis* at two temperature profiles, *Can. J. Bot.*, 46, 21, 1968.
149. **Linde-Laursen, I., Jensen, H. P., and Jorgensen, J. H.**, Resistance of *Triticale, Aegilops, and Haynaldia* species to the take-all fungus, *Gaeumannomyces graminis, Z. Pflanzenzeucht.*, 70, 100, 1973.
150. **Ling, K. C.**, *Rice Virus Diseases*, The International Rice Institute, Los Banos, Philippines, 1972.
151. **Lyrene, P. M., Dean, J. L., and James, N. I.**, Inheritance of resistance to pokkah boeng in sugarcane crosses, *Phytopathology*, 67, 689, 1977.
152. **Lyskanowska, K. and Rejman, A.**, Bacterial canker of sweet cherry *(Prunus avium)* in Poland. III. Degree of resistance of wild cherry seedlings, *Plant Dis. Rep.*, 62, 500, 1978.
153. **MacDonald, J. D., Leach, L. D., and McFarlane, J. S.**, Susceptibility of sugarbeet lines to the stalk blight pathogen *Fusarium oxysporum* f. sp. *betal.*, *Plant Dis. Rep.*, 60, 192, 1976.
154. **McFarlane, J. S.**, Elimination of downy mildew as a major sugarbeet disease in the costal valleys of California, *Plant Dis. Rep.*, 52, 297, 1968.
155. **McFarlane, J. S., Skoyen, I. O., and Lewellen, R. T.**, Development of sugarbeet breeding lines and varieties resistant to yellows, *J. Am. Soc. Sugar Beet Technol.*, 15, 347, 1969.
156. **Madden, G. D. and Malstrom, H. L.**, Pecans and hickories, in *Advances in Fruit Breeding*, Janick, J. and Moore, J. N., Eds., Purdue University Press, West Lafayette, Ind., 1975, 420.
157. **Magnus, H. A.**, Studies on resistance to leaf spot disease in barley (trans.), *Nord. Jordbrugsforsk.*, 56, 30, 1974, abstr. in *Rev. Plant Pathol.*, 55, 3533.
158. **Martin, W. J., Hernandez, T. P., and Hernandez, T. P**, Development and disease reaction of Jasper, a new soil rot-resistant sweet potato variety from Louisiana, *Plant Dis. Rep.*, 59, 388, 1975.
159. **Mathur, R. L., Mathur, B. N., Singh, R. R., Sharma, B. S., and Godara, P. S.**, Varietal reaction of grapes to *Gloeosporium ampelophagum* causing anthracnose, *Indian J. Mycol. Plant Pathol.*, 3, 179, 1974.
160. **Mayee, C. D. and Charanjit, K.**, Little leaf incidence in relation to the vector population in some varieties of brinjal, *Indian Phytopathol.*, 28, 93, 1975.
161. **Meister, C. W.**, Differential reaction of citrus species to diseases and pests at Koronivia, *Fiji Agric. J.*, 35, 75, 1973, abstr. in *Rev. Plant Pathol.*, 53, 4445.
162. **Melkonyan, M. V.**, Breeding as a means of controlling virus diseases of grapevine (trans.), *Biol. Zh. Arm.*, 28, 45, 1975, abstr. in *Plant Breed. Abstr.* 46, 4733.
163. **Mercer, P. C.**, Control of peanut pod rot in Malawi, *Plant Dis. Rep.*, 61, 55, 1977.
164. **Mikheeva, L. P.**, The susceptibility of potato sprouts to Rhizoctonia after artificial infection (trans.), *Ref. Zh.* 10.55.593, abstr. in *Plant Breed. Abstr.*, 46, 11250.
165. **Moniz, L., Mohinddin, S. G., and Patil, B. D.**, Reactions of citrus varieties and species to isolates of *Xanthomonas citri* (Hasse) Dowson, *Res. J. Mahatma Phule Agric. Univ.*, 5, 124, 1974, abstr. in *Rev. Plant Path.*, 55, 1803.
166. **Moore, J. N., Brown, E., and Sistrunk, W. A.**, "Comanche" blackberry, *HortScience*, 9, 245, 1974.
167. **Morsi, L. R., Mansour, A. A., Khalifa, M. I., Ghobrial, E., and Sabet, T. M.**, Sources of resistance to net blotch of barley, *Helminthosporium teres* (Sacc.), *Agric. Res. Rev.*, 53, 33, 1975, abstr. in *Plant Breed. Abstr.*, 47, 9406.
168. **Moseman, J. G.**, Barley diseases and their control, *U.S. Dep. Agric. Agric. Handb.*, 338, 1968.
169. **Munthe, T.**, Resistance to barley yellow dwarf virus in barley (trans.), *Nord. Jordbrugsforsk.*, 57, 1121, 1975, abstr. in *Rev. Plant Pathol.*, 55, 5692.
170. **Muramatsu, M.**, The breeding of malting barley varieting which is resistant to barley yellow mosaic (trans.), *3rd Int. Barley Genet. Symp.*, 1975, 82, Abstr., 46, 25221.
171. **Murray, G. M., Maxwell, D. P., and Smith, R. R.**, Screening Trifolium species for resistance to *Stemphylium sarcinaeforme, Plant Dis. Rep.* 60, 35, 1976.
172. **Naqvi, S. Q. A. and Mahmood, K.**, Responses of certain varieties of sugar beet *(Beta vulgaris saccharifera)* to the attack of the root-knot nematode, *Meloidogyne incognita* (Kofoid and White) Chitwood. *Labdev J. Sci. Technol. India*, 12B, 85, 1974.
173. **Neagu, M. I. and Georgiscu, M. M.**, Contribution to the improvement of resistance to *Taphrina deformans* (Berk) in peach (trans.), *Lucr. Stiint. Inst. Agron. Hortic.* 16, 19, 1976, abstr. in *Plant Breed. Abstr.*, 46, 9432.
174. **Nelson, R. R.**, Genetics of horizontal resistance to plant diseases, *Annu. Rev. Phytopathol.*, 16, 359, 1978.
175. **Nilsson, H. E.**, Breeding for resistance to take-all disease, report given at the Eucarpia Symp., Vejle, Denmark, 1972.

176. **Nowicki, B.**, The susceptibility of cultivars of cruciferous plants grown in Poland to club root, *Plasmodiophora brassicae* Wor. (trans.), *Rocz. Nauk Roln., Ser. E,* 4, 205, 1975, abstr. in *Plant Breed. Abstr.,* 46, 7598.
177. **Nyerges, P., Szabo, E., and Donko, E.**, The role of anthocyanin and phenol compounds in the resistance of grapes against Botrytis infection, *Acta Phytopathol. Acad. Sci. Hung.,* 10, 21, 1975.
178. **O'Bannon, J. H. and Esser, R. P.**, Evaluation of citrus, hybrids, and relatives as hosts of *Pratylenchus coffeae,* with comments on other hosts, *Nematologia Mediterranea,* 3, 113, 1975.
179. **O'Brian, M. J. and Rich, A. E.**, Potato diseases, *U.S. Dep. Agric. Handb.,* No. 474, 1976.
180. **Ohm, H. W. and Shaner, G. E.**, Segregation for slow leaf rusting in wheat, *Agron. Abstr.,* 67, 59, 1975.
181. **Ondrej, M.**, Determination of the susceptibility of flax varieties to fusarioses (trans.), *Len Konopi,* 12, 53, 1974, abstr. in *Rev. Plant Pathol.,* 55, 756.
182. **Ondrej, M. and Rosenberg, L.**, Phoma wilt of flax (trans.), *Len Konopi,* 13, 35, 1975, abstr. in *Rev. Plant Pathol.,* 55, 4126.
183. **Ou, S. H. and Jennings, P. R.**, Progress in the development of disease - resistant rice, *Annu. Rev. Phytopathol.,* 7, 383, 1969.
184. **Ou, S. H.**, *A Handbook of Rice Diseases in the Tropics,* International Rice Research Institute, Los Banos, Phillipines, 1973.
185. **Ourecky, D. K.**, Brambles, in *Advances in Fruit Breeding,* Janick, J. and Moore, J. N., Eds., Purdue University Press, West Lafayette, Ind., 1975.
186. **Parlevliet, J. E.**, Evaluation of the concept of horizontal resistance in the barley *Puccinia hordei* host-pathogen relationship, *Phytopathology,* 66, 494, 1976.
187. **Parlevliet, J. E. and Zadoks, J. C.**, An integrated concept of disease resistance: a new view including horizontal and vertical resistance in plants, *Euphytica,* 26, 5, 1977.
188. **Parui, N. R. and Bandyopadhyay, D. C.**, A note on the screening of rai *(Brassica juncea* Coss.) against white rust *(Albugo candida* (Pers.) Kuntzel), *Curr. Sci.,* 42, 798, 1973.
189. **Patil-Kulkarni, B. G., Puttarudrappa, A., Kajjari, N. B., and Goud, J. V.**, Breeding for rust resistance in sorghum, *Indian Phytopathol.,* 25, 166, 1972.
190. **Patterson, F. L., Gallum, R. L., and Roberts, J. J.**, Registration of Arthur wheat, *Crop Sci.,* 14, 910, 1974.
191. **Patterson, F. L. et al.**, Registration of Sullivan wheat, *Crop Sci.,* 19, 297, 1979.
192. **Petrie, G. A.**, Diseases of Brassica species in Saskatchewan, 1970—72. I. Staghead and aster yellows, *Can. Plant Dis. Surv.,* 53, 19, 1973.
193. **Petrie, G. A.**, Diseases of Brassica species in Saskatchewan, 1970—72. II. Stem, pod, and leaf spots, *Can. Plant Dis. Surv.,* 53, 83, 1973.
194. **Petrie, G. A.**, personal communication, 1979.
195. **Pittarelli, G. W. and Stavely, J. R.**, Direct hybridization of *Nicotiana repandax, N. tabacum, J. Hered.,* 66, 281, 1975.
196. **Porter, D. M. and Hammons, R. O.**, Differences in plant and pod reaction of peanut lines to infection by *Diplodia gossypina, Peanut Sci.,* 2, 23, 1975.
197. **Pospisil, B.**, Evaluation of economic qualities of breeding material resistant to flax anthracnose (trans.), *Len Konop,* 13, 73, 1975, abstr. in *Rev. Plant Pathol.,* 55, 4127.
198. **Potlarchuk, V. I., Isin, M. M., and Dernovskaya, L. I.**, Varietal resistance to wilt in some fruit crops (trans.), *Ref. Zh.,* 9.55.813, abstr. in *Plant Breed. Abstr.,* 47, 11889.
199. **Prakash, J., Saingar, D. K., Mathur, R. S., and Sundram, N. V.**, Resistance to five leaf spotting fungi in forage and grain sorghums in India, *Plant Dis. Rep.,* 59, 1979, 1975.
200. **Puttarudrappa, A., Kulkarni, B. G. P., Kajjari, N. B., and Goud, J. V.**, Inheritance of resistance to downy mildew *(Scerospora sorghi)* in sorghum, *Indian Phytopathol.,* 25, 471, 1973.
201. **Rai, R. A., Srivastava, V. S., and Agrawat, J. M.**, Reaction of barley varieties to stripe disease, *Indian Phytopathol.,* 28, 291, 1975.
202. **Raj, S. A. and Prasad, N. N.**, Reaction of groundnut to *Rhizoctonia bataticola, Indian Phytopathol.,* 28, 440, 1975.
203. **Ristanovic, M. and Festic, H.**, *Podosphaera tridactyla* (Wallr.) A. de Bary on stone fruits in greenhouse conditions (trans.), *Jugosl. Vocarstvo,* 6, 821, 1972, abstr. in *Rev. Plant Pathol.,* 53 4049.
204. **Roach, B. T.**, Resistance to yellow spot disease in Saccharum species and hybrids, in *Proc. Queensl. Soc. Sugarcane Technol. 42nd. Conf.,* Mackay, Queensland, 1975, abstr. in *Rev. Plant Pathol.,* 54, 5537.
205. **Roussel, N., Roelants, W., and Vreven, T.**, Results of comparative trials of sugar beet varieties in Belgium from 1973 to 1975 (trans.), *Publ. Trimestrielle Inst. Belge Amelior. Betterave* 44, 1976, abstr. in *Plant Breed. Abstr.,* 46, 11313.

206. **Ruppel, E. G., Hills, I. J., and Mumford, D. L.,** Epidemiological observations on the sugar beet powdery mildew epiphytotic in western U.S. in 1974, *Plant Dis. Reptr.*, 59, 283, 1975.
207. **Sadamori, S., Yoshida, Y., Tsuchiya, S., Haniuda, T., Murakami, H., Suzuki, H., and Ishizuka, S.,** On the new apple variety Akane (trans.), *Bull. Hortic. Res. Stn., C,* No. 8, 1, 1973, abstr. in *Plant Breed. Abstr.*, 44, 2693.
208. **Saharan, G. S.,** Botryosphaeria nut rot of pecan, *Plant Dis. Rep.*, 58, 1030, 1974.
209. **Saito, K., Nakayama, R., and Takeda, K.,** Seedling screening for resistance to Alternaria leaf spot in apple (trans.,) *Agric. Hortic.*, 49, 923, 1974, abstr. in *Plant Breed. Abstr.*, 46, 5654.
210. **Sams, D. W. and Bienz, D. R.,** Relative susceptibility of spinach plant introduction accessions to curly top, *HortScience,* 9, 600, 1974.
211. **Savitsky, H.,** Hybridization between *Beta vulgaris* and *B. procumbens* and transmission of nematode *(Heterodera schachtii)* resistance to sugarbeet, *Can. J. Genet. Cytol.*, 17, 197, 1974.
212. **Savitsky, H.,** Nematode *(Heterodera schachtii)* resistance and meiosis in diploid plants from interspecific *Beta vulgaris* x *B. procumbens* hybrids, *Can. J. Genet. Cytol.*, 20, 177, 1978.
213. **Saynor, M.,** The distribution of pathotypes of the cereal cyst-eelworm *Heterodera avenae* in England and Wales, *Ann. Appl. Biol.*, 81, 215, 1975.
214. **Schafer, J. F., Caldwell, R. M., Patterson, F. L., and Compton, L. E.,** Wheat leaf rust resistance combinations, *Phytopathology,* 53, 569, 1963.
215. **Schafer, J. F.,** Tolerance to plant disease., *Annu. Rev. Phytopathol.*, 9, 235, 1971.
216. **Scott, D. H. and Lawrence, F. J.,** Strawberries, in *Advances in Fruit Breeding,* Janick, J. and Moore, J. N., Eds., Purdue University Press, West Lafayette, Ind., 1975, 71.
217. **Scotts La Massese, C.,** Tests of some fruit rootstocks and varieties as hosts for *Pratylenchus vulnus* Allen and Jenson (trans.), *C. R. Seances Acad. Agric. France,* 1088, 1975, abstr. in *Plant Breed. Abstr.*, 46, 10430.
218. **Sellam, M. A. and Wilcoxson, R. D.,** Bacterial leaf blight of wheat in Minnesota, *Plant Dis. Rep.*, 60, 242, 1976.
219. **Semina, S. N. and Naletova, O. D.,** The resistance of peach varieties to *Clasterosporium carpophilum* (trans.), *Ref. Zh.* 8.55.720, abstr. in *Plant Breed. Abstr.*, 43, 4588.
220. **Sequeira, L.,** Resistance to corky root rot in lettuce, *Plant Dis. Rep.*, 54, 754, 1970.
221. **Shaner, G.,** Evaluation of slow-mildewing resistance of Knox wheat in the field, *Phytopathology,* 63, 867, 1973.
222. **Shaner, G., Finney, R. E., and Patterson, F. L.,** Expression and effectiveness of resistance in wheat to Septoria leaf blotch, *Phytopathology,* 65, 761, 1975.
223. **Shaner, G. and Finney, R. E.,** The effect of nitrogen fertilizer on the expression of slow-mildewing resistance in Knox wheat, *Phytopathology,* 67, 1051, 1977.
224. **Shaner, G., Ohm, H. W., and Finney, R. E.,** Response of susceptible and slow leaf-rusting wheats to infection by *Puccinia recondita, Phytopathology,* 68, 471, 1978.
225. **Sharp, E. L., Sally, B. K., and McNeal, F. H.,** Effect of Pyrenophora wheat leaf blight on the thousand-kernel weight of 30 spring wheat cultivars, *Plant Dis. Rep.*, 60, 135, 1976.
226. **Sharpe, R. H. and Sherman, W. B.,** 'Flordablue' blueberry, *HortScience,* 11, 64, 1976.
227. **Shchelko, L. G. and Trainina, S. I.,** Immunity characteristics of barley of various origins (trans.), *Tr. Prikl. Bot. Genet. Sel.* 53, 113, 1974, abstr. in *Plant Breed. Abstr.*, 45, 7265.
228. **Sheppard, J. W. and Petterson, J. F.,** Chlorogenic acid and Verticillium wilt of tobacco, *Can. J. Plant Sci.*, 56, 157, 1976.
229. **Shevchenko, V. N. and Toporovskaya, Yu. S.,** Resistance of sugar beet to storage rot in the early stages of root development (trans.), *Sel. Semenovod.*, 31, 80, 1975, abstr. in *Plant Breed. Abstr.*, 46, 11321.
230. **Shirko, V. N., Pivavorova, N. S., and Burenin, V. I.,** Disease resistance in beet varieties from the world collection (trans.), *Tr. Prikl. Bot. Genet. Sel.*, 52, 244, 1974, abstr. in *Plant Breed. Abstr.*, 45, 499.
231. **Schneider, Y. I. and Ilyukhina, M. K.,** The methodology of evaluating winter-wheat varieties for resistance to bacterial diseases (trans.), *Sel. Semsenovod.*, 5, 25, 1974, abstr. in *Plant Breed. Abstr.*, 45, 2610.
232. **Sievert, R. C.,** Effect of potato virus Y on cultivars and hybrids of burley tobacco, *Phytopathology,* 68, 974, 1978.
233. **Simons, M. D.,** Polygenic resistance to plant disease and its use in breeding resistant cultivars, *J. Environ. Qual.*, 1, 232, 1972.
234. **Simons, M. D. and Murphy, H. C.,** Oat diseases, in *Oats and Oat Improvement,* Coffman, F. A., Ed., American Society of Agronomy, Madison, Wis., 1961, 330.
235. **Singh, R. A.,** Varietal resistance to Rhynchosporium leaf scald of rice, *FAO F.A.O.U.N. Int. Rice Comm. Newsl.*, 24, 105, 1975.

236. **Sitterly, W. R.,** Breeding for disease resistance in cucurbits, *Annu. Rev. Phytopathol.* 10, 471, 1972.
237. **Skarbilovich, T. S.,** The susceptibility of various varieties of flax to parasitic nematodes (trans.), *Ref. Zh.*, 11.55.764, 1973, abstr. in *Plant Breed. Abstr.*, 46, 3536.
238. **Skipp, R. A. and Samborski, D. J.,** The effect of the Sr6 gene for host resistance on histological events during the development of stem rust in near-isogenic wheat lines, *Can. J. Bot.*, 52, 1107, 1974.
239. **Skovmand, B., Roelfs, A. P., and Wilcoxson, R. D.,** Relationship between slow rusting and genes for specific resistance to stem rust, *Proc. Am. Phytopathol. Soc.*, 2, 59, 1975.
240. **Smith, O. D., Boswell, T. E., and Thames, W. H.,** Lesion nematode resistance in peanuts, *Crop Sci.*, 18, 1008, 1978.
241. **Smith, P. R. and Challen, D. I.,** Aetiology of the rosette and decline disease of peach and interactions between Prunus necrotic ringspot, prune dwarf, and dark green sunken mottle viruses, *Aust. J. Agric. Res.*, 23, 1027, 1972.
242. **Soost, R. K. and Cameron, J. W.,** Citrus, in *Advances in Fruit Breeding,* Janick, J. and Moore, J. N., Eds., Purdue University Press, West Lafayette, Ind., 1975, 507.
243. **Sorensen, H. L., Hackervott, H. L., and Harvey, T.,** Registration of KS10 pest-resistant alfalfa germplasm (Reg. No. GP44), *Crop Sci.*, 15, 105, 1975.
244. **Sowell, G., Jr., Smith, D. H., and Hammons, R. O.,** Resistance of peanut plant introductions to *Cercospora arachidicola, Plant Dis. Rep.*, 60, 494, 1976.
245. **Srivastava, M. P.,** A method of grading for resistance to Myrothecium leaf spot of cotton, *Haryana Agric. Univ. J. Res.*, 5, 94, 1975, abstr. in *Plant Breed. Abstr.*, 46, 10310.
246. **Stakman, E. C. and Piemeisel, F. J.,** A new strain of *Puccinia graminis, Phytopathology,* 7, 73, 1917.
247. **Stakman, E. C., Stewart, D. M., and Loegering, W. Q.,** Identification of physiologic races of *Puccinia graminis* var. *tritici,* U.S. Department of Agriculture, Agricultural Research Service, E617, 1962, 53.
248. **Statler, G. D. and Kiesling, R. L.,** Inheritance of resistance to black point in Leeds durum wheat, *Proc. Am. Phytopathol. Soc.*, 1, 87, 1974.
249. **Stevenson, F. J. and Jones, H. A.,** Some sources of resistance in crop plants, in *Plant Diseases, The Yearbook of Agriculture,* 1953, 192.
250. **Stevenson, N. D., Brown, A. H. D., and Latter, B. D. H.,** Quantitative genetics of sugarcane. IV. Genetics of Fiji disease resistance, *Theoret. Appl. Genet.*, 42, 262, 1972, abstr. in *Rev. Plant Pathol.*, 52, 846.
251. **Stevenson, W. R.,** personal communication, 1979.
252. **Stoyanova, M. and Konotop, A.,** A study on the resistance to Lycopersicon virus 3 Smith in progenies of *Nicotiana tabacum L.* X *N. glauca* Grah interspecies hybridization, *C. R. Acad. Agric. Georgi Dimitrov,* 8, 43, 1975, abstr. in *Plant Breed. Abstr.*, 46, 1653.
253. **Svirskis, A. and Strukchinskas, M.,** Roots rots of red clover and breeding of varieties more resistant to them (trans.), N.-i *Inst. Zomledeliya,* abstr. in *Plant Breed. Abstr.*, 45, 9663, 1973.
254. Swedish Seed Association, *Sver. Utsadesfoeren. Tidskr.*, 85, 67, 1975, abstr. in *Plant Breed. Abstr.*, 46, 1125.
255. **Takemata, T. and Sakai, K.** Characteristics of sweet potatoes introduced from New Zealand (trans.), *J. Cent. Agric. Exp. Stn.*, 22, 203, 1975, abstr. in *Plant Breed. Abstr.*, 46, 1578.
256. **Tamassy, I.,** New ways in horticultural plant breeding, *Agrartud. Kozl.*, 33, 269, 1974, abstr. in *Plant Breed. Abstr.*, 47, 5748.
257. **Taylor, N. L., Gibson, P. B., and Knight, W. E.,** Genetic vulnerability and germplasm resources of the true clovers, *Crop Sci.*, 17, 632, 1977.
258. **Terenteva, N. N.,** The resistance of lucerne to root rot and wilt (trans.), *Byull. Vses. Ordena Lenina Inst. Rastenievodstva Imeni N.I. Vavilova,* 33, 64, 1973, abstr. in *Rev. Plant Pathol.*, 53, 1861.
259. **Thakur, R. P., Patel, P. N., and Verma, J. P.,** Genetical relations between reactions to bacterial leaf spot, yellow mosaic, and Cercospora leaf spot diseases in mungbean *(Vigna radiata), Euphytica,* 26, 767, 1977.
260. **Thompson, J. M. and Taylor, J.,** Genetic susceptibility to Glomerella leaf blotch in apple, *J. Hered.*, 62, 303, 1971.
261. **Thurston, H. D.,** Relationship of general resistance: late blight of potato, *Phytopathology,* 61, 620, 1971.
262. **Tiwari, D. K. and Tiwari, J. P.,** Ecophysiological adaptability of new potato varieties at Jabalpur (M.P.), *JNKVV Res. J.*, 9, 44, 1975, abstr. in *Plant Breed. Abstr.*, 47, 463.
263. **Todd, W. A., Green, R. J., Jr. and Horner, C. E.,** Registration of Murray Mitcham peppermint, *Crop Sci.*, 17, 188, 1977.
264. **Topchiiska, M. and Topchiiska, I.,** Rubbery wood virus on pear (trans.), *Ovoscharstvo,* 55, 32, 1976, abstr. in *Rev. Plant Pathol.*, 55, 4189.

265. **Ullstrup, A. J.,** The impacts of the southern corn leaf blight epidemics of 1970—1971, *Annu. Rev. Phytopathol.,* 10, 37, 1972.
266. **Ullstrup, A. J.,** Disease of corn, in *Corn and Corn Improvement,* Sprague, G. F., Ed., Agron. Monogr. No. 18, American Society of Agronomy, Madison, Wis., 1977, 391.
267. **Van der Plank, J. E.,** *Plant Diseases: Epidemics and Control,* Academic Press, New York, 1963.
268. **Vares Megino, F.,** Citrus anthracnose in Spain. II. Rate of fungal development in leaves and buds of the host (trans), *An. Inst. Nac. Invest. Agrar. Ser. Prot. Veg.,* No. 4, 239, abstr. in *Rev. Plant Pathol.,* 54, 2248.
269. **Vasilenko, T. N.,** Bacterial blight of apple and pear in the Primore region (trans)., *Ref. Zh.,* 12.55.1350, 1973, abstr. in *Plant Breed. Abstr.,* 46, 3655.
270. **Vazquez, G., Peterson, G. A., and Timian, R. G.,** Inheritance of barley stripe mosaic reaction in crosses among three barley varieties, *Crop Sci.,* 14, 429, 1974.
271. **Vlasova, E. A. and Komarova, R. A.,** A study of the main diseases of lettuce under cover in Leningrad province, *Tr. Prikl. Bot., Genet. Sel.,* 53, 185, 1974, abstr. in *Plant Breed. Abstr.,* 45, 10571.
272. **Walker, J. C.,** *Plant Pathology,* 2nd ed., McGraw-Hill, New York, 1957.
273. **Walker, J. C.,** Disease resistance in the vegetable crops. III., Bot. Rev., 31, 331, 1965.
274. **Walker, J. C.,** Use of environmental factors in screening for disease resistance, *Annu. Rev. Phytopathol.,* 3, 197, 1965.
275. **Walter, J. M.,** Hereditary resistance to disease in tomato, *Annu. Rev. Phytopathol.,* 5, 131, 1967.
276. **Watkins, J. E. and Littlefield, L. J.,** Relationship of anthesis in Waldron wheat to infection by *Claviceps purpurea, Trans. Br. Mycol. Soc.,* 66, 362, 1976.
277. **Watts, L. E.,** The response of various breeding lines of lettuce to beet western yellows virus, *Ann. Appl. Biol.,* 81, 393, 1975.
278. **Westwood, M. N., Cameron, H. R., Lombard, P. B., and Cordy, C. B.,** Effects of trunk and rootstock on decline growth and performances of pear, *J. Am. Soc. Hortic. Sci.,* 96, 147, 1971.
279. **Whitney, E. D. and Lewellen, R. T.,** Bacterial vascular necrosis and rot of sugarbeet: genetic variability and selecting for resistance, *Phytopathology,* 68, 651, 1978.
280. **Wiese, M. V.,** *Compendium of Wheat Diseases,* American Phytopathological Society, St. Paul, 1977.
281. **Williams, E. B., Janick, J., Emerson, F. H., Dayton, D. F., Morory, J. B., Hough, L. F., and Bailey, C. H.,** "Sir Prize" apple, *HortScience,* 10, 281, 1975.
282. **Yankulov, I., Alipur, Kh., and Ivanov, In.,** A study of experimentally produced autopolyploids of *Mentha piperita L., Rastenievu dp Nauki,* 11, 23, 1974, abstr. in *Plant Breed. Abstr.,* 45, 5946.
283. **Zambettakis, C., Bockelec-Morvan, A., Waliyar, F., and Rossion, J.,** Varietal differences in susceptibility of groundnut to infection by *A. Flavus* in the field and under artificial conditions (trans.), *Oleagineux,* 32, 377, 1977, abstr. in *Rev. Plant Path.,* 57, 1916.
284. **Zaumeyer, W. J. and Meiners, J. P.,** Disease resistance in beans, *Annu. Rev. Phytopathol.,* 13, 313, 1975.
285. **Zentmeyer, G. A. and Orr, H. D.,** *Avocado Root Rot,* Leaflet 2440 Division of Agricultural Science, University of California, Berkeley, 1978.
286. **Zimmer, D. E.,** personal communication, 1979.

CONSERVATION TILLAGE IN RELATION TO PLANT DISEASES

M. G. Boosalis, B. Doupnik, and G. N. Odvody

INTRODUCTION

Although conservation tillage is a relatively new concept in farming, it has been adopted in many areas of the world. It is now a well-established cultural practice in the Great Plains states of the U.S. for the production of corn, sorghum, and wheat. There are several reasons for the dramatic acreage increase utilizing conservation tillage. This cultural practice has been shown to reduce wind and water erosion, minimize pollution, increase water uptake in soil, improve soil tilth, save energy by reducing mechanical tillage, and increase crop yields over conventional tillage.[100,101,68]

Conservation tillage may include minimum tillage, no tillage, stubble-mulch, chisel-plant, or till-plant. For a detailed description of these and other methods of tillage, see References 58, 197, and 198.

Because most conservation tillage methods leave high amounts of residues on the surface of the ground, they may profoundly affect plant disease outbreaks. These residues may increase, decrease, or have little effect on the incidence and severity of diseases. It is for this reason that plant pathologists have increased research on conservation tillage in relation to plant diseases.

Fungus and bacterial diseases are the principal diseases associated with conservation- or reduced-tillage operations. One reason for the recent upsurge of some plant diseases relates to the large quantities of surface residues existing over large areas of a state or region. A continual supply of surface residue resulting from reduced-tillage systems, coupled with a monoculture system of cropping in many areas, may maintain levels of inoculum of a pathogen sufficient to initiate severe disease outbreaks.

Surface residue is not the only factor involved in plant disease epidemics. Equally important and interrelated are factors conducive to disease development, including temperature and moisture, nutrition of the host and the pathogen, viability and virulence of the pathogen, variability of the pathogen, and susceptibility or resistance of the host. These factors, in turn, may be affected by other factors such as the kind, rate and time of fertilizer application, genetic uniformity of the crop, high plant populations, narrow row spacing, irrigation, monoculture, pesticides that may directly or indirectly influence the susceptibility of the host, and the virulence, germinability, and survival of the inoculum. Many of these factors, in turn, may relate to conservation tillage.

This chapter presents six ways in which conservation tillage may affect diseases: (1) residues provide an excellent habitat for the survival, growth, and multiplication of both soilborne and foliar pathogens—residues may serve as foci for the effective dissemination of inoculum; (2) there are many factors (nutrition, fertilizer, antagonistic microorganisms, etc.) that may affect the growth and multiplication of pathogens associated with residues; (3) residues may alter the physical and chemical components of the soil, creating an environment hostile or beneficial to the pathogen or host or both; (4) in the inevitable process of residue decomposition, compounds are produced which may adversely or beneficially affect the pathogen, the host, or both; (5) the use of pesticides, particularly herbicides, with conventional-tillage systems also may affect the pathogen, the host, or both; (6) in altering the physical and chemical environment of the soil, surface residues may affect the growth and configuration of roots to make them resistant or vulnerable to soilborne diseases.

The last part of the chapter presents control measures against diseases associated with conservation tillage.

PLANT RESIDUES AS SURVIVAL HABITATS FOR PATHOGENS

It is clear that some plant pathogenic fungi and bacteria overwinter in or on residues from diseased plants of the previous season or on residues colonized by the pathogen after the crop is harvested. Results from intensive research since 1970, made throughout the cornbelt of the U.S., show unequivocably that *Helminthosporium maydis* race T, the new physiologic race incitant of southern corn leaf blight, is dependent largely on surface corn residue for winter survival.[32,61,76,114,161,179,186] Thus, the relationship of minimum or no tillage and the source of initial inoculum is obvious. Abundant corn residue in fields may be an equally important primary source of the pathogen causing northern leaf blight of corn. It was shown that this pathogen, *H. turcicum,* overwinters on corn residues.[24,128] Results from preliminary studies indicate that *H. carbonum,* causing *Helminthosporium* leaf spot, overwinters on stalks under the attached sheaths on the nodes and on the inner surface of the leaf sheath as mycelia and conidia.[24] In contrast to *H. turcicum, H. carbonum* is a good saprophyte whose overwintering mycelia produce abundant, virulent spores on the residue in the spring. Vizvary[188] showed that *Colletotrichum graminicola,* the incitant of anthracnose of corn, overwinters on cornstalk residue. Yellow leaf of corn incited by *Phyllosticta maydis* also is reported to overwinter on surface corn residues.[11] Nyvall and Kommedahl[135,136] reported that *Fusarium monoliforme* Sheld, one of the primary fungus pathogens causing stalk rot of corn, survives as thick-walled mycelia in corn residue.

A leaf spot disease of wheat, called tanspot, caused by *Pyrenophora trichostoma,* was found in North Dakota in 1967.[84] Since then, the pathogen has spread very rapidly throughout the Midwest, extending from Oklahoma to Canada. The establishment of the pathogen over a wide area in Nebraska in 1977 was probably due to favorable weather and, perhaps more importantly, to the increased wheat residue on the ground.[194] Hosford[85] showed that the pathogen overwinters on surface residue from the infected crop. Incorporating infested residue into the soil prevents the formation of the overwintering, perithecial stage of the pathogen and destroys the pathogen.

In 1969, a new bacterial disease of corn was found in Nebraska.[200] This disease, known as ''Goss's bacterial wilt and blight of corn'' and also as ''leaf freckles and wilt of corn'', is caused by the bacterium *Corynebacterium nebraskense.*[163,164] The disease is now present in most counties in Nebraska and in some of the adjacent counties in Kansas, Colorado, South Dakota, and Iowa. The disease has not yet caused great losses to Nebraska's corn crop. Nevertheless, it constitutes a potential threat to the crop because the pathogen is highly infectious and can attack the plant at all stages of development. The bacterium overwinters in corn residue and, particularly, on surface residues.[164]

An epidemic of Holcus leaf spot, caused by *Pseudomonas syringae,* occurred in east-central Nebraska in June of 1966.[196] The sudden appearance of this disease after a severe wind and rainstorm, coupled with the abundance of corn residue on the ground from the previous season's crop, suggested that the inoculum for this epidemic was of local origin. This may be the case since the bacterial pathogen overwinters in corn residue.[196] It is noteworthy that since 1966 severe local outbreaks of Holcus leaf spot occurred in southeast Nebraska and in many areas throughout the Midwest.

Other diseases whose pathogens overwinter in residues are: sooty stripe of sorghum,[137] verticillium wilt of cotton,[26] sharp eyespot of cereals,[149] Cephalosporium stripe,[29,30] Septoria glume blotch of wheat,[159] Septoria leaf blotch of wheat and barley,[115,159] take-all of wheat,[68,118] eyespot of wheat,[46] Rhizoctonia of sugar beets and other crops,[23,138] southern leaf blight of tomato,[199] brown spot of corn,[33] and bacterial diseases such as bacterial blight of soybean,[48] halo blight of beans,[127] bacterial blight of cotton,[27] tomato canker,[18,56] and many others.

The slow rate of decomposition of surface residues may relate to plant diseases in several ways. The persistence of surface residues through the winter, spring, and early summer months extends the longevity of many pathogenic fungus propagules and some pathogenic bacteria. Also, the relatively slow decomposition of surface residues provides readily available nutrients over a long period for germination, growth, reproduction, and colonization by many pathogens. In these ways, slowly decomposing surface residues may be effective pathogen reservoirs leading to disease outbreaks. Conversely, slowly decomposing residues may create a physical environment in soil that is conducive to the development of a vigorous plant which is more resistant to pathogens, particularly soilborne pathogens. Conservation-tillage operations, characterized by slowly decomposing surface residue contrasted with conventional-tillage systems, will generally increase soil moisture content and maintain a lower soil temperature near the ground surface. In areas with low rainfall and high soil temperature during the growing season, such changes in soil moisture and temperature, resulting from conservation tillage, may be beneficial to the plant and detrimental to development of some soilborne pathogens.

The relatively rapid decomposition of residues incorporated into the soil with conventional-tillage systems reduces the inoculum potential of many fungus and bacterial pathogens associated with residues. Additionally, the decomposition of residues in soil may stimulate the growth of microorganisms detrimental to pathogens associated with residue and, thus, reduce the incidence of soilborne disease. Alternatively, residue in soil may favor the development of soilborne diseases by stimulating the growth and reproduction of soil-inhabiting pathogens and by producing materials that are phytotoxic to roots. Thus, one can not generalize and state that conventional tillage, which incorporates residues into the soil, will always reduce the incidence of diseases.

There is every indication that innovative cultural practices involving reduced-tillage systems will continue to be developed and refined and, possibly, replace conventional-tillage practices of crop production in many areas. The rapid changeover to reduced-tillage systems is exemplified by many of the cereal crops. The significance of the relationship between some pathogens and abundant plant residues resulting from reduced-tillage operations signals danger with regard to some diseases. The point is that surface residues provide a most suitable habitat for the establishment, survival, growth, multiplication, and effective dissemination of inoculum of some destructive pathogens. This is particularly the case with those pathogens that cannot live in the soil per se or that cannot saprophytically colonize residues in soil. Thus, the role of residues in the development of disease epidemic must be considered and studies made in formulating control measures against diseases associated with reduced-tillage systems.

FACTORS AFFECTING SURVIVAL AND GROWTH OF PATHOGENS ASSOCIATED WITH RESIDUES

Germination of fungus propagules associated with standing residues or with residues buried, partially buried, or only contacting soil may be stimulated by nutrients or volatiles released during the decomposition of this residue. Such stimulatory agents of germination may also come from other sources, including fertilizers, root exudates, decomposition of dead soil biota, etc.[13,45]

In the absence of a suitable food base to colonize or to parasitize, the germlings (germ tubes or young mycelia developing from germinated propagules) may lyse or perish as a result of starvation or from damage caused by the antagonistic activities of microorganisms residing in residues and soil. Chinn,[41,42] working with *Helminthosporium sativum* on wheat, suggested that stimulating germination of the conidia, followed by lysis of the germ tubes of the pathogen prior to planting of wheat, is a possible means for controlling common root rot of wheat.

Results from laboratory studies showed that ponderosa pine litter extract, added to pine soil intolerant of *Fusarium oxysporum,* stimulated the germination of chlamydospores of that fungus. However, in coniferous soils amended with litter extract, germination generally resulted in the production of abnormal germ tubes that lysed rapidly before the formation of new chlamydospores.[182] The authors[182] postulated that this phenomenon accounts, in part, for the general absence of *Fusarium* spp. in coniferous soils in California.

It cannot be assumed that the germlings are doomed to die soon after they confront an unfavorable environment for their growth. However, while this may be the case for some fungi and, particularly, for those that are not natural inhabitors of the soil, the germlings of other fungi that normally reside in soil and/or residues may continue to function for some time under adverse conditions.

Park[139] observed a premature, "precocious" development of chlamydospores and conidia of *F. roseum* in response to a lytic factor produced by *Bacillus macerans* and *B. pumilus.* According to Park,[139] precocious sporulation frequently was so pronounced that no germ tube developed between the parent and the new conidia. This phenomenon of precocious sporulation was noted when the fungus and bacterial isolates were grown together in small quantities of sterile Czapek's solution or sterile soil solution or when the conidia were placed in sterile filtrates from a sand culture in which both organisms had grown. No reference was made to precocious sporulation in soil.

Conidia of *H. sativum* germinated and subsequently sporulated two ways in soil amended with potato-dextrose broth, dextrose broth, potato broth, cornmeal, or wheatmeal.[22] Precocious type of sporulation was characterized by a single conidium borne at the tips of unbranched germ tubes. With the normal type of sporulation, many conidia were borne on conidiophores developing from branched hyphae. A precocious type of conidial germination of *H. sativum* was noted in field soil adjacent to wheat roots.[24a] These two types of sporulation of the germlings is significant with regard to the survival of the pathogen. The diversity of sporulation assures survival of the pathogen whenever the germlings encounter unfavorable conditions for growth. Under unfavorable conditions, the germlings are capable of sporulating quickly and, apparently, at all stages of development. The young germ tubes also may survive and remain viable during adverse environmental conditions by changing into a thick-walled, irregular-shaped structure. The modified germ tubes will commence to grow and sporulate upon return of a favorable environment. Also, these resting germ tubes of *H. sativum* were shown to be pathogenic on wheat.[24a] The variability of spore germination of *H. sativum* in response to its environment and, particularly, to nutrition may partly explain the inconsistency of crop rotation systems and soil amendments in alleviating common root rot of wheat and other diseases.

Burke's studies[31] with buried conidia of *F. solani* f. *phaseoli,* the cause of root rot of beans, indicated a mechanism whereby the pathogen was depressed more in one soil than in another. The cessation of vegetative growth of the germ tubes in root rot soil apparently increases the inoculum potential in the form of chlamydospores over that occurring in resistant soil. It is noteworthy that lysis of the germ tubes and mycelia occurred within 6 days in both soils so that the pathogen survived only as chlamydospores which were larger and more numerous in the root rot soil. Thus, survival of the pathogen is better in soil conductive to root rot, nothwithstanding the detrimental effect of that soil environment to the growth of germ tubes and mycelia. All of the foregoing phenomena could occur with fungus propagules associated with residues— particularly with residues in contact with soil or with weathered residues containing soil.

Nutrients released from decomposition of residues may play an important role in the multiplication and survival of pathogenic fungi in soil or associated with residues. Nutrients contained in diffusates from crop residues may stimulate fungus-resting strauctures to germinate, to vegetate briefly, and to form new resting structures and, thus, prolong the survival of the pathogen in soil and on residues. Schroth[182] and Toussoun[183] showed that chlamydospores of *F. solani* f. *phaseoli* are stimulated to germinate when residues of different crops are added to the soil. The intensity of stimulation of chlamydospore germination and the subsequent vigorous growth of germ tube hyphae varied with the different residue extracts. In this regard, the most stimulatory extracts were obtained from residues during the early stages of decomposition. In soil experiments, only the chlamydospores close to residue fragments germinated. The results of this study support the contention that substances from decomposing residues in soil or on the soil surface may increase or decrease the population of propagules of pathogens and, thus, are an important factor influencing the survival of pathogens in soil or pathogens associated with residues.

The level of nutrients in soil or residues, and particularly nitrogen, may influence the survival of some soil-inhabiting fungi. The survival of two cereal root rot fungi, *Ophiobolus graminis* (Sacc.) Sacc. *(Gaeumannomyces graminis)* and *F. roseum* f. *cerealis*, is prolonged when the nitrogen supply to infested straw was not limiting. The amount of nitrogen supply to residues did not affect survival of *Cercosporella herpotrichoides* and *Curvularia ramosa*. On the other hand, the saprophytic longevity of *H. sativum* and *Phialophora radicicola* was reduced when the residue had an excess of nitrogen.[50,64,66,67,69,70]

One explanation why nitrogen prolongs the survival of *O. graminis* in straw is that nitrogen is required for the pathogen to grow into fresh cellulose substrates when those in the vicinity of existing hyphae become exhausted.[70] Without these new sources of cellulose, the pathogen would die.[66] In other words, *O. graminis* survives in wheat straw by active saprophytism of that substrate. The amount of nitrogen available to strains of *F. solani* f. sp. *phaseoli* affected the virulence as well as the survival of the pathogen. Strains of the pathogen grown on a medium with high nitrogen survived better in soil and were more virulent in producing root rot of bean than were those grown on a medium with low or no nitrogen.[119] Conversely, the longevity of *H. sativum* in straw in soil is reduced if the soluble N content in soil is high.[71] Garret[71] believes that *H. sativum* is unable to survive under these conditions because it is a poor competitor. Other straw-inhabiting fungi, such as *F. roseum* "Culmorum" and *O. graminis,* are more competitive than *H. sativum* against other soil-inhabiting microorganisms in soil with a high soluble N content and, thus, can survive longer. When soil content of soluble N is limiting, however, the longevity of survival of the straw-inhabiting fungi is proportional to their cellulolytic ability. In this regard, *H. sativum* is superior to *O. graminis* and *F. roseum* "Culmorum".[71] Crop residues may be used to control root rots caused by *Fusarium, Rhizoctonia,* and *Thielaviopsis* of bean. In this case, control is probably achieved by tying up available nitrogen in soil by microorganisms decomposing the residue.[175] The pathogens under these conditions are deprived of the needed food base and cannot attack the host.

The retention of wheat residue infected with *F. roseum* on the soil surface or the addition of fertilizer (5:20:20 NPK at 243 kg/ha) or both favored the survival of this fungus.[191] Survival of *F. roseum* is also better when residue is colonized through parasitism and saprophytic colonization may not be important in survival.[191] Applications of fertilizers caused rapid decomposition of residue and thereby reduced the population of *F. roseum* in the soil; but when the residue was retained, less decomposition and thereby less antagonism occurred, so the population was greater. The removal of residues favors survival of *F. oxysporum* over *F. roseum.*[191]

Garrett[70] believes that successful colonization and survival of fungi associated with straw is dependent largely on their ability to tolerate the fungistatic growth products of microbial origin.

The paper by Baeumer and Bakermans[12] presents an excellent summary of the literature on nutrient concentration and distribution in tilled and untilled soil. The amount of available K and P near the surface of the soil was higher in untilled than tilled soil, but the reverse was found in deeper layers. On the other hand, Mg and Ca content was lower near the surface of the soil and higher in the lower layer of the untilled soil. The pH was generally lower for the untilled soils. Zero-tilled soils generally had a lower concentration of soluble N. Whether these differences in nutrient concentration, distribution, and pH between tilled and untilled soil affect the growth and survival of soil inhabiting microorganisms, including pathogens, and plant development in relation to specific plant diseases is unknown. However, since nutrition and pH are important factors affecting the growth of microorganisms and plants, it may follow that the nutritional and pH differences in tilled and untilled soil undoubtedly have some effects on disease outbreaks.[12]

The ability of resting structures of some soil- and residue-inhabiting fungi to germinate several times enhances their survival value. This means that the resting structure of such fungi may persist and remain viable after they have unsuccessfully germinated several times. The sclerotia of *Rhizoctonia solani* germinated readily in response to various stimuli and were able to regerminate at least 10 times.[149] Furthermore, the sclerotia retained their virulence after each time they germinated. Hsu and Lockwood[88] reported that some of the original germinated conidia of *H. sativum* and *H. victoriae* on nucleopore membranes on soil were able to germinate at least five successive times when placed on an acidified agar medium following complete lysis of germ tubes on soil. This phenomenon has long-term survival value for these two pathogens that reside in soil or residue. The authors[88] stated that this was the first report of the ability of conidia of Helminthosporium to regerminate after lysis of germ tubes.

Like seeds of many weeds, survival of fungus propagules in soil or in residues may be dependent upon the heterogeneity of a given propagule population with regard to germination. That is, not all of the viable propagules of a population will germinate at any given time under a favorable environment. Only a small proportion of the sclerotial cells of the microsclerotia of *Verticillium albo-atrum* germinated in a suitable environment, leaving the remainder to germinate at later dates, thus increasing their survival value.[161]

The survival of pathogens associated with crop residues may be affected by the age of the tissue at the time that the resting propagules are produced and by the variety of the host. The vascular pathogen of melon, *F. oxysporum* f. sp. *melonis,* survived longer in older tissues of resistant melon than in younger tissues of the susceptible host which decomposed more rapidly.[15] Apparently, the decomposition of infested melon residues in the field decreases the population of the pathogen as it cannot survive long apart from the host residue.

Nelson et al.[128] showed that the survival of *H. turcicum* on maize is affected, in part, by the genotypes of the host on which the pathogen overwinters and by the race of the pathogen. In this connection, races of *H. turcicum* differ in their ability to survive on residues from different host genotypes. These findings may explain why the pathogen apparently does not overwinter on corn residue in some of the corn-producing areas. The authors[128] point out that there is no evidence to indicate that better survival abilities of the pathogens are necessarily associated with increased virulence.

Discrepancies of reports on the longevity of survival of specific fungus pathogens associated with plant residues may be due to several factors. The methodology used to determine how long fungi survive on residues is one variable factor affecting the results. A case in point is the conflicting reports on the survival of *O. graminis* in association with wheat straw. These conflicting reports are summarized by MacNish and Dodman[118] who state that, with few exceptions, studies on the survival of *O. graminis* in soil have been made with artificially colonized straw. They go on to say that Garrett[63] established the use of this type of survival unit because of the ease of standardization and convenience of preparation. Although this practice has led to uniformity of medium, Chambers and Flentje[39] showed that the virulence of the isolate used in survival studies has a marked effect on the results. Chambers[38] found that the method of sterilization of straws affects survival of *O. graminis.* Furthermore, there appears to be a discrepancy between the periods of survival of the fungus in artificially colonized straw and naturally infested material. In the longest period of survival in artificially colonized straw reported by Butler,[38] only 3% of straws contained viable *O. graminis* after 52 weeks of burial in soil. However, Fellows[57] found *O. graminis* could survive in naturally infested soil for more than 2 years. The results of survival studies of *O. graminis* in the field, made by MacNish and Dodman[118] indicate differences in climate and soil type may affect the survival of this pathogen.

For a detailed account on the nature of pathogen survival in and on plant residues, the reader is directed to the recent publication by Cook et al.[45]

EFFECT OF REDUCED TILLAGE ON PHYSICAL AND CHEMICAL ENVIRONMENT OF SOIL

Surface residues resulting from reduced-tillage systems have a substantial effect on the physical environment of the soil. Such changes, in turn, may greatly influence the outbreak of diseases. Unfortunately, very little research has been reported on the effect of residues on the physical environment of soil as it relates to plant diseases. The fact that surface residues may profoundly affect soil temperature, moisture, pH, density, and other physical factors should be of great interest to plant pathologists doing research on reduced-tillage systems in relation to plant diseases. A resume of some of the research in the area follows.

There are many excellent reports on the effect of residues on soil moisture. The effect of no-tillage vs. conventional-tillage corn plots on soil moisture was studied by Blevins et al.[21] In this study, the residual sod of bluegrass *(Poa pratensis)* was either plowed down and disked two times for conventional tillage or killed with herbicides for no tillage. All plots were sprayed with appropriate herbicides (paraquat and atrazine) and fertilized with N, P, and K. No-tillage treatments had higher volumetric moisture contents to a depth of 60 cm during most of the growing season. The greatest difference occurred in the upper 0- to 8-cm depth. Beyond a depth of 60 cm, systems of tillage had little influence on soil moisture during the growing season. The decrease in evaporation and the greater ability to store moisture under no tillage produces a greater water reserve. This can often carry the crop through periods of short-term drought and avoid the development of detrimental moisture stresses in the plant which could, in turn, be a factor in reducing losses from certain diseases such as stalk rot.

It has been well documented[49,82,146] that in the early part of the growing season, under conventional tillage, evaporation accounts for a high percentage of water loss. As the aerial portion of the plant grows, a shading effect is produced and evaporation decreases. The increased growth in greater leaf area, and transpiration becomes the major source of water loss. The mulch from the chemically killed bluegrass sod (or other sources of residue systems) significantly reduces evaporation during the early period of the growing season.

Studies by Triplett et al.[184] indicated that corn stover mulch had a beneficial effect on no-tillage corn yield due to increased water infiltration and soil moisture. Jones et al.[100,101] reported increased grain yields and improved water retentions with use of mulches.

McCalla and Army[121] found that runoff was reduced under a mulch. It would appear that more moisture could be conserved during fallow under a mulch than on bare soil. This may not always be true. Initial evaporation rates were less under a mulch than on a bare soil, but evaporation continued for a longer time under the mulch. Gardner[62] concluded that a surface mulch may have little long-range benefit over a bare soil unless lower initial evaporation rate permits greater downward percolation of water. Harris[81] reported that soil moisture at seeding time was significantly greater on wheat stubble-mulch plots than on no-residue plots in 3 of 6 years in a wheat-fallow crop rotation. Additional moisture was conserved under stubble-mulch in years with high rainfall. Nitrate-N in soil was less under stubble-mulch than under no-residue, but amounts were adequate to produce grain yields in line with existing moisture supply. Furthermore, the increased soil water made available by mulching usually overcomes any depressive effects the residue might have and good management, in most cases, can prevent a decreased yield.[83]

Baeumer and Bakermans[12] state that "soil water tensions were affected by the tillage system down to a soil depth of 220 cm". In this connection, zero-tilled soil showed a lower soil water tension compared with tilled soils with the same water content. This means that zero-tilled soils, compared with tilled soils, have a smaller resistance to water uptake by roots and a higher conductivity of soil water. Another desirable characteristic of untilled soil is that it generally has less resistance to rainwater infiltration than tilled soils. Additionally, Baeumer and Bakermans[12] state that "most of the evaporation losses in row-cropped soil should occur before the closing canopy reduces the incident radiation and thereby evaporation on the soil surface. Hence, differential gains in soil moisture content by means of zero-tillage or induced by mulch protection are to be expected mainly during the early growth stages of row crops when evaporation rates of the more or less saturated crops are high".

Allmaras et al.[2] state that "soil temperatures under surface residues and in tiller layers are governed by many factors, some of which are the receipt of solar radiation, emission of long-wave radiation, soil and air heat flow, and evaporation".

Van Wijk et al.[187] conducted uniform mulching experiments in Iowa, Minnesota, and South Carolina. They found that in the northern states, where early-season soil temperatures are near the minimum for corn growth, a decrease in soil temperature from mulching causes a decrease in corn growth. In southern states, where the soil temperature is nearer the optimum, a decrease in soil temperature from mulching has little effect and, where temperatures are above the optimum for corn growth, a decrease may be beneficial. In this connection, Burrows et al.[34] did studies in Iowa to determine whether varying amounts of corn residue mulch affected corn growth and soil temperature. They found that lower soil temperature under mulch was the primary cause in reducing growth of corn where mulched soil is compared with bare soil. As the amount of mulch is increased, corn growth, in terms of plant height and dry matter production, is decreased progressively. As little as 2240 kg/ha of chopped cornstalks reduced the growth of corn early in the season. The smaller the amount of soil cover, the greater the amount of heat retained in the soil during the growing season. In Burrows' experiments,[34] each .91 ton (metric) of mulch applied over the range from 0 to 8960 kg/ha reduced the average soil temperature during May and June at the 10-cm depth by about 0.4°C.

Wheat grown on stubble-mulched soil sometimes yields less than wheat grown on bare soil.[201] At a depth of 2.5 cm, soil with 2240 kg/ha mulch on the surface may be as much as 6°C cooler than bare soil in the spring.[99] This alone could cause yield reductions and, also, affect the development of soilborne diseases.

Smika and Ellis[173] grew hard red winter wheat with and without wheat straw on the soil surface, both in growth chambers with controlled soil temperatures and in the field where soil temperatures were measured at crown depths. Plants grown with soil temperatures below 10°C for 50 days, in growth chambers or in the field, had two or three fewer tillers/plant and 0.5 to 1.2 fewer heads/plant than plants grown with soil temperatures below 10°C for only 15 days. Soil temperature did not affect weight/head or total plant weight. Cooler temperatures reduced the concentrations of only K and Fe in plants grown in chamber at heading and N in plants grown in the field at heading. When mulched and bare soil temperatures were identical, mulch did not reduce numbers of tillers or heads/plant, plant nutrient concentration, total uptake of any nutrient, or final grain yields in plants in the growth chamber or field. Nutrient concentration of plant material was not reduced in plants grown in slowly warmed soil in the field nor by mulch in either the growth chamber or field; thus, reduced development of yield indicators or final grain yields in plants grown in slowly warmed soil could not be attributed to reduced nutrient concentration. Other investigators have reported similar results.[37,98]

Allmaras and Nelson[3] reported the average daily shallow soil temperature from May 27 to June 12 and from June 13 to June 26 under wheat residue was about 2°C lower than for its counterpart bare-plow treatment.

Soil erosion results from the action of wind or water or both. Baeumer and Bakermans[12] state that "accumulation of organic matter near the surface of untilled soil may cause significantly higher stability of soil aggregates. This, as well as the absorption of energy by impact of falling raindrops and the impedence to water flow by surface trash, can increase infiltration rates and reduce runoff, and, hence, soil losses due to water erosion".

Smika et al.[172] found rates of wheat straw mulch greater than 3360 kg/ha were needed to significantly reduce soil NO_3-N during fallow at three Great Plains locations. Smaller amounts of NO_3-N in the soil at the end of fallow with stubble-mulch did not limit grain production. Mulched soils contained more water and less NO_3-N at seeding than bare soils. At all locations, in 7 of 11 years when grain yields were higher from mulched soil than from bare soil, soil NO_3-N was available in sufficient quantity under both treatments so the final grain yields were not limited. In those years when grain yields were reduced by the presence of a straw mulch, no evidence was obtained to indicate this reduction was due to an inadequate supply of N.

Baeumer and Bakermans[12] stated, "With zero tillage, soils are loosened only locally and superficially; yet they have to bear the normal load of traffic in the field. Hence, natural consolidation and mechanical compaction will cause a denser packing of zero-tilled plots." Although untilled soil is generally denser, it also may have more structural homogeneity in space compared to conventionally tilled soils. These workers[12] state that "a relatively higher amount of smaller pores, but greater homogeneity in time as well as space are thus the dominant changes in porosity when a soil remains untilled for a long time."

Barrentine and Waddle[17] conducted research to examine the effect of mulches on cotton plant growth and development and on verticillium wilt *(Verticillium albo-atrum)* incidence in two contrasting cotton cultivars—the more susceptible cultivar "Rex-SL" and the more resistant cultivar "DPL-SL". Soil surface mulches of asphalt, asphalt and white paint, cotton burs, and nonmulches (check) were compared in field experiments involving cotton grown on verticillium wilt-infested soil to determine if resulting changes in soil temperature could influence wilt incidence. Mean soil temperature, dry weight of seedling cotton plants, and seed cotton yield at first harvest were significantly reduced by the cotton bur mulch. The mulch from cotton burs was also associated with a lowered wilt incidence in the more resistant cotton cultivar, DPL-SL. The more wilt-susceptible and earlier-maturing cotton cultivar, REX-SL, gave no mulch response as measured by wilt incidence. Apparently, the lower mean soil temperature resulting from the mulch treatment resulted in a slower growth and maturity in the more resistant cultivar, which was sufficient to depress growth and development of the fungus within the plant.

The unfavorable environmental conditions in the Great Plains involving low precipitation and high wind velocities require fallow practices as part of the cropping systems. During the fallow period, the soil is subject to wind and water erosion unless appropriate measures are taken. A wheat-fallow rotation is commonly used to reduce erosion and conserve soil moisture throughout the Great Plains. A wheat-sorghum-fallow rotation is commonly used in the central and southern Great Plains.[147] In 1958, Phillips[147] initiated research to establish a system to replace part or nearly all of the mechanical cultivation by using herbicide treatments in a wheat-sorghum-fallow rotation. During the past 15 years, Smika and Wicks[173] have refined the wheat-sorghum or corn-fallow rotation system and named it 'ecofallow'. Ecofallow is defined as a system of controlling weeds and conserving soil moisture in a crop rotation with minimum disturbance of crop residues and soil. In essence, ecofallow is a combination of herbicide fallow-minimum- or no-tillage practice using a wheat-sorghum or corn-fallow rotation. Weeds are controlled either with sweep tillage and herbicides (minimum tillage) or herbicides alone (no tillage). The ecofallow system allows the production of two crops in three years in nonirrigated areas with limited precipitation. In contrast, the conventional rotation (normally winter wheat-fallow) produces two crops in four years.

The long-term effect of ecofallow on the incidence of plant diseases is important, because certain plant pathogens can survive and even increase in crop residues left on the soil surface and in volunteer plants and weeds when reduced tillage is practiced. Data obtained in 1972, 1973, and 1974 showed ecofallow greatly decreased the amount of stalk rot of sorghum.[53] Stalk rot averaged 39% in the conventionally tilled plots, whereas the amount of stalk rot in the minimum-and no-tilled plots averaged 23% and 11%, respectively. The average grain yield for sorghum during this period was 3912, 3831, and 2719 kg/ha for minimum tillage, no tillage, and conventional tillage, respectively. This represents an average decrease in stalk rot of 41% and 72% and an average increase in sorghum grain yield of 44% and 41% with minimum tillage and no tillage, respectively, over conventional tillage. These differences were highly significant between the two ecofallow systems and the conventional tillage system. It is noteworthy that foliar diseases of wheat or sorghum have not been observed to be a problem with any of the three tillage systems.

The dramatic effects of ecofallow on stalk rot and yield of grain sorghum may offer an answer to some of the plant disease problems encountered when reduced tillage is practiced. Ecofallow differs from most reduced tillage practices in that one crop is planted into the residue of a different crop rather than into the residue of the same type of crop. Two other possible ways ecofallow reduces the amount of stalk rot and increases grain yield of sorghum is by minimizing environmental stress on the crop—particularly soil moisture and temperature. In this connection, 7.5 to 10.0 additional cm of soil water can be conserved under ecofallow. Ecofallow also significantly reduces daily high, daily fluctuation in soil temperature, and daily average soil temperature as well as "head unit" accumulation.[53a]

The ecofallow system is applicable in commercial plantings of sorghum and corn. About 3240 ha of ecofallow corn and sorghum were planted in southwestern Nebraska in 1974, 4860, 12,150, and 24,300 ha in 1975, 1976, and 1977, respectively. There is high expectation that in the near future most of the corn and sorghum in western Nebraska will be in ecofallow.

In other studies,[3,99] untilled or mulched soils were cooler than tilled soils during the growing season. Soil temperatures on mulched soil were higher than tilled soil during cooling periods in May.[3] Reduced soil temperatures prevailing in the spring and early summer with reduced tillage may retard seed germination and seedling development. This situation may be conducive to the development of damping-off seedling blight and root diseases caused by fungi favored by low soil temperatures. Thus, the importance of special distribution of residues away from the emerging plants is obvious.

RESIDUE TOXINS AND STIMULANTS

Plant residues in one form or another and in all stages of decomposition are always present in field soil. During the decomposition process of these residues, various organic compounds are produced. These organic compounds exhibit a wide range of properties and may be directly or indirectly involved in plant disease development. They may have a toxic or a stimulatory effect on plants and microorganisms[65,120,122,123,189] or no apparent effect.

Results from greenhouse and laboratory studies, made by Patrick and Koch,[143] showed that, in some instances, decomposition of plant residues often resulted in formation of phototoxic substances. Patrick et al.[143] reported that decompcsition plant residue may produce phytotoxic substances under natural conditions in the field. They report such substances are generally formed during the early stages of the decomposition process and are metabolized to nonphytotoxic substances during subsequent decomposition. Depending on the decomposition condition, substances highly toxic, nontoxic, or stimulatory to plants may be formed during the decomposition of plant residues.[143]

Effective concentrations of phytotoxic substances were found in decomposing plant residue fragments and in soil in contact with such fragments. Thus, the extent of root injury and the total effect on plants is contingent on the frequency of chance encounter of the growing root system with plant residue fragments undergoing decomposition and producing toxic substances. For example, phytotoxic injury to roots of lettuce and spinach seedlings was found only on roots in direct contact or in the immediate vicinity of decomposing residue in the soil. An exception to this would be in instances where abnormally large quantities of plant residue were incorporated into the soil.[143] This might be the case with some reduced-tillage systems which incorporate residues into the soil. In these cases, harmful effects might be widespread and severe if conditions for decomposition were favorable for production of phytotoxins.

Guenzi et al.[78] showed that the state of decomposition of residues affects the potency of the phytotoxicity of the substances derived from the residues. In this case, residues of wheat, oat, corn, and sorghum at harvest time contained water-soluble materials toxic to growth of wheat seedlings, and the order of increasing toxicity was wheat, oat, corn, and sorghum residues. Wheat and oat residues were devoid of toxic components after 8 weeks of exposure in the field. Residues of corn and sorghum required about 22 to 28 weeks of decomposition before the water-soluble portion of the residue was relatively nontoxic. Additionally, the germination of wheat seed and the growth of wheat seedlings was affected differently by the toxic substances from straw of nine wheat varieties.

Linderman[113] was able to partially purify and identify the major phytotoxic components of decomposing barley residue. These were benzoic, phenylacetic, 3-phenylpropionic (hydrocinnamic), and 4-phenyl-butyric acids.[113] These compounds did not stimulate germination of chlamydospores of *Thielaviopsis basicola* in nonsterile soil. However, root rot caused by this pathogen was more severe when cotton seedling roots were treated with the barley extract, phenylpropionic acid or benzoic acid. Nearly avirulent clones of *T. basicola* were pathogenic on roots of barley treated with phenylpropionic acid or the barley residue extract. The authors[113] hypothesize that changes in root exudates incited by the toxins may be responsible for the increased pathogenic activities of the fungus.

Additionally, Guenzi and McCalla[77] showed that the adverse effects of residues on crop growth may be associated with the phenolic acid, *p*-coumaric acid, released from crop residue during decomposition.

Carley and Watson[36] reported that aqueous extracts of 23 crop residues retarded root development in clover, radish, and wheat seedlings. Phytotoxins from barley, bean, alsike clover, carrot, green soybean, sagebrush, onion, red clover, pea, sugar beet, and potato residues were selectively toxic to the assay plants.

Boyd and Phillips[25] reported that water extracts of soil containing various crop residues of peanut, soybean, corn, sorghum, and cotton plants inhibited radicle growth of germinating peanut seeds. Extracts from soil with soybean or peanut residue significantly reduced mycelial growth of *Sclerotium rolfsii*. Toxicity was greater with the highest and lowest soil moisture levels, and extracts from soil with mature as well as immature crop residues were toxic. Some extracts reduced germination percentage of *S. rolfsii*.

Kimber[104,105] showed extracts from rotted wheat straw inhibited seedling growth of wheat and reported cold aqueous extracts of several grasses and legumes that had been rotted for periods up to 21 days inhibited the growth of wheat seedlings grown under aseptic conditions. The degree of inhibition varied from one species to another and with the rotting period. Green straws produced a higher level of toxicity than the fully matured straws. Extracts from unweathered and weathered wheat straw did not show strong phytotoxic activity against wheat seedling roots. Behmer and McCalla[20] stated that stubble-mulch practices have been widely used in the control of soil erosion by wind and water. Although this method of leaving crop residue on the surface has been effective in combating erosion, depressed plant growth and reduced grain yields often occurred in wet years. Similar growth problems were noted during unseasonably cool, wet springs in field experiments at Lincoln, Neb.

Because increased numbers of certain microorganisms occur immediately beneath a decomposing surface mulch, the production of phytotoxins under these conditions may well have been associated with the adverse effect on plant growth. In fact, a strain of *Penicillium urticae* isolated from wheat stubble-mulch was found to produce a phytotoxic substance in culture media.[134] Behmer and McCalla[20] reported that wheat seedlings grown in soil treated with alfalfa residue, inoculated with *P. urticae* and incubated for 3 to 4 weeks prior to planting developed a root curling rot evident with other residues (corn, alfalfa). Although residues alone effected a reduction in seedling growth, *P. urticae* reduced growth only in the presence of residues. It is suggested that crop residues in soil, alone or in combination with *P. urticae*, may cause depressive effects on the growth of wheat seedlings. The action of certain soil-inhabiting fungi on crop residues may induce growth retardation in plants whose root systems are exposed to degradation products of crop residues. However, these relationships have not been demonstrated in the field.

There is evidence to indicate phytotoxins are involved in the etiology of root diseases.[44,74,143,180] In addition to their direct toxic action, the phytotoxins apparently predispose roots to invasion by various low-grade pathogens.[44,143,185] Patrick and Koch[144] found that tobacco plants treated with phytotoxic solutions from decomposing rye and timothy residues increased their susceptibility to black root rot caused by *T. basicola.* There were no significant differences among 16 tobacco varieties, and disease was equally severe on roots of resistant and susceptible varieties after the toxin treatment. The effect of the toxic compounds from residues appears to be largely on the host rather than on the fungus. The results of this study indicate that phytotoxins may be important host-conditioning factors in the development of root rots.

Toussoun and Patrick[181] found that the severity of bean root diseases caused by *F. solani* f. *phaseoli, T. basicola,* and *R. solani* increased when roots were exposed to toxic, water-soluble substances obtained from decomposing barley, rye, wheat, timothy, broccoli, and broad bean residues. *R. solani* invaded only roots exposed to the toxic extract. Apparently, the main role of the extract was to predispose roots to fungus invasion. The authors[181] suggested that this is the major role of toxic residue compounds in root rot etiology in nature.

Linderman and Toussoun[113] stated that ether extracts of barley residues did not stimulate germination of chlamydospores of *T. basicola* in nonsterile soil. However, the toxins from the barley residue increased the severity of root rot caused by this pathogen when the seedling roots of cotton were treated with chlamydospore-amended nonsterile soil. The authors[113] believe that the increased activity of the fungus on toxin-treated roots may be due to changes in root exudation caused by the toxin. The roots of cotton seedlings treated with extracts from barley residue were more susceptible to a nearly avirulent clone of *T. basicola.* This suggests a plant could be rendered susceptible to an innocuous microorganism by having the roots exposed to phytotoxins produced during decomposition of barley residue.

Black root rot of tobacco[142] caused by *T. basicola,* was increased in the presence of phytotoxic residue extracts. Disease was increased on certain roots treated with crude extract from barley residue or hydrocinnamic acid.[111] The rate of proliferation of the fungus on or in the root tissue treated with residue extracts or hydrocinnamic acid was much greater than on water-treated roots. The residue decomposition product may have affected the plants' resistance mechanism(s). Hydrocinnamic acid could render resistant varieties susceptible to *T. basicola.* The increase in, or occurrence of, root lesions was due to an increased number of chlamydospores germinating to infect the roots as well as to a reduced hypersensitive reaction.[112]

Under certain conditions of plant residue decomposition, substances may be released into the soil to stimulate germination and growth of fungus pathogens in the soil. Residue extracts stimulated germination of chlamydospores of *F. solani* f. sp. *phaseoli* in soil and bean root rot caused by this pathogen increased following treatment with such extracts.[183]

Harricks[80] reported common root rot of wheat, barley, and oat, though slight, was less severe in the heavy than light rape residue areas. Heavy rape residue reduced the growth of oats more than of either wheat or barley. Loss in plant weight of the three cereals was caused more by stunting of the plants than by reduction in their number. The stunting may have been caused by the lower soil temperature in the area of the heavy rape residue or by a toxin from the rape residue or one formed during its decomposition, and not from pathogenic action or a nutritional deficiency.

Toussoun and Patrick[181] make an excellent point in stating that the differences in behavior of soilborne pathogens in relation to plant residues is contingent on the prevailing ecological environmental factors. Since toxic products were most often obtained from crop residues decomposing under cold, wet, or anaerobic conditions, it follows that conditions not conducive to the formation of phytotoxic substances may prevent the disease, whereas decomposition conditions resulting in formation of phytotoxic compounds would tend to increase root rot severity.

Lewis and Papavizas[109] reported that decomposing mature residues of rye, oat, sorghum, buckwheat, corn, barley, timothy, and soybean were more inhibitory than corresponding immature residues to chlamydospore germination of *F. solani* f. sp. *phaseoli* in vitro and in soil. There was also a reduction in the incidence of Fusarium root rot of bean in soil amended with these mature residues. Failure of chlamydospore germination in vitro and in soil was ascribed to production of a fungitoxicant from the decomposing residue as well as to fungistasis due to nutrient deficiency. In this connection, crop residues may influence soil fungistasis and either decrease or increase the severity of disease.[1]

The absence of Fusarium from forest soils is due, in part, to a detrimental type of chlamydospore germination of *F. oxysporum* caused by substances produced from the decomposing needle duff layer.[140,141,151]

It appears that these fungus stimulants from decomposing residues may be widespread, and they may activate resistant fungus structures to germinate in an otherwise unfavorable environment, which ultimately could result in the elimination of that propagule from the soil.

The distance and rate of diffusion of plant residue decomposition products into the soil solution may limit their effectiveness in activating pathogenic fungus propagules. The effective distance of diffusion of residue byproducts in their gaseous state, however, may be greater. Sclerotia of *S. rolfsii* placed on moist field soil were stimulated to germinate and grow when exposed to volatiles from alfalfa.[74] Exposing soil to volatiles for one week also increased its antagonistic potential to sclerotia placed on the soil and given a secondary exposure to the distilled vapor.

EFFECT OF PESTICIDES ON PLANT DISEASES

The success of reduced-tillage systems is dependent upon the use of pesticides for the control of weeds, insects, and fungi. Some herbicides and insecticides may increase diseases by stimulating growth of the pathogens, stimulating growth of antagonistic microorganisms, increasing exudation from the host, and by increasing the susceptibility of the host and alternate host.

The recent comprehensive and scholarly review articles by Altman and Campbell[7] and by Katan and Eshel[102] on the effect of herbicides on plant diseases detail many aspects of the subject.

Researchers report that some herbicides may increase or decrease the incidence and severity of plant diseases. Huber et al.[92,93] noted that fall application of the herbicide Diuron (3-(3,4-dichlorophenyl)-1-dimethylurea) to control weeds in winter wheat reduced the incidence and severity of root rots, eyespot and sharp eyespot. Plants from plots treated with Diuron had more tillers/plant than plants from nontreated plots when root rot was a problem. The incidence of seedling blights of barley caused by *H. sativum,* commonly associated with surface crop residue,[108] was reduced substantially with the applications of the insecticides Aldrin (1,2,3,4,10,10-hexachloro-1,4,4a,5,8,8a-hexahydro-*exo* -1,4-endo-5,8-dimethanonaphthalene), Endrin (hexachloroepoxyoctahydro-*endo,endo*-dimethanonaphthalene), or Chlordane (1,2,4,5,6,7,8,8-octochlor-2,3,3a,4,7,7a-hexahydro-4,7-methonoindane), and with soil applications of the herbicides NAP (*N*-1-naphthylphthalamic acid), DNBP (2-*sec*-Butyl-4,6-dinitrophenol), or Dalapon (2-2-dichloropropionic acid).[152] Sewell and Wilson[167] studied the effect of no tillage and normal tillage on the incidence of *Verticillium* wilt of hop. No tillage resulted in an average annual wilt of 18% compared with 25% wilt with normal tillage during 3 years when wilt incidence was high. However, when wilt incidence was low, there was slightly more wilt with no tillage. These workers believe the reduction of wilt under no tillage is due mainly to the eradication of weed host plants by the herbicide simazine (2-chloro-4,6-*bis* [ethylamino]-*s*-triazine). The effect of 2,4-dichlorophenoxyacetic acid, ethyl ester (2,4-D ester) 2,4-dichlorophenoxyacetic acid, alkanolamine (2,4-D amine), 2-methyl, 4-chlorophenoxyacetic acid, butyl ester (MCPA ester), 2-methyl, 4-chlorophenoxyacetic acid, dimethylamine (MCPA amine), and 4-chloro-2-butynyl *N*-(3-chlorophenyl) carbamate (barban) on sporulation of *Rhynchosporium secalis,* the cause of barley scald, was reported by Skoropad and Kao.[170] The results of their studies in vitro and in vivo showed that these five widely used herbicides partly or completely inhibited sporulation of *R. secalis.* It is noteworthy that the herbicides changed the form of the fungus growth in vivo from exclusively conidial-type to a mycelial-type. Moreover, the converted mycelial fungus cultures did not revert to the conidial-type after 14 transfers over a period of 140 days. Suppression of sporulation in barley leaf lesions by the five herbicides, at concentrations lower than used in the field, could reduce or prevent new infections.

Kaufman[103] reported s-triazine and phenylurea herbicides affected populations of *Fusarium* spp. in corn and soybean-cropped soil. The effect varied with the herbicide and cropping system. Diphenamid (*N,N*-dimethyl-2,2-diphenylacetamide) decreased the incidence of preemergence damping-off and increased postemergence damping-off caused by *Pythium aphanidermatum* and *R. solani* on peas. Growth of the two pathogens was decreased in synthetic media containing the herbicide. Application of trifluralin (α,α,α-trifluoro-2,6-dinitro-*N,N*-dipropyl-*p*-toluidine) at 0.56 kg/ha, followed by propachlor at 5.6 kg/ha, reduced root rot of peas caused by *Aphanomyces euteiches.*[75]

Warren et al.[192] found that an actinomycete, antagonistic to *Sclerotium rolfsii* and *R. solani,* became dominant in unsterilized soil sprayed with 2,4-D. Naito and Kohima[126] discovered a species of *Gloeosporium,* grown on media containing either 2-chloro-phenoxyacetic acid or 2-methyl-phenoxyacetic acid, produced an antibiotic substance only in the absence of these chemicals. Paraquat (1,1′-dimethyl-4,4′ bipyridinium salt), at concentrations from 12.5 to 75 μg/mℓ of modified Czapek's solution medium, reduced the mycelial growth of *S. rolfsii.*[155] Results from studies using sterilized and nonsterilized field soil suggested the herbicide prometryn [2,4-*bis*(isopropylamino)-6-methylthio-*s*-triazine] (Caparol) suppressed the growth of *F. oxysporum* f. sp. *vasinfectum* at low concentrations of the herbicide and stimulated the growth of the fungus at the higher concentrations. Prometryn apparently inhibited spore germination in soil by acting directly on the spore rather than by changing a fungistatic factor of the soil.[43]

Soil application of the herbicides trifluralin as dinitramine (N^4,N^4-diethyl-α,α,α-trifluoro-3-5-dinitrotoluene,2,4-diamine) reduced root rot of pea caused by *Aphanomyces euteiches.*[75]

Deep and Young[51] showed the incidence of crown gall caused by *Agrobacterium tumefaciens* on woody rosaceous hosts increased after preplanting treatments with the fungicide dichlone (2,3-dichloro-1,4-naphthoquinone). The increased incidence by this treatment was not due to a direct effect of the fungicide on the pathogen or to an influence on host predisposition. In this case, dichlone suppressed antagonistic microorganisms of *A. tumefaciens* and, thus, increased the effectiveness of the inoculum, leading to a higher amount of crown gall.

Ko and Farley[106] reported that pentachloronitrobenzene (PCNB) is gradually converted to pentachloroaniline (PCA) in moist soil. PCNB has a broader antimicrobial spectrum and is inhibitory to soil microorganisms at lower concentrations than is PCA. PCNB is more inhibitory to *Rhizoctonia solani* than PCA. Because PCA is very stable in soil and is also inhibitory to *R. solani,* conversion of PCNB to PCA in soil may be a factor in the long-term control of *R. solani* in field soils treated with PCNB.

The susceptibility of wheat pathogenic fungi in straw to soil fumigants was investigated by Ebbels.[54] The addition of formalin to soil in the field decreased the survival of *Cercosporella herpotrichoides* but had no effect on survival of *Ophiobolus graminis* or *F. roseum*. The fumigants D-D (Nemafene), chlorpicrin, and dazomet (tetra-hydro-3,5-dimethyl-2*H*-thiadiazine-2-thione) had no effect on survival of *C. herpotrichoides* in straw. The survival of *F. roseum* was not affected by D-D but reduced by formalin and dazomet soil treatments. In 150-mm soil columns, D-D (Nemafene) and cloropicrin were more effective in killing fungus inoculum in straw below the point of injection than above it. None of the fumigants tested eradicated *C. herpotrichoides, F. roseum,* or *O. graminis* in straws near the soil surface.

Richardson[153] showed that early blight of tomato, caused by *Alternaria solani,* increased with the application of the herbicide 2,4-D and the insecticide isodrin. On the other hand, the incidence of this disease decreased with the application of the insecticides endrin, dieldrin (hexachloro-epoxy-octahydro-*endo,exo*-dimethanonaphthalene), demeton (systox), and aldrin, and the herbicides MH (1,2-dihydropyridazine-3,6-dione), NPA, IPC (isopropyl carbanilate), and dalapon (2,2-dichloropropionic acid). Lindane (1,2,3,4,5,6-hexachlorocyclohexane), isodrin, and dalapon increased the severity of *Fusarium* wilt caused by *F. oxysporum* f. *lycopersici.* The application of herbicide in greenhouse and field studies showed an increase in the incidence of both foliar and soilborne diseases. Several investigators reported on the interaction of pathogens and herbicides to the detriment of plants.[4,102,148] Results from greenhouse and field experiments indicate *R. solani* is subject to less injury from herbicides than soil saprophytes antagonistic to *R. solani.*[5,6,130]

Cycloate at 3 ppm increased the severity of *R. solani* disease in sugarbeet.[130] Both tillam (S-propyl butylethylthiocarbamate) and pyramin [5-amino-4-chloro-2-phenyl-3-(2*H*) pyridazinone] increased damping-off caused by *R. solani* in steamed soil over that caused by the pathogen alone.[4] Altman and Ross[9] reported that the herbicides pebulate (S-propyl butylethylthiocarbamate) and pyrazone [5-amino-4-chloro-2-phenyl-3(2*H*)-pyridazinone] increased the susceptibility of sugar beets to *R. solani.* These researchers hypothesized that two herbicides affected the susceptibility of the host by maintaining it in a seedling stage for a prolonged period (4 or 5 weeks) after emergence. The incidence and severity of *R. solani* infection on snapbeans was increased by the application of the herbicides trifluralin or dinoseb (2,4-dinitro-6-*sec*-butylphenyl). The two herbicides apparently altered the structural and biochemical defenses of the host, resulting in increased disease. Increasing levels of atrazine progressively increased the susceptibility of maize hybrids to maize dwarf mosaic virus.[117] The application of the herbicide trifluralin to soil increased the incidence of disease on cotton caused by *R. solani* both in sterilized and nonsterilized soil.[130] Trifluralin in soil increased both the susceptibility of the host and the activity of the fungus.

Results from field studies indicated the herbicides pebulate and PCA [5-amino-4-chloro-2-phenyl-3-(2*H*)-pyridazinone], used for sugar beet weed control, increased the incidence of *R. solani* on sugar beets.[8,9] It was shown that greater amounts of glucose and certain ions leaked out of the plant tissues in herbicide-treated soil than from plants in untreated soil.[8,9] Thus, the increase in available nutrients for the pathogen may account for the herbicide-induced increase in *R. solani* of sugar beets. Pinckard and Standifer[148] and Standifer et al.[176] reported that trifluralin increased the incidence of seedling blight of cotton caused by *R. solani*. Trifluralin incorporated in potato-dextrose agar suppressed the growth of *R. solani* but had no effect on the virulence of the pathogen.[130] On the other hand, the virulence of *H. sativum* to wheat was increased when the pathogen was grown on potato dextrose agar with 5,000 ppm 2,4-D.[87]

Percich and Lockwood[145] studied the effect of atrazine (2-chloro-4-ethylamine-iso-propylamino-S-triazine) in soil artificially infested with *F. solani* f. sp. *pisi*, the cause of pea root rot, or with *F. roseum* f. sp. *cerealis* 'Culmorum', the cause of seedling blight of corn. They reported that the incidence of root rot of peas and seedling blight of corn was increased two- and threefold in soil containing atrazine at 30 μg/g, compared with soil infested with the pathogens but devoid of atrazine. Amiben (3-amino-2,5,dichlorobenzoic acid) and trifluralin applied to soybeans retarded seedling emergence and thus increased susceptibility to damping-off fungi.[55] Nilsson[133] stated that mesoprop increased the incidence of take-all of spring wheat caused by *O. graminis*. The incidence of this disease was increased on winter wheat by the application of the herbicide oxitril at 50 ℓ/ha.[133] Root rot of wheat and corn seedlings was increased in soil treated with picloram.[166] Atrazine increased the susceptibility of corn hybrids to maize dwarf mosaic virus.[116,117] Results from field trials[132] showed that the herbicide oxitril 4 (a.i. 300 g/kg dichloropropt + 130 g/kg of MCPA + 58 g/kg of oxynil + 38 g/kg of bromoxynil) significantly increased the amount of take-all and eyespot disease of wheat. The commercial compound Basonor (Bas-240 H; a.i. 250 g/kg of brompyrazon + 250 g/kg of tricuron) slightly increased the incidence of take-all.[132] The herbicide mesoprop (2-[4-chloro-*o* -tolyl] oxy) propionic acid) (a.i. 640 g mesoprop/liter) tended to increase incidence of take-all on spring wheat. Malformed structures produced on wheat roots by mesoprop were highly susceptible to *Gaeumannomyces graminis*.

The effects of combinations of pesticides on soil microorganisms has been studied. Bates[19] showed that fungus populations in thiram [bis(dimethylthiocarbamoyl) disulfide] atrazine combinations at field rates were about the same as in soil treated with atrazine. Bates suggested that atrazine may stimulate a specific group of soil fungi tolerant to thiram and showed that thiram-fluometuron combinations stimulated fungus populations more than did the individual pesticides. More recently, Houseworth and Tweedy[86] evaluated the effects of atrazine in combination with Captan or thiram upon the population of fungi and bacteria in soil when atrazine-resistant or -susceptible plants were grown. Results from the fungicide studies showed Captan [cis-*N*-(trichloromethyl)thio 4-cyclohexene-1-2-dicarboximide] and thiram in soil initially reduced the fungus populations, but by 108 days the populations returned to levels found in untreated soils. Bacterial populations varied inversely with the fungus populations in soils treated with either of the two fungicides. In soil planted with corn, the herbicide atrazine had no direct effect upon the populations of microorganisms, but an indirect effect by killing plants such as bean and wheat that are susceptible to atrazine. The increase in microbial populations in the atrazine-treated soil was due to the decay of dead wheat and bean plants. Atrazine in combination with captan or thiram had no synergistic or antagonistic effects upon the microbial populations.

Improved emergence of pea seedlings was reported by Richardson[154] to result from various insecticide-fungicide combination seed treatments as compared with single treatments with either insecticides or fungicides. In this study, the pea seeds were planted in sterilized soil infested with *Pythium ultimum* or *R. solani*. The pea seeds were treated singly with commercial preparations of aldrin, dieldrin, lindane, thiram, captan, or chloranil (2,3,5,6-tetrachloro-1,4-benzoquinone) or with an insecticide-fungicide combination of these materials. Richardson[154] stated that "increased pre-emergence killing following insecticide treatments without a fungicide has been reported by various workers but the cause of the reaction has not been definitely established. It is also possible that insecticides have a physiological effect on the host, predisposing it to infection. Another possibility is that the insecticide neutralizes or prevents the formation of substances by the host which would normally inhibit the pathogen."

Brodie[28] believes that the use of mixtures of selective pesticides to achieve multiple pest control on a field-scale appears most promising. When exact ratios of toxicants (pesticides) for effective control are determined, Brodie[28] believes that a "super" granule (or other suitable formulations) with a nematicide, herbicide, and fungicide might be formulated and successfully applied for use in the production of many crops. Each formulation of this type necessarily would be tailored for specific crops and pest control used under specific environmental conditions at a given location. Certainly, such innovative use of pesticides would be applicable to reduced-tillage systems of crop production.

Further research is needed to ascertain whether the current herbicides and insecticides used in reduced-tillage systems and the new pesticides, particularly herbicides, affect plant diseases. Such information is necessary to select the appropriate herbicide or insecticide which may simultaneously control several different pests such as fungi and weeds, fungi and insects, etc.

ROOT DEVELOPMENT IN RESPONSE TO RESIDUE

It is well documented that surface residues may change the physical and chemical environment of the soil. And if the changes in soil temperature, moisture, pH, aeration, structure, nutrition, etc. are of sufficient magnitude, they may, in turn, affect the growth and pattern or configuration of roots.[3,16,40,129,178,193] Unfortunately, there is a dearth of research on the development of crop roots relative to conservation tillage systems. Moreover, in the few cases where reduced tillage was shown to affect root development, it is difficult to ascribe significance to the results in relation to crop yield, root disease development, drought resistance, etc. Compounding this difficulty is the fact that we cannot define the optimal root system relative to specific cultural practices or to specific soil environmental conditions.

Root density and weight at different growth stages of the plant and at different soil layers were lower on zero-tilled soil[12] Also, in the early vegetative growth phases, the roots were shallower in the zero-tilled soil.

Choudhary and Prehar[40] studied rooting patterns, root distribution, and growth of corn under four cultural practices: (1) control (no mulch, no cultivation, no compaction); (2) post-plant cultivation (5 cm deep) by hand hoe; (3) a 2-cm mat of wheat straw mulch (approximately 6 metric tons/ha); and (4) inter-row compaction by compacting the central 40-cm inter-row space. Straw mulch and cultivation treatments increased root growth in the upper 15 cm of soil and increased the lateral spread of roots. Root development was less in the mulch plot than in the control treatment. A more symmetrical root distribution was noted in the cultivated plots. Inter-row compaction arrested lateral spread of roots in the surface layers and caused downward growth of roots. Five centimeters below the soil surface, straw mulch reduced maximum soil temperature by 2.6°C in 1970 and 2.3°C to 6.3°C in 1971. A higher moisture content in the 0- to 7.5-cm soil depth under mulch, despite more roots in this layer than the control, indicated that evaporation loss is less in mulch. Greater lateral spread and enhanced root growth in the surface layers of mulched plots were associated with reduced soil temperature and moisture losses and resulted in better plant growth and higher yield.

The effect of tillage on root growth was studied also by Allmaras and Nelson.[3] In their 1973 comprehensive review article, Allmaras and Nelson[2] stated, "Tillage and residue management have a predictable effect on soil temperature and moisture at the time of root development in annual crop species. Analysis of root growth response to these environmental factors in row crops requires, first, that the soil environmental differences between the row and interrow be recognized. This should also be the case for close-seeded crops such as wheat or peas. Moreover, this environmental difference is time-dependent with respect to crop development. Secondly, not all parts of the underground portion of plants require the same environmental conditions for optimum growth and development." Adventitious root initiation of corn was sensitive to temperatures at the upper 10 cm of soil under the row.[2] Horizontal root development of corn was stimulated by the high soil water content resulting from interrow surface residue placement in a moderate moisture year, whereas in a wet and cold year, the interrow surface residue suppressed root extension. These results indicated adventitious root initiation of corn was favored by high temperature, while root elongation and branching have a lower optimum temperature.[2] Other factors conducive to adventitious root initiation and root elongation are presented by Allmaras et al.[2]

The significance of the apparently greater lateral spread and enhanced root growth in the surface layer of soil containing surface residue in relation to soilborne diseases has not been determined. It may be conjectured that there is greater opportunity for pathogens to infect roots near the surface layer of the soil because higher populations of microorganisms, including plant pathogens, exist in the top 4 to 5 cm of soil. On the other hand, the more vigorous root system produced near the surface layer may be highly resistant to soilborne pathogens. More robust, resistant roots may result from the improved chemical and physical environment of the soil or from the stimulation of growth of microorganisms in the rhizosphere that are antagonistic to the pathogens or that enhance nutrient availability to the roots.

Ferguson and Boatwright[59] indicated that the surface residue may influence the location of the crown node of winter wheat in a greenhouse study at 19°C. There are at least two reasons why the location of the crown node is important. First, winter survival of winter wheat relates to the location of the crown node. Lawrence[107] and Dobrenz[52] believe that winter injury is associated with a high crown position (crown close to the soil surface) and, secondly, adventitious roots develop from the crown node. In the Great Plains, where winter wheat is typically planted 5.0 cm deep to reach moist soil in the fall, the crown node and the adventitious roots form near the surface of the soil during dry periods. This dry soil, coupled with low temperature, may occur frequently in the fall to prevent development of adventitious roots.[59] When adventitious roots are initiated at the crown nodes in a dry soil, they generally do not elongate and develop unless the water content of the soil is at or above 15-bar value.[59] Weaver[198] noted a root system that is well developed at the start of winter can better withstand stress conditions of alternate thawing and freezing and heaving of the soil. Thus, winter survival of the crop is enhanced by the formation of adventitious roots in the fall.

Ferguson and Boatwright[59] noted that as the surface straw increased, the crown node formed farther from the seed—closer to the soil surface—and, in some cases, above the soil surface. They reported a positive variety-straw rate interaction. Winterhardy varieties formed their crown nodes deeper in the soil compared with non-winterhardy varieties. As the soil temperature decreased or the light available to the seedling decreased, the crown node formed closer to the soil surface. There was a positive variety-temperature-light intensity interaction, and each variety reacted differently to a specific set of conditions.

Selection of winter wheat varieties well adapted to surface residue, coupled with the increased soil moisture content resulting from surface residue, may overcome the undesirable effects of surface residues which lead to high soil temperature and reduced light intensity in the fall. In this way, the crown node may be produced well below the surface of the soil and reduce root and crown rot and the accompanying winterkilling of wheat. In this connection, Nilsson[131] noted there is a wheat varietal difference in root development and summarized the literature on the subject as well as the disease aspects relating to crown and root development.

CONTROL

Because of the great benefits derived from reduced tillage with respect to wind erosion, reduced soil temperature, soil moisture conservation, absence of root damage due to cultivation, better pattern of root distribution, reduced soil compaction, and reduced cost of crop production, it is not practical to advocate the elimination of this cultural practice to eliminate surface residues for the control of plant diseases. Furthermore, some diseases such as northern leaf blight, yellow leaf blight, and southern leaf blight of corn, Goss's bacterial wilt and blight of corn, and some other diseases associated with surface residues are presently effectively controlled by resistant varieties, and with further research, other diseases associated with reduced-tillage systems may be controlled by resistant or tolerant varieties.

Control of plant diseases associated with reduced-tillage systems may be achieved by means other than resistant varieties. One way is to develop crop varieties better adapted to reduced-tillage cultural practices. In this connection, it may be feasible to develop crop varieties with marked changes in plant architecture with regard to size of the plant and foliage arrangement. Such structural plants could render the microenvironment unfavorable for the development of diseases. For example, the thicker upright canopy of leaves produced by dwarf upright varieties or varieties with an open canopy of leaves[165] may increase the amount of light intercepted by the plants, reduce the relative humidity in the immediate vicinity of leaves, lower the temperature at the soil surface, and reduce the total leaf area exposed to rain splashing off residues. All of these factors associated with a change in the architecture of the plant may adversely affect the development and spread of many pathogens associated with residues.

Ferguson and Boatwright[59] noted surface residue caused crown nodes of wheat to develop near the surface of the soil. This response of wheat to surface residues is undesirable as crown nodes near the surface of the soil are subject to winterkilling. They suggested it might be possible to overcome this problem by developing or selecting wheat cultivars that produce crown nodes well below the soil surface containing large amounts of surface residues.

Research on the biochemistry of host-parasite relations undoubtedly will lead to the development of more effective chemotherapeutic agents for the control of diseases associated with reduced tillage. A greater number of candidate chemicals will be available for testing against many destructive and refractory plant diseases. Mitchell's[124] intriguing pioneer studies showed that the excretion from roots of chemical compounds applied to the foliage bode well for the future development of chemotherapeutic agents against specific root-attacking pathogens.

The mode of translation of herbicides and other chemicals into and out of roots and the significance of this in relation to soil microorganisms is the subject of a review article entitled "Factors Affecting Root Exudation," authored by Hale, Fay, and Shay.[79] The highlights of this review article are as follows: herbicides and other growth regulators differ markedly in the rate of uptake and subsequent translocation both into and out of roots, as shown by autoradiography, isotope counting, and bioassay.[47]

Some compounds are known to be translocated not only from the shoots into roots, but secreted from roots into the culture medium. Accumulation and, perhaps, active secretion of certain of these substances are suspected. The loss from roots of trans-cinnamic acid is a classical example of the release of a naturally occurring growth regulator by plant roots. Subsequently, many other exogenous growth regulators were reported to be exuded from roots.[60,96,97,110,125]

Sullia[177] applied 100 ppm each of gibberellic acid and indole-3-acetic acid to foliage of sickle pod and rattle-box. There was a quantitative increase in the number of fungi but no qualitative change in the species present in the rhizosphere. Balasubramanian and Rangaswani[14] reported that foliar applications of nitrogen in the form of $NaNO_3$ (0.1%) increased the concentrations of amino acids exuded by roots of sorghum, sun-hemp, ragi, and tomato, but applications of P as Na_2HPO_4 reduced the concentration. Conversely, nitrogen applications reduced total sugars exuded, while P applications increased the sugar concentrations. Root exudates may serve as substrates for micro-organisms which could liberate nutrients from soil. Also, root exudates may play a direct role in liberating nutrients from soil.

Rovira[158] points out that changes in rhizosphere populations of microorganisms included by foliar application of nutrients and other substances may be of great significance in controlling plant diseases. This also has practical implications with regard to control of soilborne diseases associated with reduced-tillage systems.

Light and temperature, as they affect photosynthesis and translocation, may have a profound effect on the amounts and kinds of root exudate. The pattern of amino acids was shown to vary with temperature, light intensity, plant age, and species.[157] Also, soil water stress may affect the kinds of organic substances exuded from roots. Changes in kinds and amounts of substances exuded from roots due to the aforementioned factors certainly relate to conservation tillage as it affects the chemical environment of the soil which, in turn, affects plant diseases.

Soilborne fungus diseases associated with conservation tillage may be controlled by the kind, the amount, and the time of fertilizer application. Research done by Huber[89] and others[95] on the relationship of soil fertilization and soilborne diseases may be applicable to conservation tillage.

Spring application of ammonium sulfate controlled take-all and increased the yield of grain straw. Fall application of N for springseeded wheat increased take-all, and reduced yields of grain and straw.[89] Nitrogen application to low-fertility soil increased root growth and reduced take-all.[72,73,90,95] Smiley and Cook[174] concluded from their greenhouse studies that the lowered rhizosphere pH resulting from ammoniacal N application arrested the growth of *O. graminis*. This mechanism probably does not account for the disease control observed in fields in this study because of the relatively low rates of N applied (broadcast) to the alkaline soil (pH 8.2).

Huber[89] concluded from his studies on the effect of spring vs. fall fertilization on take-all of spring wheat in the irrigated wheat areas that soil moisture in the fall is usually adequate and nutrification is not arrested for a long time by low temperature. His[89] data indicated that where take-all is a potential hazard, fall application of N before planting spring wheat may increase the disease.

The source of nitrogen may affect the susceptibility of winter wheat to root rot.[91] Pittman and Horricks[150] studied the effect of NH_4NO_3 and $CaH_4(PO_4)_2$ fertilizers and winter wheat residue on the growth, yield, and common root rot of barley caused by *H. sativum*. The amount of residue alone did not greatly affect the severity of common root rot in two of three years. However, the severity of the disease increased as the rate of applied nitrogen was increased. The authors pointed out that plants heavily fertilized with N often mature slowly and such plants are more vulnerable to root rot.

Contrary to previous workers[91] they[150] reported an increase in both root rot severity and yield of barley with increased N application, especially on residue-free soil. Applied phosphorus generally had no apparent effect on root rot severity and yield, although the average number of fungus pathogens, particularly *Fusarium* spp. associated with the crowns of plants grown under heavy residue and high rates of P alone, was lower than that from other treatments.

Warren and Kommedahl[191] concluded from their studies on the relationship of *Fusarium* spp., fertilizer and crop residue that, because soil populations of *Fusarium* spp. remain fairly stable, the mechanism(s) of action affecting population changes may be found in the rhizosphere or rhizoplane. Changing the microbial populations of the roots by adding fertilizer or incorporating organic matter may aid in inhibiting root rot and seedling blight of wheat. They concluded that incorporation of residue, addition of fertilizer, or both are more effective in reducing wheat seedling blight and root rot than the removal of the residue.

Shipton[169] reported that nitrogen applied to wheat or barley stubble after harvest in the autumn did not reduce the incidence of take-all on barley. There was no significant difference in levels of disease between rate or form of nitrogen applied as urea/ammonium nitrate spray at 0, 12, 24, or 48 units/acre (1 unit/acre = 1.25 kg/ha) ammoniacal gas liquid spray at 48 units/acre and anhydrous ammonia injected at 40 units/acre.

Most fertilizer nitrogen used in this country is applied as ammonium nitrogen which is strongly adsorbed to soil colloids in an unleachable form. Ammonium nitrogen is converted in soil via nitrification to nitrite ions and then to nitrate ions which are available to plants and are subject to leaching or denitrification. Fall application of ammonium sulfate with the compounds 2-chloro-6-(trichloro-methyl) pyridine (nitrapyrin) causes a longer retention of nitrogen as the ammonium ion in the soil prevented movement of applied nitrogen below the 30.5-cm soil depth and increased yield of wheat by 32% to 45% compared with increases in yield without nitrapyrin.[90] In this regard, the biological activity of nitrapyrin is unique in allowing greater retention of nitrogen or the ammonium ion in the upper soil. This is explained by the fact that nitrapyrin is selectively inhibitory to *Nitrosomonas* spp. of soil bacteria which convert ammonium to nitrate. Huber and Watson[94] reported that the ammonium form of nitrogen stabilized by the addition of nitrapyrin reduced the severity of take-all of wheat. More recently, Warren et al.[190] observed that fall application of nitrapyrin with anhydrous ammonium increased corn yield on the average of 68% over the anhydrous ammonia treatment without nitrapyrin. Stalk rot incidence was reduced 60% to 90% in treatments of anhydrous ammonium containing nitrapyrin. The application of nitrapyrin with anhydrous ammonium had no effect on stalk rot of corn in Nebraska.[53a]

The form of nitrogen, the stabilization of nitrogen with nitrapyrin, and the time of application are being investigated further for the control of soilborne pathogens associated with reduced-tillage systems. This approach for the control of plant diseases may prove practical with minimum- or zero-tillage because the inclusion of one or all of the nitrogen treatments mentioned herein would not necessitate a major change of a particular reduced-tillage operation.

Integrated pest management methods, employing an integration of biological, chemical, cultural, and other techniques, will continue to become a major method of controlling many diseases of crops, including crops utilizing conservation-tillage systems, for these reasons. Many diseases currently are not controlled consistently nor economically by using only one or two methods of control; biological and chemical control (pesticides) have not been successful against many destructive diseases; integrated pest management is the only method we have for long-term solution to our urgent disease problems; and recent changes in cultural practices involving monoculture, coupled with reduced or zero tillage, dictate the use of integrated pest management methods for the control of disease. The reader is referred to an elegant and thought-provoking article on integrated pest management by Apple.[10]

The utilization of short crop rotation with reduced tillage may be an effective means of controlling plant diseases in conservation tillage systems. An example of this method of controlling disease is the "ecofallow" system, previously described in this section, for reducing losses from stalk rot of corn. Ecofallow is a combination of herbicide fallow-minimum- or no-till practice using a wheat-corn or sorghum-fallow rotation. Ecofallow reduces the incidence of stalk rot by reducing environmental soil stress conditions, particularly with regard to soil temperature and moisture and, perhaps, by preventing an increase of inoculum resulting from a monoculture system of crop production.

Chemical and biological control may be feasible to use on surface residues harboring pathogens—particularly fungus pathogens. The principal objective here is to kill the pathogen on and in the residue or to prevent the pathogen from producing reproductive structures on the residue. This might be achieved by spraying the residues with fungicides or other pesticides soon after the crop is harvested—before the fungus pathogen produces fruiting structures. Various kinds of nutrients may be applied to the residues to stimulate growth of microorganisms antagonistic to the pathogens associated with residues. Also, application of cellulitic enzymes to residues could result in biological control of residue-inhabiting pathogens.

REFERENCES

1. **Adams, P. B. and Papavizas, G. C.,** Sensitivity of *Thielaviopsis basicola* to soil fungistasis in natural and alfalfa-amended soil, *Phytopathology,* 58, 1041, 1968.
2. **Allmaras, R. R., Black, A. A., and Rickman, R. W.,** Tillage, soil environment, and root growth, *Proc. Natl. Conf. Soil Conserv. Soc. Am.,* Ankeny, Iowa, 1973, 62.
3. **Allmaras, R. R. and Nelson, W. W.,** Corn *(Zea mays L.)* root configuration as influenced by some row-interrow variants of tillage and straw mulch management, *Soil Sci. Soc. Amer. Proc.* 35, 974, 1971.
4. **Altman, J.,** Predisposition of sugarbeets to *Rhizoctonia solani* damping off with herbicides, *Phytopathology,* 59, 1015, 1969.
5. **Altman, J.,** Increased and decreased plant growth responses resulting from soil fumigation, Symposium Vol. II, Root diseases and soil-borne pathogens, University of California Press, 216, 1970.
6. **Altman, J.,** Increased glucose exudate and damping off in sugarbeets in soils treated with herbicides, *Phytopathology,* 62, 743, 1972.
7. **Altman, J., and Cambell, C.,** Effect of herbicides on plant diseases, *Annu. Rev. Phytopathol.,* 15, 361, 1977.
8. **Altman, J. and Ross, M.,** Unexpected preplant herbicide damage in sugarbeets, *Phytopathol.* 55, 1051, 1965.
9. **Altman, J., and Ross, M.,** Plant pathogens as a possible factor in unexpected preplant herbicide damage in sugarbeets, *Plant Dis. Rep.,* 51, 86, 1967.
10. **Apple, J.,** Integrated pest management: philosophy and principles, *Symp. Ecol. Agric. Prod.,* University of Tennessee, Knoxville, 1973, 89.
11. **Army, D. C., Worf, G. L., Ahrens, R. W., and Lindsey, M. F.,** Yellow leaf blight of maize in Wisconsin: its history and the reactions of inbreds and crosses to the inciting fungus *(Phyllosticta* sp.), *Plant Dis. Rep.,* 54, 281, 1970.
12. **Baeumer, K. and Bakermans, W. A. P.,** Zero-tillage, *Adv. Agron.,* 25, 77, 1973.
13. **Baker, K. F. and Cook, R. J.,** *Biological Control of Plant Pathogens,* W. H. Freeman, San Francisco, 1974, 433.

14. **Balasubramanian, A. and Rangaswami, G.,** Studies on the influence of foliar nutrient sprays on the root exudation pattern in four crop plants, *Plant Soil,* 30, 210, 1969.
15. **Banihashemi, Z. and de Zeeuw, D. J.,** Saprophytic activities of *Fusarium oxysporum* f. sp. *melonis* in soil, *Trans. Br. Mycol. Soc.,* 60, 205, 1973.
16. **Barley, K. P. and Greacen, E. L.,** Mechanical resistance as a soil factor influencing the growth of roots and underground shoots, *Adv. Agron.,* 19, 1, 1967.
17. **Barrentine, W. L. and Waddle, B. A.,** Influence of mulches on the growth and development of cotton grown on *Verticillium* infested soil, *Agron. J.,* 64, 616, 1972.
18. **Basu, P. K.,** Temperature, an important factor determining survival of *Corynebacterium michiganense* in soil, *Phytopathology,* 60, 825, 1970.
19. **Bates, J. J.,** The effect of herbicide-fungicide combinations on soil bacteria and fungi in treated soil and treated cultures, M.S. thesis, University of Missouri, Columbia, 1969.
20. **Behmer, D. E. and McCalla, T. M.,** The inhibition of seedling growth by crop residues in soil inoculated with *Penicillium urticae* Bainer, *Plant Soil,* 18, 199, 1963.
21. **Blevins, R. L., Cook, D., Phillips, S. H., and Phillips, R. E.,** Influence of no-tillage on soil moisture, *Agron. J.,* 63, 593, 1971.
22. **Boosalis, M. G.,** Precocious sporulation and longevity of conidia of *Helminthosporium sativum* in soil, *Phytopathology,* 52, 1172, 1962.
23. **Boosalis, M. G. and Scharen, A. L.,** Methods for microscopic detection of *Aphonomyces euteiches* and *Rhizoctonia solani* and for isolation of *Rhizoctonia solani* associated with plant debris, *Phytopathology,* 59, 192, 1956.
24. **Boosalis, M. G., Sumner, D. R., and Rao, A. S.,** Overwintering of conidia of *Helminthosporium turcicum* on corn residue and in soil in Nebraska, *Phytopathology,* 57, 990, 1967.

24a. **Boosalis, M. G.,** unpublished data.

25. **Boyd, H. W. and Phillips, D. V.,** Toxicity of crop residue to peanut seed and *Schlerotium rolfsii, Phytopathology,* 63, 70, 1973.
26. **Brinkerhoff, L. A.,** The influence of temperature, aeration, and soil microflora on microsclerotial development of *Verticillium albo-atrum* in abscised cotton leaves, *Phytopathology,* 59, 805, 1969.
27. **Brinkerhoff, L. A. and Fink, G. B.,** Survival and infectivity of *Xanthomonas malvacearum* in cotton plant debris and soil, *Phytopathology,* 54, 1198, 1964.
28. **Brodie, B. B.,** Use of nonselective and mixtures of selective pesticides for multiple pest control, *Phytopathology,* 60, 12, 1970.
29. **Bruehl, G. W. and Lai, P.,** Prior-colonization as a factor in the saprophytic survival of several fungi in wheat straw, *Phytopathology,* 56, 766, 1966.
30. **Bruehl, G. W. and Lai, P.,** The probable significance of saprophytic colonization of wheat straw in the field by *Cephalosporium gramineum, Phytopathology,* 58, 464, 1968.
31. **Burke, D. W.,** *Fusarium* root rot of beans and behavior of the pathogen in different soils, *Phytopathology,* 55, 1122, 1965.
32. **Burns, E. E. and Shurtleff, M. C.,** Factors affecting overwintering and epidemiology of *Helminthosporium maydis* in Illinois corn, *Phytopathology,* 62, 749, 1972.
33. **Burns, E. D. and Shurtleff, M. C.,** Observations of *Physoderma maydis* in Illinois: effects of tillage practices in field corn, *Plant Dis. Rep.,* 57, 630, 1973.
34. **Burrows, W. C. and Larson, W. E.,** Effect of amount of mulch on soil temperature and early growth of corn, *Agron. J.,* 54, 19, 1962.
35. **Butler, F. C.,** Saprophytic behaviour of some cereal root-rot fungi, IV. Saprophytic survival in soils of high and low fertility, *Ann. Appl Biol.,* 47, 28, 1959.
36. **Carley, H. E. and Watson, R. D.,** Plant phytotoxins as possible predisposing agents to root rots, *Phytopathology,* 57, 401, 1967.
37. **Case, V. W., Brady, N. C., and Lathwell, D. J.,** The influence of soil temperature and phosphorus fertilizers of different water-solubilities on the yield and phosphorus uptake by oats, *Soil Sci. Soc. Am. Proc.,* 28, 409, 1964.
38. **Chambers, S. C.,** Some factors affecting the relative importance of hosts in the survival of *Ophiobolus graminis, Aust. J. Agric. Res.,* 22, 111, 1971.
39. **Chambers, S. C. and Flentje, N. T.,** Studies on variation with *Ophiobolus graminis, Aust. J. Biol. Sci.,* 20, 941, 1967.
40. **Chaudhary, M. R. and Prihar, S. S.,** Root development and growth response of corn following mulching, cultivation, or interrow compaction, *Agron. J.,* 66, 350, 1974.
41. **Chinn, S. H. F.,** A slide technique for the study of fungi and actinomyces in soil with special reference to *Helminthosporium sativum, Can. J. Bot.,* 31, 718, 1953.
42. **Chinn, S. H. F., Ledingham, R. J., Sallans, B. J., and Simmonds, P. M.,** A mechanism for the control of the common root rot of wheat, *Phytopathology,* 43, 701, 1953.

43. **Chopra, B. K. and Rodriquez-Kabana, R.,** Influence of prometryn in soil on growth-related activities of *Fusarium oxysporium* f. sp. *vasinfectum, Phytopathology,* 60, 717, 1970.
44. **Cochrane, V. W.,** The role of plant residues in the etiology of root rot, *Phytopathology,* 38, 185, 1948.
45. **Cook, R. J., Boosalis, M. G., and Doupnik, B.,** Influence of crop residues on plant diseases, *Agron. J.,* (In press), 1978.
46. **Cox, J. and Cock, L. J.,** Survival of *Cercosporella herpotrichoides* on naturally infected straws of wheat and barley, *Plant Pathol.,* 11, 65, 1962.
47. **Crafts, A. S.,** Herbicide behaviour in the plant, in *The Physiology and Biochemistry of Herbicides* Audus, L. J., Ed., Academic Press, New York, 1964, 75.
48. **Daft, G. G. and Leben, C.,** Bacterial blight of soybeans: field overwintered *Pseudomonas glycinea* or possible primary inoculum, *Plant Dis. Rep.,* 65, 156, 1973.
49. **Davis, J. R.,** Evaporation and evapotranspiration research in the United States and other countries, Report of Evapotranspiration Committee, American Society of Agricultural Engineers, St. Joseph, Michigan, 1956, 1.
50. **Deacon, J. W.,** Behaviour of *Cercosporella herpotrichoides* and *Ophiobolus graminis* on buried wheat plant tissues, *Soil Biol. Biochem.,* 5, 339, 1973.
51. **Deep, I. W. and Young, R. A.,** The role of preplanting treatments with chemicals in increasing the incidence of crown gall, *Phytopathology,* 55, 212, 1965.
52. **Dobrenz, A. K.,** Deep-crowned barley survives winter best, *Crops Soils,* 19(6), 18, 1967.
53. **Doupnik, B., Jr., Boosalis, M. G., Wicks, G., and Smika, D.,** Ecofallow reduces stalk rot in grain sorghums, *Phytopathology,* 65, 1021, 1975.

53a. **Doupnik, B., Jr. and Boosalis, M. G.,** unpublished data.

54. **Ebbels, D. L.,** Effect of fumigants on fungi in buried wheat straw, *Trans. Br. Mycol. Soc.,* 54, 227, 1970.
55. **Espinosa, W. G., Adams, R. S., Jr., and Behrens, R.,** Interaction effects of atrazine and CDAA, linuron, amiben, and trifluralin on soybean growth, *Agron. J.,* 60, 183, 1968.
56. **Farley, J. D.,** Recovery of *Corynebacterium michiganense* from overwintered tomato stems by the excised-petiole inoculation method, *Plant Dis. Rep.,* 55, 654, 1971.
57. **Fellows, H.,** Effect of certain environmental conditions on the prevalence of *Ophiobolus graminis* in the soil, *J. Agric. Res.,* 63, 715, 1941.
58. **Fenster, C. R., Owens, H. I., and Follett, R. H.,** Conservation tillage for wheat in the Great Plains, U.S. Dep. Agric. Ext. Serv. PA1190, 1977, 32.
59. **Ferguson, H. and Boatwright, G. O.,** Effects of environmental factors on the development of the crown node and adventitious roots of winter wheat *(Triticum aestivum), Agron. J.,* 60, 258, 1968.
60. **Foy, C. L. and Hurtt, W.,** Further studies on root exudation of exogenous growth regulators in *Phaseolus vulgaris, Proc. Weed. Soc. Am.* (Abstr.), 1967, 40.
61. **Futrell, M. C. and Scott, G. E.,** Overwintering and early season development of *Helminthosporium maydis* in Mississippi, *Plant Dis. Rep.,* 55, 954, 1971.
62. **Gardner, W. R.,** Solutions of the flow equation for the drying of soils and other porous media, *Soil Sci. Soc. Am. Proc.,* 23, 183, 1959.
63. **Garett, S. D.,** Soil conditions and the take-all disease of wheat. III. Decomposition of the resting mycelium of *Ophiobolus graminis* in infected wheat stubble buried in the soil, *Ann. Appl. Biol.,* 25, 742, 1938.
64. **Garrett, S. D.,** Soil conditions and the takeall disease of wheat. VIII. Further experiments on the survival of *Ophiobolus graminis* in infected wheat stubble, *Ann. Appl. Biol.,* 31, 186, 1944.
65. **Garrett, S. D.,** *Biology of Root-infecting Fungi,* Cambridge University Press, Cambridge, England, 1956, 293.
66. **Garrett, S. D.,** Cellulose-decomposing ability of some cereal foot-rot fungi in relation to their saprophytic survival, *Trans. Br. Mycol. Soc.,* 49, 57, 1966.
67. **Garrett, S. D.,** Effect of nitrogen level on survival of *Ophiobolus graminis* in pure culture on cellulose, *Trans. Br. Mycol. Soc.,* 50, 519, 1967.
68. **Garrett, S. D.,** *Pathogenic Root-infecting Fungi,* Cambridge University Press, Cambridge, England, 1970, 1.
69. **Garrett, S. D.,** Effects of nitrogen level on survival of *Phialophora radicola* and *Cochliobolus sativus* in pure culture on cellulose, *Trans. Brit. Mycol. Soc.,* 57, 121, 1971.
70. **Garrett, S. D.,** Factors affecting saprophytic survival of six species of cereal foot-rot fungi *Trans. Br. Mycol. Soc.,* 59, 445, 1972.
71. **Garrett, S. D.,** Influence of nitrogen on cellulolysis rate and saprophytic survival in soil and some cereal-rot fungi, *Soil Biol. Biochem.,* 8, 229, 1976.

72. **Glynn, M. D.,** Effects of cultural treatments on wheat and on the incidence of eyespot, lodging, take-all and weeds, *Ann. Appl. Biol.,* 38, 665, 1951.
73. **Glynn, M. D.,** Wheat yield and soil-borne diseases, *Ann. Appl. Biol.* 40, 221, 1953.
74. **Graham, V. E. and Greenberg, L.,** The effect of salicyclic aldehyde on the infection of wheat by *Pythium arrhenomanes* Drechsler and the destruction of the aldehyde *Actinomyces erythropolis* and *Penicillium* sp., *Can. J. Res.,* 17, 52, 1939.
75. **Grau, C. R. and Reiling, T. P.,** Effect of trifluralin and dinitramine on *Aphanomyces euteiches, Phytopathology,* 67, 273, 1977.
76. **Gudauskas, R. T., Teem, D. H., To, W.-N., and Whatley, T. L.,** Overwintering of *Helminthosporium maydis* in Alabama, *Plant Dis. Rep.,* 55, 947, 1971.
77. **Guenzi, W. D. and McCalla, T. M.,** Phenolic acids in oat, wheat, sorghum, and corn residues and their phytotoxicity, *Agron. J.,* 58, 303, 1966.
78. **Guenzi, W. D., McCalla, T. M., and Norstadt, F. A.,** Presence and persistence of phytotoxic substances in wheat, oat, corn and sorghum residues, *Agron. J.,* 59, 163, 1967.
79. **Hale, M. G., Foy, C. L., and Shay, F. J.,** Factors affecting root exudation, *Adv. Agron.,* 23, 89, 1971.
80. **Harricks, J. S.,** Influence of rape residue on cereal production, *Can. J. Plant Sci.,* 49, 632, 1969.
81. **Harris, W. W.,** Effects of residue, rotation, and nitrogen fertilizer on small grain production in northwestern Kansas, *Agron. J.,* 55, 281, 1963.
82. **Harrold, L, L., Peters, D. B., Dreibelis, R. F., and McGuiness, J. L.,** Transpiration evaluation of corn grown on a plastic-covered lysimeter, *Soil Sci. Soc. Am. Proc.,* 23, 174, 1959.
83. **Hauser, V. L. and Taylor, H. M.,** Evaluation of deep tillage treatments on a slowly permeable soil, *Trans. ASAE,* 7, 134, 1964.
84. **Hosford, R. M.,** A form of *Pyrenophora trichostoma* pathogenic to wheat and other grasses, *Phytopathology,* 61, 28, 1976.
85. **Hosford, R. M., Jr.,** Fungal leaf spot diseases of wheat in North Dakota, *N. D. Agric. Exp. Stn. Bull.,* 500, 1976, 1.
86. **Houseworth, L. D. and Tweedy, B. G.,** Effect of atrazine in combination with captan or thiram upon fungal and bacterial populations in the soil, *Plant Soil,* 38, 493, 1973.
87. **Hsia, Y. and Christensen, J. J.,** Effect of 2,4-D on seedling blight of wheat caused by *Helminthosporium sativum, Phytopathology,* 41, 1011, 1951.
88. **Hsu, S. C. and Lockwood, J. L.,** Responses of fungal hyphae to soil fungistasis, *Phytopathology,* 61, 1355, 1971.
89. **Huber, D. M.,** Spring versus fall nitrogen fertilization and take-all of spring wheat, *Phytopathology,* 62, 434, 1972.
90. **Huber, D. M., Murray, G. A., and Crane, J. M.,** Inhibition of nitrification as a deterrent to nitrogen loss, *Soil, Sci. Soc. Am. Proc.,* 33, 975, 1969.
91. **Huber, D. M., Painter, C. G., McKay, H. C., and Peterson, D. L.,** Effect of nitrogen fertilization on take-all of winter wheat, *Phytopathology,* 58, 1470, 1968.
92. **Huber, D. M., Seely, C. I., and Watson, R. D.,** Effects of the herbicide diuron on foot rot of winter wheat, *Plant Dis. Rep.,* 50, 852, 1966.
93. **Huber, D. M., Seely, C. I., and Watson, R. D.,** Nonfungicidal, chemical control of foot rot of winter wheat with the herbicide diuron, *Phytopathology,* 58 (Abstr.), 1054, 1968.
94. **Huber, D. M. and Watson, R. D.,** Nitrogen form and plant disease, *Down to Earth,* 27(4), 14, 1972.
95. **Huber, D. M. and Watson, R. D.,** Nitrogen form and plant disease, *Annu. Rev. Phytopathol.,* 12, 139, 1974.
96. **Hurtt, W. and Foy, C. L.,** Excretion of foliarly applied dicamba and pichloram from roots of black Valentine beans grown in soil, sand, and culture solution, *Proc. Northeast. Weed Control Conf.,* 19 (Abstr.), 602, 1965.
97. **Hurtt, W. and Foy, C. L.,** Some factors influencing the excretion of foliarly applied dicamba and pichloram from roots of black Valentine beans, *Plant Physiol.* (Suppl. XIVIII), 40, 1965.
98. **Ippolitov, D. V. and Kolyasev, F. E.,** The effect of small changes in soil temperature on cereal yields, *Bot. Zh.,* 41, 703, 1956.
99. **Jacks, G. V., Brind, W. D., and Smith, W.,** Mulching, *Common. Bur. Soils, Tech. Comm.,* No. 49, 1955.
100. **Jones, J. N., Jr., Moody, J. E., and Lillard, J. L.,** Effects of tillage, no-tillage and mulch on soil water and plant growth, *Agron, J.,* 61, 719, 1969.
101. **Jones, J. N., Jr., Moody, J. E., Shear, G. M., Moschler, W. W., and Lillard, J. H.,** The no-tillage system for corn *(Zea mays* L.), *Agron. J.,* 60, 17, 1968.
102. **Katan, J. and Eshel, Y.,** Interactions between herbicides and plant pathogens, *Residue Rev.,* 45, 145, 1973.

103. **Kaufman, D. D.,** Effect of s-triazine and phenylurea herbicides on soil fungi in corn- and soybean-cropped soil, *Phytopathology,* 54(Abstr.), 897, 1964.

104. **Kimber, R. W. L.,** Phytotoxicity from plant residues. I. The influence of rotted wheat straw on seedling growth, *Aust. J. Agric. Res.,* 18, 361, 1967.

105. **Kimber, R. W. L.,** Phytotoxicity from plant residues. II. The effect of time rotting of straw from some grasses and legumes on the growth of wheat seedlings, *Plant Soil,* 38, 347, 1973.

106. **Ko, W. H. and Farley, J. D.,** Conversion of pentachlorobenzene to pentachloroaniline in soil and the effect of these compounds on soil microorganisms, *Phytopathology,* 59, 64, 1969.

107. **Lawrence, T. and Ashford, R.,** Mesocotyl development in Russian wild ryegrass and its effect on survival of shoot apices. *Nature (London),* 201, 727, 1964.

108. **Ledingham, R. J., Sallans, R. J., and Wenhardt, A.,** Influence of culture practice on incidence of common root rot of wheat, *Can. J. Plant Sci.,* 40, 310, 1960.

109. **Lewis, J. A. and Papavizas, G. A.,** Effect of plant residues on chlamydospore germination of *Fusarium solani* f. sp. *phaseoli* and on Fusarium root rot of beans, *Phytopathology,* 67, 925, 1977.

110. **Linder, P. J., Mitchell, J. W., and Freeman, G. D.,** Persistence and translocation of exogenous regulating compounds that exude from roots, *J. Agric. Food Chem.,* 12, 437, 1964.

111. **Linderman, R. G. and Toussoun, T. A.,** Pathogenesis of *Thielaviopsis basicola, Phytopathology,* 57 (Abstr.), 819, 1967.

112. **Linderman, R. G. and Toussoun, T. A.,** The behavior of chlamydospores and endoconidia of *Thielaviopsis basicola* in nonsterilized soil, *Phytopathology,* 57, 729, 1967.

113. **Linderman, R. G., and Toussoun, T. A.,** Predisposition to *Thielaviopsis* root rot of cotton by phytotoxins from decomposing barley residues, *Phytopathology,* 58, 1571, 1968.

114. **Litrell, R. H., and Sumner, D. S.,** Overwintering of *Helminthosporium maydis* in Georgia, *Plant Dis. Rep.,* 55, 947, 1971.

115. **Lutey, R. W. and Fezer, K. D.,** The role of infected straw in the epiphytology of Septoria leaf blotch of barley. *Phytopathology,* 50, 910, 1960.

116. **MacKenzie, D. R., Cole, H. and Ercegovich, C. D.,** Resistance breakdown in maize to maize dwarf mosaic virus after treatment with atrazine, *Phytopathology,* 58 (Abstr.), 1058, 1968.

117. **MacKenzie, D. R., Cole, H., Smith, C. B., and Ercegovich, C.,** Effects of atrazine and maize dwarf mosaic virus infection on weight and macro and micro element constituents of maize seedlings in the greenhouse, *Phytopathology,* 60, 272, 1970.

118. **MacNish, G. C. and Dodman, R. L.,** Survival of *Gauemannomyces graminis* var. *tritici* in the field, *Aust. J. Biol. Sci.,* 26, 1309, 1973.

119. **Maier, C. R.,** Influence of nitrogen nutrition on Fusarium root rot of pinto bean and on its suppression by barley straw, *Phytopathology,* 58, 620, 1968.

120. **Martin, A.,** Chemical aspects of ecology in relation to agriculture, *Can. Department of Agriculture* Publ. 1015, Ottawa, Ontario, Canada, 1968, 96.

121. **McCalla, T. M. and Army, T. J.,** Stubble mulch farming, *Adv. Agron.,* 13, 125, 1961.

122. **McCalla, T. M. and Haskins, F. A.,** Phytotoxic substances from soil microorganisms and crop residues, *Bacteriol. Rev.,* 28, 181, 1964.

123. **Miller, E. C.,** *Plant Physiology,* McGraw-Hill, New York, 1931, 900.

124. **Mitchell, John W.,** Perspectives of biochemical plant pathology: progress in research on absorption, translocation, and exudation of biologically active compounds in plants, *Conn. Agric. Exp. Stn. Bull.,* 663, 49, 1963.

125. **Mitchell, J. W., Smale, B. C., and Preston, W. H., Jr.,** New plant regulators that exude from roots, *J. Agric. Food Chem.,* 7, 841, 1959.

126. **Naito, N. and Kohima, Y.,** Fungitoxic substance production by *Gleosporium olivarum* which was grown on media containing either 2-chlorophenoxyacetic acid or 2-methylphenoxyacetic acid, *Kagama Agr. Univ. Tech. Bull.,* 9, 18, 1957.

127. **Natti, J. J.,** Overwintering survival of *Pseudomonas phaseolicola* in New York, *Phytopathology,* 57, 343, 1967.

128. **Nelson, R. R. and Sheifele, G. L.,** Factors affecting the overwintering of *Trichometasphaeria turcica* on maize, *Phytopathology,* 60, 369, 1970.

129. **Nelson, W. W. and Allmaras, R. R.,** An improved monolith method for excavating and describing roots, *Agron. J.,* 61, 751, 1969.

130. **Neubauer, R. and Avizohar-Hershenson, Z.** Effect of the herbicide, Trifluralin, on *Rhizoctonia* disease in cotton, *Phytopathology,* 63, 651, 1973.

131. **Nilsson, H. E.,** Studies of root and foot rot diseases of cereals and grasses, *Ann. Agric. College Swed.,* 35(NR3), 1969, 807.

132. **Nilsson, H. E.,** Influence of herbicides on take-all and eyespot disease of winter wheat in a field trial, *Swed. J. Agric. Res.,* 3, 115, 1973.

133. **Nilsson, H. E.**, Influence of the herbicide mecoprop in *Gaeumanomyces graminis* and take-all of spring wheat, *Swed. J. Agric. Res.*, 3, 105, 1973.
134. **Norstadt, F. A. and McCalla, T. M.**, A phytotoxic substance from a species of *Penicillium, Science,* 140, 410, 1963.
135. **Nyvall, R. and Kommedahl, T.**, Thickened hyphae as a survival mechanism in *Fusarium moniliforme, Phytopathology,* 56 (Abstr.), 893, 1966.
136. **Nyvall, R. F. and Kommedahl, T.** Individual thickened hyphae as survival structures of *Fusarium moniliforme* in corn, *Phytopathology,* 58, 1704, 1968.
137. **Odvody, G. N. and Dunkle, L. D.**, Overwintering capacity of *Ramulispora sorghi, Phytopathology,* 63, 1530, 1973.
138. **Papavizas, George C.**, Survival of root-infecting fungi in soil. VIII. Distribution of *Rhizoctonia solani* in various physical fractions of naturally and artificially infested soils, *Phytopathology,* 58, 746, 1968.
139. **Park, D.**, Effect of substrate on a microbial antagonism with reference to soil conditions, *Trans. Br. Mycol. Soc.*, 39, 239, 1956.
140. **Patrick, Z. A.**, Decomposition of plant residues in soil, formation of phytotoxins and their role in root disease, *Can. Phytopathol. Soc. Proc.*, 29 (Abstr.), 15, 1962.
141. **Patrick, Z. A. and Koch, L. W.**, Inhibition of respiration, germination, and growth by substances arising during the decomposition of certain plant residues in the soil, *Can. J. Bot.*, 36, 621, 1958.
142. **Patrick, Z. A. and Koch, L. W.**, The adverse influence of phytotoxic substances from decomposing plant residues on resistance of tobacco to black root rot, *Can. J. Bot.*, 41, 747 1963.
143. **Patrick, Z. A., Toussoun, T. A., and Synder, William C.**, Phytotoxic substances in arable soils associated with decomposition of plant residues, *Phytopathology,* 53, 152, 1963.
144. **Patrick, Z. A., Toussoun, T. A., and Thorpe, H. J.**, Germination of chlamydospores of *Thielaviopsis basicola, Phytopathology,* 55, 466, 1965.
145. **Percich, J. A. and Lockwood, J. L.**, Influence of atrazine on the severity of *Fusarium* root rot in pea and corn, *Phytopathology,* 5, 154, 1975.
146. **Peters, D. B. and Russell, M. B.**, Relative water losses by evaporation and transpiration in field corn, *Soil Sci. Am. Proc.*, 23, 170, 1959.
147. **Phillips, W. M.**, A new technique of controlling weeds in sorghum in a wheat-sorghum-fallow rotation in the Great Plains, *Weeds,* 12, 42, 1964.
148. **Pinckard, J. A. and Standifer, L. C.**, An apparent interaction between cotton herbicidal injury and seedling blight, *Plant Dis. Rep.*, 50, 172, 1966.
149. **Pitt, D.**, Studies on sharp eyespot disease of cereals. II. Viability of sclerotia: persistence of the causal fungus, *Rhizoctonia solani* Kuhn, *Ann. Appl. Biol.*, 54, 231, 1964.
150. **Pittman, U. J., and Horricks, J. S.**, Influence of crop residue and fertilizers on stand, yield, and root rot of barley in southern Alberta, *Can. J. Plant Sci.*, 52, 463, 1972.
151. **Rands, R. D. and Dopp, E.**, Influence of certain harmful soil constituents on severity of *Pythium* root rot of sugar cane, *J. Agric. Res.*, 56, 53, 1938.
152. **Richardson, L. T.**, Effect of insecticides and herbicides applied to the soil on the development of plant diseases. I. The seedling disease of barley caused by *Helminthosporium sativum* P. K. and B., *Can. J. Plant Sci.*, 37, 196, 1957.
153. **Richardson, L. T.**, Effect of insecticides and herbicides applied to the soil on the development of plant diseases. II. Early blight and Fusarium wilt of tomato, *Can. J. Plant Sci.*, 39, 30, 1959.
154. **Richardson, L. T.**, Effect of insecticide-fungicide combinations on emergence of peas and growth of damping-off fungi, *Plant Dis. Rep.*, 44, 104, 1960.
155. **Rodriquez-Kabana, R., Curl, E. A., and Funderburk, H. H., Jr.**, Effect of paraquat on growth of *Sclerotium rolfsii* in liquid culture and soil, *Phytopathology,* 57, 911, 1967.
156. **Romig, W. R. and Sasser, M.**, Herbicide predisposition of snapbeans to Rhizoctonia solani, *Phytopathology,* 62 (Abstr.), 785, 1972.
157. **Rovira, A. D.**, Root excretions in relation to the rhizophere effect, *Plant Soil,* 11, 53, 1959.
158. **Rovira, A. D.**, Plant root exudates, *Bot. Rev.*, 35, 35, 1969.
159. **Scharen, A. L.**, Environmental factors on development of glume blotch in wheat, *Phytopathology,* 54, 300, 1964.
160. **Schenck, N. C.**, Overwintering of *Helminthosporium maydis* in Florida, *Plant Dis. Rep.*, 55, 949, 1971.
161. **Schreiber, L. R. and Green, R. J.**, Comparative survival of mycelium, conidia, and microsclerotia of *Verticillium albo-atrum* in mineral soil, *Phytopathology,* 52, 288, 1962.
162. **Schroth, M. N. and Hendrix, F. F., Jr.**, Influence of nonsusceptible plants on the survival of *Fusarium solani* f. *phaseoli* in soil, *Phytopathology,* 52, 906, 1962.
163. **Schuster, M. L.**, Leaf freckles and wilt of corn incited by *Corynebacterium nebraskense* Schuster, Hoff, Manel, Lazar, *Nebr. Agric. Exp. Stn. Res. Bull.*, 270, 1975, 40.

164. **Schuster, M. L., Compton, W. A., and Hoff, B.,** Reaction of corn inbred lines to the new Nebraska leaf freckles and wilt bacterium, *Plant Dis. Rep.*, 56, 863, 1972.
165. **Schwartz, H. F., Steadman, J. R., and Coyne, D. P.,** Influence of *Phaseolus vulgaris* blossoming characteristics and canopy structure upon reaction to *Sclerotinia sclerotiorum, Phytopathology,* 68, 465, 1978.
166. **Semeniuk, G. and Tunac, J. B.,** Tarton increase of root rot severity in wheat and corn, *Proc. S. D. Acad. Sci.*, 47 (Abstr.), 346, 1968.
167. **Sewell, G. W. and Wilson, J. F.,** The influence of normal tillage and non-cultivation on *Verticillium* wilt of the hop, *Ann. Appl. Biol.*, 76, 37, 1974.
168. **Shear, G. M. and Moschler, W. W.,** Continuous corn by the no-tillage and conventional tillage methods: A six-year comparison, *Agron. J.*, 61, 524, 1969.
169. **Shipton, P. J.,** Influence of stubble treatment and autumn application of nitrogen to stubbles on the subsequent incidence of take-all and eyespot, *Plant Pathol.*, 21, 147, 1972.
170. **Skoropad, W. P. and Kao, W.,** The effect of some herbicides on sporulation of *Rhynchosporium secalis, Phytopathology,* 55, 43, 1965.
171. **Smika, D. W., Black, A. L., and Greb, B. W.,** Soil nitrate, soil water, and grain yields in a wheat-fallow rotation in the Great Plains as influenced by straw mulch, *Agron. J.*, 61, 785, 1969.
172. **Smika, D. E. and Ellis, R., Jr.,** Soil temperature and wheat mulch effects on wheat plant development and nutrient concentrations, *Agron. J.*, 63, 388, 1971.
173. **Smika, D. E. and Wicks, G. A.,** Soil water storage during fallow in the central Great Plains as influenced by tillage and herbicide treatments, *Proc. Soil Sci. Soc. Am.*, 32, 591, 1968.
174. **Smiley, R. W. and Cook, R. J.,** Influence of nitrogen fertilizers on rhizosphere pH and take-all of wheat caused by *Ophiobolus graminis, Phytopathology,* 61 (Abstr.), 711, 1971.
175. **Snyder, W. C., Schroth, M. N., and Christou, T.,** Effect of plant on root rot of bean, *Phytopathology,* 49, 755, 1959.
176. **Standifer, L. C., Jr., Melville, D. R., and Phillips, S. A.,** A possible interaction between herbicidal injury and the incidence of disease in cotton plantings, *Proc. South. Weed Control Conf.*, 19, 1966, 126.
177. **Sullia, S. B.,** Effect of foliar spray of hormones on rhizosphere microflora of leguminous weeds, *Plant Soil,* 29, 292, 1968.
178. **Taylor, H. M.,** Effect of tillage-induced soil environmental changes on root growth, *Proc. Conf. Tillage Greater Crop Production,* American Society of Agricultural Engineers, St. Joseph, Mich., 1967, 15.
179. **Thompson, D. L. and Herbert, T. T.,** Winter survival of *Helminthosporium maydis* in North Carolina, *Plant Dis. Rep.*, 55, 956, 1971.
180. **Toussoun, T. A. and Patrick, Z. A.,** Effect of substances produced in the soil during decomposition of crop residues on the pathogenesis of *Fusarium solani* f. *phaseoli, Phytopathology,* 52 (Abstr.), 30, 1962.
181. **Toussoun, T. A. and Patrick, Z. A.,** Effect of phytotoxic substances from decomposing plant residues on root rot of bean, *Phytopathology,* 53, 265, 1963.
182. **Toussoun, T. A., Menzinger, W., and Smith, R. S., Jr.,** Role of conifer litter in ecology of *Fusarium*: stimulation of germination in soil, *Phytopathology,* 59, 1396, 1969.
183. **Toussoun, T. A., Patrick, Z. A., and Snyder, W. C.,** Influence of crop residue decomposition products on the germination of *Fusarium solani* f. *phaseoli* chlamydospores in soil, *Nature (London),* 197, 1314, 1963.
184. **Triplett, G. B., Van Doren, D. M., Jr., and Schmidt, B. L.,** Effect of corn *(Zea mays* L.) stover mulch on no-tillage corn yield and water infiltration, *Agron. J.*, 60, 236, 1968.
185. **Tyner, L. E.,** Colonization of organic matter in the soil by fungi, *Phytopathology,* 51, 625, 1961.
186. **Ullstrup, A. J.,** Overwintering of race T. of *Helminthosporium maydis* in midwestern U.S., *Plant Dis. Rep.*, 55, 563, 1971.
187. **van Wijk, W. R., Larson, W. E., and Burrows, W. C.,** Soil temperature and the early growth of corn from mulched and unmulched soil, *Proc. Soil Sci. Soc. Am.*, 23, 428, 1959.
188. **Vizvary, M. A.,** Saprophytism and survival of *Colletotrichum graminicola* in soil and corn stalk residue and pathogenic behavior on corn roots, M.S. thesis, Purdue University, Lafayette, Ind., 1974.
189. **Waksman, S. A.,** *Soil Microbiology,* John Wiley & Sons, New York, 1952, 356.
190. **Warren, H. L., Huber, D. M., Nelson, D. W., and Mann, O. W.,** Stalk rot incidence and yield of corn as affected by inhibiting nitrification of fall-applied ammonium, *Phytopathology,* 67, 555, 1975.
191. **Warren, H. L. and Kommedahl, T.,** Fertilization and wheat refuse effects on *Fusarium* species associated with wheat roots in Minnesota, *Phytopathology,* 63, 103, 1973.
192. **Warren, J. R., Graham, F., and Gale, G.,** Dominance of an actinomycete in a soil microflora after 2,4-D treatment of plants, *Phytopathology,* 41, 1037, 1951.

193. **Warrick, A. W., Biggar, J. W., and Nielson, D. R.,** Simultaneous solute and water transfer for an unsaturated soil, *Water Resour. Res.*, 7, 1216, 1971.
194. **Watkins, J. E., Odvody, G. N., Boosalis, M. G., and Partridge, J. E.,** An epidemic of tanspot of wheat in Nebraska, *Plant Dis. Rep.* 62, 132, 1978.
195. **Weaver, J. E.,** *Root Development of Field Crops,* McGraw-Hill, New York, 1926, 160.
196. **Weihing, J. L. and Vidaver, A. K.,** Report on holcus leaf spot *(Pseudomonas syringae)* epidemic on corn, *Plant Dis. Rep.*, 51, 396, 1967.
197. **Wittmuss, H. D., Triplett, G. B., Jr., and Greb, B. W.,** Conservation tillage, *Proc. Natl. Conf. Soil Conserv. Soc. Am.*, Ankeny, Iowa, 1973, 5.
198. **Woodruff, N. P., Lyles, L., Siddoway, F. H., and Fryrear, D. W.,** How to control wind erosion, *U.S. Dep. Agric. USDA Agric. Inf. Bull.*, 354, 1972, 22,
199. **Worley, R. E., Morton, D. J., and Harmon, S. A.,** Reduction of southern blight on tomato by deep plowing, *Plant Dis. Rep.*, 50, 443, 1966.
200. **Wysong, D. S., Vidaver, A., Stevens, H., and Stenberg, D.,** Occurrence and spread of an undescribed species of *Corynebacterium* pathogenic on corn in the western corn belt., *Plant Dis. Rep.*, 57, 291, 1973.
201. **Zingg, A. W. and Whitfield, C. J.,** Stubble-mulch farming in the western states, *U.S. Dep. Agric. Tech. Bull.*, No. 1166, 1957.

HYPERPARASITISM FOR CONTROL OF PLANT PATHOGENS*

R. D. Lumsden

INTRODUCTION

The ability of microorganisms to cause diseases in plants is considered to be exceptional. Higher plants usually are immune or very tolerant to the effects of stressful microorganisms and succumb to them often only under conditions of adversity. A natural balance prevails, especially in nondisturbed environments. In agricultural environments this balance is often completely disrupted by the manipulation of the environmental or cultural conditions necessary to produce a crop economically. Practices such as continuous culture, dense planting, monoculture, tillage, and the use of pesticides can disrupt the natural balance in favor of the pathogenic populations, thus permitting disease to occur. We know, however, that microbial equilibria are established in the environment and that these successfully suppress or prevent disease. An example of the consequences of disturbing such equilibria is the effect of sterilization of soil on the incidence of disease. In steam-sterilized soil, if plant pathogens are introduced they can grow rampantly, and even organisms not ordinarily pathogens can cause significant disease. In contrast, nonpathogenic microorganisms in naturally suppressive soil or even in partially sterilized soil can prevent or greatly reduce plant infection by resident pathogens, unless the soil is overwhelmed by a high pathogen inoculum level.[6]

The equilibria in the environment are established as the result of antagonistic processes termed (1) competition, (2) antibiosis, including fungistasis, and (3) hyperparasitism. All of these types of antagonism occur commonly, especially in soil, and all can influence the activity of plant pathogenic microorganisms. For example, a pathogenic soil-borne fungus may be subject to: (1) fungistasis, which prevents germination of fungal spores, (2) antibiosis, which can inhibit vegetative growth, (3) competition from other organisms for colonizable organic matter in soil, upon which the fungus is dependent for an energy source to reproduce itself, and (4) parasitism by other microorganisms upon fungal mycelium and survival structures.

This last phenomenon is the subject of this review and is called hyperparasitism. Hyperparasitism is a broad term for the parasitism of one parasite on another. If the parasitism involves one fungus attacking another fungus, the term mycoparasitism generally is used. Mycoparasitism is widespread in natural environments and is well documented.[35] Most examples in this review will therefore be of mycoparasitism.

Two types of mycoparasitism are recognized.[9,35] One is biotrophic or balanced mycoparasitism. This type of parasitism is often obligate, and the parasite obtains nutrients from living cells with little or no apparent harm to the host. Apparently the loss of its ability to synthesize some nutrient has resulted in the parasite's dependence on the host for survival. The parasite may merely contact the host cell without penetration, obtaining nutrients through "buffer" or absorptive cells; it may penetrate the host with a peg from an appressorium-like swelling and form haustoria within the host cell; or it may penetrate the host hyphae and grow internally from cell to cell with no apparent effect in the early stage of development. Later the association may cause distortion of hyphae and a reduction in host sporulation as in parasitism of *Rhizopus oryzae* by *Syncephalis californica*.[28] The second type of mycoparasitism is necrotrophic or destructive mycoparasitism. The parasite destroys the host cell after, or slightly before, invasion and uses nutrients from the dying or dead host. The infection is initi-

* The literature survey for this review was completed August, 1978.

ated by hyphal contact or coiling of hyphae around fungal cells, or by direct penetration and invasion of hyphae or survival structures. Several reviews have treated mycoparasitism in detail.[7-9,11,12,16,35]

Hyperparasitism also occurs between a bacterium and a fungus;[37] it is then called mycolysis, exolysis, or bacterial necrosis. This phenomenon is not well understood, especially because of the difficulty in distinguishing this type of lysis from autolysis.[34] Fungi commonly autolyse under adverse conditions, possibly as a means of survival. The nutrients derived from the breakdown of vegetative mycelium are used in the formation of persistent survival structures.

Hyperparasitism among plant pathogens also occurs between two bacteria, a virus (bacteriophage) and a bacterium, or a virus (mycovirus) and a fungus.[24,58] Additionally, plant parasitic nematodes are parasitized by an array of microorganisms including fungi, bacteria, and protozoa.[12,36,40,48] Specific examples of control of plant disease by these agents are not numerous, but they may be significant in the balance of microorganisms in natural ecosystems.

OCCURRENCE OF HYPERPARASITISM AND ITS SIGNIFICANCE IN BIOLOGICAL CONTROL

Undoubtedly hyperparasitism is widespread in natural ecosystems and hyperparasites have adverse effects on their hosts.[24,35,40,48,58] The actual effect of hyperparasites on the balance of natural ecosystems and their importance in natural or imposed biological control of plant disease, however, have been minimally documented. Mycoparasitism has been suggested to greatly influence populations of soil-borne fungi by reducing the longevity of survival structures.[17,49] Experimental evidence recently has been shown for the occurrence of mycoparasitism in natural soils[3,4,28,49] and in aerial environments.[33,52] Examples of biological control of plant diseases resulting from mycoparasitism, however, are limited.[6,11,35,41,50] The reasons for this minimum demonstration of both natural and imposed mycoparasitic biological control can be attributed to the inadequacy of research efforts and methods and to the complexity of the ecosystems involved. Virus infection of fungi has also been reported, and in several cases the possible effects of a virus on the pathogenicity of the fungus have been studied.[24] Likewise, bacteriophages have given protection from bacterial disease symptoms.[13,58]

Progress is being made in discovering important new examples of biological control of plant disease by hyperparasites. Recent increased research efforts and the study by the U.S. Department of Agriculture[56] on the status and prospects of biological agents for pest control should give renewed interest and impetus to this alternative to traditional methods for disease control.

The need for more research has been emphasized.[6,27,56] The following examples of disease control, either natural or imposed, illustrate the great potential of hyperparasites as biological control agents. The mechanism of antagonism has not been studied extensively in each example, but in each case hyperparasitism is at least suggested as the mode of action.

Control of Wood and Tree Decay

Control of wood decay and diseases of forest trees is the most successful application of biological control to date. Hyperparasitism or "hyphal interference"[29] has been suggested as the mode of action in at least one case, but competition of antibiosis probably is also involved. The topical application of antagonistic microorganisms to cut wood surfaces suppresses entry of wood decay fungi.[19-21,43,45,46] The most successful control, applied extensively on a commercial scale, is the control of *Fomes annosus* butt rot of pine trees by *Peniophora gigantea*.[20,46] Stump inoculation of freshly cut pine trees with a conidial suspension of *P. gigantea* prevents stump infection by *F. annosus*. This pathogen can spread through the roots; when it comes into contact with roots of adjacent trees, these neighboring trees become infected and are killed. *Peniophora gigantea* is produced commercially and has been used routinely by the British Forestry Commission since 1962.[20,46]

Other examples of hyperparasitic control of wood decay include the protection of Douglas fir poles from *Poria carbonica* rot by *Scytalidium lignicola*;[45] the suppression of the silver leaf disease organism, *Stereum purpureum*, with *Trichoderma viride*;[21] and the action of *Trichoderma polysporum* against *Fomes annosus* heart rot of spruce.[45] Other wood rots that have been controlled experimentally are the control by *Trichoderma harzianum* of hymenomycete decay in maple tree wounds[43] and the suppression by *Trichoderma viride* of citrus root colonization by *Armillaria mellea*.[10,39]

Control of Airborne Pathogens

Interactions of antagonists of hyperparasites and pathogens on foliage or fruits and stems can control disease. Few examples of hyperparasitism in relation to successful foliar disease control exist, however, and these are not well understood.

Swendsrud and Calpouzos[52] attempted to control wheat leaf rust caused by *Puccinia recondita* by the use of pycnidiospores of the rust mycoparasite, *Darluca filum*. Even with varied inoculation sequences and periods of misting of wheat plants, control was not effective. In fact, with prior inoculation of leaves with *D. filum* spores and prolonged misting, uredia formation by *P. recondita* was higher than in the controls. *D. filum* was also studied in attempts to control fusiform rust (caused by *Cronartium* spp.) on oak and pine.[33] The mycoparasite infected a large percentage of rust sori in a dense stand of oak trees and effectively controlled the rust within the pure oak stand. In less dense stands, however, control was less effective. For this and other reasons biological control of this rust disease was not considered practical.[33]

More positive results were obtained with *Trichoderma pseudokoningii*, an antagonist to *Botrytis cinerea*, which causes the dry eye rot disease of apple.[54] The antagonism, thought to be due to growth inhibitors and mycoparasitism, significantly decreased the frequency of dry eye rot of artificially infected apples in the field. Natural infection was not controlled. Similar studies on the control of fruit rot of strawberry[53] with *Trichoderma* indicated reduction of disease incidence. These applications of biological control still require extensive research on the influence of environmental factors, such as temperature, on the successfulness of control.

Powdery mildew infection of greenhouse cucumber has been significantly decreased with the mycoparasite *Ampelomyces quisqualis*.[30] The mycoparasite, however, also caused small angular lesions on leaves, and sunken lesions on fruit, that detracted from market quality. These disadvantages must be overcome for mycoparasitic control agents to be used commercially.

The possibilities of biological control of chestnut blight with hypovirulence is intriguing. Although no virus-like particles have been found, *Endothia parasitica*, which causes the devastating disease on chestnut, has double-stranded RNA associated with weakened or hypovirulent strains. These strains transmit the hypovirulence to virulent strains by hyphal anastomoses.[24] Active cankers have been cured by inoculation with hypovirulent strains. Problems have also been encountered with this natural biological control, but if the technique succeeds it may offer an important new disease-control method — that is, if hypovirulent strains of other pathogens can be found.

The bacteriophage-bacterial plant pathogen interaction may also be a useful biological control mechanism. The presence of phage on peach foliage reduces *Xanthomonas pruni* infection.[13] Infection is not affected, however, if the bacterium is applied to foliage first. This and other examples of phages affecting bacterial infections[58] indicate the possibility of these agents being utilized for biological control. Many problems and difficulties will have to be overcome, however, before they can be of practical use.

Control of Soil-borne Pathogens

Recent studies have demonstrated that the survival of soil-borne pathogens and their ability to cause disease may be successfully manipulated by hyperparasites. *Trichoderma harzianum,* for example, effectively controls *Sclerotium rolfsii* in soil. The first report of its efficacy was that of Wells, Bell, and Jaworski[59] in which they demonstrated excellent control of southern blight on lupine, tomato, and peanuts. They attributed the successful biological control to the use of a supplemental food base for vigorous growth of the mycoparasite. When *Trichoderma* spp. come in contact with or proximity to hyphae of *Sclerotium rolfsii,* the hyphae are disrupted and die. The *Trichoderma* subsequently colonize the *S. rolfsii* hyphae and uses their nutrients.[60] *Trichoderma* spp. have been grown successfully in bulk and they are considered to be economically feasible for commercial application and to offer advantages over chemical control means.[5] These biological control agents also have been demonstrated to control *Rhizoctonia solani* under greenhouse conditions.[22]

White rot of onions, caused by *Sclerotium cepivorum,* recently has been successfully controlled by the mycoparasite *Coniothyrium minitans,* also in greenhouse experiments.[1] Pycnidial dust of *C. minitans* was applied to soil and as a seed dressing to protect onion seeds planted in *S. cepivorum*-infested soil. Control of the disease and appearance of the onion seedlings were more enhanced by the *C. minitans* treatments than in a standard mercury fungicide control.

Partial control of *Sclerotinia sclerotiorum* wilt of sunflower was also demonstrated with *C. minitans.*[25] In field studies, the incidence of wilt was 25% in the mycoparasite-treated plots and 43% in untreated plots. Other results[26,55] indicated that the use of *C. minitans* reduced the numbers of sclerotia of *Sclerotinia* spp. in the field soil. Use of this mycoparasite may have potential to control Sclerotinia diseases.

Another destructive mycoparasite, *Sporidesmium sclerotivorum,* has been recently isolated from *S. sclerotiorum.*[3] *Sporidesmium sclerotivorum* causes a significant decline in survival of *Sclerotinia* spp. sclerotia in soil over a period of several weeks. Its use as a biological control agent against *Sclerotinia* spp. seems promising. In addition, sclerotia of *Sclerotium cepivorum* were attacked by the mycoparasite, but those of *Macrophomina phaseolina* apparently were not.

Successful control of black root rot of pine seedlings caused by *M. phaseolina* was demonstrated by De La Cruz and Hubbell.[15] They showed an increase in seedling survival from 57% in soil artificially infested with *M. phaseolina* to 81% in soil simultaneously infested with *M. phaseolina* and an unidentified sterile basidiomycete. This increased stand compared favorably with the survival rate of 84% in the noninoculated controls. The unidentified control agent has potential economical application in forest seedling nurseries with high-density plantings.

An interesting and complex natural biological control system has been studied for a number of years with regard to the take-all disease of wheat caused by *Gaeumannomyces graminis.*[14] Outbreaks of this disease are frequently followed by a significant decline in disease that continues for several years until monoculture of wheat is interrupted. One intriguing hypothesis for the mechanism of this decline was that virus infection caused the fungus to lose virulence, thus reduced disease incidence.[24] This suggestion could not be substantiated, however.[44] Another, better-substantiated hypothesis holds that the fungus *Phialophora radicicola* competes with *G. graminis* in cereal monoculture, thus causing the decline in severity of symptoms.[14] It has also been suggested that the mycoparasite *Pythium oligandrum* may control take-all alone or in conjunction with other means of control, perhaps by altering balances between the pathogen and competing microorganisms.[14]

Limited control of damping-off diseases of seedlings has been reported also with *Pythium oligandrum*. This fungus, very weakly pathogenic to higher plants, is strongly mycoparasitic to *Pythium debaryanum* and *P. ultimum*, which cause serious damping-off in emerging sugarbeet seedlings.[57] Damage from these soil-borne pathogens was reported to be reduced when *P. oligandrum* was introduced into the soil. This mycoparasite may not have potential as a useful applied biological control agent because of its weak plant pathogenicity, but it may contribute to the decline of pathogenic fungi in soils in natural environments.

Pythium debaryanum has also been experimentally controlled by the suppressive action of lytic rhizosphere bacteria.[37] *Arthrobacter* spp. lysed mycelia of *P. debaryanum* and, when used as a seed inoculant, protected tomato seedlings from damping-off. Lytic bacteria may have substantial potential for biocontrol, especially if isolates can be obtained that lyse oospores and sporangia of *Pythium* spp. and other phytomycetes.

Several reports indicate partial control of black root rot of cucumber.[18,38,51] This commercially important greenhouse disease was reduced by adding *Trichoderma* spp.[18,51] or *Gliocladium roseum*[38,51] to soil infested with the pathogen, *Phomopsis sclerotioides*. These antagonists may help alleviate this serious problem in European greenhouse-grown cucumbers.

There has been considerable optimistic speculation on the feasibility of biological agents for control plant diseases caused by soil-inhabiting plant parasitic nematodes.[12,40] As with soil-borne fungal pathogens, however, numerous failures have occurred in attempts to control nematode diseases of plants by using parasites of nematodes. Again, future success will depend greatly on a better understanding of the ecological interactions between nematodes and their antagonists in soil. Studies on the cereal cyst nematode *(Heterodera avenae)*[31] have made considerable progress in understanding the ecology of the disease caused by this nematode. As with the decline of take-all disease caused by *Gaeumannomyces graminis*, populations of the cereal cyst nematode decrease when cereals were grown continuously. Fungal parasites, including an *Entomophthora*-like fungus and *Verticillium chlamydosporium* have been commonly associated with parasitized eggs and females of *H. avenae*. These nematophagous fungi may be responsible for the decline of this nematode. If so, this will be the first documented case of control of a cyst nematode by naturally occurring parasites.

These examples of biological control of soil-borne plant pathogens are only a few of the perhaps myriad cases of natural control. Soil-borne disease organisms are especially vulnerable to microbial pressure because of the vast and complex ecological systems in soil. Optimistically, we may look forward to isolating and exploiting many of these natural biological agents for the control of soil-borne diseases that have been so difficult or even impossible to control by conventional methods. Integrated pest management systems that include hyperparasitic control of diseases may be the answer to these difficult problems.

HYPERPARASITISM IN PEST MANAGEMENT SYSTEMS

The use of hyperparasites to control plant diseases is in its infancy. Examples of the successful application of hyperparasites to the control of plant disease are few. This does not reflect a lack of biological control through hyperparasitic action in natural ecosystems; specific documented cases of mycoparasitism[35] and other types of hyperparasitism[24,40,58] abound. What it does reflect is a lack of concentrated research efforts and commitment to the search for natural control agents; of expanded research to understand how and why these control agents work in nature; and of intensified efforts to use them in our agricultural system. Hopefully, new direction and impetus for obtaining positive results are forthcoming.[56] The interest in biological control of plant pathogens is gathering momentum.

It is becoming more and more recognized that our agriculture should not depend on single, isolated, one-shot approaches to pest control.[2] Instead, several means, such as host resistance, cultural practices, pathogen-free plant sources, minimal levels of pesticides, and multiple-use pesticides need to be employed. Biological control by hyperparasites can and should also be a part of pest management systems. In fact, perhaps the common failure of biological control agents under field conditions can be overcome by integrated or multipurpose approaches. Such use has been suggested repeatedly.[6,36,41] Coordinated use of more than one approach to control may overcome some of the difficulties previously encountered, because the various components may work best at different times and places and may compensate for each other's deficiencies. Each disease is different and will require imaginative approaches aimed at enhancing the biological system to the advantage of the hyperparasite and the disadvantage of the disease organism. We need to learn more about these control systems to better understand how they can be manipulated. It is clear, however, that the natural biological balance has been altered when disease occurs and that we must restore that balance or tip the balance in favor of the antagonists. This can be done by (1) adding large numbers of antagonists to the environment so that they successfully compete with the pathogen and inhibit its ability to cause disease; (2) applying shock treatments such as heat, chemicals, or organic soil amendments that greatly stimulate the antagonists to the disadvantage of the pathogen; (3) altering the host crop by genetic or cultural means to enhance its relationship to antagonistic microorganisms and to weaken its relationship to the plant pathogen; and (4) introducing hyperparasites into areas where they are not normally found and where they may flourish. Some of these approaches have been used successfully in the past and will undoubtedly lead to more successes in the future.

Implementation of Hyperparasitism in Pest Management

The techniques used in studying interactions of microorganisms and in applying research findings to actual disease outbreaks determine the success or failure of control efforts. It is important, therefore, to consider the methods for using hyperparasites in pest management systems.

Preparation of Inoculum

The nutrition of mycoparasites is important in determining their ability to infect their hosts.[9,11,35,59] Thus, the composition of the substrate for growth of the biocontrol agent is important. Various materials have been employed, including standard laboratory media,[20,32] natural substrates such as cereal kernels or wheat bran,[22,59] and inert materials with an organic nutrient added.[5] This later method has been incorporated into a system for growth and delivery of *Trichoderma harzianum* and was developed by Backman and Rodriguez-Kabana.[5] Diatomaceous earth granules impregnated with a molasses solution, plus potassium nitrate and potassium phosphate, was used as the growth substrate. Modifications of this technique are being experimentally applied to mass production of inocula of antagonists by solid fermentation procedures by commercial companies in both the U.S.[31a] and Europe.[33a] These endeavors by commercial companies will greatly encourage field testing of potential hyperparasites and perhaps hasten the successful application of biological control agents in pest management systems. Commercial production and marketing of biological control agents, however, pose their own special considerations and problems. These include stability of antagonist strains, safety standards needed in manufacturing and field use, and effective means of transporting living inoculum.

Delivery Systems and Quality Control

Inoculum of the biological control agent must be economically producible and easily applicable, and its viability and effectiveness must be dependable. Antagonists grown on organic media have been added directly for control,[59] preparations have been dried and ground into a dust for application,[1,5] spores in a slurry have been used to coat seeds[32] or to paint tree wound surfaces or stumps,[20,21,43,46] and spores have been added to carriers such as bark pellets or peat.[51]

Two delivery systems have proved effective on a mass scale. Spores of *Peniophora gigantea* were produced on malt agar, packaged in small packets,[46] and applied economically[23] to cut tree stumps in a water suspension. In a recently adopted improvement of the technique,[19] spores of *P. gigantea* from commercial packets are added to chainsaw lubricating oil. In this way the stumps are automatically inoculated with the antagonist at the time the trees are felled. This method compares favorably with the conventional inoculation of stumps after felling. A similar mechanical application by dipping pruning shears in spore suspensions of *T. viride,* has been suggested to control silver leaf disease.[21]

Delivery of *T. harzianum* in diatomaceous earth-molasses medium is advantageous.[5] The preparation can be dried, greatly reducing the final weight; the resulting granules are easily applied to the field in standard agricultural granule machinery; and the viability of the resulting product is not affected.[5]

The viability of a product is of utmost concern, since the hyperparasite must be highly germinable and vigorous to be effective. This concern was considered by Parker[42] who developed viability tests for *P. gigantea.* A sequential sampling scheme determined the viability and the wood-decaying ability of commercially produced packets. Symptoms of wood decay are assessed on fresh pine wood in petri dishes for suspensions of the fungus that have a shelf life of 3 to 4 months in the refrigerator. Other biocontrol agents will require careful monitoring for quality control.

Use of Hyperparasitism in Integrated Pest Management

Sophisticated pest-management regimes using hyperparasites have not been fully developed. Devising schemes to best use microorganisms for control will require extensive knowledge of the ecology of the parasite, the host, and the crop plant to be protected. It will be necessary to take a broad ecological overview of pest problems associated with the major agroecosystems and to develop unified, ecology-based approaches to control of plant pathogens. Attention must be directed to production methods and to manipulation of the biological control agents in various combinations and in different situations to gain insight into how they are affected by other factors. These factors include time of the year for application, temperature and moisture conditions for application, advantages or disadvantages of applying chemicals simultaneously, and the rate of application. Each disease will require evaluation and formulation of the most effective control measures for that particular crop. Implementation of new methods and use of old methods in the best possible combination can open up new approaches to plant disease control.

Current Status and Future Prospects

To date, few examples of integrated control of plant diseases with hyperparasites are available; however, a significant beginning has been made. Several examples are available to demonstrate that an integrated approach can be made to work effectively and economically. Two examples amenable to the integrated approach are the control of *Fomes annosus* by *Peniophora gigantea* and control of *Sclerotium rolfsii* by *Trichoderma harzianum* that have already been discussed. Integrated control of *Armillaria mellea* in citrus orchards is another example. Bliss[10] found that *A. mellea* in roots was not killed immediately by soil fumigation with carbon bisulfide, but was killed in 24 days by *Trichoderma viride* in the soil. The fumigant stimulated the activity of *T. viride* against *A. mellea* which was weakened by the fumigant. Sublethal doses of fumigant are used in orchards, where stronger doses would injure the trees. In addition, manipulation of temperature and moisture in the soil stimulated the activity of *T. viride* and suppressed *A. mellea*.[39] The combination of treatments enhanced the action of *Trichoderma* in this instance.

Recently, successful use of an antagonist in the integrated control of *Rhizoctonia solani* damping-off of radish in the greenhouse was reported.[22] *Trichoderma harzianum* protected radish seedlings from damping-off and also enhanced germination of the seeds. Protection lasted for five successive plantings. In addition, use of pentachloronitrobenzene with *T. harzianum* had an additive effect. Survival of *R. solani* inoculum in soil was most adversely affected by the combination of fungicide and antagonist. Again, this example points out the advantages of favoring the antagonist while weakening the plant pathogen. *Trichoderma harzianum* was effective both in preventing the buildup and reducing the population of *R. solani* in the greenhouse. These studies should be expanded to evaluate its effectiveness in the field.

Another interesting possibility is the multiple effect of pesticides on pest control. Although it was shown only experimentally, the herbicide atrazine (2-chloro-4-[ethylamino]-6-[isopropylamino]-S-triazine) at high rates of application adversely affected the growth of *Sclerotium rolfsii*.[47] In contrast, growth of *T. viride* was either little affected or promoted. The effect of combinations of treatments on the success of biological control deserves exploration, and the use of such combinations could prove to be an economical, environmentally sound method of controlling more than one pest. Future attempts to control disease biologically may be highly dependent on such manipulation of the environment to favor the hyperparasite over the plant pathogen to be controlled.

CONCLUSIONS

The widespread occurrence of hyperparasitism in natural ecosystems suggests that plant diseases are often naturally controlled by this form of antagonism of one microorganism or agent against another. Relatively few cases of deliberate use of this form of biological control have been documented. The few successful examples do, however, demonstrate that this form of control is feasible and can be realized as a component in our integrated arsenal of control practices.

The future development, application, and success of biological control with hyperparasites and other antagonists will require greatly expanded research efforts to discover new control agents, understand how they work, learn to mass produce them and manipulate them in the agricultural environment, and learn to use them effectively and economically.

REFERENCES

1. **Ahmed, A. H. M. and Tribe, H. T.,** Biological control of white rot of onion *(Sclerotium cepivorum)* by *Coniothyrium minitans, Plant Pathol.,* 26, 75, 1977.
2. **Apple, J. L. and Smith, R. F.,** *Integrated Pest Management,* Plenum Press, New York, 1976.
3. **Ayers, W. A. and Adams, P. B.,** Mycoparasitism of sclerotia of *Sclerotinia* and *Sclerotium* species by *Sporidesmium sclerotivorum, Can. J. Microbiol.,* 25, 17, 1979.
4. **Ayers, W. A. and Lumsden, R. D.,** Mycoparasitism of oospores of *Pythium* and *Aphanomyces* species by *Hypochytrium catenoides, Can. J. Microbiol.,* 23, 38, 1977.
5. **Backman, P. A. and Rodriguez-Kabana, R.,** A system for the growth and delivery of biological control agents to the soil, *Phytopathology,* 65, 819, 1975.
6. **Baker, K. F. and Cook, R. J.,** *Biological Control of Plant Pathogens,* W. H. Freeman & Co., San Francisco, 1974.
7. **Barnett, H. L.,** The nature of mycoparasitism by fungi, *Annu. Rev. Microbiol.,* 17, 1, 1963.
8. **Barnett, H. L.,** Mycoparasitism, *Mycologia,* 56, 1, 1964.
9. **Barnett, H. L. and Binder, F. L.,** The fungal host-parasite relationship, *Annu. Rev. Phytopathol.,* 11, 273, 1973.
10. **Bliss, D. E.,** The destruction of *Armillaria mellea* in citrus soils, *Phytopathology,* 41, 665, 1951.
11. **Boosalis, M. G.,** Hyperparasitism, *Annu. Rev. Phytopathol.,* 2, 363, 1964.
12. **Boosalis, M. G. and Mankau, R.,** Parasitism and predation of soil microorganisms, in *Ecology of Soil-borne Plant Pathogens,* Baker, K. F. and Snyder, W. C., Eds., University of California Press, Berkeley, 1965.
13. **Civerolo, E. L. and Keil, H. L.,** Inhibition of bacterial spot of peach foliage by *Xanthomonas pruni* bacteriophage, *Phytopathology,* 59, 1966, 1969.
14. **Deacon, J. W.,** Biological control of take-all by *Phialophora radicicola* Cain, *EPPO Bull.,* 6, 297, 1976.
15. **De La Cruz, R. E. and Hubbell, D. H.,** Biological control of the charcoal root rot fungus *Macrophomina phaseolina* on slash pine seedlings by a hyperparasite, *Soil Biol. Biochem.,* 7, 25, 1975.
16. **DeVay, J. E.,** Mutual relationships in fungi, *Annu. Rev. Microbiol.,* 10, 115, 1956.
17. **Drechsler, C.,** Two hyphomycetes parasitic on oospores of root-rotting oomycetes, *Phytopathology,* 28, 81, 1938.
18. **Gindrat, D., van der Hoeven, E., and Moody, A. R.,** Control of *Phomopsis sclerotioides* with *Gliocladium roseum* or *Trichoderma, Neth. J. Plant Pathol.,* 83 (Suppl. 1), 429, 1977.
19. **Greig, B. J. W.,** Inoculation of pine stumps with *Peniophora gigantea* by chainsaw felling, *Eur. J. For. Pathol.,* 6, 286, 1976.
20. **Greig, B. J. W.,** Biological control of *Fomes annosus* by *Peniophora gigantea, Eur. J. For. Pathol.,* 6, 65, 1976.
21. **Grosclaude, C., Ricard, J., and Dubos, B.,** Inoculation of *Trichoderma viride* spores via pruning shears for biological control of *Stereum purpureum* on plum tree wounds, *Plant Dis. Rep.,* 57, 25, 1973.
22. **Henis, Y., Ghaffar, A., and Baker, R.,** Integrated control of *Rhizoctonia solani* damping-off of radish: effect of successive planting; PCNB, and *Trichoderma harzianum* on pathogen and disease, *Phytopathology,* 68, 900, 1978.
23. **Hodges, C. S., Jr.,** Cost of treating stumps to prevent infection by *Fomes annosus, J. For.,* 72, 402, 1974.
24. **Hollings, M.,** Mycoviruses: viruses that infect fungi, *Adv. Virus Res.,* 22, 1, 1978.
25. **Huang, H. C.,** Biological control of Sclerotinia wilt of sunflowers, *Annu. Contrib. Manit. Agron.,* 1976, 69.
26. **Huang, H. C.,** Importance of *Coniothyrium minitans* in survival of sclerotia of *Sclerotinia sclerotiorum* in wilted sunflower, *Can. J. Bot.,* 55, 289, 1977.
27. **Huffaker, C. B.,** Biological control in the management of pests, *AgroEcosystems,* 2, 15 1975.
28. **Hunter, W. E., Duniway, J. M., and Butler, E. E.,** Influence of nutrition, temperature, moisture, and gas composition on parasitism of *Rhizopus oryzae* by *Syncephalis californica, Phytopathology,* 67, 664, 1977.
29. **Ikediugwu, F. E. O., Dennis, C., and Webster, J.,** Hyphal interference by *Peniophora gigantea* against *Heterobasidion annosum, Trans. Br. Mycol. Soc.,* 54, 307, 1970.
30. **Jarvis, W. R. and Slingsby, K.,** The control of powdery mildew of greenhouse cucumber by water sprays and *Ampelomyces quisqualis, Plant Dis. Rep.,* 61, 728, 1977.
31. **Kerry, B. R.,** Fungi and the decrease of cereal cyst-nematode populations in cereal monoculture, *EPPO Bull.,* 5, 353, 1975.

31a. **Kenny, D. S.,** Abbott Laboratories, Long Grove, Ill., personal communication.

32. **Kommedahl, T. and Windels, C. E.,** Evaluation of biological seed treatment for controlling root diseases of pea, *Phytopathology,* 68, 1087, 1978.
33. **Kuhlman, E. G., Matthews, F. R., and Tillerson, H. P.,** Efficacy of *Darluca filum* for biological control of *Cronartium fusiforme* and *C. strobilinum, Phytopathology,* 68, 507, 1978.

33a. **Lisansky, S. G.,** Tate and Lyle, Ltd., Readings, England, personal communication.

34. **Lockwood, J. L.,** *Colloquium* on soil fungistasis and lysis: summary and synthesis, in *Biology and Control of Soil-borne Plant Pathogens,* Bruehl, G. W., Ed., The American Phytopathological Society, St. Paul, Minn., 1975.
35. **Lumsden, R. D.,** Ecology of Mycoparasitism, in *The Fungal Community, its Organization and Role in the Ecosystem,* Wicklow, D. and Carroll, G., Eds., Marcel Dekker, New York, 1981, 295.
36. **Mankau, R.,** Utilization of parasites and predators in nematode pest management ecology, *Proc. Tall Timbers Conf. Ecol. Anim. Contr. Hab. Manage.,* 4, 129, 1972.
37. **Mitchell, R. and Hurwitz, E.,** Suppression of *Pythium debaryanum* by lytic rhizosphere bacteria, *Phytopathology,* 55, 156, 1965.
38. **Moody, A. R. and Gindrat, D.,** Biological control of cucumber black root rot by *Gliocladium roseum, Phytopathology,* 67, 1159, 1977.
39. **Munnecke, D. E., Wilbur, W., and Darley, E. F.,** Effect of heating or drying on *Armillaria mellea* or *Trichoderma viride* and the relation to survival of *A. mellea* in soil, *Phytopathology,* 66, 1363, 1976.
40. **Norton, D. C.,** *Ecology of Plant-Parasitic Nematodes,* John Wiley & Sons, New York, 1978.
41. **Papavizas, G. C.,** Status of applied biological control of soil-borne plant pathogens, *Soil Biol. Biochem.,* 5, 709, 1973.
42. **Parker, E. J.,** Viability tests for the biological control fungus *Peniophora gigantea* (Fr.) *Eur. J. For. Pathol.,* 7, 251, 1977.
43. **Pottle, H. W., Shigo, A. L., and Blanchard, R. O.,** Biological control of wound hymenomycetes by *Trichoderma harzianum, Plant Dis. Rep.,* 61, 687, 1977.
44. **Rawlinson, C. J., Hornby, D., Pearson, V., and Carpenter, J. M.,** Virus-like particles in the take-all fungus, *Gaeumannomyces graminis, Ann. Appl. Biol.,* 74, 197, 1973.
45. **Ricard, J.,** Experience with immunizing commensals, *Neth. J. Plant Pathol.,* 83 (Suppl. 1), 443, 1977.
46. **Rishbeth, J.,** Stump inoculation: a biological control *Fomes annosus,* in *Biology and Control of Soil-borne Plant Pathogens,* Bruehl, G. W., Ed, The American Phytopathological Society, St. Paul, Minn., 1975.
47. **Rodriguez-Kabana, R., Curl, E. A., and Funderburk, H. H.,** Effect of atrazine on growth activity of *Sclerotium rolfsii* and *Trichoderma viride* in soil, *Can. J. Microbiol.,* 14, 1283, 1968.
48. **Sayre, R. M.,** Biotic influences in soil environment, in *Plant Parasitic Nematodes,* Vol. 1, Academic Press, New York, 1971, 235.
49. **Sneh, B., Humble, S. J., and Lockwood, J. L.,** Parasitism of oospores of *Phytophthora megasperma* var. *sojae, P. cactorum, Pythium* sp., and *Aphanomyces euteiches* in soil by oomycetes, chytridiomycetes, hyphomycetes, actinomyces, and bacteria, *Phytopathology,* 67, 622, 1977.
50. **Snyder, W. C., Wallis, G. W., and Smith, S. N.,** Biological control of plant pathogens, in *Theory and Practice of Biological Control,* Huffaker, C. B. and Messenger, P. S., Eds., Academic Press, New York, 1976.
51. **Sundheim, L.,** Attempts at biological control of *Phomopsis sclerotioides* in cucumber, *Neth. J. Plant Pathol.,* 83 (Suppl. 1), 439, 1977.
52. **Swendsrud, D. P. and Calpouzos, L.,** Effect of inoculation sequence and humidity on infection of *Puccinia recondita* by the mycoparasite *Darluca filum, Phytopathology,* 62, 931, 1972.
53. **Tronsmo, A. and Dennis, C.,** The use of *Trichoderma* species to control strawberry fruit rots, *Neth. J. Plant Pathol.,* 83 (Suppl. 1), 449, 1977.
54. **Tronsmo, A. and Raa, J.,** Antagonistic action of *Trichoderma pseudokoningii* against the apple pathogen *Botrytis cinerea, Phytopathol. Z.,* 89, 216, 1977.
55. **Turner, G. J. and Tribe, H. T.,** Preliminary field plot trials on biological control of *Sclerotinia trifoliorum* by *Coniothyrium minitans, Plant Pathol.,* 24, 109, 1975.
56. U.S. Department of Agriculture, Biological Agents for Pest Control: Status and Prospects, U.S. Government Printing Office, Washington, D.C., 1978.
57. **Veselý, D.,** Potential biological control of damping-off pathogens in emerging sugar beet by *Pythium oligandrum* Drechsler, *Phytopathol. Z.,* 90, 113, 1977.
58. **Vidaver, A. K.,** Prospects for control of phytopathogenic bacteria by bacteriophages and bacteriocins, *Annu. Rev. Phytopathol.,* 14, 451, 1976.
59. **Wells, H. D., Bell, D. K., and Jaworski, C. A.,** Efficacy of *Trichoderma harzianum* as a biocontrol for *Sclerotium rolfsii, Phytopathology,* 62, 442, 1972.
60. **Wells, H. D.,** personal communication.

ENVIRONMENTAL CONTROL OF WEEDS

F. W. Slife

The plants we call weeds are well adapted to exploit the ecological niches left open in those environments altered by man. Most weeds have developed the necessary survival mechanisms through decades of selection pressure in the environment of other living organisms.

The objective of integrated weed management is to create conditions unfavorable to weeds while maintaining suitable conditions for crops or other beneficial vegetation. Weeds are most efficiently controlled by the simultaneous application of a variety of practices. These practices may be classified as preventive, physical, managerial, biological, and chemical.

PREVENTIVE WEED CONTROL

The most basic of all weed control methods is prevention, that is, those measures taken to forestall the introduction and spread of weeds. Many of our present weed species were introduced into this country with the early settlers and spread with their pursuit of agriculture. Unfortunately, we continue to introduce weeds into new areas today, in a variety of ways.

Weed Seed in Crop Seed

Some crop seeds, such as cereals, legumes and grasses, are rarely free of weed seeds. The difficulty of removing weed seed of the same size and density of the crop seed makes the problem a continuing one. Some crop seed is bought or traded among farmers without cleaning, and under these conditions weed seed is effectively spread. Fortunately, most crop seed is now purchased through seed companies that strive to produce crop seed free of weed growth. In addition, sophisticated cleaning equipment is used to remove contaminants. Farmers can prevent infestation or reinfestation of potentially serious weeds by purchasing certified crop seed or seed from reliable sources.

In the U.S., both the federal and state governments have clean-seed laws. These prohibit distribution of mislabeled or contaminated seed locally and in interstate commerce. In addition, these laws regulate seed imported into the U.S.

Farm Equipment

Weed seeds can be moved long distances by ships, airplanes, railroad cars, trucks, automobiles, and other types of transportation. Seeds may adhere to the tires or surfaces of the vehicles or they may be a part of the agricultural commodity being moved.

Although the farmer has no control over most transportation machinery, he can do much to reduce the spread of weeds by his own agricultural machinery and that of custom harvesters. All kinds of farm equipment are responsible for spreading weed seeds and vegetative organs from field to field and farm to farm. Seedbed preparation equipment scatters vegetative organs of weeds over fields. Mature weed seeds are often harvested by combines, threshing machines, choppers, hay balers, and other harvesting machines. Some of these seeds may be returned to the field or become contaminants in the harvested product.

Combines and similar harvesters should be equipped with seed cleaners, and will trap much of the weed seed during harvest. This seed can be destroyed rather than letting it fall on the ground with crop residue. Although combines are difficult to clean, all possible residue should be removed. A portable air compressor is an effective tool for blowing seeds out of harvesting machines.

Planting and tillage equipment can be cleaned easily before moving to a new location by removing vegetative plant parts and soil that clings to the equipment.

Animals

Many weeds are disseminated by weed seeds adhering to the hair or feathers of animals or by passing through the digestive tract. Managers of livestock can do much to reduce the spread of weeds through the movement of animals. Proper disposal of manure at loading and unloading points is necessary and animals should be penned or held for several days when being moved from weedy areas to less weedy ranges. Most weed seeds will pass through the digestive tracts in 3 to 4 days.

Fence Rows, Irrigation, and Drainage Ditches

Ditches are important sources of weed seeds. They not only contain aquatic weeds, but most of the weeds of cultivated areas grow on the ditch banks. Unless controlled, the water serves as an excellent method of distribution for both seeds and vegetative parts of these plants.

Dredging, draglining, or chaining has been used in the past to remove this vegetation, but in recent years aquatic herbicides have been widely used. Mowing the banks of the ditches and selectively using herbicides to aid in the establishment of desirable grass species reduces the terrestrial weed population. Fence rows provide a continuing source of weed seeds unless managed properly. Mowing or the use of selective herbicides can be used to control weeds and aid in the establishment of desirable grasses.

PHYSICAL METHODS

Weed control by physical means remains the predominant method of weed control in spite of the very rapid increase in herbicide use. Physical methods are used to destroy annual weeds and thus give the crop the competitive advantage before another crop of annual weeds appears. Repeated physical methods are used to destroy the top growth of perennial weeds to prevent seed production and exhaust the food reserve in the root system. Physical methods of weed control can be classified into manual means, machine tillage, mowing and cutting, flooding, heat, and mulching.

Manual Means of Control

Hand pulling and tillage by hand is still used extensively worldwide. On small farms in many tropical areas it is the only method of weed control. For home gardens it is used extensively because only a few herbicides are available for such use. Hand tillage is also used extensively on minor crop acreage. These specialized crops, although extremely important, may occupy only a few thousand acres of land each year. Since herbicides are rarely registered for these minor use crops, hand weeding is a necessity. Hand weeding is also used on some large acreage crops such as soybeans. It is used to remove scattered weed plants left from cultivation and herbicide treatment. This procedure prevents seed production. Hand weeding can greatly reduce the invasion of a weed species into new areas by removal of these scattered plants.

Hand pulling or hand tillage is very effective on annual or biennial weed species, but is less effective on perennials.

Machine Tillage

Tillage with implements powered by animals or machine is a common practice throughout the world. Animal-drawn tillage equipment is rapidly decreasing in the more developed agricultural regions in favor of motorized tillage.

Tillage for weed control remains the most important method of weed control in all parts of the world. Herbicides are used to supplement tillage practices and in many instances have reduced the amount of tillage. There are only a few examples where herbicides have eliminated tillage operations.

Tillage reduces or eliminates weed competition for moisture, nutrients, light, and carbon dioxide, and thereby improves crop growth. In addition, turning under crop residues improves the soil structure, adds organic matter to the soil, and reduces other pest problems. Tillage is used to alter the physical conditions of certain soils that have clay pans or other soil barriers.

With the introduction of herbicides, it has been possible to study the effect of tillage other than for weed control. These studies indicate that tillage is of little benefit other than for weed control under most soil conditions. Farmers are currently utilizing the zero tillage concept on several million acres of cropland with successful results.

Fallowing for weed control is aimed at controlling patches of perennial weeds, usually one particular species, and it is used to conserve moisture in the arid areas in the western half of the U.S. In the case of perennial weeds, the repeated removal of the top growth prevents the plant from storing manufactured food in the root system, but at the same time depletes the food reserve, since the early growth of the perennial is made at the expense of the root system. The repeated tillage breaks up rhizomes or roots into short pieces which results in more plants within a few weeks. These plants are destroyed by the next tillage which hastens the process of food depletion. The underground root system of some perennial weeds can be eliminated with one season of fallowing if it is done correctly. On the other hand, species that have deep and extensive root systems, such as field bindweed and Canada thistle, may require two full seasons of fallowing for complete elimination.

In some of the arid regions, crops are planted in alternate years because there is not enough soil moisture for growing a crop each year. When the land is not in crop, the weed growth must be destroyed in order to conserve the moisture. Historically, this has been accomplished through repeated shallow tillage, leaving the soil susceptible to wind and water erosion. In recent years much of the fallow land has not been tilled and herbicides have been used to control weeds. This practice reduces erosion, conserves more soil moisture, and appears to be more economical than the costly fallowing operation.

Mowing and Cutting

The first cutting implements were sharp rocks and their use dates back thousands of years. These eventually gave way to more complex machines of sharp metal. Manual cutting implements are still widely used throughout the world. They include the machete, scythe, hoe, and a wide variety of other implements. Manual implements are employed in areas inaccessible to power-driven equipment or where power equipment does not exist.

Mowing controls weeds in two ways. If properly timed, mowing prevents plants from producing seed, and repeated mowings also aid in the control of some perennial plants by depleting the underground food supply. Mowing is ineffective against prostrate or short weeds growing close to the ground. In some cases, prostrate or low-growing weeds are often invaders of frequently cut areas because the competition to their growth and existence has been reduced. Most weeds should be cut in the bud stage or earlier since they may produce viable seed if the heads are cut during flowering. Perennial weeds should be cut in the early bloom stage since this is usually the period in which the weed has the lowest amount of food reserves in the roots.

There are about one billion acres of pasture and grazing lands in the U.S. On parts of nearly all of this area, woody plants present some problem. On the western dryland ranges, undesirable woody plants rapidly invade under poor management systems. The delicate balance between desirable and undesirable species is very easily upset. Repeated mowing, cutting, or defoliation is widely used in the western ranges to suppress the growth of brush species. Chemicals have replaced some of the mowing and cutting procedures in recent years.

Highway rights-of-way provide an area for weed plants to grow unmolested. Some of the rural areas are not mowed or most are not mowed frequently enough to prevent seed production. Perennial weed species predominate in these areas after desirable perennial grasses are established. Timely mowing will prevent most of the weed species from spreading and give the turf species the competitive advantage. In some areas, after years of mowing, the grass sod had become so well established that mowing has been eliminated. Herbicides are used for spot treatment of perennial weeds as the need arises. This latter system is more economical than the mowing procedure, and weed growth is minimal.

Flooding

Flooding is utilized to control many species of weeds in rice fields. Unfortunately, some weed species such as barnyardgrass have nearly the same growth habits as rice and often cannot be controlled by flooding. Selective herbicides are used for those weeds in rice not controlled by flooding.

In some of the western states, flooding is used for controlling established perennial weeds such as silverleaf nightshade, camelthorn, and Russian knapweed. This is a costly procedure since dikes must be built and 6 to 10 in. of water are needed for 3 to 8 weeks for elimination. Flooding has the advantage over chemical sterilants in that no harmful soil residue is left on the treated area.

Heat

One of the more neglected methods of weed control is the use of fire and heat. In many situations, burning cannot be equaled for removal of existing uneconomic vegetation and for the removal of general noxious vegetation in a wide variety of situations. Controlled burning of rangeland in the western states is perhaps the single most important practice used to improve or increase livestock forage by eliminating competing plants or by reducing litter and stimulating growth of desirable forage plants. Wildlife habitat also can be improved by controlled burning. Burning is usually restricted to those areas that have accumulated enough litter so that the fire will continue to burn. Controlled burning also reduces the possibility of wild fires that occur frequently under those arid conditions. Controlled burning is also important in certain coniferous forests. In this case, the fire is used to keep the forest free of an undesirable understory of vegetation. Reducing the understory of vegetation greatly reduces the possibility of forest fires.

Heat is used to control weeds by steam sterilization of nursery soils in certain plant materials to kill noxious weed seed and prevent its dissemination. Another use of heat is the procedure known as flame cultivation. The flaming machine is driven through the cropland with the burners directed diagonally at the base of the crop row. Speed of travel and the temperature are adjusted such that tender weed seedlings are killed without injuring the larger crop plants. The flame does not kill by actual combustion, but by raising the temperature of the plant tissue slightly above the normal death point. Flaming has been used for many years in the cotton growing areas, but its use is declining rapidly with the use of suitable herbicides for the cotton crop.

Mulches

The use of mulches to control weeds is increasing. Nonliving material such as straw, sawdust, bark, wood chips, paper, and plastic are used to exclude light and, hence, the control of annual weeds is quite good. Most perennial weeds are not well controlled by this method since they have enough food reserve in their root system to emerge through the mulch. In the past, mulching has been confined to gardens, orchards and landscape material. Presently, however, there is much interest in using sheet materials for mulching in high intensity crop productions. For example, seedbeds are covered with strips of black polyethylene or paper, and plants of many horticultural crops and home ornamentals are transplanted through holes in the cover. Paper mulching has proven particularly effective and economical on unirrigated sugar and pineapple plantations in Hawaii. In addition to weed control, there are often additional environmental and cultural advantages to this practice.

MANAGERIAL PRACTICES

Proper management of our forests, rangeland, and intensive cropped land can do much to reduce the seriousness of the weed problem. Proper management is using as many practices as possible that will allow the crop to gain a competitive advantage over the weeds. Management practices are not static and need modification as the weed spectrum changes and as new cultural practices are introduced. Scouting fields to determine which weed species are present is a vital part of a weed management program.

The individual preventive, mechanical, biological, or chemical methods of weed control employed separately are but minor forces toward accomplishing the objective of most weed control programs. It is only when methods are integrated in such ways to reinforce one another that they become major deterrents to invading vegetation. Each manageable factor in the habitat then should be reviewed as a potential weed control method and used accordingly. The most important agent in cropland weed control is the crop itself. The vigor and competitive ability of the crop, in terms of local conditions and the weed flora present, are critical factors in determining whether or not the crop will capture and maintain a dominant position.

Some weed problems can be reduced by avoiding peak periods of weed seed germination. Delaying crop planting can be used to destroy one or more crops of early germinating summer annual weeds. In irrigated areas, irrigation can be used to stimulate weed seed germination. After weeds emerge, they are killed by shallow cultivation. The crop is then planted deep enough to utilize the soil moisture for crop seed germination. By the time irrigation is needed again, the crop has the competitive advantage. Many of our forage crops are planted in late summer to avoid the weed problem that is always present in spring seedlings.

The row spacing of crops greatly affects weed intensity. Historically, row spacing was determined by the width of animals used in planting and tilling the soil. Recently, row width has been reduced and thus the crop canopy covers the area between the rows much faster. The result is that the crop is extremely competitive to weeds that germinate later in the season.

Perhaps the most important management practice is what could be termed good husbandry. This includes using proper grazing practices, timely cultivation procedures, proper irrigation procedures, and all other practices that have a direct or indirect effect on weed intensity.

Crop rotations were one of the earliest management practices adapted. Historically, reasons given for recommending a sequence of crops were: (1) maintenance of good soil tilth and structure, (2) supply of the legume nitrogen for other crops, (3) better water infiltration and less erosion, (4) less economic risk, (5) better labor distribution, and (6) better weed, insect, and disease control. In spite of the possible advantages for rotations, the trend towards concentrating on one or two crops on a farm continues. It would appear that the economics of concentrating on fewer crops is more favorable than wide crop diversification. Abundant fertilizer and pesticides have made this system possible. From the standpoint of weed control without herbicides, growing a series of crops with divergent life cycles is extremely beneficial. Those weed species with the same life cycle as the crop tend to increase rapidly when the crop is grown continuously. For example, in a continuous corn system, summer annual weeds—both grasses and broadleaves—would tend to dominate the area. Where winter wheat is alternated with a summer row crop, the intensities of the summer annual weeds decrease because winter wheat is an excellent competitor in the early spring, and thus reduces germination of summer annual weeds and their further establishment. It should be pointed out, however, that some of the winter annual weeds with the same life cycle as wheat will become established when the land is sown to wheat.

Rotations that include several crops with different life cycles will tend to prevent any one group of weeds from becoming dominant and extremely serious. On the other hand, the same rotations will have a diversity of species simply because some can persist in one crop and not another.

Cropping systems can be altered to fight a particular weed problem. Winter wheat is frequently used to control johnsongrass infestations. Although johnsongrass is a perennial, it grows vigorously from spring to late fall. Fall planted winter grains become well established by early spring and, hence, offer severe competition to johnsongrass. These winter grains will be harvested before johnsongrass resumes its vigorous growth. After grain harvest, control measures can be initiated in that particular area.

In some areas, corn is grown for several years in the same area so that annual broadleaf weeds such as cocklebur can be treated with 2,4-D. Then land can be returned to other crops in the rotation where cocklebur control is more difficult. Vegetable crop growers frequently grow a nonvegetable crop that has good weed control for several years to reduce the weed problem before returning the land to intensive vegetable production.

Although the trend has been away from a wide diversity of crops on a farm unit, more emphasis will probably be placed on rotations in the future. For example, growing corn continuously on the same area has been possible because pesticides have been available to control pest problems. However, the developing resistance of the corn insects, particularly root worm, to insecticides indicates that corn will have to be rotated with other crops to reduce the seriousness of this problem. Crop diversification will probably receive more emphasis as integrated pest management systems develop.

Biological Control

Biological control is mentioned here since it is a part of the environmental control of weeds. It is covered in depth in other sections of this book. The use of natural enemies of weeds—parasites, predators, and pathogens—is by far the cheapest method of control and there are some striking examples of success in noncultivated areas. In a broader sense, biological control of weeds could be defined as the use of any living organism to control weeds. Some of our domestic animals, particularly sheep and goats, have long been used for weed control. Smother crops have been of value in controlling perennial weeds. Dense stands of millet, sudangrass, sweet clover, and others have been used for this purpose. Recent work on the allelopathic effects of crops on weeds indicates that in the future we may be able to develop crops that suppress weeds by this means.

Chemical Control

Chemical control is mentioned here since it has made a major impact on physical methods of weed control and management practices. Herbicides are now widely used in our forests, rangelands, and cropland areas. They are used to supplement other sound weed control practices. The wise uses of herbicides integrated into weed management programs have resulted in a level of weed control that was not possible before modern day herbicides were introduced. Weed resistance to herbicides has not become a problem as yet, and as long as herbicide treatments are changed frequently, the likelihood of weed resistance to herbicides is low. In addition to greatly reducing weed interference with our crops, herbicides have increased our total production of food, feed, and fiber by allowing new crop culture techniques to develop. Zero tillage is an example. This practice is used on several million acres and is increasing. Zero tillage reduces erosion and conserves soil moisture, and has allowed crops to be grown on sloping land without significant erosion losses. Other forms of conservation tillage are being practiced even more widely than zero tillage. The introduction and wide acceptance of conservation tillage systems would not have been possible without herbicides. Detailed information on herbicides is given elsewhere in this book.

REFERENCES

1. **Pace, C. P. and Grandfelt, C. E.,** The use of fire on Arizona rangelands, *Rocky Mt. For. Range Exp. Stn.* Publ. No. 4, Arizona Interagency Range Committee, Ft. Collins, Colo., 1977.
2. **Fryer, J. D. and Matsunaka, S.,** *Integrated Control of Weeds,* University of Tokyo Press, Tokyo, 1977.
3. **Klingman, G. C. and Ashton, F. M.,** *Weed Science Principles and Practices,* John Wiley & Sons, New York, 1975.
4. National Academy of Sciences, *Principles of Plant and Animal Pest Control,* Vol. 2, National Academy of Sciences, Washington, D. C., 1968.
5. National Academy of Sciences, *Pest Control Strategies for the Future,* National Academy of Sciences, Washington, D.C., 1972.
6. **Roberts, Daniel A.,** *Fundamentals of Plant-Pest Control,* W. Freeman, San Francisco, 1978.

ENVIRONMENTAL CONTROL OF RODENTS

David E. Davis

INTRODUCTION

This section describes the results of efforts to manage rodent populations by altering some aspect of the environment. However, the addition of poisons to the environment is excluded from this section since it is more appropriately described elsewhere. The principle followed in environmental control is to decrease some factor (food, shelter, etc.) so that the capacity of the area is lowered. Such manipulation can reduce the rodents and their damage to acceptable levels. The rodents will remain at such levels as long as the capacity is kept low. Lamentably, although many programs have altered the habitat of rodents, documentation is rare. Biological control, which is the introduction of a predator or parasite, will be included since such agents are part of the environment.

Many definitions of control exist. Here, control is considered to be achieved when a population remains below a level of economic importance without frequent (annual, monthly) treatments specially designed to affect the rodent. Thus, the annual distribution of poison is not considered satisfactory, whereas the improvement of housing (that has multiple benefits and lasts years) is considered satisfactory.

RODENTS (RODENTIA)

Rodentia, the largest order of mammals, contains some 35 families and 350 genera. More than half the individual mammals in the world are rodents. Anatomically, their characteristics are rather uniform. Most are small, terrestrial or arboreal, and lack specialization of fur or limbs except in a few cases such as porcupines and flying squirrels. The distinguishing characteristic is the presence in each jaw of two persistently growing incisor teeth, adapted to gnawing. A gap separates these teeth from the grinding molars behind.

Rodents occur throughout the world and have been introduced accidentally or purposefully on islands and continents. The damage by rodents is vast but rarely documented by numbers except for propaganda purposes.

The data on environmental control of rodents will be discussed under the major groups of rodents. Information on classification may be obtained from several texts.[1,2]

Ground Squirrels and Allies

A large division of Rodentia includes marmots, ground squirrels, chipmunks, and prairie dogs. They occur throughout the northern hemisphere of the Old and New World and are major agricultural pests in grain crops, rangeland, and orchards. They are often a nuisance in parks and they are a multiplier (not a reservoir) of plague. For decades various poisons have provided temporary control so that no environmental manipulation has been documented. It has been known for years[3] that in many species dense or tall vegetation will deter the establishment of colonies, but no experiments have been conducted.

Voles (Microtines)

A vast number of small rodents belong to the Microtine, a subfamily of the Cricetidae family of mice. Microtines occur in the northern parts of the world and generally live in open or grassy areas. Most are terrestrial and may make runways or tunnels in soil, grass, or snow. They have very specific preference for particular plants as food.[4,5] Microtines are opportunists that invade a habitat in early stages of succession and increase rapidly. They cause extensive (but undocumented) damage to fruit and forest trees, nurseries, and grain crops.

Experimental work on environmental control of voles in orchards has demonstrated results.[6-8] Cultivation of the area under apple trees reduced the population and kept it low (Figure 1) and the use of herbicides also affected the population (Figure 2). Populations in relatively clean orchards remained low (Figure 3). For another species, weed control is claimed to reduce damage to plantations of forest trees.[9]

Mice (Cricetines)

A large number of mice belong to the cricetine subfamily of the Cricetidae. Nomenclatural confusion arises because some rodents, also called mice (e.g., house mouse), do not belong to this group. The cricetines are found throughout the world, and, while most are terrestrial, some are arboreal or even aquatic. The tails of most species are short and the molar teeth are adapted to a diversified diet including insects. They live in diverse habitats and rarely cause extensive damage except to forestry plantations. The usual method of control is to use poison on seeds before distribution or during the seedling stage. Later, as succession procedes, the mice no longer present problems.

Cricetines of several species in North America and Russia are the reservoir of plague (i.e., they survive infection and maintain fleas). Control programs directed at these species have not been developed.

Rats (Muridae)

This large family of Old World rodents has become a serious urban pest as well as an agricultural pest. Rats and house mice have followed man throughout the world. Members of the Muridae may be large or small, they have teeth adapted to a diverse diet, and they inhabit marshes, fields, forests, and now cities. While many species exist, only 10 to 20 are serious pests. The damage throughout the world includes crops, buildings, and plague and other diseases. Data on numbers and damage are notoriously unreliable because the figures are inflated for propaganda purposes.

The major crops attacked by rats are grains in Europe and Russia, sugar cane in various countries, rice in southeast Asia, and smaller crops such as coconuts and fruits. Barriers are often useful; the damage by rats to cane has been reduced by plowing a strip around the field, the destruction of rice was stopped by an electric fence, and coconuts were protected by guards on the trees.[10]

Economic damage in buildings has been reduced by sanitation and rat-proofing for decades.[11] Unfortunately, because the programs usually are conducted on a massive scale, documentation is rare. The extensive programs during the Depression were not documented nor were the wartime efforts for the control of typhus fever. However, environmental control was documented in Baltimore for 1945 to 1950. The rehabilitation of housing resulted in a reduction of rats (Figure 4), the removal of fences produced a decline (Figure 5), the reduction of food, trash, and fences also produced a decline (Figure 6) and the cleanup of garbage caused a precipitous decline (Figure 7).

Some other work provides some data. Storage containers for grain can be constructed to prevent damage.[12] A "barrier" of poison bait reduced the invasion of sewers.[13] Chemosterilants, which may be classed as a means of environmental control, show promise. Use of a sterilant over a period of 4 months on a city dump stopped reproduction (zero of nine pregnant) in contrast to a reference population (six of nine pregnant). Also, few young were present.[14] Some experiments were inconclusive due

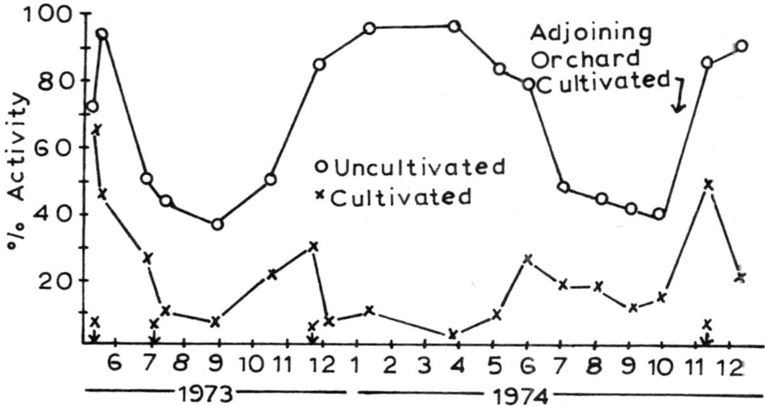

FIGURE 1. Cultivation several times a year (at arrows) reduced vole numbers (as measured by activity) in comparison with an uncultivated orchard. Cultivation was done by a special Rototiller® under the canopy of the tree, but not in the alleys.[6]

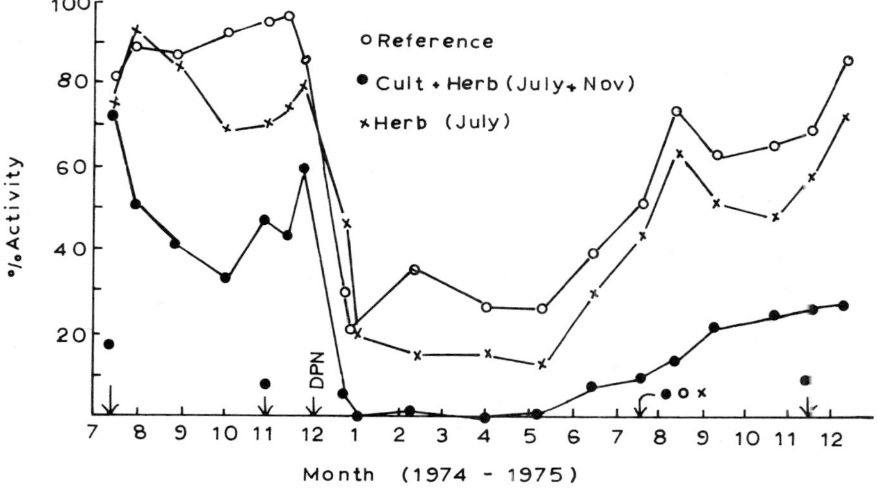

FIGURE 2. Procedures for reducing vegetation by herbicides reduced the amount of activity by voles in orchards.[6]

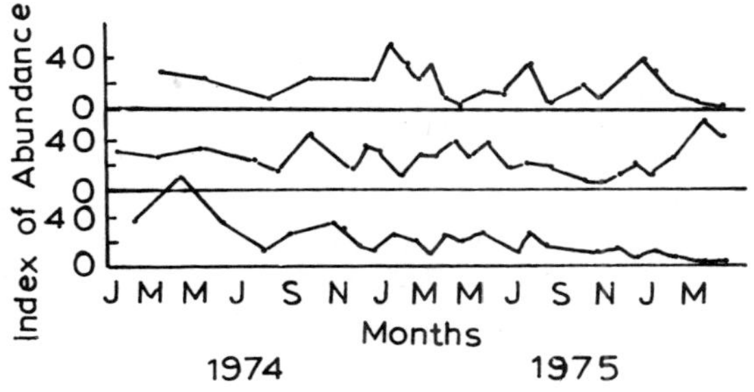

FIGURE 3. Regular mowing of grass in the alleys and killing of weeds by herbicides under the trees kept pine voles at a level below economic damage. The index of abundance is the percentage of trees that have tunnels. The percentage with live voles is, of course, lower.[8]

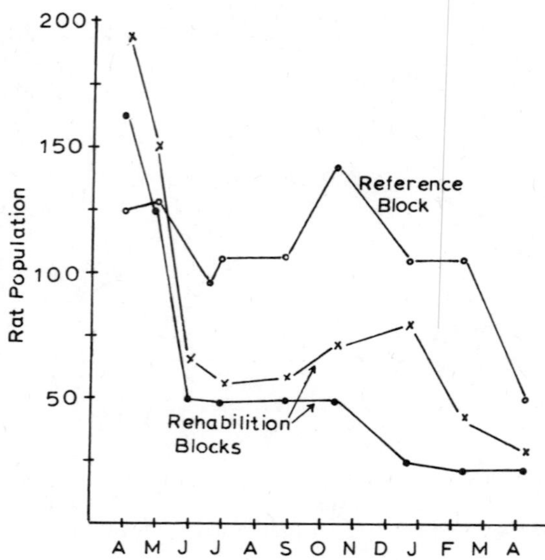

FIGURE 4. A comprehensive program for improvement of housing conditions resulted in reduction of numbers of rats.[11]

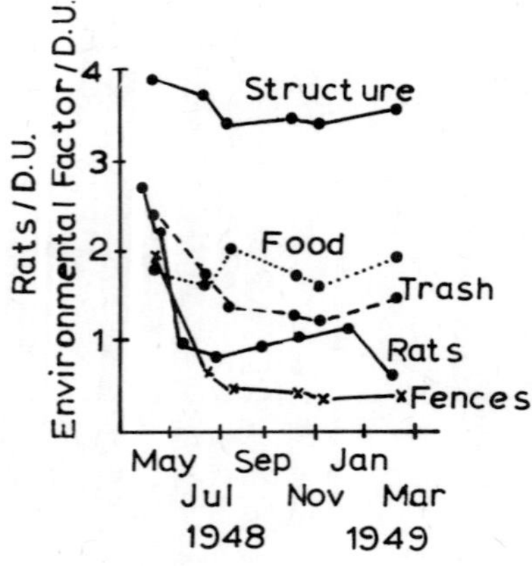

FIGURE 5. Removal of fences produced a decline of rats in residential areas even though the amount of food and trash remained the same.[11]

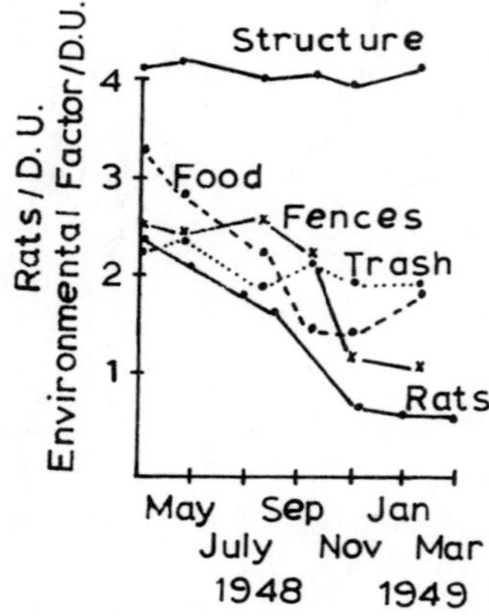

FIGURE 6. Removal of food, trash, and fences resulted in a more drastic decline than did removal only of fences.[11]

to seasonal decline in the reference area[15] but others succeeded in reducing the percentage of juvenile rats from 44 to 13% in contrast to the reference area (21 to 25%.)[16] In summary, environmental control has been sufficiently successful when practical problems of action are solved, but few cases are documented.

Biological control of rats by the introduction of cats has been moderately successful. Cats can keep a population low after reduction by poisoning[17] and can reduce a population by killing the young until some other source of food, such as squab, becomes available.[18] Attempts to control rats by mongooses have been equivocal.[19] Although mongooses may kill rats, they soon turn to more easily captured prey such as lizards and birds.

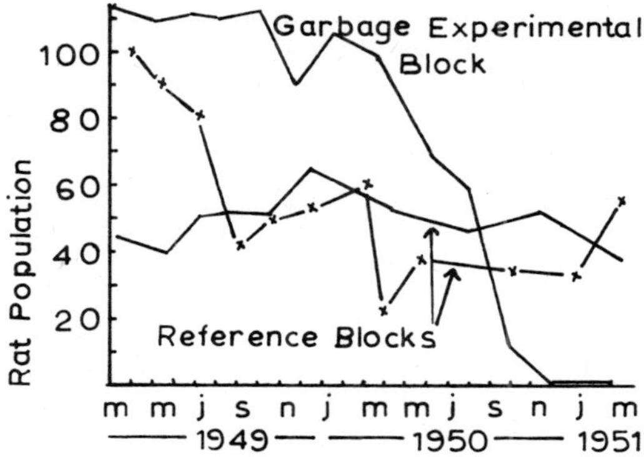

FIGURE 7. Reduction in rat population in a block followed the reduction in food (garbage) available in contrast to the blocks without reduction of food.[11]

Table 1
THE NUMBER OF GOPHERS TRAPPED BEFORE AND AFTER DESTRUCTION OF VEGETATION IN JULY 1956 IN AN EXPERIMENTAL AREA[19]

	1956	1957
Reference area	101	110
Experimental area	117	15

Other Rodents

Many species of rodents become a problem in local areas. Porcupines may damage timber in parts of North America but no environmental control has been documented. Pocket gophers damage lawns and crops in western North America. One report[21] used a herbicide to destroy forbs in pastures (Table 1). Destruction of forbs may improve rangeland for cattle, but in most areas (lawns, gardens, alfalfa fields) gophers compete directly.

Rabbits

While rabbits are not rodents, they do resemble rodents ecologically in many ways. No documentation of pure environmental control is available although fences have been used for centuries. However the most spectacular example of biological control of a vertebrate concerns the Australian and European rabbit *(Oryctolagus cuniculus)*. Documentation is abundantly available in Australia[20,22] and in England[23] to show a decline of 90%. The populations remain in many places in Australia at 1% of their level of 1950. Differences, however, exist throughout Australia. In zones where the virus is transmitted by mosquitoes, the epizootics occur regularly in spring and summer, keeping the populations so low that predators, by catching the young, regulate the population. However, in other areas the vector is a mite which increases in winter.

The epizootics are sporadic but may cause mortality of 90%.[20] In England the flea is the vector, and reduction by the myxoma virus is less. The relationships among the rabbit, the virus, and the vectors are evolving towards a balance of numbers.

REFERENCES

1. **Anderson, S. and Jones, J. K.,** *Recent Mammals of the World,* Ronald Press, New York, 1967.
2. **Walker, E. P.,** *Mammals of the World,* Vols. I and II, Johns Hopkins Press, Baltimore, 1964.
3. **Bond, R. M.,** Range rodents and plant succession, *Trans. N. Am. Wildl. Conf.,* 10, 229, 1945.
4. **Abbott, H. G.,** Tree seed preferences of mice and voles in the Northeast, *J. For.,* 60, 97, 1962.
5. **Thompson, D. Q.,** Food preferences of the meadow vole *(Microtus pennsylvanicus)* in relation to habitat affinities, *Am. Mid. Nat.,* 74, 76, 1965.
6. **Byers, R. E., Young, R. S., and Neely, R. D.,** Review of cultural and other control methods for reducing pine vole populations in apple orchards, *Proc. Vert. Pest Conf.,* 7, 242, 1976.
7. **Davis, D. E.,** Management of pine voles, *Proc. Vert. Pest Conf.,* 7, 270, 1976.
8. **Davis, D. E.,** Advances in rodent control, *Z. Angew. Zool.,* 64, 193, 1977.
9. **Radvanyi, A.** Harmful effects of small mammal populations on a tree plantation in southern Ontario, *Can. Field Nat.,* 89, 53, 1975.
10. **Strecker, R. L., Marshall, J. T., Jackson, W. B., Barbehenn, K. R., and Johnson, D. H.,** Pacific Island rat ecology; report of a study made on Ponape and adjacent islands 1955—1958, *Bernice P. Bishop Mus. Bull.,* 225, 274, 1962.
11. **Davis, D. E.,** The characteristics of rat populations, *Quart. Rev. Biol.,* 28, 373, 1953.
12. **Chaturvedi, G. C., Patel, M. J., and Madsen, C. R.,** Studies on improved storage systems, *Proc. All India Rodent Seminar* (1975), 136, 1977.
13. **Greaves, J. H., Hammond, L. E., and Bathard, A. H.,** The control of re-invasion by rats of part of a sewer network, *Ann. Appl. Biol.,* 62, 341, 1968.
14. **Gwynn, G. W.,** Field trial of a chemosterilant in wild Norway rats, *J. Wildl. Manage.,* 36, 823, 1972.
15. **Bowerman, A. M., and Brooks, J. E.,** Evaluation of U-5897 as a male chemosterilant for rat control, *J. Wildl. Manage.,* 35, 618, 1971.
16. **Marsh, R. E., and Howard, W. E.,** Evaluation of mestranol as a reproductive inhibitor of Norway rats in garbage dumps, *J. Wildl. Manage.,* 33, 133, 1969.
17. **Elton, C. S.,** The use of cats in farm rat control, *Br. J. Anim. Behav.,* 1, 151, 1953.
18. **Davis, D. E.** The use of food as a buffer in a predator-prey system. *J. Mamm.,* 38, 466, 1957.
19. **Pimentel, D.,** Biology of the Indian mongoose in Puerto Rico., *J. Mamm.,* 36, 62, 1955.
20. **Davis, D. E., Myers, K., and Hoy, J. B.,** Biological control among vertebrates, in *Theory and Practice of Biological Control,* Huffaker, C. B. and Messenger, P. A., Eds., Academic Press, New York, 1976.
21. **Keith, J. C., Hansen, R. M., and Ward, A. L.,** Effect of 2,4-D on abundance and foods of pocket gophers, *J. Wildl. Manage.,* 23, 137, 1959.
22. **Myers, K.,** The rabbit in Australia, in *Dynamics of Populations,* de Boer, P. J. and Gradwell, G. R., 1971.
23. **Mead-Briggs, A. R.,** The European rabbit, the European rabbit fleas and myxomatosis, in *Applied Biology,* Vol. II, Coaker, T. H., Ed., 1977.

ENVIRONMENTAL CONTROL OF BIRDS

Keith A. Arnold

INTRODUCTION

Problems arise between bird and human populations for three reasons: economics, health hazards, and nuisances. Bird populations are competing with humans for resources and, consequently, they increase the cost of those resources to humans. Bird populations threaten the life and health of humans and their domesticated animals. Congregations of birds cause discomfort to humans through noise, defecation, and/or sheer numbers. Often, one of the former two problems is given to provide reason for relieving nuisance situations.

The general reaction of humans to bird problems has been the extermination (or attempted extermination) of the offending birds. Even chemicals promoted as bird repellents are often diluted avicides which function by repelling the birds through sounds and behavior of dying birds.

There are several reasons why extermination of an offending bird population is not desirable. First, few means of extermination are selective. Chemical poisons are available to kill any bird which will consume a particular type of bait and which inhabit the problem area. Very few, if any, toxins are so specific in their actions that non-target species will not be affected. Wetting agents used at bird roosts are effective only under certain conditions, and then may not kill the offending species, while killing desirable species. For example, use of a wetting agent at a Tennessee roost may have actually favored the starling *(Sturnus vulgaris)*, the major agricultural pest species in the roost, as the spraying program killed large numbers of blackbirds (family Icteridae), but few starlings.[1]

Bird problems generally are restricted to a very small geographical area, but the bird populations are frequently derived from a much larger geographic region.[2-4] Thus, while exterminating large numbers of birds may temporarily relieve a localized problem, the extermination may have other consequences at other times either in the same locality or in other areas.

Further, the economics of bird-human problems are poorly understood. Where a direct loss is noticeable, our ability to measure the loss is often inadequate. At the same time, such losses are rarely balanced against the financial gains produced by these same avian populations at different times and, often, at different locations. For example, many pest species of the blackbird family (Icteridae), that consume quantities of grain during parts of the nonnesting season, feed upon large quantities of insects (including many insects harmful to man and his activities) and feed insects to their young during the nesting season.[5-7]

Finally, the extermination of birds, especially in large numbers, is repulsive to a large portion of our population. The fact that these same "pest" species may also be aesthetically pleasing to some humans complicates the resolution of bird-human problems.

The resolution of a bird-human conflict, *where needed,* is best accomplished by environmental modification. In every bird problem situation, some resource is attracting the birds. If that resource can be removed or in some way modified to be less attractive to the offending birds, then the problem will be resolved or, at least, reduced to a more tolerable level. As such, environmental modification is defined here as encompassing any activity which changes the environment in any manner without resulting in the loss of large numbers of birds or in the introduction of chemicals into the environment.

PROBLEMS AND SOLUTIONS

Although much overlap occurs among different types of bird problems, the type of environmental change used to resolve the problem will depend upon the specific situation. Such situations can be broadly categorized as: bird roosts; crops and livestock feeding; human habitations; airports; and waste disposal.

Bird Roosts

Many species of birds roost in aggregations that often number in the millions.[8,9] The mere presence of such numbers may in itself be a problem because of noise and odor. Because of fecal accumulations at roosts, health hazards are also considered a factor.[4] Finally, large numbers of birds require large amounts of food and, since these aggregations usually are consumers of seeds and grain, problems develop at livestock feeding operations and in grain fields, resulting in economic losses which can reach severe proportions locally.[10-12]

Birds seek roost sites which offer one or more advantages: space to accommodate the numbers in the aggregation; closeness to needed food resources; predator protection; and protection from climatic factors such as extreme temperatures, wind, and precipitation. No study has yet demonstrated that any one factor is the primary reason for roost site selection.

Late summer and early fall roosts of starlings, brown-headed cowbirds *(Molothrus ater)*, and grackles *(Quiscalus quiscula* and *Q. mexacanus)* in central Texas were abandoned as the numbers using the roost rapidly increased through immigration of migrants.[9] Coon reasoned that the linear space in these roost sites was less than needed for the total aggregation, including the individual distance required by each bird. He suggested that overcrowding produced constant agitation in a roost population and this resulted in abandonment of that particular site.

Among climatic factors, temperature is a likely candidate as a proximate reason for roost site selection. However, Francis reported that temperature differences within and outside a pine plantation roost were but 2°C, although there was also a lag in temperature drop.[13] However, this difference was lost under strong winds (greater than 5 m/sec). Because the blackbirds in this roost were closely aggregated and did not use the entire plantation, Francis suggested that the additional metabolic heat may be an important factor, even though it results in but a 0.5°C increase. He did find that wind was significantly reduced within the roost. In central Texas temperature is less likely to be a factor than in northern roosts, in that winter temperatures are severe for relatively short periods and the severity is much less than that of the northern roosts. Like Francis, studies in central Texas show that live oak trees *(Quercus virginianus)* used as roosts provide significant reduction in wind speeds.[16] Kendeigh states: "A breeze carries the body heat of a bird away faster, and gets in between the feathers and increases the conductive capacity of the air".[15] Logically, reduction of wind by roost cover may allow the birds to survive temperatures in their environment that otherwise might be lethal.

In contrast, communal roosts are considered as "information centers" for food-gathering bird aggregations.[16,17] A study in England suggests that birds experiencing difficulty in finding food the previous day can follow other birds to new feeding areas.[18]

Whatever the proximate reason for roost-site selection by bird aggregations, solutions to problems arising from the presence of roosts must make such a site unattractive. Currently, two such mechanisms are available.

Sound may be used to discourage birds from using a particular site. Pyrotechnics, including automatic "cannons", scare cartridges, noise bombs, and fireworks, have been successfully used in Kentucky, Tennessee, and central Texas.[19,20] In Kentucky and Tennessee, shooter access was apparently more important than the size of the bird population or roost area. The Texas roosts were linear in nature and access was not a consideration. In Texas, timing of daily firing seemed to be important, i.e., such action had to begin as the birds began to arrive to be effective; after the birds settled into the roost, they did not respond to the activities. Recent scare tactics in Texas have involved a combination of automatic "cannons" and distress recordings; these tactics have worked well on blackbirds and starlings, but failed to dislodge house sparrows *(Passer domesticus)* and mourning doves *(Zenaida macroura)* roosting in the same area.

The roost site may be made less attractive by modification of the site itself. Usually, this includes cutting of vegetation. In Texas, severe pruning of live oak roost trees results in immediate abandonment.[21] However, simple removal of sucker growth is insufficient.[22] Francis found similar results with pine plantations, although here it was removal of complete trees, rather than portions of trees.[13] Usefulness of pruning will depend upon the type of plant involved and its ability to stand such pruning, as well as the aesthetic appearance of the plant(s) after pruning.

Crops and Livestock

A wide variety of bird problems arise because of the diversity of crops and livestock feeding situations that occur. Crops include fruits, grains, and trees.[23-25] Assessing damage to crops, especially grains, is not easy. Losses include not only actual consumption, but also those devalued by partial consumption or exposed to disease through damage.[26-27] Such damage often leads to reduced sale value of the crop. Among grain crops, assessment is especially difficult where plantings are close and the grain is in the form of a small head with many seeds, such as rice and grain sorghum. Frequently, bird problems exist for very short periods of time, when the crop is in a particular growth stage.[28,29]

Livestock feeding operation problems vary, from the small farmer faced with increased costs from concentrations of a few hundred birds, to a feedlot operation where thousands of birds may result in substantial monetary losses.[30] In most problem situations losses accrue both from actual consumption of feeds and from bird defecation into the feed troughs which made remaining feeds unpalatable to livestock. In some situations, particularly with poultry-feeding operations, wild birds feeding at the troughs are responsible for transmission of diseases.[31]

No ready solutions are available to the many problem types occurring with crops and livestock. With grain crops, sound and pyrotechniques are useful only if the bird species is in the area for a relatively short time, or if the crop is vulnerable for a short period. Mobility of the soundmaking devices will produce longer-lasting results than when the noisemakers are stationary. Also, some species seem less affected by sounds, such as distress calls, than others. For example, the common crow *(Corvus brachyrhynchos)*, a serious pest on pecans and some grain crops, is apparently attracted by distress call recordings of its own species.[32]

Some crops, especially fruits, may be protected by use of coverings of nylon netting.[18] This is a tedious and expensive process and should not be considered unless the value of the crop warrants such protection. Placement of devices such as mountings of "toy" owls, hawks, and snakes generally are of little value in crop protection.

Change in planting and harvesting of crops is a methodology that deserves further consideration. Because birds that cause problems often are flocking species, it is possible to predict when such flocks will form and disperse, and when such flocks can be expected in any region. Furthermore, a number of crops, at least in some areas, have planting and harvesting latitudes of a nature that allows agriculturists to delay or advance their crops to avoid or minimize bird problems.[33] Starling depredation on winter wheat in Kentucky might be eliminated by advancing planting by 2 weeks.[34] Indeed, a number of problems result when humans attempt to secure an additional harvest of a particular crop; this is especially true in areas with extended growing seasons. A corollary would be the development of genetic strains of crops that were less susceptible to bird damage.

Livestock-feeding operations may solve their problems if it is possible to present feed in covered troughs or in buildings.[35] Such techniques will virtually eliminate blackbirds and starlings, but not house sparrows. This solution is not available at operations where large quantities of livestock are confined in small areas and a need exists for more or less continuous presentation of feed to the animals. One avenue that should be explored is the concept of "demand" feeding by livestock, similar to the use of watering devices for dairy cattle that permits the cow to dispense water by pressing a bar. Under this concept, small amounts of feed would be dispersed by some action of the individual animal. This would keep less feed material available to attract birds. This solution may not be practical in feedlots with high livestock densities.

Human Habitations and Buildings

Like crops and livestock feeding, a diverse set of problems exists around human habitation. These include not only housing for man, but numerous other structures used for human activities. The problems involve nuisance situations, consumption and/or destruction of food, damage to buildings, interference with power supplies, and potential for disease.

Perhaps no other problem area involves such a diversity of bird species. Not only are the well-known blackbirds, starlings, house sparrows, and rock doves *(Columba livia)* involved, but also owls, chimney swifts *(Chaetura pelagia)*, woodpeckers, and many songbirds.[36] A number of these species find suitable nesting places, food, and shelter around and in buildings. Rafters, window ledges, and similar structures are excellent supports for nests or places for roosting. Chimneys are used for nesting by the chimney swift and for roosting by several species of owls. Woodpeckers find wooden buildings, particularly those with cedar siding, attractive for roost/nest holes. Besides the woodpecker damage, deterioration of structures may result from bird defecation. Presence of birds around man-made structures is also encouraged by plantings of food-items and bird feeders and by the protection offered from predators.

The presence of nesting material is a source both of ectoparasites and fire hazard, especially if the nest is placed on or near electrical wiring. Birds have caused severe damage to powerlines and electrical transformers because of nesting and/or roosting. Many birds species are capable of serving as reservoirs for organisms causing disease in humans or domesticated animals.[37] This situation is very prevalent among those species closely associated with human habitations: house sparrows, starlings, and rock doves.

Around bulk food storage facilities several bird species are consumers of foods meant for human consumption or contaminate such foods. Such losses increase the cost of these foods for humans or may present health hazards. House sparrows, rock doves, and blackbirds are the common species associated with this problem.

Finally, a wide variety of nuisance situations exist. A number of birds strike windows in buildings, sometimes causing damage, but more likely to be annoying to the building occupants. Other birds, especially fruit-eating species, such as cedar waxwings *(Bombycilla cedrorum)*, congregate around human habitations after consuming large quantities of fruit, and defecate on buildings, ornamental shrubbery, sidewalks, and cars. The mere presence of birds in large numbers, actively calling and moving in a restricted area, can be a nuisance.

Solutions to these problems are as varied as the problems themselves. The buildings can be made less attractive by eliminating potential nesting or roosting sites. This is best done in the construction of the building, but can be accomplished, with careful planning, in buildings where bird problems already exist. Exposed ledges should be avoided and holes either filled in or covered with screening. Birds can be excluded from chimneys by placing screens over the chimneys when the problem birds are absent from the area; placing the screen while birds are actively using a chimney will produce other undesirable effects because of dying birds.

Planting around buildings should be made with shrubbery and trees which do not produce foods or shelter for the undesirable birds. If this cannot be avoided, then these plantings should be trimmed to minimize the attractiveness. Climbing vines and shrubbery close to chimneys, windows, and rough surfaces such as brick should be avoided.

Finally, since many birds are considered desirable around human habitations and efforts are made to attract them, undesirable birds will also be attracted. These latter birds can be discouraged by using specific types of bird feeders or by limiting the size of birds that can use a feeder by placing a protective screen around the feeder.

Airports and Aircraft

The potential for loss of human life occurs with bird strikes on aircraft and this potential is probably at its maximum near and at airports. Aircraft in the process of landing or taking off are undoubtedly most vulnerable to damage in a bird strike in that they are either decelerating or not yet at full power.[38] Even if no human life is lost, the cost in damage is great because of the large amount of investment in even the smallest aircraft. Besides the actual cost of repair, time lost because the aircraft is grounded must also be considered.[38]

Furthermore, airfields and associated buildings provide direct attractants for birds. The open spaces of landing strips and the accompanying grassy or short-vegetated areas around them attract flocks of foraging birds, while the thermals created by the landing strips attract larger, soaring birds, such as hawks, vultures, gulls, albatrosses, etc.[39] Buildings, towers, and similar structures provide roosting and nesting sites for many birds.

Like situations around human habitations, construction and modification of buildings should eliminate potential nesting and roosting sites. Plantings of trees and shrubs in open areas that do not endanger aircraft will discourage use by foraging flocks by breaking up the open areas into a more patchy environment. At the same time, contouring of areas adjacent to landing strips will reduce attraction of foraging flocks and reduce thermals. Obviously, safety for aircraft must be considered in such contouring and plantings.

In some situations, problem birds can be moved from landing strips via sound and pyrotechnique devices.[40] At times, the best solution is a brief cessation of aircraft activity; this has been done at some military fields, but is not a viable option where commercial flights are scheduled. Trained birds-of-prey have been used with some success at airfields to prevent use of problem birds. However, this requires special handling and maintenance of the trained birds, which is difficult; the birds-of-prey may be difficult to acquire and permits are required; and size of the airfield may be too great to allow this solution.[38]

In establishment of new airports, consideration should be given to potential bird problems when a site is selected. Sanitary landfills, active roost sites, and feeding sites for flocking species should be considered as prime reasons for rejecting a particular location as an airfield. Similarly, migration patterns of birds over a potential site should receive serious study.

Sanitary Landfills

These operations offer one major resource for birds: food. Landfills also create open spaces, at least temporarily, that become attractive to foraging flocks. As such, the problem is not one of the landfill, per se, but of attracting birds into an area where they will cause problems in other areas such as airports, and residential areas.

The primary solution is better operation at the land fill to cover the materials deposited, Although a 6-in. covering is standard, this is often not realized.[41] Further, covering operations frequently are minimized so that larger amounts of material are left uncovered through much of each day and sometimes through the weekend when the landfill is not operating.

Birds may also be discouraged from landfills by removing trees and shrubs from areas adjacent to the operation. Many birds use such vegetation as loafing areas while not feeding. As with airports, location of landfill operations should take into consideration activity patterns of local bird populations.

SUMMARY

The methodology of environmental manipulation as a means of resolving bird-human conflicts is in its infancy. The diversity of problems dictates that we examine each problem for a definitive solution for a particular species in a particular situation. It is unlikely that every problem can be solved by this means, but we should give consideration to resolution by environmental manipulation over mass destruction of bird populations unless an emergency dictates otherwise.

Most bird-human populations result from initiation of activities by man without consideration of potential problems or the needs of local bird populations. It may often be as easy to modify the man-made structures and/or human activities. We are just beginning to understand the needs of bird species in our environment and how they are affected by our environmental manipulations.

REFERENCES

1. **Dolbeer, R. A., Woronecki, P. P., Stickley, A. R., Jr., and White, S. B.,** Agricultural impact of a winter population of blackbirds and starlings, *Wilson Bull.,* 90, 31, 1978.
2. **Royall, W. C., Jr. and Guarino, J. L.,** Movements of starlings banded in north-central Colorado, 1960—74, *North Am. Bird Bander,* 1, 58. 1976.
3. **Coon, D. W. and Arnold, K. A.,** Origins of brown-headed cowbird populations wintering in central Texas, *North Am. Bird Bander,* 2, 7, 1977.
4. **Monroe, B. L., Jr. and Cronholm, L. S.,** Blackbird Study. Final Report, Kentucky Environmental Quality Commission, Louisville, 1977.
5. **Tenovuo, R. and Lemmetyinen, R.,** On the breeding ecology of the starling, *Sturnus vulgaris,* in the archipelago of south-western Finland, *Ornis Fennica,* 47, 159, 1970.
6. **Davis, W. R., II and Arnold, K. A.,** Food habits of the great-tailed grackle in Brazos county, Texas, *Condor,* 74, 439, 1972.
7. **Robertson, R. J., Weatherhead, P. J., Phelan, F. J. S., Holroyd, G. L., and Lester, N.,** On assessing the economic and ecological impact of winter blackbird flocks, *J. Wildl. Manage.,* 42, 53, 1978.
8. **Meanley, B. and Webb, J. S.,** Nationwide populations estimates of blackbirds and starlings, *Atl. Nat.,* 20, 189, 1965.
9. **Coon, D. W.,** Daily Movement of the Brown-headed Cowbird in South-Central Texas During Winter, Ph.D. thesis, Texas A & M University, College Station, 1974.
10. **Royall, W. C., Jr., DeCino, T. J., and Besser, J. F.,** Reduction of a starling population at a turkey farm, *Poultry Sci.,* 46, 1494, 1967.

11. **Stone, C. P., Mott, D. F., Besser, J. F., and DeGrazio, J. W.,** Bird damage to corn in the U.S. in 1970, *Wilson Bull.*, 84, 101, 1972.
12. **Glahn, J. F., Steffen, D. E., and Clark, L. S.,** A preliminary survey of bird activity in dairy feed troughs in Monroe county, Tennessee, Bird Damage Res. Rep. No. 83, Denver Wildlife Research Center, U.S. Fish and Wildlife Service, Denver, Colorado, 1978.
13. **Francis, W. J.,** Micrometeorology of a blackbird roost, *J. Wildl. Manage.*, 40, 132, 1976.
14. **Arnold, K. A.,** unpublished data, 1978—79.
15. **Kendeigh, S. C.,** The role of environment in the life of birds, *Ecol. Monogr.*, 4, 297, 1934.
16. **Zahavi, A.,** The function of pre-roost gatherings and communal roosts, *Ibis*, 113, 106, 1971.
17. **Ward, P. and Zahavi, A.,** The importance of certain assemblages of birds as "information-centers" for food-finding, *Ibis*, 115, 517, 1973.
18. **Brown, D. M., Dick, W. J. A., Johnson, C. E., Sales, D. I., and Zahavi, A.,** Pied wagtail roosting and feeding behavior, *Bird Study*, 23, 267, 1976.
19. **Mott, D. F., Bray, O. E., Grandpre, T. D., Marquess, K. L., and Williams, C. A.,** Dispersing blackbirds and starlings from winter roosts in Kentucky and Tennessee by using pyrotechnic devices, Bird Damage Res. Rep. No. 73, Denver Wildlife Research Center, U.S. Fish and Wildlife Service, Denver, Colorado, 1978.
20. **Arnold, K. A.,** personal observations.
21. **Arnold, K. A. and Garza, Jr., N. E.,** unpublished data, 1978—79.
22. **Good, H. B. and Johnson, D. M.,** Experimental tree trimming to control an urban blackbird roost, *Proc. 7th Bird Contr. Sem.*, Bowling Green, Ohio, 1976, 54.
23. **Royall, W. C., Jr. and Ferguson, E. R.,** Controlling bird and mammal damage in direct seeding loblolly pine in east Texas, *J. For.*, 60, 37, 1962.
24. **Stone, C. P.,** Bird damage to agricultural crops in the U.S. — a current summary, *Proc. 6th Bird Contr. Sem.*, Bowling Green, Ohio, 1973, 264.
25. **Brown, R. G. B.,** *Bird Damage to Fruit Crops in the Niagara Peninsula*, Canadian Wildlife Service Rep. Ser. No. 27, Ottawa, 1974.
26. **Heichel, G. H. and Washko, W. W.,** Bird damage to Connecticut corn, Conn. Agric. Exp. Stn., New Haven Bull., 761, 1976.
27. **Crase, F. T., Stone, C. P., DeHaven, R. W., and Mott, D. F.,** Bird damage to grapes in the U.S. with emphasis on California, Special Scientific Report — Wildlife, No. 197, U.S. Fish and Wildlife Service, Washington, D. C., 1976.
28. **Rogers, J. G., Jr. and Linehan, J. T.,** Some aspects of grackle feeding behavior in newly planted corn, *J. Wildl. Manage.*, 41, 444, 1977.
29. **Stickley, A. R., Otis, D. L., Bray, O. E., Heisterberg, J. F., and Grandpre, T. D.,** Survey of bird and mammal damage to ripening field corn in Kentucky and Tennessee in 1977, Bird Damage Res. Rep. No. 68, Denver Wildlife Research Center, U.S. Fish and Wildlife Sevice, Denver, Colorado, 1978.
30. **Besser, J. F., DeGrazio, J. W., and Guarino, J. L.,** Costs of wintering starlings and red-winged blackbirds at feedlots, *J. Wildl. Manage.*, 32, 179, 1968.
31. **Grimes, J. E.,** Disease transfer from wild birds to turkeys, in *Brief Reports on Agricultural Research in Texas*, Texas A & M University, College Station, 1977, 3.
32. **Frings, H., Frings, M., Jumber, J., Busnel, R.-G., Giban, J., and Gramet, P.,** Reactions of American and French species of *Corvus* and *Larus* to recorded communication signals tested reciprocally, *Ecology*, 39, 126, 1958.
33. **DeHaven, R. W.,** Blackbirds and the California rice crop, *Rice J.*, 74, 1, 1971.
34. **Dolbeer, R. A., Stickley, A. R., Jr., and Woronecki, P. P.,** Starling, *Sturnus vulgaris*, damage to sprouting wheat in Tennessee and Kentucky, U.S.A., unpublished manuscript, 1978.
35. **Arnold, K. A.,** personal observations.
36. **Arnold, K. A.,** personal observations.
37. **Shillinger, J. E.,** Diseases of wildlife and their relationship to domestic livestock, in Keeping Livestock Healthy, Part 9, U.S. Department of Agriculture, Washington, D. C., 1942.
38. **Solman, V. E. F.,** Bird control and air safety, in Studies of bird hazards to aircraft, Canadian Wildlife Service Rep. Ser. No. 14, Ottawa, 7, 1971.
39. **Munro, D. A. and Harris, R. D.,** Du danger que consituent les Oiseaux pres des Aerodromes du Canada in *Colloque Le Problems des Oiseaux sur les Aerodromes*, Institut National de la Recherche Agronomique, Paris, 1963, 173.
40. **Solman, V. E. F.,** The ecological control of bird hazards to aircraft, *Proc. 3rd Bird Contr. Sem.*, Bowling Green, Ohio, 1966, 38.
41. **Arnold, K. A.,** personal observations.

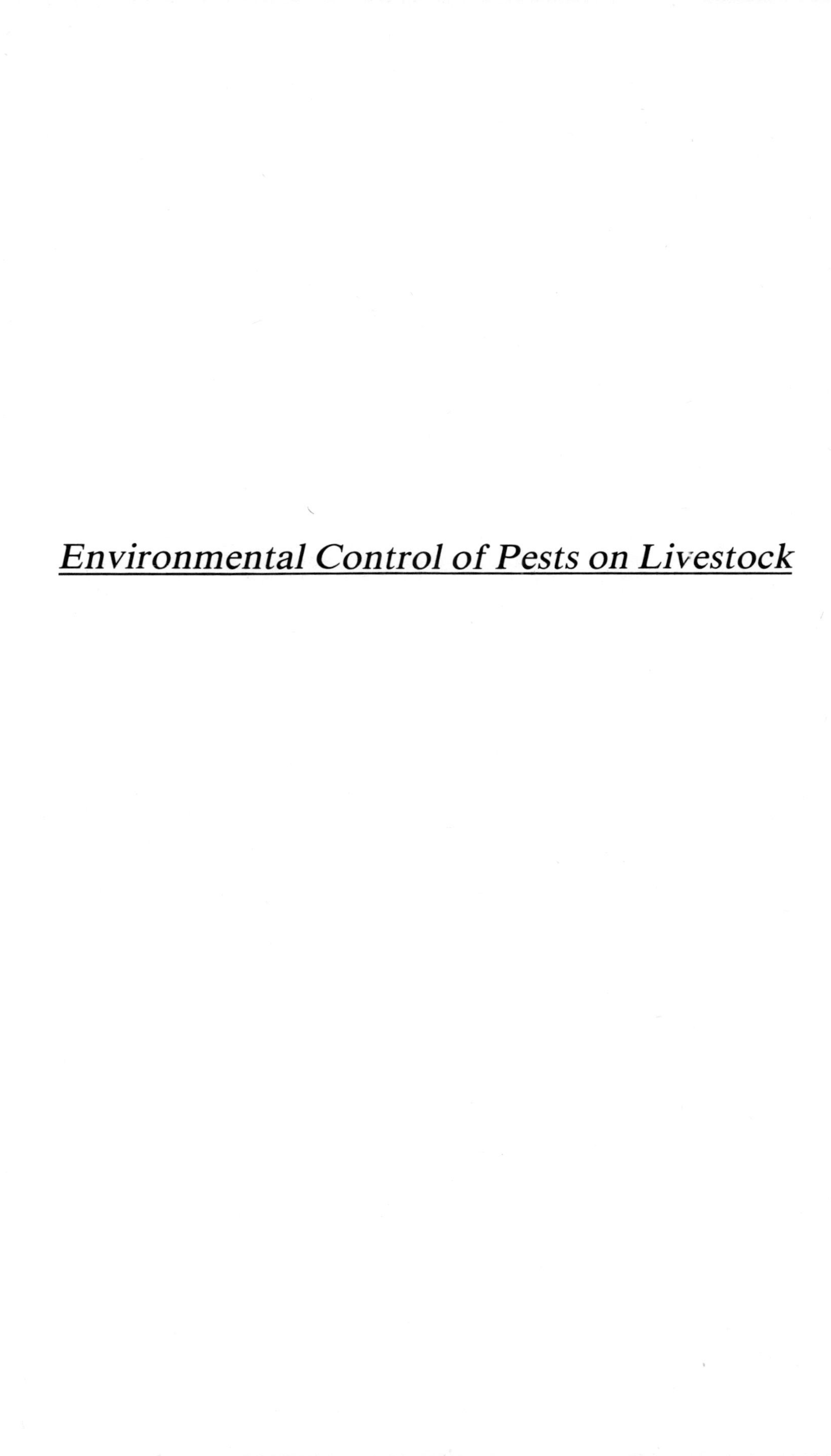

Environmental Control of Pests on Livestock

ENVIRONMENTAL CONTROL OF ARTHROPOD PESTS OF LIVESTOCK

C. Dayton Steelman

ENVIRONMENTAL CONTROL OF LIVESTOCK ARTHROPODS

During the early 1960s the concept of integrated control was broadened from the two-component system consisting of chemical and biological control to include all practices, procedures, and techniques that aid in the suppression of pest populations. Subsequently, environmental control has become recognized as that method of control wherein one or more specific components of the environment are modified to the pests' detriment, with the result that the pest population is suppressed, eliminated, or reduced below the level of the injury threshold. The effectiveness of many techniques utilized in environmental control of arthropods affecting livestock has been reported since the early 1900s. The integration of these procedures into a balanced pest management system has been neglected, primarily due to dependence on chemical control of pests since the late 1940s and because of a lack of injury and economic threshold data for the various arthropod pests of livestock.

Until recently, the philosophy of most livestock producers has been that the presence of any pests on an animal is undesirable and that they should be kept at the lowest possible population levels by regularly scheduled applications of insecticides. The dependence on chemicals as the only method for the control of arthropod pests that attack livestock has caused the development of resistance to some insecticides by several pest species, has resulted, on occasion, in the presence of unwanted residues in animals and animal food products, has caused unnecessary animal and human health hazards, and has suppressed populations of some beneficial species.

Environmental control for arthropods that attack livestock include barriers, habitat alteration of the pest species, pasture rotation or spelling, and removal, restriction, or elimination of alternate hosts.

BARRIERS

Barriers may be used to reduce or prevent contact between an arthropod pest and domestic livestock that serve as host animals. These environmental barriers are of two types: (1) areas cleared of vegetation, which serve to disrupt a specific arthropod environment to the extent that the range of dispersal to obtain a blood meal is restricted, and (2) areas containing certain types of vegetation that either direct the pest away from an area containing domestic animals or impede dispersal of the pest into these areas.

Early work in Africa showed that clearing broad areas through intense bush stopped tsetse fly movement.[1] In the Lake Victoria region of Kenya, 12- to 16-km-long cleared barriers prevented *Glossina palpalis* (R.-D.) from reinvading areas that had been cleared by insecticides.[2] The first organized barrier clearing was initiated in 1928 on the Gold Coast of Africa to control tsetse fly transmission of *Trypanasoma* species on the Eastern Cattle Route.[3] The bush was cleared at six river crossings harboring high populations of *G. palpalis* and *G. tachinoides* West along 112.63 km of this route. All trees and shrubs were cut from the banks of the river for 0.8 km on each side of the cattle crossings and 18.3 m back from the river. These barriers reduced the tsetse fly incidence at river crossings by about 90% compared to that in the surrounding bush, and resulted in a corresponding drop in bovine trypanosomiasis of 63% in the year following clearing.

Areas cleared to form barriers should be at least 3.2 km in length to be effective. Otherwise, tsetse flies move from one bush environment to another at points where no barrier exists, and further disperse into the protected zone behind the barriers.[4]

In Uganda, 6-km-wide consolidation barriers have been cleared between human settlements and bush still infested with tsetse flies. Agriculture, including cattle production, is encouraged within the barrier areas.[5] The most effective barrier has been shown to be the clearing of all shrubs and trees (sheer-clearing) from the barrier area.[6] Data are shown in Figure 1 to exemplify the barrier by clearing vegetation method of environmental control. It is obvious from these data that barrier-clearings less than 274-365 m in length are of little value, and that corresponding increases are obtained in the percent reduction of tsetse as the barrier zone is lengthened. Considerable differences occur in the efficacy of the barriers, depending upon season of the year and species of tsetse fly.[6] With respect to *G. palpalis,* the wet season decreases the effectiveness of barriers, and certain factors are believed to cause this reduction in efficacy: (1) the increase in humidity during the wet season is conducive to greater dispersal of flies and allows them to survive longer in open barrier areas, and (2) the increase in vegetative matter (high grass and dense leaf cover) causes difficulty in host finding, which results in greater host-searching flights. *Glossina tachinoides* exhibited a similar seasonal pattern of activity to that of *G. palpalis,* but it has a greater variation in its reaction to barriers because it more readily enters the barriers during the wet season. This indicates the importance of obtaining biological and ecological data for each pest species relative to the environment in which it lives.

Certain vegetative barriers impede the dispersal of some tabanids from certain types of marshland breeding habitats.[7] Through retention of natural vegetation or by strategic planting of certain species of vegetation and careful selection of planting sites, the number of tabanids reaching upland areas can be reduced. A dense wall of tall bushes and seed grasses between *Tabanus nigrovittatus* Macquart breeding grounds in salt marsh habitats and seashore areas provide an effective barrier.[8] An average of 5.6 flies per trap per day was collected behind the barrier, while 41.12 flies per trap per day were collected by traps set in the gaps of the barrier. During August, the time of peak population, some 800 flies per trap per day were collected in the gap traps, while just over 100 per trap per day were collected behind the barrier.

HABITAT ALTERATION

Vegetation Clearing

Clearing by Hand Labor

Habitat alteration by bush clearance was initiated for the control of tsetse flies as early as 1909 in Nigeria.[9] About 1.5 months after the bush was cleared, the area was relatively free of tsetse flies. The tsetse fly population tends to concentrate periodically in certain identifiable plant communities that comprise a relatively small proportion of the bush or woodland.[4] A specific microclimate is required by the tsetse fly and only the ground within the true forest or bush associated with certain plant species is adequate. In the late dry season tsetse flies are already living within 3.5°C of their critical temperature (39°C), and clearing the bush destroyed their habitat.[10, 11]

The removal of all woody growth in a fly-infested area of land is called "sheer clearing," although this procedure is seldom employed as a tsetse fly control measure.[4] Sheer-clearing is now only used to provide selected bands or strips of bush to check tsetse fly encroachment and to effect barriers of defense around human settlements

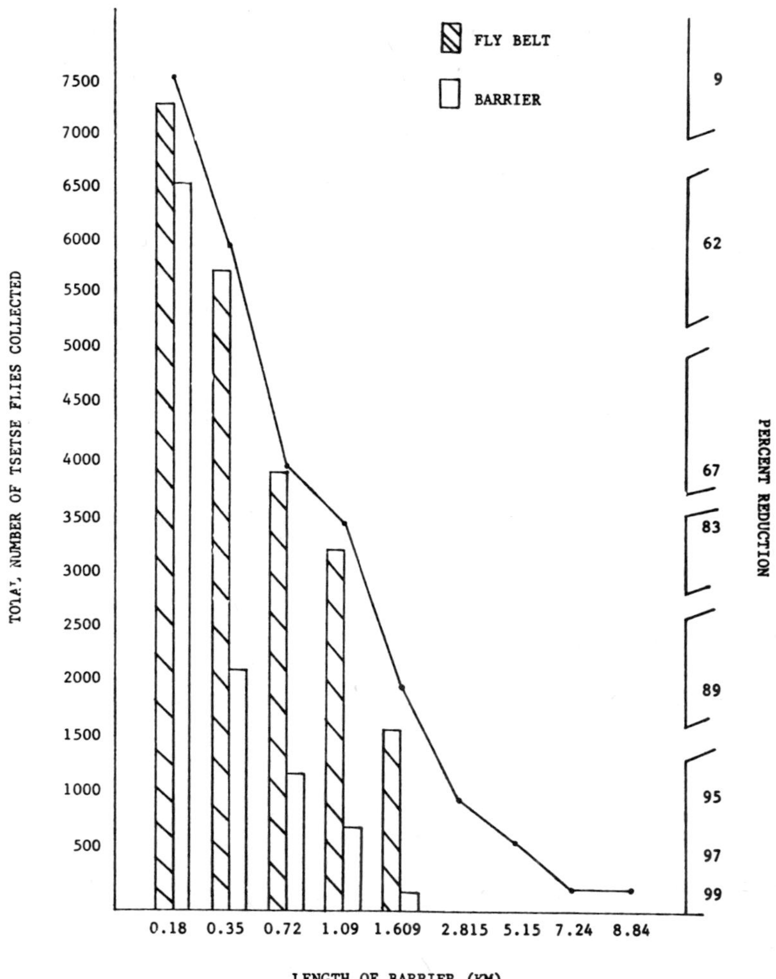

FIGURE 1. Number of adult tsetse flies collected in barriers cleared for various distances and in uncleared fly-belt areas, and percent control of tsetse using this method of environmental control.[6]

and along main routes of cattle movement. Another method or degree of vegetation clearing is called discriminative clearing, and entails the removal of a definite vegetational type consisting of different species of trees through the bush, thus limiting the vegetation to one or more species of trees. Initially cutting back 1 mi of fly-infested bush 457 m wide requires 500 man-days.[12] Partial clearing to remove only undergrowth and low-branching trees of riverine vegetation allows the entry of hot dry wind from adjoining savannahs, which destroys the moist microhabitat of the tsetse fly.[13]

Selective clearing has been used to describe the removal of certain species of trees through the bush in whatever vegetation type they occur.[12] The effectiveness of selectively clearing bush to reduce tsetse populations is exemplified in Table 1. Some 1689 km^2 of bush was "selectively cleared" between 1940 and 1945.[6] Prior to the clearing project, *G. tachinoides* populations ranged from 5600 to 6800 flies per year (catches by 20 flyboy-days per month) along the main course of the Kamba River, and from 1600 to 4000 flies per year along the upper branches of the river. Three years after

Table 1
EFFECT OF SELECTIVELY CLEARING BUSH ON NATURAL POPULATIONS OF *GLOSSINA TACHINOIDES*, 1941—45[6]

		Distance (km) into cleared bush from uncleared area							
		8.8		11.3		35.4		70.0	
Year	Uncleared area (number/year)	No.	% Reduced	No.	% Reduced	No.	% Reduced	No.	% Reduced
1941	3334	1413[a]	58	23	99	83	98	1204[a]	64
1942	7741	185	96	21	99	21	99	5	99
1943	4260	52	98	1	99	4	99	0	100
1944	5860	30	99	2	99	7	99	4	99
1945	8932	288	96	24	99	2	99	0	100

[a] Clearing only partially completed.

clearing, tsetse fly populations were reduced to 0 to 7 flies per year, while catches in the uncleared area ranged from 4260 to 8932 per year.

The various methods or types of bush clearing for tsetse fly control have been shown to be quite effective: sheer-clearing,[14-16] discriminative clearing,[15,17] partial clearing,[18-20] and selective clearing.[6]

Mechanical Clearing

Due to the high costs involved with labor, hand clearing of bush has virtually ceased to be a large-scale method for tsetse fly control.[21] Mechanical means of bush clearing have been developed in many countries. The techniques of bulldozing and chain-clearing with caterpillars to remove tsetse-infested bush are effective and economical methods.[22] Some 28,280 ha in Ankole and 32,845 ha in Bunzor have been mechanically cleared using two D-8 caterpillars towing a 91.4 m heavy chain. With this equipment, the land clearing rate ranged from 1.7 to 4 ha per tractor per hr.[23]

Chemical Clearing

The earliest trials using chemical defoliants for control of tsetse fly-infested bush were conducted during 1952 to 1954 in Kenya and Tanzania (Table 2).[24] Defoliation was achieved, but not uniformly, so that the overall reduction in shade was not sufficiently intense or prolonged enough to have more than temporary effects on the tsetse fly populations. Although few trees were killed, an increase in grass growth was observed at one treatment site as a result of the chemical applications. These results indicated that defoliation alone as a control measure for tsetse fly was not practical, but that destruction of normally fire-resistant bush could be achieved by a controlled burn following a defoliant application. The defoliation and burn destroyed the habitat of *Glossina pallidipes* Aust causing a 77% reduction in 1953 and an 84% reduction in 1954, while the unburned lakeshore habitat of *G. palpalis* was left relatively intact. This resulted in an increase in fly population numbers in the overall defoliated areas. The combined technique of chemical defoliation and burning is influenced greatly by rainfall during the period of defoliation, through its effect on the growth of grass. A low amount of rainfall during this time reduces the rate of grass growth, which in turn reduces the development of a fierce burn which is necessary to destroy the woody thicket vegetation in the following dry season.

Other efforts to control tsetse fly-infested bush with chemicals have been reported.[25-29] When 2, 4, 5, T ester was applied (1.1 kg/ha), 60 to 80% reduction of the three dominant species of thicket-bush habitat was effected and resulted in a considerable increase in grass production.[29]

Herbicidal oil applications used to eliminate specific vegetation growing in date farms resulted in almost complete absence of eye gnat *(Hippelates)* breeding (Table 3).[30] The primary requirement for *Hippelates* breeding is the presence of organic matter. Therefore, the elimination of vegetation prior to tillage effectively controls eye gnat breeding regardless of other cultural practices or conditions.

Burning

Another effective method of habitat alteration is the gradual clearing of bush to reduce favorable tsetse fly habitats by annual grass fires.[1] Burning not only destroys tsetse habitat but causes considerable reduction in the numbers of tsetse fly larvae and pupae.[18] Late-season burning conducted by various tribes at the end of the dry season is more effective for tsetse control than the early-season burning recommended by the Rhodesian government.[31]

Certain species of unpalatable plants provide efficient shelter for ticks (due to the

Table 2
EFFECTS OF VEGETATION CONTROL BY CHEMICAL DEFOLIATION ON NATURAL POPULATIONS OF *GLOSSINA PALLIDIPES* AUST. AND *G. PALPALIS* (R.-D.)[24]

	G. pallidipes			*G. palpalis*		
Year	Treated	Untreated	% Reduction	Treated	Untreated	% Reduction
1952[a]	143	134	−6.3	81	71	−12.3
1953[b]	81	350	77.0	95	55	−42.0
1954	56	351	84.0	78	22	−72.0

Note: Figures = average number of flies collected.

[a] Three applications of 2, 4, 5-T were applied to the treated area (April 2 and May 7, and July 8, 1952).

[b] Nine months after the first application of 2, 4, 5-T, between a third and one half of the vegetation in the treated area was burned.

Table 3
CONTROL OF *HIPPELATES* GNATS BY KILLING VEGETATION WITH HERBICIDAL OIL IN CALIFORNIA DATE GARDENS[30]

	Number gnats per cage			
	H. collusor		*H. hermsi*	
Soil surface	Herbicidal oil	% Reduction	Herbicidal oil	% Reduction
		1960		
No-tillage	0.10	97	0.20	99
Tillage	0.06	98	0.73	98
Control	4.5	—	49.5	—
		1961		
No-tillage	0.14	100	0	100
Tillage	0.58	99	0	100
Control	30.7	—	14.0	—

mat of debris formed by the leaf drop), in comparison to palatable plants which are grazed, and thus do not provide adequate tick habitat.[32] Field or pasture burning not only destroys ticks that are on vegetation awaiting a host animal, but also destroys the plant material that provides the tick habitat at ground level. Pasture burning also destroyed the habitats of alternate host animals such as the red hare and the elephant shrew. Larval ticks can be destroyed by burning, and by flaming the soil over large pasture areas with hydrocarbon flames operating at 1000 to 1300°C.[33,34] Heat treatments for tick control in pastures are most effective in early spring and late autumn. In areas subject to flooding, the pastures should be flamed immediately following the flood. Burning pasture vegetation in areas along the delta of the Don River, which are subject to intermediate flood and drain cycles, caused the death of *Aedes* spp. mosquito eggs that had been deposited on the soil surface.[35]

Flooding

Water has been used to control certain livestock pests through the development of specific management practices. In Africa, bush that supports tsetse flies has been cleared by flooding.[1] After holding water on land for a sufficient period of time to kill the vegetation, subsequent drying decreases the attraction of the area through the destruction of the tsetse microhabitat.

Water Impoundment

Environmental modification has been a common procedure for mosquito control, and includes methods of draining or filling marsh areas, diking and flooding with seawater, lining canals and other waterways with concrete, and the rapid flushing of water through sluggish water channels.[36] Permanent source reduction for mosquito control often is consistent with good agricultural practices.[37]

Impoundments have been used extensively to control various species of mosquitoes through water level management which destroys eggs, larvae, and pupae directly, or kills them indirectly by exposing them to natural enemies.[38] Emergent aquatic vegetation that provides a favorable habitat for *Anopheles* mosquitoes has been successfully controlled by dropping water levels in rapid steps.

The effectiveness of various water management techniques to control *Aedes, Culex,* and *Culiseta* mosquito breeding in Utah salt marsh habitats has been reported.[39-41] Permanent withdrawal of water from specific areas eliminated *Aedes* mosquito production. Modification of shoreline to maintain water depths of at least 30 cm has allowed cattails *(Typha latifolia* L.) to increase from 10 to 70% of the vegetation (replacing alkali bulrush *(Scirpus paludosus* A. Nels.) and salt grass *(Distichlis stricta* (Torr. Rybd.), which indicates that water levels are remaining constant), thereby virtually eliminating the breeding of *Aedes dorsalis* (Meigen), *Culex tarsalis* Coquillett, and *Culiseta inornata* Williston.

Techniques of water management have been utilized experimentally to control various species of tabanids.[42-45] Impoundments have eliminated *Chrysops callidus* Osten Sacken breeding for a 2-year period, and sudden changes in the water level of ponds is destructive to certain species of horse fly larvae and pupae.[42] Removal of emergent plants and projecting debris that occur in or within 3 to 3.7 m of the water-edge has stopped egg deposition in the aquatic habitat.[45] When impoundment of larvae and pupal habitats was timed to correspond to the period of the year when *Chrysops fuliginosus* Wiedemann was in larval-pupal and pupal-imaginal molts, effective control was attained (Figure 2).

Impoundment of marsh habitat has been shown to be an effective method of controlling *Culicoides* species.[46-47] After diking and removing the water from 404 ha of Florida salt marsh, an average reduction of 82% in the number of *Culicoides* larvae occurred over a 2-year period (Figure 3).[48]

Water Agitation

Agitation of water surfaces by water-sprinkling has prevented *Culex pipiens* L. breeding in poultry lagoons.[49] Mosquito populations prior to sprinkling averaged 1233 larvae of all stages, 471 pupae, and 700 egg rafts per dip. After 2 days of continuous sprinkling, averages were reduced to 100-4th instar larvae, 40 pupae, and no egg rafts. After 6 days no live mosquitoes of any stage were collected.

Flooding and Watercourse Changes

Environmental modification to control black flies (Simuliidae) is most difficult, as it requires extensive changes in watercourses. Environmental control of black flies can

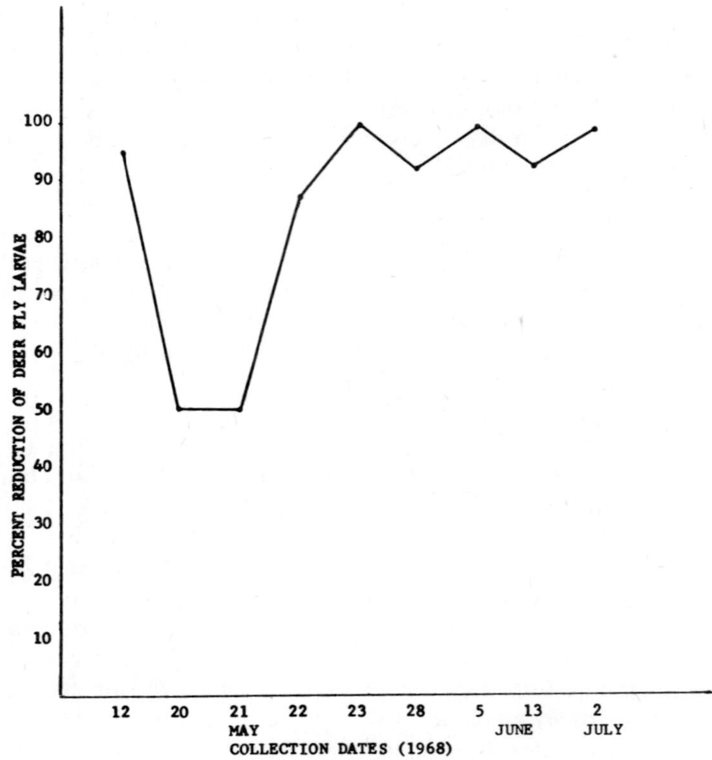

FIGURE 2. Percent reduction in number of *Chrysops fuliginosus* caused by salt marsh impoundment in 1968.[45]

be accomplished by building irrigation systems that avoid the use of areas that have rapidly moving water, and by the reduction in the amount of vegetation in flowing water courses.[36] Floods have been reported to reduce the level of infestations in streams due to the combined effects of the sudden rise in the water level and the constant disturbance of the stream bed caused by logs moving downstream in the floodwater flow.[50] After one such flood, a time period of 2 years was required before the larval population in the stream reached its pre-flood population level. Periodic deflection of streams, and the construction of numerous dams which convert streams into a system of receptacles of calm water, eliminate large black fly breeding areas.[51]

Agricultural Practices

In some situations the livestock pest habitat can be changed by the proper timing of agricultural practices so that they occur at critical times in the life cycle of pests.[52] The elimination of the damp basal vegetative mat in pastures by drainage, harrowing, and application of lime renders this habitat difficult for the sheep tick, *Ixodes ricinus* L., to survive. A consistant positive correlation exists between vegetative mat thickness and tick population density.[53] Plowing or discing pastures in early fall season has been reported to kill larval ticks that utilize the soil-cracks habitat within the pasture.[33]

Cultural methods have been shown to be effective in controlling mosquitoes breeding in Playa Lakes in West Texas.[54-55] Plowing of the lake beds prior to initial flooding resulted in reduced production of both *Culex* and *Aedes* mosquitoes (Figure 4). Rototilling was the most effective control method tested, and it was accomplished much faster than moldboard-plowing, which was the second best cultural control method evaluated.

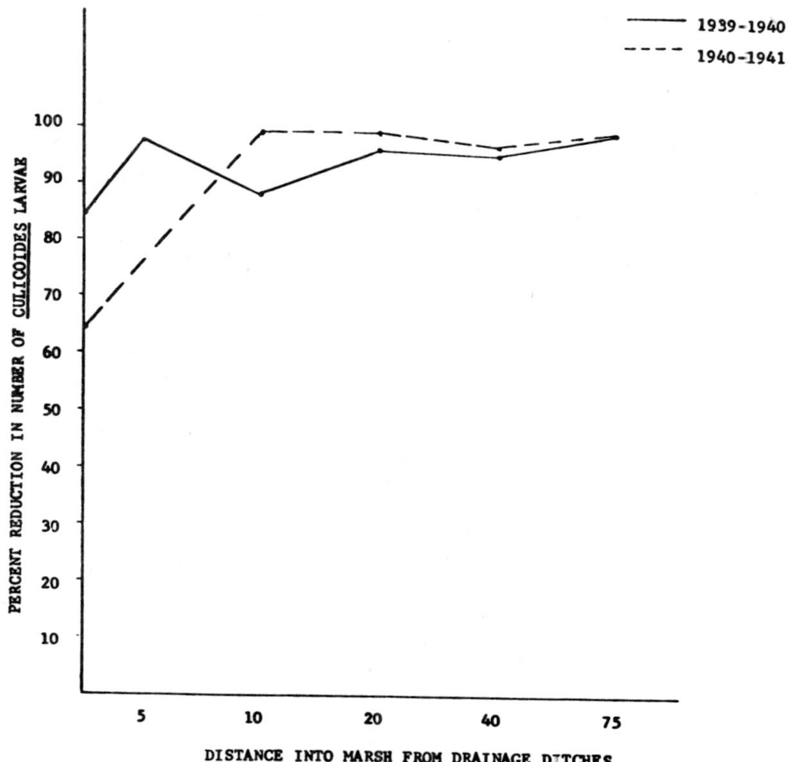

FIGURE 3. Percent reduction in numbers of *Culicoides* larvae caused by diking in Florida marsh.[48]

Non-cultivation techniques of vegetation control have been indicated as methods of choice where eye gnats *(Hippelates* spp.) are important pests of livestock.[56–59] *Hippelates* spp. gnats are either attracted to freshly-plowed soil for oviposition, or eggs deposited in uncultivated soil do not develop to adults in significant numbers. The data shown in Figure 5 indicate that greater numbers of adult gnats emerge from plowed soil than from unplowed soil.[59]

Plant Species Substitution

Replacing plant species that form mosquito breeding habitats with other species of plants that eliminate these breeding sites has been reported. In certain areas of the world, the mosquito *Mansonia annulifera* (Theob.) is dependent on the aquatic plant *Pistia stratiotes* Linn. *Salvinia auriculata* Aubet., which does not support *M. annulifera,* crowds out *P. stretiotes* rapidly when introduced into the same aquatic habitat, thereby eliminating the breeding area of this mosquito species.[60] Water-tolerant tree plantations established along lake margins shade out emergent aquatic vegetation that provides breeding habitat for anopheline mosquitoes. Baldcypress *(Taxodium distichum* Rich.) and water tupelo *(Nyssa aquatica* L.) have been reported to reduce anopheline breeding by 62%.[61]

Human Settlement

Organized concentrated human settlement has been used effectively to hold land for agricultural purposes and stop the advance of tsetse fly in many parts of Africa.[1,62] These management programs were initiated by gathering scattered communities of peo-

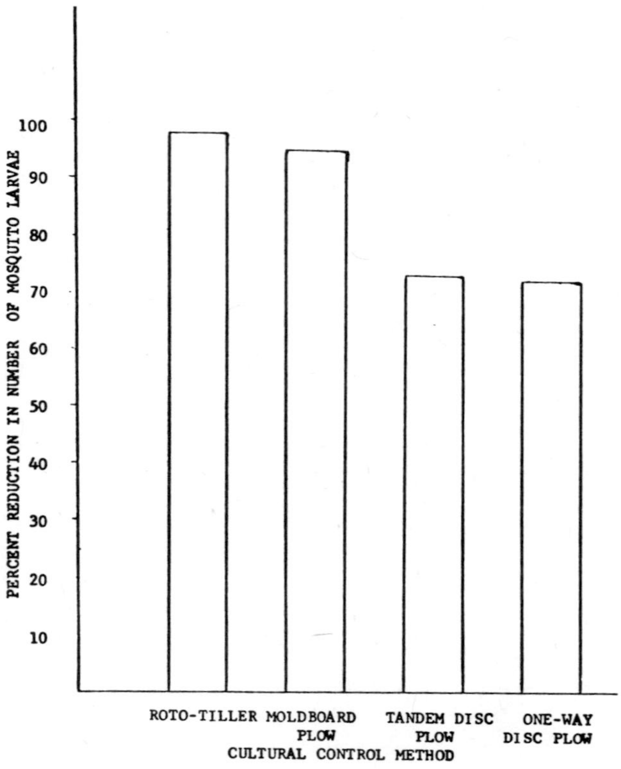

FIGURE 4. Percent reduction in numbers of mosquito larvae caused by cultural control methods.[55]

ple into settlements of 34,000 people and introducing good agricultural practices.[12,23] Villages of this size, completely surrounded by cultivation and with the husbandry of grazing cattle and goats, are reported to be tsetse fly free.[1] The policy established for this method of tsetse control is to increase the population density from 16 to 18 per 1.609 km^2 up to about 100 per 1.609 km^2, forming settlements of 2600 to 3500 people. At the higher human population density, the normal clearing of farmland suppresses fly populations by habitat destruction, and the results are dramatic. However, the settlements can only be maintained by constant vigilance.[15]

PASTURE MANAGEMENT

Rotation of Animals

The husbandry practice of alternately grazing cattle in adjacent pastures, allowing a sufficient time interval between the use of pastures to ensure that most of the ticks in the unoccupied pasture have died, is variously termed pasture spelling, pasture rotation, pasture vacating, and pasture resting.[63,64] This technique was recommended in 1932 by the U.S. Department of Agriculture for the control of cattle fever ticks.[63] Extensive research programs showing the effectiveness of this management practice in controlling ticks has been reported from the U.S.[63] the U.S.S.R.,[33,65] Africa,[66,16] and Australia.[64,67-69]

Pasture rotation has been reported to be an effective component when included in a tick control program.[67] When two herds of cattle were grazed in the same pastures continuously, the number of ticks counted on both herds was similar. When one of

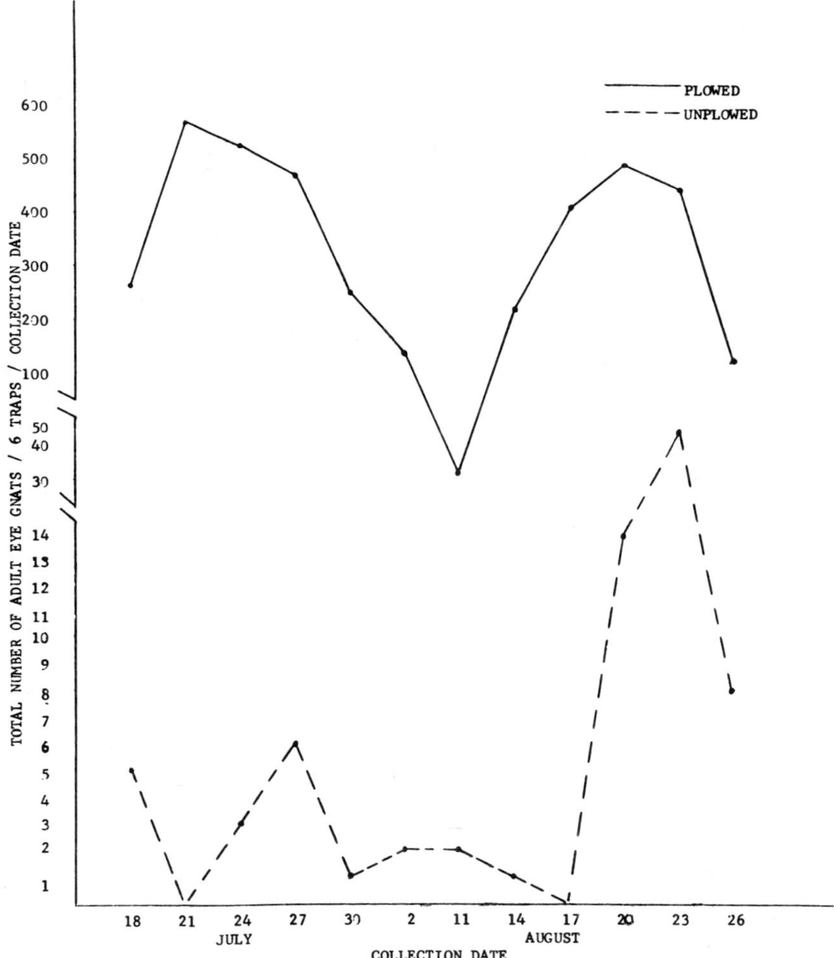

FIGURE 5. Number of eye gnats *(Hippelates* spp.) collected from plowed and unplowed fields.[59]

the herds was placed on a pasture rotation system the tick infestation was about 1/40 that of the herd maintained on the same pasture continuously (Table 4). During this 18-month study the herd that was grazed on one pasture continuously was sprayed with DDT on four occasions to control excessive tick infestations, while the rotation herd did not require treatment. The low number of ticks counted on the rotation-herd was thought to be the result of the first pasture rotation (May 1955—Table 4) when engorged ticks produced low numbers of progeny. Subsequent cattle rotations at approximately 4-month intervals prevented the tick population from rising to damaging levels.

Well-planned pasture management systems which include rotational grazing may improve the quality of the pasture by allowing an increase in the pastures' content of more palatable grasses. The more palatable grasses seem to offer less cover to the ticks as well as for the small animals which serve as alternate hosts.[66]

Pastoralism

Pastoralism is the term used in parts of Africa to describe a form of pasture rotation

Table 4
AVERAGE NUMBER OF ADULT FEMALE TICKS PER MONTH ON CATTLE GRAZED CONTINUOUSLY IN THE SAME PASTURE AND ON CATTLE ALTERNATELY GRAZED IN TWO PASTURES[67]

Months	Average number adult female ticks per one side of animal[a]	
	Continuously grazed on same pastures	Alternately grazed on two pastures
	1955	
January	18	17
February	62	35
March	8[b]	33[b]
April	101[b]	70[b]
May	109	57[e]
June	124	0
July	135[c]	0[c]
August	42	0
September	18	0[e]
October	18	0
November	27[d]	0
December	26	7
	1956	
January	74	0[e]
February	52	0
March	85	2
April	84[d]	2
May	16	5[e]
June	88[d]	0
July	17	0
August	21	0
September	3	0[e]
October	0	0
November	9	0
December	23[d]	0

[a] Approximate average number ticks per animal per month.[67]
[b] Animals sprayed to point of run-off with 0.16% arsenical preparation for tick control.
[c] Animals dusted with 1.5 oz. of 0.6% gamma BHC for tick control.
[d] Animals sprayed with 1% DDT emulsion for tick control.
[e] Pasture rotation dates.

wherein a regular movement of people and livestock occur.[16] However, pastoralism is reported to be accompanied by the lack of modern animal husbandry techniques. This method of rotation is satisfactory as long as livestock are carefully routed through the areas that have not been previously occupied by livestock.

Livestock Removal

Keeping livestock out of tick-infested areas for specific periods of time is a widely-used modification of the pasture-rotation management system. Through the removal or exclusion of the major host animals, it is possible to alter the local habitat sufficiently to reduce and perhaps even elminate the tick population in a particular area.[5,33,70-73]

In the state of Missouri (U.S.) keeping tick-infested pastures free from livestock from mid-June until the first heavy frost for 2 successive years, and occasionally thereafter, has resulted in a significant reduction in the number of ticks.[71] In Great Britian, sheep are often confined to the lowland pastures where ticks are absent until after the spring "rise" of *Ixodes ricinus* L. in the upland or hill pastures.[74] This management system prevents heavy infestation, especially where young lambs are concerned, and the sheep can then graze on the upland pastures during the summer with little danger of tick populations reaching damaging levels of infestation. On farms of Central Asian Republics where pastures are naturally tick-free, as in the mountains and deserts, these pastures are grazed in the autumn and spring, a time when the tick larvae are especially numerous elsewhere.[75]

Pasture Selection

Excessive attack by tabanids can be avoided during peak periods of activity by placing cattle in large pastures free of wooded areas, or in pastures well removed from wooded areas.[76]

Large, open pastures away from wooded areas could be reserved and used during the periods of greatest tabanid activity. The data presented in Table 5 show the number of tabanids collected from the borders of wooded areas and pastures. A statistically significant ($P < 0.05$) decrease in host-seeking activity occurred at distances of 237 m or greater from the woods' edge into open pasture areas and at approximately 18 m into the wooded area. Similar observations on areas of tabanid activity have been reported.[77,78] However, some species of tabanids have shown a preference for open areas.[79,80]

ELIMINATION OF ALTERNATE HOSTS

Game Destruction

The dependence of the tsetse fly on wild game animals, which served as hosts before large numbers of domesticated cattle were present in Africa, is evidenced by the fact that *Glossina morsitans* West. disappeared after the great rinderpest epizootic in 1896.[81] However, during the early 1920s some workers stated that game destruction was not desirable since it was expensive, and only the more noble game was killed while leaving the bush-pigs and other small animals that could also serve as hosts for the tsetse fly.[1] In Anglo-Egyptian Sudan, tribal hunters reduced the game populations to such an extent that *G. morsitans* populations were reduced.[82]

Game destruction has been used as a routine method of tsetse fly control in Southern Rhodesia, Uganda, Portuguese East Africa, Northern Rhodesia, and Bechuanaland, and many thousands of square miles have been cleared of the fly.[4,14,83-85] The policy of game elimination, combined with fierce, late dry-season burning to stop the bush from becoming more dense, was initiated about 1945, and by 1958 some 9654 km² of land had been reclaimed from the tsetse fly and most of it re-occupied by cattle. To accomplish this, some 58,000 wild animals were killed.[13]

The game destruction program in Uganda reclaimed some 7723 km² of country from the tsetse fly by the elimination of 2179 buffalo, 69 hippopotami, 10 rhinoceros and 25,163 smaller game animals, of which the bush buck composed 4901. This program was initiated in 1945 and ended in 1957 and the success was described as "no

Table 5
MEAN NUMBER OF TABANIDS COLLECTED AT VARYING DISTANCES FROM THE BORDERS OF WOODED AREAS AND PASTURES[76]

	Mean numbers of tabanids[a]							
	Into woods		From woods' edge					
	128 m	18 m	18 m	128 m	237 m	456 m	776 m	1370 m
St. Gabriel-traps	6.9 a	9.5 a	58.0 b	42.1 b	55.0 b	41.8 b	—	12.2 a
Caffarel-traps	—	—	291 c	305 c	210 d	175 d	190 d	—
Thistlewaite-traps	19 e	52 f	275 g	—	—	—	—	—
Caffarel-mare	—	—	62.5 h	—	—	20.9 i	23.8 i	—

[a] Means followed by the same letter are not significantly different at 5% level, using Duncan's new multiple range test.

other known tsetse fly control measure could have achieved so much in so short a time — nor could any other method have been successfully applied at so low a cost."[86] Over all animals, the program had a mean kill of six game animals per 1.609 km^2 at a cost of \$0.60/0.4 ha. Total cattle production in Uganda increased between the years of the program (1945 to 1957) from 2,294,000 to 3,094,000, respectively, or an increase of 800,000 cattle which at the current prices of the time were worth \$10,000,000.

An example of the effectiveness of alternate host animal elimination was shown by the results of the Shinyanga game destruction experiment in Tanganyika.[87] The primary objective was to determine if the common savanna species of tsetse fly, *G. morsitans, G. swynnertoni,* and *G. pallidipes,* could be controlled by the destruction of the hoofed game animals in a 965 km^2 area under conditions that prevailed in Tanganyika; and, if so, how long it would take and at what cost. The total number of animals recorded as shot was 8554 from 1945 to 1950, with the addition that a great number of small animals were destroyed and not reported.

The results of the program on tsetse fly populations were best shown for *G. swynnertoni* since it was collected in the largest numbers both in the experimental area and in the control (no game destruction) area (Figure 6). *Glossina morsitans* and *G. swynnertoni* were exterminated by the shooting of wild game and *G. pallidipes* were rarely collected. The experiment cost \$0.50/1.609 km^2, which was many times less than the costs that would have been required to clear the area of bush, and much less than the cost of discriminative clearing of the three habitats breeding the three species of tsetse flies. As a result of the study, game elimination was recommended only for isolated areas of manageable size.

Game-proof Fences

Fences constructed across long distances to exclude game animals have been used successfully in place of game destruction or in conjunction with game destruction programs.[88] A game fence was erected around an area of about 322 km^2 in Rhodesia after driving out the elephants and buffaloes. The favored host animals of *G. morsitans* which could not be driven out before fencing were shot. Soon after the shooting had been completed, the number of tsetse flies declined to a low level. A smaller experiment where game was excluded by fencing a 0.8 km^2 area by a 4.8 km fence significantly reduced the number of tsetse during a 2-year period.[18] The remainder of the tsetse flies inside the fenced area were reported to have starved.

The effectiveness of game fences was also indicated in the Shinyanga game destruction experiment, as a 80.4 km game fence was constructed during the study.[87] From

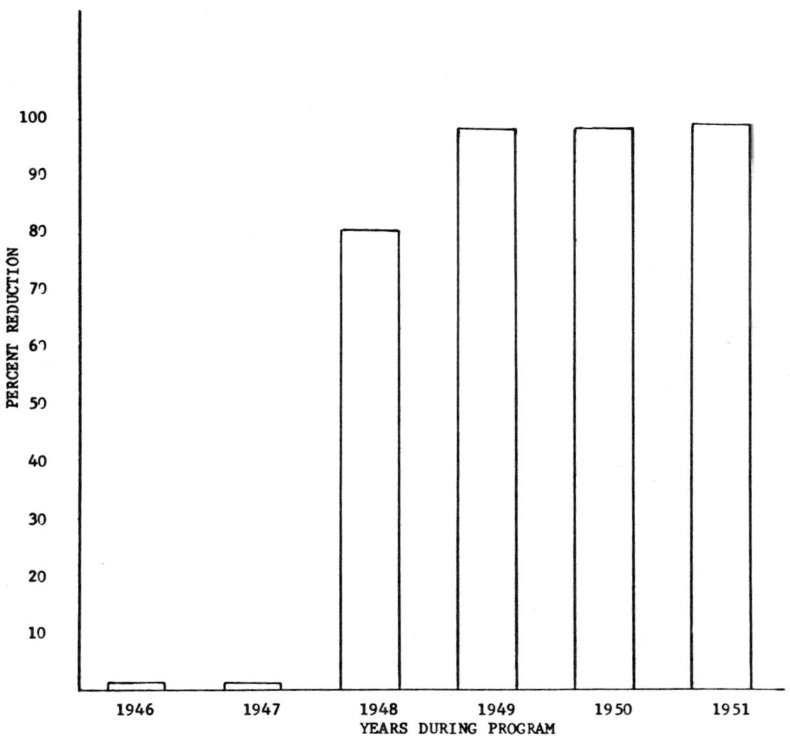

FIGURE 6. Percent reduction in the numbers of *Glossina swynnertoni* caused by game destruction in Tanganyika, 1945—1950.[87]

1945 to September 1948, about one half of the animals shot were plains game (3340 out of 6695), but after the fence was constructed, the plains species formed only a fifth of the total killed (378 out of 1859).

The methods of environmental control of arthropod pests of livestock described herein are helpful and desirable where and when they can be used. While they frequently are not completely satisfactory when used alone, they can provide an additional component for inclusion in an integrated pest management system for the control of livestock pests.

REFERENCES

1. **Swynnerton, C. F. M.,** The entomological aspects of an outbreak of sleeping sickness near Mwanza, Tanganyika Territory, *Bull. Entomol. Res.,* 13, 317, 1922.
2. **Whiteside, E. F.,** The control of animal trypanosomiasis in Kenya, *Inter-African Advisory Committee on Epizootic Diseases, Symp. Animal Trypanosomiasis,* Publ. No. 45, Committee Technology Cooperative in Africa South of the Sahara, Luanda, 1958, 81.
3. **Pomeroy, A. W. J. and Morris, K. R. S.,** The tsetse problem on the eastern cattle route in the Gold Coast, *Bull. Entomol. Res.,* 23, 501, 1932.
4. **Hocking, K. S., Lamerton, J. F., and Lewis, E. A.,** Tsetse-fly control and eradication, *Bull. W. H. O.,* 28, 811, 1963.
5. **Belshaw, D. G. R.,** An outline of resettlement policy in Uganda 1945 to 1963, in *Land Settlement and Rural Development in Eastern Africa, Kampala,* Transition Books Ltd., 14, 1970, chap. 3.

6. **Morris, K. R. S.,** The control of trypanosomiasis (of man and animals) by entomological means, *Bull. Entomol. Res.,* 37, 201, 1946.
7. **Axtell, R. C.,** Coastal Horse Flies and Deer Flies (Diptera:Tabanidae), in Marine Insects, Cheng, L., North-Holland, Amsterdam, 415, 1976.
8. **Morgan, N. O. and Lee, R. P.,** Vegetative barriers influence flight direction of saltmarsh greenheads, *Mosq. News,* 37, 263, 1977.
9. **Moiser, B.,** Notes on the haunts and habits of *Glossina tachinoides,* near Geidam Barnu Province, Northern Nigeria, *Bull. Entomol. Res.,* 3, 195, 1912.
10. **Nash, T. A. M.,** The part played by microclimates in enabling *Glossina submorsitans* and *G. tachinoides* to withstand the high temperatures of a West Africa dry season, *Bull. Entomol. Res.,* 27, 339, 1936.
11. **Nash, T. A. M.,** The ecology of the puparium of *Glossina* in Northern Nigeria, *Bull. Entomol. Res.,* 30, 259, 1939.
12. **Bax, S. N.,** A practical policy for tsetse reclamation and field experiment. II, *East Afr. Agric. J.,* 9, 83, 1943.
13. **Nash, T. A. M.,** A review of the African trypanosomiasis problem, *Trop. Dis. Bull.,* 57, 974, 1960.
14. **Buxton, P. A.,** Natural history of tsetse flies, *London School Hyg. Trop. Med.,* Memo No. 10, 816, 1955.
15. **Nash, T. A. M.,** *Africa's Bane: The Tsetse Fly,* Collins, London, 1969, 224.
16. **Jahnke, H. E.,** The economics of controlling tsetse flies and cattle trypanosomiasis in Africa, *Weltforum Verlag.,* 359, 1974.
17. **Swan, J. F. C.,** Annual report, Department of Veterinary and Tsetse Control Services for 1960, *Trans. R. Entomol. Soc. London,* 84, 1, 1961.
18. **Lloyd, L. L., Johnson, W. B., and Rawson, P. H.,** Experiments in the control of tsetse-fly, *Bull. Entomol. Res.,* 17, 423, 1926.
19. **Nash, T. A. M.,** The effect upon *Glossina* of changing the climate of the true habitat by partial clearing of vegetation, *Bull. Entomol. Res.,* 21, 201, 1940.
20. **MacLennan, K. J. R.,** Recent advances in techniques for tsetse-fly control: with special reference to Northern Nigeria, *Bull. W.H.O.,* 37, 615, 1967.
21. **Jordan, A. M.,** Recent developments in the ecology and methods of control of tsetse flies *(Glossina* spp.) (Dipt.,Glossinidae)—a review, *Bull. Entomol. Res.,* 63, 361, 1974.
22. **Glover, P. E., Trump, E. C., and Williams, R.,** Note on mechanical bush clearing, *East Afr. Agric. J.,* 25, 18, 1959.
23. **Wooff, W. R.,** Consolidation in tsetse reclamation, in *Int. Sci. Council Trypanosomiasis Res. 11th Meet.,* Organization for African Unity, Science Technology, and Research Commission, Publication Bureau, Nairobi Lagos, No. 100, 141, 1967.
24. **Fryer, J. D., Johns, D. L., and Yeo, D.,** The effects of a chemical defoliant on an isolated tsetse community and its vegetation, *Bull. Entomol. Res.,* 48, 359, 1957.
25. **Little, E. C. S. and Ivens, G. W.,** The control of bush by herbicides in tropical and subtropical grassland, *Herb. Abstr.,* 35, 1, 1965.
26. **Ivens, G. W.,** Seasonal differences in kill of two Kenya bush species after foliar herbicide treatment, *Weed Res.,* 11, 150, 1971.
27. **Pratt, D. J.,** Bush-control studies in the drier areas of Kenya. VI. Effects of fenuron (3-phenyl-1,1-dimenthylurea), *J. Appl. Ecol.,* 8, 239, 1971.
28. **Mason, G. W.,** A review of chemical control of brushweeds on agricultural land in New Zealand, *Proc. 4th Asian-Pacific Weed Sci. Soc. Conf.,* 2, 339, 1973.
29. **Ivens, G. W. and Pratt, D. J.,** Aerial application of chemicals to control bush in east Africa, *East Afr. Agric. For. J.,* 40, 278, 1975.
30. **Mulla, M. S.,** An ecological basis for the suppression of *Hippelates* eye gnats, *J. Econ. Entomol.,* 56, 768, 1963.
31. **Buxton, P. A.,** *Trypanosomiasis in East Africa,* Her Majesty's Stationary Office, London, 1948, 44.
32. **Stampa, S. and DuToit, R.,** Paralysis of stock due to the Karoo paralysis tick, *I. rubicundus, S. Afr. J. Sci.,* 54, 241, 1958.
33. **Galuzo, I. G.,** Ecological bases of the control of the vectors of the *ileriasis* of cattle, *Izv. Kazakhsk. Fil Akad. Nauk. SSSR (Ser. Zool.),* 3, 92, 1944.
34. **Krotkov, A. A. and Lavrentyev, A. I.,** Utilization of a machine-operated gas flame against pasture parasites of animals, *Tr. 2 Nauch. Konf. Parasitol. U.S.S.R., Kiev,* 37, 1960.
35. **Shumkov, M. A.,** Effect of burning vegetation on the death of *Aedes* mosquito eggs in the delta of the Don River, *Med. Parazitol. Parazit. Bolez.,* 38, 102, 1969.
36. **James, M. T. and Harwood, R. F.,** *Herm's Medical Entomology,* Macmillan, 6th ed., 484, 1969.
37. **Davis, S.,** Soil, water and crop factors that indicate mosquito production, *Mosq. News,* 21, 44, 1961.

38. **Gartrell, F. G., Barnes, W. W., and Christopher, G. S.,** Environmental impact and mosquito control water resource management projects, *Mosq. News,* 32, 337, 1972.
39. **Rees, D. M. and Anderson, J. N.,** Results of multipurpose water management studies on marshes adjacent to the Great Salt Lake, Utah, *Mosq. News,* 26, 160, 1966.
40. **Rees, D. M. and Winget, R. N.,** Water management units and shore line modifications used in mosquito control investigations, *Mosq. News,* 28, 305, 1968.
41. **Rees, D. M. and Winget, R. N.,** Mosquito control in Utah by shore line modification, *Mosq. News,* 29, 368, 1969.
42. **Schwardt, H. H.,** Horseflies of Arkansas, *Arkansas Agric. Exp. Stn. Bull.,* 332, 66, 1935.
43. **Oldroyd, H.,** *The Natural History of Flies,* Weidenfeld and Nicolson, London, 1964, 324.
44. **Kneen, F. R.,** Water management in the control of *Chrysops niger, Proc. N.J. Mosq. Exterm. Assoc.,* 55, 94, 1968.
45. **Anderson, U. F. and Kneen, F. R.,** The temporary impoundment of salt marshes for the control of coastal deer flies, *Mosq. News,* 29,239,1969.
46. **Blanton, F. S., Graham, O. H., and Keenan, C. M.,** Notes on *Culicoides furens* (Poey) at Fort Kobbe, Canal Zone, *Mosq. News,* 15, 13, 1955.
47. **Rogers, A. J.,** Effects of impounding and filling on the production of sand flies *(Culicoides)* in Florida salt marshes, *J. Econ. Entomol.,* 55, 521, 1962.
48. **Hull, J. B., Shields, S. E., and Platts, N. G.,** Diking as a measure for sand fly control in salt marshes, *J. Econ. Entomol.,* 36, 405, 1943.
49. **Schober, H.,** Agitation of water surfaces by sprinkling to prevent mosquito breeding, *Mosq. News,* 26, 144, 1966.
50. **Peterson, D. G. and Wolfe, L. S.,** The biology and control of black flies (Diptera: Simuliidae) in Canada, *Proc. 10th Int. Cong. Entomol.,* 3, 551, 1958.
51. **Dampf, A.,** Los simulidos transmisores de la oncocercosis en los Estados de Oaxaca y chiapas, *Medicina (Mexico City),* 11, 753, 1931.
52. **Markov, A.,** Protozoal blood parasites of domestic animals and their control in U.S.S.R., *Bull. Off. Inst. Epizol.,* 49, 96, 113.
53. **Milne, A.,** The ecology of the sheep tick, *Ixodes ricinus* L., *Parasitology,* 35, 186, 1944.
54. **Huddleston, E. W. and Ward, C. R.,** Multipurpose modification of Playa sinks, Completion Report, Proj. 29, Division of Water Supply and Pollution Control, U.S. Public Health Service, Lubbock, Texas, 80, 1969.
55. **Owens, J. C., Ward, C. R., Huddleston, E. W., and Ashdown, D.,** Non-chemical methods of mosquito control for Playa Lakes in West Texas, *Mosq. News,* 30, 571, 1970.
56. **Tinkham, E. R.,** The eye gnats of the Coachella Valley with notes on the 1951 larviciding program, *Proc. Calif. Mosq. Cont. Assn.,* 20, 83, 1952.
57. **Bay, E. C. and Legner, E. F.,** The prospect for the biological control of *Hippelates collusor* (Townsend) in southern California, *Proc. 31st Conf. Calif. Mosq. Cont. Assn.,* 31, 76, 1963.
58. **Mulla, M. S.,** Biology and control of *Hippelates* eye gnats, *Proc. 33rd Ann. Conf. Calif. Mosq. Cont. Assn.,* 33, 26, 1965.
59. **Gaydon, D. M. and Adkins, T. R., Jr.,** Effect of cultivation on emergence of eye gnats *(Hippelates* spp) in South Carolina, *J. Econ. Entomol.,* 62, 312, 1969.
60. **Joseph, C., Menom, M. A. U., Unnithan, K. R., and Raman, S.,** Studies on the comparative hospitability of *Salvinia auriculata* Aubet and *Pistia stratiotes* Linn. to *Mansoniodes annulifera* (Theobald) in Kerala, *J. Malariol.,* 17, 311, 1963.
61. **Smith, G. E., Pickard, E., and Hall, T. F.,** Tree plantings for mosquito control, *Mosq. News,* 29, 161, 1969.
62. **Swynnerton, C. F. M.,** The tsetse flies of East Africa, *Trans. R. Soc. London,* 84, 1, 1936.
63. **Ellenberger, W. P. and Chapin, R. M.,** Cattle-fever ticks and methods of eradication, *U.S. Dep. Agric. Farmers' Bull,* 1057, 270, 1932.
64. **Wilkinson, P. R.,** The spelling of pasture in cattle tick control, *Aust. J. Agric. Res.,* 8, 414, 1957.
65. **Galuzo, I. G.,** Ecological foundations of control measures against transmitters of haemosporidiosis of horses, *Dermacentor marginatus, Izv. Kazakhsk. Fil. Akad. Nauk SSSR (Ser. Zool.),* 3, 12, 1944.
66. **Stampa, S.,** Tick paralysis in the Karoo areas of South Africa, *Onderstepoort J. Vet. Res.,* 28, 169, 1959.
67. **Wilkinson, P. R.,** Pasture spelling as a control measure for cattle ticks in southern Queensland, *Aust. J. Agric. Res.,* 15, 822, 1964.
68. **Harley, K. L. S. and Wilkinson, P. R.,** A comparison of cattle tick control by "Conventional" acaricidal treatment, planned dipping, and pasture spelling, *Aust. J. Agric. Res.,* 15, 841, 1964.
69. **Harley, K. L. S. and Wilkinson, P. R.,** A modification of pasture spelling to reduce acaricide treatments for cattle tick control, *Aust. Vet. J.,* 47, 108, 1971.
70. **Theiler, A.,** Diseases of ticks and their eradication, *Agric. J. Un. S. Africa,* 1, 491, 1911.

71. **Portman, R. W.,** Pasture management kills lone star ticks, *Univ. Missouri Agric. Exp. Stn. Cir.* 297, 4, 1945.
72. **Wilkinson, P. R.,** Temporary destocking of pastures to aid control of the cattle tick, *Nature (London),* 176, 515, 1955.
73. **Barnett, S. F.,** The control of ticks on livestock, *F.A.O. F.A.O.U.N. Agric. Stud.,* 54, 115, 1961.
74. **Milne, A.,** The ecology of the sheep tick, *Ixodes ricinus* L., *Parasitology,* 36, 142, 1945.
75. **Galuzo, I. G.,** Ticks (Ixodiodea) of Kazakhstan and central Asian Republics, *Bull. Off. Int. Epizol.,* 49, 386, 1958.
76. **Sheppard, C. and Wilson, B. H.,** Relationship of horse fly host seeking activity to the edge of wooded areas in southern Louisiana, *Environ. Entomol.,* 7, 781, 81977.
77. **Hine, J. S.,** Second report upon the horse flies of Louisiana, *La. Agric. Exp. Stn. Bull.,* 98, 1, 1907.
78. **Duke, B. L. L.,** Studies of the biting habits of *Chrysops.* IV. The dispersonal of *Chrysops silacae* over cleared areas from the rainforest at Kumba British Cameroons, *Ann. Trop. Med. Parasitol.,* 49, 368, 1955.
79. **Jamnback, H. and Wall, W. J.,** The common salt marsh Tabanidae of Long Island, New York, *N. Y. State Mus. Sci. Ser. Bull.,* 375, 77, 1959.
80. **Thornhill, A. R. and Hays, K. L.,** Dispersal and flight activities of some species of *Tabanus* (Diptera: Tabanadae), *Environ. Entomol.,* 1, 602, 1972.
81. **Jack, R. W.,** Tsetse fly and big game in Southern Rhodesia, *Bull. Entomol. Res.,* 5, 97, 1914.
82. **Lewis, D. J.,** The tsetse fly problem in the Anglo-Egyptian Sudan, *Sudan Notes Rec.,* 30, 179, 1949.
83. **Cockbill, G. F., Pilson, R. D., and Vale, G. A.,** Observations on populations of *Glossina morsitans* Westw. and of game animals on Island 173/174, Lake Kariba, Rhodesia, *Proc. Trans. Rhod. Sci. Assoc.,* 53, 88, 1969.
84. **Cockbill, G. F.,** *Annual Report of the Branch of Tsetse and Trypanosomiasis Control Department of Veterinary Services,* Ministry of Agric., Salisbury, Rhodesia, Government Printer, 38, 1972.
85. **McKelvey, J. J., Jr.,** *Man Against Tsetse: Struggle for Africa,* Cornell University Press, Ithaca, 305, 1973.
86. **Robertson, A. G. and Bernacca, J. P.,** Game elimination as a tsetse control measure in Uganda, *East Afr. Agric. J.,* 23, 254, 1958.
87. **Potts, W. H. and Jackson, C. H. N.,** The Shinyanga game destruction experiment, *Bull. Entomol. Res.,* 43, 365, 1953.
88. **Cockbill, G. F.,** The history and significance of trypanosomiasis problems in Rhodesia, *Proc. Trans. Rhod. Sci. Assn.,* 52, 7, 1967

ENVIRONMENTAL CONTROL OF LIVESTOCK DISEASE

Leon H. Russell, Jr., Pete D. Teel, and D. E. Bay

INTRODUCTION

Environmental control of livestock diseases as discussed in this chapter concerns those diseases transmitted by arthropod vectors or produced directly through the action of arthropod parasites on their hosts. **Vectors** may be classified as either biological or mechanical transmitters of disease agents, i.e., pathogens. **Biological vectors** are those essential to disease transmission since the pathogen undergoes a required part of its cycle, either propagation or maturation, within the body of the arthropod (e.g., babesiosis transmission by *Boophilus* ticks[1]).[2] **Mechanical vectors** are not essential for the disease cycle, serving merely to amplify the transmission of the pathogen through passive short-term carriage (e.g., bovine pink eye transmission by face flies[3]).[2]

A diagrammatic representation of infectious disease transmission is presented in Figure 1. Successful transmission is dependent upon three essential components; i.e., reservoir, transfer, and new susceptible host. Disease control programs may be accomplished by implementing one or more of the following techniques: (1) reservoir neutralization, (2) transfer inhibition, and/or (3) host resistance modification.[4] Ideally, control methodologies should proceed from right to left in this scheme as permanent control, i.e., eradication, can be accomplished only through reservoir neutralization. Control measures will also be influenced by the history of each disease within a specific geographical area. Endemic diseases, i.e., those which occur at relatively constant frequencies, are more likely to be the object of control programs that emphasize inhibition of transmission and/or host resistance modification.[2,5] Epidemic diseases, i.e., those diseases occurring in excess of "normal" or endemic levels, are more susceptible to control through a combination of all three techniques, with emphasis on reservoir neutralization.[2,5]

RESERVOIR NEUTRALIZATION

The reservoir of a particular infectious disease may be defined as the geographical location or animal host in which the causative agent is perpetuated or maintained over an extended period of time.[2] Although clinical cases of disease may constitute sources of exposure, animal reservoirs frequently do not have clinical syndromes during the times which they are sources of biological vector infection or mechanical vector contamination. Animals in a subclinical "shedding" phase are commonly referred to as carriers and are often proficient sources of pathogen exposure.[5] The classification of carriers is based on the duration of the carrier state (Table 1). Following recovery from clinical disease, chronic carriers maintain the etiological agent for 6 months or longer while convalescent carriers maintain the etiological agent for shorter time periods. A healthy carrier is one which has never demonstrated clinical disease, but is a source of the etiological agent for vector transmission.[2,5] The type of carrier, and number of species constituting carriers, are both important factors influencing the ease of disease transmission.

Infectious disease agents which utilize chronic carriers as reservoirs and mechanical vectors as transmission modes are generally the most readily transmitted.[6] Anaplasmosis, for example, produces a lifetime chronic carrier state in which the causative agent can be transmitted by any number of mechanical vectors, such as tabanids and

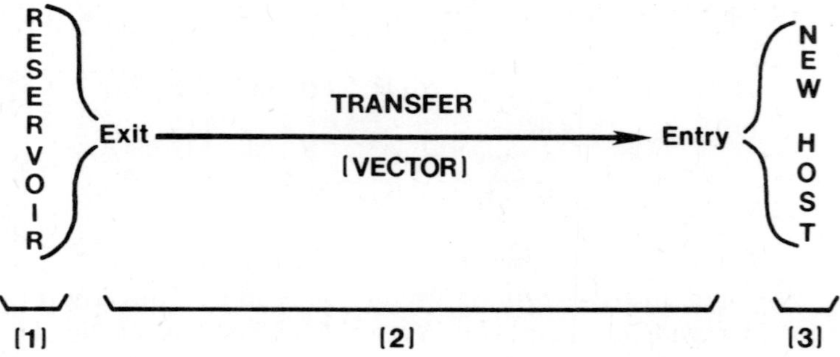

FIGURE 1. Diagrammatic representation of infectious livestock disease transmission. Phases of control: (1) reservoir neutralization, (2) transfer inhibition, and (3) host resistance modification.

Table 1
ANIMAL RESERVOIRS AND VECTORS OF SELECTED LIVESTOCK INFECTIOUS DISEASES

Disease	Etiologic agent	Principal livestock host	Vector type	Reservoir	Carrier state (duration)
African horse sickness[a]	Virus	Horse	Biological (*Culicoides* spp.)	Horse	Chronic (months)
African swine fever	Virus	Swine	Biological (*Ornithodorus* spp.)	Swine, tick	Chronic (lifetime)
Bluetongue[a]	Virus	Sheep, cattle	Biological (*Culicoides* spp.)	Sheep, cattle	Convalescent (4 months); chronic (4 years)
Encephalitides Western equine[a] Eastern equine[a] Venezuelan[a]	Virus	Horse	Biological (Mosquito spp.)	Bird	Convalescent (weeks)
Epizootic bovine abortion	Virus	Cattle	Biological (*Ornithodorus* spp.)	Cattle, tick	
Equine infectious anemia	Virus	Horse	Mechanical (stable fly—tabanid)	Horse	Chronic (lifetime)
Vesicular stomatitis	Virus	Cattle	Biological (sand fly)	Cattle, wildlife	Convalescent (months)
Anaplasmosis[a]	Rickettsia	Cattle	Mechanical (biting Diptera—tick)	Cattle	Chronic (lifetime)
Eperythrozoonoses	Rickettsia	Swine	Mechanical (sucking lice)	Swine	Chronic (lifetime)
Pink eye	Bacteria	Cattle	Mechanical (face fly)	Cattle	Chronic (1 year)
Babesiosis:	Hemo-protozoa		Biological	Tick,	Transovarian
Bovine		Cattle,	(tick ssp.)	cattle	Chronic (2 years)
Equine		horse,		horse	Chronic (lifetime)
Porcine		swine		swine	Chronic (lifetime)
Habronemiasis	Nematode	Horse	Biological (housefly—stable fly)	Horse	Chronic (lifetime)

[a] Vaccines available.

Hippelates spp., for 2 to 3 days after contamination with *Anaplasma marginale* from the carrier, or, rarely, a clinical case.[7] By comparison, the least readily transmissible diseases are those which usually utilize single reservoir host and biological vector species. Bovine babesiosis is an example of this latter type, in that the pathogen, *Babesia bigemina,* is transmitted from bovine carriers by cattle fever ticks, *Boophilus annulatus* or *B. microplus;* however, even here, transmission may be facilitated by overwintering of the vector and transovarian passage of the pathogen as a reservoir mechanism.[8]

Reservoir neutralization is usually accomplished through one of two methods, i.e., segregation or elimination. Segregation is carried out through either quarantine or isolation.[4] Quarantine involves the segregation of exposed animals; isolation refers to the segregation of diseased stock for the duration of the disease or carrier state. Elimination, generally referred to as eradication, is accomplished by either selective or complete destruction of both active and potential reservoirs.

Segregation

The physical segregation of diseased from healthy stock has long formed the basis for control programs of scabies in sheep and cattle throughout the world.[9] Psoroptic scabies was declared a quarantinable animal disease in the U.S. by the federal government in 1894 and by 1895, U.S. Department of Agriculture regulations were invoked prohibiting the interstate movement of infested sheep. Rigorous state-federal programs of quarantine and chemical treatment and cooperation by sheep raisers virtually eradicated sheep scab from much of the U.S. by 1929, although it was not until 1943 that psoroptic sheep scabies was eliminated from the western states.[10] Reinfestation of these areas commonly occurred from midwestern states, where scabies continued to exist, so that an accelerated program was initiated in 1961.[11,12] Vigorous eradication efforts resulted in the last case of sheep scabies being reported from New Jersey in 1970 and the disease officially eradicated in 1973.[8]

Simultaneously, cooperative state-federal eradication programs involving quarantine and treatment were undertaken against psoroptic cattle scabies. Progress seemed equally successful, so that the mite was considered virtually eradicated when no new cases were reported during 1950 to 1952.[13] However, an outbreak occurred in southeastern Colorado in 1954 and subsequent cases have continued almost every year since that time.[8,14] (More outbreaks of psoroptic cattle scabies occurred during the first 6 months of fiscal year 1978 than had previously been recorded by the U.S. Department of Agriculture.[14]) An excellent discussion of factors complicating the control of cattle scabies in the U.S. is presented by Meleney and Christy;[14] a breakdown of state-federal quarantine restrictions and/or proper treatment of infested stock is generally involved. As pointed out by these authors, effective suppressive measures, i.e., segregation and treatment, must be applied in an organized manner throughout the range of the disease. Pest management practices, as they are presently implemented, are not acceptable alternatives to an eradication program when one is faced with quarantinable parasites.

Eradication

Selective removal of diseased stock is generally carried out through "test and slaughter" programs as has been done with bovine tuberculosis over the past six decades in the U.S.[15] Selective slaughter is most effective against slowly spreading diseases which allow time for the identification of infected animals, i.e., reactors, thus limiting the pathogen to defined herds or areas. Selective eradication was also the basis for the enactment of legislation in New South Wales in the 1850s requiring the destruction of sheep infested with scabies.[16]

Complete elimination of the reservoir may be carried out by a herd depopulation program as has been done with foot and mouth disease in North America.[17] In this type of eradication program, both animals showing clinical symptoms and those exposed are destroyed on the premises. Depopulation is generally employed against rapidly spreading diseases or diseases newly introduced into an area (e.g., African swine fever in western Europe). A modification of reservoir elimination may be practiced against diseases both transmitted and reservoired by a biological vector. *Boophilus annulatus* is both a biological vector and a transovarian reservoir of *B. bigemina,* the causative agent of bovine babesiosis. Both the pathogen and its vector were eradicated from the U.S. in 1960 through an extensive and lengthy cattle dipping program which began in 1906.[5,8] Although no new cases of babesiosis have been recorded, *B. annulatus* and *B. microplus* have been periodically reintroduced into the U.S. from Mexico since 1968.[8] An active state-federal eradication program is currently maintained along the U.S.-Mexican border to prevent the entry and spread of these ticks.

TRANSFER INHIBITION

Many environmental methods, evolved primarily through observation and experience, presently exist to inhibit the transfer of livestock diseases from reservoirs to new hosts. The methodology employed will usually vary with the disease, geographical area, and total livestock management program. The following discussion of transfer inhibition in disease control programs is directed primarily toward vector-borne diseases but may also apply to those resulting directly from the presence or feeding of arthropods (for example, myiasis, scabies, and ixodiasis) as well as some non-vector-borne diseases. In a total livestock pest management program, most of these procedures should reduce the need for clinical treatment; however, incorporation of the two methodologies into an integrated system is often the most productive approach.

Pasture/Premise Vacation

Pasture vacation is an important environmental control concept in the elimination of cattle fever ticks, *B. annulatus* and *B. microplus.*[8] In the U.S., the presence of these ticks requires quarantine of both herd and premises. In addition to chemical control through a systematic dipping program, pasture vacation is an effective means of eradicating these potential vectors of babesiosis and anaplasmosis. Pasture vacation not only removes the host source, but also subjects larval ticks to environmental stresses of temperature and relative humidity for extended periods. Osmoregulation is a critical factor in the survival of ticks; environmental stress through exposure to adverse relative humidity and temperature is known to create an imbalance beyond which ticks cannot survive without moisture replenishment through the atmosphere or through a bloodmeal.[18-20] The Gulf Coast tick, *Amblyomma maculatum,* has a definite seasonal period during which adults feed on the ears of cattle producing significant production losses.[21-23] In pastures where numbers of Gulf Coast ticks are intolerable, pasture vacation during the adult tick period would reduce production losses and environmentally stress existing tick populations as described above.

Premise vacation as a means of environmental control is also an important factor, although often overlooked, in the control of scabies in cattle feedlots. Infested animals have intense irritation resulting in alleviation attempts through biting, scratching, and rubbing.[24] As a result, various live stages of the mite may be found off the host. When the weather is damp and cool, scab mites may survive up to 3 weeks off a host; however, direct sunlight and dryness may destroy them in 48 hr or less.[25] Although it is generally conceded that the possibility of a new host acquiring an infestation from such contaminated premises is remote, the potential does exist. Recent studies indicate that transport vehicles, enclosures, and pastures vacated for 2 weeks after occupancy by diseased sheep, and presumably cattle, are safe for reuse.[26]

Habitat Modification

Habitat modification may also be employed within a total range management program to control ticks and tickborne livestock diseases. Vegetative habitats of both the lone star, *Amblyomma americanum,* and Gulf Coast ticks have been described, and their respective survival and longevity studied in relation to environmental protection by various types of vegetation.[21,27-29] In general, areas with higher relative humidities and lower temperatures during spring and summer months supported higher tick numbers. A program in eastern Oklahoma woodlots involving habitat modification through mechanical and herbicidal brush control elevated soil temperature and reduced relative humidity and soil moisture resulting in a 90% reduction of larval lone star tick populations.[30-32] The feasibility and benefits of such programs in other geographical areas in which cattle are plagued with heavy tick burdens needs to be evaluated.

Prescribed burning for the control of undesirable vegetation and pasture improvement may also provide tick control in areas where vegetation is present and terrain conducive.[33-35] Pasture improvement methods, such as root-plowing and the introduction of improved grasses, not only alter microhabitats as described, but improve forage quality and increase the stocking capacity of the pasture.[36] These practices also influence wildlife which serve as alternate and/or intermediate hosts for ticks by altering the frequency with which certain species enter environmentally modified areas. The exact interrelationships between these environmental modifications, cattle, alternate hosts, and ticks is unknown; however, it should be noted that wildlife are an important aspect of any long-term tick control program, and careful consideration should be given to wildlife management through habitat alteration.

Animal Movement Modification

The dispersion, migration, and intermixing of animal populations presents a unique situation within a pest management program in that it increases the transfer of reservoirs and/or vectors resulting in increased frequency of vector-borne disease. As a result, control programs may become quite complicated. The type and maintenance of fencing can play a vital role in such a program. If improperly maintained, fencing allows the intermingling of herds, thus increasing the possibility of disease transmission. In addition, the type of fencing and its maintenance can be a technique for the exclusion of some species of wildlife from livestock pastures. This can be of importance where white-tailed deer are alternate hosts for tick species of concern.[37-39]

Sanitation

House flies and stable flies, the two most common dipteran species associated with commercial livestock production (e.g., feedlots, stables, and dairy barns) manifest their presence primarily through annoyance and irritation; however, either may also be an intermediate host of habronemiasis. In addition, the characteristic interrupted feeding behavior of the stable fly makes it a good mechanical vector for diseases such as equine infectious anemia.[40]

Throughout livestock production areas, both species breed commonly in animal excrement, especially that which becomes mixed with decaying vegetative matter such as hay or straw; soggy grain spilled from feed troughs also presents an excellent breeding site. Hermes[41] has calculated that a single pound of horse feces could support an average of 685 housefly larvae. High density livestock operations, such as beef cattle feedlots with capacities of 50,000 to 100,000 head, may thus become serious fly-breeding sources.

The importance of proper animal waste management as a means of environmental modification to control housefly- and stable fly-populations has long been recognized, but rarely properly employed in livestock production. Undoubtedly, a contributing factor to this reluctance has been the relief, albeit temporary, afforded by numerous synthetic organic insecticides. However, as resistance developed, especially in houseflies, it became obvious that the basic requisite for primary fly control was environmental sanitation.[42-45]

Moisture is essential to the larval development of both the housefly and stable fly. Bay[46] has shown the optimum moisture content of the larval substrate to be between 75 and 80% for both species. Not coincidentally, therefore, the most common sources of larval development in animal production areas are those associated with high degrees of available moisture. Walsh[47] found the most consistent sites of fly breeding in California feedlots to be damp areas adjacent to water troughs, corral surfaces with poor drainage, and individual cattle droppings within corrals. (The author observed 2929 adult flies to emerge from one such dropping removed to the laboratory.) Other problem areas were accumulations under feed troughs and corral fences and feed residues remaining in troughs. Physical measures, such as concrete slabs in watering areas and adequate corral surface grades to eliminate standing water, will substantially reduce fly breeding in these areas. Production in undisturbed droppings can easily be reduced by stocking pens at heavy densities to maximize tramping and thus increase desiccation.[47] Similarly, it is necessary to promptly clean recently vacated pens which are to be left idle. Feedlot construction should employ features facilitating waste removal, such as concrete aprons adjacent to feeding and watering areas and metal fencing.[47]

The removal of animal waste matter from the production arena does not necessarily eliminate the potential for fly development however. Unless the waste is properly disposed, the larval breeding site may be transferred from one location to another. Loehr[48] has concluded that no profitable method of livestock manure utilization exists, therefore, efficient low-cost methods of waste disposal are needed. Various authors have concluded that disposal onto agricultural land appears to be the most economical method of handling such wastes; however, success is dependent upon the ability to maintain high crop yields without nitrogen build-up that could contaminate ground water.[49] Removed wastes may also be composted, preferably in high, steeply sloping mounds. Through the action of facultative aerobic microorganisms, principally thermophilic, the waste substrate is broken down at temperatures in excess of those optimal to larval development. Fly development may continue to occur in the outer portions of such mounds and subsequent waste disposal must still be accomplished.

Other Environmental Control Methods

The importance of other environmental management practices (e.g., drainage, flooding, watercourse modification, and vegetative clearing) for control of arthropod disease vectors, especially aquatic breeding Diptera (mosquitoes, black flies, ceratopogonids, and tabanids), has long been recognized. For a comprehensive discussion of this methodology, the reader is referred to Steelman.[50]

HOST RESISTANCE MODIFICATION

Innate or acquired host resistance may be used to produce disease-free stock in cases where reservoir neutralization is not biologically or economically feasible and/or transfer inhibition is not highly effective in the prevention of pathogen exposure.[5,6] Artificial modification of host resistance is generally accomplished through inoculation with an inactivated or attenuated pathogen, frequently termed a vaccine. Immunization is generally the technique employed in the early phases of a disease control program especially where the occurrence of the pathogen is very common or endemic; for example, in the immunization of sheep against the virus of bluetongue.[51,52]

Cattle breeds possess varying degrees of innate resistance to many arthropods and pathogens. For example, zebu cattle *(Bos indicus)* have been demonstrated to be more resistant to mosquitoes, horn flies, ticks, and babesiosis than European breeds *(Bos taurus).*[22,23,53-55] Resistance in European-zebu crosses tends to be intermediate.[56] Selection of cattle for tick resistance to *B. microplus,* and their utilization in a tick control program, has proved effective in Australia.[57,58] Relatively little information exists concerning innate resistance of livestock to microbial pathogens; however, research has demonstrated that genetically controlled resistance does occur in chickens and mice.[59]

Other host-resistance modification techniques for susceptible livestock are chemoprophylaxis and nutrition. Chemoprophylaxis is the application of drugs to animals or their feed to prevent clinical infection, e.g., the inclusion of tetracycline antibiotics in cattle feed to control anaplasmosis.[60] Increasing the plane of nutrition may modify host resistance, resulting in decreased severity and reduced occurrence of diseases such as bluetongue in cattle.[61] Generally, however, this type of control program is successful only where there is definite clinical malnutrition such as vitamin A or protein deficiencies, e.g., infectious diarrhea in calves, or yellow fever in humans.[62,63] Host nutrition may also influence vector production. *Boophilus* ticks fed on cattle receiving a protein and energy deficient diet produced larger replete females and more progeny than those on cattle receiving a balanced ration.[64] In addition, vitamin A-deficient calves have been observed to be more susceptible to infestation by the common cattle grub, *Hypoderma lineatum.*[65]

In summary, successful control of vector-borne disease is dependent upon the number of reservoir host species as well as the number of vector species. Control programs by types of reservoir and disease cycles are presented in Table 2. Generally, environmental control is more likely to succeed when it is directed against biological rather than mechanical vectors, since the former have more intimate host interactions. Diseases exemplified by control programs B and D can be eliminated if biological vector eradication is possible; however, immunization may be needed to augment disease elimination, especially for type D programs. If vector eradication is not feasible, a control program to reduce the incidence of disease requires both vector abatement and immunization. In control program A, diseases can be eliminated primarily through reservoir elimination. Type C control programs are most proficient through a combination of vector abatement and reservoir immunization. The difficulty in controlling vector-borne diseases varies directly with the number and type of reservoir host species, as well as the duration of the carrier state. It should be emphasized that control procedures A through D are based on generalities; exceptions may occur which will require modification of these recommendations.

Table 2
CONTROL PROGRAMS FOR INFECTIOUS LIVESTOCK DISEASES BY RESERVOIR-DISEASE CYCLE

Control program	Number of reservoir host species	Type of vector transmission	Proficient disease control	
			Reduction[a]	Elimination
A	Single	Mechanical	Vector abatement	Reservoir test and slaughter
B	Single	Biological	Vector abatement	Vector eradication
C	Multiple	Mechanical	Vector abatement	Reservoir immunization
D	Multiple	Biological	Vector abatement	Reservoir immunization, vector eradication

[a] Immunization may also be necessary to reduce the incidence of disease.

REFERENCES

1. **Smith, T. and Kilborne, F. L.,** Investigations into the nature, causation, and prevention of Texas or Southern cattle fever, *U.S. Dep. Agric. Bull.*, 1893, 1.
2. **Benenson, A. S., Ed.,** *Control of Communicable Diseases in Man,* 12th ed., American Public Health Association, Washington, D.C., 1975.
3. **Cheng, T. H.,** Frequency of pink eye incidence in cattle in relation to face fly abundance, *J. Econ. Entomol.*, 60, 598, 1967.
4. **Russell, L. H.,** *Epidemiology,* Texas A & M University, College Station, 1978, 63.
5. **Schwabe, C. W., Riemann, H. P., and Franti, C. E.,** *Epidemiology in Veterinary Practice,* Lea & Febiger, Philadelphia, 1977, chap. 2.
6. **Fox, J. P., Hall, C. E., and Elveback, I. R.,** *Epidemiology, Man and Diseases,* MacMillan, New York, 1970, chap. 4, 5.
7. **Roberts, R. H. and Love, J. N.,** Infectivity of *Anaplasma marginale* after ingestion by potential insect vectors, *Am. J. Res.*, 38, 1629, 1977.
8. **Graham, O. H. and Hourrigan, J. L.,** Eradication programs for the arthropod parasites of livestock, *J. Med. Entomol.*, 13, 629, 1977.
9. **Lewis, E. A.,** Sheep scab: remedial measure reviewed, *Bull. Entomol. Res.*, 28, 11, 1937.
10. **Hall, M. C.,** Parasites and parasitic diseases of sheep, *USDA Farmers' Bull.* , 1330, 36, 1929.
11. **Kemper, H. E. and Roberts, I. H.,** Eradication of sheep scabies, *Proc. U.S. Livestock Sanit. Assoc.*, 54, 247, 1950.
12. **Anderson, R. J.,** State-Federal cooperative animal disease eradication, in *U.S. Dep. Agric. Agric. Handb.*, 167, 1960, 28.
13. **Kemper, H. E. and Peterson, H. O.,** Cattle scab and methods of control and eradication, *USDA Farmers' Bull.*, 1017, 26, 1953.
14. **Meleney, W. P. and Christy, J. E.,** Factors complicating the control of psoroptic scabies of cattle, *J. Am. Vet. Med. Assoc.*, 173, 1473, 1978.
15. **Bruner, D. W. and Gillespie, J. H.,** *Hagan's Infectious Disease of Domestic Animals,* 6th ed., Cornell University Press, Ithaca, 1973, 440.
16. **Seddon, H. R.,** Eradication of sheep scab from New South Wales, *Aust. Vet. J.*, 40, 418, 1964.
17. **Callais, J. J.,** National and international control programmes in Panama, Central and North America, *Br. Vet. J.*, 134, 10, 1978.
18. **Sauer, J. R.,** Acarine salivary glands-physiological relationships, *J. Med. Entomol.*, 14, 1, 1977.
19. **Hair, J. A., Sauer, J. R., and Durham, K. A.,** Water balance and humidity preference in three species of ticks, *J. Med. Entomol.*, 2, 37, 1975.
20. **McMullen, H. L., Sauer, J. R., and Burton, R. L.,** Possible role in uptake of water vapour by ixodid tick salivary glands, *J. Insect Physiol.*, 22, 1281, 1976.

21. **Semtner, P. J. and Hair, J. A.,** Distribution, seasonal abundance, and hosts of the Gulf Coast tick in Oklahoma, *Ann. Entomol. Soc. Amer.,* 66, 1264, 1973.
22. **Williams, R. E., Hair, J. A., and Buckner, R. G.,** Effects of the Gulf Coast tick on blood composition and weights of drylot Hereford steers, *J. Econ. Entomol.,* 70, 229, 1977.
23. **Stacey, B. R., Williams, R. E., Buckner, R. G., and Hair, J. A.,** Changes in weight and blood composition of Hereford and Brahman steers in drylot and infested with adult Gulf Coast ticks, *J. Econ. Entomol.,* 71, 967, 1978.
24. **Downing, W.,** The life history of *Psoroptes communis* var. *ovis,* with particular reference to latent or suppressed scab, *J. Comp. Path. Therap.,* 49, 163, 1936.
25. **Roberts, I. H. and Cobbett, N. G.,** Cattle scabies, in *Animal Diseases, The Yearbook of Agriculture 1956,* U.S. Government Printing Office, 1956, 292.
26. **Roberts, I. H.,** Cattle scabies, in *Great Plains Beef Cow-Calf Handbook,* Cooperative Extension Service-Great Plains States, 3255.1., 1975.
27. **Semtner, P. J., Barker, R. W., and Hair, J. A.,** The ecology and behavior of the lone star tick (Acarina:Ixodidae). II. Activity and survival in different ecological habitats, *J. Med. Entomol.,* 8, 719, 1971.
28. **Semtner, P. J., Howell, D. E., and Hair, J. A.,** The ecology and behavior of the lone star tick (Acarina:Ixodidae). I. The relationship between vegetative habitat type and tick abundance and distribution in Cherokee Co., Oklahoma, *J. Med. Entomol.,* 8, 329, 1971.
29. **Stacey, B. R.,** Season and Habitat Effect on Gulf Coast Tick Biology in Central Oklahoma, M.S. thesis, Oklahoma State University, 1977.
30. **Clymer, B. C., Howell, D. E., and Hair, J. A.,** Environmental alteration in recreational areas by mechanical and chemical treatment as a means of lone star tick control, *J. Econ. Entomol.,* 63, 504, 1970.
31. **Hoch, A. L., Barker, R. W., and Hair, J. A.,** Further observations on the control of lone star ticks (Acarina:Ixodidae) through integrated control procedures, *J. Med. Entomol.,* 8, 731, 1971.
32. **Hoch, A. L., Barker, R. W., and Hair, J. A.,** Measurement of physical parameters to determine the suitability of modified woodlots as lone star tick habitat, *J. Med. Entomol.,* 8, 725, 1971.
33. **Wright, H. A.,** Range burning, *J. Range Manage.,* 27, 5, 1974.
34. **Oefinger, R. D. and Scifres, C. J.,** Gulf cordgrass production, utilization, and nutritional value following burning, *Texas Agric. Exp. Stn. Bull.,* 1176, 19, 1977.
35. **Scifres, C. J.,** Systems for improving Macartney rose infested coastal prairie rangeland, *Texas Agric. Exp. Stn. Misc. Publ.,* 1225, 12, 1975.
36. Research Monograph: Grasses and legumes in Texas-development, production and utilization, *Texas Agric. Exp. Stn. Res. Monogr.,* 6C, 1976, 2.
37. **Patrick, C. D. and Hair, J. A.,** Seasonal abundance of lone star ticks on white-tailed deer, *Environ. Entomol.,* 6, 263, 1977.
38. **Kellogg, F. E., Kistner, T. P., Strickland, R. K., and Gerrish, R. R.,** Arthropod parasites collected from white-tailed deer, *J. Med. Entomol.,* 8, 495, 1971.
39. **Samuel, W. M. and Trainer, D. O.,** *Amblyomma* (Acarina:Ixodidae) on white-tailed deer, *Odocoileus virginianus* (Zimmermann), from south Texas with implications for theileriasis, *J. Med. Entomol.,* 7, 567, 1970.
40. **Stein, C. D., Lotze, J. C., and Mott, L. O.,** Transmission of equine infectious anemia by the stablefly, *Stomoxys calcitrans,* the horsefly, *Tabanus sulcifrons* (Macquart), and by injection of minute amounts of virus, *Am. J. Vet. Res.,* 3, 183, 1942.
41. **Hermes, W. B.,** The housefly in its relation to the public health, *Univ. Calif. Agric. Exp Stn. Bull.,* 215, 513, 1911.
42. **West, L. B.,** *The Housefly,* Comstock, Ithaca, 1951, chap. 18.
43. **Burton, J. S.,** Insect control in the food process industries. III. Flies in food factories, *Food Trade Rev.,* 28, 19, 1958.
44. **Hugh, T. H.,** 6-point program assures fly-free plant, *Food Eng.,* 31, 86, 1959.
45. **McDuffie, W. C.,** Fly control, *Soap Chem. Spec.,* 35, 82, 1959.
46. **Bay, D. E.,** unpublished data, 1979.
47. **Walsh, J. D.,** A survey of fly production in cattle feedlots in the San Joaquin Valley, Calif., *Vector Views,* 11, 33, 1964.
48. **Loehr, R. C.,** Pollution implication of animal wastes — a forward oriented review, Federal Water Pollution Control Administration, U.S. Department of the Interior, Washington, D.C., 1968, 175.
49. **Manges, H. L., Schmid, L. A., and Murphy, L. S.,** Land disposal of cattle feedlot wastes, in *Livestock Waste Management and Pollution Abation, Proc. Intern. Symp. Livestock Wastes,* American Society of Agricultural Engineers, St. Joseph, Mo., 1971, 62.

50. **Steelman, C. D.,** Environmental control of arthropod pests of livestock, in, *CRC Handbook of Pest Management,* Vol.I, Pimentel, D., Ed., CRC Press, Boca Raton, Fla., 1981.
51. **Hourrigan, J. L. and Klingspon, A. L.,** Epizootiology of bluetongue: the situation in the United States of America, *Aust. Vet. J.,* 51, 203, 1975.
52. **Metcalf, H. E. and Luedke, A. J.,** Bluetongue and related diseases, in *Proc. Vet. Prev. Med. Epid. Work Conf.,* Ames, Iowa, 1, 44, 1978.
53. **Steelman, C. D., White, T. M., and Schilling, P. E.,** Efficacy of Brahman characters in reducing weight loss of steers exposed to mosquito attack, *J. Econ. Entomol.,* 69, 499, 1976.
54. **Tugwell, P., Burns, E. C., and Turner, J. W.,** Brahman breeding as a factor affecting the attractiveness of repellency of cattle to the horn fly, *J. Econ. Entomol.,* 62, 56, 1969.
55. **Francis, J.,** Resistance of zebu and other cattle to tick infestation and babesiosis with special reference to Australia: an historical review, *Br. Vet. J.,* 122, 301, 1966.
56. **Francis, J.,** The role of exotic cattle in northern Australia, *Aust. Vet. J.,* 40, 114, 1964.
57. **Hewetson, R. W.,** Selection of cattle for resistance against *Boophilus microplus,* in *Tick-borne Diseases and their Vectors,* Wilde, J. K. H., Ed., Lewis Reprints Ltd., Tonbridge, U.K., 1978, 258.
58. **Campbell, R. S. F.,** The use of resistant cattle in the control of ticks and vector-borne diseases, in *Tick-borne Diseases and their Vectors,* Wilde, J. K. H., Ed., Lewis Reprints Ltd., Tonbridge, U.K., 1978, 251.
59. **Freeden, H. T.,** Genetic aspects of disease resistance, *Can. Vet. J.,* 4, 219, 1963.
60. **Magonigle, R. A., Simpson, J. E., and Frank, F. W.,** Efficacy of a new oxytetracycline formulation against clinical anaplasmosis, *Am. J. Vet. Res.,* 39, 1407, 1978.
61. **Golitz, J.,** Bluetongue in cattle: a review, *Can. Vet. J.,* 19, 95, 1978.
62. **Jones, T. J., Harmon, B. G., and Perry, T. W.,** Malnutrition and disease resistance, *Mod. Vet. Pract.,* 44, 47, 1963.
63. **Neumann, C. G., Jelliffe, D. B., and Jelliffe, E. F. P.,** Interaction nutrition and infection, *Clin. Pediat.,* 17, 807, 1978.
64. **Gladney, W. J., Graham, O. H., Trevino, J. L., and Ernst, S. E.,** *Boophilus annulatus:* effect of host nutrition on development of female ticks, *J. Med. Entomol.,* 10, 123, 1973.
65. **Gingrich, R. E. and Barrett, C. C.,** Effect of dietary vitamin A on the larvae of *Hypoderma lineatum* (Diptera:Oestridae), *J. Med. Entomol.,* 12, 13, 1975.

ENVIRONMENTAL CONTROL OF PREDATORS

F. H. Wagner

INTRODUCTION

Following the 1972 Executive Ban on use of toxicants on public land in the western U.S. by federal predator-control agents, and subsequent delisting of those toxicants by the Environmental Protection Agency, substantial new funds were appropriated for research on predator ecology, behavior, and alternate control measures. Pressures for the new effort arose out of the stockmen's fear that the abandonment of toxicants would allow sharp increases in predator populations and increased livestock losses.

CONTROL STRATEGIES

Prior to 1972, predator control in the western U.S. was directed primarily at coyotes *(Canis latrans)* in behalf of the sheep industry. It consisted of two broad strategies. One was the trouble-shooting type of control in which predator-control agents operated in locations where ranchers were suffering losses, and attempted to destroy the predators causing those losses. The second strategy was aimed at generalized population reduction over broad geographic areas, carried out on the assumption that livestock losses in general were a function of coyote population density. If those densities could be reduced, the livestock industry in general would experience lower losses. There were some forms of control intermediate between these two strategies, as when control agents would intensify population reduction in localized areas prior to the arrival of sheep herds for lambing. All control consisted of lethal methods, and resolved into four basic techniques: trapping, denning, shooting, and use of toxicants.

Trapping

In the early decades of this century, when institutional predator control was first getting underway,[1] trapping was used as one means of attempting population reduction; however, wide-scale trapping is labor-intensive and therefore expensive. Coyotes can be caught fairly readily in late summer and fall when large numbers of naive young are present in the population. At other times of the year, the older animals are considerably more difficult to catch, and, in general, trapping by 25 to 50 predator-control agents in a state is of questionable value as a population-reducing method. The same may not be true in the case of several hundred fur trappers in a state, but data are lacking which permit a confident statement. Trapping carries the added onus of being relatively unspecific in what is caught. It remains a tool in the control agent's kit for trouble-shooting apprehension of offending individual animals.

Denning

Denning consists of locating dens in spring when the pups are small and remain in the den when it is approached. They are dispatched, either after the den is dug out manually, or by releasing carbon-monoxide-emitting cartridges in the den and covering over the entrance. Dens are located by searching afoot, on horseback, or from the air.[2] Denning tends to be labor-intensive, and is used more for trouble-shooting control, or for the intermediate prophylactic treatment on restricted areas. Control agents contend that lambing-ground losses are often alleviated if a nearby den can be found and the young destroyed.

Shooting

Control agents have long carried firearms routinely as they have gone about their activities, and shot coyotes whenever the opportunity presented itself. At times coyotes have been called and decoyed in order to shoot them. This technique has been useful in some cases of trouble-shooting problems. Generalized population reduction has been attempted through aerial gunning, usually in winter. This approach has been greatly increased, particularly with the use of helicopters, in the years since the Executive Ban, in an effort to substitute for toxicants. Aerial gunning has also been used for trouble-shooting situations. Control agents aver that a coyote causing trouble on a lambing ground, which is too experienced to be trapped, can often be spotted from the air and shot.

Toxicants

It was the use of toxicants in which control personnel had the greatest confidence, and which doubtless was the most economical in terms of population reduction per unit cost. Several substances have been used with varying evidence of effect.

Probably the oldest substance used was strychnine impregnated into balls of suet or tallow. Probably millions of these baits have been sown over the western U.S. as they have been in Australia for dingo *(Canis familiaris)* control. Sometimes released from surface vehicles, but not infrequently sown from the air, they were often distributed in areas of known high coyote populations or chronic sheep-loss problems. No data exist to evaluate the effectiveness of strychnine poisoning, either in reducing coyote numbers or sheep losses. They have the very real disadvantage of being lethal to nontarget animals, both wild and domestic.

Three toxic agents came into use in the western U.S. in the 1940s or early 1950s. One was the "coyote-getter", originally a short pipe buried in the ground with a cartridge-ejected load of cyanide. A wool or cotton tuft at the protruding end of the tube was scented with coyote urine. The animal attempted to pull it with its teeth, and in the process triggered the charge and received the poison in its mouth. The later-model "M-44" is spring ejected. Control agents generally hold the view that coyote-getters were effective in reducing coyote populations in the 1940s when they first appeared, but that within a few years selection or habituation had taken place to where coyotes commonly avoided them and they had lost their effectiveness. No data are available to evaluate these contentions. Getters are considered to be relatively coyote specific by virtue of the scent specificity and the canine oral dexterity needed to trip them. But cases of nontarget loss are known, particularly among other canines.

The highly lethal toxin, thallium, was used to a limited extent in the 1940s, but generally was abandoned because of its high toxicity, chemical stability, strong tendency to produce secondary poisoning, and nonspecificity.

The *pièce de résistance* of the control agent's arsenal was the toxin sodium monofluoroacetate, or "1080". This substance was developed by the U.S. Fish and Wildlife Service during the 1940s, and first put into westernwide use in the latter 40s or early 50s. It was the primary agent of prophylactic population reduction in the West. The substance was injected into the circulatory system of a dead sheep, quarter of a horse, or some other carcass with the intent of permeating the flesh with the poison.

Sodium monofluoroacetate was considered to be relatively coyote-specific for three reasons: (1) laboratory development had shown it to be several times more toxic to canids than to other mammals and birds, (2) the carcasses were placed in the landscape overwinter at a time when some species (e.g., bears) were hibernating and many raptors had migrated, and (3) they were placed, one per township, at 10-km intervals. Because the coyote is a highly mobile animal, this was close enough spacing to contact most of them in a region. Foxes, skunks, badgers, weasels, and bobcats have much lower movement radii. While it was acknowledged that some individuals of these species in the immediate proximity of the baits might contact them, many in the interstices of the grid would not, and the population effects would be minimal.

Because 1080 came into general use during a short, datable interval in time; and because federal predator-control personnel kept detailed records on its use; Wagner[3] was able to derive some indication of the effectiveness of this substance in reducing coyote numbers[3] and sheep losses[4] in the 1 or 2 decades immediately following its initial use. In general, coyote populations appear to have been most affected by 1080 in the northernmost states of the West where most 1080 was used, and where coyotes are probably more hardpressed for food during the severe winters. In Wyoming and Montana, mean coyote populations may have been reduced by one half to two thirds.[3] Population reduction appears to have been less marked in the intermediate states; and in the southern states (e.g., Texas, Arizona) the evidence suggests no measurable, generalized population reduction during the period of 1080 use. In these latter states, winters are mild, 1080-baits decay rapidly, and alternate food sources are available. Hence, the toxin is less effective, and less was used.

It proved more difficult to detect changes in sheep-loss rates associated with these changes in coyote numbers.[4] There are no widespread, reliable figures on predocking (pre-tail removal) losses. But there are some data sets on postdocking losses, most extensively those of the U.S. Department of Agriculture Statistical Reporting Service (S.R.S.) and, to a lesser degree, the U.S. Forest Service. The S.R.S. data indicated no statistically real change in postdocking lamb losses, either in the northern states, where coyote reduction was most pronounced and reduction in losses was more expected, or in the southern states where no population change was measurable and no change in losses was expected. The data did suggest a statistically significant reduction from about 10 to about 8% in postdocking lamb losses in such intermediate states as Utah, Colorado, Idaho, and Nevada.

Wagner and Pattison[4] rationalized this pattern as follows. Coyote population densities in the western U.S. increase from north to south. Losses due to predation constitute a relatively small fraction of total losses in the north because coyote numbers are relatively low, and inclement weather looms as the major source of mortality. Reduction in coyote numbers by 1080 reduced a comparatively minor mortality source, and had no measurable effect on total-loss levels. (The S.R.S. data only report total, postdocking losses due to all causes. They do not partition the losses by cause.) In the southern states, there were no measurable changes in coyote densities, and no changes in postdocking losses were expected. It was in the intermediate states where coyote densities were high enough to produce significant lamb losses, and where enough 1080 was used to effect material reductions in coyote numbers, that a slight reduction in total lamb losses during the 1950s and 60s could be detected.

With the general impression that coyote densities had been reduced over the West by 1080, there was considerable interest on the part of biologists, and concern on the part of stockmen, as to what coyote populations would do after the Executive Ban in 1972. In the fall of that year, the Fish and Wildlife Service instituted a region-wide coyote census network designed to index coyote densities annually.[5,6] Through fall 1977, the expected population response had not occurred. During the first 3 years, the region-wide mean increased about 12% but in the following 2 years the index slipped back to about its 1972 level.

Failure of the population to increase has variously been attributed to (1) the fact that 1080 in reality did not have a material population effect; (2) grazing and other forms of land use have so changed the ecology, especially the coyote food base, that the populations could not return to pre-1080 numbers; (3) the newer control programs, especially the intensified aerial shooting, have compensated for the cessation of toxicant use; or (4) some combination of these. Meanwhile, many stockmen claim that coyotes have increased, and plea for increased control effort by whatever means.

Barriers

Within the quest for new control methods during the 1970s, the use of barriers has had some scrutiny. Prior to this time, there had been both opponents and advocates of the view that livestock could be effectively and economically fenced against predatory attack. Among the advocates were those who pointed to the Edwards Plateau of central Texas, a sheep- and goat-rearing area extensively cross fenced with net wire to contain these animals. The Plateau has traditionally had low coyote densities and relatively low predation losses.[7] These low coyote densities have often been attributed to the extensive, net wire fencing which somehow is assumed to have served as a barrier to coyote influx.

Several lines of evidence raise considerable uncertainty about this conclusion, however. Knowlton et al.[8] and Wagner[9] have shown that coyote densities in Texas, including those of the Plateau, are a function of food abundance in the form of rodent biomass. For as yet unexplained reasons, rodent populations tend to be low on the Plateau. Furthermore, Thompson[10] has experimentally shown the ease with which coyotes go over and through typical livestock fencing, while Shelton[11,12] has emphasized the coyote proclivity for going under fences. In addition one may ask why, even if coyotes have been barred from immigration into the Plateau in postsettlement times by fencing, they were not there prior to settlement.

Despite these uncertainties abou the effectiveness of barrier fencing, interest and research have been renewed in recent years. McKnight[13] draws the distinction between exclusion fencing — citing as an example the fencing of small garden plots to protect against cows, goats, and other herbivores — and barrier fencing designed to exclude animals from large geographic areas. The best example of the latter is the dingo fence across large portions of Australia (Figure 1).

The distinction would seem to be somewhat illusory, however. All such fences built for protection against predators are designed to prevent them from moving into an area where there are domestic animals. In a sense the barrier fence around southeastern Australia constitutes an exclusion fence around that portion of the continent, and makes the region an exclosure. Thus, barrier fencing in fact falls along a spectrum ranging from the enclosure of small stockyards, as with the Masai and Somali herdsmen of East Africa who nightly enclose their cattle in bomas constructed of thorny *Acacia* boughs, to the enclosure of southeastern Australia with a dingo fence.

What form(s) of barrier fencing along this spectrum would be feasible for a given area would appear to depend on (1) size and nature of the livestock operation, and attendant economics of construction; (2) physical capabilities of the predator and type of barrier needed to materially exclude it; and (3) absence of predators within the exclosure, or ability to maintain their numbers at densities significantly lower than outside the barrier. Let us examine some forms of barrier fencing along the spectrum.

Where livestock operations are labor intensive and on a small scale by worldwide standards, as when herds ranging in size from a few dozen to a few hundred animals are continuously attended by individuals, such herds can be driven nightly into small exclosures where they can be protected from predator risk. This is the pattern with the Masai and Somali pastoralists in protecting their animals from predators, particularly lions. A similar pattern can be found in parts of North Africa and the Middle East where small flocks of sheep and goats are continuously shepherded, and nightly enclosed in small enclosures termed ''grishas'' in North Africa.

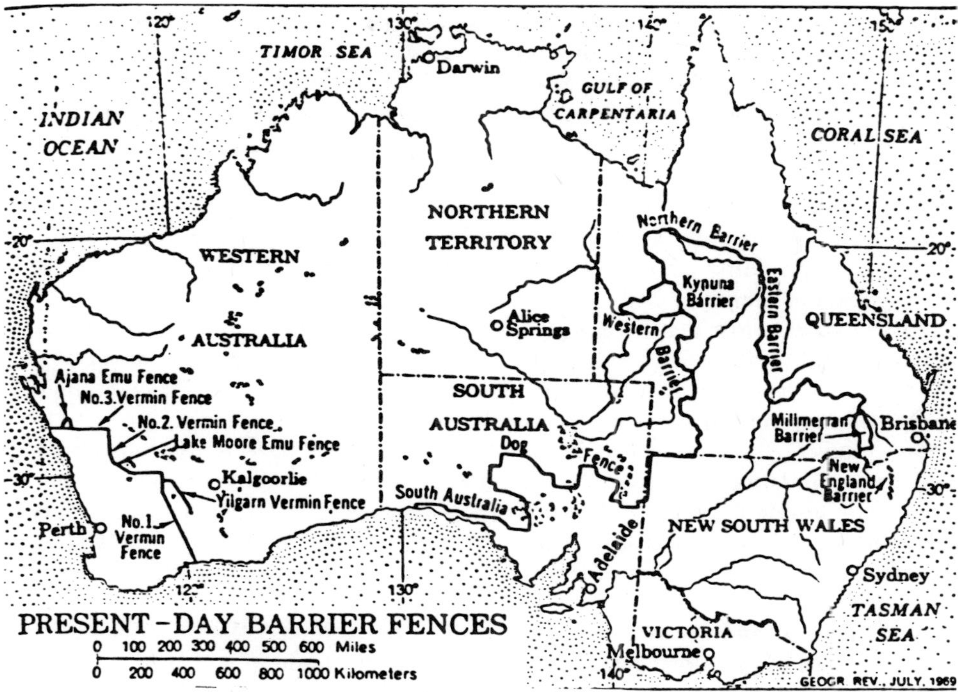

FIGURE 1. Major regional barrier fences used to exclude "vermin" in Australia as of 1969 (after McKnight[13]). Numerous fences of various types, and in various stages of repair and function, also exist around individual properties and smaller geographic areas.

The sheep-ranching practices of the western U.S. and Australia are at the other extreme of labor intensiveness and size. The ranch land of west Texas and Australia is, for the most part, privately owned and fenced into pastures ranging anywhere from 10 to 100 km^2 or more. Sheep, numbering from several hundred to a few thousand, are released into these pastures. They are rounded up periodically for lambing; for castrating, marking, and docking; for shearing; for treatment of disease and insect problems; and for shipping to market. Between these periods of activity, the animals are left to roam unattended over the large areas of landscape which the pastures constitute.

In other states of the western U.S. much of the grazing is on public land, and not so extensively fenced. Here, sheep herds may be tended by herders whose task it is to keep animals in specified grazing allotments within a national forest or area of Public Domain land. Although the herders may be present in the locale, the animals are allowed to disperse over mountainous and forested areas to feed, and are not always within view of or in close proximity to the herder.

In all of these American and Australian cases, the land is relatively pristine and inhabited with predatory species. In the U.S. the coyote is the primary animal of concern by virtue of its abundance and physical prowess as a predator. But some losses are caused by the considerably less numerous bobcats *(Lynx rufus)*, mountain lions *(Felis concolor)*, and black bear *(Ursus americanus)*; and locally by wintering concentrations of golden eagles *(Aquila chrysaetos)*. In Australia, the dingo is the major predator on sheep, although some losses are caused by the introduced European red fox *(Vulpes vulpes)*, and occasionally by feral pigs *(Sus scrofa)*.

While these large-scale operations obviously do not lend themselves to nightly confinement like the African examples, the animals can be, and are enclosed in small areas during certain key stages of the annual husbandry cycle. In the western U.S., pens or sheds are used for lambing. Ewes are enclosed in these shortly before they are expected to drop their young. Birth takes place in the pens in the presence of stockmen who assist in the process. The animals are kept in the pens for some days until acceptance of the young by the ewes is assured, and the lambs have gained sufficient strength to remain in the company of free-ranging mothers.

While the purpose of pen or shed lambing is not primarily for protection against predators, it probably achieves a measure of that protection. The presence of human activity around lambing sheds, their frequent proximity to human dwellings and the physical barriers provided by the facilities together probably afford a measure of deterrence. That they are not used more generally is probably a function of the construction costs of recent years. Most shed-lambing operations utilize facilitites which have been in operation for a considerable number of years.

Utility of small-area barrier fencing at other times of the year depends on whether husbandry operations maintain animals in limited areas for short periods and purposes other than lambing. In the southern states of the western U.S., as in Texas, sheep are allowed free access to the entire countryside throughout the year. After lambing, the animals are released to roam free over large pastures and not gathered again until docking time.

In the central and northern states, a severe winter blocks the use of mountainous terrain except in late spring and summer. Sheep are maintained on valley or desert ranges in winter. Between lambing and movement of animals onto summer range, they are often held in relatively small agricultural fields or spring, foothill pastures. There may be instances where these could be economically fenced to prevent predator attack. Gates et al.[14] describe the fencing of a 64-ha pasture with electric fencing, which is a third less expensive than regular sheep-type fencing, and which effectively bars coyote entry.

In the final analysis, the feasibility of small-area fencing will depend on whether the economic value of the animals saved from predation exceeds that of the fencing cost. During lambing, the animals are extremely vulnerable, and heavy losses in a short period of time a distinct possibility. Later, when the animals are partially or fully grown, losses proceed at a much lower rate per unit of time. At this stage, penning animals for a month or two might or might not save enough from predatory loss to offset the cost of fencing.

Above the small-area level of fencing intensity, the scale of fencing would involve whole ranches of some scores to hundreds of square kilometers, groups of ranches, or whole geographic areas. Many observers are skeptical of the potential for barrier fencing at this level for the following reasons. (1) Depending on the nature of the fence, the cost could be prohibitive. (2) Barrier fencing requires intensive maintenance in order to remain effective. Breakage by intrusion of large animals, falling trees, and floods plus crossings produced by wind-blown sand or vegetation are frequent occurrences and must be continuously guarded against and repaired. (3) Any large areas fenced today are already populated by predators which would have to be intensively controlled by lethal methods in order to be effective.

The extreme example is the Australian dingo fence designed to enclose the major sheep-growing areas of that country (Figure 1). The fence extends nearly 11,000 km in the four states of Western Australia, South Australia, New South Wales, and Queensland, more than half of this in the last state.[13] The Queensland portion includes more than 2500 km of new fence[15] built since 1954 at a cost of 1.25 million Australian pounds.[13] Extensive dingo control is carried out within the fenced region under a complex fabric of government and local stockmen's organizations, bounties, and the use of toxins and other control measures.

No systematic attempts have been made to evaluate the effectiveness with which (1) dingo are prevented from immigrating into the enclosed area, and (2) their numbers are significantly reduced by the control activities within. McKnight[13] states that fewer dingo scalps are presented for bounty in that portion of New South Wales inside the fence (Figure 1) than the adjacent portions of South Australia and Queensland outside it. He further cites the widespread impression among graziers that control and fencing are effective. Bauer[15] contends that dingo control was effective before World War II, but that with relaxation of control at that time dog numbers and sheep losses mounted. Other observers point to the greater abundance of kangaroos inside the fence than outside, the assumption being that dingo predation is lighter inside.

However, a large number of critics doubt the effectiveness of the control measures, pointing to the ubiquity of dingo inside the fenced area and the chronic disrepair of the fence. On the kangaroo question, they point out that the fence encloses the more mesic and productive vegetation area of southern and eastern Australia, and which as a result is the center of the sheep industry. Much of the area outside the fence is more arid, has different vegetation, and is used for cattle ranching. Hence there are other important variables which doubtless affect kangaroo numbers.

Despite the widespread skepticism about the feasibility of extensive barrier fencing, Shelton[11,12,16,17] has been one of the stronger, recent proponents and has conducted research on the technique for several years. Initially working on net wire fencing, he[17] provides detailed information on construction specifications and costs. He gives particular attention to what he considers to be the primary mode of coyote entry, burrowing under, and details the apron construction needed to prevent this. He also points out how costs per head of livestock can be reduced by fencing large areas, combining forces with several ranchers, and increased stocking rates.

More recently, Shelton[12,16] has explored the use of 2- and 3-wire electric fencing at the base of, and single strands at the bottom and top of, existing fences. This has presumably been done out of recognition for the cost deterrent of new, net wire fencing. He[16] presents information on design specifications, and data on the degree to which loss rates were reduced. With these results and those of Gates et al.,[14] there appears to be some promise in large-scale use of electric fencing. The matter of keeping predator populations down within the fenced areas is not discussed by these authors.

Habitat Alteration

Within the increase of predation research following the Executive Ban in 1972, some attention has been given to the question of what constitutes favorable habitat for coyotes. The desire has been to learn something about the ecological circumstances under which coyote numbers and sheep losses are high. No clear generalizations have been developed from this as yet, but one principle has emerged: an important determinant of coyote density is prey abundance.[8,9,18] Consequently, while the findings to date do not yet hold out promise that habitat alteration can be used to influence coyote numbers directly, there does appear to be some potential in indirect effects.

Any influence on the land, such as heavy grazing, which alters the vegetation in such a way as to reduce rodent and lagomorph numbers (the staple food items for coyotes) would be likely in the long run to reduce coyote populations. One such influence has been the plowing of natural vegetation and reseeding of exotic, perennial grasses over large areas of the western U.S. These seedings tend to carry much lower densities of rodents and lagomorphs than does the natural vegetation,[18] and Howard and Wolfe[19] have explored the possibility that these reduced prey numbers have led to reductions in raptor populations. No one has yet explored the possible effects on coyote numbers.

Of course, any consideration of using such vegetation manipulation as a tool for coyote control would need to weigh the other ecological effects involved. Because the long-term ecological effects of planting huge areas to monotypes of exotic grasses have not been fully evaluated, and because of the possible effects on desirable elements of the fauna, large-scale grass seeding by public agencies in the western U.S. is currently in abeyance.

Parenthetically, the progressive reduction of wild ungulate numbers in eastern and southern Africa due to shooting, competition with livestock, and habitat alteration almost certainly is causing reductions in the number of large carnivores over and above reduction from direct exploitation and control. This reduction is equally likely to ease predation pressures on livestock. The sheep and goat herds of North Africa, where once there were sizeable populations of ungulates and carnivores, experience almost no predatory loss in today's wildlife-impoverished landscape.

Other Forms of Environmental Management

The presence of a potential enemy to predators in the vicinity of livestock herds can be a deterrent to predatory attack. The Masai and Somali herdsmen tend their animals in part to ward off attacks by lions or hyenas. Davenport et al.[20] showed that summer sheep losses to predators were more than twice as high in flocks without herders as in those attended by herders in a Utah sample they investigated. In some areas of the world, dogs have been trained to accompany livestock herds and attack approaching predators. Heady[21] mentions the use of such animals in Israel.

Antifertility agents, which have had some experimentation in the U.S., bear brief mention. Experiments on reproductive inhibition in carnivorous mammals have generally centered around the use of diethylstilbesterol. Pen research has shown some success in reproductive inhibition in red foxes *(Vulpes fulva)*[22] and coyotes,[23] but attempts at using it in the field for control of free-ranging coyote populations have not been successful so far.[24]

REFERENCES

1. **Cain, S. A., Kadlec, J. A., Allen, D. L., Cooley, R. A., Hornocker, M. G., Leopold, A. S., and Wagner, F. H.,** *Predator Control — 1971/Report to the Council on Environmental Quality and the Department of the Interior by the Advisory Committee on Predator Control,* University of Michigan Press, Ann Arbor, 1, 1972.
2. **Clark, F. W.,** Influence of jackrabbit density on coyote population change, *J. Wildl. Manage.,* 36, 343, 1972.
3. **Wagner, F. H.,** Coyotes and Sheep/Some Thoughts on Ecology, Economics and Ethics, Utah State Univ. 44th Faculty Honor Lecture, Logan, 1, 1972.
4. **Wagner, F. H. and Pattison, L. G.,** Analysis of existing data on sheep loss and predator activity, in **Bennett, J. A., Bowns, J. E., Davenport, J. D., Gilbert, B. K., Nielsen, D. B., Pattison, L. G., Stoddart, L. C., Wagner, F. H., and Workman, J. P.,** *Predator Control Study,* Final Report to Four Corners Regional Comm., Utah State Univ., Logan, 18, 1973.
5. **Linhart, S. B. and Knowlton, F. F.,** Determining the relative abundance of coyotes by scent station lines, *Wildl. Soc. Bull.,* 3, 119, 1975.
6. **Roughton, R. D.,** Indices of Predator Abundance in the Western United States, Denver Wildlife Research Center, U.S. Fish and Wildlife Service, 1975.
7. **Reynolds, R. N. and Gustad, O. C.,** Analysis of Statistical Data on Sheep Losses Caused by Predation in Four Western States During 1966—69, Division of Wildlife Services Mimeographed Report, U.S. Bureau of Sport Fishing and Wildlife, Washington, D.C., 1971.
8. **Knowlton, F. F., Carley, C. J., and McBride, R. T.,** *Mammal Damage Control Research — Predators and Predator-prey,* Annu. Prog. Rep. Work Unit DF-103.9, Denver Wildlife Research Center, 1972.

9. **Wagner, F. H.,** Coyotes and Non-game Wildlife, presented at *Symp. Coyotes, Wildl. Meat Prod., 143rd Annu. Meet. Am. Assoc. Adv. Sci.,* Denver, 1977, 1.
10. **Thompson, B. C.,** Fence-crossing behavior exhibited by coyotes, *Wildl. Soc. Bull.,* 6, 14, 1978.
11. **Shelton, M.,** One approach to the predator problem — Fencing, *San Angelo Standard-Times,* p. 10E, May 6, 1973.
12. **Shelton, M.,** Electric fencing as a means of deterring coyote predation, *Ranch Mag.,* 58, 1977.
13. **McKnight, T. L.,** Barrier fencing for vermin control in Australia, *Geogr. Rev.,* 59, 330, 1969.
14. **Gates, N. L., Rich, J. E., Godtel, D. D., and Hulet, C. V.,** Development and evaluation of anti-coyote electric fencing, *J. Range Manage.,* 31, 151, 1978.
15. **Bauer, F. H.,** Queensland's dingo fence, *Aust. Geogr.,* 9, 244, 1964.
16. **Shelton, M.,** The use of electric fencing as a deterrent to coyote predation, *Texas Agric. Exp. Stn. Prog. Rep.,* in press.
17. **Shelton, M.,** Fencing as a means of protecting livestock from predation, unpublished data.
18. **Wagner, F. H.,** Some concepts in the management and control of small mammal populations, in *Populations of Small Mammals Under Natural Conditions,* Snyder, D. P., Ed., Pymatuning Laboratory of Ecology, University of Pittsburgh, Linesville, Pa., 1978.
19. **Howard, R. P. and Wolfe, M. L.,** Range improvement practices and ferruginous hawks, *J. Range Manage.,* 29, 33, 1976.
20. **Davenport, J. W., Bowns, J. E., Workman, J. P., and Nielson, D. B.,** Assessment of sheep losses, in **Bennett, J. A., Bowns, J. E., Davenport, J. D., Gilbert, B. K., Nielsen, D. B., Pattison, L. G., Stoddart, L. C., Wagner, F. H., and Workman, J. P.,** *Predator Control Study,* Final Report to Four Corners Regional Comm., Utah State Univ., Logan, 3, 1973.
21. **Heady, H. F.,** *Rangeland Management,* McGraw-Hill, New York, 162, 1975.
22. **Linhart, S. B. and Enders, R. K.,** Some effects of diethylstilbesterol on reproduction in captive red foxes, *J. Wildl. Manage.,* 28, 358, 1964.
23. **Balser, D. S.,** Management of predator populations with antifertility agents, *J. Wildl. Manage.,* 28, 352, 1964.
24. **Linhart, S. B., Brusman, H. H., and Balser, D. S.,** Field evaluation of an antifertility agent, stilbesterol, for inhibiting coyote reproduction, *Trans. North Am. Wildl. Nat. Resour. Conf.,* 33, 386, 1968.

Index

INDEX

A

B

environmental control of, 499—504
airports and aircraft, 503
crops and livestock, 501—502
human habitations and buildings, 502—503
roosts, 500—503
sanitary landfills, 504

C

E

F

G

H

I

J

K

M

O

P

Q

R

Resistance factors, plant resistance to insects, 179—180, 184—185, 188—190

S

T

U

V

W

X

Y

Z